Plant Anatomy

also by Katherine Esau
Anatomy of Seed Plants

Second Edition

Plant Anatomy

Katherine Esau
Professor of Botany, University of California

John Wiley & Sons, Inc., New York · London · Sydney

THIRD PRINTING, MARCH, 1967

Copyright © 1953, 1965 by John Wiley & Sons, Inc.

Library of Congress Catalog Card Number: 65-12713
Printed in the United States of America

29,916

Preface

The great expansion of biological research that has occurred since the publication of the first edition of this book has had a strong impact on the field of plant anatomy. The accumulation of new material was less important in this connection than the change in emphasis and in direction of interest. Of particular significance were, and still are, the increase in awareness of the unifying features in the organic world and the resulting efforts to discover the principles of structure and development common to all organisms. As community of principles is based on community of molecular structure, biological research has logically become oriented toward the molecular realm of life. This aspect of scientific development is well known and need not be discussed here. But a few words must be said about the place, in the modern scheme of things, of a fundamentally descriptive text in the anatomy of plants.

A biologist, regardless of his line of specialization, cannot afford to lose sight of the whole organism if his goal is the understanding of the organic world. Knowledge of the grosser aspects of structure is fundamental for effective teaching and research in the more specialized areas of biology. Moreover, the trend toward a reduction of emphasis on factual information in modern teaching makes a readily accessible depository of the basic information on plant structure doubly important. A rather strong evidence of the continued importance of reference works in plant anatomy is the good reception the first edition of *Plant Anatomy* enjoyed through the years of its existence.

The foregoing remarks are not intended to imply that plant anatomy has become a field which merely provides part of the background for other aspects of plant study. New approaches and techniques keep plant anatomy a live field and enable the plant anatomist to maintain the spirit of discovery and to participate effectively in the interdisciplinary search for integrated concepts of growth and morphogenesis. The time-honored comparative anatomy also continues to be a fertile field for uncovering new facts and creating new concepts on relationships and evolution of plants and plant organs.

The purpose of the book, its organization, and the type of presentation of the subject matter, as characterized in the preface to the first edition, have been maintained in this second edition. But the revision is not limited to an integration of new facts. Parts dealing with areas distinguished by active research required a reconsideration of emphasis

v

and sometimes a revision of the basic concepts and terms. Ultra-structural research, for example, has considerably modified our views on the protoplast and the interrelationship of its part and has affected the interpretation of cell-wall growth. In the study of meristems, the interest has shifted toward the relation between structure and function, particularly that in the apical meristems, and the methodology has become more sophisticated and imaginative. Use of increasingly more refined methods of study of development has resulted in remarkable insights into the factors governing growth, differentiation, and organization of plants.

Naturally, in areas of active research many conclusions are tentative and concepts controversial. Some of the interpretations might become obsolete before the book is published. This circumstance need not be a source of discouragement; instead, it should give the student a feeling for the dynamic state of science and help him to recognize fruitful areas for further research.

The enormity of modern literary scientific output is commonly recognized. In the field of plant structure, too, original papers appear in greater numbers and in many more languages than before. Then, there are all the annual reviews, the many handbooks, and the steady flow of collections of papers given at national and international symposia. The selection of references for citations in a textbook has become more difficult and the chance of omission of important papers greater. There is also the dilemma that older references cannot be deleted indiscriminately. Some continue to be the main source of certain information; others are classics that must be brought to the attention of the student.

These remarks should make it clear that the new edition is not claimed to be a "definitive" text in plant anatomy. If it will attract the student to the field or if it will provide him, as well as the more mature scientist, with an orientation he needs in his work with plants, the book will have served well.

K. E.

Santa Barbara, California
December, 1964

Preface to the First Edition

The writing of this volume was prompted by a wish to bring together, in a comprehensive form, the substance of a course in the anatomy of seed plants. The book has been planned primarily for advanced students in various branches of plant science and for teachers of plant anatomy. At the same time, an effort has been made to attract the less advanced student by presenting the subject in a straightforward style and by explaining and analyzing the basic terms and concepts.

My botanical interests, directed toward research in developmental anatomy, unavoidably color the presentation of the material. Developmental aspects are utilized to enhance the understanding of plant structure and its variability. Phylogenetic data and information on the relation between structure and function are reviewed for the same purpose, but less extensively. Consideration of historical aspects plays a minor role, despite the recognized pedagogic value of such an approach.

A large number of selected references are listed to support descriptions and interpretations, and to direct the interested person toward wider reading. Many references that seemed of less immediate import than those listed were omitted, and some pertinent ones were, no doubt, overlooked. If an author has a review paper adequately covering his own research, such a review is sometimes listed in the place of the individual publications by the same author. Among the references listed, those that were deemed strongest in supporting interpretations and conclusions have been placed in the foreground. Frequently, the descriptive matter has been substantiated by an examination of original preparations of pertinent plant material.

The organization of the subject matter in plant anatomy and the order of its presentation are challenging problems which are related to classification of cells and tissues and to matters of emphasis and attitudes in teaching. In this book, the problems of classification are not resolved and the subject matter is presented in an orthodox sequence, considering first the cell and tissue types, then the arrangement of the structural elements within the plant organs. In general, the topics are delimited and arranged in accordance with the organization developed by A. S. Foster in his *Practical Plant Anatomy.* (D. Van Nostrand Company, New York, 1949). This organization is simple and coherent and permits the development of each chapter as an organic whole.

Admittedly, some students may find the topics on meristems too complex to master at the beginning of the course. However, an early acquaintance with the structure and growth of the meristems and with phenomena of tissue differentiation should increase appreciation of the developmental interpretation which is emphasized throughout the book.

The chapters on flower, fruit, and seed were approached with a feeling of adventure. The boundary between morphology, in the sense of study of external form, and anatomy, in the sense of study of internal form, seems to be especially vague in investigations dealing with the flower and its derivatives. The study of the flower also merges with the vast field of investigations of phenomena of reproduction. It is, therefore, difficult to recognize the proper limits in a discussion of these plant parts. The chapters on the flower, the fruit, and the seed are here offered as an experiment in treatment of these topics.

Despite its length, this book does not cover the subject exhaustively. Instead of depicting numerous examples, it treats a few in detail. However, the student is made aware of the endless variability in form and structure and of the vagueness of boundaries between different types of structures. This attitude should prepare him to interpret an unfamiliar structure and to relate it to those that are known.

This book is not a generous source of new concepts and terms. Existing ones, however, are scrutinized for accuracy and usefulness. Some terms and concepts appeared to have lost their accuracy and had to be revised. There are also those that have been relegated to the domain of history because they seemed to have outlived their usefulness. The guiding principle in this evaluation was the realization that, unless terms and concepts are flexible, they fail to account for the inherent variability in the phenomena to which they refer. Readers may disagree with the treatment of some of the established notions. It is hoped, however, that the usage in this book is clear in every instance and consistent throughout.

Illustrations form an important part of a book in plant anatomy. Although a combination of quality, adequacy, and balance of illustrative material was the goal in the present undertaking, shortcomings were unavoidable. Illustrations whose source is not indicated in the legends are original. The others were copied from various research papers and occasionally from books. With few exceptions, the original drawings and the copies were made by the writer. The original photomicrographs were prepared from research material, original and borrowed, and from slides used in teaching. The slides were either purchased from various commercial concerns or prepared locally. For the sake of economy in printing the halftones had to be assembled at the end of the book in the form of plates.

In the tracing of the origin of technical terms to their Greek or Latin roots, B. D. Jackson's *A Glossary of Botanic Terms* (Duckworth, London, 1928) was the principal reference.

In conclusion I wish to express my appreciation to those who so generously gave of their time to review the manuscript or parts of it. In particular, Dr. A. S. Foster and Dr. V. I. Cheadle extended competent counsel regarding matters of organization and presentation; Dr. A. S. Crafts has advised on physiological aspects; Dr. I. W. Bailey was generous with information from research still unpublished. Valuable suggestions were offered by Dr. E. M. Gifford, Jr. and Dr. R. H. Wetmore. Thanks are also due to Dr. R. B. Wylie for reading the chapter on the leaf; to Dr. Charlotte G. Nast and to Dr. R. M. Brooks for reviewing the chapters on the flower, the fruit, and the seed; and to Dr. G. M.. Smith for the loan of his lecture notes on morphology of the angiosperm flower. Mrs. Fay V. Williams was the trusted assistant with the preparation of the manuscript. Persons who extended the courtesy of lending microscope slides, negatives, or finished illustrations are mentioned in the appropriate legends.

K. E.

Davis, California
January, 1953

Contents

Contents

Contents

Contents

General References

Aleksandrov, V. G. *Anatomiĭa rastenii.* [Anatomy of plants.] Moskva, Sovetskaĭa Nauka. 1954.

Andrews, H. N. *Studies in paleobotany.* New York, John Wiley and Sons. 1961.

Bailey, I. W. *Contributions to plant anatomy.* Waltham, Mass., Chronica Botanica Company. 1954.

Biebl, R., and H. Germ. *Praktikum der Pflanzenanatomie.* Wien, Springer-Verlag. 1950.

Boureau, E. *Anatomie végétale.* 3 vols. Paris, Presses Universitaires de France. 1954, 1956, 1957.

Brachet, J., and A. E. Mirsky, eds. *The cell. Biochemistry, physiology, morphology.* Vol. II. *Cells and their component parts.* New York, Academic Press. 1961.

Braun, H. J. *Die Organisation des Stammes von Bäumen und Sträuchern.* Stuttgart, Wissenschaftliche Verlagsgesellschaft. 1963.

Carlquist, S. *Comparative plant anatomy.* New York, Holt, Rinehart and Winston. 1961.

De Bary, A. *Comparative anatomy of the vegetative organs of the phanerogams and ferns.* Oxford, Clarendon Press. 1884.

De Robertis, E. D. P., W. W. Novinski, and F. A. Saez. *General cytology.* Philadelphia, Saunders Company. 1960.

Eames, A. J. *Morphology of vascular plants. Lower groups.* New York, McGraw-Hill Book Company. 1936.

Eames, A. J. *Morphology of the angiosperms.* New York, McGraw-Hill Book Company. 1961.

Eames, A. J., and L. H. MacDaniels. *An introduction to plant anatomy.* 2nd ed. New York, McGraw-Hill Book Company. 1947.

Esau, K. *Anatomy of seed plants.* New York, John Wiley and Sons. 1960.

Foster, A. S. *Practical plant anatomy.* 2nd ed. New York, D. Van Nostrand Company. 1949.

Foster, A. S., and E. M. Gifford, Jr. *Comparative morphology of vascular plants.* San Francisco, W. H. Freeman and Company. 1959.

Goebel, K. *Organographie der Pflanzen, insbesondere der Archegoniaten und Samenpflanzen.* 3rd ed. Jena, Gustav Fischer. 1928–33.

Goebel, K. *Organography of plants, especially of the Archegoniatae and Spermatophyta.* Part 1. *General organography.* Part 2. *Special organography.* Oxford, Clarendon Press. 1900–05.

Haberlandt, G. *Physiological plant anatomy.* London, Macmillan and Company. 1914.

Hasman, M. *Bitki anatomisi.* [Plant anatomy.] Istanbul, Matbaasi. 1955.

Hayward, H. E. *The structure of economic plants.* New York, The Macmillan Company. 1938.

Hofmann, E. *Paläohistologie der Pflanze.* Wien, Julius Springer. 1934.

Huber, B. *Grundzüge der Pflanzenanatomie.* Berlin, Springer-Verlag. 1961.

Jackson, B. D. *A glossary of botanic terms.* 4th ed. New York, Hafner Publishing Co. 1953.

Jane, F. W. *The structure of wood.* New York, The Macmillan Company. 1956.

Jeffrey, E. C. *The anatomy of woody plants.* Chicago, University of Chicago Press. 1917.

Johansen, D. A. *Plant microtechnique.* New York, McGraw-Hill Book Company. 1940.

Kaussmann, B. *Pflanzenanatomie.* Jena, Gustav Fischer. 1963.

Korsmo, E. *Anatomy of weeds.* Oslo, Grondahl. 1954.

Küster, E. *Pathologische Pflanzenanatomie.* 3rd ed. Jena, Gustav Fisher. 1925.

Küster, E. *Die Pflanzenzelle.* 3rd ed. Jena, Gustav Fischer. 1956.

Linsbauer, K., ed. *C. K. Schneiders illustriertes Handwörterbuch der Botanik.* 2nd ed. Leipzig, Wilhelm Engelmann. 1917.

Linsbauer, K., ed. *Handbuch der Pflanzenanatomie.* Band 1 et seq. Berlin, Gebrüder Borntraeger. 1922–43.

Mägdefrau, K. *Paläobiologie der Pflanzen.* 3rd ed. Jena, Gustav Fischer. 1956.

Mansfield, W. *Histology of medicinal plants.* New York, John Wiley and Sons. 1916.

Metcalfe, C. R. *Anatomy of the monocotyledons.* I. *Gramineae.* Oxford, Clarendon Press. 1960.

Metcalfe, C. R., and L. Chalk. *Anatomy of the dicotyledons.* 2 vols. Oxford, Clarendon Press. 1950.

Rauh, W. *Morphologie der Nutzpflanzen.* Heidelberg, Quelle und Meyer. 1950.

Sass, J. E. *Botanical microtechnique.* 3rd ed. Ames, Iowa State College Press. 1958.

Sinnott, E. W. *Plant morphogenesis.* New York, McGraw-Hill Book Company. 1960.

Smith, G. M. *Cryptogamic botany.* Vol. 2. *Bryophytes and Pteridophytes.* New York, McGraw-Hill Book Company. 1938.

Solereder, H. *Systematic anatomy of the dicotyledons.* Oxford, Clarendon Press. 1908.

Solereder, H., and F. J. Meyer. *Systematische Anatomie der Monokotyledonen.* Berlin, Gebrüder Borntraeger. Heft 1, 1933; Heft 3, 1928; Heft 4, 1929; Heft 6, 1930.

Stebbins, G. L., Jr. *Variation and evolution in plants.* New York, Columbia University Press. 1950.

Stover, E. L. *An introduction to the anatomy of seed plants.* Boston, D. C. Heath and Company. 1951.

Strasburger, E. *Über den Bau und die Verrichtungen der Leitungsbahnen in den Pflanzen. Histologische Beiträge.* Band 3. Jena, Gustav Fischer. 1891.

Takhtajan, A. *Die Evolution der Angiospermen.* Jena, Gustav Fischer. 1959.

Tomlinson, P. B. *Anatomy of the monocotyledons.* II. *Palmae.* Oxford, Clarendon Press. 1961.

Troll, W. *Vergleichende Morphologie der höheren Pflanzen.* Band 1. *Vegetationsorgane.* Berlin, Gebrüder Borntraeger. Heft 1, 1937; Heft 2, 1939; Heft 3, 1941–42.

Troll, W. *Praktische Einführung in die Pflanzenmorphologie.* 1. Teil: *Der vegetative Aufbau.* 2. Teil: *Die blühende Pflanze.* Jena, Gustav Fischer. 1954, 1957.

Tschirch, A. *Angewandte Pflanzenanatomie. Ein Handbuch zum Studium des anatomischen Baues der in der Pharmacie, den Gewerben, der Landwirtschaft und dem Haushalte benutzten pflanzlichen Rohstoffe.* Wien und Leipzig, Urban und Schwarzenberg. 1889.

Tschirch, A., and O. Oesterle. *Anatomischer Atlas der Pharmakognosie und Nahrungsmittelkunde.* Leipzig, H. Tauschnitz. 1900.

Wardlaw, C. W. *Phylogeny and morphogenesis.* London, Macmillan and Company. 1952.

Zimmermann, W. *Die Phylogenie der Pflanzen; ein Ueberblick über Tatsachen und Probleme.* 2nd ed. Stuttgart, Gustav Fischer. 1959.

Plant Anatomy

1

The Plant Body

THE PLANT ORGANS

The subject of this book is the structure and development of seed plants, with emphasis on the angiosperms. The complex multicellular body of a seed plant is a result of evolutionary specialization of long duration. This specialization has led to the establishment of morphologic and physiologic differences between the various parts of the plant body and has caused the development of the concept of *plant organs* (Arber, 1950; Troll, 1937). At first many organs were recognized; later their number was reduced to three: *stem, leaf,* and *root* (Eames, 1936).

The relationship of stem, leaf, and root to each other and to the plant as a whole has long been, and still is, one of the fundamental problems of plant morphology. In this connection the principal question is whether plant organs differ essentially from one another or whether they are modifications of one basic type of structure. Students of evolution postulate that the organization of the oldest land plants was extremely simple, perhaps resembling that of the leafless and rootless Devonian plants such as *Rhynia* (Foster and Gifford, 1959). If the seed plants have evolved from plants consisting of branched axes without appendages, the leaf, the stem, and the root would be closely related by phylogenetic origin (Arnold, 1947; Eames, 1936). Ontogenetically, the organs have common origin in the zygote and the resulting embryo; and in the apical meristems of the shoot, the leaf and the stem increments are formed as a unit. At maturity, too, the leaf and the stem imperceptibly merge with one another both externally and internally. The root and the stem constitute one continuous structure also and have many common features in form, anatomy, function, and method of growth.

1

The morphologic nature of the angiospermous flower is another subject of much research and discussion. One of the most widely used interpretations is that the flower is homologous with a shoot and the floral parts with leaves. Both the leaves and the floral parts are thought to have originated from branch systems. The manner and the relative time of divergence between the vegetative and floral organs thus originating is of major concern in the interpretation of the relationship between the two.

Despite the lack of absolute distinction among the various parts of the plant, the division of the plant body into morphological categories of stem, leaf, root, and flower (where present) is commonly resorted to for convenience in treating the material descriptively. Such division is also useful for the discussion of the functions of the plant and its parts.

DEVELOPMENT OF THE PLANT BODY

A vascular plant begins its existence as a morphologically simple unicellular zygote. The zygote develops into the embryo and eventually into the mature sporophyte. This development involves division, enlargement, and differentiation of cells, and an organization of cells into more or less specialized complexes, the *tissues* and *systems of tissues*. The embryo of a seed plant (fig. 1.1) has a relatively simple structure as compared with the adult plant. It has a limited number of parts—frequently only an axis bearing one or more cotyledons—and its cells and tissues are mostly at a low level of differentiation. However, the embryo has a potentiality for further growth because of the presence, at two opposite ends of the axis, of meristems (the *apical meristems*) of future shoot and root. During the development of shoot and root, subsequent to seed germination, the appearance of new apical meristems

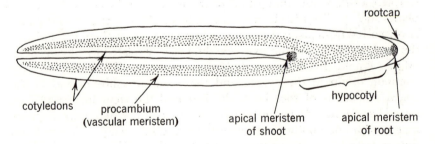

rootcap

cotyledons   procambium   apical meristem   hypocotyl
              (vascular meristem)   of shoot   apical meristem
                                                        of root

FIG. 1.1. Organization of mature embryo of *Lactuca sativa* (lettuce) in longitudinal view. (×34.)

may cause a repetitive branching of these organs. After a certain period of vegetative growth, the plant enters the reproductive stage with the development of spore-bearing structures.

The increments of plant organs originating from the apical meristems usually undergo a period of expansion in width and length. This initial growth of the successively formed roots and vegetative and reproductive shoots is commonly termed *primary growth*. The plant body formed by this growth is the *primary plant body* consisting of *primary tissues*. In most vascular cryptogams and monocotyledons, the entire life cycle of the sporophyte is completed in a primary plant body. The gymnosperms, most dicotyledons, and some monocotyledons show an increase in thickness of stem and root by means of *secondary growth*. This growth may be diffuse in that it involves cells of the ground tissue not localized in a specific region, or it is brought about by a special meristem. The secondary growth of the first type may be called *diffuse secondary growth* (Tomlinson, 1961). It is characteristic of some monocotyledons, such as palms, and of some tuberous structures. The second type is *cambial secondary growth* in that it depends on the production of cells by a cambium. The principal cambium is the *vascular cambium* producing the secondary vascular tissues. The formation of these tissues causes the increase in diameter of the stem and the root. In addition, a *cork cambium*, or *phellogen*, commonly develops in the peripheral region of the expanding axis and produces a *periderm*, a secondary tissue system assuming a protective function when the primary epidermal layer is disrupted during the secondary increase in thickness. The tissues produced by the vascular cambium and the phellogen are more or less clearly delimited from the primary tissues and may be referred to as *secondary tissues* and, in their entirety, as the *secondary body*. The products of the diffuse secondary growth are not readily separable from the primary tissues. Figure 1.2 illustrates in a diagrammatic manner the relation between primary and secondary growth in a dicotyledonous plant.

INTERNAL ORGANIZATION

The morphologic units of the multicellular plant body, the cells, are associated in various ways with each other, forming coherent masses, or tissues. In vascular plants the cells are of many different kinds, and their combinations into tissues are such that different parts of the same organ may vary considerably from one another. The arrangement of cells and tissues is not random. It is possible to recognize larger units of tissues which show topographic continuity, or physiologic similarity,

The Plant Body

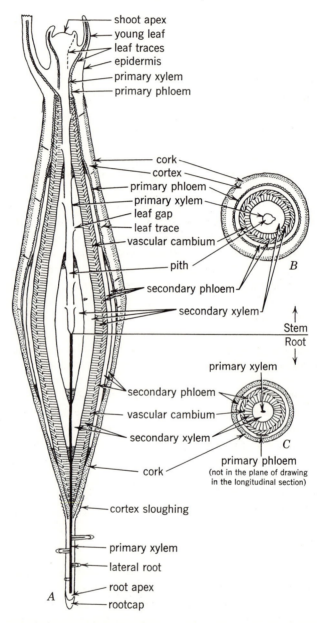

- shoot apex
- young leaf
- leaf traces
- epidermis
- primary xylem
- primary phloem

- cork
- cortex
- primary phloem
- primary xylem
- leaf gap
- leaf trace
- vascular cambium
- pith

B

- secondary phloem
- secondary xylem

Stem
Root

- primary xylem

- secondary phloem
- vascular cambium
- secondary xylem

C

primary phloem
(not in the plane of drawing
in the longitudinal section)

- cortex sloughing

- primary xylem
- lateral root
- root apex
- rootcap

A

FIG. 1.2. Relation between primary and secondary growth in a dicotyledonous plant in diagrams. *A*, longitudinal view of entire plant. *B*, transection of stem. *C*, transection of root. Thickest part of axis has three increments of secondary xylem and phloem. The usual increase in thickness of primary plant body, from the base upward, has been disregarded. (Adapted from Strasburger, *Histologische Beiträge* 3, 1891.)

or both. Such tissue units may be called tissue systems (De Bary, 1884; Foster, 1949; Haberlandt, 1914; Lundegårdh, 1922; Sachs, 1875). Thus the structural complexity of the plant body results from variation in the form and function of cells and also from differences in the manner of combination of cells into tissues and tissue systems.

Despite the long-time concern of botanists with the classification of cells, tissues, and tissue systems, agreement on the subject has not been reached. (For a critical review of the problems of such classification, see Foster, 1949, Exercise IV.) If one attempts to separate cells and tissues into distinct categories, he finds fundamental difficulties. The different kinds of cells intergrade in their characters. Living cells are capable of changing their function and structure. Cells of common origin may differ greatly from each other, and cells derived from different meristems may be basically similar. Tissues also often grade one into the other and overlap in structure and function. Cells of a certain kind may form a coherent tissue, or they may occur in groups or individually among other kinds of cells of contrasting structure and function. No one single criterion, such as structure, origin, or function of cells, or even simple topographic continuity can be applied consistently to express the complex interrelationships of plant-body cells in terms of categories of cells and tissues.

In the following discussion the principal tissues of a vascular plant are reviewed with reference to their arrangement in a dicotyledonous plant (fig. 1.3). According to Sachs' (1875) old but convenient classification based on topographic continuity of tissues, the body of a vascular plant may be pictured as composed of three systems of tissues, the *dermal*, the *vascular*, and the *fundamental* (or *ground-tissue* system). The dermal system forms the outer protective covering of the plant and is represented, in the primary plant body, by the *epidermis*. During secondary growth the epidermis may be replaced by another dermal system, the periderm, with the cork cells forming the new protective tissue. The vascular system is composed of the two principal conducting tissues, the *phloem* and the *xylem*. These tissues contain many types of cells. Some of these are peculiar to the vascular tissues; others have counterparts in the fundamental and dermal systems.

The system of fundamental tissues includes all tissues other than the dermal and the vascular. *Parenchyma* is one of the most common ground tissues. Some of the parenchyma may be modified as a thick-walled supporting tissue, the *collenchyma*. Still other modifications of parenchyma cells are found in the various secretory structures which may occur in the ground-tissue system as individual cells or as more or less extensive cell complexes. The fundamental system often contains highly specialized, mechanical elements combined in coherent masses

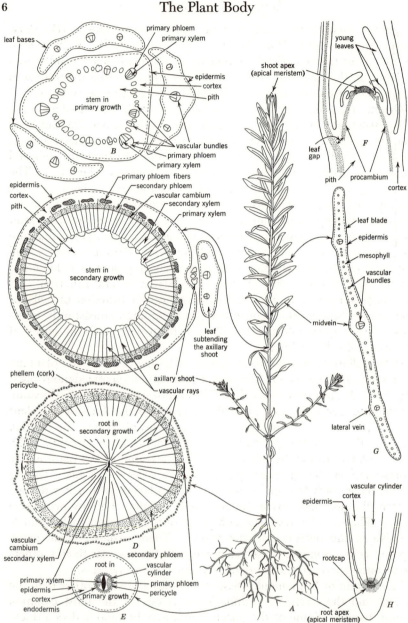

FIG. 1.3. Organization of a vascular plant. *A*, habit sketch of *Linum usitatissimum* L. (flax) in vegetative state. Transections of stem at *B*, *C*, and of root at *D*, *E*. *F*, longitudinal section of terminal part of shoot with apical meristem and developing leaves. *H*, longitudinal section of terminal part of root with apical meristem (covered by rootcap) and subjacent root regions. *G*, transection of leaf blade. (*A*, ×⅓; *B, E, F, H*, ×43; *C*, ×27; *D*, ×6; *G*, ×16. *A*, drawn by R. H. Miller.)

as *sclerenchyma* tissue or dispersed as individual sclerenchyma cells.

The three vegetative organs, stem, root, and leaf, are distinguished by the relative distribution of the vascular and ground tissues (fig. 1.3). The vascular system of the stem frequently occupies a restricted position between the epidermis and the center of the axis. Such an arrangement leaves some ground tissue—the *cortex* (bark or rind in Latin)—between the epidermis and the vascular region, and some—the *pith*—in the center of the stem (fig. 1.3*B*, *C*). In the root, pith may be absent (fig. 1.3*E*), and the cortex is commonly shed during secondary growth (fig. 1.3*D*). The arrangement of primary vascular tissues in the form of a ring of bundles in a transection of a stem (fig. 1.3*B*) is one of several possible patterns in vascular plants. In the secondary state the original structure of the primary vascular system may be obscured by an interpolation of secondary vascular tissues between the primary xylem and the primary phloem (fig. 1.3*C*). In the leaf the vascular system consists of numerous interconnected strands imbedded in the ground tissue, which in this organ is usually differentiated as photosynthetic parenchyma, the *mesophyll* (fig. 1.3*G*).

The three tissue systems of the primary body are derived from the apical meristems (fig. 1.3 *F, H*). When the derivatives of these meristems are partly differentiated, they may be classified into *protoderm, procambium,* and *ground meristem.* These are meristematic precursors of the epidermal, vascular, and fundamental tissue systems, respectively. The vascular tissue system is enlarged secondarily by secondary growth originating in the vascular cambium (fig. 1.3*C, D*). The periderm, if present, is derived from a separate meristem, the phellogen or cork cambium.

SUMMARY OF CELL TYPES AND TISSUES

The familiar types of cells and tissues of a seed plant are summarized here with no attempt to revise existing classifications or to establish a new one. The plant cells derived from a meristem acquire their distinctive characteristics through developmental changes. Some cells undergo more profound changes than others. In other words, cells become specialized to varied degrees. On the one hand, there are the relatively little specialized cells retaining living protoplasts and having the ability to change in form and function (various kinds of parenchyma cells). On the other, there are the highly specialized cells that develop thick rigid walls, become devoid of living protoplasts, and cease to be capable of structural and functional changes (various kinds of sclerenchyma and related cells). Between these two extremes are cells with varying levels of metabolic activity and different degrees of structural

and functional specialization.   The distinctions among cells and tissues
given in the following summary serve to delimit the typical structures,
but the occurrence of intermediate structures must be borne in mind
in evaluating the distinctions.

Epidermis.   Epidermal cells form a continuous layer on the surface
of the plant body in the primary state.   They show various special
characteristics related to their superficial position.   The main mass of
epidermal cells, the epidermal cells proper, vary in shape but are often
tabular.   Other epidermal cells are guard cells of the stomata and various
trichomes, including root hairs.   The epidermis may contain secretory
and sclerenchyma cells.   The principal distinctive feature of the epidermal
cells of the aerial parts of the plant is the presence of a cuticle on the
outer wall and the cutinization of some or all of the other walls.   The
epidermis gives mechanical protection and is concerned with restriction
of transpiration and with aeration.   In stems and roots having secondary
growth, the epidermis is commonly supplanted by the periderm.

Periderm.   The periderm comprises cork tissue or *phellem,* cork
cambium or phellogen, and *phelloderm.*   The phellogen occurs near the
surface of axial organs having secondary growth.   It arises in the epidermis,
the cortex, the phloem, or the root pericycle and produces phellem
toward the outside, phelloderm toward the inside.   Phelloderm may be
absent.   The cork cells are commonly tabular, are compactly arranged,
lack protoplasts at maturity, and have suberized walls.   The phelloderm
usually consists of parenchyma cells.

Parenchyma.   Parenchyma cells form continuous tissues in the cortex
of stem and root and in the leaf mesophyll.   They also occur as vertical
strands and rays in the vascular tissues.   They are primary in origin in
the cortex, the pith, and the leaf; primary or secondary in the vascular
tissues.   Parenchyma cells are characteristically living cells, capable of
growth and division.   The cells vary in shape, are often polyhedral, but
may be stellate or much elongated.   Their walls are commonly primary,
but secondary walls may be present.   Parenchyma is concerned with
photosynthesis, storage of various materials, wound healing, and origin
of adventitious structures.   Parenchyma cells may be specialized as
secretory or excretory structures.

Collenchyma.   Collenchyma cells occur in strands or continuous
cylinders near the surface of the cortex in stems and petioles, and along
the veins of foliage leaves.   Collenchyma is a living tissue closely related
to parenchyma; in fact, it is commonly regarded as a form of parenchyma

specialized as supporting tissue of young organs. The shape of cells varies from short prismatic to much elongated. The most distinctive feature is the presence of unevenly thickened primary walls.

Sclerenchyma. Sclerenchyma cells may form continuous masses, or they may occur in small groups or individually among other cells. They may develop in any or all parts of the plant body, primary and secondary. They are strengthening elements of mature plant parts. Sclerenchyma cells have thick, secondary, often lignified walls and frequently lack protoplasts at maturity. Two forms of cells are distinguished, sclereids and fibers. The sclereids vary in shape from polyhedral to elongated and often are branched. Fibers are generally long, slender cells.

Xylem. Xylem cells form a structurally and functionally complex tissue which, in association with the phloem, is continuous throughout the plant body. This tissue is concerned with conduction of water, storage of food, and support. The xylem may be primary or secondary in origin. The principal water-conducting cells are the tracheids and the vessel members. The vessel members are joined end to end into vessels. Storage occurs in the parenchymatic cells, which are arranged in vertical files and, in the secondary xylem, also in the form of rays. Mechanical cells are fibers and sclereids.

Phloem. Phloem cells form a complex tissue. The phloem tissue occurs throughout the plant body, together with the xylem, and may be primary or secondary in origin. It is concerned with support and with conduction and storage of food. The principal conducting cells are the sieve cells and sieve-tube members, both enucleate at maturity. Sieve-tube members are joined end to end into sieve tubes and are associated with parenchymatic cells, the companion cells. Other phloem parenchyma cells occur in vertical files. Secondary phloem also contains parenchyma in the form of rays. Supporting cells are fibers and sclereids.

Secretory Structures. Secretory cells—cells producing a variety of secretions—do not form clearly delimited tissues but occur within other tissues, primary or secondary, as single cells or as groups or series of cells, and also in more or less definitely organized formations on the surface of the plant. The principal secretory structures on plant surfaces are glandular epidermal cells and hairs and various glands, such as floral and extrafloral nectaries, certain hydathodes, and digestive glands. The glands are usually differentiated into secretory cells on their surfaces and nonsecretory cells supporting the secretory functionally.

Internal secretory structures are secretory cells, intercellular cavities or canals lined with secretory cells (resin ducts, oil ducts), and secretory cavities resulting from disintegration of secretory cells (oil cavities). Laticifers may be placed among the internal secretory structures. They are either single cells (nonarticulated laticifers), usually much branched, or series of cells united through partial dissolution of walls (articulated laticifers). Laticifers contain a fluid called latex, which may be rich in rubber. They are commonly multinucleate.

REFERENCES

Arber, A. *The natural philosophy of plant form.* Cambridge, Cambridge University Press. 1950.

Arnold, C. A. *An introduction to paleobotany.* New York, McGraw-Hill Book Company. 1947.

De Bary, A. *Comparative anatomy of the vegetative organs of the phanerogams and ferns.* Oxford, Clarendon Press. 1884.

Eames, A. J. *Morphology of vascular plants. Lower groups.* New York, McGraw-Hill Book Company. 1936.

Foster, A. S. *Practical plant anatomy.* 2nd ed. New York, D. Van Nostrand Company. 1949.

Foster, A. S., and E. M. Gifford, Jr. *Comparative morphology of vascular plants.* San Francisco, W. H. Freeman and Company. 1959.

Haberlandt, G. *Physiological plant anatomy.* London, Macmillan and Company. 1914.

Lundegårdh, H. *Zelle und Cytoplasma.* In: K. Linsbauer. *Handbuch der Pflanzenanatomie.* Band 1. Lief. 1 and 2. 1922.

Sachs, J. *Textbook of botany.* Oxford, Clarendon Press. 1875.

Tomlinson, P. B. *Anatomy of the monocotyledons. II. Palmae.* Oxford, Clarendon Press. 1961.

Troll, W. *Vergleichende Morphologie der höheren Pflanzen.* Band 1. *Vegetationsorgane.* Heft 1. Berlin, Gebrüder Borntraeger. 1937.

2

The Protoplast

THE CONCEPT OF THE CELL

The study of cells, the units of structure in plants and animals, constitutes the field of science called *cytology* and is treated in detail in various specialized texts and reference works (Brachet and Mirsky, 1959–1961; Guilliermond, 1941; Küster, 1956; Sharp, 1934). The differences in structure and function of cells and the diversities in their groupings bring about the differentiation, within the organismal bodies, of organs and tissues of more or less specialized nature.

The concept that the cell is the universal elementary unit of organic structure and function forms the basis of the so-called *cell theory,* whose formulation is usually connected with the names of Schleiden and Schwann, two German biologists of the early nineteenth century. The fundamental features of this concept are, however, older than the formulation of the cell theory, and many other workers have contributed to the important recognition of cells as units of living organisms (Conklin, 1940).

The word *cell* (from the Latin *cellula,* a small apartment) was introduced by the English microscopist Robert Hooke in the seventeenth century. Hooke first used the term with reference to the small units delimited by walls visible in magnified views of cork tissue. Later he recognized cells in other plant tissues and saw that the cavities of the living cells were filled with "juices" (Conklin, 1940; Matzke, 1943).

With further studies of cells the protoplasm and its inclusions received increasing attention, and the view developed that the protoplasm was the essential part of the cell, the wall not being a necessary component. In plant cells the cell wall appeared to be a secretion of the protoplast,

11

that is, it was dependent on the protoplast for its origin, and in animal cells rigid envelopes did not occur.

The substance within the cell received the name of *protoplasm* (from the Greek words for first and moulded), meaning living matter in its simplest form (Studnička, 1937; Weber, 1936). In 1880 Hanstein introduced the term *protoplast* to designate one unit of this protoplasm contained within one cell and suggested that it be used instead of the term cell; but the term persisted. Since the word cell may be related not only to the Greek *cytos* (kutos), meaning hollow place, but also to the Roman *cella*, which designated receptacles with contents (Matzke, 1943), cell is not an inappropriate designation for the protoplast with its envelopes, at least with reference to plant cells.

The component parts of the protoplast were recognized one by one. In 1831 Robert Brown became aware of the general occurrence of a clear spherical body in each cell and named it the nucleus (in Latin, kernel). In 1846 Hugo von Mohl introduced a distinction between the protoplasm and cell sap, and in 1862 Kölliker applied the name of cytoplasm to the material surrounding the nucleus. Discoveries of other details followed, first with the light microscope (Sharp, 1934), and later with the electron microscope (Mercer, 1960; Sitte, 1961; Whaley et al., 1960).

The following parts are commonly recognized in the protoplast of plant cells (figs. 2.1, 2.2). First, a group of *protoplasmic components: cytoplasm,* the general protoplasmic material enclosing the other protoplasmic bodies and nonprotoplasmic materials and containing various granules and membrane systems; *nucleus,* a body considered to be the center of synthetic and regulatory activity and the seat of hereditary units; *plastids,* bodies concerned with assimilatory metabolism, especially photosynthesis; *mitochondria,* bodies smaller than the plastids and known to be associated with respiratory activities. Second, the *nonprotoplasmic components: vacuoles* (cavities with cell sap) and various more or less solid inclusions, such as crystals, starch grains, and oil droplets. The nonprotoplasmic substances in the cytoplasm and vacuoles constitute nutritive materials or other products of metabolism and are generally classified as *ergastic materials* (from the Greek *erg,* work). The cell wall may be considered as composed of ergastic substances that do not remain in the protoplast but are deposited on its surface.

In classifying the parts of the protoplast one customarily describes the protoplasmic components as living, the nonprotoplasmic as nonliving. To draw a sharp distinction between living and nonliving constituents is obviously impossible, because the property or properties that are the

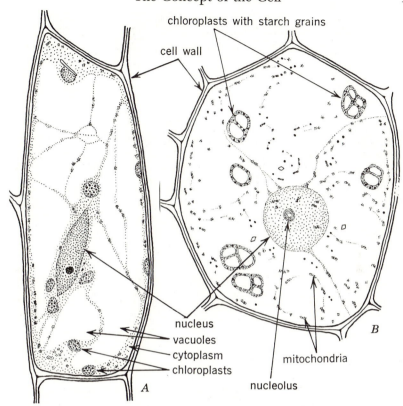

chloroplasts with starch grains

cell wall

nucleus
vacuoles
cytoplasm
chloroplasts

mitochondria

nucleolus

A

B

FIG. 2.1. Components of plant cells. *A,* cell from petiole of sugar-beet leaf. Vacuolated cytoplasm with fine and coarse granules (some of these are mitochondria), nucleus, and granular chloroplasts. *B,* starch-sheath cell from young stem of tobacco. Chloroplasts contain starch grains (left white in drawing). (Both, × 1,190.)

cause of the living state of protoplasm are not known. The individual substances composing the protoplasm, such as proteins, fats, and water, taken separately, are not alive. Yet they are in a sense alive when they are part of the protoplasm. The nonprotoplasmic substances, such as crystals, oil bodies, or starch, are lifeless even when they are imbedded in the protoplasm, but they or their component parts may become incorporated in the living protoplasm through metabolic changes. Nevertheless, it is tenable to designate nonprotoplasmic substances as nonliving when they are not obviously incorporated in the protoplasm or when they appear to be temporarily inactive.

Thus, the cell may be defined as a protoplast with or without a nonliving envelope, the cell wall, and consisting of protoplasmic components

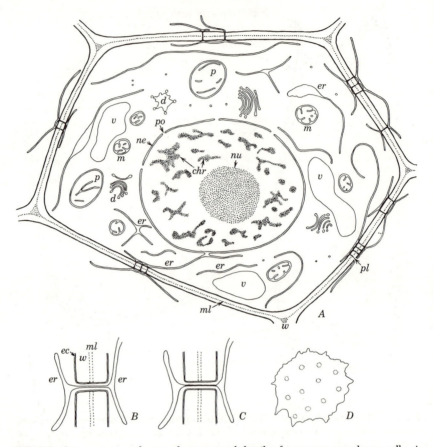

FIG. 2.2. Interpretation of some ultrastructural details of a young parenchyma cell. A, entire cell. B, C, two interpretations of structure of plasmodesmata: tubular connection of endoplasmic reticulum across plasmodesma in B; central connection solid in C. D, surface view of fragment of nuclear envelope with pores. Details: *chr*, chromatin; *d*, dictyosome; *ec*, ectoplast; *er*, endoplasmic reticulum; *m*, mitochondrion; *ml*, middle lamella; *ne*, nuclear envelope; *nu*, nucleolus; *p*, plastid; *pl*, plasmodesma; *po*, pore; *v*, vacuole; *w*, cell wall.

and of nonprotoplasmic materials, the latter intimately connected with the vital activities of the protoplast. For convenience the term cell is applied, in plants, also to the dead remains of a cell, consisting largely of cell wall.

Nuclei as discrete bodies may not be discernible in cells of some lower groups of plants, but in the higher plants nuclei are typically clearly delimited. Some cells contain more than one nucleus. Multinucleate cells are difficult to interpret with reference to the ordinary uninucleate protoplasts. They may constitute entire organisms remaining multinucleate all their lives, as certain fungi and algae. Sometimes, however, the multinucleate state is only a stage in the development of a tissue or an organ, as in the endosperm of certain angiosperms or in embryos of gymnosperms. The multinucleate state may also occur in the development of cells attaining considerable dimensions, such as fibers or laticifers. The view has been advanced that in some multinucleate structures each nucleus and the adjacent cytoplasm represent a cell, and the whole structure is an aggregation of protoplasmic units termed *coenocyte*, after the Greek words *coeno*, in common, and *cyte*, a vessel.

If one disregards the multinucleate protoplasmic masses, the concept of the cell as a unit of structure has a considerable theoretical significance because it enables one to define the structure and morphologic origin of plant tissues and organs. The value of interpreting the cell as a physiologic unit, however, may be questioned. Physiologically, the plant or animal body is not an aggregation of independent units but an organism in which the various parts are interrelated in their growth and activities. This consideration and certain others have resulted in the evolution of the *organismal theory*, which, in contrast to the cell theory, emphasizes the unity of the protoplasmic mass of the organism as a whole, rather than the division of this mass into cells (Sharp, 1934).

PROTOPLASMIC COMPONENTS

The Cytoplasm

As seen with the light microscope the cytoplasm is the visibly least differentiated part of the protoplast enclosing all the other components of it (fig. 2.1A). The electron microscope reveals membranous differentiations in the interior of the cytoplasm, notably the *endoplasmic reticulum* and the *dictyosomes* (fig. 2.2A; pl. 1A, C). Surface membranes delimit the cytoplasm from the wall (*plasma membrane, plasmalemma,* or *ectoplast*) and from the vacuole (*vacuolar membrane*, or *tonoplast*). The cytoplasm also includes granules of various sizes. Granules of 0.25 to 1 micron in diameter, containing lipids and proteins, constitute the *spherosomes* (formerly called microsomes; Perner, 1958). These granules appear to be free in the cytoplasm and are highly mobile in living cells. At the submicroscopic level, a granule about 150Å in diameter, the *ribosome*, attracts particular attention because it appears to be a globular macromolecule of ribonucleoprotein (Setterfield, 1961; Sitte, 1961) that participates in protein synthesis (Watson, 1963). Ribosomes occur free in the cytoplasm or are also associated with the endoplasmic reticulum.

The discovery of the ultrastructural membranous differentiations in the ground substance of the protoplast raises the question of appropriate use of the term cytoplasm. In this book cytoplasm is treated as a mixture composed of a ground substance in which no constant structure has been recognized thus far (*hyaloplasm*, Frey-Wyssling, 1955; Porter, 1961) and of resolvable elements of membranous and granular nature. This treatment is no more than tentative because continued technical improvements may be expected to reveal further resolvable elements in the hyaloplasm and further details in the now resolvable components of the cytoplasm.

Some of the resolvable entities in the protoplast, such as the nucleus, plastids, and mitochondria, are referred to as organelles. With the increase in knowledge of the structure and function of protoplasmic units, more and more of these units are becoming known as organelles. The endoplasmic reticulum and the dictyosomes are called sometimes membrane systems, sometimes organelles.

In living cells the cytoplasm appears as a transparent, semifluid substance. Water constitutes its basic medium and is the most abundant ingredient of the active cytoplasm (85 to 95 per cent of fresh weight; Crafts et al., 1949). Cold injury apparently results from withdrawal of water by ice formation and the consequent alteration of protein structure (Parker, 1963). Various organic and inorganic substances occur in the aqueous medium either in true solution or in a colloidal state. Salts, carbohydrates, and other water-soluble substances are in molecular and ionic dispersion. Other organic compounds, mainly proteins and fatty substances, are in a colloidal state and are also the principal components of the membranous systems present in the cytoplasm.

Studies of chemical and physical properties of the cytoplasm, including those revealed by ultraviolet microscopy and polarization optics (Frey-Wyssling, 1953) suggest the presence of a continuous but labile framework of proteins interpenetrated by the aqueous component of the system. This concept is yet to be integrated with views obtained with the electron microscope. According to one theory (Frey-Wyssling, 1957), the cytoplasm contains elementary units in the form of globular protein macromolecules. These become associated into chains forming fibrillar elements, into membranes forming boundaries and lamellated structures, and into three-dimensional porous complexes. Through interaction with each other the macromolecules play the major role in the gel $\rightleftarrows$ sol transformations characteristic of living cytoplasm. Cytoplasmic streaming is one of the external manifestations of these transformations. How the existence of cytoplasmic streaming is to be reconciled with the presence of membranous systems in the cytoplasm is an open question.

Cytoplasmic Membranes. Among the membranes mentioned above, the two surface films, ectoplast and tonoplast, have long been associated with the important physiological characteristics of the protoplast, differential permeability and capacity for active transport of substances, even against a concentration gradient (Collander, 1959). These films are difficult to recognize with the light microscope, but the electron microscope seems to confirm their morphologic identity (Mercer, 1960). They may appear as single or double lines depending on preparation and degree of resolution. The tonoplast sometimes appears thinner than the ectoplast (Falk and Sitte, 1963).

The endoplasmic reticulum is a system of membrane-bound cavities, or cisternae (Buvat, 1961; Porter, 1961). The cisternae are commonly much flattened so that their sections appear as double lines (fig. 2.2, pl. 1C). Each of these lines may be called a single membrane, the two together a double membrane or paired membranes (Weier and Thomson, 1962). The two membranes enclose an inner phase of unknown composition. The endoplasmic reticulum is thought of as possibly providing the cell with a large internal membrane surface along which enzymes are orderly distributed; and also with a system of compartments segregating the metabolites and, if the system is continuous within the cell, transporting the metabolites from one part of the cell to another.

Dictyosomes (in animal cells, components of the Golgi apparatus) are stacks of flattened sacks or cisternae, approximately circular in outline, each surrounded by vesicles (fig. 2.2A, pl. 1C). The vesicles appear to be originating from the margins of the cisternae and passing into the cytoplasm. Secretory activities are ascribed to dictyosomes, including some related to wall formation (Mollenhauer et al., 1961).

The Nucleus

The nondividing or metabolic nucleus is a spheroidal or ellipsoidal, sometimes more or less lobed, body enclosed within the cytoplasm (figs. 2.1; 2.2A; pl. 1A, B). The nucleus is bounded by a film generally spoken of as the *nuclear membrane,* or *nuclear envelope,* which has the same submicroscopic appearance of a double membrane as the endoplasmic reticulum. Moreover, the two kinds of membranes may be continuous with each other (fig. 2.2, pl. 1A). As the endoplasmic reticulum is also connected with the plasmodesmata, it appears that a continuous membrane system exists between nuclei of neighboring cells. The nuclear envelope has pores through which its contents merge with the surrounding cytoplasm (fig. 2.2A, D; pl. 1A).

The concept of identity of the nuclear envelope with the endoplasmic reticulum is supported by the submicroscopic views of mitosis (pl.

5*A, B*). In late prophase the nuclear envelope breaks up into entities indistinguishable from those of the endoplasmic reticulum. In telophase similar entities coalesce around the chromosomes and form new envelopes around the daughter nuclei. A multiplication of the endoplasmic reticulum apparently occurs between the prophase and the subsequent telophase.

Within the nuclear envelope are the matrix or *karyolymph* (nuclear sap), the reticulum composed of *chromatin,* which during nuclear division becomes aggregated into chromosomes, and the *nucleolus* or *nucleoli* (pl. 1*B*). The electron microscope has revealed no membranous differentiations within the nucleus so that the chromatin, the nucleolus, and the karyolymph are not sharply separated from one another (Sitte, 1961).

Because of the large amount of karyolymph, the nucleus may be characterized as more or less fluid. The proportion of proteins is higher in the cytoplasm than in the nucleus. One of the important chemical distinctions between the nucleus and the cytoplasm is based on the nature and the amounts of *nucleic acids* in the two parts of the protoplast. The deoxyribose nucleic acid (DNA) is characteristic of the nucleus (Mirsky and Osawa, 1961) and is considered to be the carrier of the genetic substance. The relative amount of DNA per nucleus depends on the degree of ploidy in the organism. The ribose nucleic acid (RNA) is more abundant in the cytoplasm than in the nucleus, and within the nucleus it is most characteristic of the nucleolus.

Nuclei vary in size and shape not only in different plants but also in the different tissues of the same plant (Trombetta, 1942). The differences in nuclear size may depend on the number of chromosomes, the volume of individual chromosomes, and the amount of karyolymph. Nuclei also may show diurnal fluctuations in their volume (Bünning and Schöne-Schneiderhöhn, 1957).

The nucleoli (Vincent, 1955) are typical intranuclear bodies. They usually disappear during nuclear division and then, at telophase, arise again from certain chromosomes. Each nucleus in nearly all organisms has at least one pair of chromosomes each member of which gives rise to one nucleolus. The number of nucleoli is as characteristic for a species as the number of chromosomes. As many as ten have been counted in some plants. In a given tissue the number of nucleoli seems variable because soon after the telophase the nucleoli may fuse and form a single large nucleolus before the next mitosis. The nucleoli are viscous and semisolid, denser than the karyolymph. They frequently contain vacuoles and crystal-like bodies. The ultrastructure of the nucleolus has been little investigated.

Plastids

Plastids are clearly delimited protoplasmic bodies of specialized structure and function. Lower plants may lack plastids or may contain one or two in one cell, but in the higher plants each protoplast commonly contains many plastids. The animal cell has no exact counterpart of a plastid.

Plastids are viscous bodies that may show amoeboid changes in shape (fig. 2.3B). Ultrastructurally, the plastids are seen as possessing an outer limiting membrane, usually appearing double, and, with some exceptions, a more or less elaborate system of internal membranes. Although they vary in structure and function, the plastids are interrelated through origin from similar primordial structures in meristems, and one kind of plastid may change into another.

The classification of plastids is based on the presence or absence of pigments in these bodies. Colorless plastids are called *leucoplasts;* pigmented ones, *chromoplasts.* Among the chromoplasts, green plastids termed *chloroplasts* are the most common and the most important physiologically because of their role in photosynthesis. Other chromoplasts carry pigments other than green but have no special names. Some cytologists prefer to use the term chromoplast only with reference to the colored plastids having no chlorophyll and consider chloroplasts as a separate group (Küster, 1956). Such a classification is employed in this book.

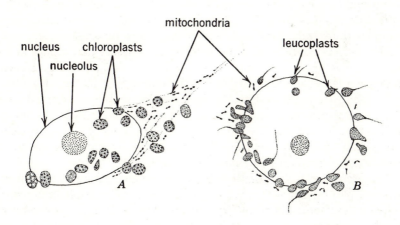

FIG. 2.3. Components of plant cells. *A*, nucleus, granular chloroplasts, and mitochondria from the petiole of sugar-beet leaf. *B*, nucleus, amoeboid leucoplasts, and mitochondria from the pith of a sugar-beet hypocotyl. (Both, ×1,110. Esau, *Jour. Agr. Res.* 69, 1944.)

Chloroplasts. These plastids have been the object of many detailed investigations before and after the development of electron microscopy (Granick, 1961; Menke, 1962). Chloroplasts are most abundant in the principal photosynthetic tissue, the mesophyll of leaves. Thirty to 40 per cent of the total leaf nitrogen may be localized in these chloroplasts. They also occur in other green parts of the plant and even in deep tissues, removed from light, as in parenchyma cells of the vascular tissues or in embryos enclosed within seed coats and fruits.

The chloroplasts of higher plants are usually disc-shaped bodies (pl. 2A), sometimes curved like a saucer. They are relatively constant in shape and size. In many plants the chloroplasts measure between 4 and 6 microns in diameter, although smaller and larger plastids may be found. In photosynthetic cells, they occur in a single layer in the cytoplasm, oriented so that one flat side is turned toward the interior of the cell, the other toward the wall. Under certain environmental conditions they are rounded off; under others they are flattened out. In the flattened state they may, in lining the wall, touch and deform each other and appear angular in outline. In some cells the chloroplasts aggregate near the nucleus (fig. 2.3A).

As seen through the light microscope the chloroplasts appear either granular (fig. 2.3A, pl. 2A) or homogeneous in structure. Studies with the electron microscope have confirmed the existence of chloroplast granules, or *grana* (pls. 2B, C; 3A). A granum is a stack of membrane-bound disc-like compartments or flattened vesicles, also called lamellae. According to some investigators (Weier, 1961), the grana are connected with each other at irregular intervals by a system of membrane-bound channels (intergrana lamellae), which may form an anastomosing fretwork. Others consider that the intergrana lamellae participate in the formation of grana (Wehrmeyer and Perner, 1962). The grana and intergrana lamellae are imbedded in the chloroplast matrix, or stroma, and the entire complex is bounded by a usually double outer membrane. The grana appear to be the principal seat of the chlorophyll. A concept has been advanced that the chlorophyll is associated with units, quantasomes, that have been recognized as granules in orderly arrangement on the surface of granum membranes (Calvin, 1962).

The grana reach their highest degree of differentiation in the chloroplasts of photosynthetic tissues of higher plants. Chloroplasts in tissues more or less removed from light have a less well-developed internal membrane system. Grana vary in structure in different groups of plants (Weier, 1963). Algal grana are plate-like, and those of *Anthoceros* and *Isoetes* form honeycomb structures. Angiosperms usually have cylindrical grana, but chloroplasts having no grana also occur (pl. 3B).

The ontogenetic development of the internal structure of the chloroplast is being related by some investigators to the presence of a so-called primary granum, or plastid center, in the young plastid (Menke, 1962). This center consists of vesicles or tubules which may be arranged in a crystalline lattice. The grana develop from elements of the primary granum. Other investigators find the primary granum only in etiolated tissues. The origin of grana from the invaginating inner layer of the outer membrane has also been described (Menke, 1962).

Chromoplasts. These plastids show varied shapes—elongated, lobed, angled, spheroidal (fig. 2.4)—and are usually yellow or orange. The pigments responsible for these colors belong to the large group of carotenoids (Zscheile, 1941). Chromoplasts with carotenoids may have the following inclusions: crystals of carotenoids (root of *Daucus*, carrot; fruit of *Lycopersicon*, tomato), microscopic and submicroscopic globules (petals of *Ranunculus*), bundles of submicroscopic filaments (fruit of *Capsicum*, pepper). The carotin in carrot chromoplasts appears first as granules but later crystallizes in the form of ribbons, plates, or spirals. Whether the mature crystals have a plastid covering is apparently not known. The development of chromoplasts with globular and fibrous inclusions from chloroplasts involves the destruction of the original grana system (Menke, 1962). Chromoplasts develop from leucoplasts also.

Leucoplasts. Leucoplasts are not a clearly defined group of plastids. They occur in mature cells that are not exposed to light, as, for example, in the pith of many stems or in underground organs. They are not well-differentiated from immature plastids of meristematic cells. The plastids of the epidermis frequently appear nonpigmented and are then classified as leucoplasts.

Leucoplasts are relatively fragile and in fresh preparations break down more readily than chloroplasts. In permanent preparations they are best preserved by the same nonacid fixatives that are used for demonstrating mitochondria. The leucoplasts often appear as small masses of protoplasm, variable and unstable in form. They commonly aggregate near the nucleus (fig. 2.3B).

Leucoplasts form starch in granules of variable sizes. If the leucoplasts are specialized as starch-storing bodies, they are called *amyloplasts*. The *elaioplasts* appear also to be leucoplasts, those concerned with the formation of lipoidal material (Walek-Czernecka and Kwiatkowska, 1961). A developmental study of these bodies in *Iris* (Faull, 1935) has suggested that they are definitely functional plastids, capable of forming starch, in addition to oil. Elaioplasts are particularly common in liverworts and monocotyledons.

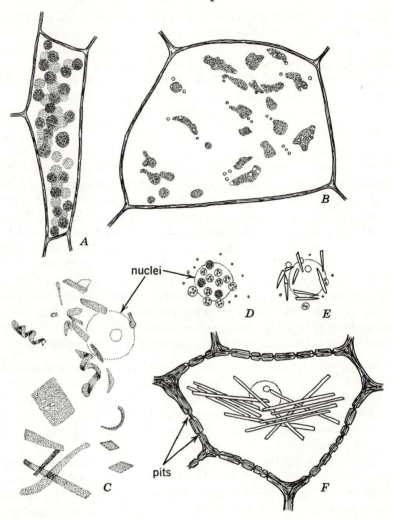

FIG. 2.4. Chromoplasts (*A, B, D*) and related pigment bodies (*C, E, F*). *A*, from petal of *Calendula*. *B*, from fruit of *Pyracantha*. *C*, from root of *Daucus* (carrot). *D, E, F*, from fruit of *Lycopersicon* (tomato). (All, ×880.)

Fats have been described as arising not only in elaioplasts but also directly in the cytoplasm (Sharp, 1934). Frequently, highly refractive granules, showing the same staining reactions as oil, occur in chromoplasts and chloroplasts. These granules are assumed to be lipid granules (Mikulska, 1960).

Origin of Plastids. Plastids are capable of multiplying by division in cells of various ages. These divisions are usually not correlated with mitoses of the nuclei. Meristems have small plastids with little or no internal structure but often with a starch granule. These plastids are regarded as plastid primordia or proplastids (Menke, 1962). If they do not contain starch, their distinction from young mitochondria may be uncertain (pl. 1A).

Mitochondria

Mitochondria are constant elements of protoplasts. They are considered to have genetic continuity and appear to divide (Weier, 1963). *Mitochondria* (singular, *mitochondrion*, from the Greek *mitos*, thread, and *chondrion*, small grain) is one of several names given to these bodies, another common one being *chondriosome* (grain-like body). The whole complement of these structures in an organism is referred to as the *chondriome*.

Through the light microscope mitochondria appear as small granules, rods, or filaments (figs. 2.1B, 2.3; pl. 4C, D). In living material they are commonly identified by means of the Janus-green stain (Hackett, 1955). They are highly sensitive to changes in environment and are easily destroyed by many ordinary cytological fixatives, especially those containing acids. Mitochondria are largely composed of protein and lipid.

At the ultrastructural level mitochondria display a membranous structure. A double membrane encloses an apparently undifferentiated matrix and a number of internal membranes attached to the outer bounding membrane (fig. 2.2A; pl. 4A, B). The internal membranes are derived from the inner layer of the outer membrane and have the form of folds (cristae), or saccules, or tubules. A high degree of differentiation of the internal membrane is characteristic of mitochondria that are highly active metabolically (De Robertis et al., 1960). Mitochondria contain some of the main oxidative enzymes and participate in the Krebs-cycle reactions.

NONPROTOPLASMIC COMPONENTS

Vacuoles

Vacuoles (in Latin *vacuus*, empty) are cavities within the cytoplasm filled with a liquid, the *cell sap*, whose composition may vary in different cells and even in the different vacuoles of the same cell. In sections of living tissue the vacuoles are colorless or pigmented; in well-fixed material they appear as clear areas bounded by the stained cytoplasm. All the vacuoles of a cell or an organism may be regarded as a system and referred to as the *vacuome*.

The principal component of cell sap is water, and in it are various substances either in true solution or in colloidal state (Crafts et al., 1949; Seifriz, 1936; Zirkle, 1937). Salts, sugars, organic acids and other soluble compounds, proteins, and even fatty substances have been identified in plant vacuoles. Tannins are commonly found. Bluish and reddish pigments of the anthocyanin type are also characteristically dissolved in the vacuolar liquid (Blank, 1958; Dangeard, 1956). The materials present in the vacuoles are classified as ergastic. They are either reserve substances that may be utilized by the protoplast for vital activities or they are by-products of metabolism. The vacuolar liquid is more or less viscous, but usually less so than the cytoplasm. The viscosity of the cell sap is probably associated with the presence of colloids, which may sometimes appear in the form of true gels (petals of *Echium vulgare*). Vacuoles containing tanniniferous compounds are often highly viscous.

Much has been learned about the nature of vacuoles by studies of cells in the living state and by staining them with noninjurious vital dyes. With regard to *p*H, two types of vacuoles have been recognized. The relatively alkaline types stain reddish orange with neutral red, and the markedly acid ones assume a bluish-magenta color with the same dye (Zirkle, 1937). The concentration of the vacuolar sap varies, and when a substance accumulates beyond its saturation point it may crystallize out. An increase in concentration may also occur through withdrawal of water, as, for example, in the drying of seeds (Sharp, 1934). Water can be withdrawn artificially from a vacuole by placing living cells into a hypertonic solution. As is well known, such treatment causes cell plasmolysis.

Vacuoles vary in shape and size in relation to the stage of development and the metabolic state of the cell. In meristematic cells vacuoles are often numerous and small. In mature cells commonly one single vacuole occupies the central part of the protoplast, whereas the cytoplasm and the other protoplasmic components are restricted to a parietal position, that is, next to the wall. Some meristematic cells as, for example, those of the vascular cambium have a very extensive vacuolar system. The occurrence of vacuoles is considered to be almost universal in plant cells, including those of meristems (Zirkle, 1937), although meristematic cells often appear to lack vacuoles as seen through the electron microscope (pl. 1A). The small vacuoles of meristematic cells increase in size through the uptake of water and gradually coalesce as the cell enlarges and ages. Thus the enlargement of a plant cell involves both an increase in the amount of its cell sap and an extension of its wall. The protoplasm may also grow in amount (Frey-Wyssling, 1953). Vacuoles are less characteristic of animal cells, and the enlargement of these cells is associated mainly with an increase in the amount of protoplasm.

Opinions vary regarding the origin of vacuoles. According to one hypothesis, certain colloidal products having a strong attraction for water become separated from the cytoplasm and by taking up large amounts of water are converted into vacuolar sap. Ultrastructurally such vacuoles are said to appear as loosened regions within the cytoplasm initially not delimited by a tonoplast (Mühlethaler, 1960). Some workers consider the vacuolar system to be permanent and self-perpetuating (Dangeard, 1956). Another view is that vacuoles arise from enlarging endoplasmic reticulum cisternae or cisternae resembling them (Buvat, 1961).

Ergastic Substances

Ergastic substances are products of metabolism. These substances may appear and disappear at different times in the life of the cell. They are reserve or waste products resulting from cellular activities and are usually simpler in structure than protoplasmic bodies. Some of the well-known ergastic substances are the visible carbohydrates, cellulose and starch; protein bodies; fats and related substances (Eckey, 1954); and mineral matter in the form of crystals. They include also many other organic substances, such as tannins, resins, gums (Howes, 1949), rubber, and alkaloids, whose nature or function or both are known imperfectly (Paech, 1950). Ergastic substances occur in the vacuoles and in the cell wall, and may be associated with the protoplasmic components of the cell.

Carbohydrates. Cellulose and starch are the principal ergastic substances of the protoplast. Cellulose is the chief component of plant cell walls, whereas starch occurs as reserve material in the protoplast itself. Both these carbohydrates are composed of long chain-like molecules, whose basic units are anhydrous glucose residues of the formula $C_6H_{10}O_5$. Both cellulose and starch have an orderly arrangement of molecules and, therefore, show optical anisotropy and double refraction. In starch granules the molecules are radially arranged with the result that in polarized light a cross pattern is seen (pl. 6A).

Glucose residues are associated with water in the two carbohydrates, starch having more of it than cellulose. In the walls of plant cells other materials beside water usually accompany cellulose (chapter 3). In their combination with water and other materials starch and cellulose show colloidal characteristics, such as the ability to imbibe water and swell, well exemplified by the formation of pastes and jellies from starch treated with hot water.

The morphologic variation of starch grains is so extensive that they may be used for the identification of seeds and other starch-containing plant parts (fig. 2.5; Küster, 1956). The following figures (in microns) exemplify their variations in size: 70 to 100 in potato, 30 to 45 in wheat, 12 to 18 in maize. Starch grains of many plants show conspicuous concentric layering because of the alternation of more and less diffractive layers. These layers are successively deposited around a point, the *hilum*, the position of which may be central in some grains, eccentric in others. Compound grains with two or more hila are characteristic of some plants. The layering is not visible in dry starch grains, but swelling in water may reveal it by dislocating the successively deposited layers (Badenhuizen, 1959). The deposition of starch in layers appears to depend particularly on fluctuations in the supply of carbohydrates.

Starch arises almost exclusively in plastids, mainly leucoplasts and chloroplasts. The chloroplasts commonly synthesize *assimilation starch* (Sharp, 1934), a temporary product which remains in the plastid as long

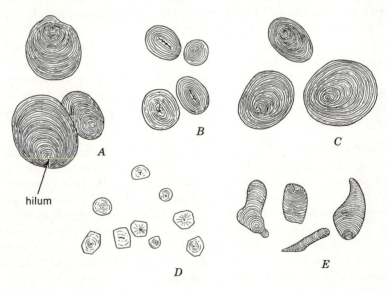

FIG. 2.5. Starch grains from following organs and plants: *A*, root of arrowroot (*Maranta*). *B*, seed of bean (*Phaseolus*). *C*, tuber of potato (*Solanum*). *D*, grain of maize (*Zea*). *E*, fruit of banana (*Musa*). (All, ×285.)

as there is an excess of carbohydrate in the cell. Leucoplasts often produce *storage starch*. One or more starch grains may arise in one plastid (fig. 2.1*B*). The starch grains contained in one plastid may remain discrete, or they may grow together into a compound grain.

Starch deposition occurs widely in the plant body, but the particularly common places of its accumulation are seeds, the parenchyma of the secondary vascular tissues in stems and roots, and the parenchyma of specialized storage organs, such as fleshy roots, tubers, rhizomes, and corms (Radley, 1954).

Proteins. Proteins are the main ingredients of the living protoplasmic bodies, but they also occur as temporarily inactive ergastic substances. Ergastic protein is known as a storage material and is found deposited in amorphous or crystalline form. Amorphous protein forms globules or shapeless masses (in gymnosperm egg cells, algae, fungi). Like starch and cellulose, crystalline protein combines crystalline and colloidal properties, and therefore the individual units of this material are spoken of as *crystalloids* (meaning crystal-like) rather than as crystals (Steffen, 1955).

A well-known amorphous ergastic protein is gluten, which is combined with starch in the endosperm of wheat. In many seeds, the embryo, the endosperm, or the perisperm contain storage protein in the form of *aleuron* grains (aleuron means wheat flour in Greek). These grains may be simple or may contain inclusions of globoids and crystalloids of protein. Cuboidal protein crystalloids occur within parenchyma cells in the peripheral regions of the potato tuber (Hölzl and Bancher, 1958).

The origin of protein inclusions was studied mainly by following the development of aleuron grains (Dangeard, 1956). Some investigators claim that cytoplasm or plastid-like bodies are concerned with the formation of these grains; others report that the ergastic protein first occurs in vacuoles; then, after withdrawal of water from these vacuoles, the remaining contents are transformed into bodies of protein nature. Ultrastructural observations support the concept of vacuolar origin of aleuron grains (Buttrose, 1963).

Fats and Related Substances. Fats and oils are widely distributed in the plant body, and they probably occur in small amounts in every plant cell. The term fat may be used to describe not only the fats proper, that is, esters of fatty acids with glycerol, but also related substances grouped under the name of lipids; and oils may be regarded as liquid fats (Seifriz, 1936). Waxes, suberin, and cutin are fatty in nature and often occur as protective substances in and on the cell wall. Phosphatids and sterols are also related to fats.

As protoplasmic inclusions, fats and oils are common reserve materials in seeds, spores, and embryos, in meristematic cells, and occasionally in differentiated tissues of the vegetative body (Sharp, 1934). They occur as solid bodies or, more frequently, as fluid droplets of various sizes either dispersed in the cytoplasm or aggregated in larger masses. Fatty substances are thought to be elaborated directly by the cytoplasm and also by elaioplasts.

The essential oils, a class of highly volatile and aromatic substances, have a widespread occurrence in plants. In certain plants, as, for example, in the conifers, they occur in all tissues; in others they may develop only in petals (rose), petals and fruit skin (orange), bark and leaves (cinnamon), or fruit (nutmeg).

Tannins. Tannins in the wide sense of the term, are a heterogenous group of phenol derivatives, usually related to glucosides. (In a restricted sense the term tannin refers to a specific category of phenolic compounds of high molecular weight.) The anhydrous derivatives of tannins, the phlobaphenes, are yellow, red, or brown amorphous substances which are very conspicuous in sectioned material. They appear as coarsely or finely granular masses, or as bodies of various sizes. In the following discussion and throughout the book, the term tannin is used in a wide sense, including the phlobaphenes and other tannin derivatives.

Tannins are particularly abundant in the leaves of many plants; in the xylem, the phloem, and the periderm of stems and roots; in unripe fruits; in the testa of seeds; and in pathological growths like galls (Küster, 1956; Sperlich, 1939). No tissue, however, appears to lack tannins entirely, and they may be identified in meristematic cells. Sometimes tannin-containing cells are conspicuously associated with vascular bundles and occur abundantly in areas where the vascular tissue terminates beneath storage tissues or secretory cells of nectaries. The monocotyledons are notably poor in tannins (Sperlich, 1939).

Tannins may be present in individual cells or in special containers termed tannin sacs. The tannin-containing cells often form connected systems. In the individual cells the tannins occur in the protoplast and may also impregnate the walls, as, for example, in cork tissue. Within the protoplast, tannins are a common ingredient of the vacuoles (Esau, 1963), or they may occur in the cytoplasm proper in the form of small droplets which eventually fuse.

With regard to their function, the tannins are interpreted as substances protecting the protoplast against desiccation, decay, or injury by animals; as reserve substances related in some undetermined manner to the starch metabolism; as substances associated with the formation and transport of sugars; as antioxidants; and as protective colloids maintaining the homogeneity of the cytoplasm.

Crystals (Frey-Wyssling, 1935; Netolitzky, 1929; Pobeguin, 1943, 1954). In contrast to animals, which normally eliminate excess inorganic materials to the exterior, plants deposit such materials almost entirely in their tissues. The inorganic deposits in plants consist mostly of calcium salts and of anhydrides of silica. Among the calcium salts the most common is calcium oxalate, which is found in the majority of plant families. Calcium oxalate occurs as mono- and trihydrate salts in many crystalline forms. There are solitary rhombohedrons or octahedrons (prismatic or bipyramidal) (fig. 2.6*C*, pl. 6*B*). The occurrence of so-called crystal sand results from the formation of many small crystals in one cell. Crystals may be united into compound structures, the druses and the sphaerites (fig. 2.6*A*, *B*; pl. 6*D*). Elongated crystals are termed styloids and raphides. The raphides are aggregated into bundles (fig. 2.6*D*, pl. 6*C*). Plants may show constant differences in the form of crystals produced and, therefore, crystals are often of systematic value (Küster, 1956).

Calcium oxalate crystals may be commonly observed in vacuoles. Some workers, however, report that crystals are formed in the cytoplasm (Küster, 1956; Netolitzky, 1929; Scott, 1941). Some oxalate crystals arise in cells that resemble adjacent, crystal-free cells. Others are formed in specialized cells, the crystal *idioblasts* (that is, cells markedly differing from other constituents of the same tissue in form, structure, or contents; from the Greek *idio*, peculiar). Still other crystals are deposited in the cell walls. Crystals may be much smaller than the cells containing them, or they may completely fill and even deform the cells. Raphides often occur in remarkably large cells (pl. 71*B*) which, at maturity, are dead structures filled with mucilage capable of swelling. Parts of the cell wall of such raphid idioblasts remain thin, and, if the

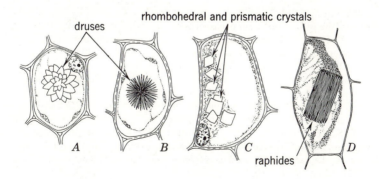

FIG. 2.6. Cells with different types of crystals. *A*, *B*, druses from cortex of *Gnetum gnemon*. *C*, prismatic and rhombohedral crystals from cortex of *Gnetum indicum*. *D*, raphides from leaf of *Vitis vinifera*. (*A–C*, ×800; *D*, ×625.)

mucilage swells, the thin wall bursts and the raphides are ejected (Cheavin, 1938). Calcium oxalate crystals may be deposited uniformly throughout the tissue, or they may be more or less restricted to certain tissue regions (for example, in cells surrounding the fiber strands in the secondary phloem of *Robinia* or in cells along the margins of the phloem rays in *Vitis*).

Calcium carbonate rarely occurs in well-formed crystals. The best-known calcium carbonate formations are the *cystoliths* (from the Greek *kustis*, a bag, and *lithos*, a stone), which are outgrowths of the cell wall impregnated with the mineral (Pireyre, 1961). They occur in ground parenchyma and in the epidermis. In the epidermis they may be formed in hairs or in special enlarged cells, the lithocysts (chapter 7).

Silica is deposited mostly in cell walls, but sometimes it forms bodies in the lumen of the cell. The Gramineae is the best-known example of a plant group having silica in both the walls and the cell lumina (Küster, 1956; Netolitzky, 1929). As discrete bodies, silica usually occurs as opal, that is, in amorphous form (Lanning et al., 1958).

REFERENCES

Badenhuizen, N. P. Chemistry and biology of the starch granule. *Protoplasmatologia* 2 B2. 1959.

Blank, F. Anthocyanins, flavones, xanthones. *Handb. der Pflanzenphysiol.* 10: 300–353. 1958.

Brachet, J., and A. E. Mirsky, eds. *The cell. Biochemistry, physiology, morphology.* 5 vols. New York, Academic Press. 1959–1961.

Bünning, E., and G. Schöne-Schneiderhöhn. Die Bedeutung der Zellkerne im Mechanismus der endogenen Tagesrhythmik. *Planta* 48:459–467. 1957.

Buttrose, M. S. Ultrastructure of the developing aleurone cells of wheat grain. *Austal. Jour. Biol. Sci.* 16:768–774. 1963.

Buvat, R. Le reticulum endoplasmique des cellules végétales. *Deut. Bot. Gesell. Ber.* 74:261–267. 1961.

Calvin, M. The path of carbon in photosynthesis. *Science* 135:879–889. 1962.

Cheavin, W. H. S. The crystals and cystoliths found in plant cells. Part I. Crystals. *Microscope, Brit. Jour. Micros. and Photomicrogr.* 2:155–158. 1938.

Collander, R. Cell membranes: Their resistance to penetration and their capacity for transport. In: F. C. Steward. *Plant physiology.* Vol. 2: *Plants in relation to water and solutes.* New York, Academic Press. 1959.

Conklin, E. G. Cell and protoplasm concepts: historical account. In: *Cell and protoplasm.* *Amer. Assoc. Adv. Sci. Publ.* 14:6–19. 1940.

Crafts, A. S., H. B. Currier, and C. R. Stocking. *Water in the physiology of plants.* Waltham, Mass., Chronica Botanica Company. 1949.

Dangeard, P. Le vacuome de la cellule végétale; morphologie. *Protoplasmatologia* 3 D1. 1956.

De Robertis, E. D. P., W. W. Nowinski, and F. A. Saez. *General cytology.* Philadelphia, Saunders Company. 1960.

Eckey, E. W. *Vegetable fats and oils.* ACS Monograph Series. New York, Reinhold. 1954.

Esau, K. Ultrastructure of differentiated cells in higher plants. *Amer. Jour. Bot.* 50:495–506. 1963.

Falk, H., and P. Sitte. Zellfeinbau bei Plasmolyse. I. Der Feinbau der *Elodea*-Blattzellen. *Protoplasma* 57:290–303. 1963.

Faull, A. F. Elaioplasts in *Iris:* a morphological study. *Arnold Arboretum Jour.* 16:225–267. 1935.

Frey-Wyssling, A. *Die Stoffausscheidung der höheren Pflanzen. Monographien aus dem Gesamtgebiet der Physiologie der Pflanzen und der Tiere.* Band 32. Berlin, Julius Springer. 1935.

Frey-Wyssling, A. *Submicroscopic morphology of protoplasm.* 2nd ed. Amsterdam, Elsevier Publishing Company. 1953.

Frey-Wyssling, A. Die submikroskopische Struktur des Cytoplasmas. *Protoplasmatologia* 2 A2. 1955.

Frey-Wyssling, A. *Macromolecules in cell structure.* Cambridge, Mass., Harvard University Press. 1957.

Granick, S. The chloroplasts: inheritance, structure, and function. In: J. Brachet and A. E. Mirsky. *The cell.* Vol. 2. New York, Academic Press. 1961.

Guilliermond, A. *The cytoplasm of the plant cell.* Waltham, Mass., Chronica Botanica Company. 1941.

Hackett, D. P. Recent studies on plant mitochondria. *Internatl. Rev. Cytol.* 4:143–196. 1955.

Hölzl, J., and E. Bancher. Über die Eiweisskristalle von *Solanum tuberosum. Österr. Bot. Ztschr.* 105:385–407. 1958.

Howes, F. N. *Vegetable gums and resins.* Waltham, Mass., Chronica Botanica Company. 1949.

Küster, E. *Die Pflanzenzelle.* 3rd ed. Jena, Gustav Fischer. 1956.

Lanning, F. C., B. W. X. Ponnaiya, and C. F. Crumpton. The chemical nature of silica in plants. *Plant Physiol.* 33:339–343. 1958.

Matzke, E. B. The concept of cells held by Hooke and Grew. *Science* 98:13–14. 1943.

Menke, W. Structure and chemistry of plastids. *Ann. Rev. Plant Physiol.* 13:27–44. 1962.

Mercer, F. The submicroscopic structure of the cell. *Ann. Rev. Plant Physiol.* 11:1–24. 1960.

Mikulska, E. Inkluzje tluszczowe w chloroplastach. [Inclusions lipidiques dans les chloroplastes.] *Soc. Bot. Polon. Acta* 29:431–455. 1960.

Mirsky, A. E., and S. Osawa. The interphase nucleus. In: J. Brachet and A. E. Mirsky. *The cell.* Vol. 2. New York, Academic Press. 1961.

Mollenhauer, H. H., W. G. Whaley, and J. H. Leech. A function of the Golgi apparatus in outer rootcap cells. *Jour. Ultrastruct. Res.* 5:193–200. 1961.

Mühlethaler, K. Die Entstehung des Vacuolensystems in Pflanzenzellen. *Internatl. Conf. Electron Micros. Verhandl.* 4:491–494. 1960.

Netolitzky, F. Die Kieselkörper. Die Kalksalze als Zellinhaltskörper. In: K. Linsbauer. *Handbuch der Pflanzenanatomie.* Band 3. Lief. 25. 1929.

Parker, J. Cold resistance in woody plants. *Bot. Rev.* 29:124–201. 1963.

Paech, K. *Biochemie und Physiologie der sekundären Pflanzenstoffe. Lehrbuch der Pflanzenphysiologie.* Band 1. Teil 2. Berlin, Springer-Verlag. 1950.

Perner, E. S. Die Sphärosomen der Pflanzenzelle. *Protoplasmatologia* 3 A2. 1958.

Pireyre, N. Contribution à l'étude morphologique, histologique et physiologique des cystoliths. *Rev. Cytol. et Biol. Vég.* 23:93–320. 1961.

Pobeguin, T. Les oxalates de calcium chez quelques Angiospermes. *Ann. des Sci. Nat.*, *Bot.* Ser. 11. 4:1–95. 1943.

Pobeguin, T. Contribution a l'étude des carbonates de calcium, précipitation du calcaire par les végétaux, comparison avec le monde animal. *Ann. des Sci. Nat.*, *Bot.* Ser. 11. 15:29–109. 1954.

Porter, K. R. The ground substance, observations from electron microscopy. In: J. Brachet and A. E. Mirsky. *The cell*. Vol. 2. New York, Academic Press. 1961.

Radley, J. A. *Starch and its derivatives*. Vol. 1. 3rd ed. New York, John Wiley and Sons. 1954.

Scott, F. M. Distribution of calcium oxalate crystals in *Ricinus communis* in relation to tissue differentiation and presence of other ergastic substances. *Bot. Gaz.* 103:225–246. 1941.

Seifriz, W. *Protoplasm*. New York, McGraw-Hill Book Company. 1936.

Setterfield, G. Structure and composition of plant-cell organelles in relation to growth and development. *Canad. Jour. Bot.* 39:469–489. 1961.

Sharp, L. W. *Introduction to cytology*. 3rd ed. New York, McGraw-Hill Book Company. 1934.

Sitte, P. Die submikroskopische Organization der Pflanzenzelle. *Deut. Bot. Gesell. Ber.* 74:177–206. 1961.

Sperlich, A. Das trophische Parenchym. B. Exkretionsgewebe. In: K. Linsbauer. *Handbuch der Pflanzenanatomie*. Band 4. Lief. 38. 1939.

Steffen, K. Einschlüsse. *Handb. der Pflanzenphysiol.* 1:401–412. 1955.

Studnička, F. K. Noch einiges über das Wort Protoplasma. *Protoplasma* 27:619–625. 1937.

Trombetta, V. V. The cytonuclear ratio. *Bot. Rev.* 8:317–336. 1942.

Vincent, W. S. Structure and chemistry of nucleoli. *Internatl. Rev. Cytol.* 4:269–298. 1955.

Walek-Czernecka, A., and M. Kwiatkowska. Elajoplasty ślazowatych. [Elaioplasts in the Malvaceae.] *Soc. Bot. Polon. Acta* 30:345-365. 1961.

Watson, J. D. Involvement of RNA in the synthesis of proteins. *Science* 140:17–26. 1963.

Weber, F. Das Wort Protoplasma. *Protoplasma* 26:109–112. 1936.

Wehrmeyer, W., and E. Perner. Der submikroskopische Bau der Grana in den Chloroplasten von *Spinacia oleracea* L. *Protoplasma* 54:573–593. 1962.

Weier, T. E. The ultramicro structure of starch-free chloroplasts of fully expanded leaves of *Nicotiana rustica*. *Amer. Jour. Bot.* 48:615–630. 1961.

Weier, T. E. Changes in the fine structure of chloroplasts and mitochondria during phylogenetic and ontogenetic development. *Amer. Jour. Bot.* 50:604–611. 1963.

Weier, T. E., and W. W. Thomson. Membranes of mesophyll cells of *Nicotiana rustica* and *Phaseolus vulgaris* with particular reference to the chloroplasts. *Amer. Jour. Bot.* 49:807–820. 1962.

Whaley, W. G., H. H. Mollenhauer, and J. H. Leech. The ultrastructure of the meristematic cell. *Amer. Jour. Bot.* 47:401–449. 1960.

Zirkle, C. The plant vacuole. *Bot. Rev.* 3:1–30. 1937.

Zscheile, F. P. Plastid pigments. *Bot. Rev.* 7:587–648. 1941.

3

The Cell Wall

The presence of nonprotoplasmic walls is considered to be the outstanding characteristic distinguishing the cells of plants from those of animals. Few plant cells lack walls, and few animal cells (these belong to the lower organisms) have nonprotoplasmic envelopes comparable to the walls of plant cells. Examples of cells without walls in plants are the motile spores in algae and fungi and the sexual cells in lower and higher plants. The sexual reproductive cells of higher plants, however, remain imbedded, throughout their existence, within the cytoplasm of other cells and some have walls of unknown composition.

The cell wall may be characterized as a nonprotoplasmic component of the protoplast because after it is formed it is removed from metabolic activities (Frey-Wyssling, 1959). In mature living cells, however, cytoplasm is present in the wall in the form of plasmodesmata. Whether during growth of the cell the relation between cytoplasm and wall is closer than at maturity is still a question (Newcomb, 1963; Wardrop, 1962). Some investigators think that the cytoplasm penetrates the growing wall, but electron-microscope views of meristematic cells indicate the presence of ectoplast delimiting the cytoplasm from the cell wall.

The cell wall largely determines the shape of the cell and the texture of the tissue (Roelofsen, 1959). Cell walls have supportive and protective functions, both as components of living cells and as remainders of no longer living cells. They assist the aerial parts of land plants in withstanding the stress of the force of gravity and protect them from desiccation. They play an important role in such activities as absorption, transpiration, translocation, and secretion (Frey-Wyssling, 1959).

The cell wall was discovered before the protoplast, and in the early history of botany it received more attention than the cell contents. Later the protoplast became the main object of study. In the present century research on cell walls has received a new impetus through the discovery of manifold industrial uses of cellulose and its derivatives and through the development of new and improved techniques in cell-wall investigations. The microchemical tests of wall materials have been further refined, and the use of polarized light, of X-rays, and of the electron microscope has become common practice in cell-wall research (Frey-Wyssling, 1959; Ott et al., 1954–1955; Roelofsen, 1959).

The term *cell wall* is now commonly used in the botanical literature written in English, but in some of the older publications in this language and in most of the German literature the term *cell membrane* is employed synonymously with cell wall.

GROSS MICROSCOPIC STRUCTURE

Classification of Wall Layers

The interpretation of the plant cell as consisting of the protoplast and the cell wall agrees with the common observation that each cell within a tissue has its own wall. The dual nature of the partitions between the contiguous protoplasts is not necessarily visible, but appropriate microchemical tests and maceration techniques reveal a noncellulosic, amorphous material between the walls of contiguous cells (Kerr and Bailey, 1934). This intercellular substance may be stained differentially or dissolved out. In the latter instance the tissue becomes macerated and falls apart into the individual cells.

The cell walls of plants vary much in thickness in relation to age and type of cell (figs. 3.1, 3.2; pl. 7). Generally, young cells have thinner walls than fully developed ones, but in some cells the wall does not thicken much after the cell ceases to grow. Whether thin or thick, the walls are of complex structure and often permit the recognition of layers variable in chemistry and structure. On the basis of development and structure, three parts are commonly recognized in plant cell walls: the intercellular substance or middle lamella, the primary wall, and the secondary wall (fig. 3.1*A*, *B*; Bailey, 1954; Wardrop, 1962). The intercellular substance occurs between the primary walls of two contiguous cells, and the secondary wall is laid over the primary, that is, next to the *lumen* (the central cavity; from the Latin, meaning light or opening) of the cell.

The *middle lamella* is amorphous and optically inactive (isotropic, pl. 7*B*). It is composed mainly of a pectic compound possibly combined with calcium (Frey-Wyssling, 1959). In woody tissues the middle

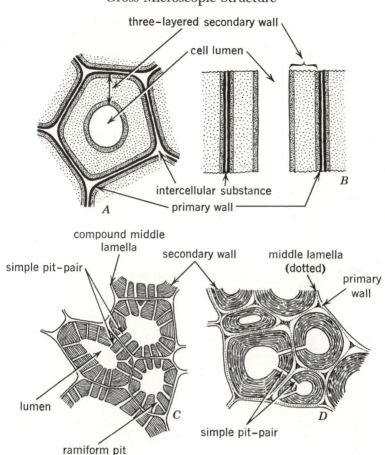

FIG. 3.1. Secondary cell walls. Common type of wall structure in cells with secondary wall layers in transverse (*A*) and longitudinal (*B*) sections. Wall layers are classified according to the concept of Kerr and Bailey (*Arnold Arboretum Jour.* 15, 1934). *C, D,* cells with secondary walls and simple pits. *C,* sclereids from a transection of *Cydonia* (quince) fruit. *D,* phloem fibers from a transection of *Nicotiana* (tobacco) stem. (*C, D,* ×560.)

lamella is commonly lignified.   Since the intercellular substance is difficult to identify microscopically, the term middle lamella has been used in the literature without much consistency.   The distinction between the intercellular lamella and the primary wall is frequently obscured during the extension growth of the cell.   In such cells as tracheids and fibers, which typically develop prominent secondary walls, the intercellular layer becomes extremely tenuous.   As a result, the two primary walls of contiguous cells and the intervening middle lamella appear as a unit, particularly when all three become strongly impregnated with lignin (fig. 3.2A). This triple structure has often been called middle lamella. The matter is still more complicated when the first layer of the secondary wall cannot be distinguished by ordinary microscopy from the primary wall, because then middle lamella, if the term is employed loosely, might refer to this compound structure consisting of five layers. The term *compound middle lamella* may be substituted if the intercellular substance is obscured, but this term would mean sometimes the three-ply, sometimes the five-ply, structure described above (Kerr and Bailey, 1934).

The *primary wall* is the first wall proper formed in a developing cell and is the only wall in many types of cells. It contains cellulose, hemicelluloses, and some pectin (Wardrop, 1962). It may become lignified.

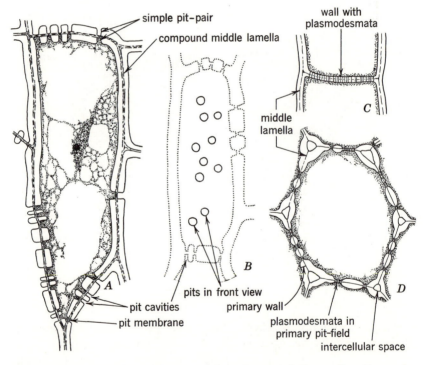

FIG. 3.2. Primary pit-fields, simple pits, and plasmodesmata. A, B, ray cells, with secondary walls (left white in drawing), from a radial section of apple wood. Simple pits and pit-pairs in sectional and face views. C, D, parenchyma cells without secondary walls from tobacco stem. Plasmodesmata dispersed throughout the wall in C and restricted to primary pit-fields in D. (A, B, ×865; C, ×420; D, ×325; C, D, adapted from Livingston, *Amer. Jour. Bot.* 22, 1935.)

Because of the presence of cellulose the primary wall is optically anisotropic (pl. 6A). Since the primary wall is initiated before the cell enlarges, it passes through a period of growth in surface area, which may
be succeeded or temporarily interrupted by a period or periods of
growth in thickness; or, possibly, the two types of growth may be
combined. Thus the primary wall may have a complex history as well
as a complex structure. If the wall is thick, it often shows conspicuous
lamination, indicating that growth in thickness has occurred by a successive deposition of layers.

Primary walls are usually associated with living protoplasts. The walls
of dividing and growing meristematic cells are primary, and so are those
of most of the cells which retain living protoplasts during the height of
their physiologic maturity. The changes that occur in primary walls
are therefore reversible. The wall may lose a thickening previously
acquired, and chemical substances may be removed or replaced by others.
Cambial walls, for example, show seasonal changes in thickness and the
thick primary walls of the endosperm in certain seeds are digested
during germination.

As its name implies, the *secondary wall* follows the primary in the
order of appearance. It consists mainly of cellulose or of varying
mixtures of cellulose and hemicelluloses. It may be modified through
deposition of lignin and various other substances. Because of its high
content of cellulose the secondary wall is strongly anisotropic (pl. 7B).
The complexity and the lack of homogeneity in structure are pronounced
in the secondary wall. The secondary walls of tracheary cells and
fibers are commonly three-layered (fig. 3.1A, B; pl. 7B), and there are
physical and chemical differences between the layers. The number of
layers may be less than three or may exceed three, and the innermost
layer sometimes consists only of a helical band.

Generally, secondary walls are laid down after the primary wall
ceases to increase in surface area. At this time, the entire cell or, in
elongating fiber cells, part of the cell (chapter 10) cease to enlarge so
that surface growth is not characteristic of the secondary wall. There
is some evidence, however, that the initial layer of secondary wall
becomes slightly extended because its deposition begins somewhat
before the increase in surface of the wall ceases (Roelofsen, 1959).

The secondary wall may be considered a supplementary wall whose
principal function is mechanical. Often the cells with secondary walls
are devoid of protoplasts at maturity (as certain fibers, tracheids, vessel
elements). Secondary walls are, in other words, most characteristic of
cells that are highly specialized and undergo irreversible changes in
their development (Bailey, 1954). But cells with active, living proto-

plasts, such as the xylem ray and xylem parenchyma cells, also may have secondary walls. Moreover, cells specialized as mechanical (sclerenchyma) elements may long retain their protoplasts, and cell division is known to occur in the presence of secondary walls (Bailey, 1961). Information is meager on the ability of protoplasts to reduce the thickness of the secondary wall or to modify its chemistry, after the cell completes its development. Delignification and dissolution of secondary walls under normal and pathological conditions have been reported in the literature (Bloch, 1941; Roelofsen, 1959).

The classification into primary and secondary walls was formulated by Kerr and Bailey (1934) and is used widely (Roelofsen, 1959; Wardrop, 1962) but not consistently. Not infrequently the later part of the primary wall, especially if the wall is conspicuously thickened, is called secondary, and the innermost layer of the secondary wall is referred to as tertiary (critique in Bailey, 1957b).

Pits

Secondary cell walls are commonly characterized by the presence of depressions or cavities varying in depth, expanse, and detailed structure. Such cavities are termed *pits*. Primary walls also have more or less conspicuous depressions. These differ from the pits in secondary walls in structure and development, and therefore the pits in the secondary walls and the depressions in the primary walls have received different designations (Wardrop, 1962): the secondary walls have *pits*, whereas the primary walls have *primary pit-fields* (Committee on Nomenclature, 1957). Thus, according to this terminology, the meristematic cells and those of their derivatives that form no secondary walls have primary pit-fields (fig. 3.2D, pl. 13B); cells with secondary walls have pits (fig. 3.2A, B).

The primary pit-fields of a meristematic cell may be so deeply depressed and so numerous that the wall appears beaded in sectional views. During the differentiation of some cells having only primary walls the primary pit-fields may be but slightly modified; in other, more specialized cells, the primary pit-fields may be considerably changed as the cell matures. In primary pit-fields the primary wall is relatively thin but continuous across the pit-field area. Furthermore, while the cell is alive, the primary pit-fields show concentrations of plasmodesmata (fig. 3.2D).

The distinguishing feature of a pit is that the secondary wall layers are completely interrupted at the pit; that is, the primary wall is not covered by secondary layers in the pit region (fig. 3.2A). Pits may be formed over primary pit-fields, one or more pit over one field. Such primary pit-fields may remain in evidence after the development of the secondary wall, or they may be obsured when, during the extension growth of the cell, the primary wall is reduced in thickness (Kerr and Bailey, 1934). Pits also arise over primary wall parts that bear no primary pit-fields, and, inversely, some primary pit-fields are completely covered by secondary wall layers. Thus there is no absolute inter-dependence between the position of the primary pit-fields in the primary wall and the development of pits in the secondary wall.

The distinction between pits and primary pit-fields has a sound morphological basis, but frequently primary and secondary walls cannot be distinguished with ordinary microscopic observation. If an uncertainty exists regarding the exact nature of a wall, neither term, pit or primary pit-field, may be applied without classifying the wall by implication. A substitute term that would include pits and primary pit-fields, however, is not available in the literature. In this book the distinction between pits and primary pit-fields is maintained whenever the nature of the wall is known. If this information is not available but the wall bears clearly circumscribed cavities, these cavities are termed pits. The adjective *pitted* is applied either to secondary walls having pits or to primary walls having primary pit-fields.

It is customary to include in the definition of the pit in a secondary wall not only the cavity but also the part of the primary wall that occurs at the bottom of the cavity (Committee on Nomenclature, 1957). Thus, fundamentally, a pit consists of a *pit cavity* and a *pit membrane*. The pit cavity is open internally to the lumen of the cell and is closed by the pit membrane along the line of junction of two cells (figs. 3.1*C, D*; 3.2*A*).

Two principal types of pits are recognized in cells with secondary walls: *simple pits* and *bordered pits*. The most fundamental difference between the two kinds of pits is that in the bordered pit the secondary wall arches over the pit cavity—this part of the wall constitutes the border—and narrows down its opening to the lumen of the cell (fig. 3.3, pl. 11*A–C*); in the simple pit no such overarching occurs (figs. 3.1*C, D*; 3.2*A*).

A pit in a wall of a given cell usually occurs opposite a complementary pit in the wall of an adjacent cell; that is, two pits are combined into a paired structure, the *pit-pair* (figs. 3.2*A*, 3.3*A*). The pit membrane is common to both pits of a pair and consists of two primary walls and a lamella of intercellular substance (fig. 3.3). Two bordered pits make up a *bordered pit-pair*, two simple pits, a *simple pit-pair*. A bordered pit may be complemented by a simple pit, the two constituting a *half-bordered pit-pair* (pl. 9*A, B*). A pit may have no complementary

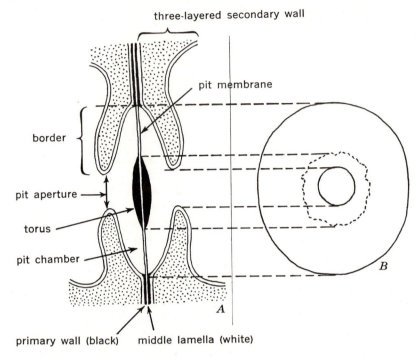

FIG. 3.3. Bordered pit-pair of *Pinus* in sectional (*A*) and face (*B*) views. Details according to the concept of Kerr and Bailey (*Arnold Arboretum Jour.* 15, 1934). The pit membrane consists of two primary walls and the intercellular lamella but is thinner than the same triple structure in the unpitted part of the wall. The torus is formed by thickening of the primary wall. In *B*, outline of torus is uneven.

structure, as, for example, when it occurs opposite an intercellular space. Such a structure is called a *blind pit*. Sometimes two or more small pits are opposed by one pit in the adjacent cell, a combination that has been named *unilaterally compound pitting*.

Simple pits may be found in certain parenchyma cells (fig. 3.2*A*, *B*; pl. 8*B*, *C*), in extraxylary fibers (fig. 3.1*D*), and in sclereids (fig. 3.1*C*). In a simple pit the cavity may be uniform in width, or it may be slightly wider or slightly narrower toward the lumen of the cell. If it narrows down toward the lumen, the simple pit intergrades with the bordered pit in its structure. The simple pits of thin walls are shallow. In thick walls the cavity of a simple pit may have the form of a canal passing from the lumen of the cell toward the pit membrane (fig. 3.1*C*, *D*). Pits may coalesce as the wall increases in thickness and give the impression of a branched canal (fig. 3.1*C*). Such pits are called *ramiform pits* (that is, pits shaped like branches, from the Latin *ramus*, branch).

Bordered pits are more complex and more variable in structure than simple pits. They occur mainly in the water-conducting and mechanical cells of the xylem, such as vessel elements, tracheids, and various fibers, but may be found in some fibers and sclereids outside the xylem also.

The part of the cavity enclosed by the overarching secondary wall, the *pit border*, is called the *pit chamber*, and the opening in the border is the *pit aperture* (fig. 3.3). The pit aperture may be circular, lenticular, or linear (figs. 3.3–3.5). The shape of the aperture may agree with the outline of the pit chamber, or it may not. Vessel elements in the angiosperms often have oval bordered pits with oval apertures (fig. 3.5B). Some tracheary cells of ferns have transversely much-elongated bordered pits with linear apertures. In the bordered pits of gymnosperms, circular, oval, or linear apertures may be associated with pit chambers and borders circular in outline (figs. 3.3, 3.4).

If the secondary wall and the border are relatively thick, the border divides the cavity into the pit chamber, the space between the pit membrane and the overarching border, and the *pit canal*, the passage from the cell lumen into the pit chamber (fig. 3.4). Such a canal has an *outer aperture* opening into the pit chamber and an *inner aperture* facing the cell lumen. The two apertures are commonly unlike in shape and size: the inner is rather large, lenticular or linear, the outer small and circular. The thicker the cell wall, the smaller and thicker is the border, the smaller the pit chamber, and the longer and narrower the inner pit aperture. With the increase in wall thickness, the inner aperture may become so long that it may reach laterally the limits of

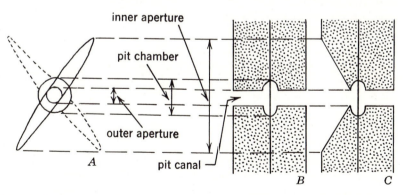

FIG. 3.4. Bordered pit-pair with extended pit apertures, flattened canals, reduced borders, and small pit cavities. *A*, face view showing: extension of pit apertures beyond the limits of pit chamber (or border); crossed arrangement of apertures of the two pits of the pair; and contrast in shape and size of inner and outer apertures. *B*, sectional view exposing pit canal along its short diameter. *C*, sectional view exposing pit canal along its long diameter.

the pit chamber and even surpass these (fig. 3.4). When the inner aperture does not extend beyond the border it is called *included*; when the long diameter of the aperture is longer than the diameter of the border the aperture is called *extended*. If the inner aperture is relatively large and linear or lenticular in outline and the outer is small and circular, the pit canal has the shape of a flattened funnel. The circular pit apertures in a bordered pit-pair appear exactly opposite each other. In a bordered pit-pair with elongated inner pit apertures the apertures may cross each other symmetrically (fig. 3.4A).

The bordered pit-pairs of conifer tracheids are particularly elaborate in their structural details (fig. 3.3; pl. 11A–C, 12A). In the large, relatively thin-walled tracheids of the early wood, these pit-pairs, as seen in face views, have large, circular or oval borders with conspicuous lenticular or circular apertures. The pit chambers are also correspondingly large, with the pit canals practically absent. The pit membrane has a thickening of primary nature, the *torus*, which is somewhat larger in diameter than the pit apertures. The thin part of the membrane surrounding the torus is called *margo* (meaning the edge, or margin; Frey-Wyssling, 1959). The pit membrane is flexible, and under certain conditions the torus occurs in a lateral position, appressed to one or the other pit aperture of the pit-pair (*aspirated* pit-pair; pl. 11C). The movements of the pit membranes and the changes in the position of the torus are reportedly influenced by pressure relations within the tracheids. Aspiration of pits that occurs in connection with heartwood formation is thought to be associated with the drying out of the central core of the wood and appearance of gases in nonconducting tracheids. The displacement of pit membranes seems to occur where a water-containing tracheid lies against one filled with gases (Harris, 1954). When the torus occurs in median position (pl. 11B), water passing through the bordered pit-pair presumably moves through the pores of the margo (Bailey, 1957c). If the torus is in lateral position, the movement of the water through the pit-pair is restricted. The torus is characteristic of the bordered pits in Gnetales and Coniferales, but may be poorly developed (pl. 13D). It is rare and sporadic in angiosperms.

In certain dicotyledons the pits of vessels develop minute outgrowths from the free surface of the secondary wall of the borders, which give the pits a sieve-like appearance. The processes are highly refractive, vary in number, shape, and size, and occur not only in the pit chambers but also on the inner surface of the secondary wall of vessels. In half-bordered pit-pairs they occur only in the bordered member of the pair. Bordered pits with such processes have been named *vestured pits* (Bailey, 1933).

The pits are variously arranged in different cells, and they are not spaced uniformly, even in a single cell. Moreover, they vary in structure within one cell. The distribution and structure of the pits within a cell depend much on the type of cells to which it is joined in a tissue. Simple pits may occur in all walls of a given cell or only in certain ones. A tracheary cell may have no pits in parts of walls joined to a fiber, large prominently bordered pits where it is connected to another tracheary cell, and much reduced borders where it is joined to a parenchyma cell. The pit-pairs between two pine tracheids have well-differentiated tori, but in the half-bordered pit-pairs which occur between tracheids and the parenchymatic members of the xylem tori are usually absent.

Pits may form definite patterns that have special names (Committee on Nomenclature, 1957). The bordered pits in tracheary cells show three main types of arrangement: scalariform, opposite, and alternate. If the pits are elongated or linear and form ladder-like series (fig. 3.5A), the arrangement is called *scalariform pitting* (from the Latin *scalaris*, pertaining to a ladder). Pits arranged in horizontal pairs or short horizontal rows characterize *opposite pitting* (fig. 3.5B). If such pits are crowded, their borders assume rectangular outlines in face view. When the pits occur in diagonal rows, the arrangement is *alternate pitting* (fig. 3.5C), and crowding gives the borders hexagonal outlines in face view. Small simple pits are often aggregated in clusters. Such an arrangement is called *sieve pitting*.

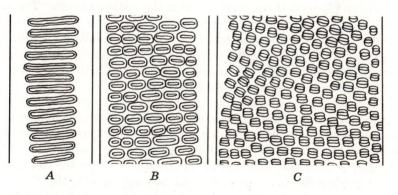

A   B   C

FIG. 3.5. Arrangement of bordered pits in vessel walls of angiosperms seen in face view. A, scalariform, *Magnolia;* B, opposite, *Liriodendron; C*, alternate, *Salix.* (All, ×375. After photomicrographs in S. J. Record, *Identification of the Timbers of Temperate North America,* John Wiley and Sons, 1934.)

Plasmodesmata

By using special techniques it is possible to demonstrate with the light microscope strand-like structures one to a few tenths of a micron in width extending from the protoplasts into the cell walls (fig. 3.2C, D; pl. 8E,F). These structures are regarded as cytoplasmic threads, the *plasmodesmata*, singular *plasmodesma* (from the Greek *desmos*, strand), interconnecting the living protoplasts of the plant body into an organic whole (Meeuse, 1957).

Plasmodesmata have been seen in red algae, hepatics, mosses, vascular cryptograms, gymnosperms, and angiosperms. They may be found throughout all living tissues, including the meristematic. Plasmodesmata, termed *ectodesmata*, have been depicted for the outer epidermal walls (Schnepf, 1959; Sievers, 1959).

Plasmodesmata either occur in groups or are distributed throughout a wall. When they are grouped, they are localized in the primary pit-fields. The relation of the plasmodesmata to primary pit-fields is characteristic: in two adjoining cells cytoplasmic processes extend into the cavities of a pair of pit-fields, and the thin wall in the pit-field is traversed by very fine threads connecting the two small masses of cytoplasm filling the depressions of the pit-fields (fig. 3.2D).

Counts of the numbers of plasmodesmata in various cells are available for the tobacco plant (Livingston, 1935). For example, the end walls (walls perpendicular to the vertical axis of the stem) in the outer cortex showed 21 to 24 threads per 100 square microns, uniformly distributed; in the side walls (walls parallel to the vertical axis of the stem) 7 to 9 threads per 100 square microns, arranged in groups. Particularly abundant plasmodesmata were found in the epidermal cells. The anticlinal walls, more or less perpendicular to the vertical axis of the organ (leaf or stem), had about 31 to 36 strands per 100 square microns; the anticlinal walls, parallel to the vertical axis of the organ, 18 to 25 strands per 100 square microns. The threads were sparse in the inner periclinal walls, and none were seen in the outer walls.

Plasmodesmata are readily seen with the electron microscope (pl. 8D). As mentioned in chapter 2, the endoplasmic reticulum appears to be connected to the plasmodesmata. Some investigators assume that tubules of this reticulum extend through the plasmodesmata (fig. 2.2B; Whaley et al., 1960), although the connection between the elements of the reticulum through a plasmodesma may appear solid (fig. 2.2C). It has been suggested that plasmodesmata arise during cell division because of persistence of endoplasmic reticulum tubules in the organizing cell plate, but they are also known to arise anew where cells form new contacts, as during cellular readjustment in tissue differentiation, in graft unions, and in junctions of tyloses (chapter 11) entering vessels from parenchyma cells. Developmental studies on *Viscum* parenchyma have shown that plasmodesmata multiply by splitting (Krull, 1960). During growth of the wall in surface area the plasmodesmata are stretched laterally and then split by interposition of wall substance. This method of growth would explain the occurrence of branched plasmodesmata.

Plasmodesmata are thought to be concerned with material transport and conduction of stimuli. They are regarded as channels permitting the movement of viruses from cell to cell, but direct proof of this assumption is lacking. The presence of plasmodesmata between the haustoria-like structures of such parasites as *Viscum, Cuscuta,* and *Orobanche* and the cells of their host plants may also be related to food and virus movement (Esau, 1948).

CHEMICAL COMPOSITION OF WALLS

The most common compound in plant cell walls is the carbohydrate cellulose. This substance received its name because it is the basic constituent of almost all cell walls in vascular plants (Ott et al., 1954–1955). It is associated with various substances, most often with other compound carbohydrates, and many walls, particularly those of woody tissues, are impregnated with lignin. The common carbohydrate constituents of the cell walls other than cellulose are hemicelluloses and pectic compounds. The fatty compounds, cutin, suberin, and waxes, occur in varying amounts in the walls of many types of cells, especially abundantly in those that are located on the periphery of the plant body. Various other organic compounds and mineral substances may be present, but they rarely play an essential part in wall structure. Water is a common constituent of cell walls and often is found in considerable amounts. Part of it occurs in microcapillaries and is relatively free; the remainder is associated with hydrophilic substances.

Cellulose is a relatively hydrophilic crystalline compound having the general empirical formula $(C_6H_{15}O_5)_n$. As a polysaccharide hexosan it is closely related to starch and its molecules are chain- or ribbon-like structures with 1,000 or more of the glucose residues held together by oxygen bridges with β-1, 4 glucosidic bonds (fig. 3.6F, G). The length of individual chains appears to vary greatly and may reach 4 microns (Frey-Wyssling, 1959).

Hemicelluloses are a heterogeneous group of polysaccharides of certain solubilities. Some of the individual members of the group are xylans, mannans, galactans, and glucans. *Pectic substances* are closely related to hemicelluloses but have different solubilities. They occur in three forms, protopectin, pectin, and pectic acid, and belong to the polyuronides, that is, polymers composed mainly of uronic acid.

Pectic compounds are amorphous colloidal substances, plastic and highly hydrophilic. The latter property suggests a possible function of maintaining a state of high hydration in the young walls. Because of the outstanding ability of pectin to jell, it is an important industrial product. As was mentioned previously, pectic compounds not only constitute the intercellular substance but also occur associated with cellulose in the other wall layers, notably the primary.

Gums and *mucilages* (or slime) should also be mentioned among the compound carbohydrates of the cell walls. These substances are related to pectic compounds and share with them the property of swelling in water. Gums appear in plants mainly as a result of physiological or pathological disturbances that induce a breakdown of walls and cell contents (gummosis or gummous degeneration). The mucilages occur in some gelatinous or mucilaginous types of cell walls. Such walls are common in the outer cell layers of plant bodies of many aquatic species and in seed coats (Frey-Wyssling, 1959).

Lignin, one of the most important wall substances, has been studied for over one hundred years, but its chemistry is still known only imperfectly (Kremers, 1959). It is a polymer of high carbon content, distinct from the carbohydrates. It consists predominantly of phenylpropanoid (C_6, C_3) units and occurs in several forms (Brown, 1961). The lignins of conifers and dicotyledons differ from one another (Gibbs, 1958). Lignin is an end product of metabolism and once formed seems to function primarily as a structural component of the cell wall. Physically it is rigid. It is the most important representative of the incrusting substances, that is, substances impregnating the wall after its initial development (Frey-Wyssling, 1959). Whether or not this process involves a removal of substances originally present in the wall is not known.

Lignin may be present in all three wall layers—the middle lamella, the primary wall, and the secondary wall. Lignification occurs in the primary wall and intercellular substance before it spreads to the secondary wall. In detail, lignification has been seen starting in the primary wall, adjacent to the corner thickenings of the middle lamella, then spreading to the intercellular layer and to the primary wall generally (Wardrop and Bland, 1959). In the secondary wall lignification was found to lag considerably behind the synthesis of cellulose and other polysaccharides. In xylem elements with secondary walls in the form of rings and helices the primary wall does not become lignified. In woody tissues the middle lamella and the primary wall are more strongly lignified than the secondary wall (Preston, 1955).

Mineral substances like silica and calcium carbonate, and diverse organic compounds like tannins, resins, fatty substances, volatile oils, and acids, as well as crystalline pigments, may impregnate walls. Silica is a common component of the walls of grasses, sedges, and horsetails. Organic compounds are frequently deposited in the xylem walls when this tissue changes from sapwood into heartwood.

The most important fatty substances are *cutin, suberin,* and *waxes.* Waxes melt readily and are easily extracted by fat solvents, whereas cutin and suberin are not meltable and show considerable insolubility in fat solvents. Suberin and cutin are closely related, highly polymerized compounds consisting of fatty acids. Suberin occurs in association with cellulose in cork cells of the periderm (chapter 14). Cutin forms a continuous layer—the cuticle—on the surface of the epidermis of all aerial parts (chapter 7). Cutin also occurs with the cellulose in the outer walls of the epidermis. These walls often show gradations from pure cellulose on the inside through layers having varying amounts of pectic compounds and fatty substances to an outermost layer of cuticle, free of cellulose and pectic compounds (Roelofsen, 1959). The phenomena of impregnation of walls with suberin and cutin are referred to as *suberization* and *cutinization,* respectively, and the formation of cuticle as *cuticularization.* Waxes are associated with suberin and cutin and may appear on the surface of the cuticle in various forms (chapter 7). Such deposition of wax is responsible for the glaucous condition (bloom) of many fruits, leaves, and stems.

Because of their chemical nature and their peripheral position in the plant body, the fatty wall substances are considered to be effective in reducing transpiration and in protecting the foliage from leaching effects of rain. The relatively hard, varnish-like cuticle, specifically, may protect against penetration of living tissues by potential parasites and against mechanical injuries.

Fatty materials are not restricted to the peripheral layers of the plant body. Suberin occurs in specialized layers like the endodermis and the exodermis (chapter 17). Inner cuticles develop in seeds during the transformation of the integuments into seed coats (chapter 20). Fatty substances, identified as cutin (Frey-Wyssling, 1959) and as suberin (Scott, 1948), occur as a coating on the mesophyll cell walls facing the internal air-space system of the leaf.

MICROSCOPIC AND SUBMICROSCOPIC STRUCTURE

The various chemical substances of cell walls combine physically and chemically with each other. Therefore, to recognize the individual compounds and their interrelations a variety of physical and chemical methods must be employed. Investigators combine observations on differential staining; differential solubilities; coarse and fine structural variations; ultrasonically disintegrated material; reaction to polarized and fluorescent light, to X-rays, and to dark-field illumination; sub-microscopic structure; refractive indices; and composition of ash (Frey-Wyssling, 1959; Roelofsen, 1959). At first the more accessible secondary wall was the main object of study, but with refinement of methods the primary wall also came to be successfully investigated. The particular significance of research on primary walls is that it yields information on the methods of growth of cell walls in surface area.

Structural Elements

The architecture of cell walls is based on cellulose. As was mentioned previously, cellulose occurs in the form of long chain molecules. These molecules are not dispersed at random in the wall but are combined into bundles of different classes of magnitude ranging from those barely discernible with the electron microscope to those visible with the light microscope. Frey-Wyssling (1959) graphically describes these structural elements and their interrelations on the basis of the secondary wall of the ramie (*Boehmeria*) fiber. One *cellulose molecule* has only 8 Å maximum width and, therefore, has not yet been resolved with the electron microscope. It may be classified as amicroscopic. Cellulose molecules are combined into an *elementary microfibril* that has a widest diameter of 100 Å and is discernible with the electron microscope. It contains 100 cellulose molecules in a transection. Both the cellulose molecules and the elementary fibrils are ribbon-like structures. Elementary fibrils form a bundle called a *microfibril*, which is 250 Å wide and contains 2,000 cellulose molecules in a transection. Electron microscope studies on cell walls are concerned mainly with this unit (fig. 3.6D, pl. 13A). Microfibrils are combined into *macrofibrils*, 0.4 micron wide and containing 500,000 cellulose molecules in transection. Finally, 2,000,000,000 cellulose molecules make up a transection of the secondary wall of the fiber.

The concept of the elementary fibril is not generally accepted but the existence of units intermediate between the microfibrils and the cellulose molecules is recognized (fig. 3.6D). From a morphological aspect the microfibril is used as the basic structural unit of the cell wall (Wardrop, 1962).

Crystallinity of Cellulose

The crystalline properties of cellulose are a result of an orderly arrangement of cellulose molecules within the fibrils. The chain molecules are combined in such a way that the glucose residues occur at uniform

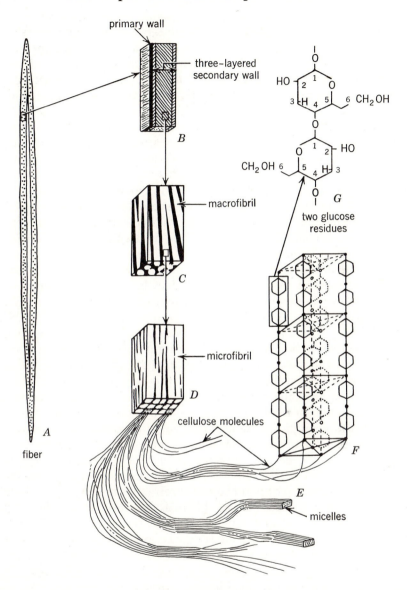

FIG. 3.6. Interpretation of wall structure. Fiber (*A*) has a three-layered secondary wall (*B*). In a fragment of the central layer of this wall (*C*) the macrofibrils (white) consist of numerous microfibrils (white in *D*) of cellulose interspersed by microporosities (black) containing noncellulosic wall materials. Microfibrils consist of bundles of cellulose molecules, partly arranged into orderly three-dimensional lattices, the micelles (*E*). Micelles are crystalline because of regular spacing of glucose residues (*F*). These residues are connected by β-1, 4-glucosidic bonds (*G*).

distances from each other and form a space lattice (fig. 3.6F). This structure has been revealed by means of X-ray studies (Frey-Wyssling, 1959). The wave lengths of the X-rays are smaller than the dimensions of cellulose molecules, and, therefore, when a beam of X-rays is allowed to impinge on a block of cellulose, a large part of the beam goes through, but part of the rays strike the atoms and groups of atoms and are scattered, or *diffracted*. The diffracted waves of light appear as reflections of the incident waves, and, when the X-ray beam strikes the crystalline material at a proper angle, the scattered waves from all points reinforce each other, and a strong beam is diffracted. For convenience, the diffracted beams are often referred to as reflections. They are reflections of the X-rays from the atoms and groups of atoms, and when caught on a photographic plate they leave a diffraction pattern. By obtaining such diffraction patterns from various sides of the same block of cellulose, one can determine the three-dimensional configuration of the molecular groups of cellulose.

Since the distances between the points on the cellulose space lattice vary in different planes, it may be said that the constituent parts of the lattice are distributed *anisotropically*, that is, arranged dissimilarly in different directions. This anisotropy is expressed in certain properties of cellulose. When, for example, cellulose is made to swell, it expands much more strongly in the direction at right angles to the long axis of the lattice or the molecular chains than in planes parallel to this axis or, when light is passed through cellulose, the light is variously affected, depending on the direction from which it strikes the lattice. In other words, cellulose shows swelling anisotropy and optical anisotropy.

Optically anistropic substances are *doubly refractive* or *birefringent*. These terms refer to the manner in which the light entering the aniso-tropic material is deflected (refracted) from its original course. When a beam of light strikes such material obliquely, the part of the beam that enters it (the other part is reflected) is refracted not as a single beam but as two beams deflected to different degrees. When the angle formed by the two refracted beams is large, the material is said to be strongly birefringent. The birefringence of a substance is readily revealed by its effect upon polarized light. As is well known, such light is interpreted as vibrating in one plane only: it is plane-polarized light. A device for using polarized light incorporates two crystalline prisms, or polaroids, one of which, the polarizer, produces polarized light, and the other, the analyzer, aids the observer to determine whether the object illuminated by the light from the polarizer has any effect upon this light. If in the absence of any object the analyzer is rotated 90 degrees with respect to the polarizer, no light will pass through the system. The two prisms are said to be crossed.

An isotropic object has no effect on polarized light, and, therefore, when it is inserted between the crossed prisms, the field seen through the microscope remains dark (middle lamellae in pl. 7B). If a doubly refractive substance is substituted for the isotropic specimen, in certain orientations it will so affect the incident light that the light will pass the analyzer and the object will appear bright (primary walls and parts of secondary walls in pl. 7B).

As was mentioned previously, birefringent material diffracts one beam of light as two. These are polarized in mutually perpendicular planes. If the doubly refractive material placed between the crossed elements of a polarizing system is so oriented that neither of its planes of polarization coincides with the plane of polarization of the polarizer, the ray coming from the polarizer is resolved into two mutually perpendicular components. The planes of these component rays are not exactly crossed with respect to the analyzer, and, therefore, both component vibrations are partially transmitted by the analyzer. The specimen appears bright on a dark background. When one or the other of the planes of polarization of the material is in alignment with the plane of polarization of the incident light, no light passes through the analyzer and the phenomenon of extinction is exhibited (central layer of secondary walls in pl. 7B). The material now does not reveal its anisotropy. In cellulose the brightest light (strongest birefringence) is seen when the light passes through it at right angles to the long axis of the molecular chains. Parallel to this position cellulose does not affect the light and remains dark between crossed prisms (Chamot and Mason, 1938).

Organization of the Wall

The crystalline aggregates of cellulose molecules are commonly called *micelles,* or *crystallites* (Wardrop, 1962). Formerly the micelles were thought to be individual units arranged in an orderly manner in an intermicellar material. This concept (and the term micelle) was introduced by Nägeli in the nineteenth century (Roelofsen, 1959). Presently, the micelles are considered to be crystalline regions in bundles of cellulose separated longitudinally by amorphous regions, that is, regions of less perfect molecular order (fig. 3.6E).

The micellar strands (or elementary fibrils) are surrounded by paracrystalline cellulose, that is, cellulose with molecules oriented in the direction of the microfibril but not arranged in a three-dimensional order. The spaces among the micellar strands constitute intermicellar spaces. Successively larger interstices occur among the microfibrils and the macrofibrils. Thus the fibrillar system of cellulose is interpenetrated by a system of capillaries of various sizes. These capillaries are filled with water and the amorphous noncellulosic components discussed under the chemical composition of the wall.

A striking demonstration of the relation between the microfibrillar and microcapillary systems can be made on heavily lignified secondary walls. In such walls it is possible to remove the lignin and leave a coherent matrix of cellulose; and, conversely, it is possible to remove the cellulose and leave a coherent matrix of lignin. The two matrices appear like positive and negative images of the original structural pattern (Bailey, 1954). Although the interstices of the cellulose matrix are filled with various substances, the cell walls are porous, as evidenced by the passage of substances through them.

The amorphous system is a continuous three-dimensional matrix in which the fibrillar system is imbedded. Views are divided regarding the question of whether or not the microfibrils of cellulose are anastomosing within the wall (Roelofsen, 1959). There is good evidence, however, that in primary walls, in which the cellulose occupies a small volume, the microfibrils are relatively widely spaced (Wardrop, 1962). The other aspect still insufficiently understood is the nature of the interaction between the components of the matrix and the cellulose in the wall (Setterfield and Bayley, 1961), but lignin is assumed to be chemically linked to polysaccharides (Brown, 1961).

Orientation of Microfibrils

As was previously stated, the degree of birefringence of wall layers, which is revealed by the polarizing microscope, is determined by the orientation of molecular chains of cellulose with reference to the beam of incident light. Since the long axes of the molecular chains and those of the microfibrils are approximately parallel, the degree of birefringence can serve to determine the orientation of the microfibrils. In addition, the fibrillar orientation may be reliably studied by observing the microscopically visible reticulations (pl. 8A) and striations (pl. 9C) and the orientation of the planes of hydrolysis caused by enzymatic activity of certain fungi (pl. 10A), also by inducing formation of crystals in the elongated porosities of the cellulose matrix, where the crystals become oriented parallel to the fibrils and are microscopically visible (Bailey, 1954, 1957a). Finally, the electron microscope reveals the microfibrils themselves.

In general the patterns formed by the microfibrils are highly variable. They vary in different woods, in wood from different parts of the tree, in different cells of the same tissue, in different layers of the same cell, and in different lamellae of the same layer. In the three-layered walls of certain vessels, tracheids, and wood fibers the fibrillar orientations of the inner and outer layers vary between transverse and helical, the helices being of comparatively low pitch, and those of the central layer fluctuate between longitudinal and relatively steeply pitched helical. In the cotton fiber the bulk of the secondary wall consists of microfibrils oriented at an angle of 45 degrees and less with respect to the longitudinal axis of the fiber (fig. 3.7; Hock, 1942). In the consecutive lamellae of the flax fiber the helices are wound in opposite directions (Anderson, 1927). In tracheary cells with annular, helical, and scalariform secondary thickenings the crystalline regions of these thickenings have a horizontal, ring-like orientation (Frey-Wyssling, 1948). Although the pitch of the helices of microfibrils varies in the secondary walls of different cells and among the layers of the same wall, within a given layer the microfibrils are usually parallel to one another and always parallel to the surface of the cell. The secondary walls may be said to have a parallel texture (fig. 3.7; pls. 10B, 12B; Frey-Wyssling, 1959).

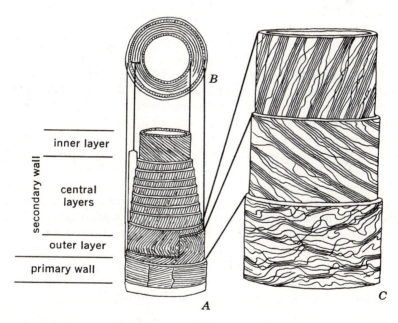

inner layer

secondary wall

central
layers

outer layer

primary wall

B

C

A

FIG. 3.7. Wall structure in the cotton fiber. A, telescoped segment and, B, transverse section of fiber, showing spatial relation of the various layers and orientation of microfibrils. C, primary wall has reticulate microfibrillar structure, the outer layer of the secondary wall combines reticulate and parallel orientation of microfibrils, and the first central layer of secondary wall has a predominantly parallel microfibrillar structure. (After Berkley, *Textile Res. Jour.* 18, 1948.)

The microfibrillar orientation in the secondary wall is reflected in that of the elongated inner aperture of bordered and simple pits: the long diameter of the aperture is parallel to the microfibrils or their aggregations (pl. 9C). Pits may cause deviations in the prevailing orientation of microfibrils. Embossed borders of the pits in tracheids of conifers, for example, show a circular orientation of microfibrils and cause a modification of the orientation in the adjacent parts of the wall (Bailey, 1954).

The microfibrillar structure of the margo of the bordered pits of conifers has been studied in considerable detail. The margo shows loosely arranged, radially oriented thin microfibrils in young cells. Subsequently, the radial strands thicken (pls. 12A, 13D) so that they may become microscopically visible (pl. 11A). This thickening is described as a gathering of thinner strands into the thicker complexes (Frey-Wyssling, 1959). It results in many varied patterns on the margo (Bailey, 1957c). Openings without membranes occur between the strands, large enough to permit carbon particles from india ink to pass (Bosshard, 1956). The torus thickens by deposition of microfibrils in circular arrangement.

The primary wall, when first formed, shows a predominantly transverse orientation of microfibrils (tube texture), but the orientation becomes more disperse as the wall increases in surface area during cell enlargement. The scattered texture of the earlier parts of the primary wall has led to the concept that primary and secondary walls may be distinguished by their microfibrillar orientation: scattered in primary walls (pl. 13A), parallel in secondary (pls. 10B, 12B). Actually, in a given wall the layers of the primary wall show an increasing degree of parallelization of microfibrils in the centripetal direction (Böhmer, 1958).

In electron micrographs of primary walls the microfibrils appear interwoven. Investigators appreciate that the microfibrillar network appears much denser in the electron microscope than it actually is because of the dehydration and the removal of most of the matrix carried out in the preparation for viewing (Frey-Wyssling, 1959). Nevertheless, the concept prevails that microfibrils are interwoven. This interpretation agrees with the common absence of lamellation in primary walls, except when they are conspicuously thickened (Wardrop, 1962).

Coarse Patterns in Walls

The presence or absence of secondary walls, the relative thickness of primary and secondary walls, and the differentiation of the secondary wall into three or more layers cause the most conspicuous variations in the appearance of walls. In addition, the secondary walls, particularly the wide central layer of the three-layered types, show diverse patterns coarser than the microfibrillar network. In cells cut at right angles to their long axes the most common patterns are: concentric layering (fig. 3.7), radial and branching lamellations, and combinations of radial and concentric lamellations. Some of these arrangements of lamellae are determined by the distribution of the noncellulosic constituents of the walls, but many specific configurations result from variations in densities and porosities in the different parts of the cellulosic matrix. In many tracheary cells and xylem fibers the denser parts of walls have more numerous fibrils per unit volume, and these are more intimately coalesced than the fibrils of the more porous parts (Bailey, 1954). In the cotton fiber the concentric lamellation has been correlated with the alternation of day and night. Every 24 hours one compact and highly birefringent and one porous and weakly anisotropic lamella are formed. If the cotton fibers are grown under continuous illumination, they have no growth rings (Hock, 1942).

Sometimes the concentric layering is caused by actual discontinuities in the cellulose matrix. In the fibers of compression wood of some gymnosperms, in the gelatinous fibers of dicotyledons, and in certain sclereids and phloem fibers, layers of truly isotropic material divide layers of cellulose (Bailey and Kerr, 1935). Some phloem fibers appear to have no cementing material between the concentric lamellae of cellulose, and the lamellae may be readily separated from one another (pl. 26A; Anderson, 1927). Tracheids in the compression wood of gymnosperms often develop a helically striate inner layer of secondary wall (pl. 9C).

A somewhat controversial subject regarding the structure of the secondary wall is the nature of the thin layer of material that has been observed lining the wall on the lumen side in many fibers and tracheary cells (Frey-Wyssling, 1959; Wardrop et al., 1959). This layer is highly resistant to sulphuric acid and often bears wart-like excrescences of submicroscopic or microscopic dimensions (pl. 13C). It is formed during the concluding stages of lignification of the secondary wall and appears to be derived from the remains of the protoplasmic contents of the cell (Liese, 1963). The wall sculpturing responsible for the vestured condition of pits in certain dicotyledons appears to be analogous to the wart structure (Côté and Day, 1963).

PROPERTIES OF WALLS

Cell walls show varying degrees of *plasticity* (property of becoming permanently deformed when subjected to changes in shape or size), *elasticity* (property of recovery of the original size and shape after deformation), and *tensile strength* in relation to their chemical composition and their microscopic and submicroscopic structure. Plasticity of walls is well illustrated by their permanent extension in certain stages of growth of cells in volume (Heyn, 1940); elasticity, by the reversible changes in volume in response to changes in turgor pressure (Frey-Wyssling, 1959). Notable tensile strength is characteristic of mechanical cells, particularly of the extraxylary fibers of monocotyledons and dicotyledons.

Some of the conspicuous differences in optical and other physical properties of walls are correlated with the orientation of the microfibrils. Thus, for example, walls or wall layers in which the microfibrils are oriented parallel to the long axis of the cell do not exhibit their anisotropy in transverse sections and do not contract longitudinally; on the contrary, walls having the microfibrils oriented at right angles to the long axis of the cell are strongly birefringent in transverse sections and contract longitudinally on drying (Bailey, 1954).

Because of its abundance in cell walls cellulose has a major influence upon their properties. Other substances add their properties or modify those imparted by the cellulose. Tensile strength is one of the remarkable characteristics of cellulose. Lignin, on the other hand, increases the resistance of walls to pressure and protects the cellulose fibrils from becoming creased (Frey-Wyssling, 1959).

FORMATION OF WALLS

Initiation of Wall during Cell Division

The process of somatic division of a protoplast into two daughter protoplasts may be separated into two stages: the division of the nucleus, or *mitosis (karyokinesis)*, and the division of the extranuclear part of the protoplast, or *cytokinesis*. In cells having cell walls the new wall is formed during cytokinesis.

The divisions of the nucleus and the cell may follow each other so closely that mitosis and cytokinesis appear as one phenomenon; or the two may be separated in time. The ordinary somatic divisions, characterizing vegetative growth from the meristems, usually show a close correlation between nuclear and cellular divisions. In contrast, the two phenomena are widely separated in the formation of pollen and endosperm in many angiosperms, and in the development of the female gametophyte and the proembryo in gymnosperms.

The partition between the new protoplasts, when first evident, is referred to as the *cell plate*. If cytokinesis follows the nuclear division immediately, the cell plate arises in the equatorial plane of a fibrous spindle, the *phragmoplast*, extending between the two groups of chromosomes that move apart during the anaphase of mitosis (pl. 4E–G). As these two groups develop into the telophase nuclei, the phragmoplast widens out in the equatorial plane and assumes the shape of a barrel. When the cell plate appears in the median part of the equatorial plane of the phragmoplast, the fibers of the phragmoplast disappear in this position but remain evident at the margins, until the cell plate appears here too.

If the diameter along which the cell is dividing is so short that the phragmoplast, after a slight widening, reaches the walls that are oriented perpendicularly to the plane of division, the phragmoplast appears to be connected to the two nuclei for the duration of cytokinesis. If, however, this diameter is longer than the original phragmoplast is wide, the phragmoplast extends laterally until it comes in contact with the cell walls, and during this extension it completely separates from the nuclei. As seen from the side, such a phragmoplast appears as two groups of fibers, disconnected from the nuclei but connected with each other by the cell plate, which follows the phragmoplast in its lateral extension (fig. 3.9A). In face views the phragmoplast has a somewhat varied appearance, depending on the shape and size of the dividing cells and also on the original position of the nucleus.

The progress of the phragmoplast and cell plate through the cell lumen is particularly striking in very long cells, for example, fusiform cambial cells, dividing longitudinally. The process of cell-plate formation in such cells is greatly extended in time and space and is clearly dissociated from the nuclear mitosis (Bailey, 1902b; chapter 6).

The phragmoplast and the mitotic spindle are proteinaceous in chemical structure (Olszewska, 1961a, b; Shimamura and Ota, 1956). The fibrous nature of the phragmoplast has been recognized in living material (Sitte, 1962) and in some electron micrographs (Sato, 1959); in others the phragmoplast was brought into relation to elements of the endoplasmic reticulum (Porter and Machado, 1960), or to the dictyosomes (Whaley and Mollenhauer, 1963), or to elements in the form of microtubules (Ledbetter and Porter, 1963). The phragmoplastic fibers appearing at the margins of the cell plate are sometimes called kinoplasmasomes, a term reflecting the old concept of the existence of a special kind of active, fibrous cytoplasm, the kinoplasm (Bailey, 1920b).

The views of cell-plate formation have been misinterpreted by some workers, and this, in turn, has led to misconceptions regarding the numbers of nuclei in ordinary somatic cells. The erroneous reports have been reviewed and corrected (Bailey, 1920a; Wareham, 1936).

Cytokinesis is not limited to meristematic cells with dense protoplasts.

Some of the meristematic cells themselves are highly vacuolated, and, furthermore, enlarging and prominently vacuolated cells of the ground tissue are known to divide actively during the growth of roots, shoots, leaves, and fruits of higher plants. In vacuolated cells the new cell plate eventually occurs in the region formerly occupied by the vacuole. One may observe, however, that during the early prophase of the nuclear division, that is, long before the beginning of cytokinesis, the nucleus comes to occupy a position corresponding to the future equatorial plate of the mitotic spindle and is surrounded by dense cytoplasm. A layer of this cytoplasm extends to the walls oriented at right angles to the future plane of division. It forms a cytoplasmic plate, for which Sinnott and Bloch (1941) coined the term *phragmosome*. The phragmosome forms a living medium in which the phragmoplast and the cell plate develop (fig. 3.8). Studies of this stage of division in the living state indicate that the amassing of cytoplasm around the nucleus before the formation of the phragmosome is associated with a cessation of the movement of particles in the streaming cytoplasm and an apparent increase in density of the cytoplasm (Jones et al., 1960). Reversion to free flow within the cytoplasm occurs only after the cytokinesis is completed.

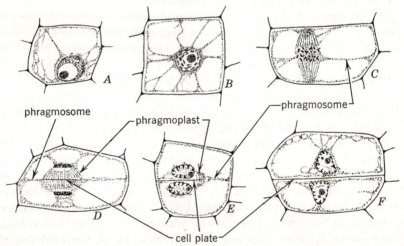

FIG. 3.8. Division of highly vacuolated cells in young pith of *Ligustrum*. *A*, cell in a nondividing state. *B*, nucleus in prophase and located in middle of cell. *C*, nucleus in early anaphase; laterally, mitotic spindle is connected to parietal cytoplasm by a cytoplasmic layer, the phragmosome. *D*, daughter nuclei in telophase; barrel-shaped spindle between nuclei is the phragmoplast; cell plate appears in its equatorial plane. *E*, cell plate intersects one of the walls of mother cell. *F*, cell division is completed and cell plate occupies former position of phragmosome. (All, ×940.)

If a cell plate is not formed immediately after nuclear division, phragmoplasts may arise later. Sometimes no phragmoplast is formed; instead, the cell divides by a process called *furrowing*. Such division has been described in lower plants and in pollen and endosperm development in higher plants. It consists of the formation of cleavage furrows within the protoplast, starting at the existing walls and advancing inward until they meet and divide the protoplast into two or more cells.

Cell-plate formation has been studied in living and fixed material and with light and electron microscopes (Becker, 1938; Porter and Machado, 1960; Sitte, 1962). It seems well substantiated that substances in semifluid state accumulate as vesicles—according to one view (Whaley and Mollenhauer, 1963), derived from dictyosomes—in the equatorial plane of the phragmoplast and cleave the protoplast in two (pl. 5C). The two new cytoplasmic surfaces become parts of the protoplasmic membranes (ectoplast, pl. 4I) of the two new cells. Pectic substances occur in the semifluid partition in the equatorial plane. These substances are regarded as forming the new middle lamella. A deposition of cellulose on both sides of this middle lamella, externally to the new protoplasmic membranes, is indicated by the appearance of double refraction, which becomes perceptible before the cell plate joins the walls of the dividing cell (Frey-Wyssling, 1959). Cellulose is deposited not only on the cell plate but around the entire daughter protoplasts (fig. 3.9A–C). Basically similar phenomena might be involved in cell division by furrowing.

Thus the partition that appears between the two sister protoplasts at cytokinesis undergoes different physical and chemical changes during the progress of cell division. There is no agreement regarding the stage of the process at which the visible partition should be called cell plate. The term has, therefore, no precise definition and merely serves, at present, as a designation for the first visible structure delimiting the two sister protoplasts from one another.

Growth of Walls

In considering the mechanism of wall growth it is necessary to differentiate between growth in surface area and growth in thickness. The former process is much more difficult to explain than the latter. Growth in thickness is particularly obvious in secondary walls but is common also in primary walls (as classified according to Kerr and Bailey, 1934). It occurs by a successive deposition of wall material, layer upon layer, that is, by a process known as *apposition*. But intercalation of new particles among those existing in the wall, that is, *intussusception*, is not necessarily excluded during thickening growth (Roelofsen, 1959).

The Cell Wall

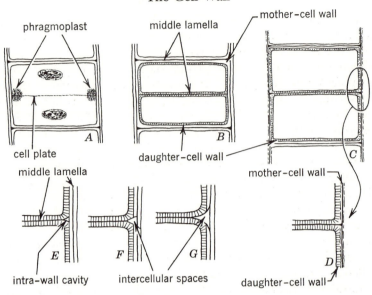

FIG. 3.9. Concepts regarding the adjustments between new and old cell walls after cell division. *A*, cell plate has been formed. *B*, two primary walls cemented by intercellular substance occupy the position of cell plate; primary daughter-cell walls have been laid down on the inside of primary mother-cell wall. *C, D*, daughter cells have expanded vertically and mother-cell wall has been stretched and ruptured opposite new wall. Thus, old and new intercellular lamellae have joined. *E–G*, establishment of continuity between old and new middle lamellae through formation of intercellular space. *E*, appearance of cavity between daughter- and mother-cell walls. *F*, dissolution of mother-cell wall next to cavity. *G*, completion of change of the intra-wall cavity into an intercellular space.

Growth of walls by apposition is usually centripetal. In other words it occurs from the outside and toward the lumen of the cell. Sometimes, however, wall growth has a centrifugal course, that is, in the direction away from the lumen. Centripetal growth is characteristic of cells forming tissues. Centrifugal growth is a specialized type of growth found in pollen grains and other spores. In such structures centrifugal growth is considered to be responsible for the formation of at least part of the exine (the outer wall). The more or less degenerated contents of the tapetal cells (chapter 18) surrounding the developing spores seem to be involved in the formation of exine (Roelofsen, 1959).

Many aspects are being considered with regard to the growth of cell walls in surface area. The question whether new material is being added to the wall during its extension is usually answered in the affirmative (Ray, 1962; Roelofsen, 1959; chapter 17). Despite the great increase in surface of the primary wall of enlarging cells, no appreciable decrease in wall thickness is observable during such growth. Moreover, accurate determinations of the amount of cell-wall material in successive stages of growth reveal considerable increase of wall material per cell. Some of the exceptions of wall extension with only negligible addition of wall material were found in the growing staminal hairs of *Tradescantia* and staminal filaments of Gramineae. Another question concerns the growth of protoplasm in cells undergoing expansion. Apparently cell walls can increase in surface without a concomitant increase in protein nitrogen in the protoplast (Matthaei, 1957).

Investigators are concerned with the question whether growth of the wall in surface involves part of a given wall or the entire wall. In ground-parenchyma tissue growth occurs, as evidenced by the uniform increase of distances among the existing pit-fields, over the entire surface of a growing wall (Wilson, 1958, 1961; Ziegenspeck, 1953). Autoradiographic studies with labeled compounds also indicate incorporation of material over the total surface of parenchyma walls (Setterfield and Bayley, 1961). Certain types of cells, however, show localized growth, as, for example, fibers and tracheids (Wardrop, 1954), in which the tips grow intrusively among other cells (chapters 4, 6), and root hairs (Dawes and Bowler, 1959), in which typically growth in length occurs at the tips.

During the over-all extension of the primary wall the primary pit-fields are not only more widely spaced but they also enlarge in area and become subdivided by the deposition of microfibrils over the pit-field (Scott et al., 1956). As previously mentioned, plasmodesmata, too, may become subdivided (Krull, 1960). During cell division, however, completely new pit-fields are added (Wilson, 1958, 1961). Thus it appears that during growth the wall maintains a characteristic density of connections with contiguous cells.

The most complex aspect of the growth of wall in surface area is the growth of the cellulosic microfibrillar system. Electron microscopists have formulated several concepts of this growth (Wardrop, 1962). According to one, for example, synthesis of wall material occurs in localized regions scattered over the wall (*mosaic growth*), in which the cytoplasm pushes apart the existing microfibrils and weaves in new ones. A more widely accepted concept is that of the *multinet growth* pattern, which visualizes an apposition of successive layers of microfibrils, with the earlier layers becoming modified in microfibrillar orientation by wall extension during cell enlargement. The structure of many primary walls seems to support this concept.

The question whether primary walls grow predominantly by apposition or by intussusception has no unequivocal answer (Roelofsen, 1959), but the preferred view is that appositional growth of microfibrils is dominant even though the microfibrils may be intertwined. On the other hand, some studies with radioactive isotopes suggest that new wall material may be deposited throughout the wall (Setterfield and Bayley, 1961). Moreover evidence has been presented (Matchett and Nance, 1962) that the distribution of isotope throughout the wall may be associated with a metabolic turnover of polysaccharides during their synthesis; that, in other words, the extension of the primary wall may be associated with a breakdown and resynthesis of the structural framework. This interpretation must be evaluated in relation to the concepts of mechanism of wall expansion, especially that considering the possibility of an increase in plasticity of the wall during growth (Setterfield and Bayley, 1961).

Studies with isotopically labeled substrates have indicated that intact glucose may be used directly in cellulose synthesis, but the mechanism of glucose polymerization has not yet been revealed (Setterfield and Bayley, 1961). It has been suggested that individual glucose residues are added to the tips of growing microfibrils and that this method of growth would explain the uniform thickness of microfibrils and absence of anastomoses.

Another complex aspect of wall growth has to do with the establishment of continuity between the new intercellular lamella and that located outside the primary wall of the mother cell. Workers visualize an extension and breakdown of the parent wall opposite the new middle lamella (fig. 3.9A–D; Priestley an Scott, 1939; Roelofsen, 1959). The formation of intercellular spaces might be associated with this phase of wall growth (fig. 3.9E–G; Martens, 1937, 1938).

FORMATION OF INTERCELLULAR SPACES

Although the cells in the meristematic tissues are generally closely packed, during tissue differentiation the close connection between walls of adjacent cells may be partly broken, with the appearance of intercellular spaces. The most common intercellular spaces result from a separation of cell walls from each other along more or less extended areas of their contact. These are the *schizogenous* intercellular spaces, so called because formerly the mechanism of their formation was thought to involve a splitting of the middle lamella (*schizo*, split; *genesis*, beginning; both from the Greek).

The origin of the schizogenous intercellular spaces is described as follows (Martens, 1937, 1938; Sifton, 1945, 1957; fig. 3.9E–G). When the new primary walls are formed between two sister protoplasts, the middle lamella between these walls comes in contact with the original mother-cell wall and not with the middle lamella cementing the mother-cell wall to that of its neighbor. A small cavity arises at the point of contact between the new middle lamella and the parent wall; then the parent wall becomes dissolved opposite the cavity. Thus the intra-wall cavity develops into an intercellular space. If a similar space is present between the mother cell and its neighbor, the new cavity and the old intercellular space may join to form one large space. In this process of space formation the intercellular substance is perhaps partly dissolved, but it does not disappear. The intercellular space remains lined by intercellular material (pl. 7A; Sifton, 1945). Certain plants, such as the submerged water plants, develop particularly large air spaces which, in the internodes, may extend like canals from node to node. These spaces are initiated like ordinary schizogenous spaces but later are enlarged by cell divisions perpendicular to the circumference of the air space (Hulbary, 1944).

Some of the schizogenous intercellular spaces form specialized structures, the secretory ducts. Examples are the resin ducts in the Coniferae (pl. 31A) and the secretory ducts in the Compositae and the Umbelliferae (Sifton, 1945). They are initiated similarly to the air spaces just described for aquatic plants. Since many cells in longitudinal or transverse series form spaces in the same position, these spaces take the form of long intercellular canals that may become connected into an intercommunicating system (De Bary, 1884). The cells lining the duct cavity are secretory and release their product into the canal.

Another type of intercellular space arises through dissolution of entire cells, which are therefore called *lysigenous* (from *lysis*, loosening, in Greek) intercellular spaces. Examples are the large air spaces in water plants and in some monocotyledonous roots (*Zea;* Sifton, 1945), and secretory cavities in *Eucalyptus, Citrus,* and *Gossypium* (De Bary, 1884; Stanford and Viehoever, 1918). In the secretory cavities the cells that break down release the secretion into the space and remain, themselves, in partly collapsed or disintegrated state around the periphery of the cavity.

REFERENCES

Anderson, D. B. A microchemical study of the structure and development of flax fibers. *Amer. Jour. Bot.* 14:187–211. 1927.

Bailey, I. W. Phragmospheres and binucleate cells. *Bot. Gaz.* 70:469–471. 1920*a.*

Bailey, I. W. The cambium and its derivative tissues. III. A reconnaissance of cytological phenomena in the cambium. *Amer. Jour. Bot.* 7:417–434. 1920*b.*

Bailey, I. W. The cambium and its derivative tissues. VIII. Structure, distribution, and diagnostic significance of vestured pits in dicotyledons. *Arnold Arboretum Jour.* 14:259–273. 1933.

Bailey, I. W. *Contributions to plant anatomy.* Waltham, Mass., Chronica Botanica Company. 1954.

Bailey, I. W. Aggregation of microfibrils and their orientations in the secondary wall of coniferous tracheids. *Amer. Jour. Bot.* 44:415–418. 1957a.

Bailey, I. W. Need for a broadened outlook in cell wall terminology. *Phytomorphology* 7:136–138. 1957b.

Bailey, I. W. Die Struktur der Tüpfelmembranen bei den Tracheiden der Koniferen. *Holz als Roh- und Werkstoff* 15:210–213. 1957c.

Bailey, I. W. Comparative anatomy of the leaf-bearing Cactaceae. II. Structure and distribution of sclerenchyma in phloem of *Pereskia, Pereskiopsis* and *Quiabentia.* *Arnold Arboretum Jour.* 42:144–150. 1961.

Bailey, I. W., and T. Kerr. The visible structure of the secondary wall and its significance in physical and chemical investigations of tracheary cells and fibers. *Arnold Arboretum Jour.* 16:273–300. 1935.

Becker, W. A. Recent investigations *in vivo* on the division of plant cells. *Bot. Rev.* 4:446–472. 1938.

Bloch, R. Wound healing in higher plants. *Bot. Rev.* 7:110–146. 1941.

Böhmer, H. Untersuchungen über das Wachstum und den Feinbau der Zellwände in der *Avena*-Koleoptile. *Planta* 50:461–497. 1958.

Bosshard, H. H. Der Feinbau des Holzes als Grundlage technologischer Fragen. *Schweiz. Ztschr. f. Forstw.* 107:81–95. 1956.

Brown, S. A. Chemistry of lignification. *Science* 134:305–313. 1961.

Chamot, E. M., and C. W. Mason. *Handbook of chemical microscopy.* Vol. 1. 2nd ed. New York, John Wiley and Sons. 1938.

Committee on Nomenclature. International Association of Wood Anatomists. International glossary of terms used in wood anatomy. *Trop. Woods* 107:1–36. 1957.

Côté, W. A., Jr., and A. C. Day. Vestured pits—fine structure and apparent relationship with warts. *Tappi* 45:906–910. 1963.

Dawes, C. J., and E. Bowler. Light and electron microscope studies of the cell wall structure of the root hairs of *Raphanus sativus.* *Amer. Jour. Bot.* 46:561–565. 1959.

De Bary, A. *Comparative anatomy of the vegetative organs of the phanerogams and ferns.* Oxford, Clarendon Press. 1884.

Esau, K. Some anatomic aspects of plant virus disease problems. II. *Bot. Rev.* 14:413–449. 1948.

Frey-Wyssling, A. *Submicroscopic morphology of protoplasm and its derivatives.* New York, Elsevier Publishing Company. 1948.

Frey-Wyssling, A. *Die pflanzliche Zellwand.* Berlin, Springer-Verlag. 1959.

Gibbs, R. D. The Mäule reaction, lignins, and the relationships between woody plants. In: K. V. Thimann. *The physiology of forest trees.* New York, Ronald Press Company. 1958.

Harris, J. M. Heartwood formation in *Pinus radiata* (D. Don). *New Phytol.* 53:517–524. 1954.

Heyn, A. N. J. The physiology of cell elongation. *Bot. Rev.* 6:515–574. 1940.

Hock, C. W. Microscopic structure of the cell wall. In: *A symposium on the structure of protoplasm. Amer. Soc. Plant Physiol. Monogr.* 1942.

Hulbary, R. L. The influence of air spaces on the three-dimensional shapes of cells in *Elodea* stems, and a comparison with pith cells of *Ailanthus. Amer. Jour. Bot.* 31:561–580. 1944.

References

Jones, L. E., A. C. Hildebrandt, A. J. Riker, and J. H. Wu. Growth of somatic tobacco cells in microculture. *Amer. Jour. Bot.* 47:468–475. 1960.

Kerr, T., and I. W. Bailey. The cambium and its derivative tissues. X. Structure, optical properties and chemical composition of the so-called middle lamella. *Arnold Arboretum Jour.* 15:327–349. 1934.

Kremers, R. E. The lignins. *Ann. Rev. Plant Physiol.* 10:185–196. 1959.

Krull, R. Untersuchungen über den Bau und die Entwicklung der Plasmodesmen im Rindenparenchym von *Viscum album. Planta* 55:598–629. 1960.

Ledbetter, M. C., and K. R. Porter. A "microtubule" in plant cell fine structure. *Jour. Cell Biol.* 19:239–250. 1963.

Liese, W. Tertiary wall and warty layer in wood cells. *Jour. Polymer Sci.* Pt.C No. 2:213–229. 1963.

Livingston, L. G. The nature and distribution of plasmodesmata in the tobacco plant. *Amer. Jour. Bot.* 22:75–87. 1935.

Martens, P. L'origine des espaces intercellulaires. *Cellule* 46:357–388. 1937.

Martens, P. Nouvelles recherches sur l'origine des espaces intercellulaires. *Bot. Centbl. Beihefte* 58:349–364. 1938.

Matchett, W. H., and J. F. Nance. Cell wall breakdown and growth in pea seedling stems. *Amer. Jour. Bot.* 49:311–319. 1962.

Matthaei, H. Vergleichende Untersuchungen des Eiweiss-Haushaltes beim Streckungswachstum von Blütenblättern und anderen Organen. *Planta* 48:468–522. 1957.

Meeuse, A. D. J. Plasmodesmata (vegetable kingdom). *Protoplasmatologia* 2 A1. 1957.

Newcomb, E. H. Cytoplasm-cell wall relationships. *Ann. Rev. Plant Physiol.* 14:43–64. 1963.

Olszewska, M. J. Recherches autoradiographiques sur la formation du phragmoplaste. *Protoplasma* 53:387–396. 1961*a*.

Olszewska, M. J. L'effect du β-mercaptoéthanol et de l'urée sur la structure du phragmoplaste. *Protoplasma* 53:397–404. 1961*b*.

Ott, E., H. M. Spurlin, and M. W. Grafflin, eds. *Cellulose and cellulose derivatives.* 3 parts. New York, Interscience. 1954–1955.

Porter, K. R., and R. D. Machado. Studies on the endoplasmic reticulum. IV. Its form and distribution during mitosis in cells of onion root tip. *Jour. Biophys. Biochem. Cytol.* 7:167–180. 1960.

Preston, R. D. Microscopic structure of plant cell walls. *Handb. der Pflanzenphysiol.* 1:722–730. 1955.

Priestley, J. H., and L. I. Scott. The formation of a new cell wall at cell division. *Leeds Phil. Lit. Soc. Proc.* 3:532–545. 1939.

Ray, P. M. Cell wall synthesis and cell elongation in oat coleoptile tissue. *Amer. Jour. Bot.* 49:928–939. 1962.

Roelofsen, P. The plant cell wall. *Handbuch der Pflanzenanatomie* Band 3. Teil 4. 1959.

Sato, S. Electron microscope studies on the mitotic figure. II. Phragmoplast and cell plate. *Cytologia* 24:98–106. 1959.

Schnepf, E. Untersuchungen über Darstellung und Bau der Ektodesmen und ihre Beeinflussbarkeit durch stoffliche Faktoren. *Planta* 52:644–708. 1959.

Scott, F. M. Internal suberization of plant tissues. *Science* 108:654–655. 1948.

Scott, F. M., K. C. Hamner, E. Baker, and E. Bowler. Electron microscope studies of cell wall growth in the onion root. *Amer. Jour. Bot.* 43:313–324. 1956.

Setterfield, G., and S. T. Bayley. Structure and physiology of cell walls. *Ann. Rev. Plant Physiol.* 12:35–62. 1961.

Shimamura, T., and T. Ota. Cytochemical studies on the mitotic spindle and the phragmoplast of plant cells. *Expt. Cell Res.* 11:346–361. 1956.

Sievers, A. Untersuchungen über die Darstellbarkeit der Ektodesmen und ihre Beeinflussung durch physikalische Faktoren. *Flora* 147:263–316. 1959.

Sifton, H. B. Air space tissue in plants. *Bot. Rev.* 11:108–143. 1945; II. 23:303–312. 1957.

Sinnott, E. W., and R. Bloch. Division in vacuolate plant cells. *Amer. Jour. Bot.* 28:225–232. 1941.

Sitte, P. Polarisationsmikroskopie der Mitose in vivo bei Staminalhaarzellen von *Tradescantia. Protoplasma* 54:560–572. 1962.

Stanford, E. E., and A. Viehoever. Chemistry and histology of the glands of the cotton plant, with notes on the occurrence of similar glands in related plants. *Jour. Agr. Res.* 13:419–436. 1918.

Wardrop, A. B. The mechanism of surface growth involved in the differentiation of fibres and tracheids. *Austral. Jour. Bot.* 2:165–175. 1954.

Wardrop, A. B. Cell wall organization in higher plants. I. The primary wall. *Bot. Rev.* 28:242–285. 1962.

Wardrop, A. B., and D. E. Bland. The process of lignification in woody plants. In: *Fourth International Congress of Biochemistry.* Vol. 2. *Biochemistry of Wood.* London, Pergamon Press. 1959.

Wardrop, A. B., W. Liese, and G. W. Davies. The nature of the wart structure in conifer tracheids. *Holzforschung* 13:115–120. 1959.

Wareham, R. T. "Phragmospheres" and the "multinucleate phase" in stem development. *Amer. Jour. Bot.* 23:591–597. 1936.

Whaley, W. G., and H. H. Mollenhauer. The Golgi apparatus and cell plate formation—a postulate. *Jour. Cell Biol.* 17:216–221. 1963.

Whaley, W. G., H. H. Mollenhauer, and J. H. Leech. The ultrastructure of the meristematic cell. *Amer. Jour. Bot.* 47:401–449. 1960.

Wilson, K. Extension growth in primary cell-walls with special reference to *Hippuris vulgaris. Ann. Bot.* 22:449–456. 1958.

Wilson, K. Pit-field distribution in relation to cell growth in dwarf peas, as affected by gibberellic acid. *Ann. Bot.* 25:363–372. 1961.

Ziegenspeck, H. Anlage und Teilung der Tüpfel des sich stark streckenden Grundgewebes im Lichte der Dichroskopie. *Phyton* (Austria) 4:300–310. 1953.

4

Meristems
and Differentiation

MERISTEMS AND GROWTH OF THE PLANT

Beginning with the division of the fertilized egg cell, the vascular plant generally produces new cells and forms new organs until it dies. In the early embryonic stages of development reproduction of cells occurs throughout the young organism, but as the embryo enlarges and develops into an independent plant the addition of new cells is gradually restricted to certain parts of the plant body, while other parts become concerned with activities other than growth. Thus portions of embryonic tissue persist in the plant throughout its life, and the adult plant is a composite of adult and juvenile tissues. These perpetually young tissues, primarily concerned with the formation of new cells, are the *meristems.*

The concentration of cell reproduction in certain parts of the plant body seems to have evolved with the phylogenetic increase in elaboration of the plant organism. In the most primitive nonvascular plants all cells are essentially alike, all take part in metabolism, photosynthesis, building of new protoplasm, and multiplication by division. With the progressive evolutionary specialization of tissues the function of cell division became largely confined to the meristems and their immediate derivatives. The presence of meristems strikingly differentiates the plant from the animal. In the plant, growth resulting from meristematic activity is possible throughout the life of the organism, whereas in the animal body the multiplication of cells mostly ceases after the organism attains adult size and the number of organs is fixed.

The term *meristem* (from the Greek *meristos,* meaning divisible) emphasizes the cell-division activity characteristic of the tissue which bears this name. Obviously, synthesis of new living substance is a fundamental part of the process of formation of new cells by division. Living tissues other than the meristems may produce new cells, but the meristems carry on such activity indefinitely, for they not only add cells to the plant body but also perpetuate themselves; that is, some of the products of division in the meristems do not develop into adult cells but remain meristematic.

Meristems are concerned with growth in the broad sense of the term, increase in mass or size or both (sometimes qualified as irreversible increase; Bloch, 1961; Whaley, 1961). Cell division may occur without an increase in size of the entity concerned (for example, formation of a gametophyte in a microspore or transformation of a multinucleate endosperm into a cellular), but usually cells enlarge before each division. Even if no cell enlargement occurs, substance is added to the system in the form of protoplasm and cell-wall material. Thus, cell reproduction is a growth process. Some authors (Haber and Foard, 1963) consider cell division as a process distinct from growth because cell division as such does not contribute to the increase in size of a structure. In this book the broad definition of growth—that including both the formation of new cells and the enlargement, or expansion, of cells—is used. In meristematic activity, growth may be approximately divided in two stages: growth with cell division and limited cell enlargement and growth without cell division and pronounced cell enlargement. The change from one to the other is more or less gradual.

Since meristems occur at the apices of all shoots and roots, main and lateral, their number in a single plant may be large. Furthermore, many plants possess additional extensive meristems, the vascular and cork cambia, concerned with the secondary increase in thickness of the axis. The combined activities of all these meristems give rise to a complex, and often large, plant body. The primary growth, initiated in the apical meristems, expands the plant body, increases its surface and its area of contact with air and soil, and eventually produces the reproductive organs. The cambia, on the other hand, aid in maintenance of the expanding body by increasing the volume of the conducting system and forming supporting and protecting cells.

Not all apical meristems present on a given plant are necessarily active. One of the well-known examples of inhibition of growth in such meristems is that dependent on the physiologic relationship between the main shoot and the lateral buds. In some plants the growth of the lateral buds is suppressed as long as the terminal shoot is actively growing. The activity of the cambia also varies in intensity, and both the apical meristems and the cambia may show seasonal fluctuations in their meristematic activity, with a slowing down or a complete cessation of cell division during the winter in the temperate zones.

MERISTEMS AND MATURE TISSUES

In the preceding discussion the meristems were described as formative tissues adding new cells to the plant body and at the same time perpetuating themselves as meristems. Thus in active meristems a part of the products of cell division remain meristematic, the *initiating cells*, and others develop into the various tissue elements, the *derivatives* of the initiating cells. In this development the derivatives gradually change, chemically, physiologically, and morphologically, and assume more or less specialized characteristics. In other words, the derivatives *differentiate* into the specific elements of the various tissue systems. The developing cell becomes different in two senses: first, it assumes characteristics that distinguish it from its meristematic precursors, and, second, it diverges from cells of similar age by following a different line of specialization.

Since the cells of vascular plants vary so much in their function and their morphologic characteristics, they also vary in details of differentiation. Moreover, different types of cells attain different degrees of differentiation as compared with their common meristematic precursors. Some diverge relatively little from the meristematic cells and retain the power of division to a high degree (various parenchyma cells); others are more thoroughly modified and lose most, or all, of their former meristematic potentialities (sieve elements, fibers, tracheary elements).

These variously differentiated cells may be considered *mature* in the sense that they have reached the degree of specialization and physiologic stability that normally characterizes them as components of certain tissues of an adult plant part. Such a concept of maturity includes the qualification that living cells may resume meristematic activities when properly stimulated. The literature contains numerous examples of fully differentiated but living cells changing morphologically and physiologically as a result of changes in the environmental conditions, such as may be induced by various stimuli (Steward and Ram, 1961), wounding (Bloch, 1941, 1952), or physiologic isolation (Gautheret, 1959). Some workers visualize a combination of the processes of *dedifferentiation* (loss of previously developed characteristics) and *redifferentiation* (development of new characteristics) in this assumption of new characteristics by a cell (Bloch, 1961).

For variable lengths of time, during the differentiation of tissues from meristems, the derivatives of meristematic cells synthesize protoplasm, enlarge, and divide. These processes of growth may persist to some degree even after the derivatives show indications of differentiation into specific kinds of cells. It is, therefore, difficult to delimit the meristem proper from its recent derivatives, and the term meristem is often used broadly to designate not only the cell complexes that show no evidence of specialization but also those whose future course of development is partly determined. The development of meristematic derivatives into mature cells is also gradual. Some activities characteristic of mature tissues (photosynthesis, starch storage) may occur while these tissues are still developing. Such overlapping of adult and juvenile characteristics makes it impossible to delimit precisely the different stages of development.

CLASSIFICATION OF MERISTEMS

Apical and Lateral Meristems

One of the most common groupings of plant meristems is based on their position in the plant body. This classification divides the formative tissues into *apical meristems,* that is, meristems located at the apices of main and lateral shoots and roots, and *lateral meristems,* that is, meristems arranged parallel with the sides of the organ in which they occur. The vascular cambium and the cork cambium (or phellogen) are lateral meristems (figs. 1.2, 1.3).

Primary and Secondary Meristems

Another classification divides the meristems into primary and secondary according to the nature of the cells that give origin to these meristems. If these cells are the direct descendants of the embryonic cells that never ceased to be concerned with growth, the meristems are called primary. If, however, the cells first differentiate and function as members of some mature tissue system, then again take up meristematic activity, the resulting meristem is called secondary. This classification of meristems is no longer popular because it is based on the obsolete concept that cells returning to a meristematic state undergo a profound readjustment—a dedifferentiation—and that they reacquire the meristematic potentialities. Although experimental studies with living tissues and cells (Gautheret, 1959) indicate that the meristematic and the histogenetic potentialities of cells are affected by their development as members of certain tissue systems, the degree of such physiologic differentiation is highly variable, and no means have been found as yet to distinguish between an acceleration of meristematic activity that had never ceased and a resumption of such activity after a period of inactivity.

The classification of meristems into primary and secondary on the basis of their origin is not employed in this book. The expressions *primary meristems* and *secondary meristems* are used only if it is necessary to indicate the relative time of origin of the meristem in a given plant or one of its organs. This classification is related to the corresponding distinction into primary and secondary parts of the plant body (chapter 1). The fundamental parts of this body, its root and stem axes, their branches and appendages, constitute the primary parts, and they originate from primary meristems. The additional tissues that may be formed after primary growth are secondary. These tissues may arise from distinct meristems—secondary meristems—or by diffuse meristematic activity, the diffuse secondary growth (Tomlinson, 1961). If this classification is correlated with the topographical classification, the apical meristems correspond to the primary meristems, the lateral to the secondary meristems.

In descriptions of the primary differentiation at the apices of root and shoot, the initiating cells and their most recent derivatives may be distinguished, under the name of *protomeristem* (Jackson, 1953), from the partly differentiated but still meristematic subjacent tissues, and the meristematic tissues are segregated according to the tissue systems that are derived from them (pl. 14A). These tissues are: the *protoderm*, which differentiates into the epidermal system; the *procambium* (also called *provascular tissue*), which gives rise to the primary vascular tissues; and the *ground meristem*, the precursor of the fundamental or ground tissue system. If the term meristem is used broadly, the protoderm, the procambium, and the ground meristem are referred to as the primary meristems (Haberlandt, 1914). In a more restricted sense, these three cell complexes constitute the partly determined primary meristematic tissues (Foster, 1949).

The terms protoderm, procambium, and ground meristem serve well for describing the pattern of differentiation in plant organs, and they are correlated with the equally simple and convenient classification of mature tissues into the three systems, epidermal, vascular, and fundamental, reviewed in the first chapter. It seems immaterial whether the protoderm, the procambium, and the ground meristem are called meristems or meristematic tissues as long as it is understood that they are tissues whose future course of development is partly determined.

Intercalary Meristems

The term *intercalary meristem* is used to designate an actively growing primary tissue region somewhat removed from the apical meristem. The word intercalary implies that the meristem is inserted between more or less differentiated tissue regions. The intercalary meristems are often grouped with the apical and the lateral meristems on the basis of position. Such grouping is not to be recommended. The intercalary growth regions contain differentiated tissue elements and eventually are completely transformed into mature tissues. Thus, they deserve the appellation of meristems only if the term is used in its wide connotation, and as meristems they are not of the same rank as the apical and lateral meristems.

The best-known examples of intercalary meristems are those found in internodes and leaf sheaths of many monocotyledons, particularly grasses (fig. 4.1; Artschwager, 1948; Lehmann, 1906; Prat, 1935), and in *Equisetum* (Golub and Wetmore, 1948). The relation between the apical meristem and the intercalary meristem is well understood in grasses (Sharman, 1942). The youngest portion of the shoot originating from the apical meristem has no internodes as such. These develop through division and expansion of cells at the bases of leaf insertions. The superposed leaf insertions or nodes thus are separated from each other by intercalary growth. The intercalated portions are the internodes. At first the cells divide throughout the young internode, but later the meristematic activity becomes confined to a more or less restricted region, frequently located at the base of the internode (fig. 4.1). The leaf elongates similarly, and in it, too, cell division gradually becomes confined to the lowermost region of the sheath. After the internodes and the leaf sheaths complete their elongation, their basal parts retain, for an extended time, the potentiality for further growth, although fully differentiated vascular and supporting cells are present in them. These potentially meristematic regions occur at the *joints*, or *pulvini* (sing. *pulvinus*, in Latin, a cushion), characterized by a thickened region in the leaf sheath (pl. 59C) or the stem. The pulvini exhibit their meristematic potentiality when the culm rises after lodging by curving away from the ground (pl. 59D). This curvature results from enlargement and division of cells located on the lower side of the lodged culm. Such growth in a pulvinus is not unlimited. As the plant ages, the pulvinus becomes mature throughout and loses its meristematic potentialities.

Being inserted between mature tissue regions, an intercalary meristem would interrupt the continuity of the vascular tissues and would weaken the structure of leaf and stem if it were completely undifferentiated. It has been shown in many monocotyledonous stems (Buchholz, 1920; Lehmann, 1906) and in the *Arachis* gynophore—an organ which elongates through the meristematic activity at the base of the ovary and forces the fruit, the peanut, beneath the ground (Jacobs, 1947)—that

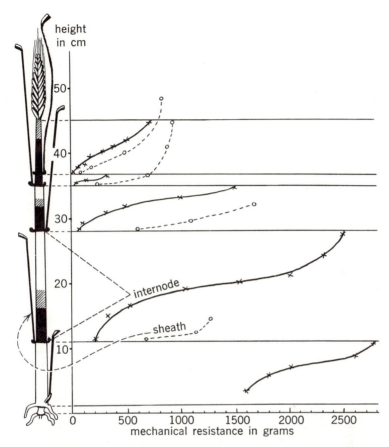

FIG. 4.1. Distribution of growth regions in a culm of rye plant. Plant has five internodes and a spike. Leaf sheaths are represented as extending upward from each node and terminating where leaf blades (shown only in part) diverge from them. Youngest tissue in internodes (intercalary meristems) is represented in black, somewhat older tissue is hatched, most mature is left white. Curves to the right indicate mechanical resistance of internodal tissues (solid lines) and of sheaths (broken lines) at the various levels of shoot. Resistance was equated with pressure, expressed in grams, necessary to make a transverse cut through internode or sheath. (After Prat, *Ann. des Sci. Nat., Bot.* 17, 1935.)

the intercalary meristems have vascular tissues while they are actively growing. The pulvini of grasses, which are active in growth only under certain conditions, have vascular and supporting cells of such a nature that they are capable of some extension and do not greatly hinder the elongation, if it is taking place (Artschwager, 1948; Lehmann, 1906).

Growth by means of intercalary meristems is not an unusual or a specialized phenomenon. Basically, all vegetative shoots articulated into nodes and internodes elongate in the manner described for the grasses: the nodes bearing leaf primordia are produced in close succession at the shoot apex and become separated one from the other by the development of internodes (pl. 14A). This phenomenon varies in intensity, timing, and degree of localization of the actively dividing region. In rosette types of plants the earlier-formed internodes fail to elongate, whereas the later-formed ones may elongate suddenly and rapidly in preparation for flowering. Obviously, the elongation of internodes contributes more to the total length of the shoot than the direct productions by the apical meristem. The activity of the internodal intercalary meristems is one of the many expressions of the primary growth that is responsible for the final form and size of plant organs. Leaves, flowers, and fruits show cell division for some time after they are initiated at the apex, and their prolonged increase in size may be regarded as an intercalary growth, less localized than that found in some internodes.

CYTOLOGIC CHARACTERISTICS OF MERISTEMS

Meristems show variable cytologic structure and are not fundamentally different from mature living tissues. During active divisions meristematic cells generally lack ergastic inclusions, and their plastids are in proplastid stages. They have a smaller amount of endoplasmic reticulum and a less elaborate internal structure in their mitochondria than the metabolically highly active parenchyma cells have. They are, in other words, relatively undifferentiated. But cork cambium may have chloroplasts, the ray initials of the vascular cambium may include starch and tannins, and the meristems of embryos usually contain various storage materials.

The degree of vacuolation of the meristematic cells varies greatly. Apical meristems contain cells with dense protoplasts (pl. 16A, B). If vacuoles are present, they are small and dispersed in the cytoplasm (Zirkle, 1932). Many plants, notably cryptogams and gymnosperms, have some conspicuously vacuolated cells in the apical meristems (pls. 16C, D; 17A; chapter 5), and the initials in the vascular cambium may be as highly vacuolated as plant-hair cells (pls. 21, 22; Bailey, 1930). In general, the larger the meristematic cell, the greater the relative amount of vacuolar material (Zirkle, 1932).

Meristematic cells are usually described as having large nuclei. But the ratio between the size of the cell and the size of the nucleus—the cytonuclear ratio—varies considerably (Trombetta, 1942). In general, the nuclei of large meristematic cells are relatively smaller in proportion to the size of the cell than those of the smaller cells. The sizes of entire meristematic cells and their shapes are also variable characteristics. One extreme is the small, nearly isodiametric cell common in apical meristems, the other, the long, narrow, fusiform initial in the vascular cambium. No less striking are the differences in wall thickness. Although commonly the meristematic cells have thin walls (pl. 17B), certain zones of the apical meristems may have thick primary walls (pl. 17A) with conspicuous primary pit-fields; and the cambial initials sometimes develop remarkably thick walls with deeply depressed primary pit-fields. Intercellular spaces are generally absent in the meristems, but they may appear very early among the still dividing derivatives (this feature is particularly well illustrated by the roots; chapter 5). One should expect biochemical differences between meristematic and nonmeristematic cells, but the biochemical approach to meristem characterization has not been intensively explored and the available information suggests considerable variation among similar meristems in different groups of plants (Steward et al., 1955). In relation to their high rate of metabolic activity, the meristematic tissues give a particularly strong reaction for peroxidase (Van Fleet, 1959). The enzyme occurs in the tissues before and during the division stage and declines after the divisions are completed.

The foregoing statements show that it would be erroneous to emphasize a certain set of characteristics as being typical of the meristematic cells. Nevertheless, the lack of conspicuous vacuolation is most commonly encountered in the meristematic tissues, and small, essentially isodiametric cells with thin walls occur in the meristems more usually than in other kinds of tissues. In recognition of the variability in the characteristics of the meristems, the term *eumeristem,* that is, true meristem, has been suggested for the designation of a meristem composed of small, approximately isodiametric cells, that have thin walls, and are rich in cytoplasm (Kaplan, 1937). This term, used judiciously, with the understanding that in a morphologic or physiologic sense a "typical meristematic cell" does not exist, often serves well for descriptive purposes.

GROWTH PATTERNS IN MERISTEMS

Meristems and meristematic tissues show varied arrangements of cells resulting from varied patterns of cell division and associated cell enlargement. Apical meristems having only one initiating cell (*Equisetum* and many ferns, fig. 5.1) have an orderly distribution of cells. In the higher plants the sequence of cell divisions in the apices is less precise, but it is not random either, for an apical meristem grows as an organized whole and the divisions and enlargement of individual cells are related to the internal distribution of growth and to the external form of the apex (Wardlaw, 1952; Wetmore and Wardlaw, 1951). These correlative influences bring about a differentiation of distinctive zones in the meristems. In some parts of the meristem the cells may divide sluggishly and attain considerable dimensions; in others they may divide frequently and remain small (pl. 17A). Some cell complexes divide in various planes (volume growth), others only by walls at right angles to the surface of the meristem (*anticlinal* divisions, surface growth).

The lateral meristems are particularly distinguished by divisions parallel with the nearest surface of the organ (*periclinal* divisions), which result in establishment of rows of cells parallel with the radii of the axes (radial seriation or alignment) and an increase in the thickness of the organ. Radial alignment is so characteristic of the immediate derivatives of the vascular cambium (pl. 21), and those of the phellogen (pl. 65), that it is often regarded as a certain indication of secondary growth. Radial seriation of cells, however, may originate during various stages of primary growth (Esau, 1943).

In cylindrical bodies, such as stems and roots, the term *tangential* division (or tangential longitudinal) is commonly used instead of periclinal division; the anticlinal division is *radial* (or radial longitudinal) if it occurs parallel with the radius of the cylinder and *transverse* if the new wall is laid down at right angles to the longitudinal axis of the cylinder.

Organs arising at the same apical meristem may subsequently assume varied forms because the still-meristematic derivatives of the apical meristems (primary meristems in the wide sense) often exhibit distinct patterns of growth. Indeed, some of these growth patterns are so characteristic that the meristematic tissues showing them have received special names. These are: the *mass meristem* (or block meristem), the *rib meristem* (or file meristem), and the *plate meristem* (Schüepp, 1926). The mass meristem grows by divisions in all planes and produces bodies that are isodiametric or spheroidal or have no definite shape. The best examples of such growth are found in reproductive organs during the formation of spores, sperms (in lower vascular plants), and endosperm, and in young embryos of some plants. The rib meristem (pl. 16C, 17C) gives rise to a complex of parallel longitudinal files ("ribs") of cells by divisions at right angles to the longitudinal axis of the cell row and also to the longitudinal axis of the plant organ. This pattern of growth is characteristic of cylindrical plant parts and is well illustrated by the development of the cortex of the root and of the pith and the cortex of the stem. The plate meristem shows chiefly anticlinal divisions so that the number of layers originally established in the young organ does not increase any further, and a plate-like structure is produced. The result of growth by a plate meristem is exemplified by the flat blades of angiosperm leaves (pl. 74). The plate meristem and the rib meristem are growth forms that occur mainly in the ground meristem. They determine the two basic forms of the plant body, the thin spreading lamina (blade) of the leaf-like organs, on the one hand, and the elongated cylindrical structures found in the root, the stem, the petiole, and the vein rib of the leaf, on the other.

DIFFERENTIATION

Concept

In a preceding part of this chapter differentiation was interpreted, at the cellular level, as the development of the derivatives of meristems into the elements of the various tissue systems of the adult plant body. In this sense differentiation comprises the many interrelated processes of chemical, physiological, and morphological nature which bring about the specialization of cells. Since the degree and kind of specialization vary in different cells, cellular differentiation ultimately results in the histological diversity characteristic of the bodies of higher plants.

Tissues that have completed their development are differentiated or mature tissues. Frequently the term differentiated is used to express not only the attainment of a certain state of development but also the occurrence of variations in structure and function resulting from developmental changes within a given cell, tissue, tissue system, or organ. One might say, for example, that certain walls of the sieve-tube elements are differentiated into sieve plates; that the xylem tissue is differentiated into tracheary elements, fibers, and parenchyma, and the vascular tissue system into xylem and phloem; or that the plant body is differentiated into root, stem, and leaf. In this sense it is acceptable to speak of differentiation in the meristem itself if the latter shows variations in the nature of the component cells.

The variation in the degree of specialization of developing cells has been previously emphasized. Many cells are so strongly modified during differentiation that they ultimately reach an irreversible state. Such a state is usually associated with a profound alteration of the protoplast or its complete disappearance. The cell thus loses the capacity to dedifferentiate and to resume meristematic activity.

Cellular Basis of Differentiation

During the differentiation of tissues, histologic diversity results from changes in the characteristics of individual cells and from readjustments in the intercellular relationships. The common alterations in the contents of differentiating cells have been mentioned in chapter 2 and need but a brief recapitulation here: conspicuous increase in the amount of vacuolar sap if the meristematic cell itself is not yet highly vacuolated; accumulation of various ergastic substances; development of plastids from the proplastids; and acquisition of color by the plastids. In highly specialized cells the protoplast or parts of it may disappear.

A nuclear phenomenon frequently encountered in cells issuing from the meristematic state is endomitotic polyploidy or endopolyploidy, that is, polyploidy resulting from nuclear division which is not followed by cell division (Partanen, 1959; Tschermak-Woess, 1956). Polyploidy has been observed in all kinds of tissues but in some tissues the phenomenon appears to be more common than in others. It is widespread in parenchyma tissues storing food and water but is less frequent in photosynthetic parenchyma and the epidermis. Polyploidization is one of the numerous features of cell differentiation and is associated with increases in nuclear volume and in DNA content (Clowes, 1961; List, 1963).

The changes in wall structure during the development of a cell have been considered in chapter 3. The increase in thickness, primary or secondary, often produces striking differences among cells. The chemistry of walls may change appreciably through lignification, suberization, or silicification. In certain cell types, such as the vessel elements, parts of the wall are removed.

One of the first gross differences that appear among the developing cells is the unequal increase in size. Some cells continue to divide with small increase in size; others cease dividing and enlarge considerably. Examples of differential growth in size are found in the elongation of the procambial cells in contrast to the lack of similar elongation of cells in the adjacent pith and cortex; or of the elements of the first sieve tubes in contrast to that of the adjacent parenchyma cells (fig. 4.2A). Size differences between two adjacent cells may also result from unequal divisions. In some plants, for example, root hairs develop from cells which are the smaller of two sister cells formed by the division of protodermal cells (fig. 4.2B, C; chapter 7).

The increase in size of a cell may be relatively uniform, but frequently the cell enlarges more in one direction than in another and thereby assumes a new form. Some cells are strikingly different in shape from their meristematic precursors (primary phloem fibers, branched sclereids, laticiferous cells); many, however, become modified in a less spectacular manner, with a change in the number of facets but a retention of the general shape (Hulbary, 1944).

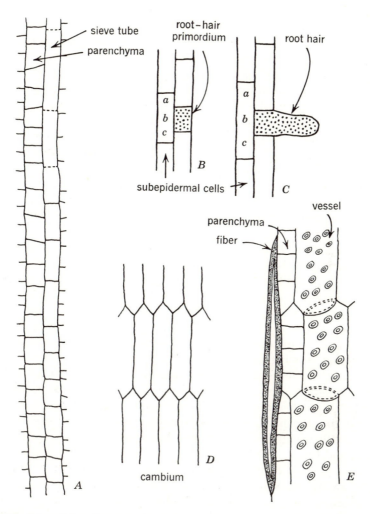

FIG. 4.2. Intercellular adjustments during tissue differentiation. *A,* series of cells from root tip of tobacco. Parenchyma cells continued dividing; phloem cells ceased dividing and began elongating. *B, C,* development of root hair from smaller of two sister cells resulting from transverse division of protodermal cell. *C,* root-hair cell is extended at a right angle to the root but not in direction of elongation of root. Apparently, in cell adjacent to root hair, wall parts *a* and *c* continued to elongate, whereas part *b* ceased to elongate after root-hair primordium was formed. *D, E,* cambium and xylem that could develop from such cambium, both in tangential sections. *E* shows results of following developmental changes in cambial derivatives: parenchyma cells were formed by transverse divisions of such derivatives; vessel elements expanded laterally; fiber elongated by apical intrusive growth.

The predominant cell arrangement in a tissue may be determined early by the growth form of its meristem (rib meristem, plate meristem). The relative position of walls in adjacent cell rows also gives a distinctive appearance to a tissue (Sinnott, 1960). Most commonly new walls alternate with the old ones in the adjacent cell row (fig. 4.2A), but in some tissues (cork, cortex of certain roots) a new wall is formed opposite the point of insertion of a previous one in the adjacent row.

The enlargement and the change in the shape of cells in a differentiating tissue are accompanied by various readjustments in the relation of the cells to each other. One of the most familiar phenomena is the appearance of intercellular spaces along the line of union of three or more cells (chapter 3). In some instances the development of intercellular spaces does not change the general arrangement of the cells, in others it profoundly modifies the appearance of the developing tissue (Hulbary, 1944).

With regard to the growth of walls during the differentiation of a tissue, two possibilities are recognized: (1) the growth of walls of adjacent cells is so adjusted that no separation of the walls occurs; (2) a separation of the walls occurs, and the growing cell comes to occupy the space formed through the separation. The first method of growth, sometimes called *symplastic growth* (Priestley, 1930), is common in organs expanding during their primary growth. Whether all cells in a complex are still dividing, or whether some have ceased to divide and are elongating (fig. 4.2A), the walls of adjacent cells appear to grow in unison, for there is no evidence of separation or buckling of walls. In this coordinated growth it is possible that part of a given wall is expanding and another is not if the two parts are associated with walls of two cells, one of which is still growing while the other has ceased to do so (fig. 4.2B, C; Sinnott and Bloch, 1939).

The second type of intercellular adjustment, that involving an intrusion of cells among others, is called *intrusive growth* (Sinnott ond Bloch, 1939) or *interpositional growth* (Schoch-Bodmer, 1945). The occurrence of such growth in the elongation of cambial initials, of primary and secondary fibers (fig. 4.2D, E), of tracheids, and of certain other cells has been well established by careful observations (Bailey, 1944; Bannan, 1956; Bannan and Whalley, 1950; Schoch-Bodmer and Huber, 1951, 1952). One of the most spectacular examples of elongation by intrusive growth is found in certain woody Liliaceae in which the secondary tracheids may become 15 to 40 times longer than the original meristematic cells (Cheadle, 1937). The elongating cells grow at their apices (*apical intrusive growth*), usually at both ends. The intercellular material seems to change in front of the advancing tip, and the primary walls of the adjacent cells become separated from each other in the same manner as they do during the formation of intercellular spaces. The common assumption is that, if plasmodesmata are present in front of the advancing tip, they are ruptured. This phenomenon has not been actually observed, but the separation of members of pairs of primary pit-fields has been noted (Neeff, 1914). Pit-pairs later appear between pairs of cells that come in contact through intrusive growth (Bannan, 1950; Bannan and Whalley, 1950). Intrusive growth also occurs in connection with the lateral expansion of cells attaining considerable width (vessel members, fig. 4.2E; chapter 11).

Early botanists assumed the occurrence of a gliding or sliding growth in the process of adjustment among differentially elongating or laterally expanding cells. The concept of gliding growth implies that a large part of a wall of a given cell expands in area and glides over the walls of other cells with which the enlarging cell is in contact before the growth begins (Krabbe, 1886; Neeff, 1914). Intrusive growth, on the contrary, is visualized as a localized extension of a wall without severance of contacts between the enlarging cell and its neighbors. Whether such localized extension involves some gliding of the new wall part over the walls of cells with which new contacts are established (Bannan, 1951) or whether the new wall is apposed along the outer surfaces of the cells that are being pushed apart (Schoch-Bodmer, 1945) is still problematical. Certain intercellular readjustments are best explained by an assumption of severance of contacts and gliding of walls (Bannan, 1951; Neeff, 1914), but intrusive growth appears to be by far the more common phenomenon. Some workers have attempted to explain all intercellular adjustments by symplastic growth (Meeuse, 1942) despite the availability of strong evidence supporting the concept of intrusive growth.

Causal Aspects of Differentiation

Growth and differentiation, which occur during the ontogeny (development of an individual) of the plant, are coordinated and regulated in such a way that the resultant plant assumes a specific form; in other words, the developing plant exhibits the phenomenon of morphogenesis (origin of form; from Greek words for shape and beginning). The term morphogenesis can be used not only with reference to the development of the external form but also to that of the internal organization. Moreover, the phenomenon of morphogenesis is revealed at various levels of organization and one can speak of the morphogenesis of individual plant organs, tissues, cells, and even components of cells. Many investigators treat the study of morphogenesis as causal morphology; that is, they seek to uncover the external and internal factors that regulate growth and differentiation, and to explain the mode of action of these factors (Wardlaw, 1952; Wetmore, 1959). (Some authors use the word morphogenesis itself for denoting the study of morphogenesis; cf. Sinnott, 1960.) These inquiries have resulted in a large body of data on possible mechanisms controlling the establishment of the external form and the histologic patterns in the internal structure of the plant (Bünning, 1953; Konarev, 1959; Sinnott, 1960; Wardlaw, 1952, 1955).

Studies in morphogenesis include observations on plants developing normally and those whose development is subjected to experimental modifications of various kinds. Examples of experimental treatments are use of chemicals, surgery, exposure to radiation, to selected day lengths or temperatures, and to mechanical stimuli. Tissue-culture methods play a particularly important role because they enable one to determine the requirements for growth of specific cells and to isolate the individual factors of growth more precisely than in work with intact plants.

Studies in morphogenesis reveal the existence of controlling mechanisms which effect the development of the plant as an integrated and organized system, that is, an organism (Erickson, 1959). Although the characteristics of the plant are primarily determined by genes, a long and complex series of processes intervenes between the primary action of genes and their ultimate effect on the morphologic character. An array of regulatory substances is produced in special tissues and these may exercise control over the responses of cells so that similar primary genic effects may result in different final expressions. The relationship is further complicated by the modifying effects of the environment to which the plant is exposed throughout its development. The various stimuli and effects and the actions of genes and enzymes have a chemical basis and must be explained at the molecular level. But a full interpretation of growth and morphogenesis will not result unless levels of organization above the molecular are also understood (Steward and Ram, 1961).

Meristematic Potentialities of Cells. One of the foremost questions in morphogenetic considerations is that concerning the developmental potency of individual cells which are members of the organized plant body. Within the plant, meristematic and mature cells are distributed in characteristic patterns. The prevailing concept is that cells assume their specific characteristics and functions in relation to their position in the plant body. This positional relation is an expression of the integrational control of differentiation of individual cells within the plant. Tissue culture provides a means of releasing the cells from the controlling mechanisms and thus testing their potentialities for growth.

As was mentioned before, some cells undergo such a high degree of specialization during differentiation that they lose their potentiality for growth. This course of events is best exemplified by cells in which the protoplasts are much altered at maturity or are absent. The presence of an active protoplast, however, does not assure that a given cell has not undergone irreversible changes. Studies on cultured tissues and on phenomena of regeneration and wound repair suggest that living cells may become limited in their meristematic potentialities (Bloch, 1941, 1944; Gautheret, 1959; Steward and Ram, 1961). At the same time, development of new techniques of tissue culture often results in successful culturing of tissues that previously seemed to have lost their potency for further growth. But the fact that special conditions and stimulants are necessary for evoking this growth is in itself an evidence of a limitation of ability to resume meristematic activity.

The technique of culturing cells in free-floating or dissociated state yields particularly instructive information with regard to the potentialities of cells released from the control of the whole organism. In cultures of phloem-parenchyma cells from carrot root (Steward, 1964), the cells first developed into randomly proliferating masses, then produced a more orderly type of growth: nodules with centrally located xylem were formed. Such nodules eventually produced roots and then shoots opposite the roots. The resulting plantlets assumed the characteristics of young carrot plants. It appeared as though the formative process of the embryo in the ovule was repeated in the tissue culture, with the nodule acting as a zygote (Steward and Shantz, 1959).

The experiment indicates that the potential toward organized growth is present in individual cells, but it also suggests that the potential is realized only under a proper balance of factors that promote growth and differentiation. If these factors are not regulated—if, for example, there is excess of nutrients—an unorganized tumorous growth occurs. It is conceivable that formation of a nodule removes the central cells from excess nutrients and thus establishes a regulatory mechanism and makes possible an organized growth (Steward et al., 1958).

Another experiment has revealed that the potency of cells with regard to organized development is less restricted in younger than in older tissues. In suspensions of cells from proliferated embryos of the carrot, many cells produced embryo-like forms that recapitulated the developmental stages of the normal embryo and became viable plants (Steward, 1964).

Internal Factors of Differentiation. Among the internal factors of differentiation, polarity, gradients, inductive effects, and mutual incompatibility of regions of vigorous growth are prominently considered in the literature on morphogenesis. Polarity refers to the orientation of activities in space. Though apparently initially induced by directing external factors (Bünning, 1952; Sinnott, 1960), polarity manifests itself early in the life of the plant and is evident in the bipolar development of the embryo from the zygote. It is later expressed in the external and internal organization of the plant into the root and the shoot and is also evident in various phenomena at the cellular level. Transplantation experiments (Gulline, 1960) and tissue culture studies (Wetmore and Sorokin, 1955) indicate that polarity is exhibited not only by the plant as a whole but also by its parts, even if these parts are isolated from the plant.

An illustration of polar behavior of individual cells within the plant body is the unequal division resulting in physiologically, and often also morphologically, different daughter cells. Unequal divisions, for example, occur in the epidermis of certain roots. After an unequal division only the smaller of the two products of a division produces a root hair (fig. 4.2B, C). Before the division, the cytoplasm appears to accumulate at the apical end of the cell (end toward the root apex) and the nucleus migrates in this direction. The nucleus divides; the cell plate is formed and separates the small future root-hair cell from the longer epidermal cell that develops no root hair (Sinnott, 1960). Biochemical differences between the two kinds of cells also become evident (Avers and Grimm, 1959). The common thought is that unequal divisions depend on a polarization of cytoplasm, because there is no evidence of unequal distribution of chromosomal material (Stebbins and Jain, 1960).

Polarity is related to the phenomenon of gradients, since the differences between the two poles of the plant axis appear in graded series. There are physiologic gradients, for example, those expressed in the rates of metabolic processes, in concentration of auxins, and in concentration of sugar in the conducting system; there are also gradients in anatomic differentiation and in the development of the external features (Prat, 1948, 1951). The plant axis shows many transitional anatomic and histologic characteristics in the transition from the root to the stem (chapter 17); the differentiation of the derivatives of meristems, in general, occurs in graded series, and adjacent but different tissues may show different gradients. Externally, graduated development is seen in the change in the form of the successive leaves along the axis, from the usually smaller and simple juvenile form to the larger and more elaborate adult form. Subsequently, after the reproductive stage is induced, smaller leaves are gradually produced, the series becoming completed with inflorescence bracts which support subdivisions of the inflorescence or the individual flowers.

The existence of inductive effects is frequently deduced from patterns of development in which similar structures appear side by side, one structure preceding the other in development. Common examples are the inception of interfascicular cambial divisions next to the previously established fascicular cambium in stems beginning secondary growth and the origin of vascular and cork cambia in wound healing and grafting (chapter 15). Studies on induction of divisions and differentiation of vascular elements in callus tissue into which a shoot tip is grafted indicate that hormonal factors and sugar concentrations are involved in these kinds of inductions (Wetmore and Rier, 1963; Wetmore and Sorokin, 1955).

A phenomenon readily interpreted as an induction effected by an individual cell within the plant body may be observed in the differentiation of stomata in the monocotyledons (Stebbins and Jain, 1960; Stebbins and Shah, 1960). In the formation of subsidiary cells of the guard cells, the divisions of the epidermal cells next to the guard-cell precursor appear to be governed by this precursor. Moreover, the sequences and results of divisions may be interpreted as indicating that with regard to the mechanism of induction the guard-cell precursors are highly independent from other cells and even from environmental conditions.

The mutual incompatibility of regions of vigorous cytoplasmic synthesis is seen as a factor determining the distribution of cells and cell complexes in characteristic patterns (Bünning, 1952, 1953). Distribution of leaf primordia at apices, of stomata in a dicotyledon leaf, and of rays in secondary vascular tissues are cited as examples of such patterns. Another utilization of the idea of incompatibility between growing regions is made in the available-space concept regarding leaf inception at the shoot apex (Wardlaw, 1952). Experiments on surgical isolation of the sites of prospective and young leaf primordia seem to indicate the existence of inhibitory effects of the older on the younger leaf primordia. A new primordium arises in the space farthest removed from the influence emanating from the physiologic field of the older leaf, that is, in the next available space.

This brief review clearly indicates that internal factors modify the potentialities of a cell during its differentiation and that the modifications may be induced by cells in both distant and proximal positions with regard to the developing cell. Both inductive and repressive stimuli may be recognized and the effects of the internal factors are difficult to separate from those of the external. But all observations give evidence of the intrinsic tendency of the plant toward regulated, organized growth.

REFERENCES

Artschwager, E. Anatomy and morphology of the vegetative organs of *Sorghum vulgare*. *U.S. Dept. Agr. Tech. Bul.* 957. 1948.

Avers, C. J., and R. B. Grimm. Comparative enzyme differentiations in grass roots. II. Peroxidase. *Jour. Expt. Bot.* 10:341–344. 1959.

Bailey, I. W. The cambium and its derivative tissues. V. A reconnaissance of the vacuome in living cells. *Ztschr. f. Zellforsch. u. Mikros. Anat.* 10:651–682. 1930.

Bailey, I. W. The development of vessels in angiosperms and its significance in morphological research. *Amer. Jour. Bot.* 31:421–428. 1944.

Bannan, M. W. The frequency of anticlinal divisions in fusiform cambial cells of *Chamaecyparis*. *Amer. Jour. Bot.* 37:511–519. 1950.

Bannan, M. W. The reduction of fusiform cambial cells in *Chamaecyparis* and *Thuja*. *Canad. Jour. Bot.* 29:57–67. 1951.

Bannan, M. W. Some aspects of the elongation of fusiform cambial cells in *Thuja occidentalis* L. *Canad. Jour. Bot.* 34:175–196. 1956.

Bannan, M. W., and B. E. Whalley. The elongation of fusiform cambial cells in *Chamaecyparis*. *Canad. Jour. Res. Sect. C., Bot. Sci.* 28:341–355. 1950.

Bloch, R. Wound healing in higher plants. *Bot. Rev.* 7:110–146. 1941; II. 18:655–679. 1952.

Bloch, R. Developmental potency, differentiation and pattern in meristems of *Monstera deliciosa*. *Amer. Jour. Bot.* 31:71–77. 1944.

Bloch, R. General survey. *Handb. der Pflanzenphysiol.* 14:1–14. 1961.

Buchholz, M. Über die Wasserleitungsbahnen in den interkalaren Wachstumszonen monokotyler Sprosse. *Flora* 14:119–186. 1920.

Bünning, E. Morphogenesis in plants. In: *Survey of Biological Progress* 2:105–140. New York, Academic Press. 1952.

Bünning, E. *Entwicklungs- und Bewegungsphysiologie der Pflanze.* 3rd ed. Berlin, Springer-Verlag. 1953.

Cheadle, V. I. Secondary growth by means of a thickening ring in certain monocotyledons. *Bot. Gaz.* 98:535–555. 1937.

Clowes, F. A. L. *Apical Meristems.* Botanical Monographs Vol. 2. Oxford, Blackwell. 1961.

Erickson, R. O. Integration of plant growth processes. *Amer. Nat.* 93:225–236. 1959.

Esau, K. Origin and development of primary vascular tissues in seed plants. *Bot. Rev.* 9:125–206. 1943.

Foster, A. S. *Practical plant anatomy.* 2nd ed. New York, D. Van Nostrand Company. 1949.

Gautheret, R. J. *La culture des tissus végétaux. Techniques et realisations.* Paris, Masson et Cie. 1959.

Golub, S. J., and R. H. Wetmore. Studies of development in the vegetative shoot of *Equisetum arvense* L. I. The shoot apex. *Amer. Jour. Bot.* 35:755–767. 1948.

Gulline, H. F. Experimental morphogenesis in adventitious buds in flax. *Austral. Jour. Bot.* 8:1–10. 1960.

Haber, A. H., and D. E. Foard. Nonessentiality of concurrent cell division for degree of polarization of leaf growth. II. Evidence from untreated plants and from chemically induced changes of the degree of polarization. *Amer. Jour. Bot.* 50:937–944. 1963.

Haberlandt, G. *Physiological plant anatomy.* London, Macmillan and Company. 1914.

Hulbary, R. L. The influence of air spaces on the three-dimensional shapes of cells in *Elodea* stems, and a comparison with pith cells of *Ailanthus*. *Amer. Jour. Bot.* 31:561–580. 1944.

Jackson, B. D. *A glossary of botanic terms.* 4th ed. New York, Hafner Publishing Co. 1953.

Jacobs, W. P. The development of the gynophore of the peanut plant, *Arachis hypogaea* L. I. The distribution of mitoses, the region of greatest elongation, and the maintenance of vascular continuity in the intercalary meristem. *Amer. Jour. Bot.* 34:361–370. 1947.

Kaplan, R. Über die Bildung der Stele aus dem Urmeristem von Pteridophyten und Spermatophyten. *Planta* 27:224–268. 1937.

Konarev, V. G. *Nukleinovye kisloty i morfogenez rastenii. [Nucleic acids and morphogenesis of plants.]* Moskva, Gosudarstvennoe Izdatel'stvo "Vysshaia Shkola". 1959.

Krabbe, G. *Das gleitende Wachsthum bei der Gewebebildung der Gefässpflanzen.* Berlin, Gebrüder Borntraeger. 1886.

Lehmann, E. Zur Kenntnis der Grassgelenke. *Deut. Bot. Gesell. Ber.* 24:185–189. 1906.

List, A., Jr. Some observations on DNA content and cell and nuclear volume growth in the developing xylem cells of certain higher plants. *Amer. Jour. Bot.* 50:320–329. 1963.

Meeuse, A. D. J. A study of intercellular relationships among vegetable cells with special reference to "sliding growth" and to cell shape. *Rec. des Trav. Bot. Néerland.* 38:18–140. 1942.

Neeff, F. Über Zellumlagerung. Ein Beitrag zur experimentellen Anatomie. *Ztschr. f. Bot.* 6:465–547. 1914.

Partanen, C. R. Quantitative chromosomal changes and differentiation in plants. In: D. Rudnick. *Developmental cytology.* New York, Ronald Press Company. 1959.

Prat, H. Recherches sur la structure et le mode de croissance des chaumes. *Ann. des Sci. Nat., Bot.* Ser. 10. 17:81–145. 1935.

Prat, H. Histo-physiological gradients and plant organogenesis. *Bot. Rev.* 14:603–643. 1948; II. 17:693–746. 1951.

Priestley, J. H. Studies in the physiology of cambial activity. II. The concept of sliding growth. *New Phytol.* 29:96–140. 1930.

Schoch-Bodmer, H. Interpositionswachstum, symplastisches und gleitendes Wachstum. *Schweiz. Bot. Gesell. Ber.* 55:313–319. 1945.

Schoch-Bodmer, H., and P. Huber. Das Spitzenwachstum der Bastfasern bei *Linum usitatissimum* und *Linum perenne. Schweiz. Bot. Gesell. Ber.* 61:377–404. 1951.

Schoch-Bodmer, H., and P. Huber. Local apical growth and forking in secondary fibers. *Leeds Phil. and Lit. Soc. Proc.* 6:25–32. 1952.

Schüepp, O. Meristeme. In: K. Linsbauer. *Handbuch der Pflanzenanatomie.* Band 4. Lief. 16. 1926.

Sharman, B. C. Developmental anatomy of the shoot of *Zea mays* L. *Ann. Bot.* 6:245–282. 1942.

Sinnott, E. W. *Plant morphogenesis.* New York, McGraw-Hill Book Company. 1960.

Sinnott, E. W., and R. Bloch. Changes in intercellular relationships during the growth and differentiation of living plant tissues. *Amer. Jour. Bot.* 26:625–634. 1939.

Stebbins, G. L., and S. K. Jain. Developmental studies of cell differentiation in the epidermis of monocotyledons. I. *Allium, Rhoeo,* and *Commelina. Devlpmt. Biol.* 2:409–426. 1960.

Stebbins, G. L., and S. S. Shah. Developmental studies of cell differentiation in the epidermis of monocotyledons. II. Cytological features of stomatal development in the Gramineae. *Devlpmt. Biol.* 2:477–500. 1960.

Steward, F. C., with M. O. Mapes, A. E. Kent, and R. D. Holsten. Growth and development of cultured plant cells. *Science* 143:20–27. 1964.

Steward, F. C., and H. Y. M. Ram. Determining factors in cell growth: some implications for morphogenesis in plants. *Advances in Morphogenesis* 1:189–265. 1961.

Steward, F. C., and E. M. Shantz. Biochemistry and morphogenesis: knowledge derived from plant tissue cultures. In: *Fourth International Congress of Biochemistry.* Vol. 6. *Biochemistry of Morphogenesis.* London, Pergamon Press. 1959.

Steward, F. C., M. O. Mapes, and K. Mears. Growth and organized development of cultured cells. II. Organization in cultures grown from freely suspended cells. *Amer. Jour. Bot.* 45:705–708. 1958.

Steward, F. C., R. H. Wetmore, and J. K. Pollard. The nitrogenous components of the shoot apex of *Adiantum pedatum. Amer. Jour. Bot.* 42:946–948. 1955.

Tomlinson, P. B. *Anatomy of the monocotyledons.* II. *Palmae.* Oxford, Clarendon Press. 1961.

Trombetta, V. V. The cytonuclear ratio. *Bot. Rev.* 8:317–336. 1942.

Tschermak-Woess, E. Karyologische Pflanzenanatomie. *Protoplasma* 46:798–834. 1956.

Van Fleet, D. S. Analysis of the histochemical localization of peroxidase related to the differentiation of plant tissues. *Canad. Jour. Bot.* 37:449–458. 1959.

Wardlaw, C. W. *Phylogeny and morphogenesis.* London, Macmillan and Company. 1952.

Wardlaw, C. W. *Embryogenesis in plants.* New York, John Wiley and Sons. 1955.

Wetmore, R. H. Morphogenesis in plants—a new approach. *Amer. Scientist* 47:326–340. 1959.

Wetmore, R. H., and J. P. Rier. Experimental induction of vascular tissues in callus of angiosperms. *Amer. Jour. Bot.* 50:418–430. 1963.

Wetmore, R. H., and S. Sorokin. On the differentiation of xylem. *Arnold Arboretum Jour.* 36:305–317. 1955.

Wetmore, R. H., and C. W. Wardlaw. Experimental morphogenesis in vascular plants. *Ann. Rev. Plant Physiol.* 2:269–292. 1951.

Whaley, W. G. Growth as a general process. *Handb. der Pflanzenphysiol.* 14:71–112. 1961.

Zirkle, C. Vacuoles in primary meristems. *Ztschr. f. Zellforsch. u. Mikros. Anat.* 16:26–47. 1932.

5

Apical Meristems

The profuse and inconsistent terminology in the voluminous literature on apical meristems (Clowes, 1961a; Gifford, 1954; Guttenberg, 1960, 1961) reflects the complexity of the subject matter. Most commonly the term *apical meristem* is used in a wider sense than merely with reference to the initials and their immediate derivatives; the term also includes variable lengths of shoot or root proximal to the apex. Yet, when determinations of the dimensions of the apices of shoots are made, only the part above the youngest leaf primordium, or youngest node, is measured. Usually *shoot apex* and *root apex* are employed as synonyms of apical meristem.

The wide meaning of apical meristem is adopted in the discussions to follow, but when it is important to differentiate the most distal part of this meristem the term protomeristem is used in the sense indicated on page 71: it refers to the least determined part of the meristem and includes the initials and their most recent derivatives. The delimitation of this protomeristem is arbitrary, but the term is useful for referring to the distal part of the apical meristem, which is given much attention in the literature. The promeristem of Clowes (1961a) includes only the initials and thus does not coincide with the protomeristem. Johnson and Tolbert's (1960) metrameristem, on the other hand, refers to the same group of cells as the protomeristem.

Apical meristem and its synonyms are appropriate substitutions for the somewhat inaccurate term growing point (Foster, 1949). Growth in the sense of cell division, which is so characteristic of the meristematic state, is not restricted to the so-called growing point but occurs abundantly—and is even more intense—at some distance from the apical meristem. Similarly, growth in the sense of increase in size of cells, tissues, and organs is most pronounced not in the apical meristem but in its derivatives.

INITIALS AND DERIVATIVES

An initial, or initiating cell (p. 69), is a cell that divides into two sister cells one of which remains in the meristem and the other is added to the meristematic tissues that eventually differentiate into the various tissues characteristic of the plant. The cell remaining in the apical meristem functions as an initial like its precursor. Investigators visualize the involvement of polarity, and a consequent cytologic differentiation, in the division into an initial and derivative; at the same time they agree that the status of a cell as an initial depends on its position in the protomeristem, and that the initial may be displaced by another cell and then differentiate into a body cell.

The inference about the existence of apical initials is generally based on microscopic views and on theoretical considerations, but there is also experimental evidence regarding this matter. By treatments with colchicine it is possible to change the number of chromosomes in individual cells. When cells occupying the position of initials in the shoot apex are thus affected, the change becomes detectable and is perpetuated developmentally in more or less extended parts of the plant body that develop after the treatment, and the alterations may be traced directly to the cells in the apical meristem. These cells thus fit the definition of initials. Changes in growth may cause a shift in the relative position of the modified cells in the apical meristem so that an initial ceases to act as such (Bain and Dermen, 1944). This observation supports the concept that a cell is an initial, not because of its inherent characteristics but only because of its particular position in the meristem.

The number of initials in root and shoot apices is variable. In many vascular cryptogams a single initial cell occurs at the apex (fig. 5.1); in other lower vascular plants, as well as in the higher, several initials are present. The single initial is morphologically rather distinct from its derivatives and is customarily spoken of as the *apical cell*. If the initials are more or less numerous, they are called *apical initials*, although considered semantically it would be appropriate to call them apical cells also. The recognition of apical initials under the microscope, in contrast to that of the single apical cells, is uncertain (pls. 16, 17).

The apical initials may occur in one or more tiers. If there is only one tier, all cells of a plant body are ultimately derived from it. In the

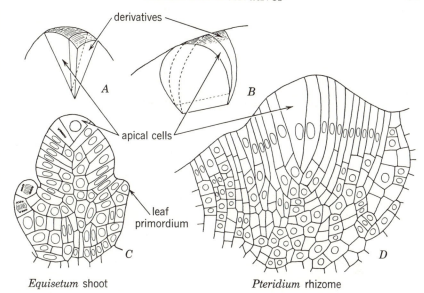

Equisetum shoot   Pteridium rhizome

FIG. 5.1. Apical cells in shoots and rhizomes. *A, B,* two forms of apical cell, pyramidal (*A*) and lenticular (*B*). Cells are cut off from three faces in the pyramidal cell, from two in the lenticular. In each drawing a derivative cell is shown attached to right side of apical cell. *C, D,* apical cells in longitudinal sections of shoot (*C*) and rhizome (*D*). In *C,* apical cells in leaf primordia; one of these (left) is dividing. (*A, B,* adapted from Schüepp, *Handbuch der Pflanzenanatomie* 4, 1926; *C, D,* ×230.)

alternative situation, different parts of a plant body are derived from different groups of initials. The existence of more than one independent layer of initials in certain plants has been clearly demonstrated in the previously mentioned experiments with colchicine. The treatment may induce polyploidy in one or more superficial layers of the apical meristem (fig. 5.2) and thus convert the plant into a cytochimera (Clowes, 1961*a*; Dermen, 1953, 1960). Induced and spontaneous chimeras showed that polyploidy could be perpetuated ontogenetically if any one of the three superficial layers in the apical meristem were polyploid, and that these three layers behaved independently in the transmission of their characteristic chromosome numbers. These plants obviously had three tiers of initials, that is, three self-propagating layers.

Induced polyploidy has also served to demonstrate the presence of more than one initial cell in each tier. In addition to periclinal chimeras, sectorial polyploidy was observed in *Vaccinium* (Bain and Dermen, 1944). The restriction of polyploidy to individual sectors of the stem is possible only if the initials occur in groups, with each component cell capable of becoming polyploid independently of the others.

Apical Meristems

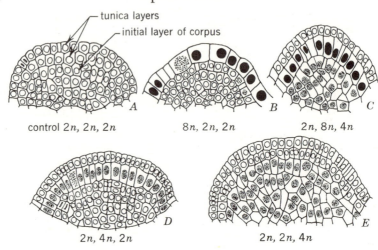

tunica layers
initial layer of corpus

control 2n, 2n, 2n   8n, 2n, 2n   2n, 8n, 4n

2n, 4n, 2n   2n, 2n, 4n

FIG. 5.2. Shoot apices of *Datura* from a diploid plant (A) and from several periclinal cyto-chimeras. Chromosomal combinations are indicated by values given below each drawing. The first figure of each group of three refers to first tunica layer; the second, to second tunica layer; the third, to initial layer of corpus. Octoploid cells are the largest, and their nuclei are shown in black for emphasis; tetraploid cells are somewhat smaller, and their nuclei are stippled; diploid cells are the smallest, and their nuclei are shown by circles. Chromosomal characteristics of tunica layers are perpetuated only in these layers and their derivatives (anticlinal divisions in tunica); those of the initial layer of the corpus are immediately transmitted to the subjacent layers (divisions in various planes). (Adapted from Satina et al., *Amer. Jour. Bot.* 27, 1940.)

EVOLUTION OF THE CONCEPT OF APICAL ORGANIZATION

As has been discussed by several writers (Foster, 1939, 1941; Romberg, 1963; Schüepp, 1926; Sifton, 1944; Wardlaw, 1945), the view concerning the number, the arrangement, and the activity of the initial cells and their recent derivatives in the apical meristems has undergone profound changes since the shoot apex was first recognized by Wolff (1759) as an undeveloped region from which growth of the plant proceeded.

The discovery of the apical cell in cryptogams led to the concept that such cells exist in phanerogams as well. The apical cell was interpreted as a constant structural and functional unit of apical meristems governing the whole process of growth. Subsequent researches refuted the assumption of a universal occurrence of single apical cells and replaced it by a concept of independent origin of different parts of the plant body. The *apical-cell theory* was superseded by the *histogen theory*.

The histogen theory was developed by Hanstein (1868, 1870) on the basis of extensive studies of angiosperm shoot apices and embryos. Its basic theses are, first, that the main body of the plant arises, not from superficial cells but from a mass of meristem of considerable depth, and, second, that this mass consists of three parts, the *histogens*, which may be differentiated by their origin and course of development. The outermost, the *dermatogen* (from the Greek words meaning skin and to bring forth), is the primordial epidermis; the second, the *periblem* (from the Greek, clothing), gives rise to the cortex; the third, the *plerome* (from the Greek, that which fills), constitutes the entire inner mass of the axis. The dermatogen and the periblem form mantle-like layers covering the plerome. The dermatogen, each layer of the periblem, and the plerome begin with one or several initials distributed in superposed tiers in the most distal part of the apical meristem.

Hanstein's "dermatogen" is not equivalent to Haberlandt's (1914) "protoderm." The protoderm refers to the outermost layer of the apical meristem, regardless of whether this layer arises from independent initials or not, and regardless of whether it gives rise to the epidermis only or to some subepidermal tissue also. In many apices the epidermis does originate from an independent layer in the apical meristem; in such apices the protoderm and dermatogen may coincide. The plerome and periblem in the sense of Hanstein are discernible in many roots but are seldom delimited in shoots. Thus the subdivision into dermatogen, plerome, and periblem has no universal application. But Hanstein's histogen theory is criticized chiefly because it contains an assumption that the destinies of the different regions of the plant body are determined by the discrete origin of these regions in the apical meristem. The prevalent view is that histogenesis and organogenesis have no obligate relationship to the segmentation and layering of cells in the apical meristems.

A modified use of histogen, meaning an already determined but still meristematic tissue, is being advocated by Guttenberg (1960). He places the histogen initials into lower levels of the apical meristem than Hanstein does and visualizes separate initials for procambium, pith, and cortex. Actually, in the shoot the ground meristem of the cortex adds cells to the procambium down to levels where vascular elements begin to differentiate. The delimitation between vascular and nonvascular tissues is not established in the apical meristem (Esau, 1943).

The apical-cell and the histogen theories have been developed with reference to both the root apex and the shoot apex. The third theory of apical structure, the *tunica-corpus theory* of Schmidt (1924), was an outcome of observations on angiosperm shoot apices. According to this theory, two tissue zones occur in the apical meristem: the *tunica*, consisting of one or more peripheral layers of cells, and the *corpus*, a mass of cells overarched by the tunica (fig. 5.6, pl. 16*A–C*). The demarcation between these two zones results from the contrasting modes of cell division in the tunica and the corpus. The layers of the tunica show anticlinal divisions; that is, they are undergoing surface growth. The corpus cells divide in various planes, and the whole mass grows in volume. Each layer of the tunica arises from a small group of separate initials, and the corpus has its own initials located beneath those of the tunica. In other words, the number of tiers of initials is equal to the number of tunica layers plus one, the tier of corpus initials. In contrast to the histogen theory, the tunica-corpus theory does not imply any relation between the configuration of the cells at the apex and histogenesis below the apex. Although the epidermis usually arises from the outermost tunica layer, which thus coincides with Hanstein's dermatogen, the underlying tissues may have their origin in the tunica or the corpus or both, depending on plant species and the number of tunica layers.

Interest in the tunica-corpus theory was strongly stimulated by the work of Foster and his students (Foster, 1939, 1941; Gifford, 1954) and has dominated studies on shoot meristems for two decades. As more plants came to be examined, the concept underwent some modifications, especially with regard to the strictness of definition of the tunica. According to one view, tunica should include only those layers that never show any periclinal divisions in the median position, that is, above the level of origin of leaf primordia (Jentsch, 1957). If the apex contains additional parallel layers that periodically divide periclinally, these layers are assigned to the corpus and the latter is characterized as being stratified. Other workers treat the tunica more loosely and describe it as fluctuating in number of layers: one or more of the inner layers of tunica may divide periclinally and thus become part of the corpus (Clowes, 1961*a*). The term *mantle* has been proposed for tunica in the loose sense; it overarches a body of cells called the *core* (Popham and Chan, 1950). Still others reject the tunica-corpus concept entirely because it does not relate the apical activity to the origin of tissues (Guttenberg, 1960). Nevertheless, the tunica-corpus theory remains useful for characterizing growth in the shoot apex of angiosperms. It is used in this book with the assumption that during the vegetative growth the tunica has a characteristic number of layers, which may be attained in steps during the development of the plant and may change during the transition to the reproductive stage; and that the corpus may vary between stratified and nonstratified configurations.

As was mentioned before, the tunica-corpus concept was developed with reference to the angiosperms; it proved to be largely unsuitable for the characterization of the apical meristem of gymnosperms (Foster, 1941, 1949; Johnson, 1951). Shoot apices of only a few gymnosperms have an independently propagating layer that could be interpreted as tunica; in others, the outermost layer divides periclinally and thus is ontogenetically related to the subjacent tissue. Studies of gymnosperm apices, stimulated by Foster (1941), led to the recognition of a zonation based not only on planes of division but also on cytologic and histologic differentiation and degree of meristematic activity of component cell complexes (figs. 5.3, 5.4; pl. 17A). Similar cyto-histologic zonation has since been observed in many angiosperms (Clowes, 1961a). The concept of zonation in Foster's sense has considerably advanced the understanding of growth in shoot apices. It has also related the apical organization to that of the underlying derivative shoot parts without reintroducing a formalized concept of histogen initials. But efforts to bring about such reintroduction are not lacking (Bartels, 1960, 1961; Guttenberg, 1960; Kalbe, 1962).

The cytologic zones that may be recognized in apical meristems vary in degree of differentiation and in details of grouping of cells. As a

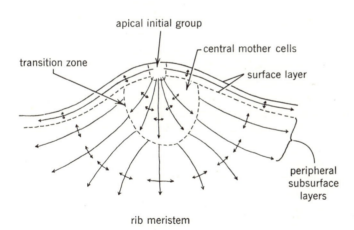

FIG. 5.3. Zones and their mode of growth in shoot tip of *Ginkgo biloba* as seen in longitudinal section. Arrows indicate prevailing direction of growth. Apical initial group contributes to surface layer by anticlinal divisions. It also adds cells by periclinal divisions to central mother-cell group. Growth in volume by cell enlargement and occasional divisions in various planes characterize the central mother-cell zone. Outermost products of divisions in this zone become displaced toward transition zone where they divide by walls periclinal with reference to mother-cell zone. Derivatives of these divisions form peripheral subsurface layers and prospective pith, the rib-meristem zone. (After Foster, *Torrey Bot. Club Bul.* 65, 1938.)

Apical Meristems

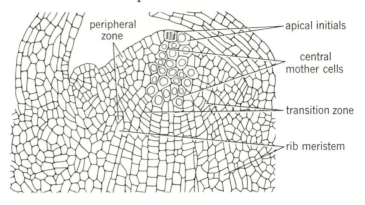

FIG. 5.4. Shoot tip of *Pinus strobus* in longitudinal view. Apical initials contribute cells to surface layer by anticlinal divisions and to central mother-cell zone by periclinal divisions. The mother-cell zone (cells with nuclei) contributes cells to the transition zone composed of actively dividing cells arranged in series radiating from the mother-cell zone. Products of these divisions form the rib meristem and the subsurface layers of peripheral zone. (×150. Slide by A. R. Spurr.)

result, the pertinent terminology is constantly growing and changing. Succinctly, the zonation may be characterized by dividing the apical meristem into a *distal axial zone* terminating the axis and two zones derived from it. One of these, the *proximal axial zone,* or the *inner zone,* appears directly below the distal zone, is centrally located in the apex, and usually becomes the pith after additional meristematic activity has occurred. The other, the *peripheral zone,* or *outer zone,* encircles the other zones. The peripheral zone is also called flank meristem in the literature because of a common tendency to describe structures as seen in sections in two dimensions.

The peripheral zone typically is the most meristematic of all three zones, has the densest protoplasts and smallest cell dimensions. It may be described as a eumeristem (p. 75). Leaf primordia and the procambium arise here, as well as the cortical ground tissue. The inner zone early shows its destiny—differentiation into the vacuolated pith— by being cytologically less dense than the outer zone. Depending on the manner of growth of the shoot, especially the degree of elongation of future internodes, the inner zone assumes more or less definitely the characteristics of a rib meristem. The distal zone is somewhat variable in appearance. All of it, or only its proximal part, may be considerably vacuolated. The term protomeristem is applicable to the distal zone in the sense that it contains the initials and their most recent derivatives.

The derivatives of the distal zone, the outer zone and the inner zone, may merge imperceptibly with the distal zone or they may be delimited from the distal zone by an additional, *transitional zone,* often compared to the cambium because of orderly seriation of cells resulting from divisions periclinal with reference to the distal zone. The transitional zone is composed of particularly actively dividing derivatives of the distal zone. The presence of the transition zone apparently depends on the rate of growth in the shoot apex, and the zone shows fluctuations in its distinctness in the same kind of apex (Philipson, 1954).

The next development in the interpretation of apical meristem resulted from the efforts of Buvat and his students to obtain a unified concept of growth of this meristem (Buvat, 1955a; Clowes, 1961a). Meristematic activity drew the chief attention in this work. Counts of mitoses, and cytological, histochemical, and ultrastructural studies served to formulate the theory that the distal zone of the apical meristem is relatively inert during vegetative growth and that the real initial zone is the peripheral one, where leaf primordia arise. The distal zone received the appellation of waiting meristem (*méristème d'attente*), because it was said to be waiting for the change from vegetative to reproductive stage before taking up meristematic activity. The peripheral zone became the initiating ring (*anneau initial*) and the inner zone the medullary (pith) meristem (*méristème medullaire*). The concept of the inactive distal zone in the apical meristem was extended from the shoots of the angiosperms to those of the gymnosperms (Camefort, 1956; who called the distal zone *zone apicale*) and the lower vascular plants (Buvat, 1955b), and to the roots (Buvat and Genèves, 1951; Buvat and Liard, 1953). The concept was later somewhat modified in that variations in degree of inactivity of the distal zone in relation to the size of the apex and its stage of development came to be recognized (Catesson, 1953; Lance, 1957; Loiseau, 1959).

The revision of the concept of apical initials by the French workers stimulated a considerable amount of research in other countries and led to refinement of techniques for determining the degree of meristematic activity in the apical meristem (Clowes, 1961a). Extensive counts of mitotic figures (Edgar, 1961; Hagemann, 1956; Hara, 1962; Jacobs and Morrow, 1961; Popham, 1958); studies of cell patterns in fixed (Paolillo and Gifford, 1961) and living apices (Ball, 1960; Newman, 1956); histochemical studies (Gifford and Tepper, 1962b); use of labeled compounds to determine the location of synthesis of DNA, RNA, and protein (Clowes, 1961a; Gifford, 1960); experimental injury of the meristem (Clowes, 1961b; Davidson, 1961; Wardlaw, 1957); and theoretical discussions (Cutter, 1959) served to evaluate the concept of the inactive distal zone in the apical meristem. Most of the investigators outside

France consider that the apparent scarcity of divisions in the distal cells of the shoot does not justify regarding these cells as being inconsequential in the construction of the shoot; these cells are the ultimate source of all other cells of the shoot and hence are the initials. This interpretation is used in the description of the shoot apices in the subsequent sections of the present chapter.

With regard to the root apices, the occurrence of an inactive center in the meristem found confirmation in many studies, which resulted in the development of the concept of *quiescent center* by Clowes (1961*a*). This center is described as a nonmeristematic group of cells roughly hemispherical in shape and surrounded by actively dividing cells, the initials, or the promeristem. The center becomes quiescent during the development of the root, either the main root (taproot) or the lateral root, after the architectural pattern of the apex is established, and it remains capable of resuming meristematic activity. There is apparently a range of development of the quiescent center. The center may be larger in large roots and smaller, or even absent, in small roots.

The origin of the architectural pattern in roots and shoots beginning with the embryo has been studied in a number of species. The subject is reviewed by Guttenberg (1960, 1961). The pattern is organized gradually in terminal apices of epicotyls, in lateral shoots, in radicles of embryos or seedlings, and in lateral and adventitious roots. Moreover, the distribution of the meristematic activity in the apical meristem changes with the development of shoot or root.

Apical meristems receive much attention in connection with studies of causal relations in morphogenesis. Many efforts have been directed toward determining the role of the apical meristem in the development of form and internal organization of plant organs (Clowes, 1961*a*; Cutter, 1959; Gifford, 1954). Some studies have been concerned with the determination of the arrangement of leaves (phyllotaxy, chapter 15) and their bilateral symmetry (chapter 16), others with the determination of the vascular patterns in roots (chapter 17) and shoots (chapter 15). Workers also consider the question whether the apex is a self-determining and dominant center of development controlling the growth of the parts derived from it or whether it is a plastic region operating under the control of stimuli sent to it from the mature subjacent tissues.

Results of experimental studies involving cultures of isolated shoot and root tips and partial isolation of apical meristems and leaf primordia by operations on growing plants have been interpreted as indicating a high degree of independence of the apical meristem. The culture studies have shown that apical meristems of roots are capable of form-ing vascularized roots and that the tissue pattern in the root is a product of apical activity (Torrey, 1955). Apical meristems of shoots including the youngest leaf primordia may grow into entire plants, whereas the subjacent regions form only vascularized masses of cells (Ball, 1946). Operations on shoot apices show that the apex may continue to grow and form primordia after its procambial connection with the subjacent region is severed (Ball, 1948; Snow and Snow, 1947; Wardlaw, 1947). Some experimental work indicates a considerable degree of resistance of the apical meristem to disturbances that may be caused by environmental conditions, such as variations in light, temperature, and nutrient conditions (Thomson and Miller, 1962).

VEGETATIVE SHOOT APEX

Vegetative shoot apices vary in shape, size, cytologic zonation, and meristematic activity. The shoot apices of conifers are commonly relatively narrow and conical in form (fig. 5.4); in *Ginkgo* (fig. 5.3, pl. 17A) and in the cycads they are rather broad and flat. The apical meristem of some monocotyledons (grasses, *Elodea*) and dicotyledons (*Hippuris*) is narrow and elongated, with the distal zone much elevated above the youngest node (pl. 17B). In many dicotyledons the distal zone barely rises above the leaf primordia (fig. 5.6), or even appears sunken (pl. 18A; Gifford, 1950). In some plants the axis increases in width close to the apex and the peripheral region bearing the leaf primordia becomes elevated above the apical meristem leaving the latter in a pit-like depression (pl. 18B; Ball, 1941; rosette type of dicotyledons, Rauh and Rappert, 1954). Examples of widths of apices measured in microns at insertion of the youngest leaf primordia are: 280, *Equisetum hiemale;* 1,000, *Dryopteris dilatata;* 2,000–3,300, *Cycas revoluta;* 280, *Pinus mugo;* 140, *Taxus baccata;* 400, *Ginkgo biloba;* 288, *Washingtonia filifera;* 130, *Zea mays;* 500, *Nuphar lutea* (Clowes, 1961a). The shape and size of the apex change during the development of a plant from embryo to reproduction, between initiation of successive leaves, and in relation to seasonal changes. An example of change in width during growth is available for *Phoenix canariensis* (Ball, 1941). The diameter in microns was found to be 80 in the embryo, 140 in the seedling, and 528 in the adult plant.

Attempts to classify apical structure of shoots have resulted in the assigning of shoot apices to several types (Johnson, 1951; Popham, 1951), but these classifications are subject to criticism on the grounds that they do not reflect fundamental differences in structure and are not helpful in making the behavior of meristems better understood (Clowes, 1961a; Newman, 1961). The simple classification into three types

(Newman, 1961) on the basis of whether there is only one initial (many vascular cryptogams), or several initials in one cell layer (most gymnosperms), or several initials in more than one layer (most angiosperms) is convenient for descriptive purposes; but the basic pattern of growth in these three kinds of apices is the same: they all consist of a distally located initiating zone (protomeristem) and two derivative zones (the outer and the inner), in which organogenesis and histogenesis begin.

Vascular Cryptogams

In the lower Tracheophyta growth at the apex proceeds from either one or few apical initial cells. These cells are often conspicuous because of their large size and relatively high degree of vacuolation. Most commonly the single apical cell is pyramidal (tetrahedral) in shape. The base of this pyramid is turned toward the free surface, the other three sides downward (fig. 5.1A). New cells are cut off roughly parallel to these three sides. In apices with a tetrahedral apical cell, the derivative cells often form an orderly pattern (fig. 5.1C), which apparently is initiated by the orderliness of divisions of the apical cells: the successive divisions follow one another in acropetal sequence along a helix. Tetrahedral apical cells are found in *Equisetum* and most leptosporangiate ferns. The eusporangiate ferns may have one or more initials. In *Botrychium*, for example, the apex bears a superficial layer of prismatic cells among which an apical cell is occasionally recognized (Bierhorst, 1958). Some investigators suggest that, in the ferns, the apex with several initials represents a more primitive evolutionary stage than the one with a single apical cell (Wardlaw, 1945). The opposite view, that an apex with a multicellular initial layer could evolve through loss of genetic provision for a single apical cell, has also been expressed (Bierhorst, 1958).

Single apical cells may be three-sided, with two sides along which new cells are cut off (fig. 5.1B). Such apical cells are characteristic of bilaterally symmetrical shoots, as in the water ferns *Salvinia* and *Azolla*. The flattened rhizome apex of *Pteridium* also bears a three-sided apical cell (fig. 5.1D; Gottlieb and Steeves, 1961).

In Lycopsida, single apical cells and groups of initial cells have been described (Härtel, 1938; Schüepp, 1926). The Isoetaceae appear to have a relatively poorly defined group of initial cells (Bhambie, 1957; Rauh and Falk, 1959). In *Psilotum nudum* a more or less distinct apical cell was observed in both the gametophyte and the sporophyte (Bierhorst, 1953, 1954).

Gymnosperms

As was mentioned previously, the cytologic zones in the apical meristem were first recognized by the study of a gymnosperm, namely *Ginkgo* (fig. 5.3; pl. 17A; Foster, 1938). The zonation found in the apex of this genus has served as a basis for the interpretation of shoot apices in other gymnosperms. In *Ginkgo*, the protomeristem has been divided in two cell groups, the apical surface initials, from which all other cells of the apex are ultimately derived, and the subjacent group of cells originating from the surface initials and termed *mother cells*. Cell division is sluggish in the interior of the mother-cell group but is active on its periphery. The products of the divisions along the periphery of the mother-cell group combine with derivatives resulting from anticlinal divisions of the apical initials. All together these lateral derivatives form a mantle-like peripheral zone of densely staining and relatively small cells which appear less differentiated (eumeristem) than the mother cells and also less so than the cells of the initiating zone. The derivatives produced at the base of the mother-cell zone become pith cells, and usually they pass through a rib-meristem form of growth. During active growth, a cup-shaped region of orderly dividing cells, the transitional zone, delimits the mother-cell group and may extend to the surface of the apical dome. The peripheral mantle of cells is the seat of origin of the leaf primordia and of the epidermis, the cortex, and the vascular tissues of the axis. Part of the pith may arise from the peripheral zone.

The details of the structural pattern just reviewed vary in the different groups of gymnosperms. The cycads have very wide apices with a large number of surface cells contributing derivatives to the deeper layers by periclinal divisions. Foster (1941, 1943) interprets this extended surface layer and its immediate derivatives as the initiation zone; others seek to confine the initials to a relatively small number of surface cells (Clowes, 1961a; Guttenberg, 1961). The periclinal derivatives of the surface layer converge toward the mother-cell zone, a pattern apparently characteristic of cycads. In other seed plants the cell layers typically diverge from the point of initiation. The convergent pattern results from numerous anticlinal divisions in the surface cells and their recent derivatives—evidence of surface growth through a tissue of some depth. This growth appears to be associated with the large width of the apex. The mother-cell group is relatively indistinct in cycads. The extensive peripheral zone arises from the immediate derivatives of the surface initials and from the mother cells. The rib meristem is more or less pronounced in the inner zone beneath the mother-cell zone.

Most conifers have periclinally dividing apical initials in the surface

layer (pl. 19). A contrasting organization, with a cell layer dividing almost exclusively by anticlinal walls, has been described in *Araucaria,
Cupressus, Thujopsis* (Guttenberg, 1961), and *Agathis* (Jackman, 1960).
In these plants the apices have been interpreted as having a tunica-
corpus organization. The mother-cell group may be well differentiated
in conifers, and a transitional zone may be present (fig. 5.4). In conifers
with narrow apices mother cells are few and may or may not be enlarged
and vacuolated. In such apices a small mother-cell group, three or four
cells in depth, is abruptly succeeded below by highly vacuolated pith
cells without the interposition of a rib meristem; and the peripheral zone
is also only a few cells wide (pl. 19A).

Coniferous shoot apices have been studied with regard to seasonal
variations in structure (Parke, 1959; Sacher, 1954; Singh, 1961). The
basic zonation does not change, but the height of the apical dome above
the youngest node is greater during growth than during rest. Because
of this difference, the zones are differently distributed in the two kinds of
apices with regard to the youngest node: the rib meristem occurs below
this node in resting apices (fig. 5.5A) and partly above it in active apices
(fig. 5.5B). This observation calls attention to a terminological problem.
If the apical meristem is defined strictly, as the part of the apex above
the youngest node, it must be interpreted as varying in its composition
during different growth phases (Parke, 1959).

The Gnetales commonly have a definite separation into a surface
layer and an inner core derived from its own initials. Therefore, the
shoot apices of *Ephedra* and *Gnetum* have been described as having a
tunica-corpus pattern of growth (Johnson, 1951). The tunica is uniseriate, and the corpus is comparable to the central mother-cell zone in its
morphology and manner of division. The shoot apex of *Welwitschia*
produces only one pair of foliage leaves and does not possess a distinct
zonation. Periclinal divisions have been observed in the surface layer
(Rodin, 1953).

The data on the shoot apices in the gymnosperms have been used to
suggest possible trends in the evolution of apical structure in this group
of plants (Foster, 1941, 1943; Johnson, 1944). The large apex of the
cycads with its extensive initiation zone, massive core of mother cells,
and generally diversified growth zones is probably primitive. Evolutionary advancement seems to have involved a refinement of the
meristem in the sense that it became simpler, with less diversity in
growth zones and, at the same time, with a more precise separation of
zones of surface and volume growth, each derived from independent
initials.

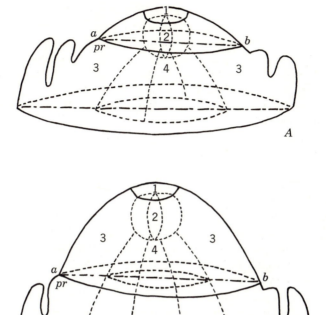

FIG. 5.5. Zonation in the shoot tip of *Abies concolor* during resting (*A*) and growth (*B*) stages. Zones are: 1, apical initials; 2, mother cells; 3, peripheral meristem; 4, central or rib meristem. Plane *ab* delimits the shoot apex above the youngest primordium (*pr*). The shoot apex, or apical meristem, differs structurally in the two shoot tips. (From Parke, *Amer. Jour. Bot.* 46, 1959.)

Angiosperms

The main features of the tunica-corpus organization of the angiospermous shoot apex have been discussed in a foregoing part of this chapter. One to five layers of tunica have been reported for dicotyledons, with two represented in the largest number of species; one to four layers for monocotyledons, with one and two predominating (Gifford, 1954; Hara, 1958; Jentsch, 1960; Thielke, 1954, 1957). An absence of tunica-corpus organization, with the outermost layer dividing periclinally, has also been observed (*Saccharum*, Thielke, 1962). To delimit the tunica from the corpus is not a simple matter. The number of parallel periclinal layers in the shoot apex may vary during the ontogeny

of the plant (Gifford and Tepper, 1962b) and under the influence of seasonal growth changes (Hara, 1962). There may also be periodic changes
in stratification in relation to the initiation of leaves (Sussex, 1955). As
was mentioned previously, some workers treat such changes as variations in the thickness of the tunica; others interpret them as reflections
of variations in the stratification of the corpus.

According to Guttenberg (1960), the tunica could consist of no more
than two layers, which he terms dermatogen and subdermatogen. The
subdermatogen sometimes lacks its own initials, a condition corresponding to a single tunica layer configuration. Beneath the two outer layers
is the central mother-cell complex, which may or may not be stratified.
Its derivatives, through intermediary meristems, are the pith, the vascular tissue, and most of the cortex. The crucial evidence for the two-
layered tunica is said to be the unbroken continuity of the dermatogen
and subdermatogen in the emerging axillary bud. It seems that this
scheme, as well as Guttenberg's concept of histogens, implies a high
degree of uniformity in the relation between apical structure and origin
of tissues below.

The analysis of apical meristems in terms of tunica and corpus is
usually combined with that based on cytological zonation (Gifford and
Tepper, 1962b; Johnson and Tolbert, 1960; Millington and Fisk, 1956;
Senghas, 1956, 1957; Smith, 1963). The characteristics of the central
mother-cell group—relatively large, light-staining cells—are sometimes
limited to the corpus or part of it; sometimes they also appear in the tunica
layers. Thus there may be a uniformly light-staining distal zone (frequently called central zone), or there may be a light-staining core
covered by a more densely staining layer or layers. The participation
of tunica and corpus in the formation of the peripheral and inner zones
depends on the relative proportions of tunica and corpus in the apex.
The degree of distinctness of zonation varies in the angiosperms, as it
does in the gymnosperms, and is usually better expressed in larger
apices. As was reviewed before, the studies of zonation may include
determinations of meristematic activity, especially with reference to the
concept of the inactive distal zone.

ORIGIN OF LEAVES

In this chapter only those features of leaf origin are considered that
are related to the structure and activity of the apical meristem. A leaf
is initiated by periclinal divisions in a small group of cells in the peripheral zone of an apical meristem. According to the concept of the
initiating ring (p. 97), the leaves arise in this ring in positions related

to the leaf arrangement. Successive sectors of the ring are visualized as being partly used up in the formation of leaves. Cell division restores each sector above a newly formed primordium so that the ring moves upward and the leaves arise at successively higher levels (Bersillon, 1956).

In the dicotyledons the first periclinal divisions initiating the leaves occur most frequently in the subsurface layer and are followed by similar divisions in the third layer and by anticlinal divisions in the surface layer (Guttenberg, 1960). In some monocotyledons the surface layer also undergoes periclinal divisions and gives rise to some or most of the internal tissue of the leaf in addition to the epidermis (pl. 17B; Guttenberg, 1960). Since the initiation of leaves in angiosperms follows a relatively consistent pattern, whereas the depth of the tunica is variable, the tunica and the corpus are variously concerned with leaf formation, depending on their quantitative relationship in a given apex.

In the gymnosperms the leaves arise in the peripheral zone. The surface layer may contribute cells to the internal tissue of the primordium by periclinal and other divisions. According to Guttenberg (1961), such activity of the protoderm is characteristic of those gymnosperms in which no independent surface layer is present in the apical meristem. In the vascular cryptogams the leaves arise from either single superficial cells or groups of such cells, one of which enlarges and becomes the conspicuous apical cell of the primordium (fig. 5.1C; Härtel, 1938; Sifton, 1944).

The divisions initiating a leaf primordium cause the formation of a lateral prominence on the side of the shoot apex (fig. 5.6D, pl. 16A). This prominence constitutes the leaf base or the so-called *leaf buttress* (Foster, 1936). Subsequently the leaf grows upward from the buttress (chapter 16). The level at which a leaf buttress appears, in relation to the initiating region of the apical meristem, varies in different species. In some species the apical meristem has the form of a relatively high cone, with the divisions that initiate the leaf appearing low on its sides (chapter 16, pl. 17B). In others the apical meristem is less prominently elevated above the youngest leaf buttress (fig. 5.6D). In still others it appears practically at one level with such a buttress (pl. 16A), or in a depression below it. Depending on the level at which a leaf primordium is initiated, the shoot apex may or may not change in shape and structure during the period between the initiation of two successive leaf primordia (or pairs or whorls of primordia in plants with an opposite or whorled leaf arrangement). Such a period has been designated *plastochron* (Schmidt, 1924).

The term plastochron was originally formulated in a rather general sense for a time interval between two successive similar events occurring in a series of similar periodically repeated events (Askenasy, 1880). In this sense the term may be applied to the time interval between a variety of corresponding stages in the development of successive leaves, for example, the initiation of periclinal divisions in the sites of origin of primordia, the beginning of apical growth of a primordium, or the initiation of lamina. Plastochron may be used also with reference to the development of internodes and of axillary buds, to stages of vascularization of the shoot, and to the development of floral parts. With reference to the development of the plant as a whole, plastochron is applicable for indicating the age of the plant. A refinement of such use is provided by the formula of Erickson and Michelini (1957) for calculating the plastochron index. In this formula, as developed for *Xanthium*, a leaf 10 mm long is used as reference, so that, if the plant has n leaves, it is n plastochrons old when leaf n is 10 mm long. For characterizing leaf development, this index proved to be more useful than the chronological age. Fresh weight, dry weight, chlorophyll synthesis, and oxygen uptake of individual developing leaves had a straight-line relationship to the plastochronic stage of growth of the leaf (Michelini, 1958).

Successive plastochrons may be of equal duration, at least during part of vegetative growth of genetically uniform material growing in a controlled environment (Stein and Stein, 1960). The stage of development of the plant and environmental conditions are known to affect the length of the plastochrons. In *Zea mays*, for example, the successive plastochrons in the embryo lengthen from 3.5 to 13.5 days, whereas those in the seedling shorten from 3.6 to 0.5 days (Abbe and Phinney, 1951; Abbe and Stein, 1954). In *Lonicera nitida* the duration of plastochrons varied from 1.5 to 5.5 days apparently in relation to changing temperature (Edgar, 1961). The rate of production of leaves is also affected by light (Mohr and Pinnig, 1962).

The changes in the morphology of the shoot apex occurring during one plastochron may be referred to as plastochronic changes. Such changes are graphically illustrated in fig. 5.6 showing a shoot apex of a plant with a decussate (opposite, with the alternate leaf pairs at right angles to each other) leaf arrangement. Before the initiation of a new leaf primordium the apical meristem appears as a small rounded mound (fig. 5.6A). It gradually widens (fig. 5.6B, C). Then leaf buttresses are initiated on its sides (fig. 5.6D). While the new leaf primordia grow upward from the buttresses, the apical meristem again assumes the appearance of a small mound (fig. 5.6E). In some plants growth of leaves overshadows that of the apex. The divisions initiating the leaves encroach upon the distal zone so that the latter appears to be almost exhausted during each plastochron and, as a consequence, the position

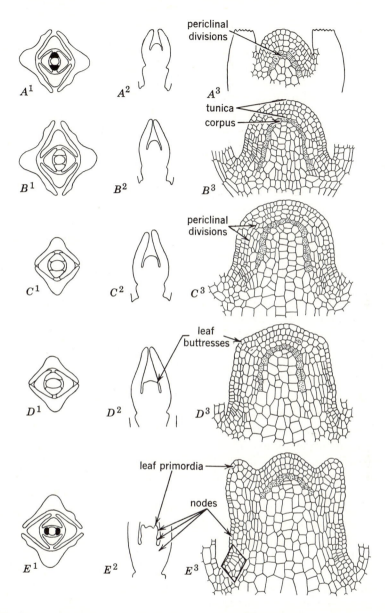

FIG. 5.6. Leaf initiation in shoot tip of *Hypericum uralum*. Changes in shape and histology of shoot apex through approximately one plastochron, starting with early stage of leaf pair shown in black in A^1, and ending shortly after emergence of leaf pair shown in black in E^1. Sections are transverse in A^1–E^1, longitudinal in A^2–E^2 and A^3–E^3. Leaves are in pairs at each node, in decussate arrangement. Protrusions in axis below the leaves in A^2–E^2 are leaf bases of next lower pair of leaves. A^3–E^3, stippling indicates outer-boundary cells of corpus and their immediate derivatives. E^3, four-sided figure, presumptive place of origin of axillary bud. (Adapted from Zimmermann, *Jahrb. f. Wiss. Bot.* 68, 1928.)

107

of this zone oscillates around the apex of the axis (Catesson, 1953; Hagemann, 1960). The other extreme is illustrated by shoots with long, slender tips in which leaves arise considerably below the distal zone and cause no plastochronic changes in the apex (Jentsch, 1960).

If the shoot apex undergoes plastochronic changes in size, both its volume and surface area change. To designate these changes, the expressions *minimal-area* and *maximal-area phases*—now abbreviated to minimal and maximal phases—were introduced (Schmidt, 1924). When the leaves are in a decussate arrangement the maximal phase is attained by a symmetrical distribution of periclinal divisions on two sides of the apical meristem. Thus two diameters of the apex, which cross one another at right angles, elongate alternately in successive plastochrons (fig. 5.6). In shoots with a helical leaf arrangement the divisions alternate in different sectors around the circumference of the apical meristem and thus the enlargement of the apex in the maximal phase is asymmetrical (pls. 52, 53; Hara, 1962). Because of the lack of delimitation between the emerging leaf primordium and the stem, the determination of the maximal stage is problematical. There is disagreement whether the foliar buttress should or should not be included in the measurement of width. The best compromise appears to be to identify the maximal phase with the first divisions initiating a leaf, before the cells resulting from these divisions begin to enlarge and thus to affect the outline of the shoot apex (Gifford, 1954).

Plastochronic changes in the apical meristem may also affect the cytologic zonation (Popham and Chan, 1950), the degree of stratification of the corpus (Soma, 1958; Sussex, 1955), and the distribution of mitoses (Edgar, 1961; Gifford, 1954; Paolillo and Gifford, 1961).

The plastochronic changes may follow a regular sequence through the successive plastochrons. In the embryo and seedling of maize, for example, both the minimum and the maximum plastochronic sizes of the apex were found to increase from plastochron 1 to plastochron 14 (latest observed). This enlargement involved an increase in the number of cells, but the size of cells remained constant (Abbe et al., 1951; Abbe and Stein, 1954). The rate of this increase, calculated as increments per unit material, were deceleratory during embryogeny and acceleratory during seedling development.

A considerable amount of research has been carried out on factors determining the emergence of leaf primordia in their characteristic arrangement, or phyllotaxy, and their development into bilateral structures. To detect the causal relationships in leaf initiation, workers use experimental methods, such as application of growth-regulating substances to the apices and the making of incisions designed to affect the development of a leaf. According to the concept of the origin of leaves in the initiating ring, the existing primordia determine the position of the new leaves. Leaf primordia arise in contact with one another along two or more helices, each of which terminates in the initiating ring with a putative generative center, which induces cell division leading to the emergence of a new leaf (Buvat, 1955a). According to the opposite and more prevalent view, a leaf arises in a locus that is spatially removed from inhibitions exercised by the distal part of the apical meristem and the adjacent youngest leaf primordia (Wetmore, 1956). This field-effect concept has been developed mainly through experimentation with ferns (Cutter and Voeller, 1959). Positions of leaves have been changed by incisions isolating potential leaf sites. Such isolations sometimes resulted in the development of a centric leaf or a bud in the place of a dorsiventral leaf, observations suggesting that the dorsiventral symmetry is imposed by the physiological environment. The dorsiventral symmetry, however, becomes fixed in older primordia. As a result, older primordia grown *in vitro* develop into dorsiventral leaves, whereas younger primordia become centric structures.

ORIGIN OF BRANCHES

In the lower vascular plants, such as *Psilotum*, *Lycopodium*, and *Selaginella*, branching occurs at the apex, without reference to the leaves. It is described as *dichotomous* when the original apical meristem undergoes a median division into equal parts and as *monopodial* when a branch arises laterally at the apical meristem (Sifton, 1944). In seed plants, branches commonly are formed in close association with the leaves—they appear to originate in the axils of the leaves—and in their nascent state they are referred to as *axillary buds*. Judged from most investigations the term axillary is somewhat inaccurate because the buds generally arise on the stem (figs. 5.6E^3, 5.7) but become displaced closer to the leaf base, or even onto the leaf itself, by subsequent growth readjustments. Such relationship has been observed in ferns (Wardlaw, 1943), dicotyledons (Garrison, 1949, 1955; Gifford, 1951; Koch, 1893), and Gramineae (Evans and Grover, 1940; Sharman, 1945). In the grasses, the lack of developmental relation between the bud and the subtending (axillant) leaf is particularly clear. The bud originates close to the leaf located above it (fig. 5.8A). Later the bud becomes separated from this leaf by the interpolation of an internode between it and the leaf. A rather similar origin of the lateral buds has been observed in other monocotyledons (*Tradescantia*, Guttenberg, 1960; *Musa*, Barker and Steward, 1962a). In the conifers, bud development resembles that in the dicotyledons (Guttenberg, 1961).

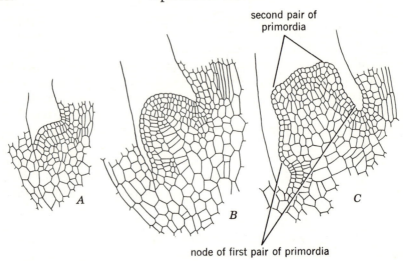

second pair of
primordia

node of first pair of primordia

FIG. 5.7. Origin of axillary bud in *Hypericum uralum.* It is formed by derivatives of three outer layers of tunica of main shoot. Two outer layers divide anticlinally and maintain their individuality as two outer layers of tunica of bud (*A–C*). Third layer divides periclinally and otherwise and gives rise to third and fourth layers of tunica and to corpus of bud. Third tunica layer is evident in bud in *C*; the fourth appears later. *C*, second pair of leaf primordia is being initiated. First pair is orientated in a plane perpendicular to surface of drawing. (Adapted from Zimmermann, *Jahrb. f. Wiss. Bot.* 68, 1928.)

Axillary buds are commonly initiated somewhat later than the leaves subtending them, frequently in the second plastochron (Seeliger, 1954; Sussex, 1955). Therefore, it is not always clear whether the meristem of the axillary bud is derived directly from the apical meristem of the main shoot or whether it originates from partly differentiated tissue of the internode. Both situations probably occur, because plants vary with regard to the number of plastochrons intervening between the origin of the leaf and that of its axillary bud (Philipson, 1949; Sifton, 1944).

The initiation of the bud in higher vascular plants is characterized by a combination of anticlinal divisions, in one or more of the superficial layers of the young axis, and of various divisions, sometimes predominantly periclinal, in the deeper layers (figs. 5.7, 5.8). This coordinated growth in surface area and in volume at greater depth causes the bud to protrude above the surface of the axis. Sometimes the divisions initiating a bud are quite regular and result in the formation of a series of curved layers approximately parallel to each other (fig. 5.8*C*). Because of this configuration the early bud meristem has been named shell

zone (Clowes, 1961a; Guttenberg, 1961). Depending on the quantitative relationships between the tunica and the corpus in the shoot apices
of angiosperms, the derivatives of the two zones variously participate in
the formation of the axillary bud meristem and not necessarily in the
same proportions as in the formation of the leaves of the same plant,
because the buds frequently arise in deeper layers than the leaves (Guttenberg, 1960). Epidermal origin of axillary buds has also been reported
(Champagnat, 1961). If the axillary bud develops into a shoot, its
apical meristem is gradually organized—commonly duplicating the pattern found in the parent shoot apex—and proceeds with the formation
of leaves (figs. 5.7, 5.8).

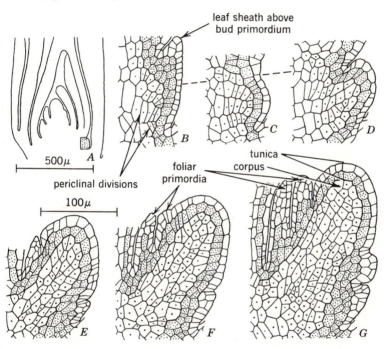

FIG. 5.8. Development of lateral bud in *Agropyron repens* (quackgrass). Median longitudinal sections. *A*, low-power view of shoot tip with several leaf primordia. Stippled part
indicates position of bud. It is formed by derivatives of the two-layered tunica and the corpus. *B–G*, derivatives of second layer of tunica are stippled, and those of corpus are indicated by a single dot in each cell. Bud is initiated by periclinal divisions in corpus
derivatives (*B, C*). Anticlinal divisions occur in tunica derivatives. Bud emerges above surface of stem (*D*). By rib-meristem growth corpus derivatives elongate the core of bud
(*E–G*). They also organize its corpus. Tunica derivatives remain in a biseriate arrangement
at apex of bud and form its two-layered tunica (*E, G*). Leaf primordia arise on bud (*E–G*).
(Adapted from Sharman, *Bot. Gaz.* 106, 1945.)

Buds that arise without connection with the apical meristem from more or less mature tissues are classified as adventitious buds (Mac-Daniels, 1953; Priestley and Swingle, 1929). No clear ontogenetic distinction exists between adventitious and axillary buds, because axillary buds also may originate in more or less differentiated parenchyma some distance from the apex. Adventitious buds occur on stems, roots, and leaves on intact plants and on isolated cuttings or leaves. In cuttings the buds usually are initiated in callus tissue, which develops before the buds. Adventitious buds may originate more or less deeply in the tissue or in the epidermis (Champagnat, 1961; Link and Eggers, 1946).

Axillary buds are described as arising *exogenously*, that is, in relatively superficial tissues. This appears to be an entirely appropriate interpretation when the origin of such buds is compared with that of lateral roots (pl. 15B), which are initiated deeply in the parent axis (*endogenous* origin). The adventitious buds may be exogenous or endogenous (Priestley and Swingle, 1929; Thompson, 1943–44).

Many physiologic studies have been carried out on initiation of both axillary and adventitious buds. The phenomenon is evidently a complex one and involves interactions of numerous determinants (Audus, 1959). Growth-regulating substances play a part but probably in a characteristic balance with a number of specific metabolites, and apparently the different stages of bud development depend on different sets of conditions.

REPRODUCTIVE SHOOT APEX

In the reproductive state in angiosperms, floral apices replace the vegetative either directly or, more frequently, through the development of an inflorescence (fig. 5.9). Flowers are borne on a wide variety of inflorescences. The structural modification that occurs in the apical meristem during the transition to the reproductive stage may become recognizable in the inflorescence apex. Thus a discussion on the reproductive apex in angiosperms should include reference to both the inflorescence and the floral apical meristem.

The change to the reproductive stage may be early detectable by the modified growth habit of the shoot. When the flowers are borne on axillary-branch inflorescences an accelerated production of axillary buds is one of the earliest indications of approaching flowering (Barker and Steward, 1962b; Hagemann, 1963; Rauh and Reznik, 1951, 1953). Concomitantly, the nature of the foliar organs subtending the axillary buds changes: they develop as bracts more or less distinct from the foliage leaves. Growth relationships appear to change. During the

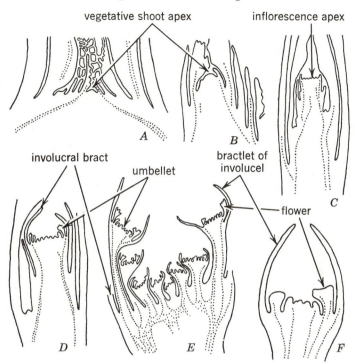

FIG. 5.9. Transformation of apical meristem during shift from vegetative growth to development of flowers in *Daucus carota*. The inflorescence is a compound umbel. It consists of axis bearing several small umbels (umbellets) in an umbellate arrangement. *A*, vegetative shoot apex at base of rosette of leaves. *B*, shoot apex at approach of reproductive stage raised above ground level by internodal elongation. *C*, *D*, flattened inflorescence (umbel) apices producing bracts and primordia of umbellets. *E*, compound umbel in a young state. Apex of each umbellet assumes an appearance similar to that of apex of umbel and produces bractlets and primordia of flowers. *F*, each flower develops a flattened apex and forms floral organs. (*A–E*, ×13; *F*, ×46. After Borthwick et al., *Amer. Jour. Bot.* 18, 1931.)

vegetative stage the growth of foliar primordia is emphasized; during the reproductive stage the axillary buds appear earlier and grow more vigorously than the subtending bract primordia (Bersillon, 1958).

The second feature frequently revealing the beginning of the reproductive stage is the sudden increase in the elongation of the internodes (Stein and Stein, 1960). This change is particularly striking in plants that have no elongated axis during the vegetative stage, as in many grasses (Bonnett, 1936; pl. 92) and rosette types of plants (Vaughan, 1955).

Histologically and cytologically the reproductive meristem differs from the vegetative in varying degrees. It may retain the same quantitative relationship between the tunica and the corpus as was present in the vegetative apex (pl. 90A, B), or the number of discrete surface layers may be reduced or augmented (Guttenberg, 1960; Philipson, 1949). The most frequently described change is expressed in the distribution of the eumeristematic and the more highly vacuolated cells (fig. 5.10). In many species the apex of the inflorescence or the flower shows a uniform, densely staining, small-celled mantle-like zone of one or more layers enclosing a lighter-staining, larger-celled core; and such an apex may be flatter and wider than the vegetative. The mantle does not necessarily coincide with the tunica; part of the corpus may be included in it (Philipson, 1949). This type of configuration is an expression of determination of growth and of a shift in its direction. Axis elongation becomes limited, and therefore the characteristic activity of the corpus, resulting in the formation of the rib meristem, is discontinued. The cells of the central tissue enlarge and vacuolate prominently and the meristematic activity becomes restricted to the mantle zone. This activity is concerned not with the elongation of the shoot and the maintenance of the initial region of the apical meristem, but only with the production of floral organs.

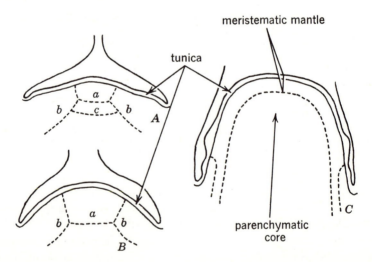

FIG. 5.10. Modification in zonation of vegetative apex during change into inflorescence apex in *Succisa pratensis. A*, apex in spring when it is forming foliage leaves; *B, C*, two stages of development of inflorescence. Details: *a*, central zone of large cells; *b*, peripheral zone; *c*, rib meristem; *a* and parts of *b* and *c* constitute the corpus. Initiation of inflorescence is associated with cessation of growth in length and disappearance of rib meristem (*B*). Later, central and peripheral zones are reorganized to form, together with the tunica, a meristematic mantle overarching a parenchymatic core (*C*). (After Philipson, *Ann. Bot.* 11, 1947.)

Some inflorescence apices retain, at least for a time, the cytological zonation of the vegetative apex (Bersillon, 1958; Vaughan, 1955). The distinctness of zonation in the reproductive apex is probably related to its degree of determinateness; indeterminate inflorescences such as those of Cruciferae have a persisting apical zonation; in the more determinate type as that of Compositae the zonation disappears in the inflorescence (Popham and Chan, 1952). Up to a certain stage, the apex of the flower itself may show a zonation of the vegetative type (Vaughan, 1955).

In the absence of internodal elongation in the axis of the flower the floral parts appear in close succession spatially and temporally. The wide meristematic surface accommodates simultaneously many centers of cell proliferation and the plastochronic rhythm that characterized the vegetative growth may become indistinguishable (Bersillon, 1956; Rauh and Reznik, 1951; Sunderland, 1961). If, however, the flower is less determinate and its apex has a prolonged meristematic activity—common features of flowers with numerous free parts—plastochronic fluctuations in size and configuration of the apex may be retained during floral ontogeny (Tucker, 1960).

Cytologic studies on transition of the apex to reproductive state have shown that mitotic activity is increased and changed in its distribution at this time (Gifford and Tepper, 1961; Jacobs and Raghavan, 1962; Sunderland, 1961). In *Xanthium* the stimulation to increased cell division in the apex was observed within 24 hours after a single inductive dark period, before any other change was detectable (Thomas, 1963). In connection with the appearance of the mantle-like eumeristem, the distinction between the highly active peripheral zone and the less active distal zone commonly seen in vegetative apices is effaced. Correspondingly, the staining that indicates presence of DNA becomes more uniform than in the vegetative stage when the distal cells stain lightly (Gifford and Tepper, 1962a). RNA and protein are uniformly distributed in both kinds of apices but both increase in concentration in the reproductive stage.

As was mentioned previously, the proponents of the waiting-meristem concept consider that the distal part of the apical meristem, which is said to be inactive during the vegetative stage, becomes active during the development of the flower (Buvat, 1955a). The initiating ring still produces the sepals but may disappear immediately after this event. The formerly inactive zone assumes two roles. The upper part is sporogenous and becomes the meristem initiating the floral parts, the lower is the receptacular meristem, which produces the axis of the flower (or the inflorescence). Thus this concept implies a functional discontinuity between the vegetative and the reproductive apical meristems and, therefore, is in agreement with the well-known view of Grégoire (1938) that the flower and the vegetative shoot are not related structures and that their meristems are fundamentally different (see reviews by Foster, 1939, and Philipson, 1949).

The concept that the reproductive apex results from a more or less extensive reorganization of the vegetative apex is the prevalent one and is accepted for both the angiosperms and the gymnosperms (Gifford and Wetmore, 1957; Wetmore et al., 1959). It is adopted in this book. The two kinds of meristem intergrade through intermediate forms and the existing differences are not fundamental; they are related to the different modes of growth of the vegetative and reproductive axes. The absence of discontinuity between the two kinds of growth is emphasized by Hillman (1962) in his review of the physiology of flowering. He suggests that floral induction represents not a sudden change in the condition of the shoot but is a process with many intermediate stages. The ontogenetic development of the reproductive apex from the vegetative is in agreement with this concept.

The change from the vegetative to the flowering state not only affects the apical meristems concerned with flower production but alters, physiologically and morphologically, other parts of the plant as well (Melchers and Lang, 1948; Philipson, 1949). This change is associated with a shift in the balance between meristematic activity and cell maturation in favor of the latter. It usually signifies the end of growth at the given apical meristem because of the determinate nature of the flower, and in annual plants it means the end of growth and the approach of death of the entire plant. The change is not irreversible, however, and may be interrupted or prevented by subjecting the plant to influences that favor vegetative growth. Even such a typical characteristic of the flower as determinate growth is not fixed, and the floral meristem occasionally resumes vegetative growth after the floral parts have been formed (Thompson, 1943-44). Thus the visible change from the vegetative to the reproductive meristem is a reflection of a physiologic change in the plant, and it may be discussed in terms of the ripeness-to-flower concept (Hillman, 1962).

ROOT APEX

In contrast to the apical meristem of the shoot, that of the root produces cells not only toward the axis but also away from it, for it initiates the rootcap. Because of the presence of the rootcap the distal part of the apical meristem of the root is not terminal but subterminal in position, in the sense that it is located beneath the rootcap (pl. 15A). The root apex further differs from the shoot meristem in that it forms no lateral appendages comparable to the leaves, and no branches. The root branches are usually initiated beyond the region of most active growth and arise endogenously (pl. 15B; chapter 17). Because of the absence of leaves, the root apex shows no periodic changes in shape and structure such as commonly occur in shoot apices in relation to leaf initiation. The root also produces no nodes and internodes, and, therefore, grows more uniformly in length than the shoot, in which the internodes elongate much more than the nodes. The rib-meristem type of growth is characteristic of the elongating root cortex (fig. 5.15, pl. 17C; Wagner, 1937).

The distal part of the apical meristem of the root, like that of the shoot, may be termed protomeristem and, as such, contrasted with the subjacent primary meristematic tissues. The young root axis is more or less clearly separated into the future central cylinder (plerome) and cortex (periblem). In their meristematic state the tissues of these two regions consist of procambium and ground meristem, respectively. The term procambium may be applied to the entire central cylinder of the root if this cylinder eventually differentiates into a solid vascular core. Many roots, however, have a pith-like region in the center. This region is regarded sometimes as potentially vascular and therefore procambial in its meristematic state, sometimes as ground tissue similar to that of the pith in stems and differentiating from a ground meristem (chapter 17). The term protoderm, if used to designate the surface layer regardless of its developmental relation to other tissues, may be applied to the outer layer of the young root. Usually the root protoderm does not arise from a separate layer of the protomeristem. It has a common origin with either the cortex or the rootcap.

Apical meristems of roots are analyzed on the basis of three concepts. The first is fundamentally Hanstein's histogen concept since it includes the assumption that a precise relation may exist between the initials in the distal zone and the tissue regions of the root. The second, the previously mentioned quiescent-center concept of Clowes (1961a), is a modification of the histogen concept. It places the initials of the tissue regions outside the distal region—the minimal constructional center of Clowes (1961a)—which is interpreted as being inactive. The third is the little-exploited body-cap (Körper-Kappe) concept of Schüepp (1917), which is comparable to the tunica-corpus concept, since it characterizes the root apex by reference to planes of division in its parts. The three concepts are not mutually exclusive. The histogen and the body-cap concepts deal with different aspects of apical activity, and the quiescent-center concept includes the postulate that the cell pattern in the distal zone is not meaningless but reflects the past history of meristem activity when the root meristem was being organized, either in embryogeny or during the origin of the lateral root.

The cellular configuration of the distal zone has been the subject of many studies and has served for the establishment of the so-called types (Schüepp, 1926) and for discussions of the phylogeny of apical organization of roots (Voronin, 1956). The principal configurations are depicted in figure 5.11, in which the distal zone is represented as containing the initials (shown in black). In the lower vascular plants all tissues are derived either from a single apical cell (Equisetaceae, Polypodiaceae; figs. 5.11A, 5.12A) or from several initials arranged in one tier (Marattiaceae). These plants usually have the same apical structure in both the root and the shoot. In some gymnosperms and angiosperms all tissue regions of the root or all except the central cylinder appear to arise from a common meristematic group of cells; in others one or more of these regions can be traced to separate initials. Guttenberg (1960) classifies the two kinds of organization as the open and the closed, respectively. He considers that both originate from a closed type present in the embryonic root or the primordium of the lateral or adventitious root. During later elongation of the root the closed pattern may be retained or replaced by an open one. In all the events of organization of the root meristem, central or connecting cells (Verbindungszellen) play the major role as initials. In position, they are periblem initials (Guttenberg, 1960).

The structure based on a single apical cell lends itself to a study of segmentation patterns among the derivatives of the apical meristem (fig. 5.12A; Clowes, 1961a). Since the root is normally radially symmetrical, the apical cell is tetrahedral. It cuts off cells either on four faces of the tetrahedron and thus produces the tissues of the root and the rootcap (*Marsilea*); or the rootcap has its own initials (*Azolla*). A root organization characterized by a precise segmentation of derivatives of the initial zone resembling that in the roots of ferns has been found in a monocotyledon, *Cyperus* (Kadej, 1963).

An analysis of divisions in the derivatives of the apical cell illustrates the body-cap concept (fig. 5.12B). The longitudinal rows of cells so prominent in roots radiate from the apical cell and many of them divide in two. Where they do so, a cell divides transversely; then one of the two new cells divides longitudinally and each daughter cell of this division becomes the source of a new row. The combination of the transverse and the longitudinal divisions results in an approximately T-(or Y-)shaped wall pattern and, therefore, such divisions of cell rows have been named T divisions. The direction of the top stroke of the T

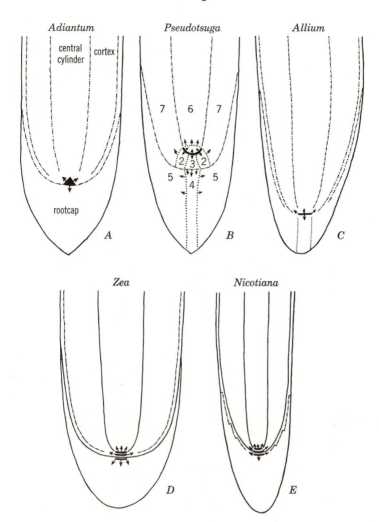

FIG. 5.11. Organisation of distal region of apical meristem of root. (*C–E*, based on the classical histogen concept.) *A*, a single apical cell (black triangle) is source of all parts of root and root-cap. *B*, initial zone (black arc) initiates mother-cell zones of the various root parts as follows: 1 (below 6, not marked) of central cylinder (6); 2, of cortex (7); 3, of column of rootcap (4). Longitudinal divisions on periphery of column give cells to peripheral part of rootcap (5). (Adapted from Allen, *Amer. Jour. Bot.* 34, 1947.) *C*, distal region with poorly individual-ized initials is source of central cylinder, cortex, and column. *D*, three tiers of initials in the initial zone. First is related to central cylinder; second, to cortex; third, to rootcap. The epidermis differentiates from the outermost layer of the cortex. *E*, three tiers of initials, first related to central cylinder; second, to cortex; third, to root-cap. The epidermis originates from the rootcap by periclinal divisions.

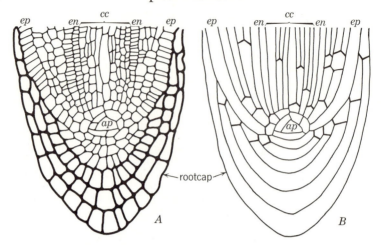

FIG. 5.12. Root tip of *Dennstaedtia*, a fern. *A*, organization of root apex with an apical cell (*ap*) and *B*, interpretation of sequences of doubling of cell layers—T (or Y) divisions— derived from apical cell. The orientation of the T distinguishes body (within primordial epidermis, *ep*) from cap (rootcap). In body, the top stroke of T is pointing toward apex, in cap, in opposite direction (toward base of root). Details: *ap*, apical cell; *cc*, central cylinder; *en*, endodermis; *ep*, epidermis. (×180. *A*, after List, *Amer. Jour. Bot.* 50, 1963.)

varies in different root parts. In the cap it is directed toward the base of the root, in the body toward the apex. The body and the cap are not sharply delimited if both arise from the same apical cell (*Marsilea*); the presence of separate rootcap initials causes the presence of a clear boundary between the cap and the body (*Azolla;* Clowes, 1961a).

The two kinds of multicellular protomeristem of angiosperms, the closed and the open in the sense of Guttenberg (1960), require separate consideration. The closed pattern is often characterized by the presence of three tiers of initials. One tier appears at the apex of the central cylinder, the second terminates the cortex, the third gives rise to the rootcap. The three-tiered meristems may be grouped according to the origin of the epidermis (rhizodermis of some authors; chapters 7, 17). In one group, the epidermis has common origin with the rootcap and becomes distinct as such after a series of T divisions along the periphery of the root (figs. 5.11*E*, 5.14*C*, 5.15*A*, pl. 20*A*). In the second, the epidermis and cortex have common initials, whereas the rootcap arises from its own initials that constitute the rootcap meristem, or *calyptrogen* (from the Greek *calyptra*, veil, and *genos*, offspring; figs. 5.11*D*, 5.13*A*, 5.15*B*). If the rootcap and the epidermis have common origin, the cell layer concerned is called *dermatocalyptrogen* (Guttenberg, 1960).

Roots with a dermatocalyptrogen are common in dicotyledons (representatives of Rosaceae, Solanaceae, Cruciferae, Scrophulariaceae, and Compositae; Schüepp, 1926), but occur also in monocotyledons (Palmae; Pillai and Pillai, 1961*b*; Schüepp, 1926). Roots with a calyptrogen are characteristic of monocotyledons (Gramineae, Zingiberaceae, some Palmae; Guttenberg, 1960; Hagemann, 1957; Pillai et al., 1961). Sometimes the epidermis appears to terminate in the distal zone with its own initials (Shimabuku, 1960). In some aquatic monocotyledons (*Hydrocharis, Lemna, Pistia*) the epidermis is regularly independent from the cortex and the rootcap.

An analysis of root meristems on the basis of the body-cap concept reveals the difference in origin of the epidermis. In the root with a calyptrogen the cap includes only the rootcap (fig. 5.14*A*), in one with a dermatocalyptrogen the cap extends into the epidermis (fig. 5.14*C*). The body-cap configuration shows other variations that elucidate patterns

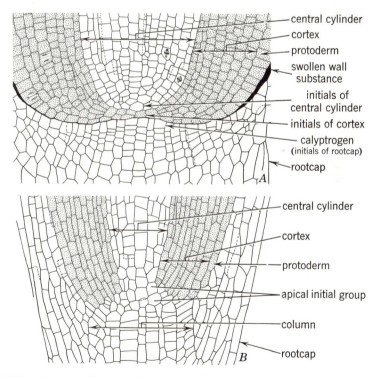

FIG. 5.13. Root tips of monocotyledons in median longitudinal sections. *A, Zea mays; B, Allium sativum.* The cells labeled initials are those concerned with the first organization of the root. Excepting the calyptrogen they may be quiescent during later growth. (Both, ×200.)

of growth of roots. In some roots the central core of the rootcap is distinct from the peripheral part in having few or no longitudinal divisions. If conspicuous enough, such a core is referred to as columella (fig. 5.14; Clowes, 1961a). The few T divisions that occur in the columella may be oriented according to the body pattern; then only the peripheral parts of the rootcap show the cap pattern.

Apices that lack a clear differentiation of initials (figs. 5.13B, 5.14B, pl. 20B)—the open type according to Guttenberg (1960)—are difficult to analyze. One common interpretation is that such roots have a *transversal meristem* without any boundaries with reference to the derivative regions of the root (Popham, 1955). The other view is that the central cylinder has its own initials in this type of meristem (Clowes, 1961a; Wilcox, 1962). Analyses of body-cap configurations indicate that the limits between the two regions may be indefinite and may change during the growth of the root (Clowes, 1961a). A reinterpretation of meristems with indefinite boundaries in the distal zone was given by Allen (1947) for *Pseudotsuga taxifolia* and by Clowes (1961a) for *Fagus sylvatica*. In *Pseudotsuga*, two kinds of initials are recognized: the "permanent" (black arc in figure 5.11B), which remain in their position indefinitely, and the "temporary" (fig. 5.11B, zones 1, 2, and 3), which

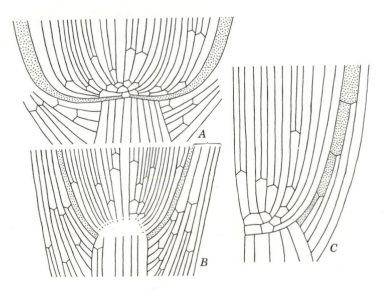

FIG. 5.14. Interpretation of root apices of *Zea* (A), *Allium* (B), and *Nicotiana* (C) in terms of body-cap concept. Body, upper stroke of T points toward apex; cap, upper stroke of T points toward base of root. Protoderm is stippled. It is part of body in A, and probably in B; part of cap in C.

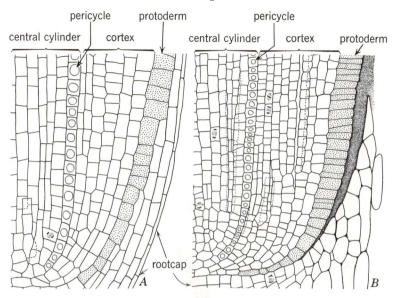

FIG. 5.15. Longitudinal sections of root tips of *Nicotiana tabacum* (A) and *Zea mays* (B), illustrating two contrasting methods of origin of epidermis. A, epidermis separates from the rootcap by periclinal divisions. B, epidermis arose from same initials as the cortex through periclinal division in a recent derivative of a cortical initial. Densely stippled area in B, gelatinized wall between rootcap and protoderm. (A, ×285; B, ×210.)

give origin to the various root regions and are from time to time replaced by derivatives from the permanent initials. *Fagus sylvatica* has a basically similar apical organization, but apparently the initials of the various regions are more independent than those of *Pseudotsuga*. Moreover, Clowes reports that the distal region enclosed by the cup-shaped group of initials is quiescent.

Apical meristems with no separate initials for the root regions have been described in dicotyledons (representatives of Proteaceae, Casuarinaceae, Leguminosae, of some Ranalean and Amentiferous families; Schüepp, 1926), monocotyledons (representatives of Musaceae, Palmae; Pillai and Pillai, 1961a, b), and some gymnosperms (Guttenberg, 1961; Wilcox, 1954). In a group of conifers (Pillai, 1964), the apex is interpreted as having (1) common initials for the central cylinder and the columella and (2) a common initiating zone for the cortex and the peripheral part of the rootcap. The initiating zone 2 surrounds the initials 1 and their recent derivatives.

The quiescent-center concept has been studied and discussed by Clowes with consistency and imagination. After various studies on normally developing roots and on experimentally treated ones, or roots that were fed labeled compounds involved in DNA synthesis, Clowes (1961*a*) concluded that the inactive state of the distal zone—the zone containing the initials according to the classical histogen theory—is a general phenomenon in roots. Whereas the classical concept assumes that the number of initials is small (Guttenberg, 1960), the quiescent-center concept indicates a large number of initials. Clowes recognizes that occasional divisions do occur in the center and that it may become active when the previously acting initials are damaged, as, for example, by radiation. The quiescent center is a reservoir of cells relatively resistant to damage because of their inactivity (Davidson, 1961; Clowes, 1961*a*, 1963). They may be sites of auxin synthesis and the source of diploid cells for replacement of polyploid and aneuploid cells that may accumulate during somatic differentiation. Finally, they are the permanent source of the active initials which themselves are not permanent, as evidenced by the fluctuations in the size of the quiescent center. Thus the role of this center may be far more important than its relative inactivity would indicate (Clowes, 1961*a*).

The tips of growing roots are frequently used in studies of growth (Clowes, 1961*a*). The zone of actively dividing cells in growing roots extends to a considerable distance from the apex, in *Zea*, for example, through 8 to 10 millimeters with a maximum at the 4 millimeter level (Erickson and Sax, 1956). The distribution of meristematic activity differs in the various root regions (chapter 17); however, the data obtained for mitotic frequency vary probably mainly in relation to methods of analysis (Clowes, 1961*a*). At the same level of the root, the processes of cell division, cell enlargement, and cell maturation overlap not only in the different tissue regions but also in the different cells of the same tissue region, and even in individual cells. The meristematic cortex vacuolates and develops intercellular spaces close to the apex, where the central-cylinder meristem still appears dense. In the central cylinder the precursors of the innermost xylem vessels cease dividing, enlarge, and vacuolate considerably in advance of the other vascular precursors (pl. 82A), and the first sieve tubes mature in the part of the root where cell division is still in progress (chapter 17). In individual cells, division, elongation, and vacuolation are combined.

REFERENCES

Abbe, E. C., and B. O. Phinney. The growth of the shoot apex in maize: external features. *Amer. Jour. Bot.* 38:737–743. 1951.

Abbe, E. C., B. O. Phinney, and D. F. Baer. The growth of the shoot apex in maize: internal features. *Amer. Jour. Bot.* 38:744–751. 1951.

References

Abbe, E. C., and O. L. Stein. The growth of the shoot apex in maize: embryogeny. *Amer. Jour. Bot.* 41:285–293. 1954.

Allen, G. S. Embryogeny and the development of the apical meristems of *Pseudotsuga.* III. Development of the apical meristems. *Amer. Jour. Bot.* 34:204–211. 1947.

Askenasy, E. Über eine neue Methode, um die Vertheilung der Wachsthumsintensität in wachsenden Theilen zu bestimmen. *Naturhist. Medic. Ver. Heidelberg, Verhandl. N. S.* 2:70–153. 1880.

Audus, L. J. Correlations. *Linn. Soc. London Jour., Bot.* 56:177–187. 1959.

Bain, H. F., and H. Dermen. Sectorial polyploidy and phyllotaxy in the cranberry (*Vaccinium macrocarpon* Ait.). *Amer. Jour. Bot.* 31:581–587. 1944.

Ball, E. The development of the shoot apex and of the primary thickening meristem in *Phoenix canariensis* Chaub., with comparisons to *Washingtonia filifera* Wats. and *Trachycarpus excelsa* Wendl. *Amer. Jour. Bot.* 28:820–832. 1941.

Ball, E. Development in sterile culture of stem tips and subjacent regions of *Tropaeolum majus* L. and *Lupinus albus* L. *Amer. Jour. Bot.* 33:301–318. 1946.

Ball, E. Differentiation in the primary shoots of *Lupinus albus* L. and *Tropaeolum majus* L. *Soc. Expt. Biol. Symposia.* No. 2. *Growth.* 1948:246–262. 1948.

Ball, E. Cell divisions in living shoot apices. *Phytomorphology* 10:377–396. 1960.

Barker, W. G., and F. C. Steward. Growth and development of the banana plant. I. The growing regions of the vegetative shoot. *Ann. Bot.* 26:389–411. 1962*a*. II. The transition from the vegetative to the floral shoot in *Musa acuminata* cv. Gros Michel. *Ann. Bot.* 26:413–423. 1962*b*.

Bartels, F. Zur Entwicklung der Keimpflanze von *Epilobium hirsutum.* II. Die im Vegetationspunkt während eines Plastochrons ablaufenden Zellteilungen. *Flora* 149:206–224. 1960.

Bartels, F. Zur Entwicklung der Keimpflanze von *Epilobium hirsutum.* IV. Der Nachweis eines Scheitelzellenwachstums. *Flora* 150:552–571. 1961.

Bersillon, G. Recherches sur les Papavéracées. Contribution à l'étude du développement des Dicotylédones herbacées. *Ann. des Sci. Nat., Bot.* Ser. 11. 16:225–447. 1956.

Bersillon, G. L'inflorescence terminale de *Reseda lutea* L.; ses rapports avec la pousse végétative. *Rev. de Cytol. et de Biol. Vég.* 19:185–197. 1958.

Bhambie, S. Studies in pteridophytes. I. The shoot apex of *Isoetes coromandeliana* L. *Indian Bot. Soc. Jour.* 36:491–502. 1957.

Bierhorst, D. W. Structure and development of the gametophyte of *Psilotum nudum.* *Amer. Jour. Bot.* 40:649–658. 1953.

Bierhorst, D. W. The subterranean sporophytic axes of *Psilotum nudum.* *Amer. Jour. Bot.* 41:732–739. 1954.

Bierhorst, D. W. Observations on the gametophytes of *Botrychium virginianum* and *B. dissectum.* *Amer. Jour. Bot.* 45:1–9. 1958.

Bonnett, O. T. The development of the wheat spike. *Jour. Agr. Res.* 53:445–451. 1936.

Buvat, R. Le méristème apical de la tige. *Ann. Biol.* 31:595–656. 1955*a*.

Buvat, R. Sur la structure et le fonctionnement du point végétatif de la *Selaginella caulescens* Spring., var. *amoena.* *Acad. des Sci. Compt. Rend.* 241:1833–1836. 1955*b*.

Buvat, R., and L. Genèves. Sur l'inexistance des initiales axiales dans la racine d'*Allium cepa* L. (Liliacées). *Acad. des Sci. Compt. Rend.* 232:1579–1581. 1951.

Buvat, R., and O. Liard. Nouvelle constatation de l'inertie des soi-disant initiales axiales dans le méristème radiculaire de *Triticum vulgare.* *Acad. des Sci. Compt. Rend.* 236:1193–1195. 1953.

Camefort, H. Étude de la structure du point végétatif et des variations phyllotaxiques chez quelques gymnospermes. *Ann. des Sci. Nat., Bot.* Ser. 11. 17:1–185. 1956.

Catesson, A. M. Structure, évolution et fonctionnement du point végétatif d'une Monocotylédone: *Luzula pedemontana* Boiss. et Reut. (Joncacées). *Ann. des Sci. Nat., Bot.* Ser. 11. 14:253–291. 1953.

Champagnat, M.   Recherches de morphologie descriptive et experimentale sur le genre *Linaria. Ann. des Sci. Nat., Bot.* Ser. 12. 22:1–170. 1961.

Clowes, F. A. L.   *Apical meristems.* Botanical Monographs Vol. 2.   Oxford, Blackwell. 1961a.

Clowes, F. A. L.   Effects of β-radiation on meristems. *Expt. Cell Res.* 25:529–534. 1961b.

Clowes, F. A. L.   X-irradiation of root meristems.   *Ann. Bot.* 27:343–364. 1963.

Cutter, E. G.   On a theory of phyllotaxis and histogenesis. *Biol. Rev.* 34:243–263. 1959.

Cutter, E. G., and B. R. Voeller.   Changes in leaf arrangement in individual fern apices. *Linn. Soc. London Jour., Bot.* 56:225–236. 1959.

Davidson, D.   Mechanisms of reorganization and cell repopulation in meristems in roots of *Vicia faba* following irradiation and colchicine. *Chromosoma* 12:484–504. 1961.

Dermen, H.   Periclinal cytochimeras and origin of tissues in stem and leaf of peach. *Amer. Jour. Bot.* 40:154–168. 1953.

Dermen, H.   Nature of plant sports. *Amer. Hort. Mag.* 39:123–173. 1960.

Edgar, E.   *Fluctuations of mitotic index in the shoot apex of Lonicera nitida.*   Univ. Canterbury Publ. 1.   1961.

Erickson, R. O., and F. J. Michelini.   The plastochron index. *Amer. Jour. Bot.* 44:297–305. 1957.

Erickson, R. O., and K. B. Sax.   Rates of cell division and cell elongation in the growth of the primary root of *Zea mays. Amer. Phil. Soc. Proc.* 100:499–514. 1956.

Esau, K.   Origin and development of primary vascular tissues in seed plants. *Bot. Rev.* 9:125–206. 1943.

Evans, M. W., and F. O. Grover.   Developmental morphology of the growing point of the shoot and the inflorescence in grasses. *Jour. Agr. Res.* 61:481–520. 1940.

Foster, A. S.   Leaf differentiation in angiosperms. *Bot. Rev.* 2:349–372. 1936.

Foster, A. S.   Structure and growth of the shoot apex in *Ginkgo biloba. Torrey Bot. Club Bul.* 65:531–556. 1938.

Foster, A. S.   Problems of structure, growth and evolution in the shoot apex of seed plants. *Bot. Rev.* 5:454–470. 1939.

Foster, A. S.   Comparative studies on the structure of the shoot apex in seed plants. *Torrey Bot. Club Bul.* 68:339–350. 1941.

Foster, A. S.   Zonal structure and growth of the shoot apex in *Microcycas calocoma* (Miq.) A. Dc. *Amer. Jour. Bot.* 30:56–73. 1943.

Foster, A. S.   *Practical plant anatomy.*   2nd ed.   New York, D. Van Nostrand Company. 1949.

Garrison, R.   Origin and development of axillary buds: *Syringa vulgaris* L. *Amer. Jour. Bot.* 36:205–213. 1949.

Garrison, R.   Studies in the development of axillary buds. *Amer. Jour. Bot.* 42:257–266. 1955.

Gifford, E. M., Jr.   The structure and development of the shoot apex in certain woody Ranales. *Amer. Jour. Bot.* 37:595–611. 1950.

Gifford, E. M., Jr.   Ontogeny of the vegetative axillary bud in *Drimys Winteri* var. *chilensis. Amer. Jour. Bot.* 38:234–243. 1951.

Gifford, E. M., Jr.   The shoot apex in angiosperms. *Bot. Rev.* 20:477–529. 1954.

Gifford, E. M., Jr.   Incorporation of H^3-thymidine into shoot and root apices of *Ceratopteris thalictroides. Amer. Jour. Bot.* 47:834–837. 1960.

Gifford, E. M., Jr., and H. B. Tepper.   Ontogeny of the inflorescence in *Chenopodium album. Amer. Jour. Bot.* 48:657–667. 1961.

Gifford, E. M., Jr., and H. B. Tepper.   Histochemical and autoradiographic studies of floral induction in *Chenopodium album. Amer. Jour. Bot.* 49:706–714. 1962a.

Gifford, E. M., Jr., and H. B. Tepper. Ontogenetic and histochemical changes in the vegetative shoot tip of *Chenopodium album*. *Amer. Jour. Bot.* 49:902–911. 1962*b*.

Gifford, E. M., Jr., and R. H. Wetmore. Apical meristems of vegetative shoots and strobili in certain gymnosperms. *Natl. Acad. Sci. Proc.* 43:571–576. 1957.

Gottlieb, J. E., and T. A. Steeves. Development of the braken fern, *Pteridium aquilinum* (L.) Kuhn.—III. Ontogenetic changes in the shoot apex and in the pattern of differentiation. *Phytomorphology* 11:230–242. 1961.

Grégoire, V. La morphogénèse et l'autonomie morphologique de l'appareil floral. I. Le carpelle. *Cellule* 47:287–452. 1938.

Guttenberg, H. von. Grundzüge der Histogenese höherer Pflanzen. I. Die Angiospermen. In: *Handbuch der Pflanzenanatomie*. Band 8. Teil 3. Berlin, Gebrüder Borntraeger. 1960. II. Die Gymnospermen. In: *Handbuch der Pflanzenanatomie*. Band 8. Teil 4. Berlin, Gebrüder Borntraeger. 1961.

Haberlandt, G. Physiological plant anatomy. London, Macmillan and Company. 1914.

Hagemann, R. Untersuchungen über die Mitosenhäufigkeit in Gerstenwurzeln. *Kulturpflanze* 4:46–82. 1956.

Hagemann, R. Anatomische Untersuchungen an Gerstenwurzeln. *Kulturpflanze* 5:75–107. 1957.

Hagemann, W. Kritische Untersuchungen über die Organisation des Sprossscheitels dikotyler Pflanzen. *Österr. Bot. Ztschr.* 107:366–402. 1960.

Hagemann, W. Weitere Untersuchungen zur Organization des Sprossscheitelmeristems; der Vegetationspunkt traubiger Floreszenzen. *Bot. Jahrb.* 82:273–315. 1963.

Hanstein, J. Die Scheitelzellgruppe im Vegetationspunkt der Phanerogamen. *Festschr. Niederrhein. Gesell. Natur- und Heilkunde* 1868:109–134. 1868.

Hanstein, J. Die Entwickelung des Keimes der Monokotylen und der Dikotylen. *Bot. Abhandl.* 1(1):1–112. 1870.

Hara, N. Structure of the vegetative shoot apex and development of the leaf in the Ericaceae and their allies. *Univ. Tokyo, Jour. Fac. Sci. Sec.* III, *Bot.* 7:367–450. 1958.

Hara, N. Structure and seasonal activity of the vegetative shoot apex of *Daphne pseudomezereum*. *Bot. Gaz.* 124:30–42. 1962.

Härtel, K. Studien an Vegetationspunkten einheimischer Lycopodien. *Beitr. z. Biol. der Pflanz.* 25:125–168. 1938.

Hillman, W. S. *The physiology of flowering*. New York, Holt, Rinehart and Winston. 1962.

Jackman, V. H. The shoot apex of some New Zealand gymnosperms. *Phytomorphology* 10:145–157. 1960.

Jacobs, W. P., and I. B. Morrow. A quantitative study of mitotic figures in relation to development in the apical meristem of vegetative shoots of *Coleus*. *Devlpmt. Biol.* 3:569–587. 1961.

Jacobs, W. P., and V. Raghavan. Studies on the floral histogenesis and physiology of *Perilla*—I. Quantitative analysis of flowering in *P. frutescens* (L.) Britt. *Phytomorphology* 12:144–167. 1962.

Jentsch, R. Untersuchungen an den Sprossvegetationspunkten einiger Saxifragaceen. *Flora* 144:251–289. 1957.

Jentsch, R. Zur Kenntnis des Sprossvegetationspunktes von *Hippuris* und *Myriophyllum*. *Flora* 149:307–319. 1960.

Johnson, M. A. On the shoot apex of the cycads. *Torreya* 44:52–58. 1944.

Johnson, M. A. The shoot apex in gymnosperms. *Phytomorphology* 1:188–204. 1951.

Johnson, M. A., and R. J. Tolbert. The shoot apex of *Bombax*. *Torrey Bot. Club Bul.* 87:173–186. 1960.

Kadej, F. Interpretation of the pattern of the cell arrangement in the root apical meristem of *Cyperus gracilis* L. var. *alternifolius*. *Soc. Bot. Polon. Acta* 32:295–301. 1963.

Kalbe, L.   Histogenetische Untersuchungen an Sprossvegetationspunkten dikotyler Holz-pflanzen. *Flora* 152:279–314. 1962.

Koch, L.   Die vegetative Verzweigung der höheren Gewächse. *Jahrb. f. Wiss. Bot.* 25:380–488. 1893.

Lance, A.   Recherches cytologiques sur l'évolution de quelques méristèmes apicaux et sur ses variations provoquées par traitments photopériodiques. *Ann. des Sci. Nat., Bot.* Ser. 11. 18:91–421. 1957.

Link, G. K. K., and V. Eggers.   Mode, site, and time of initiation of hypocotyledonary bud primordia in *Linum usitatissimum* L.   *Bot. Gaz.* 107:441–454.   1946.

Loiseau, J. E.   Observation and expérimentation sur la phyllotaxie et le fonctionnement du sommet végétatif chez quelques Balsaminacées. *Ann. des Sci. Nat., Bot.* Ser. 11. 20:1–24. 1959.

MacDaniels, L. H.   Anatomical basis of so-called adventitious buds in apple. *New York Agric. Expt. Sta. Mem.* 325. 1953.

Melchers, G., and A. Lang.   Die Physiologie der Blütenbildung. *Biol. Zentbl.* 67:105–174. 1948.

Michelini, F. J.   The plastochron index in developmental studies of *Xanthium italicum* Moretti. *Amer. Jour. Bot.* 45:525–533. 1958.

Millington, W. F., and E. L. Fisk.   Shoot development in *Xanthium pennsylvanicum*. I. The vegetative plant. *Amer. Jour. Bot.* 43:655–665. 1956.

Mohr, H., and E. Pinnig.   Der Einfluss des Lichtes auf die Bildung von Blattprimordien am Vegetationskegel der Keimlinge von *Sinapis alba* A. *Planta* 58:569–579. 1962.

Newman, I. V.   Pattern in meristems of vascular plants—1. Cell partitions in living apices and in the cambial zone in relation to the concepts of initial cells and apical cells. *Phytomorphology* 6:1–19. 1956. II. A review of shoot apical meristems of gymnosperms, with comments on apical biology and taxonomy, and a statement of some fundamental concepts. *Linn. Soc. New South Wales Proc.* 86:9–59. 1961.

Paolillo, D. J., Jr., and E. M. Gifford, Jr.   Plastochronic changes and the concept of apical initials in *Ephedra altissima. Amer. Jour. Bot.* 48:8–16. 1961.

Parke, R. V.   Growth periodicity and the shoot tip of *Abies concolor. Amer. Jour. Bot.* 46:110–118. 1959.

Philipson, W. R.   The ontogeny of the shoot apex in dicotyledons. *Biol. Rev.* 24:21–50. 1949.

Philipson, W. R.   Organization of the shoot apex in dicotyledons. *Phytomorphology* 4:70–75. 1954.

Pillai, A.   Root apical organization in gymnosperms—some conifers. *Torrey Bot. Club Bul.* 91:1–13. 1964.

Pillai, S. K., and A. Pillai.   Root apical organization in monocotyledons—Musaceae. *Indian Bot. Soc. Jour.* 40:444–455. 1961*a*.

Pillai, S. K., and A. Pillai.   Root apical organization in monocotyledons—Palmae. *Indian Acad. Sci. Proc., Sec. B.* 54:218–233. 1961*b*.

Pillai, S. K., A. Pillai, and S. Sachedeva.   Root apical organization in monocotyledons—Zingiberaceae. *Indian Acad. Sci. Proc., Sec. B.* 53:240–256. 1961.

Popham, R. A.   Principal types of vegetative shoot apex organization in vascular plants. *Ohio Jour. Sci.* 51:249–270. 1951.

Popham, R. A.   Zonation of the primary and lateral root apices of *Pisum sativum. Amer. Jour. Bot.* 42:267–273. 1955.

Popham, R. A.   Cytogenesis and zonation in the shoot apex of *Chrysanthemum morifolium. Amer. Jour. Bot.* 45:198–206. 1958.

Popham, R. A., and A. P. Chan.   Zonation in the vegetative stem tip of *Chrysanthemum morifolium* Bailey. *Amer. Jour. Bot.* 37:476–484. 1950.

References   129

Popham, R. A., and A. P. Chan. Origin and development of the receptacle of *Chrysanthemum morifolium*. *Amer. Jour. Bot.* 39:329–339. 1952.

Priestley, J. H., and C. F. Swingle. Vegetative propagation from the standpoint of plant anatomy . *U.S. Dept. Agr. Tech. Bul.* 151. 1929.

Rauh, W., and H. Falk. *Stylites* E. Amstutz, eine neue Isoëtacee aus den Hochanden Perus. 2. Teil. Zur Anatomie des Stammes mit besonderer Berücksichtigung der Verdickungsprozesse. *Heidelberg. Akad. der Wiss., Math.-Nat. Kl. Sitzber.* 2 Abh. 1959.

Rauh, W., and F. Rappert. Über das Vorkommen und die Histogenese von Scheitelgruben bei krautigen Dikotylen, mit besonderer Berücksichtigung der Ganz- und Halbrosettenpflanzen. *Planta* 43:325–360. 1954.

Rauh, W., and H. Reznik. Histogenetische Untersuchungen an Blüten- und Infloreszenzachsen. I. Teil. Die Histogenese becherförmiger Blüten- und Inflorszenzachsen, sowie der Blütenachsen einiger Rosoideen. *Heidelberg. Akad. der Wiss., Math.-Nat. Kl. Sitzber.* 3 Abh. 1951. II. Die Histogenese der Achsen köpfchenförmiger Infloreszenzen. *Beitr. z. Biol. der Pflanz.* 29:233–296. 1953.

Rodin, R. J. Seedling morphology of *Welwitschia*. *Amer. Jour. Bot.* 40:371–378. 1953.

Romberg, J. A. Meristems, growth, and development in woody plants. *U.S. Dept. Agr. For. Serv. Tech. Bul.* 1293. 1963.

Sacher, J. A. Structure and seasonal activity of the shoot apices of *Pinus lambertiana* and *Pinus ponderosa*. *Amer. Jour. Bot.* 41:749–759. 1954.

Schmidt, A. Histologische Studien an phanerogamen Vegetationspunkten. *Bot. Arch.* 8:345–404. 1924.

Schüepp, O. Untersuchungen über Wachstum und Formwechsel von Vegetationspunkten. *Jahrb. f. Wiss. Bot.* 57:17–79. 1917.

Schüepp, O. Meristeme. In: K. Linsbauer. *Handbuch der Pflanzenanatomie*. Band 4. Lief. 16. 1926.

Seeliger, I. Studien am Sprossvegetationskegel von *Thujopsis dolabrata* (L.f.) Sieb. et Zucc. *Flora* 142:183–212. 1954.

Senghas, K. Histogenetische Studien an Sprossvegetationspunkten dicotyler Pflanzen. I. Bau und Histogenese des Spross-Scheitelmeristems einiger Cruciferen. *Beitr. z. Biol. der Pflanz.* 33:85–113. 1956. II. Gestalt und Architektonik des ruhenden, embryonalen Vegetationspunktes. *Beitr. z. Biol. der Pflanz.* 33:325–370. 1957.

Sharman, B. C. Leaf and bud initiation in the Gramineae. *Bot. Gaz.* 106:269–289. 1945.

Shimabuku, K. Observation on the apical meristem of rice roots. *Bot. Mag.* [Tokyo] 73:22–28. 1960.

Sifton, H. B. Developmental morphology of vascular plants. *New Phytol.* 43:87–129. 1944.

Singh, H. Seasonal variations in the shoot apex of *Cephalotaxus drupacea* Sieb. et Zucc. *Phytomorphology* 11:146–153. 1961.

Smith, C. A. Shoot apices in the family Moraceae with a seasonal study of *Maclura pomifera* (Raf.) Schneid. *Torrey Bot. Club Bul.* 90:237–258. 1963.

Snow, M., and R. Snow. On the determination of leaves. *New Phytol.* 46:5–19. 1947.

Soma, K. Morphogenesis in the shoot apex of *Euphorbia lathyrus* L. *Univ. Tokyo Jour. Fac. Sci. Sec. III, Bot.* 7:199–256. 1958.

Stein, D. B., and O. L. Stein. The growth of the stem tip of *Kalanchoë* cv. "Brilliant Star." *Amer. Jour. Bot.* 47:132–140. 1960.

Sunderland, N. Cell division and expansion in the growth of the shoot apex. *Jour. Expt. Bot.* 12:446–457. 1961.

Sussex, I. M. Morphogenesis in *Solanum tuberosum* L.: apical structure and developmental pattern of the juvenile shoot. *Phytomorphology* 5:253–273. 1955.

Thielke, C. Die histologische Struktur des Sprossvegetationskegels einiger Commelinaceen unter Berücksichtigung panaschierter Formen. *Planta* 44:18–74. 1954.

Thielke, C. Über die Differenzierungsvorgänge bei Cyperaceen. I. Der Bau des vegetativen Vegetationskegels und die Anfangsstadien der Blattentwicklung. *Planta* 48:564–577. 1957.

Thielke, C. Histologische Untersuchungen am Sprossscheitel von *Saccharum*. II. Mitteilung. Der Sprossscheitel von *Saccharum sinense*. *Planta* 58:175–192. 1962.

Thomas, R. G. Floral induction and the stimulation of cell division in *Xanthium*. *Science* 140:54–56. 1963.

Thompson, J. McLean. Towards a modern physiological interpretation of flowering. *Linn. Soc. London, Proc.* 156:46–68. 1943–44.

Thomson, B. F., and P. M. Miller. The role of light in histogenesis and differentiation in the shoot of *Pisum sativum*. I. The apical region. *Amer. Jour. Bot.* 49:303–310. 1962.

Torrey, J. G. On the determination of vascular patterns during tissue differentiation in excised pea roots. *Amer. Jour. Bot.* 42:183–198. 1955.

Tucker, S. C. Ontogeny of the floral apex of *Michelia fuscata*. *Amer. Jour. Bot.* 47:266–277. 1960.

Vaughan, J. G. The morphology and growth of the vegetative and reproductive apices of *Arabidopsis thaliana* (L.) Heynh., *Capsella bursa-pastoris* (L.) Medic. and *Anagallis arvensis* L. *Linn. Soc. London Jour., Bot.* 55:279–301. 1955.

Voronin, N. S. Ob evoliutsii korneï rasteniï. [On evolution of roots of plants.] *Moskov. Obshch. Isp. Prirody, Otd. Biol., Biul.* 61:47–58. 1956.

Wagner, N. Wachstum und Teilung der Meristemzellen in Wurzelspitzen. *Planta* 27:550–582. 1937.

Wardlaw, C. W. Experimental and analytical studies of pteridophytes. II. Experimental observations on the development of buds in *Onoclea sensibilis* and in species of *Dryopteris*. *Ann. Bot.* 7:357–377. 1943.

Wardlaw, C. W. The shoot apex in pteridophytes. *Biol. Rev.* 20:100–114. 1945.

Wardlaw, C. W. Experimental investigations of the shoot apex of *Dryopteris aristata* Druce. *Roy. Soc. London, Phil. Trans.* 232:343–384. 1947.

Wardlaw, C. W. The reactivity of the apical meristem as ascertained by cytological and other techniques. *New Phytol.* 56:221–229. 1957.

Wetmore, R. H. Growth and development in the shoot system of plants. In: *Cellular Mechanisms in Differentiation and Growth*. Princeton, N.J., Princeton University Press. 1956.

Wetmore, R. H., E. M. Gifford, Jr., and M. C. Green. Development of vegetative and floral buds. In: *Photoperiodism and Related Phenomena in Plants and Animals*. Washington, Amer. Assoc. Adv. Sci. 1959.

Wilcox, H. Primary organization of active and dormant roots of noble fir, *Abies procera*. *Amer. Jour. Bot.* 41:812–821. 1954.

Wilcox, H. Growth studies of the root of incense cedar, *Libocedrus decurrens*. I. The origin and development of primary tissues. *Amer. Jour. Bot.* 49:221–236. 1962.

Wolff, C. F. *Theoria generationis*. Leipzig, Wilhelm Engelman. 1759.

6

The Vascular Cambium

LOCATION IN THE PLANT

The vascular cambium is the lateral meristem that forms the secondary vascular tissues. It is located between the xylem and the phloem (fig. 1.3, pl. 21) and, in stems and roots, commonly has the shape of a cylinder. When the secondary vascular tissues of an axis are in discrete strands, the cambium may remain restricted to these strands in the form of strips (*Cucurbita*, pl. 63C). It also appears in strips in most petioles and leaf veins that show secondary growth.

CELL TYPES

The tissues derived from apical meristems contain many cell types which differ strikingly from the meristematic cells in shape and size. In contrast, there is a general resemblance between the cambium cells and their derivatives, and the shape and arrangement of cells in the secondary xylem and the secondary phloem are foreshadowed in the shape and arrangement of the cambial cells (pl. 21; chapters 11, 12).

The vascular cambium contains two types of cells: elongated cells with tapering ends, the *fusiform initials* (that is, spindle-shaped initials), and nearly isodiametric, relatively small cells, the *ray initials* (figs. 6.1, 6.2; pl. 22). The exact shape of the fusiform initials of *Pinus silvestris* has been determined as that of long, pointed, tangentially flattened cells with an average of 18 faces (Dodd, 1948). The fusiform initials give rise to all the cells of xylem and phloem that are arranged with their long axes parallel to the long axis of the organ in which they occur; in other words, they form the longitudinal or axial systems of xylem and

131

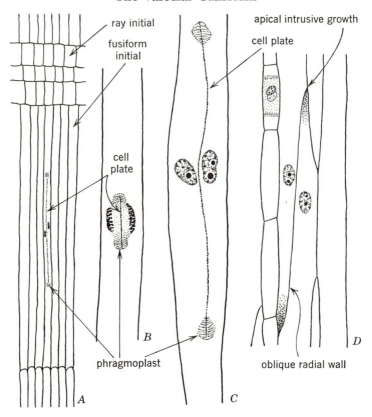

ray initial

fusiform
initial

apical intrusive growth

cell plate

cell
plate

cell
plate

B

phragmoplast

oblique radial wall

D

A

C

FIG. 6.1. Cytokinesis in vascular cambium of *Nicotiana tabacum* as seen in radial (*A–C*) and tangential (*D*) sections of stem. *A–C*, tangential divisions in side view. *B*, early stage of division; *C*, later stage. *D*, ray initial in tangential division, with the cell plate appearing in surface view, and an oblique radial wall recently formed in fusiform initial. Densely stippled areas in *D*, apices of the two new cells which were growing by apical intrusive growth, one downward, the other upward. (*A*, ×120; *B*, *C*, ×600; *D*, ×300.)

phloem (chapters 11, 12). Examples of elements in these systems are tracheids, fibers, and xylem-parenchyma cells in the xylem; sieve cells, fibers, and phloem-parenchyma cells in the phloem. The ray initials give origin to the ray cells, that is, elements of the transverse or ray system of the xylem and the phloem.

Table 6.1 gives information on the comparative characteristics of the two kinds of initials in *Pinus strobus*. These initials differ from each other most notably in length and volume, the fusiform cells being much larger than the ray initials. However, in one dimension, the radial, the ray initials surpass the fusiform. In the 60-year-old stem both kinds of initials are larger than in the 1-year-old stem. The initials are uninucleate, and, although the nuclei of the fusiform initials may be markedly larger than those of the ray initials, their volumes do not increase in proportion to the cell volumes so that the ratio of nuclear volume to cell volume is much smaller in the fusiform cells (table 6.1, last column).

The fusiform initials show a wide range of variation in their dimensions and volume (Bailey, 1920*a*). Some of these variations depend on plant species. The following figures, expressed in millimeters, exemplify differences in the lengths of fusiform initials in several plants: *Pinus strobus*, 3.20; *Ginkgo*, 2.20; *Myristica*, 1.31; *Pyrus*, 0.53; *Populus*, 0.49; *Fraxinus*, 0.29; *Robinia*, 0.17 (Bailey, 1920*a*). Fusiform initials vary in

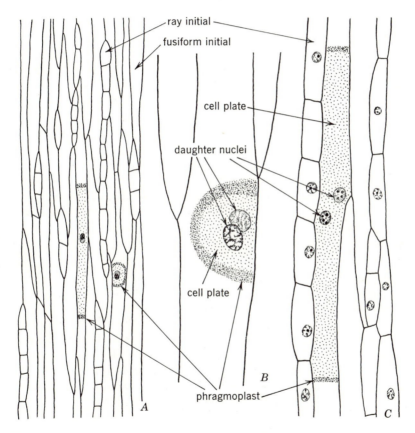

FIG. 6.2. Cytokinesis in vascular cambium of *Nicotiana tabacum* as seen in tangential sections of stem. Tangential divisions in fusiform initials. *A–C* partly formed cell plates in surface view. *B*, cell plate has reached one of the radial longitudinal walls of the parent cell. *C*, cell plate has reached both radial walls. (*A*, ×120; *B*, ×600; *C*, ×300.)

Table 6.1 Dimensions of Cambial Initials of *Pinus strobus*
(Adapted from Bailey, 1920*b*)

Age of axis in years	Kind of initial	Diameters in microns			Volume in microns^3	Ratio between volumes of nucleus and cell
		Vertical	Radial	Tangential		
1	Ray	22.9	17.8	13.8	5,000	1:14
1	Fusiform	870.0	4.3	16.0	60,000	1:60
60	Ray	24.8	26.6	17.0	10,000	1:12
60	Fusiform	4000.0	6.2	42.4	1,000,000	1:286

length within species, partly in relation to growth conditions. They also show length modifications associated with developmental phenomena in a single plant. Generally, the length of fusiform initials increases with the age of the axis, but after this length reaches a certain maximum it remains relatively stable (table 6.1; Bailey, 1920*a*; Bannan, 1960*b*; Bosshard, 1951). The changes in the size of fusiform initials bring about similar changes in the secondary xylem and phloem cells derived from these initials. The ultimate size of these cells, however, depends only partly on that of the cambial initials, because changes in size also occur during the differentiation of cells (chapter 4).

The cambial cells are highly vacuolated (pls. 21*B*, 22; Bailey, 1930). Their walls have primary pit-fields with plasmodesmata. The radial walls are thicker than the tangential walls particularly during dormancy, and their primary pit-fields are deeply depressed.

CELL ARRANGEMENT

During active growth in the cambium, the initials and their immediate derivatives form a zone of similar unexpanded meristematic cells, the *cambial zone* (pl. 21*A*). As seen in transections the cells in the cambial zone are arranged in radial series. On either side of the cambial zone, cambial derivatives expand and gradually assume the characteristics of the various xylem and phloem cells. Experiments with strips of bark partially detached from the stem indicate that mutual pressure of tissues is important in controlling the orderly pattern of differentiation of the cambial products (Brown and Sax, 1962). The prevailing concept is that the initials are arranged in one layer, one cell in thickness. In a strict sense, only the initials constitute the cambium (Bailey, 1943), but frequently the term is used with reference to the cambial zone, because it is difficult to distinguish the initials from their recent derivatives (pl. 21*B*; Bannan, 1955).

In tangential views the arrangement of cambial cells shows two basic patterns. In one, the fusiform initials occur in horizontal tiers with the ends of the cells of one tier appearing at approximately the same level (pl. 22B). Such meristem is called *storied* or *stratified cambium*. It is characteristic of plants with short fusiform initials. In the second type, the fusiform initials are not arranged in horizontal tiers, and their ends overlap (pl. 22A). This type is termed *nonstoried* or *nonstratified cambium*. It is common in plants with long fusiform initials. Intergrading types of arrangement occur in different plants. The nonstratified type is considered to be phylogenetically more primitive than the stratified. The former is found in fossil pteridophytes, in fossil and living gymnosperms, and in structurally primitive dicotyledons; the latter, in highly specialized dicotyledons (Bailey, 1923). In primitive cambia the initials vary in length more than in the more specialized meristems.

CELL DIVISION

The phloem and the xylem are formed by tangential (periclinal) divisions of cambial initials. The vascular tissues are laid down in two opposite directions, the xylem cells toward the interior of the axis, the phloem cells toward its periphery. The consistent tangential orientation of the planes of division during the formation of vascular tissues determines the arrangement of cambial derivatives in radial rows (pl. 65). Such radial seriation may persist in the developing xylem and phloem (fig. 6.3A), or it may be disturbed through various kinds of growth readjustments during the differentiation of these tissues (xylem in pl. 21A).

Tangential divisions that occur during the formation of xylem and phloem cells are not limited to the initials but are encountered also in varied numbers of derivatives, sometimes several times within the progeny of the same derivative (fig. 6.3B; Bannan, 1955, 1957; Evert, 1963). During the winter rest, xylem and phloem cells mature more or less close to the initials; sometimes only one cambial layer is left between the mature xylem and phloem elements. But some vascular tissue, frequently only phloem, may overwinter in an immature state in the cambial zone.

As the xylem cylinder increases in thickness by secondary growth, the cambial cylinder enlarges in circumference. The principal cause of this enlargement is the increase in the number of cambial cells in a tangential direction, followed by a tangential expansion of these cells. In stratified cambia the increase in the number of fusiform initials occurs by radial (anticlinal) longitudinal divisions. In nonstratified cambia, however, the fusiform initials divide by more or less oblique anticlinal walls (the so-called pseudotransverse walls), and then the resulting cells

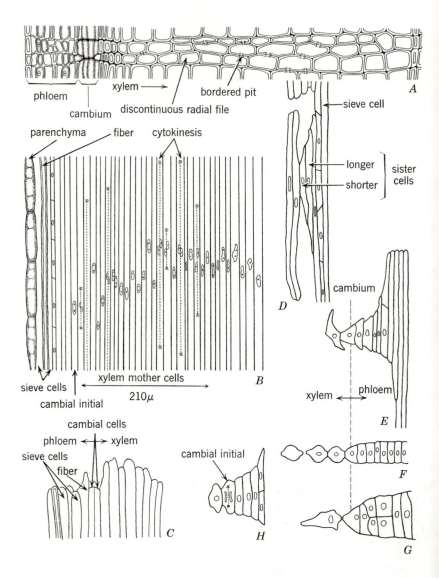

FIG. 6.3. Vascular cambium of *Thuja occidentalis*. A, transverse section showing relation of xylem and phloem to cambium. The discontinuous radial file is represented in the xylem and the phloem but not in the cambium—loss of fusiform initial. *B–H,* radial sections. *B,* wide zone of periclinally dividing xylem mother cells. *C,* differences in length of cells in cambial region. *D,* early stage in shortening of cambial cells by asymmetric periclinal division. *E,* earlier and, *F–H,* later stages in shortening of fusiform initials to dimensions of ray initials. (After Bannan, *Canad. Jour. Bot.* 31, 1953; 33, 1955.)

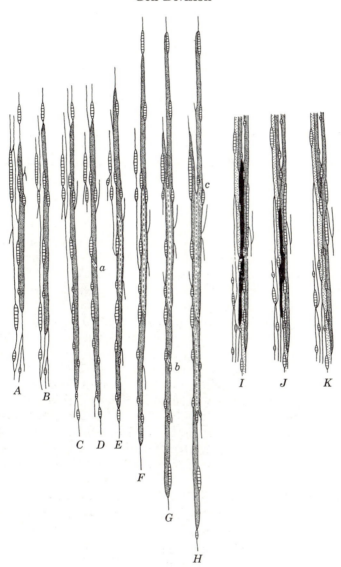

FIG. 6.4. Two series (*A–H, I–K*) of tangential sections through secondary phloem of *Taxus baccata*, illustrating developmental changes in vascular cambium that produced this phloem. In both series, cells to the left (*A, I*) are farthest from cambium. *A–H*, cambial initial, which gave origin to the stippled cells, elongated (*A–C*), then divided (*D* at *a*). The resulting sister cells elongated (*E–F*), then each divided, the lower at *b* in *G*, the upper at *c* in *H*. *I–K*, stages in disappearance of initial that gave rise to cells shown in black. (Adapted from Klinken, *Biblioth. Bot.* 19, 1914.)

elongate at their apices (apical intrusive growth; figs. 6.1D, 6.4D–H) until each cell is as long as, or longer than, the mother cell. Some investigators use the term *multiplicative* divisions with reference to the anticlinal divisions that increase the number of initials, as contrasted with the *additive*, periclinal, divisions that contribute cells to the xylem and the phloem (Bannan, 1956; Duff and Nolan, 1957).

In the longitudinal divisions of the cambial initials and their derivatives, cytokinesis is a process extended in time and space. The cell plate is initiated between the two new nuclei and then spreads through the entire length of the cell, preceded by the phragmoplast fibers (figs. 6.1, 6.2).

DEVELOPMENTAL CHANGES

Detailed studies of the vascular cambium of conifers have shown that the increase in circumference of the meristem is accompanied by profound changes in size, number, and arrangement of cells. Table 6.2 illustrates some of the quantitative changes that occur in the nonstratified cambium of a pine stem as the stem increases in girth. Both the fusiform and the ray initials are greatly multiplied in number. The fusiform initials enlarge notably along their tangential diameters, whereas the ray initials become only slightly larger in this dimension. There is also a remarkable increase in the length of the fusiform initials. The increase in number of the fusiform initials, as seen in transections, results from intrusive apical elongation (fig. 6.4A–C) following the multiplicative oblique radial divisions (fig. 6.4D–H). Since the rays in the pine are mostly uniseriate (one cell in width), the increase in the number of ray initials, as shown for the older stem in table 6.2, is a result, not of divisions of the existing ray initials, but of the addition of new ray initials.

New ray initials arise from fusiform initials or their segments. These additions maintain a relative constancy in the ratio between the rays and the axial components during the increase in the circumference of the vascular cylinder (Braun, 1955). New rays have fewer cells than older. A ray may be one cell wide and one cell high in the beginning; later, the initial divides or more initials are added to the first. The ray thus increases in height and may increase in width if multiseriate rays are characteristic of the plant. Some investigators report that new ray initials may be cut off the apices or cut out of the sides of fusiform initials (Braun, 1955; Evert, 1961). In a herbaceous species of *Hibiscus* rays were found to be derived by transverse divisions of one of the two fusiform cells resulting from an anticlinal division of a fusiform initial

Table 6.2 Difference between a 1-Year-Old and a 60-Year-Old Stem of
Pinus strobus in Circumference of Cambium and Size and Number of Initials
(Adapted from Bailey, 1923)

	Age of stem	
Item	1 year	60 years
Radius of woody cylinder	2,000 microns	200,000 microns
Circumference of cambium	12,566 microns	1,256,640 microns
Average length of fusiform initials	870 microns	4,000 microns
Average tangential diameter of fusiform initials	16 microns	42 microns
Number of fusiform initials in cross section of stem	724	23,100
Average tangential diameter of ray initials	14 microns	17 microns
Number of ray initials in cross section of stem	70	8,796

(Cumbie, 1963). Studies on certain conifers (Bannan, 1951, 1953, 1956) and *Liriodendron* (Cheadle and Esau, 1964) show that ray initiation in these plants is usually a complicated process involving subdivision of fusiform initials, elimination of some products of these divisions from the initial layer (also termed loss of initials), and transformation of others into ray initials.

Rays may increase in width and height by fusion of two or more groups of ray initials (Braun, 1955). Apparently such fusions result from changes in the intervening fusiform initials, loss of some, division and conversion to ray cells of others. The reverse process, division, or splitting, of rays occurs also, probably mainly as a result of intrusive growth of fusiform initials through a group of ray initials. Splitting resulting from elongation of ray initials into fusiform initials is probably less common.

The phenomenon of loss of initials has been studied extensively in the conifers (Bannan, 1951–1962; Forward and Nolan, 1962; Hejnowicz, 1961); less so in the dicotyledons (Cheadle and Esau, 1964; Evert, 1961). The method employed is commonly that of following the changes in radial files of cells in the xylem or the phloem as seen in serial tangential sections and reconstructing from these changes the past events in the cambium. Transections are used for confirmation since they reveal loss of initials by discontinuities in the radial files of cells (fig. 6.3A).

The loss of fusiform initials is usually gradual. Before a cell is eliminated from the initial layer, its precursors fail to enlarge normally—

possibly even diminishing in size through loss of turgor—and become abnormal in shape. Periclinal divisions separate such cells into smaller and larger derivatives, the smaller of which remains in the initial layer (fig. 6.3D, H). Thus, gradually the cell in the initial position is reduced in size, particularly in length (fig. 6.3E–G). Some of the short initials are lost from the initial layer by maturing into xylem or phloem elements; others become ray initials with or without further divisions. Ray initials, too, may disappear from the cambium. The space released by a declining initial is filled by the intrusive growth of the surviving initials (fig. 6.4I–K).

The elimination of fusiform initials is associated with the anticlinal divisions giving rise to new initials. These divisions apparently would result in an overproduction of initials if they were not accompanied by the extensive loss of cells. The loss appears to be related to vigor of growth. In *Thuja occidentalis* the survival rate was found to be 20 per cent when the annual xylem increment was 3 millimeters wide, whereas at the lowest growth rates the rate of loss and that of new production were almost equal (Bannan, 1960a). The accommodation to the increase in girth probably occurred through elongation of cells. In *Pyrus communis* the loss was calculated to be 50 per cent among the newly formed fusiform initials (Evert, 1961). Among some 300 radial files of cells of the axial system of *Liriodendron* phloem, examined in serial sections through a layer of tissue about 400 microns in radial depth, the loss of their initials by maturation and by conversion into rays nearly equaled the addition of new tiers by anticlinal divisions of fusiform initials (Cheadle and Esau, 1964). Considerable evidence indicates that in both conifers and dicotyledons longer initials tend to survive and that extensive contact of these initials with the rays increases the chance of survival (Bannan, 1956, 1963; Bannan and Bayly, 1956; Cheadle and Esau, 1964; Evert, 1961).

As was mentioned previously, anticlinal divisions are followed by intrusive elongation of the resulting cells. The direction of this elongation may be polar. In *Thuja*, for example, it was found to be considerably greater in the downward than in the upward direction (Bannan, 1956). Although the intrusive growth occurs at the tips of cells, apparently the newly formed wall continues to expand so that a slip between this wall and those with which it comes in contact is not ruled out. In such a method of growth a distinction between intrusive growth and gliding growth can hardly be made (Bannan, 1956). The tips of growing cells have thin walls and contain cytoplasmic accumulations (pl. 22A).

The walls formed during anticlinal divisions in relatively long fusiform initials show various degrees of inclination but, as seen in a tangential section of the cambium, tend to be oriented in the same direction (Bannan, 1956). In other words, the overlapping tips of the elongating fusiform initials are similarly oriented with reference to one another throughout the section. Hejnowicz (1961) suggests that this unidirectional orientation of growing cells combined with the frequent loss of initials may be causally related to the spiral arrangement of the cambial cells and of the derived vascular cells.

All the studies reviewed above deal with the vascular cambium of stems. In *Larix europea* the cambium of the root was found to have more limited elimination of initials, weaker intrusive growth, and more variable orientation of the anticlinal walls than that of a stem of similar age (Hejnowicz, 1961).

Loss of fusiform cambial initials might be less typical of herbaceous than of woody species. In *Hibiscus lasiocarpus*, a perennial herb, the loss of such initials was limited to that associated with ray formation (Cumbie, 1963).

SEASONAL ACTIVITY

The secondary growth originating in the vascular cambium is closely connected with the activities of the primary parts of the plant body and shows fluctuations in relation to the changes in the physiologic state of the plant. Annual or biennial herbaceous plants commonly have a regular sequence of vegetative stage, reproductive stage, somatic death, and seed dispersal. Between the vegetative and reproductive stages, the plant body may attain various dimensions and its vascular tissues may be increased in amount by secondary growth. This growth ceases, however, during the transition to the reproductive stage, since the cambial activity appears to be closely associated with the vegetative stage (Wilton and Roberts, 1936). In perennial plants there is a repetition of vegetative and reproductive phases, without somatic death of the whole individual. As is well known, in woody species growing in temperate regions periods of growth and reproduction alternate with periods of relative inactivity during the winter. The seasonal periodicity finds its expression in the cambial activity also. Production of new cells by the vascular cambium slows down or ceases entirely during the rest period, and the vascular tissues mature more or less closely to the initial layer.

In the spring the winter rest period is succeeded by a reactivation of the cambium. From the anatomic aspect, reactivation may be divided into two stages: (1) expansion of the cambial cells in the radial direction ("swelling" of the cambium) and (2) initiation of cell division. The radial enlargement is accompanied by a weakening of the radial walls so that a slight external force applied to the stem will cause these walls to break. The separation of the bark from the wood resulting from such a break is commonly called slipping of the bark . Slippage may also be induced later, during cell division and tissue differentiation in the cambial zone. At this time, however, the break occurs most commonly through the young xylem where the tracheary elements have attained their maximum diameters but are still without secondary walls (Bailey, 1943; Evert, 1960, 1961). In evergreen species, as in *Citrus*, the histologic aspects of slippage appear to be less definite (Schneider, 1952).

The cell divisions occurring during the second stage of reactivation are the additive periclinal divisions. Information on the exact sequence of these divisions is meager, especially with regard to the timing of xylem- and phloem-cell formation (Evert, 1960). In *Thuja occidentalis* (Bannan, 1955) periclinal divisions were found to be concentrated first in the xylem mother cells (fig. 6.3*B*); then they occurred in the initial layer. Formation of phloem cells began when the divisions in the xylem mother-cell zone were at maximum and continued till cambial activity was terminated. In *Pyrus* (Evert, 1960, 1963; fig. 6.5) additive cambial divisions began when the overwintering phloem mother cells were differentiating. Most of the first new cells were added to the phloem. Xylem cells were formed later, when a considerable amount of differentiated phloem was already present. In both conifers and dicotyledons, the annual xylem increment is characteristically wider than the corresponding phloem increment.

The resumption of cambial activity in the spring has often been found to be related to the new primary growth from buds (Fraser, 1962; Ladefoged, 1952). In many dicotyledons cambial activity of the stem begins beneath the emerging new shoots and spreads from here basipetally toward the main branches, the trunk, and the root. As an example, the data obtained with *Acer pseudo-platanus*, growing in England, may be cited (Cockerham, 1930). In this tree 9 to 10 weeks elapsed between the inception of xylem differentiation in the twigs (late in April) and that in the roots (early in July). Activity ceased in the same order. The formation of xylem stopped in the twigs in late July, in the roots, in late September. Thus, 8 to 9 weeks elapsed between the cessation of cambial activity in the branches and that in the roots. *Acer* exemplifies cambial behavior in dicotyledons with diffuse-porous wood (with vessels of similar width distributed throughout annual increment; pl. 32). The degree of development of the bud associated with cambial reactivation is variable; the bud may be still closed, or just opening, or obviously growing (Ladefoged, 1952). Many conifers

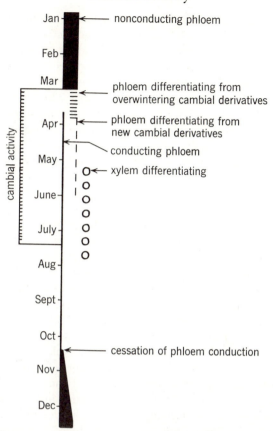

FIG. 6.5. Secondary growth during one year in a branch of pear (*Pyrus communis*). Times of differentiation are given for phloem derived from overwintering cambial cells and from new cambial cells. New xylem is derived only from new cambial cells. Data on conducting xylem are not included. (Adapted from Evert, *Calif. Univ., Publs., Bot.* 32, 1960.)

and the dicotyledons with ring-porous type of wood (characterized by aggregation of numerous wide vessels in the early wood; pl. 33) show an early rapid spread of cambial reactivation throughout the trunk in the presence of little or no bud growth (Ladefoged, 1952; Messeri, 1948; Wareing, 1951). Cessation of cambial activity follows approximately the same order as the reactivation (Fraser, 1962). The inception of cambial reactivation beneath the new shoots and its basipetal progress explains why, in a dicotyledon, the portion of a twig that might be left in pruning above the uppermost bud dries up and forms a "snag" (Wray, 1934).

The initial stimulation of cambial activity has been repeatedly related to the transport of growth substances in basipetal direction from the growing buds (Samish, 1954). The maintenance of cambial activity, however, appears to be independent of the extension growth of the new shoot (Münch, 1937). In *Robinia pseudoacacia* continuation of cambial activity was found to be dependent on exposure of the leaves to long-day conditions (Wareing and Roberts, 1956). The vascular cambium can be stimulated into activity by wounding also, possibly in relation to formation of wound hormones as a result of injury (Brown, 1937).

Intensity and amount of cambial activity varies in different seasons. Some of these variations are induced by environmental conditions, whereas others depend on an inherent rhythm of growth. At the beginning of secondary growth in pine trees, for example, the mean width of annual increments first rises then falls from one season to the next in the same internode. Other variations in ring width that may be present are overshadowed by the basic pattern (Duff and Nolan, 1953). A common feature is the decrease in the rate of growth in thickness with increasing age of the tree (Bannan, 1960b).

In some studies the multiplicative anticlinal divisions in the initial layer were found occurring toward the end of the growth season when the cambial zone was of minimum width (Bannan, 1957, 1962; Evert, 1961). Through the years, these divisions may occur more or less frequently in the same initial position. In *Thuja* the intervals between successive divisions were 1 to 8 years with an average of 3.7 years, and the frequency was reduced with increased age of the tree (Bannan, 1956, 1960b). The elongation of the new surviving initials resulting from anticlinal divisions begins directly after the divisions and continues for several years. In *Thuja* this elongation follows a familiar growth pattern in that it occurs rapidly at first, then at a decreasing rate.

The restriction of anticlinal divisions to the last part of the season of growth is not a constant feature. In *Picea* (Bannan, 1963), these divisions occurred throughout the growth period during the earlier years of growth of the stem but became limited to the terminal part of the season in later years, when narrower annual rings were being produced.

REFERENCES

Bailey, I. W. The cambium and its derivative tissues. II. Size variations of cambial initials in gymnosperms and angiosperms. *Amer. Jour. Bot.* 7:355–367. 1920a. III. A reconnaissance of cytological phenomena in the cambium. *Amer. Jour. Bot.* 7:417–434. 1920b. IV. The increase in girth of the cambium. *Amer. Jour. Bot.* 10:499–509. 1923. V. A reconnaissance of the vacuome in living cells. *Ztschr. f. Zellforsch. u. Micros. Anat.* 10:651–682. 1930.

Bailey, I. W. Some misleading terminologies in the literature of "plant tissue culture." *Science* 98:539. 1943.

Bannan, M. W. The annual cycle of size changes in the fusiform cambial cells of *Chamaecyparis* and *Thuja*. *Canad. Jour. Bot.* 29:421–437. 1951.

Bannan, M. W. Further observations on the reduction of fusiform cambial cells in *Thuja occidentalis* L. *Canad. Jour. Bot.* 31:63–74. 1953.

Bannan, M. W. The vascular cambium and radial growth in *Thuja occidentalis* L. *Canad. Jour. Bot.* 33:113–138. 1955.

Bannan, M. W. Some aspects of the elongation of fusiform cambial cells in *Thuja occidentalis* L. *Canad. Jour. Bot.* 34:175–196. 1956.

Bannan, M. W. The relative frequency of the different types of anticlinal divisions in conifer cambium. *Canad. Jour. Bot.* 35:875–884. 1957.

Bannan, M. W. Cambial behavior with reference to cell length and ring width in *Thuja occidentalis* L. *Canad. Jour. Bot.* 38:177–183. 1960a.

Bannan, M. W. Ontogenetic trends in conifer cambium with respect to frequency of anticlinal division and cell length. *Canad. Jour. Bot.* 38:795–802. 1960b.

Bannan, M. W. Cambial behavior with reference to cell length and ring width in *Pinus strobus* L. *Canad. Jour. Bot.* 40:1057–1062. 1962.

Bannan, M. W. Cambial behavior with reference to cell length and ring width in *Picea*. *Canad. Jour. Bot.* 41:811–822. 1963.

Bannan, M. W., and I. L. Bayly. Cell size and survival in conifer cambium. *Canad. Jour. Bot.* 34:769–776. 1956.

Bosshard, H. H. Variabilität der Elemente des Eschenholzes in Funktion der Kambiumtätigkeit. *Schweiz. Ztschr. f. Forstwes.* 12:648–665. 1951.

Braun, H. J. Beiträge zur Entwicklungsgeschichte der Markstrahlen. *Bot. Studien* Heft 4:73–131. 1955.

Brown, A. B. Activity of the vascular cambium in relation to wounding in the balsam poplar, *Populus balsamifera* L. *Canad. Jour. Res. Sect. C., Bot. Sci.* 15:7–31. 1937.

Brown, C. L., and K. Sax. The influence of pressure on the differentiation of secondary tissues. *Amer. Jour. Bot.* 49:683–691. 1962.

Cheadle, V. I., and K. Esau. Secondary phloem of *Liriodendron tulipifera*. *Calif. Univ., Publs., Bot.* 36:143–252. 1964.

Cockerham, G. Some observations on cambial activity and seasonal starch content in sycamore (*Acer pseudo-platanus*). *Leeds Phil. Lit. Soc. Proc.* 2:64–80. 1930.

Cumbie, B. G. The vascular cambium and xylem development in *Hibiscus lasiocarpus*. *Amer. Jour. Bot.* 50:944–951. 1963.

Dodd, J. D. On the shapes of cells in the cambial zone of *Pinus silvestris* L. *Amer. Jour. Bot.* 35:666–682. 1948.

Duff, G. H., and N. J. Nolan. Growth and morphogenesis in the Canadian forest species. I. The controls of cambial and apical activity in *Pinus resinosa* Ait. *Canad. Jour. Bot.* 31:471–513. 1953. II. Specific increments and their relation to the quantity and activity of growth in *Pinus resinosa* Ait. *Canad. Jour. Bot.* 35:527–572. 1957.

Evert, R. F. Phloem structure in *Pyrus communis* L. and its seasonal changes. *Calif. Univ., Publs., Bot.* 32:127–194. 1960.

Evert, R. F. Some aspects of cambial development in *Pyrus communis*. *Amer. Jour. Bot.* 48:479–488. 1961.

Evert, R. F. The cambium and seasonal development of the phloem of *Pyrus malus*. *Amer. Jour. Bot.* 50:149–159. 1963.

Forward, D. F., and N. J. Nolan. Growth and morphogenesis in Canadian forest species. VI. The significance of specific increment of cambial area in *Pinus resinosa* Ait. *Canad. Jour. Bot.* 40:95–111. 1962.

Fraser, D. A.   Apical and radial growth of white spruce [*Picea glauca* (Moench) Voss] at Chalk river, Ontario, Canada. *Canad. Jour. Bot.* 40:659–668. 1962.

Hejnowicz, Z.   Anticlinal divisions, intrusive growth, and loss of fusiform initials in non-storied cambium. *Soc. Bot. Polon. Acta* 30:729–748. 1961.

Ladefoged, K.   The periodicity of wood formation. *Kgl. Dansk. Vidensk. Selsk. Biol. Scr.* 7:1–98. 1952.

Messeri, A.   L'evoluzione della cerchia legnosa in *Pinus halepensis* Mill. in Bari. *Nuovo Giorn. Bot. Ital.* 55:111–132. 1948.

Münch, E.   Regelung des Dickenwachstums und der Stammform durch das Längenwachstum bei Nadelbäumen. *Deut. Bot. Gesell. Ber.* 55:109–113. 1937.

Samish, R. M.   Dormancy in woody plants. *Ann. Rev. Plant Physiol.* 5:183–204. 1954.

Schneider, H.   The phloem of the sweet orange tree trunk and seasonal production of xylem and phloem. *Hilgardia* 21:331–366. 1952.

Wareing, P. F.   Growth studies in woody species. IV. The initiation of cambial activity in ring-porous species. *Physiol. Plantarum* 4:546–562. 1951.

Wareing, P. F., and D. L. Roberts.   Photoperiodic control of cambial activity in *Robinia pseudoacacia* L. *New Phytol.* 55:356–366. 1956.

Wilton, O. C., and R. H. Roberts.   Anatomical structure of stems in relation to the production of flowers. *Bot. Gaz.* 98:45–64. 1936.

Wray, E. M.   The structural changes in a woody twig after summer pruning. *Leeds Phil. Lit. Soc. Proc.* 2:560–570. 1934.

7

The Epidermis

The term *epidermis* designates the outermost layer of cells on the primary plant body. The word is derived from two words of Greek origin, *epi*, upon, and *derma*, skin. Through the history of development of plant morphology the concept of the epidermis has undergone changes, and there is still no complete uniformity in the application of the term. This surface system of cells varies in composition, function, and origin and, therefore, does not lend itself to a precise definition based on any one criterion. In this book the term epidermis is used in a broad morphologic-topographic sense. It refers to the outermost layer of cells of all parts of the primary plant body—stems, roots, leaves, flowers, fruits, and seeds. It is considered to be absent on the rootcap and not differentiated as such on the apical meristems.

The inclusion of the surface layer of the root in the concept of epidermis is contrary to the view that the root epidermis belongs to a separate category of tissue and should have its own name, *rhizodermis* or *epiblem* (Linsbauer, 1930). The epidermis of the root differs from that of the shoot in origin, function, and structure, and, therefore, the emphasis of some workers upon the distinctness of the two parts of the epidermis is justified. At the same time, the proper definition of the root epidermis is inseparably connected with the problem of the morphologic relation between root and shoot (Allen, 1947). As long as there is no commonly accepted concept on this relation, it seems most convenient to use the term epidermis in its broadest sense to mean the primary surface tissue of the entire plant.

The normal functions of the epidermis of the aerial plant parts are

147

considered to be restriction of transpiration, mechanical protection, gaseous exchange through stomata, and storage of water and metabolic products. Some accessory functions, however, may predominate to such an extent that the epidermis assumes characteristics not typical of this tissue. In this category of functions are included photosynthesis, secretion, absorption (other than that of the root epidermis), and possibly also the perception of stimuli and causal association with the movement of plant parts. Some of the functions of the epidermis appear to be related to certain specialized anatomic characteristics (Linsbauer, 1930).

The meristematic potentialities of the epidermis merit a brief mention. In general, this tissue is relatively passive with regard to meristematic activities (Linsbauer, 1930). Nevertheless, the epidermis is known to resume such activity during the normal course of development (formation of phellogen, chapter 14) and after injuries to the plant (Gulline, 1960; Linsbauer, 1930; McVeigh, 1938).

ORIGIN AND DURATION

The origin of the epidermis is discussed in chapter 5. Briefly, the epidermis of the shoot arises from the outermost cell layer of the apical meristem, either from independent initials or jointly with the subjacent tissue layers. If the shoot apex shows a segregation into zones of surface and volume growth, that is, into a tunica and a corpus, the epidermis originates from the outermost layer of tunica. Such a layer of cells fits the definition of Hanstein's *dermatogen* (chapter 5) since its course of development into the epidermis begins in an independent initial region. In plants showing less precise zonations in the apical meristem, as most of the gymnosperms do, the epidermis does not have separate initials. It is a product of the lateral derivatives of the apical initials, which divide both anticlinally and periclinally and are the ultimate sources of the superficial as well as the interior cells of the plant body. In plants with single apical cells the epidermis also has common origin with the deeper lying tissues. In roots the epidermis may be related developmentally to the rootcap or to the cortex.

When the epidermis does not arise from separate initials, it becomes distinct at various distances from the apical meristem, depending on the architecture of the meristem. Haberlandt's term *protoderm* (chapter 5) designates such primordial epidermis, as well as the epidermis arising from separate initials. This term was coined as a morphologic-topographic designation, with no reference to the origin of the tissue. In this book protoderm is used to indicate the undifferentiated epidermis, regardless of its origin.

Organs having little or no secondary growth usually retain the epidermis as long as they exist. An exception is exemplified by some woody monocotyledons that have no secondary addition to the vascular system but develop a special kind of periderm replacing the epidermis. In stems and roots of gymnosperms and dicotyledons and of arborescent monocotyledons having secondary growth, the epidermis varies in longevity, depending on the time of formation of the periderm. Ordinarily, the periderm arises in the first year of growth of woody stems and roots, but numerous tree species produce no periderm until their axes are many times thicker than they were at the completion of primary growth. In such plants the epidermis, as well as the underlying cortex, continues to grow and thus keeps pace with the increasing circumference of the vascular cylinder. The individual cells enlarge tangentially and divide radially. An example of such prolonged growth is found in stems of a maple (*Acer striatum*) in which trunks about 20 years old may attain a thickness of about 20 cm and still remain clothed with the original epidermis (De Bary, 1884). The cells of such an old epidermis are not more than twice as wide tangentially as the epidermal cells in an axis 5 mm in thickness. This size relation clearly shows that the epidermal cells are dividing continuously while the stem increases in thickness. Another example is *Cercidium torreyanum*, a tree leafless most of the time but having a green bark and a persistent epidermis (Roth, 1963).

STRUCTURE

Composition

In relation to the multiplicity of its functions the epidermis contains a wide variety of cell types. The ground mass of tissue consists of the epidermal cells proper, which may be regarded as the least specialized members of the system. Dispersed among these cells are the guard cells of the stomata and sometimes other specialized cells. The epidermis may produce a variety of appendages, the trichomes, such as hairs and more complex structures. Trichomes with a specific function, the root hairs, develop from the epidermal cells of the roots.

Epidermal Cells

Morphology and Arrangement. Mature epidermal cells are commonly described as being tabular in shape because of their relatively small extent in depth, that is, in the direction at right angles to the surface of the organ (pl. 23C). Deviating types, cells which are much deeper than they are wide, also occur, for example, in the palisade-like epidermis of many seeds (chapter 10). In surface view the epidermal cells may be nearly isodiametric (fig. 7.1B) or elongated (fig. 7.1A). The three-dimensional shape of the epidermal cells of *Aloe aristata* and *Anacharis densa* (Matzke, 1947, 1948) approaches that of a tetrakaidecahedron cut in half. The form of epidermal cells is sometimes related to differences in position on the plant organ. Elongated epidermal cells are often found on structures which themselves are elongated, such as stems, petioles, vein ribs of leaves, and leaves of most monocotyledons. Elongated epidermal cells also occur near some hairs and stomata. Frequently epidermal cells are shallow above the strands of subepidermal sclerenchyma. In leaves the epidermal layers on the two surfaces may be dissimilar in shape and size of cells and in thickness of walls and cuticle.

In many leaves and petals the epidermal cells have wavy anticlinal walls (figs. 7.1C–E, pl. 23B), and the undulations may be present in the entire depth of the walls or only in their outermost parts. The cause of this waviness has been the subject of much study and speculation in the literature (Linsbauer, 1930). One of the explanations of the phenomenon relates the undulations to the development of stresses during the differentiation of the leaf (Avery, 1933). Another concept is that the waviness is caused by the method of hardening of the differentiating cuticle (Watson, 1942). The waviness of the walls is variable, depending on the location in the leaf or petal. Often the undulations occur only on the lower side of a leaf or are more pronounced here than on the upper side. The waviness is also affected by environmental conditions prevailing during leaf development (Linsbauer, 1930; Watson, 1942). The outer wall of an epidermal cell may be flat or convex, or it may bear one or more localized raised areas.

Some epidermal cells greatly deviate from the main mass of cells. Certain Gramineae, Gymnospermae, Dicotyledoneae, and lower vascular plants (*Adiantum, Selaginella*) contain fiber-like epidermal cells (Linsbauer, 1930). The longest epidermal fibers—up to 2 mm—were described in Stylidaceae. In Gramineae such fibers may be over 300 microns in length. Certain Cruciferae contain sac-like secretory cells (myrosin cells, chapter 13) scattered in the epidermis. In Acanthaceae, Cucurbitaceae, Moraceae (fig. 7.13C), and Urticaceae epidermal cells may develop cystoliths. Some of these cystolith-containing cells (the lithocysts) are specialized epidermal cells; others appear to be reduced trichomes (Linsbauer, 1930).

Sometimes the entire epidermis consists of highly specialized cells. Thus in certain seeds and scales the epidermis is composed of a solid layer of sclereids (chapter 10). The epidermis of the Polypodiaceae is differentiated as a photosynthetic tissue (Meyer, 1962; Wylie, 1948). The epidermal cells project into extensive intercellular spaces and contain chloroplasts.

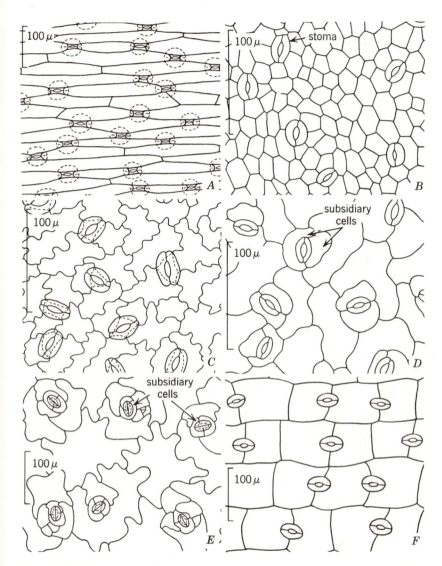

FIG. 7.1. Surface views of abaxial foliar epidermis. *A, Iris,* sunken stomata in longitudinal rows. *B, Vitis,* dispersed stomata. *C, Capsicum,* raised stomata. *B, C,* without subsidiary cells, anomocytic. *D–F,* with subsidiary cells. *D, Vigna,* paracytic. *E, Sedum,* a variant of anisocytic. *F, Dianthus,* diacytic. Wavy anticlinal walls in *C, E.* (*B, C,* courtesy of E. F. Artschwager.)

The Epidermis

The epidermal cells are arranged compactly, with rare breaks in their continuity other than those represented by the stomatal pores. Intercellular spaces occur in the epidermis of petals, but they appear to be closed on the outside by the cuticle.

Epidermis in Gramineae. The morphologic variability of the epidermis of Gramineae is used extensively for taxonomic purposes and for discussions on the evolution of this group of plants (Davies, 1959; Metcalfe, 1960; Tateoka, 1957). The gramineous epidermis typically contains long cells and two kinds of short cells, silica cells and cork cells (fig. 7.2A). The short cells frequently occur together in pairs. The silica cells are almost filled with SiO_2 which solidifies into bodies

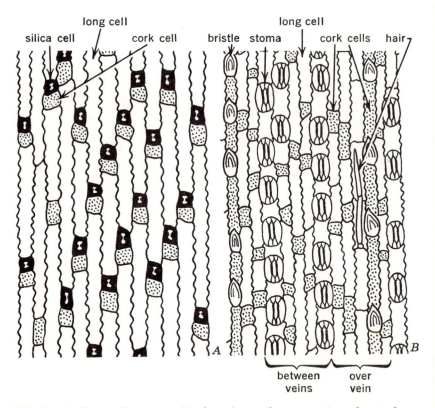

FIG. 7.2. Epidermis of sugarcane (*Saccharum*) in surface view. *A*, epidermis of stem showing alternation of long cells with pairs of short cells: cork cells and silica cells. *B*, lower epidermis from a leaf blade, showing distribution of stomata in relation to various kinds of epidermal cells. (*A*, ×500; *B*, ×320. Adapted from Artschwager, *Jour. Agr. Res.* 60, 1940.)

of various shapes. The cork cells have suberized walls and often contain solid organic material. They are also silicified. The silica in epidermal cells of oat has been identified as opal (Baker, 1960). In some parts of the plant the short cells develop protrusions above the surface of the leaf in the form of papillae, bristles, spines, or hairs. The epidermal cells of Gramineae are arranged in parallel rows, and the composition of these rows varies in different parts of the plant (Prat, 1948, 1951). The inner face of the leaf sheath at its base, for example, has a homogeneous epidermis composed of long cells only. Elsewhere in the leaves combinations of the different types of cells may be found. Rows containing long cells and stomata occur over the assimilatory tissue; only elongated cells or such cells combined with cork cells or bristles or with mixed pairs of short cells follow the veins (fig. 7.2B). In the stem, too, the composition of the epidermis varies, depending on the level on the internode and on the position of the internode in the plant.

The Gramineae and other monocotyledons possess still another peculiar type of epidermal cell, the bulliform cell. The bulliform cells, literally "cells shaped like bubbles," are large, thin-walled, highly vacuolated cells which occur in all monocotyledonous orders except the Helobiae (Linsbauer, 1930; Metcalfe, 1960). Bulliform cells either cover the entire upper surface of the blade or are restricted to grooves between the veins. In the latter situation they form bands, usually several cells wide, arranged parallel with the veins. In transections through such a band the cells often form a fan-like pattern, for the median cells are usually the largest and are somewhat wedge-shaped (pl. 70A). Bulliform cells may occur on both sides of the leaf. They are not necessarily restricted to the epidermis but are sometimes accompanied by similar cells in the subjacent mesophyll.

Bulliform cells are poor in solid contents. They are mainly water-containing cells, with little or no chlorophyll. Tannins and crystals are rarely found in these cells. Their radial walls are thin, but the outer wall may be as thick or thicker than those of the adjacent ordinary epidermal cells. The walls are of cellulose and pectic substances. The outer walls are cutinized and also bear a cuticle (Burström, 1942). Bulliform cells may accumulate silica (Parry and Smithson, 1958).

According to one view, the bulliform cells are concerned with the unrolling of the developing leaves. Their sudden and rapid expansion during a certain stage of leaf development is assumed to bring about the unfolding of the blade; hence, the term expansion cells, often applied to these cells. Another concept is that, by changes in turgor, these cells play a role in the hygroscopic opening and closing movements of mature leaves; hence, the alternative term motor cells. Still other workers doubt that the cells have any other function than that of water storage (Linsbauer, 1930). Studies on the unfolding and the hygroscopic movements of leaves of certain grasses have shown that the bulliform cells are not actively or specifically concerned with these phenomena (Burström, 1942; Shields, 1951).

Contents. In general the contents of the epidermal cells have been incompletely investigated, but since these cells possess living protoplasts they may be expected to include a variety of substances, depending on the degree of their specialization. The epidermal plastids are usually not definitely differentiated as chloroplasts, but in many plants they appear to contain chlorophyll as determined by tests for fluorescence and reduction of silver nitrate (Mikulska, 1959*a*, *b*). Starch also may occur in the epidermal plastids. Some ferns, water plants, and a number of higher vascular land plants, particularly those of shady habitats, contain well-developed chloroplasts in the epidermis (Linsbauer, 1930; Meyer, 1962). The cell sap of the epidermal cells may contain anthocyanin, as in many flowers, leaves of the purple beech and the red cabbage, and stems and petioles of *Ricinus*. Under the electron microscope the epidermal cells of *Allium* bulbs show structures similar to those found in parenchyma cells (Drawert and Mix, 1963).

Wall Structure. The epidermal walls of different plants and of different plant parts vary markedly in thickness. In the thinner-walled epidermis the outer wall is frequently the thickest (pl. 23A). Epidermis with exceedingly thick walls is found in coniferous leaves (fig. 7.4; pl. 79; Linsbauer, 1930; Marco, 1939). The wall thickening is uneven and so massive in some species that it almost obliterates the lumina of the cells. These walls are probably secondary. Thick secondary walls occur in the epidermal cells differentiated as sclereids in seed coats and scales (chapter 10).

The radial and the inner tangential walls frequently show primary pit-fields. The outer wall also may have thin places and markings resembling primary pit-fields (Linsbauer, 1930). Plasmodesmata have been described not only in the radial and the inner tangential walls but also in the outer walls, where they are called ectodesmata (Sievers, 1959). Although the ectodesmata do not pass through the cuticle they are thought to be the pathways for substances that are discharged through the cuticle (Franke, 1961).

The epidermal cells of leaves and petals in some plants show internal ridges that resemble folds (Marco, 1939). The ridges apparently consist of two wall layers cemented together by intercellular material.

The two layers may split apart with the formation of a schizogenous intercellular space. In such instances a ridge appears as a loop in transection.

Cuticle. The restriction of transpiration by the epidermis largely results from the presence of the fatty substance *cutin* as an impregnation of the cell walls (cutinization, incrustation with cutin) and as a separate layer, the *cuticle* (cuticularization, adcrustation of cutin, Frey-Wyssling and Mühlethaler, 1959), on the outer surface of the cells (pl. 23A). The cuticle covers all parts of the shoot. It occurs on the floral parts, on nectaries, and on ordinary and glandular trichomes. Some authors report the presence of a cuticle on the apical meristem (Priestley, 1943) and in the absorbing region of the root including the root hairs (Scott et al., 1958). A contrasting report states that no cuticle was found on the youngest leaf primordia of certain angiosperms (Bolliger, 1959). The cuticle can be removed from plant parts as an unbroken layer (pl. 24A), an indication that it is continuous.

Cutin has also been identified on the free surfaces of the leaf-mesophyll cells and on the inner walls of the epidermis where these are exposed to the internal air spaces (chapter 3). The inner layer of cutin is continuous with the external cuticle through the stomatal apertures, whose bounding cells, the guard cells, are covered with a cuticle on their free surfaces.

The cuticle attains variable thickness in different plants. Environmental conditions and other unknown factors affect its development. The surface of the cuticle may be smooth, or it may have various protrusions, ridges, or cracks. The origin of the complicated relief pattern in the cuticle of floral parts (pl. 24A) has been ascribed to the effects of cell growth (Priestley, 1943). The cuticle of the tomato fruit contains a yellow pigment, probably a flavonoid, whose development depends on the same light conditions that regulate flowering and seed germination in certain plants (Piringer and Heinze, 1954).

The cutinized part of the outer epidermal wall beneath the cuticle has a complicated structure (chapter 3). In plants with thick outer walls it consists of many lamellae of cutinized cellulose alternating with layers rich in pectin; and a layer of pectin is often recognized between the cuticle and the cuticular layer (Sitte, 1957).

Wax, oil, resin, salts in crystalline form (*Cressa cretica, Tamarix, Frankenia*), and caoutchouc (*Eucalyptus*) occur as surface deposits on the aerial plant parts (Linsbauer, 1930). The structure of the wax deposits has been studied by means of the electron microscope (pl. 24B; Juniper, 1959, 1960; Schiefferstein and Loomis, 1956, 1959). The wax deposits are crystalline (Kreger, 1958) and may have the form of granules, rods, often with hooked ends, network of tubes, isolated tufts, and more or less homogeneous layers. An exceptionally thick layer of wax (up to 5mm) occurs in *Klopstockia cerifera*, the wax palm of the Andes (Kreger, 1958). The carnauba wax is derived from the leaves of the Brasilian wax palm, *Copernicia cerifera*. Wax affects the wettability of the foliage because it prevents the contact of liquid with the leaf surface. Its structure and development are, therefore, of considerable interest in connection with research on agricultural sprays. Wax evidently passes through the cuticle, but the latter shows no pores that could be interpreted as pathways for its discharge. Wax is commonly present also within the cuticle and the cutinized layers beneath. Older leaves have mainly subcuticular wax accumulations which appear to be of greater ecological significance than the surface deposits.

The stratified appearance of the cutinized outer epidermal wall and the increase in the proportion of cutin toward the periphery suggest that the fatty substances do migrate outwards. Some workers think that this movement occurs through ectodesmata (Scott et al., 1958). Others have found lipoidal droplets throughout the young wall, which later apparently diffuse to the surface (Bolliger, 1959).

The development of the initial cuticle is interpreted as a flooding with a cutin precursor, or procutin (probably unsaturated fatty acids, Frey-Wyssling and Mühlethaler, 1959), analogous to a drying oil, and a subsequent hardening of this material through polymerization under the influence of the air oxygen. According to a study of the cuticle of the apple fruit (Huelin, 1959), however, cutin formation is conceived as a process controlled by enzyme action rather than by spontaneous oxidation. The hardening of the cuticle presumably terminates further extrusion of wax and cutin precursor and, therefore, these substances accumulate beneath the cuticle (Schiefferstein and Loomis, 1956). This subcuticular deposition would occur in the cellulose-containing part of the wall or it could also increase the thickness of the cuticle. Some evidence suggests that the margins of the upper epidermal wall of leaves continue to grow and retain an immature cuticle for some time. The greater susceptibility of young leaves to herbicides is ascribed to the presence of these immature, permeable zones in the cuticle (Schiefferstein and Loomis, 1956, 1959).

Cutin is semihydrophilous because some of its polar groups remain free during polymerization. This feature explains the moderate swelling of the cuticle in water and the occurrence of cuticular transpiration (Frey-Wyssling and Mühlethaler, 1959). It is possible to demonstrate excretion of water through the cuticle and its aggregation into droplets;

this occurs without any evidence of submicroscopic pores in the cuticle (Bancher et al., 1960).

Cutin is highly inert and resistant to oxidizing maceration methods. It does not decay since apparently no microorganisms possess cutin-degrading enzymes (Frey-Wyssling and Mühlethaler, 1959). Because of its chemical stability the cuticle is preserved as such in fossil material and is very useful in identification of fossil species (Dilcher, 1963; Harris, 1956).

The cuticle occurs not only over the surface of the epidermal cells but also often as rib-like projections in the anticlinal walls (chapter 9). Such ribs develop relatively late in the life of an organ. One of the explanations of these ribs is that, when new cells are produced by anticlinal divisions during the development of the epidermis, each of these cells extends tangentially and produces its own complete wall, while the parent wall becomes stretched and torn (Priestley, 1943). Thus, the outer cutinized layers accumulate as interrupted lamellae of cellulose together with pectic substances and cutin. The interruptions occur over the anticlinal walls and are filled with cutin deposits. The stretching and tearing of the outer cellulose lamellae and their permeation with cutin eventually make it difficult to distinguish the cuticle and the cutinized layers from one another (pl. 23C, D) without special treatment.

Most plants produce only epidermal cuticular layers, even if the periderm is formed late in the life of the organ and the epidermis continues to grow. In some exceptional plants like the Viscoideae and *Menispermum* (pl. 23D) cuticular layers are also formed among cortical cells in successively deeper regions of the cortex (Damm, 1902).

Other Wall Substances. Among other common wall substances, lignin is a relatively infrequent component of the epidermal walls in angiosperms. If present, it is sometimes generally distributed, sometimes restricted to a part of the outer wall. Lignification of epidermal cells is comparatively common in the lower vascular plants. It occurs also in the Cycadaceae, Cyperaceae and Juncaceae, and in a few dicotyledons (*Eucalyptus, Quercus, Laurus nobilis, Nerium oleander*), and specifically in the needles of conifers, the rhizomes of Gramineae, and the leaves of Gramineae outside the sclerenchyma strands. Many plants deposit silica in epidermal cells (*Equisetum*, ferns, Gramineae, numerous Cyperaceae, palms, and certain dicotyledons; Linsbauer, 1930).

In some dicotyledonous families (Malvaceae, Rutaceae, Loganiaceae, Gentianaceae, Euphorbiaceae) mucilaginous modifications of walls occur in individual epidermal cells, or in groups of cells; sometimes most epidermal cells are more or less mucilaginous, as, for example, in seeds (Linsbauer, 1930).

Stomata

The stomata are apertures in the epidermis, each bounded by two guard cells (fig. 7.1; pl. 23A, B). In Greek, stoma means mouth, and the term is often used with reference to the stomatal pore only. In this book, the term stoma includes the guard cells and the pore between them.

By changes in their shape, the guard cells control the size of the stomatal aperture. The aperture leads into a substomatal intercellular space, the substomatal chamber, which is continuous with the intercellular spaces in the mesophyll. In many plants two or more of the cells adjacent to the guard cells appear to be associated functionally with them and are morphologically distinct from the other epidermal cells. Such cells are called *subsidiary*, or *accessory*, *cells* (figs. 7.1D, E; 7.5).

The stomata are most common on green aerial parts of plants, particularly the leaves. The aerial parts of some chlorophyll-free land plants (*Monotropa, Neottia*) and roots have no stomata as a rule, but rhizomes have such structures (De Bary, 1884). They occur on some submerged aquatic plants, and not on others. The variously colored petals of the flowers often have stomata, sometimes nonfunctional. Stomata are also found on stamens and gynoecia. In green leaves they occur either on both surfaces (amphistomatic leaf) or on one only, either the upper (epistomatic leaf) or more commonly the lower (hypostomatic leaf). The density of stomata has been established as 100 to 300 per square millimeter for leaves of many species (Stålfelt, 1956).

In leaves with parallel veins, such as those of monocotyledons and some dicotyledons, and in the needles of conifers the stomata are arranged in parallel rows (figs. 7.2B, 7.4A; pl. 77D). The substomatal chambers in each row are coalesced, and the mesophyll cells bounding these chambers form an arch over (or beneath) the intercellular canal (fig. 7.4B, pl. 79A). In netted-veined leaves the stomata are scattered (fig. 7.1B–E).

Guard cells may occur at the same level as the adjacent epidermal cells, or they may protrude above or be sunken below the surface of the epidermis (figs. 7.1, 7.3, 7.4, 7.6). In some plants stomata are restricted to the epidermis that lines depressions in the leaf, the stomatal crypts. Epidermal hairs may also be prominently developed in such crypts (chapter 16).

The guard cells are generally crescent-shaped with blunt ends (kidney-shaped) in surface view (figs. 7.1, 7.3D, 7.6C) and often have ledges of wall material on the upper and lower sides. In sectional views such ledges appear like horns (fig. 7.3E, F, H). Sometimes a ledge occurs only on the upper side (fig. 7.3A, G, I), or none is present (fig. 7.4B, E). If two ledges are present, the upper delimits the front cavity above the stomatal pore, and the lower encloses the back cavity between the pore and the substomatal chamber (fig. 7.3F). The ledges are more or less heavily cutinized (Bondesson, 1952).

An outstanding characteristic of stomata is the unevenly thickened

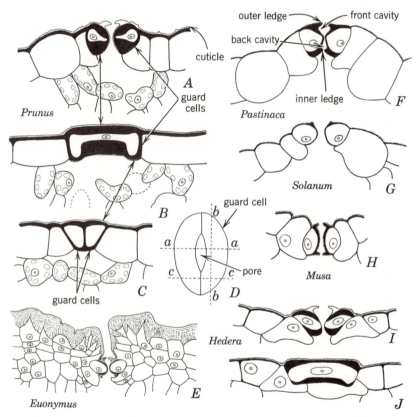

FIG. 7.3. Stomata in abaxial epidermis of foliage leaves. A–C, stomata and some associated cells from peach leaf sectioned along planes indicated in D by the broken lines aa, bb, and cc. E–I, stomata from various leaves cut along the plane aa. J, one guard cell of ivy cut along the plane bb. The stomata are raised in A, F, G. They are slightly raised in I, slightly sunken in H, and deeply sunken in E. The horn-like protrusions in the various guard cells are sectional views of ledges. Some stomata have two ledges (E, F, H); others only one (A, G, I). Ledges are cuticular in A, E, I. The Euonymus leaf has a thick cuticle; epidermal cells are partly occluded with cutin. (A–D, F–J, ×605; E, ×242.)

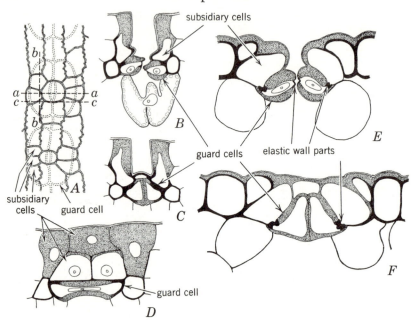

FIG. 7.4. Stomata of conifer leaves. *A*, surface view of epidermis with two deeply sunken stomata from *Pinus merkusii*. Guard cells are overarched by subsidiary and other epidermal cells. Stomata and some associated cells of *Pinus* (*B–D*), and *Sequoia* (*E*, *F*). The broken lines in *A* indicate the planes along which the sections of stomata were made in *B–F*: *aa*, *B*, *E*; *bb*, *D*; *cc*, *C*, *F*. (*A*, ×182; *B–D*, ×308; *E*, *F*, ×588. *A*, adapted from Abagon, *Philippine Univ. Nat. and Appl. Sci. Bul.* 6, 1938.)

walls of the guard cells (figs. 7.3, 7.4). This feature appears to be related to the changes in shape and volume (and the concomitant changes in the size of stomatal aperture) which are operated by turgor changes in the guard cells. In many species, the posture of the guard cells appears to be determined by turgor difference between the guard cells and the subsidiary cells (Heath, 1959).

The primary cause of changes in turgor of the guard cells is not definitely established (Heath, 1959; Ketellapper, 1963). The immediate cause appears to be the condensation and hydrolysis of starch in their chloroplasts. Photosynthesis alone is not sufficient to explain the rapid changes in osmotic pressure associated with the closing and opening mechanism. Moreover, the chloroplasts may not be well differentiated in guard cells (Brown and Johnson, 1962). Among the environmental factors carbon dioxide concentration appears to play a major role in size changes of the stomatal pore (Ketellapper, 1963).

Judged by the polymorphism of the guard cells, the mechanisms responsible for opening and closing of the stomata are varied (Stålfelt, 1956). In one common type, the change in the shape of the guard cells occurs because the wall that is turned away from the stomatal aperture, the so-called back wall, is thin and apparently elastic (fig. 7.3A, E–I; pl. 23A). When the turgor increases, the thin wall bulges away from the aperture, while the front wall (facing the pore) becomes straight or concave. The whole cell appears to bend away from the aperture, and the aperture increases in size. Reversed changes occur under decreased turgor.

Another distinct type of stomatal mechanism is illustrated by the guard cells of Gramineae and Cyperaceae. These cells are bulbous at two ends and straight in the middle (fig. 7.5). The middle part has a strongly but unevenly thickened wall; the bulbous ends have thin walls, and the wall between the bulbous ends of two adjacent cells may be incomplete so that the protoplasts of the two guard cells are partially confluent (Brown and Johnson, 1962). Increase in turgor causes a swelling of the bulbous ends and the consequent separation of the straight median portions from each other (compare A and B in fig. 7.5). The nucleus in a gramineous guard cell is extended and simulates the shape of the cell lumen. It has two enlarged ends connected by a thin thread-like middle part.

Coniferous stomata are sunken and appear as though suspended from the subsidiary cells arching over them (fig. 7.4). In their median parts the guard cells are elliptical in section and have narrow lumina (fig.

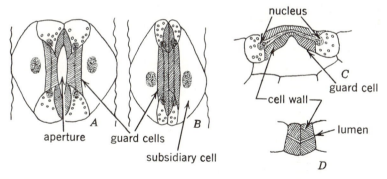

FIG. 7.5. Stomata of sugarcane (*Saccharum*). *A, B* stomata seen from outer surface in open (*A*) and closed (*B*) states. *C*, longitudinal section of one guard cell. The nucleus is much extended and appears as two masses connected by a thin thread. *D*, transection through the central portion of two guard cells from a closed stoma. Hatched parts, thick walls; circles in *A–C*, chloroplasts. (After Flint and Moreland, *Amer. Jour. Bot.* 33, 1946.)

The Epidermis

7.4*B*, *E*). At their ends they have wider lumina and are triangular in section (fig. 7.4*C*, *F*). The characteristic feature of these guard cells is that their walls, and those of the subsidiary cells, are partly lignified and partly free of lignin. This combination of more and less rigid wall parts, the manner of connection with the subsidiary cells, and the presence of thin wall parts in the subsidiary cells are features that appear to be involved in the working of the coniferous stomata (Florin, 1931). In *Equisetum* the guard cells occur beneath two subsidiary cells in which the walls that are in contact with the guard cells have conspicuous wall thickenings (Hauke, 1957).

The front cavities of stomata in conifers and some angiosperms are often occluded with finely granular or alveolar material, probably cutic-

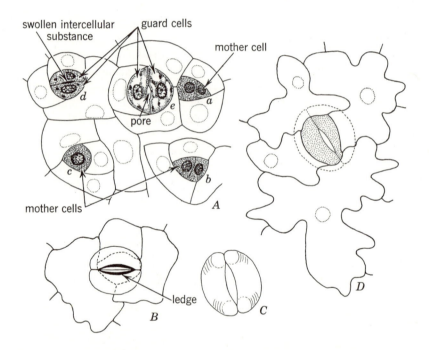

FIG. 7.6. Stomata of *Nicotiana* (tobacco) in surface views. *A*, developmental stages: *a*, *b*, soon after division that resulted in formation of stoma mother cell; *c*, stoma mother cell has enlarged; *d*, stoma mother cell has divided in two guard cells, still completely joined, but with swollen intercellular substance in position of future pore; *e*, young stoma with pore between guard cells. *B*, mature stoma seen from outer side of adaxial epidermis. *D*, similar stoma seen from inner side of abaxial epidermis. The guard cells are raised and thus appear above the epidermal cells in *B* and below them in *D*. *C*, guard cells as they appear from the inner side of the epidermis. (*A*, ×620; *B–D*, ×490.)

ular in nature (Bondesson, 1952; Turrell, 1947). The stomata may be occluded on the inner side by parenchyma cells, the so-called obturating cells, that extend into the substomatal chamber (Villaça and Ferri, 1954).

The structure of the guard-cell walls is comparable to that of the other epidermal cells of the same leaves. They are usually cutinized in the outer layers and covered with a cuticle. As previously mentioned, the cuticle extends through the stomatal aperture into the substomatal chamber where it joins the inner layer of cutin. (The cuticle is reported to be absent on the thin wall facing the pore in *Citrus;* Turrell, 1947.) Guard cells show lignification, at least in parts of their walls, in vascular cryptogams, gymnosperms, Gramineae, Cyperaceae, and certain dicotyledons (Kaufman, 1927). Ultrastructurally, a longitudinal orientation of microfibrils has been recognized in guard cells of the *Avena* coleoptile (Setterfield, 1957). According to some studies, no plasmodesmata occur between the guard cells and the adjacent epidermal cells (Brown and Johnson, 1962; Ketellapper, 1963); according to others, plasmodesmatal connections do occur between the guard cells and the subsidiary cells (Sievers, 1959).

Development. Stomata arise through differential divisions in the protoderm. After several divisions of a given protodermal cell, one of the products of these divisions becomes the immediate precursor of the guard cells. This is the stoma or guard-cell mother cell (figs. 7.6A, 7.7A), that eventually divides into the two guard cells. These enlarge and assume the characteristic crescentic shape. The area which becomes the pore shows a lenticular mass of pectic material just before the walls separate from each other (fig. 7.6A). This appearance probably results from the swelling of the intercellular material preceding its dissolution. The mother cells of the guard cells occur at the same level as the adjacent epidermal cells. If the mature stoma is raised above or sunken below the surface of the epidermis, the change in position occurs during the development of the stoma through mutual cellular readjustments within the epidermis and between the epidermis and the mesophyll (fig. 7.7). Even in the coniferous leaves, in which the guard cells are so deeply sunken, the stoma mother cells are at one level with the other epidermal cells (Cross, 1942). More or less conspicuous intercellular spaces occur in the mesophyll during the initiation of stomata (fig. 7.7A–C). A large intercellular space, the substomatal chamber, develops during the maturation of the stoma (fig. 7.7E–G).

The sequence of divisions preceding the formation of stomata varies in different species so that the guard cells and the subsidiary cells may be unrelated or more or less closely related. In Gramineae, for example, the immediate guard-cell precursor arises in one row of cells, the

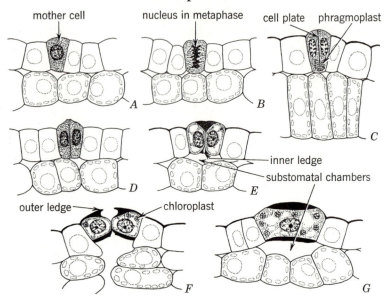

FIG. 7.7. Development of stoma of *Nicotiana* (tobacco) leaf as seen in sections. *C,* from adaxial epidermis with some palisade cells; others from abaxial epidermis. *A–C,* stoma mother cell before and during division in two guard cells. *D,* young guard cells with thin walls. *E,* guard cells have extended laterally and have begun to thicken their walls. Inner ledge and substomatal chamber have been formed. *F,* mature guard cells with upper and lower ledges and unevenly thickened walls. *G,* one mature guard cell cut parallel with its long axis and at right angles to the leaf surface. (All, ×490.)

subsidiary cells in two adjacent rows (fig. 7.8; Stebbins and Shah, 1960). In *Drimys,* a protodermal cell (the primary mother cell of stoma) gives rise, after two divisions, to a guard-cell mother cell and two subsidiary cells (Bondesson, 1952). The grass stoma exemplifies the *perigene* (from the Greek for about and offspring) type of subsidiary cells, that is, cells not arising from the primary mother cell (Florin, 1958). The subsidiary cells in *Drimys* do arise from the primary mother cell and are called *mesogene* (from the Greek for in the middle and offspring). The same stoma may have both mesogene and perigene subsidiary cells, as in *Trochodendron* (Bondesson, 1952). The classification into mesogene and perigene requires developmental studies because the mature pattern does not necessarily reveal the ontogenetic relationship of cells; and the distinction between the two kinds of cells may not be significant physiologically.

A discussion of cell relationships brings up the question as to how early in the ontogeny of a stoma cytologic differentiation in the protodermal cell indicates the inception of this ontogeny. The guard-cell precursor has been repeatedly described as being distinguishable by the density of its cytoplasm, and developmental studies indicate that this feature results from cytoplasmic polarization—an accumulation of cytoplasm in one end of the cell—before the primary mother cell of the stoma divides (Bünning and Biegert, 1953; Stebbins and Shah, 1960). An asymmetric division occurs across the gradient resulting from the polarization and gives rise to a small guard-cell precursor and a larger, less specialized epidermal cell. A polarization-asymmetry sequence is possibly operative even earlier, during the formation of the primary mother cell of the stoma (Stebbins and Shah, 1960). In *Populus pyramidalis* the guard-cell precursor appears hypertrophied and its neighbors undergo an accelerated vacuolation (Meyer, 1959).

The nuclei of the guard-cell precursors are denser than those of their sister cells. Apparently this differentiation occurs gradually, through one or more preceding generations of cells (Resch, 1952). The sequences suggest an establishment of a center with highly specialized cells—the guard cells—surrounded by less specialized cells with gradation in the degree of specialization. The distinctness of guard cells in this scheme is indicated by their inability to respond with division in wounded tissue (Resch, 1952).

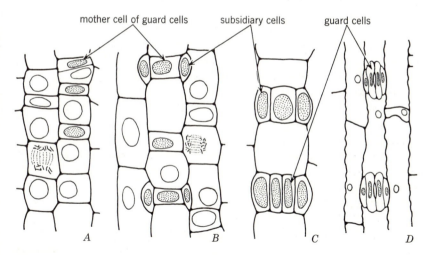

mother cell of guard cells   subsidiary cells   guard cells

A   B   C   D

FIG. 7.8. Development of stomata in oat (*Avena*). A, mother cells of guard cells formed by unequal divisions of protodermal cells. B, subsidiary cells are carved out of protodermal cells adjacent to mother cells. C, one mother cell has divided in two guard cells. D, mature stomata. (A–C, ×262; D, ×93; after photographs in Bonnett, *Univ. Illinois Agr. Expt. Sta. Bul.* 672, 1961.)

In a given leaf, the stomata arise not all at once but in succession through a considerable period of leaf growth. Two principal patterns may be distinguished in the development of stomata in the leaf as a whole (Ziegenspeck, 1944). In leaves with parallel veins, having the stomata in longitudinal rows, the developmental stages of the stomata are observable in sequence in the successively more differentiated portions of the leaf. (This sequence is basipetal, that is, from the tip of the leaf downward; chapter 16.) In the netted-veined leaves the different developmental stages are mixed in mosaic fashion so that mature stomata occur side by side with immature ones. The first pattern is characteristic of most monocotyledons and of a few dicotyledons (*Tragopogon, Thesium*); the second, of most dicotyledons and a few monocotyledons (Araceae, Smilacoideae, Taccaceae, Dioscoreaceae). Both developmental patterns are found among the vascular cryptogams.

Classification. The mode of development of stomata and their spatial relation to neighboring cells are considered with reference to problems of classification and phylogeny in the angiosperms and the conifers. The classifications refer to stomatal types but actually they are based on the relation between the stomata and the subsidiary cells.

In the gymnosperms Florin (1931, 1951, 1958) distinguishes two main types of stomatal complexes, the *haplocheilic* (simple lipped), in which the subsidiary cells are perigene, and the *syndetocheilic* (compound lipped), in which the subsidiary cells are mesogene. The haplocheilic type is highly variable in details and is regarded as the more primitive of the two. Although these categories were established by ontogenetic studies, the classification is used with regard to fossils by relying on the mature patterns characterizing the two types.

In dicotyledons the use of the mature pattern formed by the stomata and neighboring cells is suggested for the establishment of the typology (Metcalfe and Chalk, 1950). Four main types have been proposed: *anomocytic* (irregular-celled, formerly ranunculaceous), no subsidiary cells are present (fig. 7.1*B*); *anisocytic* (unequal-celled, formerly cruciferous), three subsidiary cells, one distinctly smaller than the other two, surround the stoma (fig. 7.1*E*, variant of anisocytic); *paracytic* (parallel-celled, formerly rubiaceous), one or more subsidiary cells occur on either side of the stoma parallel with its long axis (fig. 7.1*D*); *diacytic* (cross-celled, formerly caryophyllaceous), two subsidiary cells enclose the stoma, their common wall at right angles to the long axis of the stoma (fig. 7.1*F*). There are variations within these types and some probably merit separate designations as, for example, the *actinocytic* with the subsidiary cells arranged along the radii of a circle.

In monocotyledons, four categories of stomatal complexes have been described (Stebbins and Kush, 1961); two of these have four or more subsidiary cells surrounding the guard cells (*Rhoeo, Commelina*), one has two subsidiary cells (Gramineae), and one has none (*Allium*). The types with many subsidiary cells are regarded as more primitive, the other two as independently derived by reduction in the number of subsidiary cells.

Trichomes on Aerial Plant Parts

Trichomes (a word of Greek origin, meaning a growth of hair) are epidermal appendages of diverse form, structure, and functions (figs. 7.9–7.12; Uphof, 1962). They are represented by protective, supporting, and glandular hairs, by scales, various papillae, and the absorbing hairs of the roots.

Trichomes are usually distinguished from the so-called emergences (prickles, De Bary, 1844) on the basis that the emergences are formed from both epidermal and subepidermal tissues. The distinction between such emergences and trichomes is not sharp, however, because some plant hairs are raised upon a base originating by division of subepidermal cells. Trichomes also intergrade with nontrichomatous epidermal cells having protrusions in the form of papillae and with cells differentiated as "water vesicles."

Trichomes may occur on all parts of a plant. Either they persist throughout the life of an organ, or they are ephemeral. Some persisting hairs remain alive; others become devoid of protoplasts and are retained in dry state. The epidermal trichomes usually develop early in relation to the growth of the organ.

Trichomes may show wide variations within families and the smaller plant groups, and even in the same plant (fig. 7.10D, E). On the other hand, there is sometimes considerable uniformity in trichomes within a plant group. Plant-hair types have been successfully used in the classification of genera and even of species in certain families and in the recognition of interspecific hybrids (Cowan, 1950; Heintzelmann and Howard, 1948; Hummel and Staesche, 1962; Metcalfe and Chalk, 1950; Rollins, 1944).

Trichomes may be classified into different morphological categories (Foster, 1949). One common type is referred to as *hair*. Structurally, hairs may be subdivided into unicellular and multicellular. The unicellular hairs may be unbranched (fig. 7.10D, F) or branched (fig. 7.9G, H). Multicellular hairs may consist of a single row of cells (figs. 7.9I, 7.10A) or of several layers (fig. 7.9J). Some multicellular hairs are branched in dendroid (tree-like) manner (fig. 7.9D); others have the branches oriented largely in one plane (stellate hairs, fig. 7.9E). Commonly a multicellular hair can be divided into a foot, which is imbedded in the epidermis, and a body projecting above the surface (fig. 7.10B). The cells surrounding the foot are sometimes morphologically distinct from other epidermal cells.

Another common type of trichome is the *scale*, also called *peltate hair* (from the Latin *peltatus*, target-shaped or shield-like, and attached by its lower surface). A scale consists of a discoid plate of cells, often borne on a stalk or attached directly to the foot (figs. 7.9A, B, and 7.10G, H).

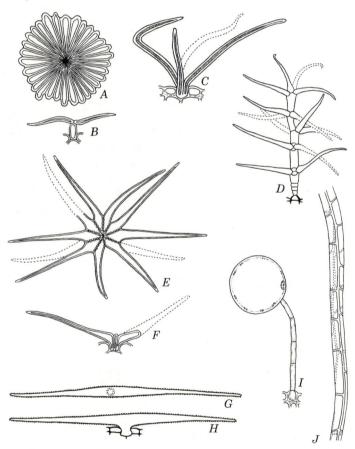

FIG. 7.9. Trichomes. A, B, peltate scale of *Olea* in surface (A) and side (B) views. C, tufted hair of *Quercus*. D, branched, candelabra hair of *Platanus*. E, F, stellate hair of *Sida* in surface (E) and side (F) views. G, H, two-armed, T-shaped unicellular hair of *Lobularia* in surface (G) and side (H) views. I, vesiculate hair of *Chenopodium*. J, part of multicellular shaggy hair of *Portulaca*. (A–C, I, ×180; D–H, J, ×90.)

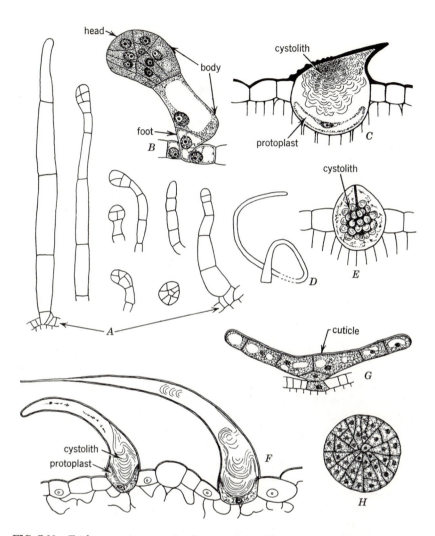

FIG. 7.10. Trichomes. *A*, group of ordinary and glandular (with multicellular heads) hairs of *Nicotiana* (tobacco). *B*, enlarged view of glandular hair of tobacco, showing characteristic density of contents of glandular head. *C*, hooked hair with cystolith of *Humulus*. *D*, long coiled unicellular hair, and *E*, short bristle with cystolith of *Boehmeria*. *F*, hooked hairs with cystoliths of *Cannabis*. *G*, *H*, glandular peltate trichome of *Humulus* seen in sectional (*G*) and surface (*H*) views. (*H* from younger trichome than *G*.) (*A*, *F*, ×100; *B*, *D*, *E*, ×310; *C*, *G*, ×245; *H*, ×490.)

The Epidermis

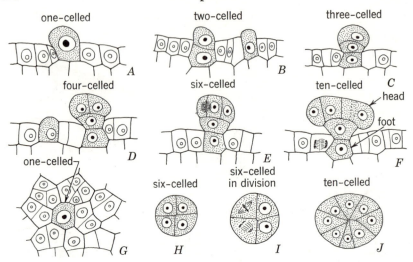

FIG. 7.11. Development of glandular trichomes (stippled cells) of *Ligustrum* as seen in sectional (*A–F*) and surface (*G–J*) views. (×490.)

Unicellular, multicellular, and peltate hairs may be glandular. Some of the simple multicellular glandular hairs consist of a stalk and a unicellular or multicellular head (fig. 7.10*B*). The head constitutes the secretory part of the hair. In a peltate glandular trichome the discoid plate is composed of glandular cells (fig. 7.10*G, H*). Some glandular trichomes consist of a multicellular core of cells covered with a palisade-like layer of secretory cells (chapter 13).

A trichome is initiated as a protuberance from an epidermal cell. The protuberance elongates, and if it develops into a multicellular structure, various divisions may follow the initial elongation (fig. 7.11).

The cell walls of trichomes are commonly of cellulose and are covered with a cuticle. They may be lignified. Plant hairs often produce thick secondary walls as, for instance, the cotton seed hairs (Anderson and Kerr, 1938) or the "climber hairs" of *Humulus* (Franz, 1935). The walls of trichomes are sometimes impregnated with silica or calcium carbonate (Beyrich, 1943). Cell contents are varied in relation to function; the most complex are probably those of the glandular cells. Chloroplasts are often present, though they may be small and not persisting. Plant hair cells, other than the glandular, are characteristically highly vacuolated. Cystoliths and other crystals may develop in hairs (fig. 7.10*C, E, F*).

Cotton seed hairs, commonly known as cotton fibers, are extremely long epidermal hairs with thick secondary walls of almost pure cellulose (Berkley, 1948). They are formed from the protoderm of the ovule during flowering and continue to arise for about 10 days after anthesis (Anderson and Kerr, 1938). The elongation lasts for 15 to 20 days, and the hairs become ½ to 2½ in. long, depending on the variety of cotton. A number of other plants produce commercially usable hairs on the seeds or other parts of the fruit (Dewey, 1943; Pearson, 1948).

Root Hairs

Root hairs are tubular structures constituting direct lateral extensions of the cells that originate them. In a study involving 37 species in 20 families, the root hairs were found to vary between 5 and 17 microns in diameter and between 80 and 1,500 microns in length (Dittmer, 1949). They are highly vacuolate and contain the nucleus in their parietal cytoplasm. The root hairs are rarely branched (Linsbauer, 1930). The adventitious roots of *Kalanchoë* growing in air have multicellular root hairs, whereas the same kind of roots growing in soil have unicellular ones (Popham and Henry, 1955). Root hairs are typical of roots, but under certain conditions they may develop on other plant parts (Haccius and Troll, 1961).

The ability of root hairs to absorb water has been demonstrated experimentally. The same experiments showed that the hairless epidermal cells also absorb water with a range of velocity comparable to that of cells bearing root hairs (Rosene, 1954).

The principal function of the root hairs is considered to be the extension of the absorbing surface of the root, and from this standpoint the available information pertaining to their numbers and surface area in a rye plant is of interest (rounded off figures from Dittmer, 1937). In this plant the 13,800,000 roots had a surface area of 2,500 sq ft. Living root hairs numbered 14 billion and had a total surface area of 4,300 sq ft. Thus the combined surface area of the roots and root hairs was 6,800 sq ft, and it was packed in less than 2 cu ft of soil. This total surface was 130 times that exposed to the outside air by the aerial parts of the same plant. If the surface of the mesophyll cells of a foliage leaf facing the intercellular spaces was taken into consideration, the root surface was still 22 times that of the transpiration area of the top. Relating these figures to the absorptive capacity of root hairs, Rosene (1955) calculated that a small number of the total root hairs can obtain all the water necessary for transpiration and growth of the plant.

Wall Structure. Although there is general agreement that the chief components of root hair walls are cellulose and pectic substances, the matter of distribution of these substances is still controversial. According to one view, the pectic substances occur as a matrix in the cellulosic microfibrillar system (Ekdahl, 1953); according to the other, calcium pectate forms a separate layer outside the cellulosic part of the wall (Cormack, 1962). An ultrastructural study of root hairs (Belford and Preston, 1961) suggests that the outer layer of the wall consists of randomly oriented microfibrils imbedded in an amorphous matrix, probably composed of pectins and hemicelluloses. The inner layer consists of axially oriented cellulose microfibrils associated with little or no amorphous material. Another study (Dawes and Bowler, 1959) recognizes, from outside in, a layer of mucilage, a cuticle, a layer of pectin, and a layer of cellulose and pectin. Environmental condition may induce the formation of callose within the root hairs (Lerch, 1960).

Development. The root hairs develop acropetally, that is, toward the apex of the root, and apparently never originate among preexisting hairs. Because of the acropetal sequence of initiation, in most seedling taproots the root hairs show a uniform gradation in size, beginning with those nearest the apex and going back to those of mature length. The hairs are initiated in the part of the root located behind the zone of most active cell division, but where the longitudinal extension of the epidermal cells may still be in progress (Cormack, 1949). Usually the root hair emerges as a small papilla at or near the apical end of a cell. If the cell continues to elongate, after the appearance of the papilla, the root hair eventually occurs some distance from this end; otherwise its position remains terminal. The root hair grows at the tip where the cellulose microfibrils have random orientation. In the basal part of the hair, where elongation has ceased, apposition of the parallel-oriented microfibrils occurs. The growing tip contains dense cytoplasm (Sievers, 1963). Some workers see the nucleus in constant position near the tip in growing hairs (Bouet, 1954); others report a continuous movement of the nucleus (Kawata and Ishihara, 1962).

The factors affecting root hair development are a subject of controversy. According to Cormack's (1962) theory, a gradual hardening of the wall through calcification of the pectic layer arrests the growth of the hair at its proximal end and confines it to the soft region of the distal end. Ekdahl (1953), on the other hand, ascribes the hardening mainly to the formation of new cellulose microfibrils.

At the ultrastructural level, evidence has been presented that dictyosomes may be concerned with the formation of root hair wall (Sievers, 1963). Vesicles, separating from the dictyosomes and developing dense contents, appear to be transported toward the wall, especially at the tip.

In some plants the root epidermis shows a morphologic differentiation into hair-forming cells (*trichoblasts*) and cells that do not form hairs (fig. 7.12). This differentiation varies in degree of expression (Cormack, 1949) but is sufficiently characteristic of many genera of grasses that it may be used for the study of relationships in this family (Row and Reeder, 1957). In general, the root hair-forming cells are shorter than the others (fig. 7.12C, D). When strongly expressed, this difference may be visible at the origin of a trichoblast (fig. 7.12A, B). The precursory protodermal cell divides into a longer and a shorter cell, the shorter being characterized by denser cytoplasm and a lower rate of elongation than the longer cell (Avers, 1957). Recently formed trichoblasts are also distinguished from their sister cells by more intense enzymatic activity (Avers, 1958; Cormack, 1962) and greater amounts of RNA (Kawata and Ishihara, 1961). Significantly, the physiologic specialization of the trichoblast is observed before its maximum elongation; in fact, it appears to be initiated through polarization phenomena at the time of the asymmetric division producing the trichoblast (Avers, 1963).

In plants having a homogeneous root epidermis all epidermal cells are potential trichomatous cells, but not all necessarily produce root hairs. The nontrichomatous cells of a root with a heterogenous epidermis may be induced to form root hairs by suitable changes in the environment, and conversely the potentially trichomatous cells may be prevented from developing such structures (Cormack, 1949).

Root hairs are short-lived. Their longevity is commonly measured in days (Linsbauer, 1930). Old root hairs collapse, and the walls of the epidermal cells, if the cells are not sloughed off, become suberized and lignified. Persisting root hairs, however, have been observed in a number of plant species (Cormack, 1949). Such hairs become thick-walled and are then probably not concerned with absorption.

MULTIPLE EPIDERMIS

One or more layers of cells beneath the epidermis in leaf, stem, and root may be morphologically and physiologically distinct from the deeper-lying ground tissue. The older literature designates all distinctly characterized subepidermal layers as *hypodermis* (from the Greek *hypo*, below, and *derma*, skin; De Bary, 1884; Guttenberg, 1943). The specialized subsurface tissue may be part of the ground tissue, or it may be derived from the protoderm by periclinal divisions. The recognition of the latter possibility has prompted workers to separate the hypodermis originating in the ground tissue from the subsurface layers of protodermal origin by introducing the concept of *multiple* or

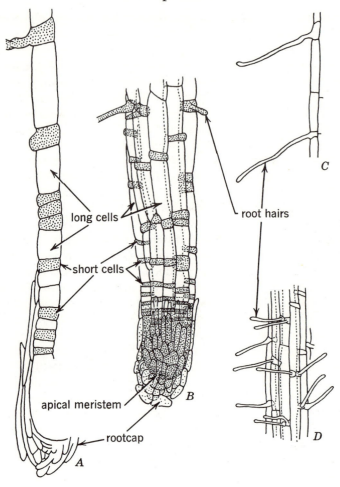

FIG. 7.12. Root-hair development from protodermal cells (short cells, or trichoblasts). *A, C, Cyperus; B, D, Anigozanthos.* (*A, B,* ×240; *D,* ×175. Redrawn from Leavitt, *Boston Soc. Nat. Hist. Proc.* 31 1904.)

multiseriate epidermis (Linsbauer, 1930). A study of mature structures rarely permits the identification of the tissue either as multiple epidermis or as a combination of epidermis and a hypodermis. The origin of the subsurface layers can be properly revealed only by developmental studies.

The outermost layer of a multiple epidermis resembles the ordinary uniseriate epidermis in having a cuticle. The inner layers are commonly differentiated as water-storage tissue lacking chlorophyll (Linsbauer,

1930). The multiple epidermis varies in thickness from 2 to 16 layers of cells (De Bary, 1884). Sometimes only individual cells of the epidermis undergo periclinal divisions. Representatives with multiple epidermis may be found among Moraceae (fig. 7.13; most species of *Ficus*), Pittosporaceae, Piperaceae (*Peperomia*), Begoniaceae, Malvaceae, Monocotyledoneae (palms, orchids), ferns, and others (Linsbauer, 1930). The *velamen* (from the Latin word for covering) of the aerial and terrestrial roots of orchids is a multiple epidermis (or rhizodermis; Engard, 1944; Linsbauer, 1930).

The periclinal divisions initiating the multiple epidermis in leaves occur at different stages of leaf growth but usually when a leaf is several internodes below the apex (Linsbauer, 1930). In *Ficus*, for example, the leaf has a uniseriate epidermis until the stipules are shed (Pfitzer, 1872). Then periclinal divisions occur in the epidermis (fig. 7.13A). Similar divisions are repeated in the outer row of daughter cells, sometimes once, sometimes twice (fig. 7.13B). During the expansion of the

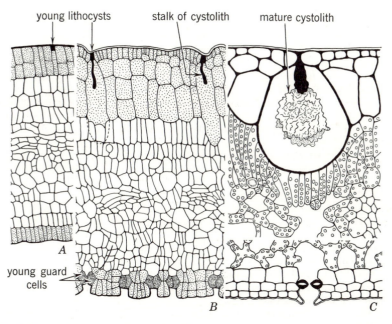

FIG. 7.13. Multiple epidermis (on both leaf surfaces) in transections of *Ficus elastica* leaves in three stages of development. Epidermis stippled in *A, B,* with thick walls in *C.* Part of leaf omitted in *C.* Cystolith development: *A,* wall thickens in lithocyst; *B,* cellulose stalk appears; *C,* calcium carbonate is deposited on stalk. Unlike other epidermal cells, lithocyst undergoes no periclinal divisions. (*A,* ×207; *B,* ×163; *C,* ×234.)

leaf, anticlinal divisions occur also, and, since these divisions are not synchronized in the different layers, the ontogenetic relation between these layers becomes more or less obscured (fig. 7.13*B*, *C*). The inner cells expand more than the outer. The outer cells remain particularly small because they expand less and undergo more numerous anticlinal divisions than the inner. The cystolith-containing cells characteristic of *Ficus* leaves do not divide but keep pace with the increasing depth of the epidermis and even overtake it by expansion and intrusion into the mesophyll (fig. 7.13; Ajello, 1941; Pfitzer, 1872). In some plants (*Peperomia*) the cells of the multiple epidermis remain arranged in radial rows and clearly reveal their common origin (Linsbauer, 1930).

REFERENCES

Ajello, L. Cytology and cellular interrelations of cystolith formation in *Ficus elastica*. *Amer. Jour. Bot.* 28:589–594. 1941.

Allen, G. S. Embryogeny and the development of the apical meristems of *Pseudotsuga*. II. Late embryogeny. *Amer. Jour. Bot.* 34:73–80. 1947.

Anderson, D. B., and T. Kerr. Growth and structure of cotton fiber. *Indus. and Engin. Chem.* 30:48–54. 1938.

Avers, C. J. An analysis of difference of growth rate of trichoblasts and hairless cells in the root epidermis of *Phleum pratense*. *Amer. Jour. Bot.* 44:686–690. 1957.

Avers, C. J. Histochemical localization of enzyme activity in the root epidermis of *Phleum pratense*. *Amer. Jour. Bot.* 45:609–613. 1958.

Avers, C. J. Fine structure studies of *Phleum* root meristem cells. II. Mitotic asymmetry and cellular differentiation. *Amer. Jour. Bot.* 50:140–148. 1963.

Avery, G. S., Jr. Structure and development of the tobacco leaf. *Amer. Jour. Bot.* 20:565–592. 1933.

Baker, G. Hook-shaped opal phytoliths in the epidermal cells of oats. *Austral. Jour. Bot.* 8:69–74. 1960.

Bancher, E., J. Hölzl, and J. Klima. Licht- und elektronenmikroskopische Beobachtungen an der Kutikula der Zwiebelschuppe von *Allium cepa*. *Protoplasma* 52:247–259. 1960.

Belford, D. S., and R. D. Preston. The structure and growth of root hairs. *Jour. Expt. Bot.* 12:157–168. 1961.

Berkley, E. E. Cotton, a versatile textile fiber. *Textile Res. Jour.* 18:71–88. 1948.

Beyrich, H. Über die Membranverkieselung einiger Pflanzenhaare. *Flora* 36:313–324. 1943.

Bolliger, R. Entwicklung und Struktur der Epidermisaussenwand bei einigen Angiospermenblättern. *Jour. Ultrastruct. Res.* 3:105–130. 1959.

Bondesson, W. Entwicklungsgeschichte und Bau der Spaltöffnungen bei den Gattungen *Trochodendron* Sieb. et Zucc., *Tetracentron* Oliv. und *Drimys* J. R. et G. Forst. *Acta Horti Bergiani* 16:169–218. 1952.

Bouet, M. Études cytologiques sur le developpement des poils absorbants. *Rev. Cytol. et Biol. Vég.* 15:261–305. 1954.

Brown, W. V., and S. C. Johnson. The fine structure of the grass guard cell. *Amer. Jour. Bot.* 49:110–115. 1962.

Bünning, E., and F. Biegert. Die Bildung der Spaltöffnungsinitialen bei *Allium Cepa*. *Ztschr. f. Bot.* 41:17–39. 1953.

Burström, H. Über die Entfaltung und Einrollen eines mesophilen Grassblattes. *Bot. Notiser* 1942:351–362. 1942.

Cormack, R. G. H. The development of root hairs in angiosperms. *Bot. Rev.* 15:583–612. 1949. II. *Bot. Rev.* 28:446–464. 1962.

Cowan, J. M. *The Rhododendron leaf: a study of the epidermal appendages.* Edinburgh, Oliver and Boyd. 1950.

Cross, G. L. Structure of the apical meristem and development of the foliage leaves of *Cunninghamia lanceolata. Amer. Jour. Bot.* 29:288–301. 1942.

Damm, O. Über den Bau, die Entwicklungsgeschichte und die mechanischen Eigenschaften mehrjähriger Epidermen bei den Dicotyledonen. *Bot. Centbl. Beihefte* 7:219–260. 1902.

Davies, I. The use of epidermal characteristics for the identification of grasses in the leafy stage. *Jour. Brit. Grassl. Soc.* 14:7–16. 1959.

Dawes, C. J., and E. Bowler. Light and electron microscope studies of the cell wall structure of the root hairs of *Raphanus sativus. Amer. Jour. Bot.* 46:561–565. 1959.

De Bary, A. *Comparative anatomy of the vegetative organs of the phanerogams and ferns.* Oxford, Clarendon Press. 1884.

Dewey, L. H. Fiber production in the western hemisphere. *U.S. Dept. Agr. Misc. Pub.* 518. 1943.

Dilcher, D. L. Cuticular analysis of Eocene leaves of *Ocotea obtusifolia. Amer. Jour. Bot.* 50:1–8. 1963.

Dittmer, H. J. A quantitative study of the roots and root hairs of a winter rye plant (*Secale cereale*). *Amer. Jour. Bot.* 24:417–420. 1937.

Dittmer, H. J. Root hair variations in plant species. *Amer. Jour. Bot.* 36:152–155. 1949.

Drawert, H., and M. Mix. Elektronenmikroskopische Studien au den Oberepidermiszellen der Schuppenblätter von *Allium cepa* L. *Protoplasma* 57:270–289. 1963.

Ekdahl, I. Studies on the growth and the osmotic conditions of root hairs. *Symb. Bot. Upsal.* 11(6):5–83. 1953.

Engard, C. J. Morphological identity of the velamen and exodermis in orchids. *Bot. Gaz.* 105:457–462. 1944.

Florin, R. Untersuchungen zur Stammesgeschichte der Coniferales und Cordaitales. *Svenska Vetensk. Akad. Handl.* Ser. 3. 10:1–588. 1931.

Florin, R. Evolution in cordaites and conifers. *Acta Horti Bergiani* 15:285–388. 1951.

Florin, R. Notes on the systematics of the Podocarpaceae. *Acta Horti Bergiani* 17:403–411. 1958.

Foster, A. S. *Practical plant anatomy.* 2nd ed. New York, D. Van Nostrand Company. 1949.

Franke, W. Ectodesmata and foliar absorption. *Amer. Jour. Bot.* 48:683–691. 1961.

Franz, H. Beiträge zur Kenntnis des Dickenwachstums der Membranen. (Untersuchungen an den Haaren von *Humulus lupulus.*) *Flora* 29:287–308. 1935.

Frey-Wyssling, A., and K. Mühlethaler. Über das submikroskopische Geschehen bei der Kutinisierung pflanzlicher Zellwände. *Naturf. Gesell in Zürich, Vrtljschr.* 104:294–299. 1959.

Gulline, H. F. Experimental morphogenesis in adventitious buds in flax. *Austral. Jour. Bot.* 8:1–10. 1960.

Guttenberg, H. von. Die Bewegungsgewebe. In: K. Linsbauer. *Handbuch der Pflanzenanatomie.* Band 5. Lief. 42. 1943.

Haccius, B., and W. Troll. Über die sogenannten Wurzelhaare an den Keimpflanzen von *Drosera-* und *Cuscuta*-Arten. *Beitr. z. Biol. der Pflanz.* 36:139–157. 1961.

Harris, T. M. The fossil plant cuticle. *Endeavour* 15:210–214. 1956.

Hauke, R. L. The stomatal apparatus of *Equisetum*. *Torrey Bot. Club Bul.* 84:178–181. 1957.

Heath, O. V. S. The water relations of stomatal cells and the mechanisms of stomatal movement. In: F. C. Steward. *Plant Physiology*. Vol. 2. New York, Academic Press. 1959.

Heintzelmann, C. E., Jr., and R. A. Howard. The comparative morphology of the Icacinaceae. V. The pubescence and the crystals. *Amer. Jour. Bot.* 35:42–52. 1948.

Huelin, F. E. Studies in the natural coating of apples. IV. The nature of cutin. *Austral. Jour. Biol. Sci.* 12:175–180. 1959.

Hummel, K., and K. Staesche. Die Verbreitung der Haartypen in den natürlichen Verwandschaftsgruppen. In: *Handbuch der Pflanzenanatomie* Band 4. Teil. 5. 1962.

Juniper, B. E. The surfaces of plants. *Endeavour* 18:20–25. 1959.

Juniper, B. E. Growth, development, and effect of environment on the ultrastructure of plant surfaces. *Linn. Soc. London, Jour., Bot.* 56:413–420. 1960.

Kaufman, K. Anatomie und Physiologie der Spaltöffnungsapparate mit verholzten Schliesszellmembranen. *Planta* 3:27–59. 1927.

Kawata, S., and K. Ishihara. Studies on the distribution of ribonucleic acid in the epidermis of the crown roots of the rice plant. *Crop Sci. Soc. Japan, Proc.* 29:387–391. 1961.

Kawata, S., and K. Ishihara. Relationships between root-hair elongation, and the movement of nuclei and cytoplasmic granules in root hairs of rice plants. *Crop Sci. Soc. Japan, Proc.* 300:161–168. 1962.

Ketellapper, H. J. Stomatal physiology. *Ann. Rev. Plant Physiol.* 14:249–270. 1963.

Kreger, D. R. Wax. *Handb. der Pflanzenphysiol.* 10:249–269. 1958.

Lerch, G. Untersuchungen über Wurzelkallose. *Bot. Studien* No. 11:1–111. 1960.

Linsbauer, K. Die Epidermis. In: K. Linsbauer. *Handbuch der Pflanzenanatomie*. Band 4. Lief. 27. 1930.

Marco, H. F. The anatomy of spruce needles. *Jour. Agr. Res.* 58:357–368. 1939.

Matzke, E. B. The three-dimensional shape of epidermal cells of *Aloe aristata*. *Amer. Jour. Bot.* 34:182–195. 1947.

Matzke, E. B. The three-dimensional shape of epidermal cells of the apical meristem of *Anacharis densa* (*Elodea*). *Amer. Jour. Bot.* 35:323–332. 1948.

McVeigh, I. Regeneration in *Crassula multicava*. *Amer. Jour. Bot.* 25:7–11. 1938.

Metcalfe, C. R. *Anatomy of the monocotyledons*. I. *Gramineae*. Oxford, Clarendon Press. 1960.

Metcalfe, C. R., and L. Chalk. *Anatomy of the dicotyledons*. 2 vols. Oxford, Clarendon Press. 1950.

Meyer, F. J. Das trophische Parenchym. A. Assimilationsgewebe. In: *Handbuch der Pflanzenanatomie*. Band 4. Teil. 7A. 1962.

Meyer, J. Le caractère précocement idioblastique des initiales stomatiques du pétiole de *Populus pyramidalis* Rozier. *Protoplasma* 51:313–319. 1959.

Mikulska, E. Chloroplastes dans l'épiderme des feuilles des Monocotylédones. *Soc. Sci. et Let. Lodz Cl. III Sci. Math. et Nat. Bul.* 10:1–8. 1959a.

Mikulska, E. Chloroplasty w skórce liści roślin dwuliściennykh. [Sur l'existence des chloroplastes dans l'épiderme des feuilles des Dicotylédones.] *Soc. Bot. Polon. Acta* 28:143–173. 1959b.

Parry, D. W., and F. Smithson. Silicification of bulliform cells in grasses. *Nature* 181:1549–1550. 1958.

Pearson, N. L. Observations on seed and seed hair growth in *Asclepias syriaca* L. *Amer. Jour. Bot.* 35:27–36. 1948.

References 179

Pfitzer, E. Beiträge zur Kenntnis der Hautgewebe der Pflanzen. III. Über die mehrschichtige Epidermis und das Hypoderma. *Jahrb. f. Wiss. Bot.* 8:16–74. 1872.

Piringer, A. A., and P. H. Heinze. Effect of light on the formation of a pigment in the tomato fruit cuticle. *Plant Physiol.* 29:467–472. 1954.

Popham, R. A., and R. D. Henry. Multicellular root hairs on adventitious roots of *Kalanchoe fedtschenkoi. Ohio Jour. Sci.* 55:301–307. 1955.

Prat, H. Histo-physiological gradients and plant organogenesis. Part I. *Bot. Rev.* 14:603–643. 1948. Part II. *Bot. Rev.* 17:693–746. 1951.

Priestley, J. H. The cuticle in angiosperms. *Bot. Rev.* 9:593–616. 1943.

Resch, A. Untersuchungen über Kerndifferenzierung in peripheren Zellschichten der Sprossachse einiger Blütenpflanzen. *Chromosoma* 5:296–316. 1952.

Rollins, R. C. Evidence for natural hybridity between guayule (*Parthenium argentatum*) and mariola (*Parthenium incanum*). *Amer. Jour. Bot.* 31:93–99. 1944.

Rosene, H. F. A comparative study of the rates of water influx into the hairless epidermal surface and the root hairs of onion roots. *Physiol. Plant* 7:676–686. 1954.

Rosene, H. F. The water absorptive capacity of winter rye root-hairs. *New Phytol.* 54:95–97. 1955.

Roth, I. Entwicklung der ausdauernden Epidermis sowie der primären Rinde des Stammes von *Cercidium torreyanum* in Laufe des sekundären Dickenwachstums. *Österr. Bot. Ztschr.* 110:1–19. 1963.

Row, H. C., and J. R. Reeder. Root-hair development as evidence of relationships among genera of Gramineae. *Amer. Jour. Bot.* 44:596–601. 1957.

Schiefferstein, R. H., and W. E. Loomis. Wax deposits on leaf surfaces. *Plant Physiol.* 31:240–247. 1956.

Schiefferstein, R. H., and W. E. Loomis. Development of cuticular layers in angiosperm leaves. *Amer. Jour. Bot.* 46:625–635. 1959.

Scott, F. M., K. C. Hamner, E. Baker, and E. Bowler. Electron microscope studies of the epidermis of *Allium cepa. Amer. Jour. Bot.* 45:449–461. 1958.

Setterfield, G. Fine structure of guard-cell walls in *Avena* coleoptile. *Canad. Jour Bot.* 35:791–793. 1957.

Shields, L. M. The involution mechanism in leaves of certain xeric grasses. *Phytomorphology* 1:225–241. 1951.

Sievers, A. Untersuchungen über die Darstellbarkeit der Ektodesmen und ihre Beeinflussung durch physikalische Faktoren. *Flora* 147:263–316. 1959.

Sievers, A. Beteiligung des Golgi-Apparates bei der Bildung der Zellwand von Wurzelhaaren. *Protoplasma* 56:188–192. 1963.

Sitte, P. Morphologie des Cutins und des Sporopollenins. In: E. Treiber. *Die Chemie der Pflanzenzellwand.* Berlin, Springer-Verlag. 1957.

Stålfelt, M. G. Die stomatäre Transpiration und die Physiologie der Spaltöffnungen. *Handb. der Pflanzenphysiol.* 3:350–426. 1956.

Stebbins, G. L., and G. S. Kush. Variation in the organization of the stomatal complex in the leaf epidermis of monocotyledons and its bearing on their phylogeny. *Amer. Jour. Bot.* 48:51–59. 1961.

Stebbins, G. L., and S. S. Shah. Developmental studies of cell differentiation in the epidermis of monocotyledons. II. Cytological features of stomatal development in the Gramineae. *Devlpmt. Biol.* 2:477–500. 1960.

Tateoka, T. Miscellaneous papers on the phylogeny of Poaceae (10). Proposition of a new phylogenetic system of Poaceae. *Jour. Jap. Bot.* 32:275–287. 1957.

Turrell, F. M. Citrus leaf stomata: structure, composition, and pore size in relation to penetration of liquids. *Bot. Gaz.* 108:476–483. 1947.

Uphof, J. C. T.   Plant hairs.   In: *Handbuch der Pflanzenanatomie.*   Band 4.   Teil 5.   1962.

Villaça, H., and M. G. Ferri.   On the morphology of the stomata in *Eucalyptus tereticornis, Ouratea spectabilis* and *Cedrela fissilis. São Paulo Univ. Faculd. de Filos., Ciên. e Let., Bot.* 11:33–51.   1954.

Watson, R. W.   The effect of cuticular hardening on the form of epidermal cells.   *New Phytol.* 41:223–229.   1942.

Wylie, R. B.   The dominant role of the epidermis in leaves of *Adiantum. Amer. Jour. Bot.* 35:465–473.   1948.

Ziegenspeck, H.   Vergleichende Untersuchungen der Entwicklung der Spaltöffnungen von Monokotyledonen und Dikotyledonen in Lichte der Polariskopie und Dichroskopie. *Protoplasma* 38:197–224.   1944.

8

Parenchyma

CONCEPT

The term *parenchyma* refers to a tissue composed of living cells variable in their morphology and physiology, but generally having thin walls and a polyhedral shape (pl. 25*A*), and concerned with vegetative activities of the plant. The individual cells of such a tissue are *parenchyma cells*. The word parenchyma is derived from the Greek *para*, beside, and *en-chein*, to pour, a combination of words that express the ancient concept of parenchyma as a semiliquid substance "poured beside" other tissues which are formed earlier and are more solid.

Parenchyma is often spoken of as the fundamental or ground tissue. It fits this definition from morphological as well as physiological aspects. In the plant body as a whole or in its organs parenchyma appears as a ground substance in which other tissues, notably the vascular, are imbedded. It is the foundation of the plant in the sense that the apical meristems and the reproductive cells are parenchymatic in nature. Furthermore, parenchyma cells are involved in phenomena of wound healing and regeneration. Phylogenetically, parenchyma is also the precursor of other tissues, as evidenced by the structure of the most primitive multicellular plants whose bodies consist of parenchyma only.

This tissue is the principal seat of such essential activities of the plant as photosynthesis, assimilation, respiration, storage, secretion, excretion— in short, activities depending on the presence of living protoplasts. Parenchyma cells which occur in the xylem and phloem tissues appear to play an important role in connection with the movement of water in the nonliving tracheary elements and with the transport of food in the sieve elements whose protoplasts lack nuclei.

181

Developmentally, parenchyma cells are relatively undifferentiated. They are unspecialized morphologically and physiologically, compared with such cells as sieve elements, tracheids, or fibers, since, in contrast to these three examples of cell categories, parenchyma cells may change functions or combine several different ones. However, parenchyma cells may also be distinctly specialized, for example, with reference to photosynthesis, storage of specific substances, or deposition of materials which are in excess in the plant body. Whether they are specialized or not, parenchyma cells are highly complex physiologically because they possess living protoplasts.

It was pointed out in chapter 4 that living cells are not fixed in their characteristics and that they possess, in varying degrees, the ability to resume meristematic activity. Parenchyma constitutes the principal category of tissue showing such developmental plasticity resulting from its relatively low level of differentiation. Apparently the potentiality to divide may be retained by parenchyma cells for many years, as evidenced by callus development from pith cells (medullary sheath, chapter 15) of an approximately 50 year-old stem of *Tilia* (Barker, 1953). This development, however, was possible only after removal of the pith tissue, by tissue-culture techniques, from the correlative inhibitions to which the cells are subjected in the plant.

DELIMITATION

Parenchyma cells may occur in extensive continuous masses as parenchyma tissue. They may also be associated with other types of cells in morphologically heterogeneous tissues. The pith and the cortex of stems and roots, the photosynthetic tissue, or mesophyll, of leaves, the flesh of succulent fruits, the endosperm of seeds are all examples of plant parts consisting largely or entirely of parenchyma. As components of heterogeneous tissues parenchyma cells form the vascular rays and the vertical files of living cells in the xylem and the phloem (chapters 11, 12). Sometimes an essentially parenchymatic tissue contains parenchymatic or nonparenchymatic cells or groups of cells, morphologically or physiologically distinct from the main mass of cells in the tissue. Sclereids, for example, may be found in the leaf mesophyll and in the pith and cortical parenchyma (chapter 10). Laticifers occur in various parenchymatic regions of plants containing latex (chapter 13). Sieve tubes traverse the cortical parenchyma of certain plants (chapter 12).

The variable structure of parenchyma tissue and the distribution of parenchyma cells in the plant body clearly illustrate the problems involved in the proper definition and classification of a tissue. On the one hand, parenchyma may fit the most restricted definition of a tissue as a group of cells having a common origin, essentially the same structure, and the same function. On the other hand, the homogeneity of a parenchyma tissue may be broken by the presence of varying numbers of nonparenchymatic cells, or parenchyma cells may occur as one of many cell categories in a heterogeneous tissue.

Thus the spatial delimitation of the parenchyma as a tissue is not precise in the plant body. Furthermore, parenchyma cells may intergrade with cells that are distinctly nonparenchymatic. Parenchyma cells may be more or less elongated and have thick walls, a combination of characters suggesting specialization with regard to support. A certain category of parenchyma cells is so distinctly differentiated as a supporting tissue that it is designated by the special name of collenchyma (chapter 9). Parenchyma cells may develop relatively thick lignified walls and assume some of the characteristics of sclerenchyma cells (chapter 10). Tannin may be found in ordinary parenchyma cells and also in cells basically parenchymatic but of such distinct form (vesicles, sacs, or tubes) that they are designated as idioblasts. Similarly, certain secretory cells differ from other parenchyma cells mainly in their function; others are so much modified that they are commonly treated as a special category of elements (laticifers; chapter 13).

The present chapter is restricted to a consideration of parenchyma concerned with the most ordinary vegetative activities of the plant, excluding the meristematic. The parenchyma cells of the xylem and the phloem are described in chapters dealing with these two tissues, and the general characteristics of the protoplasts of parenchyma cells are discussed in chapter 2.

STRUCTURE

Cell Contents

The variation in contents of parenchyma cells is intimately related to the activities of these cells (De Bary, 1884; Haberlandt, 1914; Meyer, 1962; Netolitzky, 1935; Sperlich, 1939). The cells of the photosynthetic parenchyma have variable numbers of chloroplasts. At certain times during the day chloroplasts may contain assimilation starch. Because of its high content in chlorophyll the photosynthetic parenchyma is sometimes called *chlorenchyma*. The most distinctly specialized chlorenchyma is represented by the leaf mesophyll (pl. 72), but chloroplasts occur also in the cortex (pl. 23A) and sometimes more deeply in the stem, even in the pith. Cells not concerned with photosynthesis have no chloroplasts or have chloroplasts with a weakly differentiated internal lamellar system (chapter 2). Cells devoid of chloroplasts may have leucoplasts. Actively synthesizing cells are commonly conspicuously vacuolated.

Many different food substances are synthesized and stored by parenchyma cells. The same protoplast may store one or more kinds of substances. These substances may be dissolved in the vacuolar sap, or they may be discrete solid or fluid bodies in the cytoplasm (fig. 8.1C). Bodies or masses of material may consist of such ergastic substances as starch grains, granules and crystalloids of protein, and globules of fats and oils. The cell sap may be a depository of sugars and other soluble carbohydrates and of nitrogenous substances in the form of amides and proteins. Some examples of plant organs and their storage products may be cited by reference to Netolitzky (1935). Amides, proteins, and sugar are dissolved in the cell sap of the fleshy beet root and the bulb scales of the onion. The parenchyma of the potato tuber and of the rhizomes of many other plants contains amides and proteins in the cell

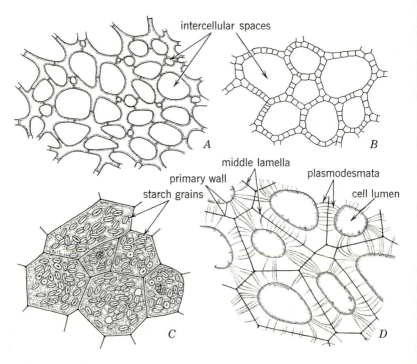

FIG. 8.1. Parenchyma tissue. A, aerenchyma with armed cells and prominent intercellular spaces from a lacuna in a leaf of *Canna*. B, aerenchyma from a transection of petiole of *Zantedeschia*. Endosperm parenchyma of *Secale* (rye), C, and *Diospyros* (persimmon), D. (A, ×90; B, ×24; C, ×180; D, ×620.)

ap and starch in the cytoplasm. Protein granules and starch grains occur in the cytoplasm of parenchyma cells in the cotyledons of the bean, the pea, and the lentil; protein granules and oil occur in the endosperm of *Ricinus* and the cotyledons of *Glycine* (soy bean). The most widely distributed storage product is starch. It occurs in parenchyma of the cortex and the pith; in parenchyma of the vascular tissues, that is, xylem and phloem parenchyma and ray parenchyma; and in parenchyma of fleshy leaves (bulb scales), rhizomes, tubers, fruits, cotyledons, and the endosperm of seeds (fig. 8.1C). In the leaves, starch predominates as the storage carbohydrate in the dicotyledons, sugars in the monocotyledons (Wanner, 1958).

The physiological activity of protoplasts varies in different kinds of storage parenchyma. In stems and roots of woody species starch accumulation undergoes seasonal fluctuations. It is deposited at one time and is removed at another. Such periodic changes indicate that the storage cells have active protoplasts. The specialized storage organs, such as tubers, bulbs, and rhizomes, may serve for storage only once, their protoplasts dying after the removal of the reserves to the growing organs. During the development of storage tissues the cells can divide in the presence of starch (Bradbury, 1953).

In seeds the living protoplasts are directly concerned with the accumulation of storage products, but their relation to the subsequent mobilization of the storage material is not always clear. Storage cotyledons that, during the growth of the seedling, emerge above the surface of the ground (epigeous germination) and become green evidently have active protoplasts capable of taking part in photosynthesis after the removal of storage products. In contrast, cotyledons that remain below the surface of the ground during germination (hypogeous germination) usually die after they release the food reserves to the growing parts of the seedling. In both types of cotyledons the storage cells themselves probably control the mobilization of the stored food. There is some evidence, however, that the epidermis of cotyledons may be the seat of production of enzymes that digest the food materials (Netolitzky, 1935). The protoplasts of the endosperm of some seeds are reported as being actively engaged in the process of dissolution of starch and other reserves. In other seeds the endosperm protoplasts die after the accumulation of the storage material is completed (Müller, 1943). In such seeds the digestion of the food reserves is initiated and regulated by the enzymic activity of the embryo, alone or in conjunction with parts of the endosperm. In the Gramineae, for example, the digestion of starch is brought about by the scutellum of the embryo and by the outermost layer of the endosperm, the aleuron layer (chapter 20).

Water is abundant in all active vacuolated parenchyma cells so that the parenchyma plays a major role as a water reservoir. In a study of species of bamboo, the variations in moisture content of the different parts of culms were found to be clearly associated with the proportions of parenchyma cells in the tissue system (Liese and Grover, 1961).

Parenchyma may be rather specialized as a water-storage tissue. Many succulent plants, such as the Cactaceae, *Aloe, Agave,* and *Mesembryanthemum,* contain in their photosynthetic organs chlorophyll-free parenchyma cells full of water. This water tissue consists of living cells of particularly large size and usually with thin walls. The cells are often in rows and may be elongated like palisade cells. Each has a thin layer of parietal cytoplasm, a nucleus, and a large vacuole with watery or somewhat mucilaginous contents. The mucilages seem to increase the capacity of the cells to absorb and to retain water and may occur in the protoplasts and in the walls.

In the underground storage organs there is usually no separate water-storing tissue, but the cells containing starch and other food materials have a high water content. Potato tubers may start shoot growth in air and provide the growing parts with moisture for the initial growth (Netolitzky, 1935). A high water content is characteristic not only of underground storage organs, such as tubers and bulbs, but also of buds and of fleshy enlargements on aerial stems. In all these structures the storage of water is combined with that of the ergastic substances serving as food reserves.

Many parenchyma cells accumulate phenol derivatives, including tannins. Tanniniferous cells may form a connected system in the plant body, or they may occur singly or in groups. In leaves they are often distributed in continuous zones with no relation to the structural characteristics of the cells occurring in these zones. In stems there may be a concentric zonation of tannin cells. Frequently tanniniferous cells are conspicuous in the outermost zone of the pith, the so-called medullary sheath. Cells containing tannins may accompany the vascular bundles or be included within them. Tannins commonly accumulate in cells located near injuries or infections.

As visible deposits, tannins occur in the vacuoles. The metabolism of carbohydrates and tannins are interrelated, and, according to some studies, starch and tannins mutually exclude each other, except when both are produced in large amounts (Sperlich, 1939). Tannin-containing cells may be as readily stimulated to growth and division as cells free of tannins. Cells with tannins, for example, may divide in callus cultures (Ball, 1950), initiate phellogen, and produce tyloses—proliferations of parenchyma cells into lumina of vessels (pl. 37A–C)—or divide with the rest of the ground-parenchyma cells during stem elongation (Bloch, 1948).

Parenchyma cells also store mineral substances and form the different kinds of crystals described in chapter 2. Some cells giving rise to crystals retain their protoplasts; others die after the development of crystals.

Cell Walls

Chlorenchyma and many kinds of storage cells commonly have thin primary walls. Some storage parenchyma, however, develops remarkably thick walls (Bailey, 1938). The carbohydrates deposited in these walls, notably the hemicelluloses (chapter 20), are regarded by some workers as reserve materials (Netolitzky, 1935). Thick walls occur, for example, in the endosperm of date palm (*Phoenix dactylifera*), persimmon (*Diospyros;* fig. 8.1*D*), *Asparagus,* and *Coffea arabica.* They become thinner during germination. Although the thick-walled storage cells appear to have living protoplasts, the removal of the walls is not necessarily an independent activity of such protoplasts but may be regulated by the embryo (Netolitzky, 1935).

Relatively thick and often lignified secondary walls also occur in parenchyma cells, particularly in those of the secondary xylem.

Cell Arrangement

Mature parenchyma tissue either is closely packed or is permeated by a more or less prominent air-space system. The storage parenchyma of fleshy axial organs or fruits has abundant intercellular spaces. In contrast, the endosperm of most seeds contains none or only small intercellular spaces (fig. 8.1*C*). During germination, however, the cells are gradually separated from each other (Netolitzky, 1935). This structural peculiarity seems to support the notion, mentioned previously, that the mobilization of the reserves in the endosperm is stimulated and regulated, not by the activity of the storing cells themselves, but by the embryo and perhaps also by the peripheral layers of the endosperm.

Chlorenchyma is a familiar example of a tissue having a well-developed aerating system. This structural detail is particularly characteristic of the leaf mesophyll, where the proportion of air by volume may be between 77 and 713 parts in 1,000 (Sifton, 1945). Intercellular spaces are abundant in the photosynthetic parenchyma of stems too. In general, they characterize such parenchyma in all groups of land plants from the liverworts and mosses to the angiosperms. Parenchyma removed from light, as in the pith and roots, also has more or less prominent intercellular spaces. On the basis of studies on penetrability of plant organs by gases under pressure, the concept has been introduced that plants possess two kinds of intercellular-space systems, continuous and discontinuous (Redies, 1962).

The intercellular spaces of vascular plants commonly arise schizogenously or lysigenously (chapter 3). The schizogenous method may give rise to very large spaces, particularly if the cells divide with reference to these spaces (Hulbary, 1944). In stems and leaves of *Elodea* and other monocotyledons cells divide parallel to the longitudinal axis of the stem or petiole and perpendicular to the surface of the initial air spaces so that these spaces become bounded by an increasingly large number of cells (fig. 8.1B). Other large air spaces may arise lysigenously or *rhexigenously* (by mechanical rupture; from the Greek *rhexis*, a rending). For example, cortical cells break down in the roots of some Gramineae, Cyperaceae, and other families (chapter 17) leaving large lacunae arranged radially or tangentially (Sifton, 1945, 1957). Parenchyma tissue with large and abundant intercellular spaces is termed *aerenchyma*.

Air spaces reach a particularly high development in the aquatic angiosperms, both in individual size and in combined volume (Sifton, 1945, 1957). In these plants, the aerenchyma forms an elaborate system that appears to be continuous from the leaf to the root. The significance of the development of aerenchyma in aquatic plants has been much discussed in the literature. The continuity of the system through the plant suggests a provision for aeration. The air also gives buoyancy to the plant. But these roles may be incidental to those that are determined by the primary requirement met with in an aquatic environment: a structure that for a given diameter provides strength with the least possible amount of tissue (Williams and Barber, 1961). A honeycomb structure answers this double requirement.

Cell Shape

Parenchyma cells are commonly described as having a polyhedral shape, with the various diameters differing relatively little from each other (pl. 25A, B), but they vary considerably even in the same plant (Mia, 1962). Many kinds of parenchyma cells are more or less elongated and merge imperceptibly with the so-called prosenchyma cells (elongated cells with tapering ends) in shape and dimensions. Furthermore, parenchyma cells in the mesophyll and other plant parts may be variously lobed, folded, and armed (fig. 8.1A; pl. 79; chapter 16; Geesteranus, 1941).

Parenchyma cells have served as an object for intensive studies on cell shape, involving special techniques for isolating cells, constructing models of cells, and subjecting the models to statistical analyses (Marvin, 1939; Matzke, 1946; Matzke and Duffy, 1955, 1956). These studies show that, in general, parenchyma cells in relatively homogeneous complexes, with small intercellular spaces or none, have the shape of polyhedra with an approximate average of 14 faces. A geometrically perfect, 14-sided polyhedron with 8 hexagonal and 6 quadrilateral faces has been named the orthic tetrakaidecahedron. This ideal figure is extremely rare among plant cells but is approximated more commonly and more closely than the 12-sided figure with 12 rhombic faces (the rhombic dodecahedron) which the early botanists considered to be the fundamental form of cells in undifferentiated parenchyma. From the beginnings of botany, cells in tissues were regarded as assuming shapes giving the greatest economy of space—minimum surface with maximum volume—and consequently they were interpreted as potential spheres that had a polyhedral shape because of mutual contact and pressure. The rhombic dodecahedron was at first assumed to fit this interpretation best; later, it was found that the orthic tetrakaidecahedron satisfies more conditions in liquid films and also has a greater economy in the surface-to-volume relationship. The rare occurrence of the ideal tetrakaide-cahedron among plant cells is understandable. Even in the most homogeneous tissues cells are not of equal volumes, and are not regularly spaced.

The approximation to a 14-sided figure was observed, for example, in parenchyma of various vegetative parts in monocotyledons and dicotyledons, of the carpel vesicles in citrus, and of petioles of a fern (Matzke and Duffy, 1955). Presence of intercellular spaces, particularly of large ones, reduces the number of contacts (Hulbary, 1944). If a tissue contains large and small cells, the number of faces is correlated with size. The small cells have fewer than 14 facets, the large cells more than 14. In *Elodea* cells, the number of faces rises to almost 17 during preparation for cell division, but each new cell has at first less than 13 faces (Matzke and Duffy, 1956).

By studies of nonliving systems attempts were made to determine some of the possible factors that influence cell shape. In one system —lead shot placed in metal cylinders and subjected to pressure—pressure was the main factor determining the shape (Marvin, 1939; Matzke, 1939). In the other—soap foam bubbles placed in a container and left to adjust themselves—the surface tension played the major role (Matzke, 1946; Matzke and Nestler, 1946). Plant cells occupied an intermediate position between lead shot and soap bubbles in the characteristics of the three-dimensional configurations. These observations suggest that pressure and surface tension both might play a part in shaping the cells. But there must be other factors as well.

The forces operating in the growth of folded cells (arm-palisade cells) or of cells with internal ridges, as in *Pinus* mesophyll (pl. 79; Küster, 1956; Meyer, 1962), are obscure. In the ontogeny of stellately armed parenchyma cells (fig. 8.1*A*) lateral tension appears to be one of the factors determining the final shape (Geesteranus, 1941). Ultrastructural studies of growing stellate cells of *Juncus* indicate that the arms are elongating through their entire length and that the growth of the cell wall is of the multinet type (Houwink and Roelofsen, 1954). Certain developmental phenomena, such as increase in length of cells and division of cells, violate the principle of least surfaces (Matzke and Nestler, 1946); and in cell division the usual position of a new wall indicates no relation to surface tension phenomena (Sinnott and Bloch, 1941).

ORIGIN

The parenchyma tissue of the primary plant body, that is, the parenchyma of the cortex and the pith, of the mesophyll of leaves, and of the flower parts, differentiates from the ground meristem. The parenchyma associated with the primary and secondary vascular tissues is formed by the procambium and the vascular cambium, respectively. Parenchyma may also arise from the phellogen in the form of phelloderm, and it may be increased in amount by diffuse secondary growth.

REFERENCES

Bailey, I. W.   Cell wall structure of higher plants.   *Indus. and Engin. Chem.* 30:40–47. 1938.

Ball, E.   Differentiation in a callus culture of *Sequoia sempervirens*.   *Growth* 14:295–325. 1950.

Barker, W. G.   Proliferative capacity of the medullary sheath region in the stem of *Tilia americana*.   *Amer. Jour. Bot.* 40:773–778.   1953.

Bloch, R.   The development of the secretory cells of *Ricinus* and the problem of cellular differentiation.   *Growth* 12:271–284.   1948.

Bradbury, D.   Division of starch-containing cells.   *Amer. Jour. Bot.* 40:286–288.   1953.

De Bary, A.   *Comparative anatomy of the vegetative organs of the phanerogams and ferns.* Oxford, Clarendon Press.   1884.

Geesteranus, R. A. M.   On the development of the stellate form of the pith cells of *Juncus* species.   *Nederl. Akad. van Wetenschap. Proc.* 44:489–501, 648–653.   1941.

Haberlandt, G.   *Physiological plant anatomy.*   London, Macmillan and Company.   1914.

Houwink, A. L., and P. A. Roelofsen.   Fibrillar architecture of growing plant cell walls. *Acta Bot. Neerland.* 3:385–395.   1954.

Hulbary, R. L.   The influence of air spaces on the three-dimensional shapes of cells in *Elodea* stems, and a comparison with pith cells of *Ailanthus*.   *Amer. Jour. Bot.* 31:561–580. 1944.

Küster, E.   *Die Pflanzenzelle.*   Jena, Gustav Fischer.   3rd ed.   1956.

Liese, W., and P. N. Grover.   Untersuchungen über den Wassergehalt von indischen Bambushalmen.   *Deut. Bot. Gesell. Ber.* 74:105–117.   1961.

References

Marvin, J. W. The shape of compressed lead shot and its relation to cell shape. *Amer. Jour. Bot.* 26:280–288. 1939.

Matzke, E. B. Volume-shape relationships in lead shot and their bearing on cell shapes. *Amer. Jour. Bot.* 26:288–295. 1939.

Matzke, E. B. The three-dimensional shape of bubbles of foam—an analysis of the role of surface forces in three-dimensional cell shape determination. *Amer. Jour. Bot.* 33:58–80. 1946.

Matzke, E. B., and R. M. Duffy. The three-dimensional shape of interphase cells within the apical meristem of *Anacharis densa*. *Amer. Jour. Bot.* 42:937–945. 1955.

Matzke, E. B., and R. M. Duffy. Progressive three-dimensional shape changes of dividing cells within the apical meristem of *Anacharis densa*. *Amer. Jour. Bot.* 43:205–225. 1956.

Matzke, E. B., and J. Nestler. Volume-shape relationships in variant foams. A further study of the role of surface forces in three-dimensional cell shape determination. *Amer Jour. Bot.* 33:130–144. 1946.

Meyer, F. J. Das trophische Parenchym. A. Assimilationsgewebe. In: *Handbuch der Pflanzenanatomie*. Band 4. Teil 7A. 1962.

Mia, A. J. Polymorphic parenchymatous cells of *Rauwolfia vomitora* Afzl. *Texas Jour. Sci.* 14:305–318. 1962.

Müller, D. Tote Speichergewebe in lebenden Samen. *Planta* 33:721–727. 1943.

Netolitzky, F. Das trophische Parenchym. C. Speichergewebe. In: K. Linsbauer. *Handbuch der Pflanzenanatomie*. Band 4. Lief. 31. 1935.

Redies, H. Über "homobare" und "heterobare" Interzellularensysteme in höheren Pflanzen. *Beitr. z. Biol. der Pflanz.* 37:411–445. 1962.

Sifton, H. B. Air-space tissue in plants. *Bot. Rev.* 11:108–143. 1945. II. *Bot. Rev.* 23:303–312. 1957.

Sinnott, E. W., and R. Bloch. The relative position of cell walls in developing plant tissues. *Amer. Jour. Bot.* 28:607–617. 1941.

Sperlich, A. Das trophische Parenchym. B. Exkretionsgewebe. In: K. Linsbauer. *Handbuch der Pflanzenanatomie*. Band 4. Lief. 38. 1939.

Wanner, H. Die Speicherung von Kohlenhydraten im Blatt. *Handb. der Pflanzenphysiol.* 6:841–854. 1958.

Williams, W. T., and D. A. Barber. The functional significance of aerenchyma in plants. *Soc. Expt. Biol. Symp.* 15:132–144. 1961.

9

Collenchyma

Collenchyma is a living tissue composed of more or less elongated cells with thick primary nonlignified walls. The structure and the arrangement of collenchyma cells in the plant body indicate that the primary function of the tissue is support. Morphologically, collenchyma is a simple tissue, for it consists of one type of cell.

The presence of living protoplasts denotes a close physiologic similarity between collenchyma and parenchyma cells. In form and structure the two types of cells also intergrade. Collenchyma cells are commonly longer and narrower than parenchyma cells, but some collenchyma cells are short and, on the other hand, some parenchyma cells are considerably elongated. Where collenchyma and parenchyma lie next to each other, they frequently intergrade through transitional types of cells. The resemblance to parenchyma is further stressed by the common occurrence of chloroplasts in collenchyma and by the ability of this tissue to undergo reversible changes in wall thickness and to engage in meristematic activity. In view of these similarities and in view of the structural and functional variability of parenchyma (chapter 8), collenchyma is commonly interpreted as a thick-walled kind of parenchyma structurally specialized as a supporting tissue. The terms parenchyma and collenchyma are also related, but in the latter the first part of the word, derived from the Greek word *colla*, glue, refers to the thick glistening wall characteristic of collenchyma.

POSITION IN PLANT

Collenchyma is the typical supporting tissue, first, of growing organs and, second, of those mature herbaceous organs that are only slightly

192

modified by secondary growth or lack such growth completely. It is the first supporting tissue in stems, leaves, and floral parts, and it is the main supporting tissue in many mature, dicotyledonous leaves and some green stems. Collenchyma may occur in root cortex (Guttenberg, 1940), particularly if the root is exposed to light (Van Fleet, 1950). It is absent in stems and leaves of many of the monocotyledons that early develop sclerenchyma (Falkenberg, 1876; Giltay, 1882).

Collenchyma characteristically occurs in peripheral position in stems and leaves (fig. 9.1). It may be present immediately beneath the epidermis, or it may be separated from the epidermis by one or more layers of parenchyma. If it is located next to the epidermis, the inner tangential walls of the epidermis may be thickened like the walls of collenchyma. Sometimes the entire epidermal cells are collenchymatous. In its subepidermal position, collenchyma occurs in the form of continuous or somewhat discontinuous cylinders (fig. 9.1A, C) or in discrete strands (fig. 9.1D–F). In stems and petioles with protruding ribs, collenchyma is particularly well developed in the ribs. In leaves it may differentiate on one or both sides of the veins (fig. 9.1B) and along the margins of the leaf blade.

In many plants the elongated parenchyma cells of the outermost part of the phloem develop thick walls after the sieve elements are obliterated, and the tissue ceases to be concerned with conduction. The resulting structure is commonly called the bundle cap. The parenchyma on the inner periphery of the xylem may be similarly differentiated. If the entire bundle is enclosed by thick-walled elongated cells, it is said to have a bundle sheath. The bundle caps and the bundle sheaths sometimes have thickened primary walls, sometimes lignified secondary walls. The tissues forming these caps and sheaths are often interpreted as collenchyma when they have nonlignified primary walls (Duchaigne, 1955), and as sclerenchyma when they have secondary walls. The comparative characteristics of subepidermal collenchyma, on the one hand, and of nonlignified caps and sheaths, on the other, are imperfectly known. In a study of celery grown in a boron-deficient medium the collenchyma walls were found to be thinner than normal, while the walls of parenchyma cells of the phloem forming the bundle caps and of the ground parenchyma were thicker than normal (Spurr, 1957). In a comparison of the strength of collenchyma and of bundle-cap tissue from the same petioles of celery, the collenchyma strands proved to be much stronger (Esau, 1936). In this book collenchyma refers only to the supporting tissue in peripheral positions in the plant. If bundle caps and sheaths resemble collenchyma, they are referred to as *collenchymatous*, an adjective implying similarity to collenchyma, but not necessarily morphologic identity.

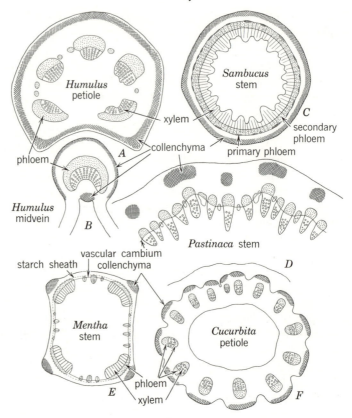

FIG. 9.1. Distribution of collenchyma (crosshatched) and vascular tissues in various plant parts. Transections. (*A, B,* ×19; *C–F,* ×9.5.)

STRUCTURE

Cell Shape

Collenchyma cells may vary in length, but typically they are considerably elongated—cells 2 mm in length have been recorded—and resemble fibers in having tapering ends (Haberlandt, 1914; Majumdar, 1941). The shorter collenchyma cells are prismatic like many parenchyma cells. Both kinds of cells are polygonal in transection. Collenchyma cells may vary in shape and size in the same strand. These variations are related to the origin of the cells. A collenchyma strand is formed by a series of longitudinal divisions, which spread from a central point toward the periphery of the future strand. The longitudinal divisions are followed by elongation of the resulting cells, so that the first, that is, the innermost, cells start to elongate earlier than the more peripheral ones and attain a greater length.

The development of collenchyma was studied in considerable detail in the umbellifer *Heracleum* (Majumdar, 1941; Majumdar and Preston, 1941). In this plant the elongation of a collenchyma cell either follows immediately the last longitudinal division of a mother cell or is preceded by one or rarely more transverse divisions. In macerated preparations the products of the latest transverse divisions often remain together, enclosed by the common mother-cell wall. Such cell complexes resemble septate fibers (chapter 10). When transverse divisions occur before elongation, the shape of the cells is affected. The ends formed by transverse divisions may be slightly oblique or almost transverse. Without such divisions the cells taper at both ends. Peripheral cells of a collenchyma bundle are short, and their end walls taper little.

Cell Wall

The structure of the cell wall is the most distinctive feature of collenchyma cells. The thickenings are deposited unevenly in a manner somewhat variable in different groups of plants. As seen in transections, a common form of collenchyma shows the main deposits of wall material in the corners where several cells are joined together (*Ficus, Vitis, Ampelopsis, Polygonum, Beta, Rumex, Boehmeria, Morus, Cannabis, Begonia, Pellionia;* fig. 9.2*B*; pl. 25*B*). The degree of restriction of wall thickenings to the angles varies in relation to the amount of wall thickening present on other wall parts. If the general wall thickening becomes massive, the thickening in the corners is obscured and the lumen of the cell assumes, in transections, a circular outline, instead of an angular one. Such developmental modification is observed, for example, in the Umbelliferae (Esau, 1936; Majumdar, 1941). In another form of collenchyma the thickening occurs chiefly on the tangential walls (*Sambucus, Sanguisorba, Rheum, Eupatorium;* fig. 9.2*A*). Still another form is characterized by intercellular spaces and the development of the collenchymatic thickenings on the walls facing the intercellular spaces (Compositae, *Salvia, Brunella, Malva, Althaea;* fig. 9.2*C*). The three forms of collenchyma have been named by Müller (1890) angular (Eckencollenchym), lamellar (Plattencollenchym), and lacunar (Lückencollenchym), respectively. The word lamellar has reference to the plate-like arrangement of the thickenings; the lacunar, to the presence of intercellular spaces. In collenchymatous bundle caps and sheaths the cell-wall thickening is sometimes most emphasized in the corners. More commonly, however, the thickening is either relatively even over the entire wall, or is uneven without being restricted to the corners or the tangential walls.

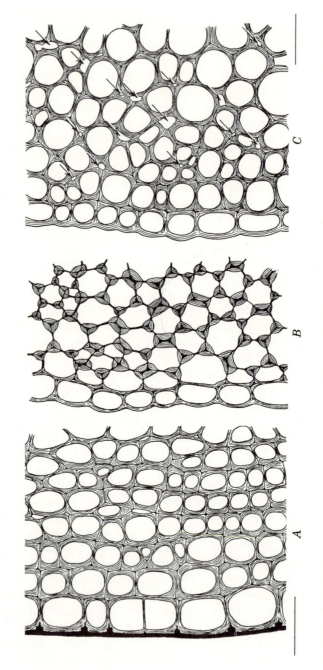

FIG. 9.2. Collenchyma in stem transections. In all drawings the epidermal layer is to the left. A, *Sambucus*; thickenings mainly on tangential walls (lamellar collenchyma). B, *Cucurbita*; thickenings in the angles (angular collenchyma). C, *Lactuca*; numerous intercellular spaces (indicated by arrows) and the most prominent thickenings located next to these spaces (lacunar collenchyma). Thick cuticle (shown in black) in A. (All, ×320.)

In longitudinal sections collenchyma shows thin and thick wall portions, depending on the direction of the cut with reference to the thickenings (pl. 25C). The nearly transverse end walls are usually thin, whereas the pointed ends may appear solid because of accumulation of wall material (Majumdar, 1941). Primary pit-fields occur in collenchyma cells, in both the thinner and the thicker parts of the walls.

The walls of collenchyma consist mainly of cellulose and pectic compounds and contain much water (Anderson, 1927; Cohn, 1892; Majumdar and Preston, 1941). In some species collenchyma walls appear to have an alternation of layers rich in cellulose and poor in pectic compounds with layers that are rich in pectic compounds and poor in cellulose (Czaja, 1961). Ultrastructurally, the thickenings of collenchyma in celery petioles show an alternation of layers of longitudinally oriented microfibrils and noncellulosic material (Beer and Setterfield, 1958). According to a study with polarization optics, cellulose forms transverse as well as longitudinal lamellations (Czaja, 1961).

Collenchyma walls may contain over 60 per cent water, based on fresh weight; over 200 per cent, based on dry weight (Cohn, 1892). Heating destroys the ability of walls to absorb water. When collenchyma walls lose their water under the influence of dehydrating agents they shrink visibly. This shrinkage is greatest in a radial direction and smallest in a vertical direction.

The characteristic thickening of collenchyma walls begins to develop before the extension of the cell is completed. Apparently the successive layers are formed around the entire cell, but the individual layers are thicker where the wall is most massive in final state (Majumdar and Preston, 1941). In the electron microscope a merging of layers of microfibrils in the thinner wall parts was recognized (Beer and Setterfield, 1958).

As was previously mentioned, collenchyma may or may not have intercellular spaces. In the absence of spaces the corners where several cells meet frequently show prominent accumulations of pectic substances. These accumulations may not fill the space completely but protrude into it in the form of warts or coralloid structures (Carlquist, 1956; Duchaigne, 1955). Similar formations may occur in parenchyma tissue (Kisser, 1928).

The wall thickness in collenchyma is increased if during development the plants are exposed to motion by wind (Walker, 1960). Apparently inhibition of cell elongation occurs at the same time. The wall thickenings of collenchyma are sometimes removed, as, for instance, when a phellogen arises in this tissue or when collenchyma cells respond to injuries with wound-healing reactions. Loss of wall material in collenchyma was also induced experimentally by etiolation (Walker, 1960).

The occurrence of simultaneous increase in thickness and surface area of collenchyma walls, that is, the increase of wall thickening during the elongation of cells is a noteworthy phenomenon. Because of this development the term thickened primary wall has been applied to the collenchyma wall (Majumdar and Preston, 1941). Ultrastructurally, too, the collenchyma has been interpreted as primary (Beer and Setterfield, 1958).

Collenchyma walls may become modified in older plant parts. In woody species with secondary growth, collenchyma follows, at least for a time, the increase in circumference of the axis by active growth with retention of the original characteristics. In some plants (*Tilia, Acer, Aesculus*) collenchyma cells enlarge and their walls become thinner (De Bary, 1884). Apparently it is not known whether this reduction in thickness results from a removal of wall material or from stretching and dehydration. Collenchyma may develop lignified secondary walls. It thus becomes changed into sclerenchyma (Duchaingne, 1955; Funk, 1912; Went, 1924).

Cell Contents

As was stated previously, collenchyma cells have living protoplasts at maturity. Chloroplasts occur in variable numbers. They are most numerous in collenchyma which approaches parenchyma in form. Collenchyma consisting of long, narrow cells—the most highly specialized type—contains only a few small chloroplasts or none. Tannins may be present in collenchyma cells.

STRUCTURE IN RELATION TO FUNCTION

Collenchyma appears to be a mechanical tissue particularly adapted for support of growing organs. Thick walls and close packing make it a strong tissue. At the same time, the peculiarities of growth and the structure of its walls permit adjustments to elongation of the organ without loss of strength. As has been previously stressed, collenchyma cells are capable of increasing simultaneously the surface and the thickness of the walls and, therefore, can develop thick walls while the organ is still elongating.

Collenchyma tissue combines considerable tensile strength with flexibility and plasticity. Measurements of the strength of collenchyma have been made by determining the weight necessary to break a strand of the tissue dissected out from the organ (Ambronn, 1881; Curtis,

1938; Esau, 1936). The values thus obtained were in some instances recalculated per unit area of strand and used to express the tensile strength of the tissue. Such values obviously give a measure, not of the tensile strength of the wall proper, but of the tissue as a whole. Nevertheless, they provide useful information about the strength of a tissue, since in the plant body the mechanical effect of a tissue is determined not only by the nature of the walls but also by the shape and arrangement of the cells.

A comparison of collenchyma with fibers is particularly interesting. Collenchyma has been found capable of supporting 10 to 12 kg per mm^2, fiber strands, 15 to 20 kg per mm^2 (Ambronn, 1881). The fiber strands regain their original length even after they have been subjected to a tension of 15 to 20 kg per mm^2, whereas collenchyma remains permanently extended after it has been made to support 1½ to 2 kg per mm^2. In other words, fibers are elastic, and collenchyma is plastic. If fibers were to differentiate in growing organs, they would hinder tissue elongation because of their tendency to regain their original length when stretched. Collenchyma, on the other hand, could be expected to respond with a plastic change in length under the same conditions.

The importance of the plasticity of collenchyma walls for the internal adjustment of growing tissues is emphasized by the observation that much of the elongation of the internodes occurs after the collenchyma cells have thickened their walls. In one particular study on *Heracleum* plants (Majumdar, 1941; Majumdar and Preston, 1941) thick-walled collenchyma cells were found in young internodes that were several times shorter than the extended internodes of the same axes. The individual collenchyma cells in the young internodes were markedly shorter than those in the extended ones.

The plasticity of collenchyma changes with age. Old tissue is harder and more brittle than young (Curtis, 1938). As mentioned previously, in some plants collenchyma may finally become sclerified. Hardened collenchyma occurs in plant parts that have ceased to elongate.

ORIGIN

Collenchyma has been variously reported as originating jointly with the vascular tissues from the procambium (Ambronn, 1881; Haberlandt, 1914; Majumdar, 1941) and separately from these tissues in the ground meristem (Ambronn, 1881; Esau, 1936; Haberlandt, 1914; Wisselingh, 1882). This disagreement results from a difference in the interpretation of histogenetic events. Although it is proper to speak of a differentiation of the derivatives of the apical meristems into protoderm, procambium, and ground meristem, these meristems usually become delimited from each other gradually, particularly in shoots. The protoderm may be distinguished close to the initial region and may even have its own initials (chapter 5), but the procambium of stems and leaves is built up by longitudinal divisions involving an increasingly larger number of cells in the meristem that also gives rise to ground tissues (chapter 15). Thus, at first it is impossible to designate a certain part of the meristem as giving rise to the procambium, another to the ground meristem. It may be said, therefore, that the cortical collenchyma and the procambium originate in a common meristem. The final delimitation of the pro-cambium occurs later in some plants than in others, and consequently the ontogenetic relationship between the cortex and the procambium appears close in some plants (Umbelliferae, Piperaceae, Araceae) and distant in others (Labiatae, *Clematis, Aristolochia,* certain Cucurbitaceae, *Chenopodium,* Compositae; Ambronn, 1881).

The development of collenchyma in the Umbelliferae clearly illustrates the lack of separation between cortex and procambium in the early stages of their development (Esau, 1936). In mature petioles of celery, collenchyma strands appear near the periphery in the ribs, separated by cortical parenchyma from the vascular bundles (fig. 9.3). In early ontogeny longitudinal divisions occur in the peripheral part of the petiole. Some of these initiate the procambium, others form the cortex. Subsequently, the procambium becomes distinct from the cortex because its cells are smaller in transverse diameters and greater in length. A secretory duct differentiates outside the procambium. After the procambium appears the cells between the duct and the protoderm

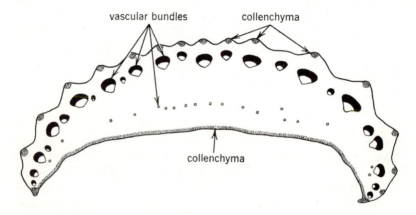

FIG. 9.3. Transection of celery petiole showing distribution of collenchyma and vascular bundles. Collenchyma occurs in strands in the ribs on the abaxial side and as a continuous layer on the adaxial side. (×16.)

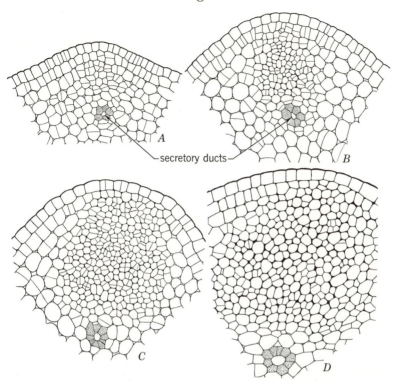

secretory ducts

FIG. 9.4. Development of collenchyma. Transections of celery petioles in different stages of development. *A*, longitudinal divisions initiated between secretory duct and epidermis. *B, C,* further divisions and appearance of thickenings in angles, probably resulting from accumulation of intercellular material. *D*, divisions completed and primary wall thickenings developing. (×302; Esau, *Hilgardia* 10, 1936.)

—ground meristem cells—undergo a series of divisions giving rise to the collenchyma (fig. 9.4).

Collenchyma that differentiates early in a given organ becomes highly specialized in its morphology, whereas that formed later is more parenchymatous. This difference is also reflected in the nature of the meristem giving rise to the different kinds of collenchyma. The more specialized collenchyma has its origin in a procambium-like meristem, the less specialized in a parenchymatic ground meristem. Since Haberlandt (1914) extended the concept of procambium to include meristems giving rise to all elongated cells of the primary body, he called the collenchyma meristem having elongated cells procambium, a usage not followed in this book.

REFERENCES

Ambronn, H. Über die Entwickelungsgeschichte und die mechanischen Eigenschaften des Collenchyms. Ein Beitrag zur Kenntnis des mechanischen Gewebesystems. *Jahrb. f. Wiss. Bot.* 12:473–541. 1881.

Anderson, D. Über die Struktur der Kollenchymzellwand auf Grund mikrochemischer Untersuchungen. *Akad. der. Wiss. Wien, Math.-Nat. Kl.* 136:429–440. 1927.

Beer, M., and G. Setterfield. Fine structure in thickened primary walls of collenchyma cells of celery petioles. *Amer. Jour. Bot.* 45:571–580. 1958.

Carlquist, S. On the occurrence of intercellular pectic warts in Compositae. *Amer. Jour. Bot.* 43:425–429. 1956.

Cohn, J. Beiträge zur Physiologie des Collenchyms. *Jahrb. f. Wiss. Bot.* 24:145–172. 1892.

Curtis, D. S. Determination of stringiness in celery. *Cornell Univ. Agric. Expt. Sta. Mem.* 212. 1938.

Czaja, A. T. Neue Untersuchungen über die Struktur der partiellen Wandverdickungen von faserförmigen Kollenchymzellen. *Planta* 56:109–124. 1961.

De Bary, A. *Comparative anatomy of the vegetative organs of the phanerogams and ferns.* Oxford, Clarendon Press. 1884.

Duchaigne, A. Les divers types de collenchymes chez les Dicotylédones: leur ontogénie et leur lignification. *Ann. des Sci. Nat., Bot.* Ser. 11. 16:455–479. 1955.

Esau, K. Ontogeny and structure of collenchyma and of vascular tissues in celery petioles. *Hilgardia* 10:431–476. 1936.

Falkenberg, P. *Vergleichende Untersuchungen über den Bau der Vegetationsorgane der Monokotyledonen.* Stuttgart, Ferdinand Enke. 1876.

Funk, G. Beiträge zur Kenntnis der mechanischen Gewebesysteme in Stengel und Blatt der Umbelliferen. *Bot. Centbl. Beihefte.* 29:219–297. 1912.

Giltay, E. Sur le collenchyme. *Arch. Néerland. des Sci. Exact. et Nat.* 17:432–459. 1882.

Guttenberg, H. von. Der primäre Bau der Angiospermenwurzel. In: K. Linsbauer. *Handbuch der Pflanzenanatomie.* Band 8. Lief. 39. 1940.

Haberlandt, G. *Physiological plant anatomy.* London, Macmillan and Company. 1914.

Kisser, J. Untersuchungen über das Vorkommen und die Verbreitung von Pektinwarzen. *Jahrb. f. Wiss. Bot.* 68:206–232. 1928.

Majumdar, G. P. The collenchyma of *Heracleum Sphondylium* L. *Leeds Phil. Lit. Soc. Proc.* 4:25–41. 1941.

Majumdar, G. P., and R. D. Preston. The fine structure of collenchyma cells in *Heracleum Sphondylium* L. *Roy. Soc. London, Proc.* Ser. B. 130:201–217. 1941.

Müller, C. Ein Beitrag zur Kenntnis der Formen des Collenchyms. *Deut. Bot. Gesell. Ber.* 8:150–166. 1890.

Spurr, A. R. The effect of boron on cell wall structure in celery. *Amer. Jour. Bot.* 44:637–650. 1957.

Van Fleet, D. S. A comparison of histochemical and anatomical characteristics of the hypodermis with the endodermis in vascular plants. *Amer. Jour. Bot.* 37:721–725. 1950.

Walker, W. S. The effect of mechanical stimulation and etiolation on the collenchyma of *Datura stramonium*. *Amer. Jour. Bot.* 47:717–724. 1960.

Went, F. A. F. C. Sur la transformation du collenchyme en sclérenchyme chez les Podostémonacées. *Rec. des. Trav. Bot. Néerland.* 21:513–526. 1924.

Wisselingh, C. van. Contribution à la connaissance du collenchyme. *Arch. Néerland. des Sci. Exact. et Nat.* 17:23–58. 1882.

10

Sclerenchyma

The term *sclerenchyma* refers to complexes of thick-walled cells, often lignified, whose principal function is mechanical. These cells are supposed to enable plant organs to withstand various strains, such as may result from stretching, bending, weight, and pressure, without undue damage to the thin-walled softer cells. The word is derived from the Greek and is a combination of *sclerous*, hard, and *enchyma*, an infusion (chapter 8); it emphasizes the hardness of sclerenchyma walls. The individual cells of sclerenchyma are termed *sclerenchyma cells;* collectively, sclerenchyma cells form the sclerenchyma tissue. In terms of the entire mechanical system of a plant, collenchyma and sclerenchyma are combined in the physiological concept of *stereome* (Foster, 1949; Haberlandt, 1914). The plastic, highly hydrated primary walls of collenchyma, however, distinguish this tissue from sclerenchyma with its hard, elastic secondary walls.

Sclerenchyma cells show much variation in form, structure, origin, and development, and the different types of cells are intergrading. A division of such a graded series of forms into a limited number of categories can be only arbitrary, and the value of such a division depends on the clearness of definition of the criteria upon which it is based. Judged by the variety of systems that have been proposed for the classification of sclerenchyma cells (Foster, 1944; Tobler, 1957), precise criteria for the separation of the forms have not been established.

Most commonly, the sclerenchyma cells are grouped into fibers and sclereids. Fibers are described as long cells, sclereids as relatively short cells. Sclereids, however, may grade from short to conspicuously

203

elongated, not only in different plants but also in the same individual. The fibers, similarly, may be shorter or longer. Although the pitting is generally more conspicuous in the walls of sclereids than in those of fibers, this difference is not constant either. Sometimes the origin of the two categories of cells is considered to be the distinguishing characteristic: sclereids are said to arise through secondary sclerosis of parenchyma cells, fibers from meristematic cells that are early determined as fibers. But there are sclereids that differentiate from cells early individualized as sclereids (*Camellia*, Foster, 1944; *Monstera*, Bloch, 1946), and in certain plants phloic parenchyma cells differentiate into fibers when the tissue becomes old and ceases to be concerned with conduction (chapter 12). When it is difficult to assign sclerenchyma cells to one or the other category, the compound term *fiber-sclereid* may be used.

Sclerenchyma cells frequently possess no living protoplasts at maturity. This characteristic, combined with the presence of secondary walls, distinguishes sclerenchyma from parenchyma and collenchyma. But ground parenchyma cells may develop secondary walls (*sclerotic parenchyma*, Bailey and Swamy, 1949) and fibers and sclereids may retain their protoplasts at maturity. Thus, parenchyma and sclerenchyma are not sharply delimited from one another.

FIBERS

Occurrence and Arrangement in Plant

Fibers occur in separate strands or cylinders in the cortex and the phloem, as sheaths or bundle caps associated with the vascular bundles, or in groups or scattered in the xylem and the phloem. In the stems of monocotyledons and dicotyledons the fibers are arranged in several characteristic patterns (De Bary, 1884; Haberlandt, 1914; Schwendener, 1874; Tobler, 1957). In many Gramineae the fibers form a system having the shape of a ribbed hollow cylinder, with the ribs connected to the epidermis (fig. 10.1*A*, pl. 63*D*). In *Zea, Saccharum, Andropogon, Sorghum* (fig. 10.1*B*), and other related genera the vascular bundles have prominent sheaths of fibers (pl. 57*B*) and the peripheral bundles may be irregularly fused with each other or united by sclerified parenchyma into a sclerenchymatic cylinder. The hypodermal parenchyma may be strongly sclerified (Magee, 1948). A hypodermis containing long fibers, some over 1 mm long, has been recorded in *Zea mays* (Murdy, 1960). In the palms the central cylinder is demarcated by a sclerotic zone that may be several inches wide (Tomlinson, 1961). It consists of vascular bundles with massive, radially extended fibrous sheaths. The associated ground parenchyma also becomes sclerotic. In addition, fiber strands occur in the cortex and a few in the central cylinder. Other patterns may be found in the monocotyledons and patterns may vary at different levels of the stem in the same plant (Murdy, 1960). Fibers may be prominent in the leaves of monocotyledons (fig. 10.1*E*). Here they form sheaths enclosing the vascular bundles, or strands extending between the epidermis and the vascular bundles (pl. 70*C*), or subepidermal strands not associated with the vascular bundles.

In stems of dicotyledons, fibers frequently occur in the outermost part of the primary phloem, forming more or less extensive anastomosing strands or tangential plates (fig. 10.1*C, F*). In some plants no other than the peripheral fibers (primary phloem fibers) occur in the phloem (*Alnus, Betula, Linum, Nerium*). Others develop fibers in the secondary phloem also, few (*Nicotiana, Ulmus, Boehmeria*) or many (*Clematis, Juglans, Magnolia, Quercus, Robinia, Tilia, Vitis;* pl. 44*A*). Some dicotyledons have complete cylinders of fibers, either close to the vascular tissues (*Geranium, Pelargonium, Lonicera,* some Saxifragaceae, Caryophyllaceae, Berberidaceae, Primulaceae) or at a distance from them, but still located to the inside of the innermost layer of the cortex (fig. 10.1*H*; pls. 55, 63*C*; *Aristolochia, Cucurbita*). In dicotyledonous stems without secondary growth the isolated vascular bundles may be accompanied by fiber strands on both their inner and outer sides (*Polygonum, Rheum, Senecio*). Plants having phloem internal to the xylem may have fibers associated with this phloem (*Nicotiana*). Finally, a highly characteristic location for fibers in the angiosperms is the primary and the secondary xylem where they have varied arrangements (chapter 11). Roots show a distribution of fibers similar to that of the stems and may have fibers in the primary (fig. 10.1*D*) and in the secondary body. Gymnosperms usually have no fibers in the primary phloem, but many have them in the secondary phloem. Cortical fibers are sometimes present in stems (fig. 10.1*G*).

Classification

Fibers are divided into two large groups, *xylary fibers* and fibers of the various tissue systems outside the xylem, the *extraxylary fibers*. The developmental and topographic relationships of xylary fibers are usually quite precise. These fibers develop from the same meristematic tissues as the other xylem cells and constitute an integral part of the xylem. The assigning of the extraxylary fibers to their proper tissue systems is much less simple and direct. Some of the extraxylary fibers are as definitely related to the phloem as the xylary fibers are to the xylem; in many others the developmental relationship is less clear. The

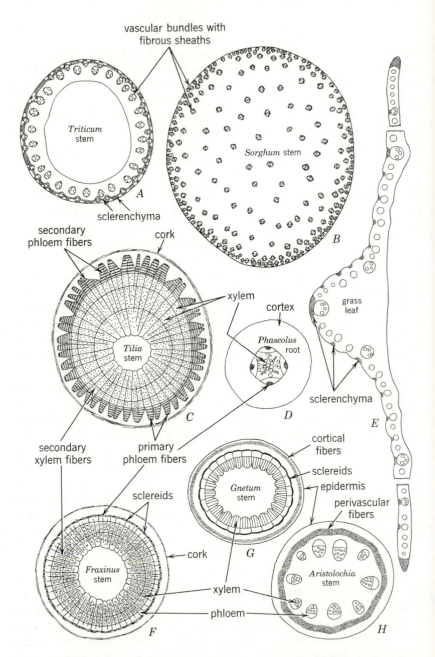

FIG. 10.1. Transections of various plant organs showing distribution of sclerenchyma (stippled), mainly fibers, and of vascular tissues. *A, Triticum* stem, sclerenchyma ensheaths vascular bundles and forms layers in peripheral part of stem. *B, Sorghum* stem, sclerenchyma in fibrous sheaths about vascular bundles. *C, Tilia* stem, fibers in primary and secondary phloem and in secondary xylem. *D, Phaseolus* root, fibers in primary phloem. *E,* grass leaf,

206

fibers that form continuous cylinders in monocotyledonous stems arise in the ground tissue, at varying distances from the epidermis (fig. 10.1A). They might be classified as cortical fibers except that vascular bundles occur among them and that the limits of the cortex in the monocotyledons are generally vague. The fibers forming sheaths around the vascular bundles in the monocotyledons arise partly from the same procambium as the vascular cells, partly from the ground tissue. The fibers of some vine types of stem, such as *Aristolochia* and *Cucurbita*, occur inside the layer of cells characterized by abundant starch accumulation, the starch sheath, which is commonly interpreted as the innermost layer of the cortex (chapter 15). Thus these fibers are part of the vascular cylinder but are not related to the phloem on a developmental basis (Blyth, 1958).

The fibers located on the outer periphery of the vascular cylinder are frequently classified as pericyclic fibers. The pericycle is interpreted as a tissue separate from the vascular, both topographically and developmentally (chapter 15). But in the stems of the majority of dicotyledons investigated ontogenetically, the phloem abuts on the cortex and there is no distinct tissue between the cortex and the phloem that could be termed pericycle in the usual sense of the word (Blyth, 1958; Kundu and Sen, 1961; fig. 10.2; pl. 27). Nevertheless, in much of the literature the primary phloem fibers are called pericyclic fibers because the developmental relation of these fibers to the phloem is being disregarded (Metcalfe and Chalk, 1950) or not recognized. It would be desirable to assign all the extraxylary fibers to the tissue systems to which they belong by origin, but such a classification requires developmental studies and also a proper reevaluation of the concept of the pericycle.

The extraxylary fibers are sometimes combined into a group termed *bast fibers* (Foster, 1949). The word bast was originally applied to fiber strands obtained from the extracambial region of dicotyledonous stems (Haberlandt, 1914). In its etymology "bast" is related to the verb "to bind." It was adopted as a name for the fiber strands because they were used for binding. The bast of the extracambial region of dicotyledonous stems is constituted, in most instances, of fibers of the phloem. In its development the concept of bast followed a double course. In one direction it was extended to cover the extraxylary fibers in various other arrangements than those in the dicotyledonous stems; in the other, it became a specific term for the phloem and was widened to include all

---

sclerenchyma in strands beneath abaxial epidermis and along margins of blade. *F, Fraxinus* stem, fibers in primary phloem and secondary xylem; phloem fibers alternate with sclereids. *G, Gnetum gnemon* stem, fibers in cortex and sclereids in perivascular position. *H, Aristolochia* stem, cylinder of fibers inside starch sheath in perivascular position. (*A, G,* ×12.5; *B, C, F,* ×6; *D,* ×8.5; *E,* ×26; *H,* ×11.5.)

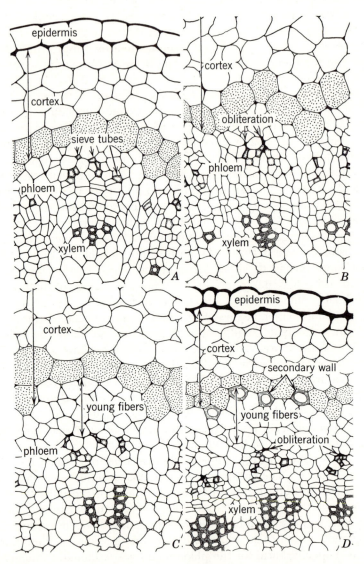

FIG. 10.2. Development of primary phloem fibers in *Linum perenne* L. *A*, first primary sieve tubes are mature. *B*, *C*, new sieve tubes differentiate while older ones are obliterated. *D*, cells remaining after obliteration of sieve tubes begin to develop secondary walls characteristic of flax fibers. (*A–C*, ×620; *D*, ×330.)

the cells of this tissue. In the second usage, the nonsclerified conducting and parenchymatic elements of the phloem received the name "soft bast," the fibers, "hard bast" (Haberlandt, 1914). The term bast fibers is still used for phloem fibers in references dealing with the economic use of plant fibers (Harris, 1954).

In the present book, the term extraxylary fibers is used for fibers not included in the xylem, and these are grouped as follows: *phloic* or *phloem* fibers, fibers originating in primary or secondary phloem; *cortical fibers*, fibers originating in the cortex; *perivascular fibers*, fibers located on the periphery of the vascular cylinder inside the innermost cortical layer but apparently not originating in the phloem. The term perivascular has been used in the literature (Van Fleet, 1948) in a similar descriptive topographic sense.

The xylem or wood fibers have a common origin, but they are morphologically heterogeneous. They intergrade with the imperforate tracheary elements (the tracheids) and with the parenchyma cells, and certain xylem fibers resemble phloem fibers. Wood fibers are subdivided into two main categories, the *fiber-tracheids* and the *libriform fibers* (Committee on Nomenclature, 1957). The fiber-tracheids are transitional between the tracheids and the extreme, or most specialized, libriform fibers. The libriform fibers resemble phloem fibers, hence, the name libriform. It is derived from *liber*, a Latin term for "inner bark," that is, phloem. Some of the xylem fibers form transverse partitions late in their development and are spoken of as *septate fibers*. Phloem fibers also may be septate.

Structure

Extraxylary Fibers. Although a long spindle-like shape is considered typical of extraxylary fibers (and fibers in general), these elements may vary in length, and their ends may be blunt, rather than tapering, or branched. Generally, primary extraxylary fibers are longer than the secondary. The bast fibers of commerce (various extraxylary fibers) vary from a fraction of a millimeter to about ½ meter (primary phloem fibers of ramie, *Boehmeria nivea*, Aldaba, 1927).

The cell walls of the extraxylary fibers are frequently very thick. In the phloem fibers of flax (*Linum usitatissimum*) the secondary thickening may amount to 90 per cent of the area of the cell in cross section (fig. 10.3). The pits are simple or slightly bordered. Some extraxylary fibers have lignified walls, others nonlignified. Flax, hemp, and ramie fibers have little or no lignin and their secondary wall consists of 75 to over 90 per cent cellulose (Harris, 1954). Some extraxylary fibers, notably those of the monocotyledons, are strongly lignified.

Concentric lamellations may be observed in extraxylary fibers with or without treatment with swelling reagents. In flax fibers the individual lamellae vary in thickness from 0.1 to 0.2 micron and the cellulosic layers are alternately strongly and weakly birefringent and vary in density of staining, probably a reflection of varying densities of the cellulosic matrix in the successive lamellae (Hock, 1942). In certain types of extraxylary fibers the lamellation appears to result from an alternation of cellulosic and noncellulosic layers (Bailey, 1938). The orientation of the cellulosic microfibrils has also received attention and has been found to vary in fibers from different plants (Hock, 1942; Preston, 1943).

Xylary Fibers. Wood fibers typically have lignified secondary walls. They vary in size, shape, thickness of wall, and type and abundance of pitting (chapter 11). The variations in structural details and the corresponding divisions of this group of fibers into categories are best understood if they are considered with reference to the possible evolution of the xylem fibers. Phylogenetically, these fibers are considered as

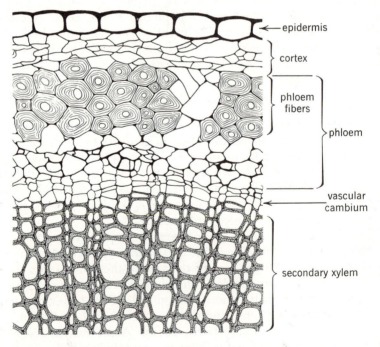

FIG. 10.3. Transection of stem of *Linum usitatissimum* showing position of primary phloem fibers. (×320.)

having been derived from an imperforate xylem cell combining the functions of water conduction with that of support, that is, a tracheid. A good indication that fibers and tracheids are phylogenetically related is the occurrence of almost imperceptible gradations between these two cell types in certain angiosperms such as the oak. The gradations suggest the following principal changes during the evolution of the fiber from a tracheid: increase in wall thickness, decrease in length, and reduction in the size of bordered pits (fig. 11.1). In the extreme condition the pit is simple or nearly so. Of these characteristics, wall thickness and particularly the nature of pitting are used to differentiate between the two main categories of wood fibers, fiber-tracheids, and libriform fibers (Committee on Nomenclature, 1957). Even these criteria do not permit a circumscribed characterization of each category for use in the identification of elements in different species. The limits of the categories are best decided upon by comparing the elements of a given species among themselves (Bailey, 1936). First the tracheid is identified by the resemblance of its pitting to that of the vessel members in the same tissue. Then the limits for the fiber-tracheids are established by the identification of cells with pits whose borders are reduced as compared with those in the tracheids. Finally the cells with simple pits, or essentially so, are classified as libriform fibers (chapter 11).

Commonly, the thickness of wall increases in the sequence of tracheid, fiber-tracheid, libriform fiber. This increase results in an increase in the length of the pit canal. In the fiber-tracheid the pit canals lead into small but evident pit chambers, and the inner apertures are lenticular to slit-like and usually extended beyond the outlines of the border. The libriform fibers also have long slit-like canals, but their pit chambers are much reduced or absent, an expression of the extreme reduction of borders. The inner apertures of the pit-pairs in the fiber-tracheids and libriform fibers are often crossed with each other (chapter 3).

The phylogenetic decrease in length during the evolution of a fiber from a primitive tracheid is a concomitant of a decrease in length of cambial fusiform initials. In a given sample of wood, however, the tracheids are usually shorter and the fibers longer, with the libriform fibers attaining the greatest length. The fibers become longer than the associated tracheids because they undergo a more extensive apical elongation during tissue differentiation.

Fiber-tracheids and libriform fibers may both be septate (Bailey, 1936). The septa are true walls, but they are formed after the deposition of the secondary layers on the longitudinal walls of the element. The formation of septa in the fiber-tracheids of *Hypericum* has been found to involve a regular mitosis followed by cytokinesis (Vestal and Vestal, 1940).

Septate and nonseptate fibers may retain living protoplasts in the sapwood and serve for storage of starch, oils, and other reserve materials (Bailey, 1957; Fahn and Leshem, 1963). Thus the living fibers intergrade with xylary parenchyma cells in function. The retention of protoplasts by fibers is an evolutionary advance (Bailey, 1953) and is associated with a reduction or an elimination in amount of axial parenchyma in the xylem (Money et al., 1950).

In the reaction wood of dicotyledons (tension wood, chapter 11), the fibers—either fiber-tracheids or libriform fibers—are frequently of the *gelatinous* type (pl. 10C; Committee on Nomenclature, 1957). The name gelatinous refers to the appearance of a layer in the secondary wall which has a peculiar cellulosic structure and frequently lacks lignin. The cellulosic matrix has a coarse texture and in some species has been found to be highly crystalline, with the micelles oriented axially (Dadswell et al., 1958). The wall is highly hygroscopic and undergoes striking changes in volume on drying (Bailey and Kerr, 1937).

Origin and Development

Remarks made in the beginning of the chapter indicate that fibers arise from various meristems. Fibers of the xylem and the phloem are derived from procambium or cambium. In the cambium, the fibers arise from the fusiform initials. Extraxylary fibers other than those of the phloem arise from the ground meristem, but the cells that eventually become fibers early cease to divide transversely and elongate (Meeuse, 1938). In some Cyperaceae fibers are epidermal in origin (Thielke, 1957). Protodermal cells divide periclinally and anticlinally and the derivatives differentiate into fibers, except for the outermost, which assume ordinary epidermal characteristics. In plants having fibrous bundle sheaths, part of the fibers may be derived from the procambium and part from the ground meristem (Esau, 1943a; Sinnott and Bloch, 1943). In the shoots of some monocotyledons the proportion of bundle-sheath fibers in a vascular bundle may be very high, or the bundles may consist of fibers only (De Bary, 1884). Since such fibrous bundles may be connected with vascular bundles and since there are bundles with various proportions of fibers and vascular elements, the fibrous bundles probably should be considered as originating from the procambium.

From the developmental standpoint the attainment of great length by fibers is of particular interest. Fibers originating during primary growth have a different kind of development than those formed in the secondary tissues. Primary fibers are initiated before the organ has elongated, and they can reach considerable length by elongating while the associated cells are still dividing. To this symplastic growth may be added apical intrusive growth (chapter 4). In contrast, secondary fibers originate in the part of the organ that has ceased to elongate, and they can increase in length only by intrusive growth (chapters 4, 6). This difference in the method of growth explains why in the same stem primary phloem fibers may attain greater lengths than the secondary. In *Cannabis* (hemp), for example, the primary phloem fibers were found to average about 13 mm, the secondary about 2 mm (Kundu, 1942).

The growth of primary extraxylary fibers in unison with the other tissues in the growing organ causes longer fibers to occur in longer organs. For example, the adult length of primary phloem fibers in *Cannabis* and *Boehmeria* is correlated with the adult length of the internodes (Kundu, 1942; Kundu and Sen, 1961). Similarly, the longest phloem fibers in flax occur in the longest stems (Tammes, 1907). In *Sanseviera*, *Agave*, and *Musa* the average length of extraxylary fibers depends on the length of the part of the leaf from which the fibers are obtained (Meeuse, 1938).

The great length attained by some primary extraxylary fibers cannot be explained, however, on the basis of elongation by symplastic growth only. In *Sanseviera*, *Agave*, and *Musa* the fibers become 40 to 70 times longer than the meristematic cells from which they arise (Meeuse, 1938). In *Luffa* the elongation of the fibers of the fruit at first keeps pace exactly with the increase in size of the fruit itself, but, after the fibers attain about 200 microns in length, their rate of extension becomes greater than that of the fruit as a whole (Sinnott and Bloch, 1943). Thus the fibers appear to have an independent growth in addition to that correlated with the other tissues. Microscopic observations support this assumption (Kundu, 1942; Kundu and Sen, 1961; Schoch-Bodmer and Huber, 1951; Sinnott and Bloch, 1943). The apices of the fibers long remain thin walled and rich in cytoplasm. They may be serrated and forked because of adjustments to the outlines of adjacent cells. Furthermore, the number of fibers, as determined in transections of stems, gradually increases although longitudinal divisions do not occur. All these observations indicate that the apices of the fibers elongate and intrude among the associated cells. Since this growth occurs in a stem that is still elongating, intrusive growth is probably followed by symplastic growth of the new three-ply wall system formed by the apposition of the new wall of the fiber apex to that of another cell. In flax the phloem fibers have been found growing at both apices, upward and downward, and the length of the stem in which this apical growth of fibers was taking place was estimated to be about 19 mm (Schoch-

Bodmer and Huber, 1945; 1951). Although the secondary phloem fibers fail to attain the same lengths as the primary, they commonly become longer than the cambial initials by apical intrusive growth (Kundu, 1942; Schoch-Bodmer, 1960).

Apical elongation is well substantiated with reference to the secondary xylem fibers (Schoch-Bodmer, 1960; chapter 4). Table 10.1 illustrates the result of such growth by comparing the lengths of fiber-tracheids with those of cambial cells in certain woods. Frequently the occurrence of intrusive growth in secondary xylary fibers may be recognized in the mature shape of the cells. The cells consist of a wider median part corresponding to the unelongated cambial cell and two slender ends that originated during intrusive growth. Pits are limited to the median part in such fibers (Schoch-Bodmer, 1960).

When the extraxylary fibers begin to develop they cease dividing. The nuclei, however, may continue to divide so that the developing fibers become multinucleate. This phenomenon is particularly characteristic of the very long primary phloem fibers (literature in Esau, 1943b). In the same plants the primary phloem fibers may be multinucleate and the shorter, secondary phloem fibers uninucleate (Kundu, 1942; Kundu and Sen, 1961).

The prolonged growth in length of the primary phloem fibers results in a highly complicated method of secondary wall development. As has been explained in chapter 3, the deposition of secondary walls begins after the primary wall completes its increase in surface. While the primary fibers elongate by symplastic growth in correlation with the

Table 10.1 Comparison of Lengths of Fiber-Tracheids and Cambial Cells in Certain Dicotyledonous Trees (Based on data from Bailey, *Amer. Jour. Bot.* 7, 1920, and Forsaith, *Syracuse Univ. Tech. Publ.* 18, 1926)

Species of Tree	Length in millimeters		Ratio of length of fiber-tracheid to cambial cell × 100
	Cambial cell	Fiber-tracheid	
Liquidambar Styraciflua L. Red gum	0.70	0.96	136
Betula populifolia Marsh. Gray birch	0.94	1.31	140
Quercus alba L. White oak	0.53	1.00	189
Carya ovata (Mill.) C. Koch Hickory	0.52	1.30	250
Fraxinus americana L. White ash	0.29	0.96	330
Ulmus americana L. White elm	0.35	1.53	436
Robinia Pseudo-Acacia L. Black locust	0.17	0.87	510

surrounding cells, they remain thin walled. Presumably at this stage the entire fiber wall is increasing its surface. Later, during the apical-growth stage, the apices of the cells remain thin, whereas the median portions of the cells, which have completed their elongation, begin to form secondary walls. This secondary thickening of the primary phloem fibers was studied in particular detail in *Linum* and *Boehmeria* (Aldaba, 1927; Anderson, 1927). In these two plants the secondary wall of the fibers develops in the form of distinct lamellae, each being tubular in shape and growing from the base upward. (Presumably in the earlier stage of the process there is also a downward growth of the tubule while the lower end of the fiber is still elongating. It is conceivable that this end would cease growing first, because it is imbedded in maturing tissues, whereas the upper end is advancing into a still-growing tissue.) Thus, several telescoping hyaline tubules arise successively, each tubule being longer than the immediately following one (fig. 10.4). After the cell ceases to grow at the apex, some of the successively formed layers reach the apex, others stop their growth at lower levels, while new layers arise above them and complete the wall thickening in the upper cell parts. This partial interruption of wall growth results in the formation of compartments, which may be in communication with each other. Apparently the deposition of secondary walls in primary fibers may continue long past the stage at which the cell completes its elongation. In flax and in hemp the fibers of the phloem in the adult plant parts are reported to possess living protoplasts and to continue depositing secondary wall layers (Kundu, 1942; Tammes, 1907).

One of the striking features observed in the growth of the secondary wall in primary phloem fibers is that this wall is not cemented to the primary wall and the successive layers of the secondary wall also appear to be distinct, at least while the cell is not yet mature (Aldaba, 1927; Anderson, 1927; Kundu, 1942). In sectioned material the secondary wall of an immature fiber commonly appears detached from the primary and is often separated into two or more layers which are more or less infolded (pl. 26A). This infolding and wrinkling might be an artifact, but it also is taken as an indication that the secondary wall layers are in a loose and relaxed state during their formation (Anderson, 1927; Kundu, 1942).

Economic Fibers

Fiber plants have been utilized economically since ancient times. Flax is known to have been cultivated by man as early as 3,000 years B.C. in Europe and Egypt, and hemp at approximately the same time

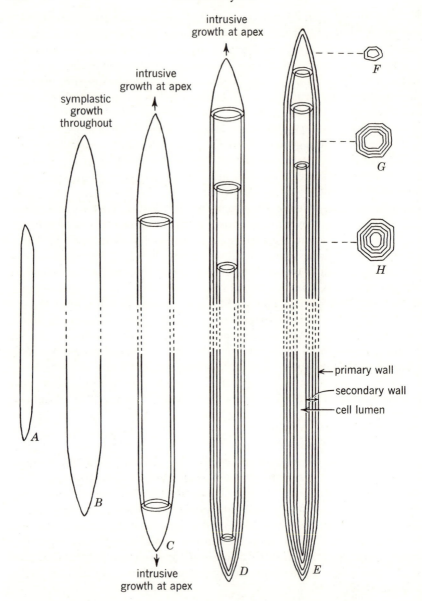

intrusive
growth at apex

intrusive
growth at apex

symplastic
growth
throughout

F

G

H

primary wall

secondary wall

cell lumen

A

B

C

intrusive
growth at apex

D

E

FIG. 10.4. Interpretation of growth and differentiation of primary phloem fibers. *A*, young fiber, narrow and short. *B*, fiber has increased in width and length by symplastic growth. *C*, median part of fiber has reached final length and has formed the first layer of secondary wall; apices are elongating by intrusive growth. *D*, apical growth has been completed at lower end. Successive lamellae of secondary wall, tubular in structure, are deposited one upon the

in China (Ash, 1948; Dewey, 1943). In the technical field, the term fiber usually does not have the strict botanical connotation of individual cells of a certain category of sclerenchyma. In plants in which the commercial fibers originate in the phloem (flax, hemp, ramie, jute), the term fiber denotes a fiber strand. The fibers obtained from monocotyledonous leaves commonly represent vascular bundles together with the associated fibers (pl. 70C). Raffia consists of leaf segments of *Raphia* palm; rattan, of stems of *Calamus* palm. The epidermal hairs of cotton seed and of the kapok seed pod are also termed fibers. In still other plants the vascular system of the root (*Muhlenbergia*) or of the entire plant (*Tillandsia*) are used as fibers.

Commercial fibers are separated into hard fibers and soft fibers. The hard fibers are monocotyledonous leaf fibers with heavily lignified walls, hard and stiff in texture. Examples of plants yielding such fibers and ranges of lengths of these fibers in millimeters, according to Harris (1954), are *Agave* species (henequen and sisal, 0.8–8.0), *Musa textilis* (abaca, 2–12), *Yucca*, and *Phormium tenax* (New Zealand hemp, 2–15; pl. 70C). The soft fibers, that is, bast fibers, may be lignified or free of lignin, but all are soft and flexible. Here are included the phloem fibers of such plants as *Linum usitatissimum* (flax, 9–70), *Cannabis sativa* (hemp, 5–55), *Corchorus capsularis* (jute, 0.8–6.0). *Boehmeria nivea* (ramie, 50–250), and *Hibiscus cannabinus* (kenaf). The seed hairs of *Gossypium* (cotton) range 16–30 mm in length.

The length of the fiber strand depends on the length of the organ from which the fibers are obtained and on the degree of anastomosing of the strands within the plant. The vascular bundles and fiber strands of the monocotyledonous leaves commonly have a long, straight course with rather small, weak cross anastomoses uniting the bundles with each other. The phloem fiber strands of the dicotyledons, on the other hand, form a network in which the individual strands have no identity as such. It is assumed that the shape and length of the fiber cells, their degree of overlapping, and their connection with each other are factors in the development of strength in fiber strands.

In the preparation of commercial fiber the plants are subjected to a process of partial maceration called *retting* (technical form of the word rotting). In this process the plant material is exposed to a decomposing action by bacteria and fungi. These are allowed to act on the plant

---

other and successively closer to apices of cell. *E*, growth in length has been completed at both ends; layers of secondary wall have reached lower end of cell, but upper end is not fully mature. *F–H*, transections of oldest fiber (*E*) taken at levels with different numbers of layers of secondary wall.

parts until the tissues surrounding the fibers are so softened that the fibers can be easily freed mechanically (Ash, 1948). In the early stages of retting only the intercellular material is affected by pectic enzymes; later the primary wall may be attacked also. An effort is made to discontinue the retting process before the fiber strands are macerated into individual cells. Lignification of cell walls, which usually involves the intercellular substance also, interferes with retting (Anderson, 1927).

SCLEREIDS

Occurrence and Arrangement in Plant

Sclereids are widely distributed in the plant body (De Bary, 1884; Haberlandt, 1914). The cortex and the pith of gymnosperms and dicotyledons often contain sclereids, arranged singly or in groups. Sclereids are also common components of the xylem and the phloem, where they may intergrade with fibers. In many plants the interfascicular parenchyma cells located between the strands of primary phloem fibers develop lignified secondary walls and differentiate into sclereids which, together with the fibers, form a continuous sclerenchyma cylinder on the outer periphery of the vascular system. The plants in which a continuous sclerenchyma cylinder is present in the primary state may show a disruption of this cylinder when the vascular system surrounded by the sclerenchyma increases in circumference through secondary growth. The breaks in the sclerenchyma cylinder are filled with parenchyma cells which later may differentiate into sclereids (*Aristolochia*, pl. 55*B*).

Many species of plants, particularly in the tropics, contain sclereids in the leaves (Foster, 1944, 1945; Kitamura, 1956; Rao, 1957). The leaf sclereids may be few to abundant. In some leaves the mesophyll is completely permeated by sclereids (pl. 26*B*; Arzee, 1953*a*). In certain species the leaf sclereids occur at the ends of vascular bundles (pl. 73*B*, Foster, 1947, 1955). Sclereids are also common in fruits and seeds. In fruits they sometimes are dispersed in the soft flesh singly or in groups (*Pyrus, Cydonia, Vaccinium;* Yarbrough and Morrow, 1947). In solid layers they constitute hard coverings in the form of shells of nuts or of endocarps of stone fruits (chapter 19). The hardness and strength of the seed coat often result from the presence of abundant sclereids (fig. 10.5; Netolitzky, 1926; Zimmerman, 1936). Solid layers of sclereids occur in the epidermis of some protective scales (fig. 10.6).

Classification

Sclereids vary widely in the shape, size, and characteristics of their walls. It was inevitable, therefore, that an extensive terminology be

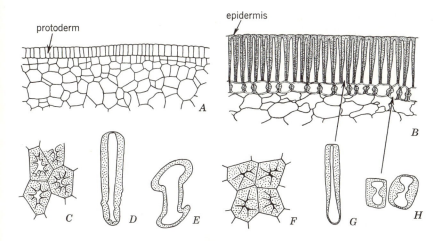

FIG. 10.5. Sclereids of leguminous seed coats. *A, B,* outer parts of *Phaseolus* seed coat from transections of seeds in two stages of development. *B,* epidermis, a solid layer of macrosclereids. Subepidermal sclereids have most of the wall thickenings localized on anticlinal walls. *C–E,* sclereids of *Pisum, F–H,* of *Phaseolus. C, F,* groups of epidermal sclereids seen from the surface. *D, G,* epidermal sclereids; *E, H,* subepidermal sclereids. (*A, B,* ×225; *C, F,* ×550; *D, E, G, H,* ×280.)

developed in the course of study of these elements (Foster, 1949). Commonly the following categories of sclereids are recognized: *brachysclereids,* stone cells, short, roughly isodiametric sclereids, resembling parenchyma cells in shape, widely distributed in cortex, phloem, and pith of stems, and in the flesh of fruits (chapter 3); *macrosclereids,* elongated rod-like cells, exemplified by sclereids forming the palisade-like epidermal layer of leguminous seeds (fig. 10.5*B–D, F, G*); *osteosclereids,* bone-shaped sclereids (that is, columnar cells enlarged at the ends; fig. 10.5*E*), as those present in leaves of many dicotyledons and in seed coats; *astrosclereids,* literally star-sclereids, cells ramified to varied degrees and often found in the leaves of dicotyledons (fig. 10.7*A*); *filiform* sclereids, long, slender cells resembling fibers (pl. 26*B*); and *trichosclereids,* branched, thin-walled sclereids resembling plant hairs, with branches extending into intercellular spaces (Bloch, 1946; Gaudet, 1960; Nicolson, 1960). This classification is rather arbitrary and does not cover all the forms of sclereids known (Bailey, 1961). Its usefulness is further limited by the polymorphism of each of the established categories and by the existence of transitions among the categories. Nevertheless, sclereid forms may be characteristic of species and may therefore be of taxonomic value (Barua and Dutta, 1959).

Sclerenchyma

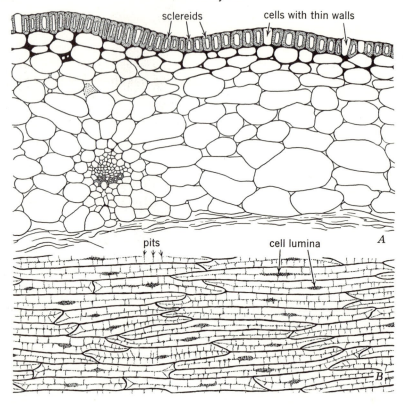

FIG. 10.6. Epidermal sclereids in a protective bulb scale of *Allium sativum* (garlic). *A*, section of scale, with sclereid walls stippled. *B*, surface view of scale showing the solid layer of epidermal sclereids overlapping each other. (Both, ×99. From Mann, *Hilgardia* 21, 1952.)

Structure

The secondary walls of the sclereids vary in thickness and are typically lignified. If the walls are relatively thin, the sclereids cannot be definitely separated from sclerotic parenchyma. The thick-walled forms, on the other hand, may strongly contrast with the parenchyma cells. In many sclereids the lumina are almost filled with massive wall deposits, and the secondary wall shows prominent pits, often with ramiform canal-like cavities. The pits are commonly simple, but sometimes the secondary wall slightly overarches a small pit chamber. The secondary wall often appears concentrically lamellated in ordinary and polarized light. The lamellation may be the result of an alternation of isotropic layers with those composed of cellulose (Bailey and Kerr, 1935).

Crystals are imbedded in the secondary wall of the sclereids in certain species (Bailey and Nast, 1948).

In some sclereids the deposition of the secondary walls is uneven. In the macrosclereids of the seed coats of the Leguminosae, for example, most of the secondary deposit is localized on the lateral walls in the end of the cell turned toward the surface of the seed (fig. 10.5B). Furthermore, this thickening is laid down in the form of longitudinal ridges arranged vertically or helically and constricting the lumen of the cell in such a way that it appears star shaped in sections cut at right angles to the long axis of the cell (fig. 10.5C, F). As was mentioned before, sclereids may either retain their protoplasts on reaching maturity or become dead cells.

Origin and Development

Sclereids arise either through a belated sclerosis of apparently ordinary parenchyma cells (secondary sclerosis) or directly from cells that are early individualized as sclereid primordia. The sclerification of cells in

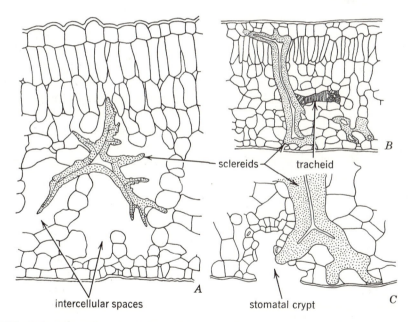

intercellular spaces   stomatal crypt

FIG. 10.7. Foliar sclereids. *A*, branched form in *Trochodendron* leaf blade. *B*, columnar, with horizontal branches above and below, in *Mouriria* leaf; sclereid is in contact with terminal tracheid of small vascular bundle. *C*, portion of sclereid as in *B*; processes extend to cuticle and one has penetrated between two guard cells into stomatal crypt. (*A*, ×155; *B*, ×115; *C*, ×333. After Foster, *Amer. Jour. Bot.* 32, 1945; 34, 1947.)

the phloem may occur after the tissue ceases to function in conduction. The *Camellia* leaf sclereids begin their development during the final stage of enlargement of the leaf (Foster, 1944). The primordia of the sclereids in the *Mouriria* leaf, on the other hand, are clearly evident before the intercellular spaces appear in the mesophyll and while the small veins are still entirely procambial (Foster, 1947). The trichosclereids of the air roots of *Monstera* develop from cells early set aside by unequal divisions in the rib meristem of the cortex (Bloch, 1946). In one and the same organ, sclereids may arise over an extended period of time, as in *Trochodendron* leaves (Foster, 1945).

Within the vascular tissues the sclereids develop from derivatives of the procambial and the cambial cells. Stone cells imbedded in the cork originate from the phellogen. Macrosclereids of the seed coats are of protodermal origin (fig. 10.5*A*, *B*; Reeve, 1946). Many sclereids differentiate from ground-parenchyma cells or, if they are individualized early, from ground-meristem cells. In some leaves the parenchyma cells developing into sclereids are part of the spongy mesophyll (Foster, 1945). In the olive leaf, the filiform sclereids originate in both palisade and spongy parenchyma cells and enlarge several hundredfold, whereas the neighboring parenchyma cells only double or triple their size (Arzee, 1953*b*). The sclereids of *Mouriria*, which are located at the terminations of the vascular bundles in the mesophyll are, from the time of their origin, in contact with procambial cells, and both the sclereids and the procambium arise from the same layer in the ground meristem (Foster, 1947).

If the sclereids resemble parenchyma cells, their development involves no striking changes in shape, compared with that of the adjacent parenchyma cells. The principal change is the development of the secondary wall. The sclereids which assume shapes strikingly different from those of the associated parenchyma cells show considerable independence in their growth. They invade intercellular spaces, intrude among other cells, and may penetrate the epidermis, sometimes between the guard cells (fig. 10.7*B*, *C*; Foster, 1947, 1955). They become very much larger than the initial cells, and assume extraordinary, often grotesque shapes.

The casual relationships in the development of the sclereids constitute a challenging problem for a student of histogenesis. Auxin levels influence the development of sclereids, tending to suppress it at high levels (Al-Talib and Torrey, 1961). In some plants sclereid growth seems to be highly individualistic and uncoordinated with the growth of the other cells (Foster, 1944, 1945). In others, the origin and the development of sclereids appear to be a part of the growth pattern of the cell complex as a whole (Bloch, 1946; Foster, 1947, 1955). Surgical experiments on *Camellia* leaves suggest that position may play the major role in inducing sclereid development (Foard, 1959). In some plants sclereids grow and branch in a relatively compact tissue (*Mouriria,* Foster, 1947); in others they begin development in a lacunose tissue and grow mainly by sending out protrusions into intercellular spaces (*Monstera,* Bloch, 1946; *Nymphaea,* Gaudet, 1960).

The mechanics of growth of the sclereids can best be explained as a combination of symplastic growth in the early stages of their development, when they still grow in unison with the adjacent cells, and apical intrusive growth in the later stages, when they elongate by penetrating into intercellular spaces and intruding among other cells (Arzee, 1953*b*; Foster, 1947).

REFERENCES

Aldaba, V. C. The structure and development of the cell wall in plants. I. Bast fibers of *Boehmeria* and *Linum. Amer. Jour. Bot.* 14:16–24. 1927.

Al-Talib, K. H., and J. G. Torrey. Sclereid distribution in the leaves of *Pseudotsuga* under natural and experimental conditions. *Amer. Jour. Bot.* 48:71–79. 1961.

Anderson, D. B. A microchemical study of the structure and development of flax fibers. *Amer. Jour. Bot.* 14:187–211. 1927.

Arzee, T. Morphology and ontogeny of foliar sclereids in *Olea europaea.* I. Distribution and structure. *Amer. Jour. Bot.* 40:680–687. 1953*a*. II. Ontogeny. *Amer. Jour. Bot.* 40:745–752. 1953*b*.

Ash, A. L. Hemp—production and utilization. *Econ. Bot.* 2:158–169. 1948.

Bailey, I. W. The problem of differentiating and classifying tracheids, fiber-tracheids, and libriform wood fibers. *Trop. Woods* 45:18–23. 1936.

Bailey, I. W. Cell wall structure of higher plants. *Indus. and Engin. Chem.* 30:40–47. 1938.

Bailey, I. W. Evolution of the tracheary tissue in land plants. *Amer. Jour. Bot.* 40:4–8. 1953.

Bailey, I. W. The potentialities and limitations of wood anatomy in the study of the phylogeny and classification of angiosperms. *Arnold Arboretum Jour.* 38:243–254. 1957.

Bailey, I. W. Comparative anatomy of the leaf-bearing Cactaceae. II. Structure and distribution of sclerenchyma in phloem of *Pereskia, Pereskiopsis* and *Quiabentia. Arnold Arboretum Jour.* 42:144–150. 1961.

Bailey, I. W., and T. Kerr. The visible structure of the secondary wall and its significance in physical and chemical investigations of tracheary cells and fibers. *Arnold Arboretum Jour.* 16:273–300. 1935.

Bailey, I. W., and T. Kerr. The structural variability of the secondary wall as revealed by "lignin" residues. *Arnold Arboretum Jour.* 18:261–272. 1937.

Bailey, I. W., and C. G. Nast. Morphology and relationships of *Illicium, Schizandra,* and *Kadsura.* I. Stem and leaf. *Arnold Arboretum Jour.* 29:77–89. 1948.

Bailey, I. W., and B. G. L. Swamy. The morphology and relationships of *Austrobaileya. Arnold Arboretum Jour.* 30:211–226. 1949.

Barua, P. K., and A. C. Dutta. Leaf sclereids in the taxonomy of *Thea* camellias—II. *Camellia sinensis* L. *Phytomorphology* 9:372–382. 1959.

Bloch, R. Differentiation and pattern in *Monstera deliciosa*. The idioblastic development of the trichosclereids in the air roots. *Amer. Jour. Bot.* 33:544–551. 1946.

Blyth, A. Origin of primary extraxylary stem fibers in the dicotyledons. *Calif. Univ., Publs., Bot.* 30:145–232. 1958.

Committee on Nomenclature. International Association of Wood Anatomists. *International glossary of terms used in wood anatomy. Trop. Woods* 107:1–36. 1957.

Dadswell, H. E., A. B. Wardrop, and A. J. Watson. The morphology, chemistry and pulp characteristics of reaction wood. In: *Fundamentals of Papermaking Fibres.* British Paper and Board Makers' Association. 1958.

De Bary, A. *Comparative anatomy of the vegetative organs of the phanerogams and ferns.* Oxford, Clarendon Press. 1884.

Dewey, L. H. Fiber production in the western hemisphere. *U.S. Dept. Agr. Misc. Publ.* 518. 1943.

Esau, K. Ontogeny of the vascular bundle in *Zea Mays. Hilgardia* 15:327–368. 1943*a*.

Esau, K. Vascular differentiation in the vegetative shoot of *Linum.* III. The origin of the bast fibers. *Amer. Jour. Bot.* 30:579–586. 1943*b*.

Fahn, A., and B. Leshem. Wood fibers with living protoplasts. *New Phytol.* 62:91–98. 1963.

Foard, D. E. Pattern and control of sclereid formation in the leaf of *Camellia japonica. Nature* 184:1663–1664. 1959.

Foster, A. S. Structure and development of sclereids in the petiole of *Camellia japonica* L. *Torrey Bot. Club Bul.* 71:302–326. 1944.

Foster, A. S. Origin and development of sclereids in the foliage leaf of *Trochodendron aralioides* Sieb. and Zucc. *Amer. Jour. Bot.* 32:456–468. 1945.

Foster, A. S. Structure and ontogeny of the terminal sclereids in the leaf of *Mouriria Huberi* Cogn. *Amer. Jour. Bot.* 34:501–514. 1947.

Foster, A. S. *Practical plant anatomy.* 2nd ed. New York, D. Van Nostrand Company. 1949.

Foster, A. S. Structure and ontogeny of terminal sclereids in *Boronia serrulata. Amer. Jour. Bot.* 42:551–560. 1955.

Gaudet, J. Ontogeny of foliar sclereids in *Nymphaea odorata. Amer. Jour. Bot.* 47:525–532. 1960.

Haberlandt, G. *Physiological plant anatomy.* London, Macmillan and Company. 1914.

Harris, M., ed. *Handbook of textile fibers.* Washington, Harris Research Laboratories. 1954.

Hock, C. W. Microscopic structure of flax and related fibers. *U.S. Natl. Bur. Standards Jour. Res.* 29:41–50. 1942.

Kitamura, R. Development of the foliar sclereids in *Sciadopitys verticillata* Sieb. et Zucc. *Bot. Mag.* [*Tokyo*] 69:519–523. 1956.

Kundu, B. C. The anatomy of two Indian fibre plants, *Cannabis* and *Corchorus* with special reference to the fibre distribution and development. *Indian Bot. Soc. Jour.* 21:93–128. 1942.

Kundu, B. C., and S. Sen. Origin and development of fibres of ramie (*Boehmeria nivea* Gaud.) *Natl. Inst. Sci. India,* Proc. 26, B (Suppl.):190–198. 1961.

Magee, J. A. Histological structure of the stem of *Zea mays* in relation to stiffness of stalk. *Iowa State Col. Jour. Sci.* 22:257–268. 1948.

Meeuse, A. D. J. Development and growth of the sclerenchyma fibres and some remarks on the development of tracheids in some monocotyledons. *Rec. des Trav. Bot. Néerland.* 35:288–321. 1938.

Metcalfe, C. R., and L. Chalk. *Anatomy of the dicotyledons.* 2 vols. Oxford, Clarendon Press. 1950.

Money, L. L., I. W. Bailey, and B. G. L. Swamy. The morphology and relationships of the Monimiaceae. *Arnold Arboretum Jour.* 31:372–404. 1950.

Murdy, W. H. The strengthening system in the stem of maize. *Mo. Bot. Gard. Ann.* 67:205–226. 1960.

Netolitzky, F. Anatomie der Angiospermensamen. In: K. Linsbauer. *Handbuch der Pflanzenanatomie.* Band 2. Lief. 14. 1926.

Nicolson, D. H. The occurrence of trichosclereids in the Monsteroideae (Araceae). *Amer. Jour. Bot.* 47:598–602. 1960.

Preston, R. D. The fine structure of the walls of phloem fibres. *Chron. Bot.* 7:414–416. 1943.

Rao, T. A. Comparative morphology and ontogeny of foliar sclereids in seed plants—I. *Memecylon* L. *Phytomorphology* 7:306–330. 1957.

Reeve, R. M. Ontogeny of the sclereids in the integument of *Pisum sativum* L. *Amer. Jour. Bot.* 33:806–816. 1946.

Schoch-Bodmer, H. Spitzenwachstum und Tüpfelverteilung bei secundären Fasern von *Sparmannia. Ztschr. des Schweiz. Forstver. Beih.* 30:107–113. 1960.

Schoch-Bodmer, H., and P. Huber. Das Spitzenwachstum der Fasern bei *Linum perenne* L. *Experientia* 1:327–328. 1945.

Schoch-Bodmer, H., and P. Huber. Das Spitzenwachstum der Bastfasern bei *Linum usitatissimum* und *Linum perenne. Schweiz. Bot. Gesell. Ber.* 61:377–404. 1951.

Schwendener, S. *Das mechanische Princip im anatomischen Bau der Monokotylen mit vergleichenden Ausblicken auf die übrigen Pflanzenklassen.* Leipzig, Wilhelm Engelmann. 1874.

Sinnott, E. W., and R. Bloch. Development of the fibrous net in the fruit of various races of *Luffa cylindrica. Bot. Gaz.* 105:90–99. 1943.

Tammes, T. Der Flachsstengel. Eine statistisch-anatomische Monographie. *Natuurk. Verhand. v. d. Holland Maatsch. d. Wetenschappen t. Haarlem.* Derde Verzameling. Deel VI. Vierde Stuk. 1907.

Thielke, C. Über Differenzierungsvorgänge bei Cyperaceen. II. Entstehung von epidermalen Faserbündeln in der Scheide von *Carex. Planta* 49:33–46. 1957.

Tobler, F. Die mechanischen Elemente und das mechanische System. In: K. Linsbauer. *Handbuch der Pflanzenanatomie.* 2nd ed. Band 4. Teil 6. 1957.

Tomlinson, P. B. *Anatomy of the monocotyledons.* II. *Palmae.* Oxford, Clarendon Press. 1961.

Van Fleet, D. S. Cortical patterns and gradients in vascular plants. *Amer. Jour. Bot.* 35:219–227. 1948.

Vestal, P. A., and M. R. Vestal. The formation of septa in the fiber tracheids of *Hypericum Androsemum* L. *Harvard Univ. Bot. Mus. Leaflet* 8:169–188. 1940.

Yarbrough, J. A., and E. B. Morrow. Stone cells in *Vaccinium. Amer. Soc. Hort. Sci. Proc.* 50:224–228. 1947.

Zimmerman, K. Zur physiologischen Anatomie der Leguminosentesta. *Landw. Vers. Sta.* 127:1–56. 1936.

11

Xylem

The vascular system of the plant is composed of xylem, the principal water-conducting tissue, and phloem, the food-conducting tissue. As components of the vascular system xylem and phloem are called *vascular tissues*. Sometimes the two together are spoken of as the *vascular tissue*. The term *xylem* was introduced by Nägeli (1858) and is derived from the Greek *xylos*, meaning wood.

The physiologic and phylogenetic importance of the vascular system and its prominence among the structural elements of the plant body has led to a taxonomic segregation of plants having such a system into one group, the so-called *vascular plants*, or *Tracheophyta* (Cheadle, 1956). This group consists of the Psilopsida, Lycopsida, Sphenopsida, and Pteropsida (ferns, gymnosperms, and angiosperms).

The terms vascular plants and Tracheophyta refer to the characteristic elements of the xylem, the vessels and the tracheary elements in general. Because of its enduring rigid walls the xylem is more conspicuous than the phloem, is better preserved in fossils (pl. 29), and may be studied with greater ease. It is this tissue, therefore, rather than the phloem, that serves for the identification of vascular plants.

Structurally, the xylem is a complex tissue, for it consists of several different types of cells, living and nonliving. The most characteristic components are the tracheary elements, which conduct water. Some of the tracheary elements combine the function of conduction with that of support. The xylem also commonly contains specialized supporting elements, the fibers, and living parenchymatic cells concerned with various vital activities. Fibers may retain their protoplasts in the

226

conducting xylem and thus combine vital functions, as starch storage, with the mechanical one of support. In certain groups of plants, the xylem includes laticifers. Sclereids differentiated from parenchymatic elements may be present.

The common association of fibers with other xylem and phloem elements brought about the introduction of the term "fibrovascular tissue" with reference to the xylem and the phloem. This term is rarely employed now.

CLASSIFICATION

The first xylem differentiates during the early ontogeny of a plant—in the embryo or the post-embryonic stage—and as the plant grows, new xylem continuously develops from the derivatives of the apical meristems. As a result of this growth, the primary plant body, which is eventually formed by the activity of the apical meristems, is permeated by a continuous xylem system (together with the accompanying phloem system) whose pattern varies in different kinds of plants. The xylem differentiating in the primary plant body is the *primary xylem*. The immediate precursor of this xylem is the *procambium* (chapter 4).

If the plant is of such a nature that, after the completion of primary growth, it forms secondary tissues through the activity of the *vascular cambium* (chapter 6), the xylem produced by this meristem constitutes the *secondary xylem* (pl. 28).

The histologic characteristics of the two kinds of xylem are given later in this chapter. Depending on the nature of the plant, the primary xylem is more or less distinct from the secondary, but in many respects the two kinds of xylem intergrade with each other (Esau, 1943). Therefore, to be useful the classification into primary and secondary xylem must be conceived broadly, relating the two components of the xylem tissue to the development of the plant as a whole, in a manner outlined in the preceding paragraphs and chapter 4.

ELEMENTS OF THE XYLEM

Tracheary Elements

Tracheids and Vessel Members. The term tracheary element is derived from "trachea," a name originally applied to certain primary xylem elements resembling insect tracheae (Esau, 1961). Two fundamental types of tracheary elements occur in the xylem, *tracheids* and *vessel members* (or *vessel elements;* figs. 11.1, 11.2D–F, 11.8, 11.9). In the mature state both are more or less elongated cells (some vessel members

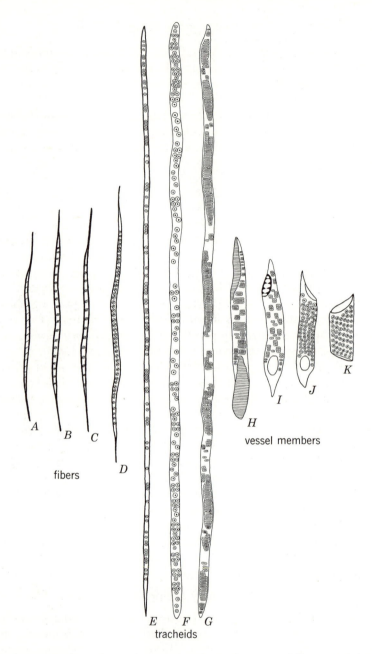

fibers

tracheids

vessel members

A B C D E F G H I J K

FIG. 11.1. Main lines of specialization of tracheary elements and fibers. *E–G*, long tracheids from primitive woods. (*G*, reduced in scale.) *E*, *F*, circular bordered pits; *G*, elongated bordered pits in scalariform arrangement. *D–A*, evolution of fibers: decrease in length, reduction in size of pit borders, and change in shape and size of pit apertures. *H–K*,

may be drum-shaped; fig. 11.9, pl. 36A) having lignified secondary walls and containing no protoplasts. They differ from each other in that the tracheids are imperforate cells having only pit-pairs on their common walls, whereas the vessel members are perforated in certain areas of union with other vessel members. Thus the vessel members are joined into long continuous tubes, the *vessels* (pl. 35B; sometimes called *tracheae*). Sap moving through these structures passes freely from element to element through the perforations, whereas in the tracheids it traverses the walls, particularly the thin pit membranes (Stamm, 1946).

The perforations of vessel members commonly occur on the end walls, but they may be present on the lateral walls too. The wall area bearing the perforation is called the *perforation plate* (Committee on Nomenclature, 1957). A perforation plate (fig. 11.2) may have a single perforation (*simple perforation plate*) or several (*multiple perforation plate*). The openings in a multiple perforation plate are arranged in a parallel series (*scalariform perforation plate*, from the Latin *scalaris*, ladder), or in a reticulate manner (*reticulate perforation plate*, from the Latin *rete*, net), or as a group of approximately circular holes (*ephedroid perforation plate*, as in *Ephedra*, fig. 11.8).

Each vessel (that is, a series of vessel members joined end to end) is limited in length, and the vessels in a series are connected to each other by imperforate walls in the same manner as tracheids. Water and aqueous solutions pass through these imperforate walls, but such substances as mercury and gases fail to do so. The exact length of vessels is difficult to determine. Some data suggest that individual vessels may be from 2 to 15 ft long, but in species with especially wide vessels in the early wood (ring-porous wood) the vessels appear to extend through the entire height of the tree (Greenidge, 1952; Handley, 1936).

Formation of a Vessel. A vessel originates ontogenetically from a longitudinal series of meristematic cells. These are procambial cells in the primary xylem, cambial derivatives in the secondary. The primordial vessel members may or may not elongate before they develop secondary walls, but they usually expand laterally (pl. 36A). After this growth is completed, secondary wall layers are deposited in a pattern characteristic for the given type of vessel element. The portions of the primary wall that later are perforated are not covered by secondary wall material. Nevertheless, they commonly become thicker, as compared

evolution of vessel members: decrease in length, reduction in inclination of end walls, change from scalariform to simple perforation plates, and from opposite to alternate pit arrangement. (After Bailey and Tupper, *Amer. Acad. Arts and Sci. Proc.* 54, 1918.)

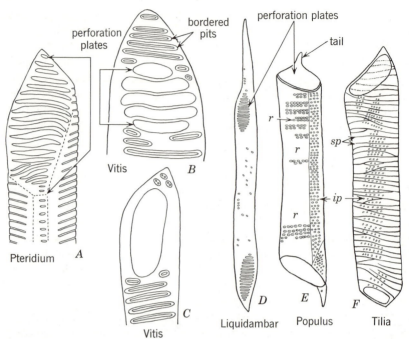

FIG. 11.2.   A–C, end walls of vessel members with perforations: A, B, scalariform, C, simple.   D–F, complete vessel members.   D, scalariform perforation plates.   E, simple perforation plates, intervessel pitting (*ip*), and ray contact areas (*r*).   F, simple perforation plates, intervessel pitting (*ip*), and spiral thickenings (*sp*).   (A, ×255; B, C, ×480; D, E, ×80, F, ×140.   D–F, after photomicrographs in Carpenter and Leney, *Coll. For. Syracuse Tech. Publ.* 74, 1952.)

with the rest of the primary wall (fig. 11.3, pl. 36C).   This thickening results, not from a deposition of additional wall substance, but from swelling of the intercellular substance.   In such walls the cellulose layers appear to remain extremely thin, whereas the intercellular pectic lamella visibly increases in thickness (Esau and Hewitt, 1940).   The swollen regions of the primary wall break down (fig. 11.3D, pl. 36D) but only after the secondary walls, where they occur, are fully formed and lignified.

The exact process of the removal of the cell wall during the perforation is not known.   According to one assumption, both cellulosic and noncellulosic components are removed by the action of the protoplast of the cell (Roelofsen, 1959); according to another, only the noncellulosic components are removed whereas the cellulosic microfibrillar network is pushed from the original position to the margins of the perforation (Frey-Wyssling, 1959). A related controversial question is whether or not plant cells contain cellulase necessary for degrading the cellulose. (The notion of a completely mechanical removal of the end wall, by tearing, during a supposed sudden expansion of the differentiating vessel is based on erroneous interpretations of microscopic views. See Esau and Hewitt, 1940.)

Typically the protoplast dies after the perforation is formed. According to ultrastructural investigations, the remnants of the dead protoplasts form a lining along the walls of tracheary elements (Scott et al., 1960). This lining has also been described as the warty layer (chapter 3; Liese, 1956).

Structure of Secondary Wall. The secondary walls of tracheary elements develop in a wide variety of patterns. Generally, in the first-formed part of the primary xylem a more limited area of the primary wall is covered by secondary wall layers than in the later-formed primary xylem and in the secondary xylem. Beginning with the earliest primary xylem, the secondary thickenings are deposited in the successive elements as rings, continuous helices, then as networks (figs. 11.4, 11.5). Such secondary thickenings are called, respectively, *annular, spiral* or *helical,* and *reticulate.* When the meshes of the net are rather distinctly elongated transversely, the thickening is called *scalariform-reticulate.* Tracheary elements with a still more extensive secondary wall constitute *pitted* elements (figs. 11.4; 11.5G, H). In

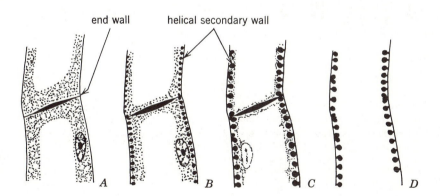

FIG. 11.3. Development of perforation plates in vessel members of celery. *A,* end wall thickened by swelling of intercellular material. *B–C,* thickened end wall and helical secondary wall thickening on lateral walls. *D,* end wall disintegrated, vessel member mature. Protoplast degenerating in *C,* absent in *D.* (× 800.)

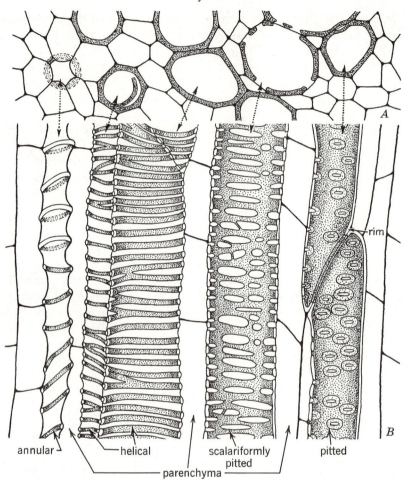

FIG. 11.4. Parts of primary tracheary elements and associated parenchyma cells from stem of *Aristolochia*. Transverse (A) and longitudinal (B) sections. In both, the earliest part of xylem is to the left. Element with annular thickenings is partly extended and adjacent parenchyma cells are bulging slightly into its lumen. Helically thickened elements have a few interconnections among the coils of the helix. Wide element with helical thickenings in B shows, above, a junction between two superposed elements. (×512.)

these, the secondary wall is interrupted only in the pit areas (and in the perforation plates of the vessel members). Pitted elements are characteristic of the latest primary xylem and of the secondary xylem. Comparative studies on fossils indicate that annular and spiral thickenings are more ancient than the pitted (Henes, 1959).

The details of the secondary wall sculpture in elements with annular, helical, and reticulate thickenings vary in different species of plants, and not all three patterns are necessarily present in a given plant. Furthermore, there may be gradual transitions among the different types, or combinations of more than one type of thickening in the same longitudinal series of elements or even in the same individual element (fig. 11.5C). Rings and helices vary in thickness. Some helices are grooved on their inner surface, occasionally so deeply that the helix appears double. Sometimes more than one helix is present in one element. The rings and the helices are more or less firmly attached to the primary wall (Badenhuizen, 1954). In many plants the thickening is connected with the primary wall by means of a narrow band. In sectional views the portion of the ring or the helix projecting over the narrow base appears like a border of a bordered pit (fig. 11.3D).

The different types of pitting encountered in tracheary cells have

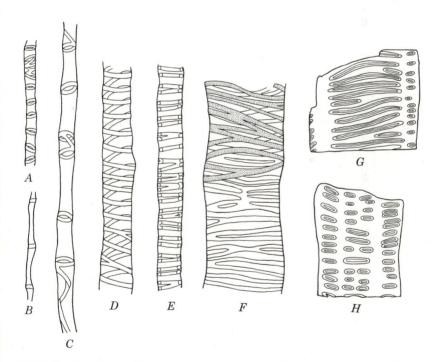

FIG. 11.5. Secondary wall structure in primary tracheary elements. *A–E, Hedera helix; F, Blechnum* (a fern); *G, H, Osmunda* (a fern). Thickenings are: annular, *A*; annular extended, *B*; annular, transitional to helical, *C*; helical, *D, E*; reticulate, *F*; scalariformly pitted, *G*; oppositely pitted, *H*. (All, ×600. After Bierhorst, *Phytomorphology* 10, 1960.)

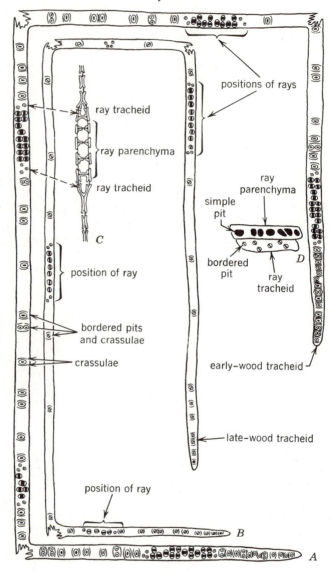

FIG. 11.6. Elements from secondary xylem of *Pinus*. *A*, early-wood and *B*, late-wood tracheids. Radial walls in face views. *C*, ray in transverse section as seen in tangential section of wood. *D*, two ray cells as seen in a radial section of wood. Tracheids in *A*, *B*, show contact areas with rays. Small pits in these areas connect axial tracheids with ray tracheids. Large pits with partial borders connect ray parenchyma cells with axial tracheids. Elsewhere tracheids have pits with full borders. (All, ×100. *A*, *B*, *D*, adapted from Forsaith, *Syracuse Univ. Tech. Publ.* 18, 1926, courtesy New York State College of Forestry.)

been described in chapter 3. Briefly, the pits are commonly bordered. The pit membranes have a torus in certain gymnosperms. If the bordered pits are elongated transversely and arranged in vertical series, the pattern is called *scalariform pitting* (figs. 11.1*G*, 11.2*A*). (Such wall thickening is often difficult to distinguish from the scalariform-reticulate). The circular or oval bordered pits are arranged in horizontal (*opposite pitting*) or oblique (*alternate pitting*) series. Helical thickening bands may develop on the surface of the pitted secondary wall without covering the pits (fig. 11.2*F*).

The pits on the wall of a given tracheary element are rarely all exactly alike (figs. 11.2, 11.6, 11.9), because the development of a pit is more or less affected by the nature of the other member of the pair of pits joining two cells. Distinctly bordered pit-pairs commonly occur on walls between two tracheary elements (*intervascular pitting*). There may be no pit-pairs or only a few small ones between a tracheary element and a fiber. Pit-pairs between tracheary elements and parenchyma cells are simple, half-bordered (with the border on the tracheary side, pl. 9*A*, *B*), or bordered.

The ontogenetic series of primary tracheary elements, beginning with the elements having annular thickenings and ending with those having pitted walls (sometimes with the omission of one or another type), occurs among vascular plants from the lowest to the highest levels on the phylogenetic scale (Bierhorst, 1960). In Ginkgoales, Coniferales, Gnetales, and Ophioglossaceae the helical and reticulate thickenings are combined with circular bordered pits of the type characteristic of the secondary tracheary elements of these plants (fig. 11.7*E*, *F*); scalariformly pitted elements are completely omitted (Bailey, 1925, 1944*b*; Bierhorst, 1960).

Phylogenetic Specialization. The xylem occupies a unique position among plant tissues in that the study of its anatomy has come to play such an important role with reference to taxonomy and phylogeny. The lines of specialization of the various structural features have been better established for the xylem than for any other single tissue. Many examples may be cited of the use that has been made of xylem anatomy to clarify taxonomic affinities (literature in Bailey, 1954; Carlquist, 1961; Metcalfe and Chalk, 1950). Among the individual features of the xylem the structure of the tracheary elements has been investigated especially thoroughly. Extensive comparative studies employing statistical methods and conducted with notable consistency have properly evaluated the variations in the morphology of the tracheary elements and explained their significance (Bailey, 1953, 1957*a*; Cheadle, 1953, 1956).

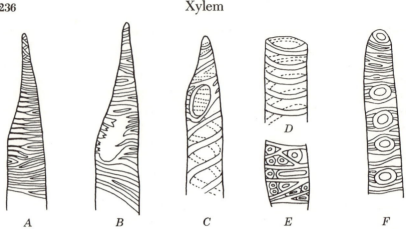

A   B   C   E   F

FIG. 11.7. Details of primary xylem elements. *A–D*, ends of dicotyledonous helically thick-ened vessel members with the following variations in perforation plates; *A*, scalariform; *B*, simple in transition from scalariform; *C*, simple, with rim; *D*, simple, with rim, on truncated end. Views in *A–C* may be used to illustrate evolutionary sequence in development of a simple perforation plate in helically thickened primary tracheary elements. *E, F*, parts of tra-cheary elements of *Ophioglossum* (*E*) and *Gnetum* (*F*) with combinations of reticulate and helical secondary thickenings and bordered pits. (*A–D, F*, after Bailey, *Amer. Jour. Bot.* 31, 1944; *E*, after Bierhorst, *Phytomorphology* 10, 1960.)

The tracheid is a more primitive element than the vessel member. The tracheid is the only kind of element found in the fossil seed plants, the pteridosperms (Andrews, 1961), and in most of the living lower vascular plants and gymnosperms (Jeffrey, 1917). Vessel members have evolved from tracheids and occur in the Gnetales; the dicotyledons, except representatives of the lowest taxonomic groups; the monocotyle-dons; certain ferns (Duerden, 1940; White, 1963*b*); *Selaginella* of the Lycopodiaceae (Duerden, 1934); and *Equisetum* (Bierhorst, 1958).

Vessels arose independently, through parallel evolution, in the six groups of plants named above. In the dicotyledons specialization of tracheids into vessel members occurred first in the secondary xylem and then gradually proceeded into the primary xylem, beginning with the latest part of this tissue (Bailey, 1944*b*). In monocotyledons (Cheadle, 1943*a, b*, 1944, 1955; Fahn, 1954*a, b*) vessels do not occur in the secondary xylem (few monocotyledons develop such tissue), and in the primary xylem they evolved first in the ontogenetically latest part, then in the earlier parts of this tissue. Monocotyledonous vessels appeared first in the root and later in stems, inflorescence axes, and leaves, in that order. The organographic origin of vessels in dicotyledons has been explored less completely, but in the secondary wood the evolution of vessels in root and stem appear to be synchronized (Bailey, 1944*b*). In *Pteridium*, in *Selaginella*, and in the secondary xylem of the dicotyledons the vessel members arose from tracheids with scalariform bordered pitting, in the Gnetales from tracheids that had circular bordered pits of the coniferous type (Bailey, 1944*b*, 1949). The vessel members of the primary xylem of angiosperms evolved not only from scalariformly pitted tracheids but also from tracheids with reticulate and helical secondary thickenings (fig. 11.7*A–D*; Bailey, 1944*b*; Cheadle, 1956). (The evolution of tracheary elements with annular thickenings has not been sufficiently investigated.) The perforation plate in vessel members derived from scalariformly pitted tracheids evolved in steps from a part of a wall bearing several bordered pits. At first only the pit membranes disappeared, then the borders ceased to develop, and finally the bars between individual openings were eliminated. Thus, a pitted wall part became a scalariform perforation plate, which later evolved into a simple perforation plate bearing a single opening. Concomitantly with these changes the vessel members gradually acquired definite end walls of decreasing degree of inclination, in contrast to the tapering ends of tracheids (fig. 11.1).

Structures representing the successive stages in the evolution of vessels of secondary xylem of dicotyledons are preserved in the existing representatives of this group of plants. Thus they are readily accessible for study and are well understood (Bailey, 1953; Cheadle, 1956). Surveys of vessel members in a broad and representative sampling of dicotyledons reveal that specialization has proceeded from long, narrow elements with tapering ends to short, wide ones having slightly inclined and transverse end walls which are almost completely eliminated by perforation (fig. 11.1). The phylogenetic shortening of vessel members is a particularly consistent characteristic and has occurred in all Tracheophyta that have developed vessels (Bailey, 1944*b*).

The pitting on the longitudinal walls of the vessel members of dicotyledons has also undergone evolutionary changes. In intervessel pitting, bordered pit-pairs in scalariform series have been replaced by circular bordered pit-pairs, first in opposite and ultimately in alternate series (fig. 11.1). The pit-pairs in walls between vessels and parenchyma have changed from fully bordered, to half-bordered, and finally to entirely simple (Frost, 1931).

Although vessels evolved in angiosperms, imperforate tracheary elements have also been retained, and they too have undergone phylogenetic modifications (fig. 11.1). The tracheids became shorter and developed a pitting essentially similar (the pits may be somewhat reduced) to that in the associated vessel members. The tracheids shortened much less than did the vessel members, and they generally did not increase in width.

In the ferns the shortening of the tracheids appears to be a less consistent characteristic than in the angiosperms (White, 1963a). The correlation between tracheid length and evolutionary divergence is obscured in this plant group by the variability in length of tracheids induced by various factors.

The different trends of specialization of tracheary elements discussed in the preceding paragraphs are not necessarily closely correlated within specific groups of plants. Some of these trends may be accelerated, others retarded, so that the more and the less highly specialized characters occur in combinations. Moreover, plants may secondarily acquire characteristics that appear primitive because of evolutionary loss. Vessels, for example, may be lost through nondevelopment of perforations in potential vessel members. In aquatic plants, parasites, and succulents, vessels may fail to develop concomitantly with a reduction of vascular tissue. These vesselless plants are highly specialized as contrasted with the primitively vesselless dicotyledons exemplified by *Trochodendron, Tetracentron, Drimys, Pseudowintera* and others (Bailey, 1953; Cheadle, 1956; Lemesle, 1956). In some families, as for example the Cactaceae and Compositae, evolutionary degeneration of vessel members involved a decrease in width of cells and nondevelopment of perforations (Bailey, 1957b; Carlquist, 1961). The resulting nonperforate cells, having the same kind of pitting as associated vessel members, are referred to as *vascular tracheids.* Another deviating trend in specialization may be the development of perforation plates of a reticulate type in an otherwise phylogenetically highly advanced family such as the Compositae (Carlquist, 1961).

Despite these inconsistencies, the main trends of vessel specialization in angiosperms are so reliably established that they play a significant role in the determination of specialization of other structures in the xylem. Furthermore, they may be used in the classification and identification of angiosperms, as well as in considerations of their origin (Bailey, 1957a; Carlquist, 1961).

Fibers

The xylary fibers are treated in chapter 10. To repeat briefly, fibers have thicker walls and reduced pit borders as compared with tracheids from which they have evolved (fig. 11.1). The two main types of xylem fiber, the fiber-tracheids and the libriform fibers, intergrade with each other and also with the tracheids. Because of this lack of clear separation between fibers and tracheids, the two kinds of elements are sometimes grouped together under the term "imperforate tracheary elements" (Bailey and Tupper, 1918). Like tracheids, fibers have undergone a phylogenetic shortening with increase in specialization of the xylem (fig. 11.1), but they are usually longer than the tracheids of the same plant because of more intensive apical intrusive growth. Fiber-tracheids have bordered pits with less developed borders than tracheids, whereas libriform fibers have simple or almost simple pits. Fibers are most highly specialized as supporting elements in those woods that have the most specialized vessel members (fig. 11.9), whereas such fibers are lacking in woods with tracheid-like vessel members (fig. 11.8). A further evolutionary advance results in the retention of protoplasts by fibers (Money et al., 1950).

Parenchyma Cells

Living parenchyma cells occur in both the primary and the secondary xylem. In the latter they are commonly present in two forms: *axial parenchyma*, derived, together with the tracheary elements and fibers, from the fusiform cambial initials, and *ray parenchyma*, formed by the ray initials of the cambium (fig. 11.11). The axial parenchyma cells may be as long as the fusiform initials (*fusiform parenchyma cells*, fig. 11.8), or they may be several times shorter, if a fusiform derivative divides transversely before differentiation into parenchyma (*parenchyma strand*, fig. 11.11). Parenchyma strands occur more commonly than parenchyma cells.

The ray parenchyma cells vary in shape but two fundamental forms may be distinguished, cells with their longest axes oriented radially (*procumbent ray cells*) and cells with their longest axes oriented vertically (*upright ray cells*). Ray cells that appear square in radial sections of the wood are called *square ray cells*, a modification of the upright type. In dicotyledons, the axial vessels may be connected across the rays by elements that have the shape of ray cells but are differentiated as vessel members. These cells are sometimes called *perforated ray cells* (Carlquist, 1960).

The ray cells and the axial parenchyma cells of the secondary xylem may or may not have secondary walls. If a secondary wall is present, the pit-pairs between the parenchyma cells and the tracheary elements may be bordered, half-bordered, or simple. Only simple pit-pairs occur between parenchyma cells. In the primary wall of parenchyma cells, the microfibrils are oriented approximately transversely to the longitudinal cell axis, in the secondary wall they form helices inclined between 30 and 60° to the cell axis (Wardrop and Dadswell, 1952).

The parenchyma cells of the xylem have a variety of contents. They

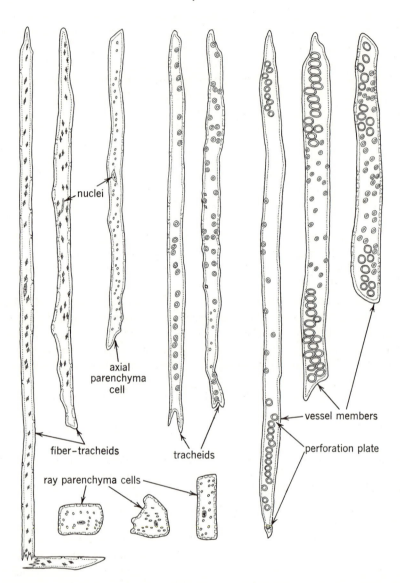

FIG. 11.8. Isolated elements from secondary xylem of *Ephedra californica* (Gnetales). Primitive wood with relatively little morphologic differentiation among elements of axial system. Typical fibers are absent. Axial and ray parenchyma cells have secondary walls with simple pits. Fiber-tracheids have living contents and pits with reduced borders. Tracheids have pits with large borders. Vessel members are slender, elongated, and have ephedroid perforation plates. (×155.)

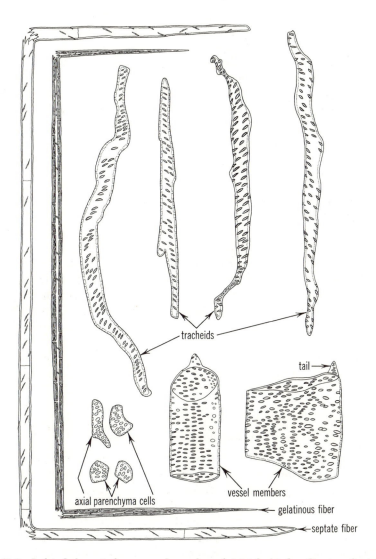

FIG. 11.9. Isolated elements from secondary xylem of *Aristolochia brasiliensis,* a dicotyle-donous vine. Specialized wood with elements of axial system diverse in form. Fibers are libriform, with reduced pit borders. Some are thin-walled and septate; others have thick gelatinous walls. Tracheids are elongated and irregular in shape, with slightly bordered pits. Vessel members are short and have simple perforations. Pits connecting vessel members with other tracheary elements are slightly bordered; others are simple. Axial parenchyma cells are irregular in shape and have simple pits. Ray parenchyma cells are not shown. They are relatively large, with thin primary walls. (×130.)

are particularly noted for storage of food reserves in the form of starch
or fat. Generally, the starch reserves accumulate toward the end of a
growth season and are depleted, not necessarily completely, during the
cambial activity of the following season. Tannins, crystals, and various
other substances may occur in these cells. Crystals vary in type and
assume characteristic distributional patterns in some families (Chattaway,
1955). In herbaceous plants and young twigs of woody plants chlorophyll often occurs in the xylary parenchyma cells, particularly the ray
cells (Gundersen and Friis, 1956).

Tyloses. In many plants the axial and the ray parenchyma cells
develop protrusions that enter tracheary cells when these become
inactive or the xylem tissue is injured (pl. 37A–C). Such outgrowths
from parenchyma cells are called *tyloses* (singular *tylose* or *tylosis*).
Tylose development occurs through the pit-pairs connecting the parenchyma cells with the tracheary elements.

Tyloses are sometimes so numerous that they completely fill the
lumen of the tracheid or vessel element (pls. 33, 37A). The nucleus of
the originating parenchyma cell and part of the cytoplasm appear in the
tylose. In the mature state tyloses may remain thin walled or develop
secondary walls which become lignified. In the primary wall the cellulosic microfibrils form a network similar to that of the primary walls
of parenchyma cells (Nečesaný, 1955). Tyloses may become subdivided
(Gertz, 1916). Sometimes they develop into sclereids.

PRIMARY XYLEM

Protoxylem and Metaxylem

When the primary xylem is studied in detail, some developmental and
structural differences usually may be observed between the earlier and
the later formed parts of this tissue. These two parts have been named
protoxylem and *metaxylem* (*proto* and *meta* are derived from the Greek
and mean first and beyond, respectively). Originally, the protoxylem
and the metaxylem were defined with regard to the relative time
of appearance of the two tissues. Later the consideration of the
morphologic differences was superimposed over the initial concept
(Bugnon, 1925; Esau, 1943). No single distinction is entirely satisfactory
because developmental details vary in different plants and commonly
the two parts of the primary xylem merge with one another imperceptibly.
In this book the terms protoxylem and metaxylem are used broadly to
characterize the basic pattern of xylem initiation in the shoot and in the
root. Temporal and positional relations are given the principal
consideration.

Protoxylem is the tissue which appears at the beginning of vascular differentiation and occupies a characteristic position in the primary vascular system of a plant or plant organ. Thus, for example, in higher vascular plants it is restricted to the largest vascular bundles in a given transection of a stem and occurs nearest the pith (endarch xylem, chapter 15), whereas in the transection of a root it appears at the outer extremities of the xylem system, that is, farthest from the center (exarch xylem, chapter 17). Ordinarily a stem, leaf, or root passes through a period of elongation after its initiation by the apical meristem. In the stem and leaf the protoxylem usually matures before these organs undergo intensive elongation. The metaxylem, which appears after the protoxylem, is in the process of differentiation while the shoot is elongating, and matures after this elongation is finished. In the root the protoxylem frequently matures beyond the region of major elongation. This relationship appears to be determined by the restriction of the elongation in the root to a shorter distance than in the stem. It may be modified in strongly elongating roots (Scherer, 1904).

In the discussion of the secondary wall of the tracheary elements, the ontogenetic sequence from annular through helical and reticulate to pitted wall sculpture was pointed out. The protoxylem elements commonly have annular and spiral thickenings, sometimes also reticulate. The metaxylem may have spiral, reticulate, and pitted secondary walls. (The Committee on Nomenclature, 1957, limits the metaxylem to the tissue with pitted tracheary elements.) The protoxylem elements, at least the first ones, are narrower than the metaxylem elements, but there may be a gradual transition in the size of cells between the two parts of the primary xylem.

If the protoxylem matures before the organ has elongated, as is typical of the shoot, its mature, nonliving tracheary elements are unable to keep pace with the extension of the surrounding tissue and are, therefore, stretched and frequently completely destroyed. During this stretching the primary wall is presumably torn, whereas the secondary wall is distorted. The rings are separated from one another and tilted, and the helices are extended (fig. 11.5C). Since the metaxylem matures after the organ completes its growth in length, its elements are not destroyed. But in the earliest metaxylem the secondary walls may be somewhat stretched during differentiation. In plants having no secondary growth, the metaxylem constitutes the only water-conducting tissue of the mature plant. In the presence of appreciable amounts of secondary growth the metaxylem usually becomes nonfunctioning, although its tracheary elements appear to remain intact. Sometimes they are filled with tyloses.

The protoxylem usually contains relatively few tracheary elements

(tracheids or vessel elements) and a considerable proportion of parenchyma cells. The latter either remain thin walled after the obliteration
of the tracheary elements or become lignified, with or without the
development of secondary walls. The metaxylem is, as a rule, a more
complex tissue than the protoxylem, and its tracheary elements are
generally wider. These elements may be differentiated into tracheids
or vessel members, and they are accompanied by parenchyma and
frequently also by fibers. The relatively high proportion of cells with
lignified secondary walls makes the metaxylem appear more compact
than the protoxylem.

Secondary Wall Structure and Xylem Development

The wall character of the primary xylem elements is influenced by the
amount of elongation of the organ in which they differentiate. The
normal proportion of the easily extensible elements with annular and
helical thickenings in the primary xylem may be affected by changes in
the character of elongation of the plant. Thus, if the elongation of a
plant organ is decelerated or inhibited (for example, by regulation of
light or use of X-rays), pitted elements instead of the extensible types
appear close to the apical meristem (Goodwin, 1942; Koernike, 1905;
Smith and Kersten, 1942). Among naturally growing roots those that
elongate much have a larger proportion of extensible forms of xylem
cells than roots showing little elongation (Scherer, 1904).

The causal relationships between the cessation of elongation and the
development of pitted elements are obscure (Goodwin, 1942; Stafford,
1948). Judged from the details of development of secondary walls seen
with the light microscope, the pattern of such walls is foreshadowed in
the cytoplasm. Before the secondary wall thickening is present the cytoplasm becomes denser along the parts of the wall that later are covered
by the secondary thickenings (Sinnott and Bloch, 1945). According to
one electron-microscope study, the dense cytoplasm contains numerous
mitochondria, dictyosomes, and vesicles of various sizes (Hepler and
Newcomb, 1963). If such cells are plasmolyzed and the protoplast
withdraws from the wall, the pattern is seen to be located in the outer
part of the protoplast, rather than on the wall (Crüger, 1855). These
observations do not support the concept that the secondary thickening
of the extensible primary xylem elements is laid down as a continuous
layer that is later pulled apart into rings, helices, or reticulae (Smith and
Kersten, 1942).

Observations on the relation between the secondary wall structure in
the primary xylem and the elongation of the growing plant show that,
in distinguishing between protoxylem and metaxylem, too much emphasis

upon wall sculpture would detract from the value of these terms. The relative time of maturation within the vascular system gives the most consistent basis for classification of the primary xylem into protoxylem and metaxylem (Esau, 1943; Goodwin, 1942).

SECONDARY XYLEM

Distinction from Primary Xylem

Like the division of the primary xylem into protoxylem and metaxylem, the classification into primary and secondary xylem is problematical. This classification, too, is of little value unless it is conceived in relation to the growth of the plant or of a plant organ as a whole (chapters 1, 4). Briefly, the primary xylem is the xylem differentiating in conjunction with the growth of the primary plant body and derived from the procambium. The secondary xylem is a part of the accessory secondary body superimposed over the primary and formed by the vascular cambium.

The secondary xylem is formed by a relatively complex meristem, the cambium, consisting of fusiform and ray initials, and is, therefore, composed of two systems, the *axial* (vertical) and the *ray* (horizontal) systems (figs. 11.10, 11.11), an architecture not characteristic of the primary xylem. In the dicotyledons the secondary xylem is commonly more complex than the primary in having a wider variety of component cells. The sculpture of the secondary walls of the primary and secondary tracheary elements has been considered earlier in this chapter. The elements of the late part of the metaxylem may intergrade with the secondary elements, since both may be similarly pitted.

Frequently the arrangement of cells, as seen in transverse sections, is stressed as a criterion for distinguishing the primary from the secondary xylem. The procambium and the primary xylem are said to have a haphazard cell arrangement, and the cambium and the secondary xylem, an orderly arrangement, with the cells aligned parallel with the radii of the secondary body. This distinction is highly unreliable, for in many plants the primary xylem shows just as definite radial seriation of cells as the secondary (Esau, 1943; chapter 15).

In many woody dicotyledons the length of tracheary cells reliably separates the primary from the secondary xylem (Bailey, 1944*b*). Although the helically thickened tracheary elements are generally longer than the pitted elements of the same primary xylem, these pitted elements are still considerably longer than the first secondary tracheary elements. Indeed, this difference may be so conspicuous that one can speak of a nonconformity between the two parts of the xylem (Bailey,

1944*b*). The apparent break in the continuity of development may be caused, not only by the elongation of the metaxylem cells during their differentiation and lack of a comparable elongation of the cambial derivatives but also by possible transverse divisions of the procambial cells just before the initiation of cambial activity. In the gymnosperms, too, the last primary xylem elements are longer than the first secondary elements (Bailey, 1920).

The change from longer to shorter tracheary cells at the beginning of secondary growth is one of the steps in the establishment of mature characteristics of the secondary xylem. Various other changes accompany this step, for example, those involving the pitting, the ray structure, and the distribution of axial parenchyma. By these changes, the secondary xylem eventually attains the evolutionary level characteristic of the species. Since the evolutionary specialization of the xylem progresses from the secondary to the primary xylem, in a given species the latter may be less advanced, or more juvenile, with regard to the evolutionary specialization. It appears that dicotyledons which are not truly woody —even if they possess secondary growth—show a protraction of juvenile characteristics into their secondary xylem (paedomorphosis, Carlquist, 1962). One of the expressions of this juvenility is a gradual, instead of a sudden, change in length of tracheary elements.

Basic Structure

Axial and Ray Systems. The arrangement of cells into the vertical, or axial, system, on the one hand, and the transverse, or ray system, on the other, constitutes one of the conspicuous characteristics of the secondary wood (figs. 11.10, 11.11). The rays and the axial system are arranged as two interpenetrating systems closely integrated with each other in origin, structure, and function. In conducting xylem the rays most commonly consist of living cells. The vertical system contains, depending on the species of plant, one or more of the different kinds of nonliving tracheary elements, fibers, and parenchyma cells. The living cells of the rays and those of the axial system are interconnected with each other, so that one can speak of a continuous system of living cells permeating the wood. Moreover, this system often is connected, through the rays, with the living cells of the pith, the phloem, and the cortex.

Since the longitudinal axis of the vertical system is parallel with the longitudinal axis of the stem or the root in which the xylem occurs, transverse and longitudinal sections of the organ coincide with the same kind of sections of the vertical system. The rays, on the contrary, have their longitudinal axes parallel with the radii of the approximately cylindrical bodies of stem and root and their branches. Consequently, transverse and radial longitudinal sections of the plant axis show the rays in longitudinal section, whereas tangential longitudinal sections expose the rays in their transverse section. If statements are made that xylem is sectioned transversely and longitudinally (radially or tangentially), the plane of sectioning is referred to the organ as a whole and, therefore, also to the axial system of the xylem.

The rays are characterized as having length, width, and height. The length is measured between the cambium and the innermost end of the ray. The width of the ray corresponds to its tangential extent and is commonly expressed in the number of cells in this direction. The height of a ray is its extent in the direction parallel with the longitudinal axis of stem or root.

The rays vary much in their dimensions in different plants and may be of more than one size in the same plant. If a ray is one cell wide, it is termed *uniseriate* (pls. 31, 32). The contrasting type is the *multiseriate* ray (pl. 33C), which may be a few cells to many cells wide (if the ray is two cells in width, it is called *biseriate*). A multiseriate ray, as seen in a tangential section of the xylem, tapers toward the upper and lower margins, where it is commonly uniseriate. Thus a wide ray appears lenticular or fusiform in its transectional outline. Both kinds of ray may be low or high. Although the height and width of rays often undergo considerable change through the successive layers of secondary xylem, the kind and extent of change induced are characteristic of a given species. The length of a ray, on the other hand, is an indefinite characteristic for three reasons: first, new rays are constantly initiated as the axis increases in circumference; second, some rays are discontinued; and, third, the length of the ray is affected by the vigor of growth.

Storied and Nonstoried Woods. In chapter 6 a distinction was made between storied, or stratified, and nonstoried, or nonstratified, cambia, with reference to the arrangement of the fusiform initials in tangential sections. Nonstoried cambia produce nonstoried woods (figs. 11.10, 11.11, pl. 31–33). The xylem derived from a storied cambium may be storied (pl. 35A, B) or only partly so, if the original stratification is obscured by changes during the differentiation of the xylem. One of the most common of such changes is the elongation of the elements in the vertical system. Tracheids, fiber-tracheids, and libriform fibers generally become longer than the fusiform cambial cells from which they were derived (chapter 6). The apices of these elements extend by intrusive growth beyond the limits of their own horizontal tier and thus partly efface its demarcation from the tiers above and below. A relatively indistinct stratification may be carried over by the xylem also from the cambium itself, since cambia show varied degrees of stratification. The stratification of a wood may be expressed only in the arrangement of the axial cells or also in that of the rays (Boureau, 1957; Cozzo, 1954). The degree of stratification may change during the development of the successive increments of xylem. The storied condition is associated with short fusiform initials and short vessel members and is, therefore, an advanced phylogenetic feature.

Growth Layers. The activity of the cambium is commonly periodic in the temperate regions, and the xylem produced during one growth period constitutes a *growth layer* (figs. 11.10, 11.11, pls. 31–33, 34A, B). In transverse sections of stems and roots such layers are referred to as *growth rings.* If the growth is definitely seasonal and occurs once during a season, the growth layer and the growth ring may be called the *annual layer* and the *annual ring,* respectively. If, however, the seasonal growth is interrupted by adverse climatic conditions, diseases, or other agents, and is later resumed, a second growth layer will be visible in the wood added during one season. Such an additional layer is called a *false annual ring* and the annual growth increment consisting of two or more growth rings is termed a *multiple annual ring.*

The growth rings are of varied degrees of distinctness, depending on the species of wood and also on growing conditions. The cause of the visibility of the growth layers in a section of wood is the structural difference between the xylem produced in the early and the late parts of the growth season. The *early wood* is less dense than the *late wood* and has generally larger cells and, proportionally, a smaller amount of wall substance per unit volume. In the temperate zone the early wood and the late wood are commonly called "spring wood" and "summer wood," respectively. The early wood of a given season merges more or less gradually with the late wood of the same season, but the division line between the late wood of one season and the early wood of the following season is ordinarily sharp.

The factors determining the change from the early-wood characteristics to those of the late wood are of continued interest to tree physiologists (Studhalter, 1955). One of the principal regulating influences appears to be the availability of auxin contributed by the growing shoots. In studies on *Pinus resinosa* the production of early-wood type was found to be associated with active extension growth, either a natural or one induced by photoperiodic treatment (Larson, 1960, 1662).

Growth rings occur in deciduous and evergreen trees. Furthermore, they are not confined to the temperate zone, with its striking contrast between the season of growth and the season of dormancy or rest, but may be present in subtropical and tropical woods. In the tropical species growth rings are often formed only under certain environmental conditions, whereas in many plants of the northern hemisphere zonation is produced under all conditions of growth (Bailey, 1944a). Growth rings have been recorded in a monocotyledon also (Chamberlain, 1921). The width of rings, that is, vigor of growth, is easily influenced by the external environment and is, therefore, variable (Glock et al., 1960; Trendelenburg, 1955). In straight parts of a tree growing under uniform conditions, the rings show orderly concentric arrangement. But many agencies of mechanical, chemical, and physiological nature may cause eccentric growth, sometimes of such a pronounced degree that part of the growth layers do not continue around the circumference of the axis.

Sapwood and Heartwood. The elements of the secondary xylem are variously specialized in relation to their function. The tracheary elements and the fibers that are concerned, respectively, with movement of water and support typically become devoid of protoplasts before their principal contribution to the physiological activity of the plant begins. The living cells, which store and translocate food (parenchyma cells and certain fibers), are alive at the height of xylem activity. Eventually, the living cells die. This stage is preceded by numerous changes in the wood that visibly differentiate the active *sapwood* from the inactive *heartwood* (Harris, 1954; Trendelenburg, 1955).

Many of the differences between sapwood and heartwood are chemical. With increasing age, the wood loses water and stored food substances and becomes infiltrated with various organic compounds, such as oils, gums, resins, tannins, and aromatic and coloring materials. Some of these substances impregnate the walls; others enter the cell lumina also. The development of color in the heartwood is a slow process dependent on the oxidation of phenols, which, in turn, follows the disappearance of starch and an apparent breakdown in the enzymatic control over the activities of living cells (Frey-Wyssling and Bosshard, 1959). In many woods tyloses develop in the tracheary cells (Chattaway, 1949). In gymnosperm xylem the pit membranes having tori may become fixed so that the tori are appressed to the borders and close the apertures (aspirated pit-pairs, chapter 3) and may be incrusted with lignin-like and other substances (Krahmer and Côté, 1963). The aspiration of bordered pits appears to be related to processes causing the drying of the central core of the wood (Harris, 1954). These various changes do not affect the strength of the wood but make it more durable than the sapwood, less easily attacked by decay organisms, and less penetrable to various liquids (including artificial preservatives).

The proportion of sapwood and heartwood and the degree of visible and actual differences between the two is highly variable in different species and in different conditions of growth. Some trees have no clearly differentiated heartwood (*Populus, Salix, Picea, Abies*), others have thin sapwood (*Robinia, Morus, Taxus*), and still others have a thick sapwood (*Acer, Fraxinus, Fagus*). In some species the sapwood is early converted into heartwood; in others it shows greater longevity . The development of heartwood sometimes results from a pathological state.

Reaction Wood. The reaction wood (Dadswell et al., 1958; Sinnott, 1952) is a type of wood produced on lower sides of branches and leaning and crooked stems of conifers (*compression wood*) and on the upper sides of the same kinds of axial parts in dicotyledons (*tension wood*). In the reaction wood, the fibers and tracheids have a rounded appearance, include intercellular spaces among them, and are shorter than normal. In the tracheids of the compression wood, the inner layer of the secondary wall is absent, the outer is wider than normal, and the middle layer shows many radial discontinuities. The wall is heavily lignified. The fibers of the tension wood are the so-called gelatinous fibers (pl. 10C; chapter 10). The gelatinous layer is high in cellulose, unlignified, and can be detected by its lack of fluorescence after a staining with fluorochromes (Siebers, 1960) and by its dark appearance with phase contrast (Jutte and Isings, 1955) and porous structure at the ultrastructural level (Côté and Day, 1962). Tension wood also shows a reduction in the number of vessels (Scurfield and Wardrop, 1962). The exact nature of the stimulus inducing the development of reaction wood is not known, but a strong correlation has been recorded in *Populus deltoides* between the proportion of gelatinous fibers and the degree of experimentally produced lean of the tree (Berlyn, 1961).

Gymnosperm Wood

The xylem of gymnosperms is generally simpler and more homogeneous than that of angiosperms (figs. 11.10, 11.11; pls. 31, 33). The chief distinction between the two kinds of wood is the absence of vessels in the gymnosperms (except in Gnetales; fig. 11.8) and their presence in most angiosperms. A further outstanding peculiarity of gymnosperm wood is the relatively small amount of parenchyma, particularly axial parenchyma (Jane, 1956).

The xylem of the Coniferales has been extensively studied, beginning

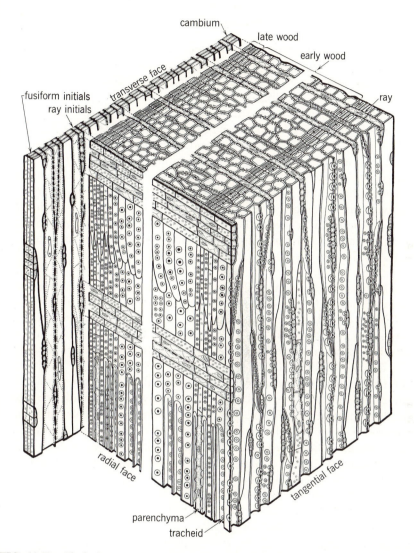

FIG. 11.10.   Block diagram of cambium and secondary xylem of *Thuja occidentalis* L. (White Cedar).   Conifer wood.   The axial system is composed of tracheids and small amount of parenchyma.   The ray system consists of low, uniseriate rays composed of parenchyma cells.   (Courtesy of I. W. Bailey.   Drawn by Mrs. J. P. Rogerson under the supervision of L. G. Livingston.   Redrawn.)

with the classical investigations of Sanio (1872–74) and continuing into modern times when the research was extended into the realm of ultrastructure (Bailey, 1954; Greguss, 1955; Wardrop and Dadswell, 1953).

Axial System. In gymnosperm xylem, the axial system consists mostly or entirely of tracheids. The late-wood tracheids develop relatively thick walls and pits with reduced borders, so that they may be classified as fiber-tracheids, but libriform fibers do not occur. The tracheids are long cells—they vary in length from 0.5 to 11 mm (Bailey and Tupper, 1918)—with their ends overlapping those of other tracheids (figs. 11.6, 11.10; pl. 31). The individual tracheids are regarded as having basically 14 sides with frequent increases in the number of faces to 18 and even 22 because of the incurved tips (Lewis, 1935). Although the fusiform initials from which these cells arise are wedge shaped at their ends, showing their pointed faces in the tangential sections and their blunt ends in the radial sections, the tracheid ends are more or less modified, because the cells undergo apical growth and adjust the shape of their ends to that of the spaces they invade. These ends may be forked (fig. 11.8).

The tracheids of extant gymnosperms are interconnected by circular or oval bordered pit-pairs in single, opposite (wide-lumened early-wood tracheids of Taxodiaceae and Pinaceae), or alternate (Araucariaceae) arrangement (figs. 11.6, 11.10). Some studies have shown that the number of pits on each tracheid may vary from approximately 50 to 300 (Stamm, 1946). The pit-pairs are most abundant on the ends where the tracheids overlap each other. In general, the pits are confined to the radial facets of the cells. Only the late-wood tracheids are pitted on their tangential walls (pl. 9D). In the bordered pit-pairs of gymnosperms more or less distinctly developed tori are present on the pit membranes in *Ginkgo*, Coniferales (pl. 12A), *Ephedra*. According to ultrastructural studies, tori are absent in *Gnetum, Welwitschia, Cycas revoluta*, and *Encephalartos* (Eicke, 1957, 1962; Eicke and Metzner-Küster, 1961; but see Bierhorst, 1960, on *Welwitschia*).

The movement of water through the tracheid system of conifers depends on the distribution of pits and the orientation of tracheids (slope of grain). In a study with injection of chemicals into conifer trunks, five patterns of movement were recognized in different species (Vité and Rudinsky, 1959). Two of these were spiral and only one entirely straight, sectorial.

The tracheids characteristically show thickenings of intercellular material and primary wall along the upper and lower margins of the pit-pairs (fig. 11.6, pl. 30A). These thickenings are called *crassulae* (from the Latin, little thickenings; Committee on Nomenclature, 1957). Still another not uncommon wall sculpture is represented by the *trabeculae*, small bars extending across the lumina of the tracheids from one tangential wall to the other. Tracheids with trabeculae commonly occur in long radial series of cells. Helical thickenings on pitted walls have been observed in the tracheids of *Pseudotsuga, Taxus, Cephalotaxus, Torreya*, and in some species of *Picea* (Phillips, 1948). Occasionally literature refers to perforated tracheids. These appear to be aberrant forms of no phylogenetic significance (Bannan, 1958).

Each tracheid is in contact with one or more rays. The proportion of the length of the tracheid wall that is joined to ray cells has been calculated as varying from 0.072 to 0.288 in different conifers (Stamm, 1931).

Where present, the axial parenchyma of the Coniferales is commonly distributed throughout the growth ring and occurs in long strands formed by transverse divisions of the mostly long fusiform cambial derivatives. Parenchyma is conspicuous in many Podocarpaceae, Taxodiaceae, and Cupressaceae; it is scanty in the Pinaceae; absent in Araucariaceae and Taxaceae (Phillips, 1948). In *Pinus*, axial parenchyma occurs only in the epithelium of the resin ducts (pl. 31). Secondary walls occur in axial parenchyma cells in the Abietoideae (*Picea, Pinus, Pseudotsuga, Cedrus, Keteleeria, Abies*).

Structure of Rays. The rays of gymnosperms are composed either of parenchyma cells alone (fig. 11.10), or of parenchyma cells and tracheids (pl. 30*B*). Ray tracheids are distinguished from ray parenchyma cells chiefly by their bordered pits and lack of protoplasts. They occur regularly in all Pinaceae, except *Abies, Keteleeria*, and *Pseudolarix*, occasionally in *Sequoia* and most Cupressaceae (Phillips, 1948).

Ray tracheids have lignified secondary walls. In some conifers these walls are thick and sculptured, with projections in the form of teeth or bands extending across the lumen of the cell. The ray parenchyma cells have living protoplasts in the sapwood and often darkly colored resinous deposits in the heartwood. They have only primary walls in Taxodiaceae, Araucariaceae, Taxaceae, Podocarpaceae, Cupressaceae, and Cephalotaxaceae (although the microfibrillar orientation of ray-cell walls of *Podocarpus amara* and *Tsuga canadensis* are interpreted as those typical of secondary walls; Wardrop and Dadswell, 1953), and have also secondary walls in Abietoideae (Bailey and Faull, 1934).

The rays of conifers are for the most part only one cell wide and from 1 to 20, sometimes up to 50, cells high. Ray tracheids may occur singly or in series, at the margins of a ray or interspersed among the layers of parenchyma cells. The presence of a resin duct in a ray makes the ray more than one cell wide except at the upper and lower limits (*fusiform ray*).

The ray cells having secondary walls are pitted with each other and also with the tracheids of the axial system. The pit pairs between the ray parenchyma cells and the axial tracheids are particularly distinctive. They are usually half-bordered, with the border being on the side of the tracheid (pl. 9*A*, *B*). The shape of these pit-pairs, their number, and their distribution on the rectangular facets of a wall, where a ray cell is in contact with an axial tracheid (the so-called *cross field*), are important features with regard to phylogeny and classification within smaller groups (Record, 1934).

Resin Ducts. Certain gymnosperms develop resin ducts in the axial system or in both the axial and ray systems (Pinaceae). Typically, resin ducts arise as schizogenous intercellular spaces by separation of paren- chyma cells from each other. After some divisions these cells form the lining, or the *epithelium,* of the resin ducts and excrete resin (chapter 13). In *Pinus* the epithelial cells are thin-walled, remain active for several years, and produce abundant resin. In *Pinus elliottii* the size and number of horizontal resin ducts per unit area of wood were found to be decreasing with the increase in age of the tree. Eventually the number became stabilized (Mergen and Echols, 1955). In *Abies* and *Tsuga* the epithelial cells have thick lignified walls and most of them die during the year of origin. These genera produce little resin. Eventually a resin duct may become closed by enlarging epithelial cells. These tylosis-like intrusions are called *tylosoids* (Record, 1934). They differ from tyloses in that they do not grow through pits.

Some workers make a distinction between resin ducts that are normal and those that are traumatic (from the Greek *trauma,* a wound), that is, arise in response to injury. Normal ducts are elongated and occur singly; traumatic ducts are cyst-like and occur in tangential series (Phillips, 1948). Other investigators consider all resin ducts in the wood traumatic (Bannan, 1936; Thomson and Sifton, 1925). The association of resin ducts with injuries has been observed under natural conditions and in controlled experiments (Bailey and Faull, 1934; Bannan, 1936; Thomson and Sifton, 1925). The phenomena that induce the develop- ment of traumatic resin ducts are numerous. Some of these are forma- tion of open and pressure wounds and injuries by frost and wind. Different groups of conifers are not alike in their response to injuries.

The variations in development and activity of the resin ducts suggest a phylogenetic series of increasing sensitivity to injury from the subtribe Abieteae to the subtribe Pineae (Bannan, 1936).

Angiosperm Wood

The designation angiosperm wood commonly refers to the secondary xylem of dicotyledons. The woody monocotyledons having secondary growth do not form a solid and homogeneous body of secondary xylem and are not a commercial source of wood (Record, 1934).

The secondary xylem of the dicotyledons is generally more complex than the wood of most gymnosperms since its elements are more varied in kind, size, form, and arrangement. The most complex dicotyledonous woods, such as that of oak, may contain vessel members, tracheids, fiber-tracheids, libriform fibers, axial xylem parenchyma, and rays of different sizes. Certain dicotyledonous woods are, however, less complicated in structure. Many Juglandaceae, for example, contain only fiber-tracheids among the imperforate nonliving cells (Heimsch and Wetmore, 1939). In the absence of vessels, the xylem of certain primitive dicotyledons appears so similar to the gymnosperm wood that it has been erroneously interpreted as being of the coniferous type (critique in Bailey, 1944a).

Because of the complexity of structure of dicotyledonous woods many characters may be used in their identification (Greguss, 1945; Kribs, 1950; Record, 1934; Record and Hess, 1943). Some of the major features are presence or absence of vessels and their distribution in the tissue; types of rays; distribution of axial parenchyma; presence of storied or nonstoried structure; types of perforation plates in the vessels.

Distribution of Vessels. The arrangement of vessels in dicotyledonous woods shows two main patterns. When the vessels have essentially equal diameters and are uniformly distributed through a growth ring, the wood is called *diffuse porous* (fig. 11.11; pl. 32; *Acer, Betula, Liriodendron*). (The word porous refers to the appearance of the vessels in transections. They seem like holes or pores in the section of the wood.) Woods with vessels of unequal diameters and with the largest vessels localized in the early wood are called *ring porous* because of the ringlike arrangement of the large vessels in transections of the xylem cylinder (pls. 33C, 34B; *Castanea, Fraxinus, Robinia,* and certain species of *Quercus*). Between these two extremes, various intergrades occur (pl. 34A). Moreover, in a given species the distribution of vessels may vary in relation to environmental conditions and may change with increasing age of a tree.

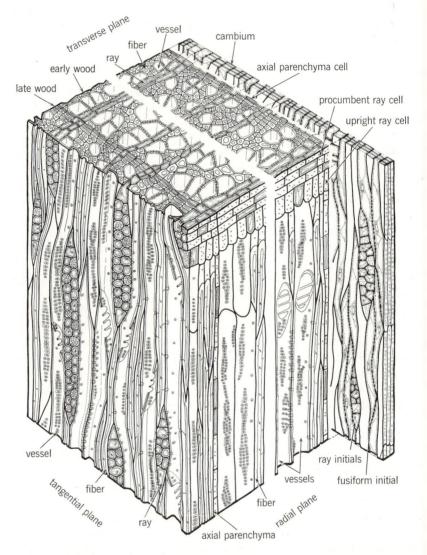

FIG. 11.11. Block diagram of cambium and secondary xylem of *Liriodendron tulipifera* L. (Tulip Tree). Dicotyledon wood. The axial system consists of vessel members with bordered pits in opposite arrangement and inclined end walls with scalariform perforation plates; fiber-tracheids with slightly bordered pits; and parenchyma strands in terminal position. The ray system contains heterocellular rays (marginal cells are upright, others procumbent), uniseriate and biseriate, of various heights. (Courtesy of I. W. Bailey. Drawn by Mrs. J. P. Rogerson under the supervision of L. G. Livingston. Redrawn.)

The ring-porous condition appears to be highly specialized and occurs in comparatively few woods, nearly all being species of the north temperate zone. Some wood anatomists consider the zone containing the large pores—the pore zone—as an additional tissue without an equivalent in the diffuse-porous woods (Studhalter, 1955). The vessels in ring-porous wood are longer than in the diffuse-porous kind (Handley, 1936).

The physiologic aspects also indicate the specialized nature of ring-porous wood. It conducts water almost entirely in the outermost growth increment or part of it (Kozlowski and Winget, 1963) and has a flow of water that is about ten times faster than that of diffuse-porous wood (Huber, 1935). Trees with ring-porous wood appear to produce their early-wood vessel system rapidly, whereas species with diffuse-porous wood form their new xylem slowly. A frequent accompaniment of the ring-porous condition is an early development of tyloses in the large early-wood vessels. It indicates that these highly specialized vessels are conducting for a short time only.

Within the major types of distributional patterns, the individual vessels, as seen in transverse sections, may be isolated from each other or they may occur in clusters of various sizes and shapes. The isolated vessels are circular or oval in outline; the clustered ones are flattened along the lines of contact with other vessels (pl. 32C).

Although the vessels may appear isolated in transections of wood, in the three dimensional aspect they are interconnected in various planes (fig. 11.12). Studies of conduction in different species by means of radioactive phosphorus and dyes indicate that in some species vessels are interconnected only within the growth increments, in others also between the growth increments (Braun, 1963). The vessels and other tracheary elements are also in contact with living cells, either with axial parenchyma or with ray cells or with both.

Distribution of Axial Parenchyma. The amount of axial parenchyma in dicotyledonous woods varies from very small or none to very large and the axial parenchyma shows diverse but intergrading patterns of distribution. Two basic types of distribution are distinguished (Committee on Nomenclature, 1957). In the *apotracheal* type (pl. 34D) the position of the parenchyma is independent of that of the vessels (although the two may be touching each other); in the *paratracheal* (pl. 34C) the two kinds of element are associated with one another. In the word apotracheal, *apo* means from in Greek and expresses, in this instance, independence from; in paratracheal, *para* means beside in Greek. In each distributional type subordinate variations are recognized. Apo-

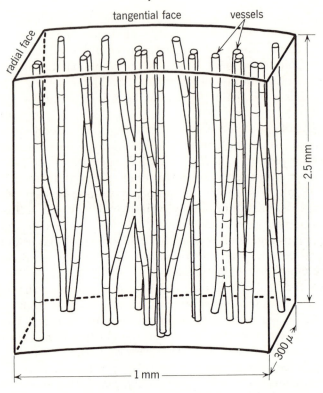

FIG. 11.12. Network of vessels in *Populus* wood with lateral connections between vessels in both radial and tangential planes. The horizontal dimensions are represented on a larger scale than the vertical. The delimitations of vessel members are approximate. (Adapted from Braun, *Ztschr. f. Bot.* 47, 1959.)

tracheal parenchyma may be *diffuse,* that is, dispersed throughout the growth ring, or *banded,* appearing in bands, or *marginal* (Carlquist, 1961), that is, restricted either to the end of a seasonal increment (*terminal parenchyma*) or to the beginning of one (*initial parenchyma*). Paratracheal parenchyma may be *scanty; vasicentric,* surrounding the vessels; *aliform,* vasicentric with wing-like tangential extensions; and *confluent,* coalesced aliform forming irregular tangential or diagonal bands. The phylogenetic sequence among the distributional types of wood parenchyma is from the diffuse arrangement to the other apotracheal and the paratracheal types (Bailey, 1957a).

It was pointed out in the discussion on fibers (chapter 10) that these tissue elements may function as starch-storage cells and that the retention of protoplasts by fibers is an evolutionary advance. This development appears to be associated with an evolutionary elimination of axial parenchyma or its reduction to scanty paratracheal or terminal (Money et al., 1950). The living fibers commonly become septate. Where such fibers are abundant they show apotracheal and paratracheal patterns of distribution similar to those shown by axial parenchyma (Spackman and Swamy, 1949).

Structure of Rays. The dicotyledons typically contain only parenchyma cells in the rays. The two main types of ray parenchyma cells, the procumbent and the upright, occur in various combinations. The ray is *homocellular* if it consists of only procumbent or only upright cells, *heterocellular* if it contains both morphological cell types (fig. 11.11; pl. 32; Committee on Nomenclature, 1957). Uniseriate and multiseriate rays both may be either homocellular or heterocellular.

The entire ray system may consist of either homocellular or heterocellular types or combinations of the two (Carlquist, 1961; Jane, 1956). On this basis the ray system is classified into *homogeneous*, rays all homocellular (all procumbent cells), or *heterogeneous*, rays all heterocellular or some are homocellular, others heterocellular. Under each of these categories further subdivisions are made with regard to whether the rays are all uniseriate, all multiseriate, or whether the two kinds are combined. Finally, the heterogeneous ray systems are subdivided into the third level of categories based on distribution of procumbent and upright cells in the component rays.

The variation in ray structure in different plant species has resulted from divergences during the evolution of the xylem (Bailey, 1957a; Kribs, 1935). Plants with primitive xylem have a combination of two kinds of ray, conspicuously high-celled (with cells vertically elongated) uniseriate rays and heterocellular multiseriate rays. During evolution, multiseriate rays have been either increased or decreased in size and number. Uniseriate rays have undergone a reduction in height and number. One or the other or both kinds of ray have been eliminated in certain evolutionary lines. Thus examples of specialized ray structure may be a combination of large multiseriate rays with small uniseriate rays (*Quercus*, pl. 33); or the presence of only one kind of ray, either multiseriate or uniseriate (pl. 32); or the complete absence of rays. Specialization has also affected the cellular composition of the rays and resulted in the development of homocellular rays from heterocellular.

The more advanced ray structure often appears only in the later increments of the xylem, with the earlier-formed secondary xylem having a primitive structure. In such instances the process of phylogenetic modification may be determined by comparing successive tangential sections through the wood and noting the changes that a particular ray underwent after its inception and during its continued growth within the consecutive growth layers. In this way the progressive modification in the ontogeny of a given ray may be revealed (Barghoorn, 1940, 1941; Shimaji, 1962). The significant implication of such changes in ray structure is that the ontogenetic stages in the xylem of the same individual plant represent different levels of phylogenetic specialization.

The evolutionary changes that may be recognized in serial sections of wood are reflections of ontogenetic changes in the cambium (chapter 6). Ray initials may be displaced by fusiform initials in the cambium, either within the group of initials or on its margins. If the displacement occurs within a group of ray initials, the ray appears to be split by the fusiform initials into two or more parts. Usually such separation of a ray into parts occurs through intrusion of a fusiform initial, by apical intrusive growth, into a group of ray initials. But the ray may also be broken up into parts by the change of some of the ray initials into fusiform initials (pl. 35D). The last method often modifies large multiseriate rays into structures resembling aggregations of small multiseriate rays. Actual aggregations accompanied by partial fusions may also occur. Rays may increase in size by fusion with one another or by radial divisions of ray initials. Fusions of rays are brought about by elimination from the cambium of fusiform initials intervening between groups of ray initials.

Secretory Canals. Intercellular canals similar to the resin ducts of the gymnosperms occur in dicotyledonous woods (Record, 1934). These are arbitrarily distinguished from the resin ducts of conifers by the designation *gum ducts*, although they may contain various substances, including resins, oils, gums, and mucilages (Stern, 1954). The gum ducts occur in the axial and ray systems and originate by schizogeny or lysigeny or by a combination of the two methods. Frequently they have no differentiated epithelium. Instead of long canals, relatively small cavities may develop. These are comparable to the resin cysts of the gymnosperms and are called gum cysts.

Many of the secretory ducts are undoubtedly traumatic in origin, and the agents inducing their formation are as varied as those responsible for the development of the resin ducts in the gymnosperms. Gum ducts often develop in association with *gummosis*, a degeneration of cells resulting in the formation of complex and variable substances commonly referred to as gum. Most investigators agree that gum is derived from decomposition of carbohydrates, particularly starch, but also of those occurring in the cell walls. Hence gummosis results in depletion of starch in cells, but it may also bring about a breakdown of the cell walls. The gum may collect in the gum ducts or in various xylem cells, including vessel members. Plants frequently respond with gummosis to disease infection, to injury by insects, and to physiological disturbances (Esau, 1948).

Differentiation in the Secondary Xylem

The derivatives that arise on the inner face of the cambium through tangential divisions of the cambial initials undergo complex changes during their development into the various elements of the xylem (figs. 11.10, 11.11). The basic distinction in form and orientation between the elements of the axial and the ray systems is determined by the structure of the cambium itself, since the cambium is composed of fusiform and ray initials. Also, all the changes in the relative proportions between these two systems—for example, the addition or the elimination of rays—originate in the cambium.

The derivatives of the ray initials undergo relatively little change during differentiation. Generally, the ray cells remain parenchymatic —some with primary walls, others with secondary walls—and their contents may not change much, since the ray initials themselves often contain such substances as starch and tannins. Ray cells enlarge radially as they emerge from the cambium, but the distinction between the upright and the procumbent cells is apparent in the cambium. A profound change occurs in the ray tracheids of gymnosperms, for these cells develop secondary walls with bordered pits and become devoid of protoplasts.

The ontogenetic changes in the axial system vary with the type of cell and may result in striking contrasts between the cambial cells and their derivatives. The cells developing into vessel members elongate slightly, if at all (fig. 4.2E), but they expand laterally, often so strongly that their ultimate width exceeds their height. Short, wide vessel members are characteristic of highly specialized xylem. In many species of dicotyledons the vessel members expand in their median parts but not at the ends, which overlap those of the vertically adjacent elements. These ends are ultimately not occupied by the perforation and appear like elongated wall processes, tails, with or without pits (figs. 11.2D, E, 11.9).

Expansion of the vessel members affects the arrangement and the shape of adjacent cells. These cells become crowded out of their original position and cease to reflect the radial seriation present in the cambial zone. The rays, too, may be deflected from their original positions. The cells in the immediate vicinity of an expanding vessel enlarge parallel with the surface of the vessel and assume a flattened appearance. But often these cells do not keep pace with the increase of the circumference of the vessel and become partly or completely separated from each other. As a result, the expanding vessel element comes in contact with new cells. The expansion of a vessel member can be pictured as a phenomenon involving both symplastic and intrusive growth. As long as the cells next to the vessel element expand in unison with it, the common walls of the various cells undergo symplastic growth. During separation of adjacent cells, the vessel-member wall intrudes between the walls of the other cells.

The separation of the cells located next to an expanding vessel causes the development of cells having odd, irregular shapes. Some remain partially attached to each other and, as the vessel member continues to enlarge, these connections extend into long tubular structures (pl. 36B). The parenchyma cells and the tracheids that are thus affected by developmental adjustments have received the names *disjunctive parenchyma cells* and *disjunctive tracheids*, respectively (Record, 1934). These cells are modified growth forms of the xylem parenchyma cells and the tracheids of the axial system.

In contrast to the vessel members, the tracheids and the fibers show relatively little increase in width but often elongate much during their differentiation. The degree of elongation of these elements in the different groups of plants varies widely. In the conifers, for example, the cambial initials themselves are very long, and their derivatives elongate only slightly. In the dicotyledons, on the contrary, the tracheids and the fibers become considerably longer than the meristematic cells. If the xylem contains tracheids, fiber-tracheids, and fibers, the fibers elongate most, although the tracheids attain the largest volume because of their greater width. The elongation occurs through apical intrusive growth. In the extreme storied woods, there may be little or no elongation of any kind of element (pl. 35B; Record, 1934).

Woods containing no vessels retain a rather symmetric arrangement of cells, because in the absence of strongly expanding cells the original radial seriation characteristic of the cambial region is not much disturbed. There is some change in alignment resulting from apical intrusive growth of the axial tracheids.

Vessel elements, tracheids, fiber-tracheids, and libriform fibers develop secondary walls and the end walls of the vessel members become perforated. Ultimately the protoplasts disintegrate in those cells that are nonliving in the mature state.

The fusiform meristematic cells that differentiate into the axial parenchyma typically do not elongate. If a parenchyma strand is formed, the fusiform cell divides transversely. No such divisions occur during the development of a fusiform parenchyma cell. In some plants the parenchyma cells develop secondary walls but do not die until the heartwood is formed. The parenchyma cells associated with resin and gum ducts in the vertical system arise like axial parenchyma cells by transverse divisions of fusiform cambial cells.

The elongation of certain cells in the xylem just discussed occurs among the derivatives of the cambial cells. Another type of elongation occurs as a result of the elongation of the fusiform cambial initials mentioned in chapter 6. Because of this phenomenon the tracheids of conifers increase in length from year to year until a maximum is attained at an advanced age of the tree (Dinwoodie, 1961). There is, secondly, a seasonal variation in length. If the multiplicative anticlinal divisions of the fusiform initials, which reduce the length of the cells, occur at the end of a seasonal growth, the early-wood tracheids are on the average shorter than those of the late wood (Chalk and Ortiz, 1961). A yearly increase in length was observed also in fibers of nonstoried dicotyledonous woods (Bosshard, 1951; Hejnowicz and Hejnowicz, 1958). As was explained in chapter 6, the increase in length of the fusiform initials in conifers and dicotyledons with nonstoried wood occurs by intrusive growth following the oblique anticlinal divisions. In storied cambia the multiplicative divisions are radial anticlinal, which do not materially change the length of the initials. This relationship is reflected in the constancy of length of parenchyma strands and vessel members in storied woods (Chalk et al., 1955). The fibers in such woods elongate independently of the length of the initials and this elongation may show an increase over the years (Hejnowicz and Hejnowicz, 1959).

Strength of Wood in Relation to Structure

The composition of the xylem tissue and the structure and the arrangement of the component elements determine the physical properties of woods and their suitability for commercial uses (Forsaith, 1926; Record, 1934). A consideration of the effect of structure upon one of the most important characteristics, strength, enhances the understanding of xylem histology. The word strength is used here in a broad sense, referring collectively to properties enabling the wood to resist different forces or loads. These properties are manifold and are not necessarily closely correlated, so that a given wood may be strong with reference to one kind of force and weak with reference to another.

Probably the most important single characteristic that gives an indication of the strength of wood is its specific gravity. In an absolutely dry wood specific gravity depends on the volume of the wall material and its chemistry. The specific gravity of the wall substance as such has been calculated to be between 1.40 and 1.62, but because of variable proportions of walls in the different woods their specific gravity may be as low at 0.04 and as high as 1.46 (Record, 1934). The strength that might be predicted from specific gravity, however, is often considerably modified by histologic structure.

It is particularly instructive to compare the effects of the different types of xylem elements upon the strength of the wood. Because of their length, thickness of walls, and sparse pits, libriform fibers and fiber-tracheids are chiefly responsible for the strength of the dicotyledonous woods. (The weakening effect of pits upon the walls has been demonstrated experimentally; Forsaith, 1926.) These types of cells are particularly influential when aggregated in dense masses. The importance of fibers as mechanical cells is clearly indicated by the close correlation often observed between fiber volume, specific gravity, and strength of woods (Forsaith, 1926).

Along with the strong fibers, dicotyledonous woods also contain elements that are relatively weak. Among these the vessels are especially notable because their diameters are large in proportion to the volume of the wall. Obviously, their number and distribution influence their weakening effect. For example, ring-porous woods with their aggregation of very large vessels in a localized region are less resistant to certain stresses than woods with more evenly distributed vessels.

Axial xylem parenchyma may influence the strength of a wood if it is abundant. In some dicotyledons it may occupy as much as 23 per cent of the total volume of the xylem (Forsaith, 1926). Apparently the distribution of parenchyma is of as much importance as its total volume, and it might be expected to reduce the resistance to certain forces, if it occurs in wide bands in recurring zonations.

The relation of rays to the strength of woods is complicated by the circumstance that woods with a greater volume of ray tissue often are highly specialized and have a large volume of heavy-walled fibers giving them a high specific gravity. If two species of woods are of the same specific gravity but have a different volume of ray tissue, the wood with the larger amount of this tissue is weaker (Forsaith, 1926).

Gymnosperm woods do not have such weak elements as the vessels of angiosperms and possess only a relatively small volume of parenchyma cells. On the other hand, they do not have such strong elements as the fibers of dicotyledonous woods. In general, gymnosperm woods vary in strength and hardness. (The terms softwood for gymnosperm wood and hardwood for angiosperm wood are misnomers because there are soft and hard woods in both taxa; Record, 1934.) The homogeneous structure of gymnosperm woods, with the predominance of long elements, makes them easily workable and particularly suitable for paper making.

The late wood is generally stronger than the early wood because of the larger volume of wall material. Variation in the width of the growth rings affects the strength of different woods in different ways. Reduction in the width of a ring of a conifer lowers the proportion of the thin-walled, large-celled early-wood type. Within certain limits, therefore, coniferous wood with narrow rings is stronger than wood with wide rings. In dicotyledons, on the contrary, reduction in width of rings occurs mainly at the expense of the late wood. Therefore, hardwoods with wider rings are stronger. These relations hold, of course, as long as no uncommon reduction of wall thickness accompanies the development of wide rings. The change of sapwood to heartwood does not increase the strength of the wood.

REFERENCES

Andrews, H. N., Jr. *Studies in Paleobotany*. New York, John Wiley and Sons. 1961.

Badenhuizen, N. P. Some observations on removable spirals in *Scilla ovatifolia* Bak. *Protoplasma* 43:429–440. 1954.

Bailey, I. W. The cambium and its derivative tissues. II. Size variations of cambial initials in gymnosperms and angiosperms. *Amer. Jour. Bot.* 7:355–367. 1920.

Bailey, I. W. Some salient lines of specialization in tracheary pitting. I. Gymnospermae. *Ann. Bot.* 39:587–598. 1925.

Bailey, I. W. The comparative morphology of the Winteraceae. III. Wood. *Arnold Arboretum Jour.* 25:97–103. 1944a.

Bailey, I. W. The development of vessels in angiosperms and its significance in morphological research. *Amer. Jour. Bot.* 31:421–428. 1944b.

Bailey, I. W. Origin of the angiosperms: need for a broadened outlook. *Arnold Arboretum Jour.* 30:64–70. 1949.

Bailey, I. W. Evolution of the tracheary tissue of land plants. *Amer. Jour. Bot.* 40:4–8. 1953.

Bailey, I. W. *Contributions to plant anatomy*. Waltham, Mass., Chronica Botanica Company. 1954.

Bailey, I. W. The potentialities and limitations of wood anatomy in the study of the phylogeny and classification of angiosperms. *Arnold Arboretum Jour.* 38:243–254. 1957a.

Bailey, I. W. Additional notes on the vesselless dicotyledon, *Amborella trichopoda* Baill. *Arnold Arboretum Jour.* 38:374–378. 1957b.

Bailey, I. W., and A. F. Faull. The cambium and its derivative tissues. IX. Structural variability in the redwood *Sequoia sempervirens*, and its significance in the identification of the fossil woods. *Arnold Arboretum Jour.* 15:233–254. 1934.

Bailey, I. W., and W. W. Tupper. Size variation in tracheary cells. I. A comparison between the secondary xylems of vascular cryptogams, gymnosperms and angiosperms. *Amer. Acad. Arts and Sci. Proc.* 54:149–204. 1918.

Bannan, M. W. Vertical resin ducts in the secondary wood of the Abietineae. *New Phytol.* 35:11–46. 1936.

Bannan, M. W. An occurrence of perforated tracheids in *Thuja occidentalis* L. *New Phytol.* 57:132–134. 1958.

Barghoorn, E. S., Jr. The ontogenetic development and phylogenetic specialization of rays in the xylem of dicotyledons. I. The primitive ray structure. *Amer. Jour. Bot.* 27:918–928. 1940. II. Modification of the multiseriate and uniseriate rays. *Amer. Jour. Bot.* 28:273–282. 1941.

Berlyn, G. P. Factors affecting the incidence of reaction tissue in *Populus deltoides* Bartr. *Iowa State Jour. Sci.* 35:367–424. 1961.

Bierhorst, D. W. Vessels in *Equisetum*. *Amer. Jour. Bot.* 45:534–537. 1958.

Bierhorst, D. W. Observations on tracheary elements. *Phytomorphology* 10:249–305. 1960.

Bosshard, H. H. Variabilität der Elemente des Eschenholzes in Funktion der Kambium-tätigkeit. *Schweiz. Ztschr. f. Forstw.* 12:648–665. 1951.

Boureau, E. *Anatomie végétale.* Vol. 3. Paris, Presses Universitaires de France. 1957.

Braun, H. J. Die Organization des Hydrosystems im Stammholz der Bäume und Sträucher. *Deut. Bot. Gesell. Ber.* 75:401–410. 1963.

Bugnon, P. Origine, évolution et valeur des concepts de protoxylème et de metaxylème. *Soc. Linn. de Normandie, Bul.* Ser. 7. 7:123–151. 1925.

Carlquist, S. Wood anatomy of Astereae (Compositae). *Trop. Woods* 113:54–84. 1960.

Carlquist, S. *Comparative plant anatomy.* New York, Holt, Rinehart and Winston. 1961.

Carlquist, S. A theory of paedomorphosis in plants. *Phytomorphology* 12:30–45. 1962.

Chalk, L., E. B. Marstrand, and J. P. De C. Walsh. Fibre length in storeyed hardwoods. *Acta Bot. Neerl.* 4:339–347. 1955.

Chalk, L., and M. Ortiz C. Variation in tracheid length within the ring in *Pinus radiata* D. Don. *Forestry* 34:119–124. 1961.

Chamberlain, C. J. Growth rings in a monocotyl. *Bot. Gaz.* 72:293–304. 1921.

Chattaway, M. M. The development of tyloses and secretion of gum in heartwood formation. *Austral Jour. Sci. Res. B, Biol. Sci.* 2:227–240. 1949.

Chattaway, M. M. Crystals in woody tissues; Part I. *Trop. Woods* 102:55–74. 1955.

Cheadle, V. I. The origin and certain trends of specialization of the vessel in the Monocotyledoneae. *Amer. Jour. Bot.* 30:11–17. 1943a.

Cheadle, V. I. Vessel specialization in the late metaxylem of the various organs in the Monocotyledoneae. *Amer. Jour. Bot.* 30:484–490. 1943b.

Cheadle, V. I. Specialization of vessels within the xylem of each organ in the Monocotyledoneae. *Amer. Jour. Bot.* 31:81–92. 1944.

Cheadle, V. I. Independent origin of vessels in the monocotyledons and dicotyledons. *Phytomorphology* 3:23–44. 1953.

Cheadle, V. I. The taxonomic use of specialization of vessels in the metaxylem of Gramineae, Cyperaceae, Juncaceae, and Restionaceae. *Arnold Arboretum Jour.* 36:141–157. 1955.

Cheadle, V. I. Research on xylem and phloem—progress in fifty years. *Amer. Jour. Bot.* 43:719–731. 1956.

Committee on Nomenclature, International Association of Wood Anatomists. International glossary of terms used in wood anatomy. *Trop. Woods* 107:1–36. 1957.

Côté, W. A., Jr., and A. C. Day. The G layer in gelatinous fibers—electron microscope studies. *Forest Prod. Jour.* 12:333–338. 1962.

Cozzo, D. Filogenia de los tipos de estructura leñosa estratificada. *Rev. Argentina Agron.* 21:196–214. 1954.

Crüger, H. Zur Entwickelungsgeschichte der Zellenwand. *Bot. Ztg.* 13:601–613, 617–629. 1855.

Dadswell, H. E., A. B. Wardrop, and A. J. Watson. The morphology, chemistry and pulp characteristics of reaction wood. In: *Fundamentals of Papermaking Fibres*. British Paper and Board Makers' Association. 1958.

Dinwoodie, J. M. Tracheid and fibre length in timber: a review of literature. *Forestry* 34: 125–144. 1961.

Duerden, H. On the occurrence of vessels in *Selaginella*. *Ann. Bot.* 48:459–465. 1934.

Duerden, H. On the xylem elements of certain ferns. *Ann. Bot.* 4:523–531. 1940.

Eicke, R. Elektronenmikroskopische Untersuchungen an Gymnospermenhölzern als Beitrag zur Phylogenie der Gnetales. *Bot. Jahrb.* 77:193–217. 1957.

Eicke, R. Die Bedeutung der Feinstrukturen des Holzes von *Welwitschia mirabilis* für die Phylogenie der Chlamydospermen. *Bot. Jahrb.* 81:252–260. 1962.

Eicke, R., and I. Metzner-Küster. Feinbauuntersuchungen an den Tracheiden von Cycadeen I. *Cycas revoluta* und *Encephalartos* spec. *Deut. Bot. Gesell. Ber.* 74:99–104. 1961.

Esau, K. Origin and development of primary vascular tissues in seed plants. *Bot. Rev.* 9:125–206. 1943.

Esau, K. Anatomic effects of the viruses of Pierce's disease and phony peach. *Hilgardia* 18:423–482. 1948.

Esau, K. *Plants, viruses, and insects*. Cambridge, Mass., Harvard University Press. 1961.

Esau, K., and W. B. Hewitt. Structure of end walls in differentiating vessels. *Hilgardia* 13:229–244. 1940.

Fahn, A. Metaxylem elements in some families of the Monocotyledoneae. *New Phytol.* 53:530–540. 1954a.

Fahn, A. The anatomical structure of the Xanthorrhoeaceae Dumort. *Linn. Soc. London, Jour., Bot.* 55:158–184. 1954b.

Forsaith, C. C. The technology of New York State timbers. *N.Y. State Col. Forestry, Syracuse Univ., Tech. Pub.* 18. Vol. 26. 1926.

Frey-Wyssling, A. *Die pflanzliche Zellwand*. Berlin, Springer-Verlag. 1959.

Frey-Wyssling, A., and H. H. Bosshard. Cytology of the ray cells in sapwood and heartwood. *Holzforschung* 13:129–137. 1959.

Frost, F. H. Specialization in secondary xylem of dicotyledons. III. Specialization of lateral wall of vessel segment. *Bot. Gaz.* 91:88–96. 1931.

Gertz, O. Untersuchungen über septierte Thyllen nebst anderen Beiträgen zu einer Monographie der Thyllenfrage. *Lunds Univ. Arsskr.* N. F. Avd. 2. Vol. 12. No. 12. 1916.

Glock, W. S., R. A. Studhalter, and S. R. Agerter. Classification and multiplicity of growth layers in the branches of trees. *Smithsonian Misc. Publs.* 140:1–292. 1960.

Goodwin, R. H. On the development of xylary elements in the first internode of *Avena* in dark and light. *Amer. Jour. Bot.* 29:818–828. 1942.

Greenidge, K. N. H. An approach to the study of vessel length in hardwood species. *Amer. Jour. Bot.* 39:570–574. 1952.

Greguss, P. *Bestimmung der mitteleuropäischen Laubhölzer und Sträucher auf xylotomischer Grundlage*. Budapest, Hungarian Museum of Natural History. 1945.

Greguss, P. *Identification of living gymnosperms on the basis of xylotomy*. Budapest, Akademiai Kiado. 1955.

Gundersen, K., and J. Friis. Chlorophyll i marv og ved hos løvfaeldende traeer [Chlorophyll in pith and xylem of deciduous trees.] *Bot. Tidsskr.* 53:60–66. 1956.

Handley, W. R. C. Some observations on the problem of vessel length determination in woody dicotyledons. *New Phytol.* 35:456–471. 1936.

Harris, J. M. Heartwood formation in *Pinus radiata* (D. Don). *New Phytol.* 53:517–524. 1954.

Heimsch, C., Jr., and R. H. Wetmore. The significance of wood anatomy in the taxonomy of the Juglandaceae. *Amer. Jour. Bot.* 26:651–660. 1939.

Hejnowicz, A., and Z. Hejnowicz. Variations in length of vessel members and fibres in the trunk of *Populus tremula* L. *Soc. Bot. Polon. Acta* 27:131–159. 1958.

Hejnowicz, A., and Z. Hejnowicz. Variations of length of vessel members and fibers in the trunk of *Robinia pseudoacacia*. *Soc. Bot. Polon. Acta* 28:453–460. 1959.

Henes, E. Fossile Wandstrukturen. In: *Handbuch der Pflanzenanatomie*. Band 3. Teil 5. 1959.

Hepler, P. K., and E. H. Newcomb. The fine structure of young tracheary xylem elements arising by differentiation of parenchyma in wounded *Coleus* stem. *Jour. Expt. Bot.* 14:496–503. 1963.

Huber, B. Die physiologische Bedeutung der Ring- und Zerstreutporigkeit. *Deut. Bot. Gesell. Ber.* 53:711–719. 1935.

Jane, F. W. *The structure of wood.* New York, The Macmillan Company. 1956.

Jeffrey, E. C. *The anatomy of woody plants.* Chicago, University of Chicago Press. 1917.

Jutte, S. M., and J. Isings. The determination of tension wood in ash with the aid of the phase-contrast microscope. *Experientia* 11:386–390. 1955.

Koernike, M. Über die Wirkung von Röntgen- und Radiumstrahlen auf die Pflanzen. *Deut. Bot. Gesell. Ber.* 23:404–415. 1905.

Kozlowski, T. T., and C. H. Winget. Patterns of water movement in forest trees. *Bot. Gaz.* 124:301–311. 1963.

Krahmer, R. L., and W. A. Côté, Jr. Changes in coniferous wood cells associated with heartwood formation. *Tappi* 46:42–49. 1963.

Kribs, D. A. Salient lines of structural specialization in the wood rays of dicotyledons. *Bot. Gaz.* 96:547–557. 1935.

Kribs, D. A. *Commercial foreign woods on the American market.* Lithoprint. Ann Arbor, Michigan, Edwards Brothers Inc. 1950.

Larson, P. R. A physiological consideration of the springwood-summerwood transition in red pine. *Forest Sci.* 6:110–122. 1960.

Larson, P. R. The indirect effect of photoperiod on tracheid diameter in *Pinus resinosa*. *Amer. Jour. Bot.* 49:132–137. 1962.

Lemesle, R. Les éléments du xyléme dans les Angiospermes à caractères primitifs. *Soc. Bot. de France Bul.* 103:629–677. 1956.

Lewis, F. T. The shape of the tracheids in the pine. *Amer. Jour. Bot.* 22:741–762. 1935.

Liese, W. Zur systematischen Bedeutung der submikroskopischen Warzenstruktur bei der Gattung *Pinus* L. *Holz als Roh- und Werkstoff* 14:417–424. 1956.

Mergen, F., and R. M. Echols. Number and size of radial resin ducts in slash pine. *Science* 121:306–307. 1955.

Metcalfe, C. R., and L. Chalk. *Anatomy of the dicotyledons.* 2 vols. Oxford, Clarendon Press. 1950.

Money, L. L., I. W. Bailey, and B. G. L. Swamy. The morphology and relationships of the Monimiaceae. *Arnold Arboretum Jour.* 31:372–404. 1950.

Nägeli, C. W. Das Wachsthum des Stammes und der Wurzel bei den Gefässpflanzen und die Anordnung der Gefässstränge im Stengel. *Beiträge z. Wiss. Bot.* Heft 1:1–56. 1858.

Nečesaný, V. Elektronenmikroskopische Untersuchung der Tyllen und der Kernstoffe der Rotbuche *Fagus sylvatica* L. *Bot. Tidsskr.* 52:48–55. 1955.

Phillips, E. W. J. The identification of softwoods by their microscopic structure. London, Dept. Sci. and Indust. Res. *Forest Prod. Res. Bul.* 22. 1948.

Record, S. J. *Identification of the timbers of temperate North America.* New York, John Wiley and Sons. 1934.

Record, S. J., and R. W. Hess. *Timbers of the New World.* New Haven, Yale University Press. 1943.

Roelofsen, P. A. The plant cell-wall. In: *Handbuch der Pflanzenanatomie*. Band 3. Teil 4. 1959.

Sanio, K. Über die Grösse der Holzzellen bei der gemeinen Kiefer (*Pinus silvestris*). *Jahrb. f. Wiss. Bot.* 8:401–420. 1872.

Sanio, K. Anatomie der gemeinen Kiefer (*Pinus silvestris* L.) 2. Entwickelungsgeschichte der Holzzellen. *Jahrb. f. Wiss. Bot.* 9:50–126. 1873–74.

Scherer, P. E. Studien über Gefässbündeltypen und Gefässformen. *Bot. Centbl. Beihefte* 16:67–110. 1904.

Scott, F. M., V. Sjaholm, and E. Bowler. Light and electron microscope studies on the primary xylem of *Ricinus communis*. *Amer. Jour. Bot.* 47:162–173. 1960.

Scurfield, G., and A. B. Wardrop. The nature of reaction wood. VI. The reaction anatomy of seedlings of woody perennials. *Austral. Jour. Bot.* 10:93–105. 1962.

Shimaji, K. Anatomical studies on the phylogenetic interrelationship of the genera in the Fagaceae. *Tokyo Univ. Forests Bul.* 57. 1962.

Siebers, A. M. The detection of tension wood with fluorescent dyes. *Stain Technol.* 35:247–251. 1960.

Sinnott, E. W. Reaction wood and the regulation of tree form. *Amer. Jour. Bot.* 39:69–78. 1952.

Sinnott, E. W., and R. Bloch. The cytoplasmic basis of intercellular patterns in vascular differentiation. *Amer. Jour. Bot.* 32:151–156. 1945.

Smith, G. F., and H. Kersten. The relation between xylem thickenings in primary roots of *Vicia faba* seedlings and elongation, as shown by soft X-ray irradiation. *Torrey Bot. Club Bul.* 69:221–234. 1942.

Spackman, W., and B. G. L. Swamy. The nature and occurrence of septate fibers in dicotyledons. Abst. *Amer. Jour. Bot.* 36:804. 1949.

Stafford, H. A. Studies on the growth and xylary development of *Phleum pratense* seedlings in darkness and in light. *Amer. Jour. Bot.* 35:706–715. 1948.

Stamm, A. J. A new method for determining the proportion of the length of a tracheid that is in contact with rays. *Bot. Gaz.* 92:101–107. 1931.

Stamm, A. J. Passage of liquids, vapors, and dissolved materials through softwoods. *U.S. Dept. Agric. Tech. Bul.* 929. 1946.

Stern, W. A suggested classification for intercellular spaces. *Torrey Bot. Club Bul.* 81:234–235. 1954.

Studhalter, R. A. Tree growth. I. Some historical chapters. *Bot. Rev.* 21:1–72. 1955.

Thomson, R. G., and H. B. Sifton. Resin canals in the Canadian spruce (*Picea canadensis* (Mill.) B. S. P.)—an anatomical study, especially in relation to traumatic effects and their bearing on phylogeny. *Roy. Soc. London, Phil. Trans.* Ser. B. 214:63–111. 1925.

Trendelenburg, R. *Das Holz als Rohstoff*. 2nd ed. Revised by Mayer-Wegelin. München, Carl Hauser. 1955.

Vité, J. P., and J. A. Rudinsky. The water-conducting systems in conifers and their importance to the distribution of trunk injected chemicals. *Boyce Thompson Inst. Contrib.* 20:27–38. 1959.

Wardrop, A. B., and H. E. Dadswell. The cell wall structure of xylem parenchyma. *Austral. Jour. Sci. Res. Ser. B, Biol. Sci.* 5:223–236. 1952.

Wardrop, A. B., and H. E. Dadswell. The development of the conifer tracheid. *Holzforschung* 7:33–39. 1953.

White, R. A. Tracheary elements of the ferns. I. Factors which influence tracheid length; correlation of length with evolutionary divergence. *Amer. Jour. Bot.* 50:447–455. 1963a. II. Morphology of tracheary elements; conclusions. *Amer. Jour. Bot.* 50:514–522. 1963b.

12

Phloem

CONCEPT

The phloem is the principal food-conducting tissue of the vascular plants. The phloem and the xylem are, as a rule, spatially associated with each other (fig. 12.1, pl. 28) and together constitute the vascular system of the plant body. Like the xylem, the phloem is composed of several different kinds of cells, concerned with different functions, and therefore it, too, exemplifies a morphologically and physiologically complex tissue.

The basic components of the phloem are the sieve elements, several kinds of parenchyma cells, fibers, and sclereids. In plants possessing a laticiferous system, elements of this system may be found in the phloem also. Various idioblasts, specialized morphologically and physiologically, are encountered in the phloem. In this chapter only the principal components of the phloem are considered in detail.

Information on the structure of the phloem is relatively incomplete, and knowledge of the phylogeny of this tissue is fragmentary. Various circumstances are responsible for this deficiency. The phloem shows unique cytologic characteristics which require exacting techniques for investigation. The phloem cells, other than the fibers and the sclereids, do not develop such rigid, persisting walls as the xylem elements, and, after the phloem ceases to act as a conducting tissue, it becomes much modified, functionally and structurally. The most pronounced changes occur in the conducting elements themselves, for they become disorganized to various degrees. Thus, in contrast to the xylem, for which a large body of comparative data is available (chapter 11), the phloem early loses its original nature and appearance, and in such a state it can-

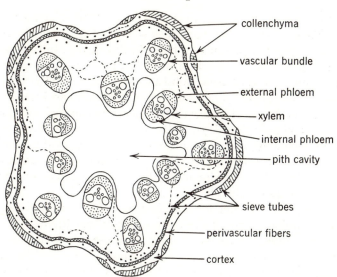

FIG. 12.1. Transection of *Cucurbita* stem. Herbaceous vine with discrete vascular bundles, each having external and internal phloem (bicollateral bundles). Vascular region is delimited on the outside by sclerenchyma (perivascular fibers). Cortex is composed of parenchyma and collenchyma. There is an epidermis. A cavity has replaced the pith. Small strands of extrafascicular sieve tubes and companion cells traverse parenchyma of vascular region and cortex. (×8.)

not be properly studied with regard to its structural details. The lack of firmness is also related to the generally poor preservation of phloem in fossils (Andrews, 1961; exceptionally well preserved phloem was found in *Tetraxylopteris*, pl. 29*B*; Beck, 1957). In much of the commercial utilization of xylem this tissue is not separated into its individual components, and the proper evaluation of its economic qualities requires the knowledge of its structure as a tissue. The commercial importance of the phloem, on the other hand, is determined largely by its content of fibers and such organic substances as tannins, spices, latex, and drugs, all products that are separated or extracted from the tissue. Thus the commercial utilization of the products of the phloem is not a strong stimulus for the study of the phloem as a tissue.

Historically, too, the significance of the xylem as a conducting tissue was recognized earlier than that of the phloem (Esau, 1961). In the phloem, the fibers attracted the first attention and, as was outlined in chapter 10, the tissue received the name of *bast*—a word related to the verb to bind—because the fibers formed in it were used for binding. After the sieve element was discovered by Hartig in 1837, the true nature of the tissue was gradually revealed. In 1858 Nägeli gave it the name of *phloem* (derived from the Greek word for bark), thus eliminating the emphasis upon the presence of fibers in this tissue. In time phloem became the generally accepted term for the food-conducting tissue of vascular plants. Nevertheless, substitute terms are still being used, particularly in German (*Leptom, Siebteil, Cribralteil*) and French (*liber, tissu criblé*) literatures. The term *leptom* deserves special mention. It refers, since Haberlandt (1914), to the soft-walled conducting part of the phloem, including sieve elements, companion cells, and parenchyma cells. The parallel term for the xylem is *hadrom*, which refers to the conducting part of the xylem including the tracheary and parenchymatic elements but excluding the fibers.

Sometimes, with reference to stems and roots, it is convenient to treat as a unit the phloem and all the tissues located outside it. The non-technical term *bark* is employed for this purpose. In stems and roots possessing only primary tissues bark most commonly refers to the primary phloem and the cortex. In axes in a secondary state of growth it may include the primary and the secondary phloem, various amounts of cortex, and the periderm (chapter 14).

From the researches on phloem conducted since the discovery of the sieve element and periodically reviewed from different points of view by various authors (Crafts, 1961; De Bary, 1884; Strasburger, 1891; Perrot, 1899; Schmidt, 1917; Huber, 1937; Esau, 1939, 1950, 1961; Swanson, 1959), the concept has become established that the main characteristic of the phloem is the presence of highly specialized cells, the sieve elements, which together with the accompanying parenchymatic members of the tissue are concerned with the translocation of elaborated food materials, and that the structural peculiarities of the sieve elements are related to their function. Furthermore, the fibers, if present, came to be regarded as components of the phloem tissue, just as wood fibers are components of the xylem tissue.

CLASSIFICATION

Like the xylem, the phloem is classified as primary or secondary on the basis of its time of appearance in relation to the development of the plant or the organ as a whole. The primary phloem is initiated in the embryo, is constantly added to during the development of the primary plant body, and completes its differentiation when the primary plant body is fully formed. Like the primary xylem, the primary phloem differentiates from procambium. If a dicotyledon or a gymnosperm plant has secondary growth, the vascular cambium that forms secondary xylem toward the interior of the stem or root produces secondary phloem in the opposite direction, that is, toward the periphery of the stem or root.

Although most commonly the phloem occupies a position external to the xylem in the axis, or abaxial in the leaves and leaf-like organs, certain ferns and many dicotyledonous families (Apocynaceae, Asclepiadaceae, Convolvulaceae, Cucurbitaceae, Myrtaceae, Solanaceae, Compositae) have a part of the phloem located on the opposite side of the xylem as well (figs. 12.1, 15.1*B*, pl. 38*A*). The two parts of the phloem are called *external* and *internal* phloem, respectively. They may be termed also *abaxial* (that is, away from the axis) and *adaxial* (that is, toward the axis) phloem. In the leaves these terms refer the position of the phloem to the stem, or axis, to which the leaf is attached. In the stems and roots the axis of reference would be an imaginary one, passing longitudinally through the center of the organ. (In the root, internal phloem occurs at levels where a pith is present.)

The term internal phloem replaces *intraxylary phloem* (Committee on Nomenclature, 1957). The latter term is sometimes confused with *interxylary phloem* referring to phloem strands or layers included in the secondary xylem of certain dicotyledons, that is, to the *included phloem*. The included phloem is called *concentric* when it appears in layers alternating with xylem layers, *foraminate* when it appears in strands surrounded by xylem tissue (chapters 15, 17; anomalous growth).

In dicotyledons the internal phloem is initiated somewhat later than the external. Nevertheless, it constitutes a part of the primary phloem system. It resembles the external primary phloem in development, composition, and structure and arrangement of cells (Esau, 1939). Generally, it is not increased in amount by cambial activity (Jean, 1926).

ELEMENTS OF THE PHLOEM

Sieve Elements

Parallel with the classification of the tracheary elements into the phylogenetically primitive tracheids and the more advanced vessel members, the conducting elements of the phloem, called collectively *sieve elements*, may be segregated into the less specialized *sieve cells* (fig. 12.7, pl. 42) and the more specialized *sieve-tube members* (or *sieve-tube elements;* fig. 12.8, pl. 43). The term *sieve tube* designates a longitudinal series of sieve-tube members, just as the term vessel denotes a longitudinal series of vessel members. In both classifications the characteristics of the wall structures—pits and perforation plates in the

tracheary elements, *sieve areas* and *sieve plates* in the sieve elements—
serve to distinguish the elements of the two kinds of categories.

Sieve Areas and Sieve Plates. The morphologic specialization of the
sieve elements is expressed in the development of the sieve areas on
their walls and in the peculiar modifications of their protoplasts. The
sieve areas (the term implies a resemblance to a sieve) are wall areas
with clusters of pores, through which the adjacent sieve elements are
interconnected by strand-like prolongations of their protoplasts (figs.
12.2, 12.3; pls. 38C, D; 39, 40). Thus the sieve areas are comparable
to the primary pit-fields with plasmodesmata that occur in primary walls
of living parenchyma cells. In fact, the sieve areas are specialized
primary pit-fields. The diameter of the pores in the sieve areas ranges
from a fraction of a micron to 15 and probably more in some dicotyledons
(Esau and Cheadle, 1959). Accordingly, the strand-like contents of

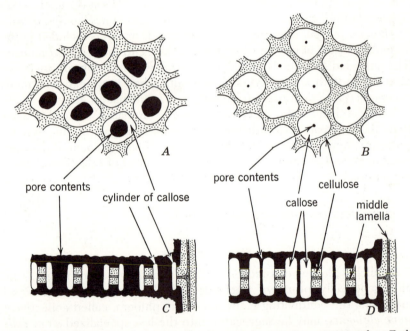

FIG. 12.2. Interpretation of structure of a sieve area in an angiosperm sieve tube. Each
drawing represents a part of a sieve area with several pores. Surface views in *A, B,* sectional
views in *C, D.* The protoplasmic contents of the sieve elements covering the sieve areas in
C and *D* are shown in black; so are also the strands connecting these contents across the sieve
areas. *A, C,* illustrate younger sieve areas; *B, D,* older sieve areas. In *B, D,* the amount of
callose lining the pores is larger and the pores narrower than in *A, C.*

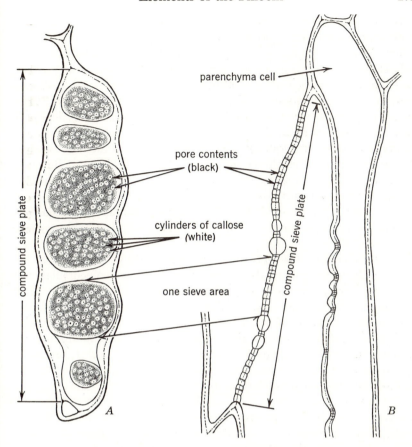

parenchyma cell

pore contents
(black)

cylinders of callose
(white)

one sieve area

compound sieve plate

compound sieve plate

A

B

FIG. 12.3. Compound sieve plate of *Nicotiana* (tobacco) in surface view (*A*) and in longitudinal section (*B*). In each sieve area, numerous pores are each lined with callose. Sieve
areas occur in depressions of sieve-plate wall. Depressions with plasmodesmata occur in wall
between sieve element and parenchyma cell in *B*. (*A*, ×1,070; *B*, ×930. After Esau,
Hilgardia 11, 1938.)

the pores range from the size of plasmodesmata to considerably larger
(pl. 38*C*, *D*).

In sectioned material the sieve-area strands are commonly seen
associated with the carbohydrate *callose*, a polymer of glucose residues
united into spirally wound chains in β-1-3 linkages (Kessler, 1958; in
contrast, cellulose occurs as straight chains of glucose residues in β-1-4
linkages; chapter 3). Callose stains a clear blue with anilin blue and
resorcin blue and, in small amounts, may be detected by its characteristic fluorescence after treatment with dilute anilin blue (Currier, 1957).

In sieve elements that are regarded as mature and conducting, the amounts of callose are relatively small (Esau, 1961; Ullrich, 1962). The callose lines the pores (fig. 12.2A, C; pl. 38C, D), constricting them only slightly, and it may form a thin layer on the surface of the sieve area (on the bars between the pores) as well. In view of the evidence that, in response to injury, callose is deposited on walls in a variety of living cells (chiefly in relation to plasmodesmata) and that this deposition may be extremely rapid (Currier, 1957; Eschrich, 1956), some investigators have raised the question whether callose is present in conducting sieve elements in the intact plant (Eschrich, 1963). But even if callose were absent in active phloem before the plant is sampled, the usual selective deposition and distribution of sieve-element callose in sectioned material are so characteristic that callose may be successfully used as a diagnostic feature of the conducting cells of the phloem (Esau et al., 1953).

The term *callose* (Mangin, 1890) was preceded by *callus*, a word first used by Hanstein (1864) with reference to the massive accumulation of callose on sieve areas of old sieve elements. The use of callus for callose has been largely abandoned. Callose is a word that parallels cellulose in indicating a carbohydrate, whereas callus has, with regard to plants, an entirely different earlier meaning. It refers to proliferations of parenchyma cells associated with wound-healing and regeneration phenomena. Callus in this sense is also the material used extensively in tissue-culture research (chapter 4).

The wall in a sieve area is obviously a double structure, because it consists of two layers of primary wall, each belonging to one of the two contiguous cells, cemented together by intercellular substance (fig. 12.3B). With reference to the secondary walls the term pit-pair is employed to designate a combination of two pits opposing each other in a wall between two cells (chapter 3). No similarly precise terminology has been introduced with reference to the sieve areas. Therefore, the term sieve area sometimes indicates a paired structure, sometimes one-half of a paired structure. This usage corresponds to the similarly flexible application of the term wall either to the wall of a given cell or to the paired walls of two contiguous cells.

In surface view, a sieve area appears like a depression in a wall with a number of dots—the transections of the pore contents—each surrounded with a ring of callose (figs. 12.2A, B, 12.3A; pls. 38C, D; 40A). In sectional views also the sieve areas are recognized as thin places in the wall with the pore contents and associated callose traversing the wall from one cell lumen to the other (figs. 12.2C, D; 12.3B; pl. 40B).

In meristematic cells the future sieve areas resemble primary pit-fields. The less specialized types of sieve areas, that is, areas with relatively small pores, differ at maturity from primary pit-fields in the greater conspicuousness of the pore contents and the usual presence of callose. Probably some enlargement of pores occurs during their differentiation. In the more highly specialized sieve areas, pore formation follows a complex sequence as seen with the electron microscope (pl. 39; Esau et al., 1962). The future pore site is at first occupied by a single plasmodesma. Sheets of endoplasmic reticulum and platelets of callose become localized on the opposing surfaces of each pore site, with the ectoplast interposed between the endoplasmic reticulum and the callose. The sheets and the platelets increase in diameter until they become as wide as the future pores. Eventually the two opposing platelets at each pore site fuse because of the disappearance of the original separating wall. A hole appears in the middle of the fused platelets and enlarges centrifugally. Thus, from its inception, a pore appears to be lined with callose.

The removal of the wall material between the callose platelets may concern only noncellulosic substances; the cellulosic microfibrils could be mechanically displaced toward the margins of the pores (Frey-Wyssling and Müller, 1957). The bars between the pores become thickened probably in part through a deposition of additional wall material and in part through the displacement of the microfibrils from the pore sites.

As the sieve element ages, the amount of callose in the sieve area is augmented (figs. 12.2B, D; 12.4). Its mass increases within the pores and constricts the protoplasmic strands. Callose is also deposited in increasing amounts on the surface of the sieve area. Consequently the sieve areas cease to appear as depressions in the wall. Instead, they become thickened regions of the wall, for the callose eventually projects above the surface of the wall (fig. 12.4E–G; pl. 40C). When the element reaches the end of its activity, the sieve areas are blocked with bulging masses of callose which may or may not be traversed by tenuous strands (fig. 12.4G). If there are several sieve areas close together, the adjacent callose masses may fuse. Since such extensive accumulation of callose usually indicates cessation of activity of the sieve element, the mass of callose at this stage of development is called *definitive callose* (Lecomte, 1889).

When the protoplast of the inactive sieve element completely disorganizes, the sieve-area strands disappear. The definitive callose commonly separates from the sieve area and also disappears (fig. 12.4H). The sieve area freed of callose represents a thin portion of cellulose wall with open perforations. Thus, the sieve-like structure of the sieve area becomes clearly evident only after the element ceases to function.

Like the pits in the tracheary elements, the sieve areas occur in

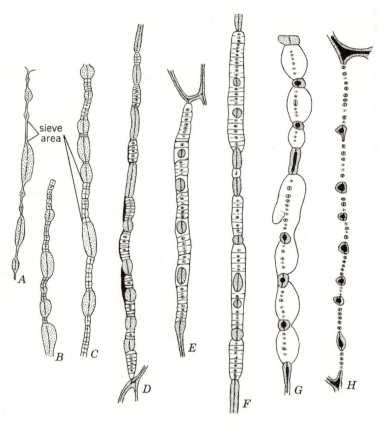

FIG. 12.4. Development of compound sieve plate of *Nicotiana*. *A*, sieve areas still in stage of primary pit-fields in cambial wall (presumably traversed by plasmodesmata in living state). *B–D*, accumulation of callose (white) and resulting thickening of sieve areas. Pore contents become readily visible during this development (black lines traversing the sieve areas). *E, F*, increase in amount of callose. *G*, most massive callose accumulations (definitive callose). Pore contents have disappeared with death of protoplasts. *H*, old sieve plate with no callose and with open pores from nonfunctional sieve tube. (All, ×860. After Esau, *Hilgardia* 11, 1938.)

various numbers and are variously distributed in sieve elements of different plants. As mentioned, they also show unequal degrees of specialization; that is, they vary in the size of pores. In some plants, the sieve areas of a given cell are all alike in the degree of specialization (fig. 12.7); in others, some of these structures possess distinctly larger pores than the rest (fig. 12.8). The sieve areas with the larger pores are commonly localized on certain walls of the sieve elements, most often on the end walls. The wall parts bearing the highly specialized sieve areas are called *sieve plates*. If a sieve plate consists of a single sieve area, it is a *simple sieve plate* (pl. 38C, D). Many sieve areas, arranged in scalariform, reticulate, or any other manner, constitute a *compound sieve plate* (fig. 12.3, pl. 40A–C). Sieve elements having sieve plates on their end walls usually bear less differentiated sieve areas on their lateral walls (fig. 12.8). In some species the sieve areas of the sieve plates and those on the lateral walls are sharply differentiated from each other by the size of their pores; in others, the two intergrade with one another through intermediate forms (Esau, 1950).

Sieve Cells and Sieve-Tube Members. The two types of sieve elements, the sieve cells and the sieve-tube members (or sieve-tube elements), differ in the degree of differentiation of their sieve areas and in the distribution of these areas on the walls. A *sieve cell* is an element with relatively unspecialized sieve areas, not strikingly differentiated from one another and consequently with no wall parts that can be conveniently distinguished from others as sieve plates (fig. 12.7). Sieve cells are commonly long and slender, and they taper at their ends or have steeply inclined end walls. In the tissue they overlap each other, and the sieve areas are usually particularly numerous on these ends.

Sieve-tube members are sieve elements in which some of the sieve areas are more highly specialized than others and are localized in the form of sieve plates (fig. 12.8). The sieve plates occur mainly on end walls which vary from much inclined to transverse. Sieve-tube members are usually disposed end to end in long series, the common wall parts bearing the sieve plates. These series of sieve-tube members are *sieve tubes.* The walls of laterally adjacent sieve tubes bear sieve areas of a lower degree of specialization than those of the sieve plates, but sometimes sieve plates occur on these walls also.

Phylogenetic Specialization. The lack of comprehensive data on the comparative anatomy of the phloem of vascular plants makes it impossible to present a precise picture of evolution of the phloem elements comparable to that given for the xylem elements in chapter 11.

The lower vascular plants and the gymnosperms generally have sieve cells as defined in this book, whereas most angiosperms have sieve-tube members. The evolutionary changes of sieve-tube members have been studied most comprehensively in the late primary phloem (metaphloem) of the monocotyledons (Cheadle, 1948; Cheadle and associate, 1941, 1948). These sieve-tube members show the following trends in evolutionary specialization: a progressive localization of highly specialized sieve areas on the end walls; a gradual change in the orientation of these end walls from very oblique to transverse; a step-wise change from compound to simple sieve plates; and a progressive decrease in conspicuousness of the sieve areas on the side walls. The specialization of sieve-tube members in the secondary phloem of dicotyledons probably has progressed along similar lines (Esau et al., 1953). In addition to the phylogenetic enlargement of pores on the end wall in dicotyledons, an increase in the per cent of transverse area occupied by the sieve-area strands appears to have occurred (Esau and Cheadle, 1959). The increase of specialization of the end-wall sieve areas, as contrasted with those on the side walls, suggests an emphasis on longitudinal penetrability of the conducting system. In the monocotyledons the specialization of sieve elements has progressed from the leaf toward the root, that is, in the direction opposite to that in which the evolution of the tracheary elements has occurred (chapter 11). Comparable information is not available for the phloem of dicotyledons.

The phylogenetic specialization of sieve elements shows some parallelism with that of tracheary elements. The sieve cells with their overlapping ends and lack of sieve plates may be compared with the tracheids that are connected with each other through bordered pits. In the sieve elements specialization has resulted in an enlargement of pores, in the tracheary elements, in the formation of perforations. The application of the term perforation plate to the open wall in the vessel members parallels the use of the term sieve plate with reference to the wall of a sieve-tube member bearing sieve areas with the largest pores. In both kinds of elements there has been a change in the orientation of the end walls from oblique to transverse, and as in the vessel member multiperforate end walls were replaced by those with simple perforations, in the sieve-tube member the compound sieve plate was succeeded by the simple sieve plate. The phylogenetic decrease in length, so well established for the vessel members, and so clearly related to the decrease in the length of the cambial initials, is less direct and consistent in the evolution of the sieve elements. In the phloem, the conducting elements did decrease in length in relation to the shortening of the cambial initials, but in many species an ontogenetic decrease in length by transverse divisions in phloem initials obscures the length relation between the sieve elements and the cambial cells (Esau and Cheadle, 1955; Zahur, 1959). The phylogenetic significance of the shortening by divisions is uncertain (Carlquist, 1961). Similarly problematic is the physiologic significance of the introduction of additional sieve plates, by the divisions, in the path of movement of assimilates. In many dicotyledons, moreover, the phloem initials also divide longitudinally with the result that the potential width of the conduit is reduced.

Wall Structure. The walls of the sieve elements are cellulosic. No well authenticated evidence for their lignification is available. The walls vary in thickness. In many species, a distinct thickening, called *nacreous* or *nacré* (meaning lustrous) thickening, is present (Esau and Cheadle, 1958). It gives a positive reaction to tests for cellulose and pectins. It is not exceptionally highly hydrated but may shrink as the cell ages. The nacreous wall may be so thick as to occlude the lumen, but it does not cover the sieve areas.

In the absence of nacreous thickening the wall of the sieve element is regarded as primary by the classification based on light microscopy. A proper classification of the nacreous wall has not been made. In one group of conifers, the Abietineae, the wall thickening of sieve elements has been interpreted as a true secondary wall (pl. 26C, D; Abbe and Crafts, 1939).

Protoplast and Function of the Cell. The interpretation of the function of the sieve element depends on the proper understanding of the nature of its contents. Although physiologic research reveals that organic solutes move largely in the phloem (Biddulph and Biddulph, 1959; Zimmermann, 1961), the proof that the sieve element is the main conduit of this movement is rather indirect (Esau, 1961). Movement of fluorescent dyes has been seen in sieve elements but the relation of this phenomenon to the translocation of assimilates is not clear (Esau et al., 1957). Some autoradiographic studies have indicated transport of radioactive materials in the phloem, but they have not demonstrated unequivocally a specific involvement of sieve elements in this movement. The strongest support for the concept of the sieve element as the conduit in the phloem is found in studies of the exudation phenomenon —release of fluid from cut or punctured phloem—especially by the use of insect mouth parts (stylets). Under the microscope, it is possible to ascertain that the exudate is derived from the sieve elements. On the other hand, it has also been demonstrated that an aphid feeding on phloem inserts its stylets into an individual sieve element and excretes, in the form of honeydew, material resembling the phloem exudate (Zimmermann, 1961). Use of aphid stylets—after cutting away the aphid anaesthetized during feeding—as a micropipette for tapping a sieve element has provided a large amount of information on the sieve element as a conducting cell (Hill, 1962; Peel and Weatherley, 1962; Weatherley, 1962; Weatherley et al., 1959; Ziegler and Mittler, 1959). These studies and others have established that the contents of the sieve element are under positive pressure (approximately 30 atmospheres), that the transported sugar is mainly sucrose (and related oligosaccharides), that the concentration of sugar may exceed 20 per cent, that the movement is rapid (frequently about 100 cm per hour), and that the physiologic activity of the sieve element is intimately connected with that of the associated parenchyma cells. The structure of the sieve-element protoplast has not yet been fully correlated with the physiologic characteristics of the cell.

The well-known property of the sieve-element protoplast is its lack of a nucleus at functional maturity. The loss of the nucleus occurs during the differentiation of the cell (fig. 12.5). In the meristematic state the sieve element resembles other procambial or cambial cells in having a more or less vacuolated protoplast with a conspicuous nucleus. Later the nucleus disorganizes and disappears as a discrete body (enucleate state). In some plants, scattered among unrelated families, the nucleolus (or nucleoli, if more than one are present) is extruded from the nucleus before the latter finally disorganizes (fig. 12.6G, H; pl. 41A, B). The extruded nucleoli persist in the sieve elements while these exist as intact cells (Esau, 1947).

In the dicotyledons the sieve elements commonly contain variable amounts of a relatively viscous substance, the so-called *slime*, which consists mainly of proteins. In the mature state the slime is dispersed in the vacuolar sap. The slime readily aggregates when the phloem is processed for microscopic observation in live or killed state and becomes displaced toward the sieve areas, chiefly those of the sieve plates (pl. 38B). The protoplast often contracts in injured cells (pl. 41C). The slime accumulation on a sieve area is called *slime plug* and its presence is regarded as an indication that the cell has been injured. Slime plugs appear to stop the exudation of contents from the cut phloem in the first stages of wound reaction. Later, the sieve areas become plugged by wound callose.

The slime originates in the cytoplasm in the form of discrete bodies, the *slime bodies* (fig. 12.5; pl. 41D). These bodies may be spheroidal, or spindle-shaped, or variously twisted and coiled. They occur singly or in multiples in one element. They absorb cytoplasmic stains and are, therefore, easily demonstrated under the microscope. During the differentiation of the sieve-tube member, the slime bodies lose their sharp outlines, become more fluid, sometimes fuse with one another, and eventually become dispersed in the vacuolar contents which are no longer delimited by a tonoplast (fig. 12.5G). Under the electron microscope the dispersed slime may show fibrous structure (Engleman, 1963). In some Leguminosae (pl. 43; *Robinia*) structures interpreted as slime bodies do not disperse. In these plants, the slime body as such appears to form the slime plug in injured cells (Resch, 1954). The slime may appear in the form of strands connected to one or both sieve plates and continuous with the pore contents.

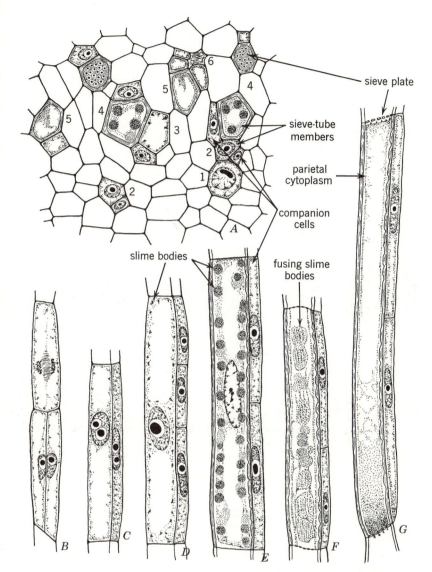

FIG. 12.5. Differentiation of sieve-tube members in primary phloem of *Cucurbita*. *A*, transection with details: (1) cell before division; (2) after division into sieve-tube member and companion cell; (3) slime bodies have appeared in sieve-element protoplast; (4) slime bodies of maximal size and thick wall in sieve element; (5) slime bodies dispersed; (6) sieve element partly obliterated. Longitudinal sections: *B*, cells in division (above) and after division (below) into sieve-tube member and precursor of companion cells. *C*, young sieve element and precursor of companion cells. *D*, sieve-tube member with small slime bodies; precursor of companion cells has formed three companion cells. *E*, slime bodies of maximal size, nucleus highly vacuolated, thick walls in sieve element. *F*, slime bodies partly fused and nucleus absent. *G*, mature sieve element with dispersed slime (somewhat denser below). In *G*, protoplast connected with the lower sieve plate but partly withdrawn from the upper. Pits in sieve-element walls facing companion cells in *E–G*. (All, ×730.)

In many arborescent species the sieve-tube protoplasts are thin in consistency and, when injured, form rather small slime plugs. The monocotyledons, the gymnosperms, and the still lower vascular plants have watery contents in the sieve elements, with small amounts of slime.

Sieve elements of many species contain small plastids (pl. 41A, B) that elaborate a form of starch commonly giving a red staining reaction when treated with iodine. In injured sectioned material the starch grains are released by the plastids and move with the slime toward the sieve areas (fig. 12.6F). The grains usually have the shape of discs with a lightly stained center. The plastid may contain one to several granules.

The nuclear degeneration in the developing sieve element indicates a profound change in the condition of the protoplast. It is associated with other seemingly disorganizational changes, some of which are detectable only at the ultrastructural level. In young cells a tonoplast delimits the vacuole; at maturity no tonoplast is present and thus the boundary between the parietal cytoplasm and vacuole disappears (Esau and Cheadle, 1962a). Accordingly, the vital stain neutral red, which is taken up selectively by vacuoles of living cells, ceases to accumulate in the sieve element. Despite the absence of tonoplast, however, the sieve element continues to be plasmolyzable (pl. 41B, E; Currier et al., 1955; Kollmann, 1960). The endoplasmic reticulum, present in the usual form of sacs in the nucleate stage, may break up into vesicles later; and the mitochondria may become devoid of internal membranes (Duloy et al., 1961; Esau and Cheadle, 1962b). The dictyosomes disappear completely. Finally, the cell has a parietal layer, apparently composed of the ectoplast and vesicles of the endoplasmic reticulum. More or less modified mitochondria and plastids, if present in a given species, also occupy the parietal position. The center of the cell is filled with a mixture of vacuolar sap and of disorganized cytoplasmic matter, chiefly slime in the dicotyledons. Since no tonoplast is present the term vacuole ceases to be appropriate with reference to the mature sieve element. The changes in the maturing sieve element resemble those occurring in tracheary elements in which the protoplasts are completely eliminated at maturity (Esau et al., 1963).

The disorganizational changes in the developing sieve element do not receive a uniform evaluation from different students of translocation. The proponents of the concept of diffusional, or molecular, movement, that is, movement of molecules independently of the solvent (water), assume that the mature sieve element has an active protoplast that provides the energy necessary for moving the solute. The proponents of the mass- or pressure-flow hypothesis, on the other hand, assume that the denaturing of the sieve-element protoplast creates a conduit in which the solute is carried by a passive mass movement with the solvent along a concentration gradient. The energy required to maintain the gradient is provided by the nucleate cells associated with the sieve element in the tissue. These cells secrete sugar into the sieve elements at places of its synthesis (mesophyll or storage tissue where starch is being hydrolyzed into sugar) and remove it from the conduits where the food is used for growth or is stored. Thus, concentration gradients are established between the sources and the sinks of the carbohydrates.

Since the movement occurs from cell to cell, the nature of the connections between superposed sieve elements is as important as that of the protoplast for interpreting the mechanism of translocation. The study of the pore contents in the sieve areas, as well as that of the protoplast as a whole, is greatly hindered by the sensitivity of the sieve element to injury. In sections that may have well-preserved parenchyma cells, the sieve elements are likely to show displaced contents and sieve areas more or less completely plugged by slime or callose depending on the treatment. The most direct cause of the displacement of contents is the positive pressure in the mature sieve element which induces a surge of sap toward the incision. The unilateral flow in response to cutting causes the denser components of the protoplasts to accumulate on the sieve areas near the cut. The accumulations are located on the sides of the sieve areas facing away from the wound surface, thus giving the impression that the dense material is filtered out from the sap flowing through the sieve areas. If cuts are made at two ends of a phloem strand, the plugs face one way (pl. 38B) at one end of the section, and another way at the opposite end, and may appear at both ends of the elements in the median part of the section.

The other obstacle in the successful observation of the pore contents, particularly in living state, is their small size. Electron microscopy based on material prepared with special care to minimize injury—but still killed and dehydrated material—indicates that in the sieve plates of dicotyledons with relatively large pores, the contents of the pores resemble those of the cells themselves; that is, they are filled with the mixture of vacuolar sap and disorganized cytoplasmic derivatives delimited by the ectoplast from the pore wall (pl. 39D; Esau and Cheadle, 1961). Thus, no differentially permeable membrane separates the protoplasts. This kind of structure would be compatible with the concept of mass movement from cell to cell, except that the state and the role of the slime in this system continues to be an enigma.

The sieve-area strands of smaller diameters have not been much studied at the ultrastructural level. In one conifer (*Metasequoia*) these strands are described as composed of the ectoplast and numerous tubules of the endoplasmic reticulum (Kollmann and Schumacher, 1962, 1963). The continuity of the endoplasmic reticulum through pores in the wall is also suggested for plasmodesmata (chapter 3). Possibly pores in sieve areas of different degrees of specialization differ in their contents and in degree of resemblance to plasmodesmata.

Companion Cells

The sieve-tube members of monocotyledons and dicotyledons are commonly associated with specialized parenchyma cells called *companion cells*. These cells arise from the same meristematic cell as the associated sieve-tube member, so that the two kinds of elements are closely related in their ontogeny (fig. 12.5). In the formation of the companion cells the meristematic precursor of the sieve-tube member divides longitudinally one or more times. One of the resulting cells, usually distinguished by being larger, differentiates into a sieve-tube member. The others become companion cells, with or without some transverse or other divisions preceding their differentiation. The number of companion cells associated with a given sieve-tube member varies from one to several in different species and may also be variable in the same plant (fig. 12.6A–C; Cheadle and Esau, 1958; Zahur, 1959). Companion cells also vary in size. Some are as long as the sieve-tube member with which they are related; others are shorter than the sieve-tube member. The companion cells of a given sieve-tube element may occur on various sides of this element, or they may form continuous longitudinal series on one side of it (fig. 12.6H). In some herbaceous dicotyledons and in many monocotyledons having little or no phloem parenchyma, the companion cells of the superposed series of sieve-tube members form continuous longitudinal series (Strasburger, 1891), but in other plants the companion cells of different elements are commonly not in contact with each other.

The wall between the companion cell and sieve element is either uniformly thin or has obviously depressed areas, primary pit-fields (fig. 12.6D, E). Under the electron microscope plasmodesmata are evident in these walls, frequently branched on the companion-cell side (Esau and Cheadle, 1962b). In macerated material the companion cells commonly remain attached to the sieve-tube element. In older sieve elements callose may be present on the pit-fields connecting them with the companion cells.

In contrast to the sieve element, the companion cell retains its nucleus at maturity (fig. 12.5). At the height of activity its protoplast may stain more heavily than that of ordinary parenchyma cells, and it is noteworthy that this chromaticity increases after the companion cell develops

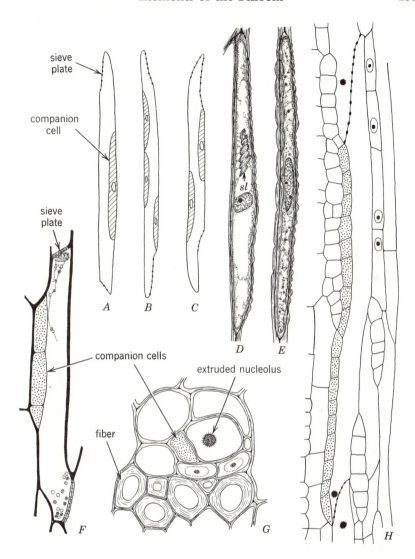

FIG. 12.6. Companion cells. *A–C,* sieve-tube elements of *Vitis;* companion cells hatched. *D, E, Vitis* companion cells, one still young and containing a slime body (*sl* in *D*), the other mature, with slime body dispersed (*E*). The sieve-tube member would have been to the right of each companion cell. *F,* sieve-tube element of *Daucus* (carrot), with companion cells stippled. Small bodies near sieve plates are plastids with starch, large body is slime. *G, H,* phloem sections of *Eucalyptus,* transverse (*G*) and longitudinal (*H*). Companion cells stippled. Extruded nucleoli in lumina of sieve elements. (*A–C,* ×100; *D, E, G,* ×850; *F,* ×450; *H,* ×300. Esau, *A–E, Hilgardia* 18, 1948; *F, Hilgardia* 13, 1940; *G, H, Amer. Jour. Bot.* 34, 1947.)

beyond the meristematic state. The deep staining of the companion cells is possibly caused by a substance similar to the sieve-tube slime. In some species (*Vitis, Robinia, Pyrus*) the companion cells develop the same kind of slime bodies as the sieve tubes (fig. 12.6D) and the chromaticity of the companion-cell protoplast increases after the dispersal of these bodies (Esau, 1947, 1948). The dense type of companion cells also has a small vacuome. The companion cells apparently form no starch, but may have leucoplasts or chloroplasts. They retain at maturity numerous mitochondria rich in internal membranes, dictyosomes, and endoplasmic reticulum (Esau and Cheadle, 1961, 1962b). The sieve-tube elements and their companion cells appear to be closely associated not only ontogenetically and morphologically but also physiologically: when the sieve-tube protoplasts are disorganized at the end of its activity, the associated companion cells die also. An instance of sclerification of companion cells in old phloem has been recorded in *Tilia* (Evert, 1963c).

Although the companion cells are considered to be characteristic components of the angiosperm phloem, no comprehensive comparative studies have been made on their distribution in this group of plants. There is evidence that they may be absent in the primitive woody dicotyledons (Bailey and Swamy, 1949). Companion cells are frequently lacking in the earliest part of the primary phloem (protophloem) of the angiosperms, a tissue that functions for only a short time (Esau, 1939). Detailed studies on secondary phloem sometimes reveal that a few individual sieve-tube members have no companion cells, probably an insignificant aberration (Cheadle and Esau, 1958; Evert, 1960, 1963a).

The sieve cells of gymnosperms (fig. 12.7, pl. 42) and vascular cryptogams have no companion cells. In conifers and *Ginkgo* certain ray and phloem parenchyma cells are closely associated morphologically and physiologically with the sieve cells (Esau et al., 1953; Grillos and Smith, 1959; Srivastava, 1963a, b). These parenchyma cells have received the name *albuminous cells*, because in sections they frequently stain deeply with cytoplasmic stains, as though they are particularly rich in proteinaceous materials (Strasburger, 1891). When albuminous cells occur in rays, they are usually located at the margins of rays and constitute the erect ray cells, which are taller and of smaller transverse diameters than the procumbent ray cells. Albuminous cells included among axial parenchyma cells appear to be mostly members of declining tiers (Srivastava, 1963b). The walls of the sieve cells facing the albuminous cells have conspicuous sieve areas. Typically, albuminous cells contain no starch. Albuminous cells die when the sieve cells are disorganized. Thus the relation between albuminous cells and sieve cells resembles that between companion cells and sieve-tube members of the angiosperms, except that typically there is no direct ontogenetic relation between albuminous cells and sieve cells (Srivastava, 1963*b*).

Parenchyma Cells

The phloem contains variable numbers of parenchyma cells other than companion and albuminous cells. These are concerned with many of the activities characteristic of living parenchyma cells, such as storage of starch, fat, and other organic food materials, and accumulations of tannins and resins.

Some parenchyma cells may arise from the same mother cells as the sieve elements (but before the companion cells are formed). Parenchyma cells, especially those ontogenetically related to the sieve elements, may die at the end of the functioning period of the associated sieve elements. Parenchyma cells, thus, may intergrade with companion cells in their relation to the sieve elements (Cheadle and Esau, 1958; Evert, 1963*b*; Srivastava and Bailey, 1962). Possibly the companion cells themselves vary in degree of specialization in the same plant (Resch, 1954).

The parenchyma cells of the primary phloem are elongated and are oriented, like the sieve elements, with their long axes parallel with the longitudinal extent of the vascular tissue. In the secondary phloem, parenchyma occurs in two systems, the axial and the ray system (figs. 12.7–12.10). The parenchyma of the axial system is called *axial phloem parenchyma*, a term corresponding to the term axial xylem parenchyma for the secondary xylem (chapter 11). The *ray parenchyma* constitutes the phloem rays.

The secondary phloem parenchyma occurs mainly in two basic forms. The cells may be either comparable to the fusiform cambial cells in length or considerably shorter because of transverse divisions in the fusiform derivatives that give rise to them. In conformity with the terminology used for the xylem, the long parenchyma cells are *fusiform parenchyma cells*, and a series of short ones derived from one fusiform cell is a *parenchyma strand*. They ray cells may be elongated in the radial direction (fig. 12.7, *procumbent cells*), or vertically (fig. 12.8, *erect cells*). Erect cells commonly occur at ray margins.

In the active phloem, the phloem parenchyma and the ray cells apparently have primary unlignified walls. After the tissue ceases to conduct, the parenchyma cells may remain relatively unchanged, or they may become sclerified. In many plants a phellogen eventually arises in the phloem (chapter 14). It is formed by the phloem parenchyma and the ray parenchyma.

The walls of both kinds of parenchyma cells have numerous primary pit-fields interconnecting axial parenchyma cells and ray cells with one another and among themselves. Pit-fields also occur between parenchyma cells and companion cells and between parenchyma cells and sieve elements. Usually the pit-field on the sieve element side is called a sieve area since it develops callose.

Fibers

The fundamental structure of phloem fibers, their origin, and their development were considered in chapter 10 (see also Esau, 1950). Fibers occur in both primary and secondary phloem. Those of the primary commonly develop in organs that are still elongating. By a combination of symplastic and apical intrusive growth, the primary fibers may become very long. The secondary phloem fibers arise from fusiform cambial cells as components of the axial system. These fibers may elongate by apical intrusive growth, but, as a rule, they remain conspicuously shorter than the primary fibers of the same plant. The primary and the secondary phloem fibers develop secondary walls after they complete their elongation, although the beginning of secondary-wall deposition may occur while the cell is still elongating at its apices (chapter 10). In some plants the fibers are typically lignified; in others they are not. The pits in their walls are usually simple, but may be slightly bordered. Septate and gelatinous fibers also occur in the phloem. In some plant species the secondary phloem fibers mature in the conducting phloem and appear to be highly specialized as mechanical elements (*Tilia*). In other species they have primary walls and active protoplasts in the functioning phloem and differentiate as fibers only after the sieve elements cease to function (*Prunus*, pl. 44B; *Parthenium*). Some workers consider such fibers to be sclerotic phloem parenchyma cells, or sclereids, and not true fibers (Holdheide, 1951). When a sclerenchyma cell has characteristics intermediate between fibers and sclereids it may be called fiber-sclereid (Evert, 1963a). Phloem fibers, like xylem fibers, may remain alive and store starch (septate fibers in *Vitis*, pl. 44A).

PRIMARY PHLOEM

In conformity with the classification of the primary xylem into protoxylem and metaxylem, the primary phloem may be divided into *protophloem* and *metaphloem*. These terms have evolved in relation to the parallel terminology for the primary xylem (chapter 11).

Protophloem

The protophloem constitutes the conducting tissue of the actively growing parts of the plant and contains sieve elements possessing the usual specialized characteristics of such elements, that is, highly vacuolate, enucleate protoplasts and walls bearing sieve areas. There is some doubt regarding the morphologic nature of the first phloem elements in the gymnosperms and since no sieve areas have been recognized in them they are referred to as *precursory phloem* cells (pl. 54*A*; Esau, 1950; Smith, 1958). In the angiosperms, sieve elements have been observed in the protophloem of roots, stems, and leaves in woody and herbaceous species (Esau, 1939, 1950). These elements appear to be sieve-tube members, but they may lack companion cells. They are elongated but have narrow transverse diameters, and their sieve areas are revealed only in good preparations and at high magnifications. The recognition of these elements is frequently facilitated by their somewhat thickened walls, which readily absorb cellulose stains (pl. 45*A*), and by the scarcity of stainable contents in their lumina. The light staining of contents often makes the sieve elements particularly conspicuous among the adjacent protophloem cells still possessing dense protoplasts.

The sieve tubes of the protophloem apparently function for a brief period only. In rapidly elongating organs they are destroyed (fig. 10.2*B*, pl. 45*B*), soon after maturation, by the effects of elongation of the surrounding cells. Being enucleate cells they are unable to keep pace with this growth by active elongation and are passively stretched. Often the surrounding cells crush both the partly stretched elements and their companion cells, if such are present. The remnants of the crushed cells may later disappear completely. This phenomenon of effacement of the sieve elements is commonly called obliteration.

In many dicotyledons the cells remaining in the protophloem after the sieve tubes are obliterated differentiate into fibers (Blyth, 1958; Léger, 1897). Certain vine types of stems, which possess a sclerenchyma cylinder outside the vascular strands (*Aristolochia, Cucurbita;* figs. 10.1*H*, 12.1), form no fibers in the protophloem. In the leaf blades and the petioles of dicotyledons the protophloem cells remaining after the destruction of the sieve tubes often differentiate into long cells with collenchymatically thickened unlignified walls (chapter 9). The strands of these cells appear, in transverse sections, like bundle caps delimiting the vascular bundles on their abaxial sides. This type of transformation of the protophloem in leaves is widely distributed and occurs also in those species that have protophloem fibers in the stems (Esau, 1950). As was pointed out in chapter 10, the profound change that the protophloem undergoes during the early stages of development of an organ obscures the original nature of the tissue and may lead to the erroneous assumption that this tissue is distinct from the rest of the phloem and constitutes part of the so-called pericycle (Blyth, 1958).

Metaphloem

Since the metaphloem matures after the growth in length of the surrounding tissues is completed, it is retained as a conducting tissue longer than the protophloem. Some herbaceous dicotyledons, most monocotyledons, and many lower vascular plants produce no secondary tissues and depend entirely on the metaphloem for food conduction after their primary bodies are fully developed. In woody and herbaceous species having cambial secondary growth the metaphloem sieve elements become inactive after the secondary conducting elements differentiate. In such plants the metaphloem sieve elements may be partly crushed or completely obliterated.

The absence of secondary growth in persisting plants, such as ferns, bamboo, and palms, raises the question whether these plants have sieve elements which, despite their enucleate protoplasts, remain functional for many years. The scanty references to this subject (Esau, 1939) suggest that such protracted longevity might occur.

The sieve elements of the metaphloem (pl. 45C) are commonly longer and wider than those of the protophloem, and their sieve areas are more distinct. In the angiosperms investigated thus far, these elements are sieve-tube members. Companion cells and phloem parenchyma are typically present in the metaphloem of the dicotyledons. In the monocotyledons, the sieve tubes and companion cells often form strands containing no phloem parenchyma cells among them, although such cells may be present on the periphery of the strands (Cheadle and Uhl, 1948). In such phloem the sieve elements and companion cells form a regular pattern, a feature that is considered to be phylogenetically advanced (Carlquist, 1961). A monocotyledonous type of metaphloem, without phloem parenchyma cells among the sieve tubes, may be found in herbaceous dicotyledons (Ranunculaceae, chapter 15).

According to the literature, the metaphloem of dicotyledons usually lacks fibers (Esau, 1950). If primary phloem fibers occur in dicotyledons, they arise in the protophloem, but not in the metaphloem, even if such elements are later formed in the secondary phloem. In herbaceous species the old metaphloem may become strongly sclerified. Whether the cells undergoing such sclerification should be classified as fibers or as sclerotic phloem parenchyma has not been determined. In monocotyledons sclerenchyma encloses the vascular bundles as bundle sheaths and may also be present in the metaphloem (Cheadle and Uhl, 1948).

The delimitation between protophloem and metaphloem is sometimes rather clear, as, for example, in the aerial parts of monocotyledons having only sieve tubes in the protophloem and distinct companion cells associated with the sieve tubes in the metaphloem (pl. 57B). In dicotyledons the two tissues usually merge gradually, and their delimitation must be based on a developmental study.

In plants having secondary phloem the distinction between this tissue and the metaphloem may be quite uncertain. The delimitation of the two tissues is particularly difficult if radial seriation of cells occurs in both tissues. An exception has been found in *Prunus*, in which the last cells initiated on the phloem side by the procambium mature as large parenchyma cells and sharply delimit the primary from the secondary phloem (chapter 15; Schneider, 1945). In general, the developmental relations between the two parts of the phloem have not been sufficiently investigated. No data are available on the relative lengths of primary and secondary sieve elements comparable to those assembled for the tracheary elements, which prove that the last metaxylem cells are distinctly longer than the first secondary elements (chapter 11).

SECONDARY PHLOEM

Basic Structure

The arrangement of cells in the secondary phloem parallels that in the secondary xylem. A vertical or axial system of cells, derived from the fusiform initials of the cambium, is interpenetrated by the transverse or ray system derived from the ray initials (figs. 12.7–12.10; pls. 42, 43). The principal components of the axial system are sieve elements (either sieve cells or sieve-tube members, the latter usually with companion cells), phloem parenchyma, and phloem fibers. Those of the transverse system are ray parenchyma cells.

Storied, nonstoried, and intermediate types of arrangement of phloem cells may be found in different species of plants. As in the xylem, the type of arrangement is determined, first, by the morphology of the cambium (that is, whether it is stratified or not) and, second, by the degree of elongation of the various elements of the axial system during tissue differentiation.

Many woody species of dicotyledons show a division of the secondary phloem into seasonal growth increments (Holdheide, 1951), although this division is less clear than in the secondary xylem. The growth layers in the phloem are distinguishable if the cells of the early phloem expand more strongly than those of the late phloem (fig. 12.9B, pl. 44A; Artschwager, 1950; Holdheide, 1951). In *Pyrus malus* a band of future fiber-sclereids and crystalliferous cells overwinter in meristematic state near the cambium and when mature can serve as a marker for delimiting the successive growth layers (Evert, 1963*b*). The collapse of the sieve elements in the nonconducting region of the phloem and the concomitant modifications in some other cells—notably the enlargement of the parenchyma cells—contribute toward obscuring the structural differences that might exist in the different parts of a growth layer at its inception (pl. 44*B*). Many gymnosperms and angiosperms form fibers in tangential bands in the secondary phloem (figs. 12.7, 12.8). The number of these bands is not necessarily constant from season to season and cannot be safely used to determine the age of the phloem tissue.

The phloem rays are continuous with the xylem rays since both arise from a common group of ray initials in the cambium. (Compare figs. 12.7 and 12.8 with figs. 11.10 and 11.11.) The phloem ray and the xylem ray together constitute the vascular ray. Near the cambium the phloem and the xylem rays having common origin are usually the same in height and width. However, the older part of the phloem ray, which is displaced outward by the expansion of the secondary body, may increase in width, sometimes very considerably (Holdheide, 1951; pl. 28*A*). Before the phloem rays become dilated their variations in form and size are similar to those of xylem rays in the same species. Phloem rays are uniseriate, biseriate, or multiseriate; they vary in height; and small and large rays may be present in the same species. The rays may be composed of one kind of cell (fig. 12.7), or they may contain both kinds of cell, procumbent and erect (fig. 12.8). Phloem rays do not attain the same lengths as xylem rays, because the vascular cambium produces less phloem than xylem and also because commonly the outer portions of the phloem are sloughed off through the activity of the phellogen.

Conifer Phloem

In conifers the phloem parallels the xylem in the relative simplicity of its structure (fig. 12.7). The axial system contains sieve cells, parenchyma cells, and frequently fibers. The rays are mostly uniseriate and contain parenchyma only or parenchyma and albuminous cells. The cell arrangement is nonstoried. The expansion of cells during differentiation is uniform, the apical elongation slight, and, therefore, the radial seriation of cells, which originates in the cambium, is retained in the mature tissue (pl. 42*C*). In general, conifer phloem seems to show relatively little developmental disturbance in the cell arrangement which it inherits from the cambium.

The sieve cells of conifers are slender, elongated elements comparable

FIG. 12.7. Block diagram of secondary phloem and cambium of *Thuja occidentalis* (White Cedar), a conifer. (Courtesy of I. W. Bailey. Drawn by Mrs. J. P. Rogerson under the supervision of L. G. Livingston. Redrawn.)

to the fusiform initials from which they are derived. They overlap each other at their ends, and each is in contact with several rays. The sieve areas are particularly abundant on the ends which overlap those of other sieve cells. Elsewhere the sieve areas are generally restricted to the radial walls (Abbe and Crafts, 1939; Strasburger, 1891). The strands in the sieve areas are somewhat larger than plasmodesmata (Kollmann and Schumacher, 1962). Within a given sieve area the protoplasmic strands are aggregated into groups, and the callose associated with the strands in one group appears to fuse into one structure. In other words, several connecting strands seemingly traverse one callose cylinder.

The phloem parenchyma cells occur chiefly in longitudinal strands (fig. 12.7). They store starch at certain times of the year but are particularly conspicuous when they contain resinous and tanniniferous inclusions (pl. 42). Crystals also are commonly deposited in the parenchyma. In Abietineae, the axial parenchyma cells occur, often in tangential bands, among the sieve cells (pl. 26C, D; Srivastava, 1963b). In representatives of Taxaceae, Taxodiaceae, and Cupressaceae, phloem parenchyma cells alternate, in tangential bands, with sieve cells and fibers (fig. 12.7). In several genera there is an orderly sequence (with some variations) of fibers, sieve cells, phloem parenchyma, sieve cells, fibers. Abietineae have no fibers but apparently develop secondary walls in the sieve cells, whereas Taxaceae, Taxodiaceae, and Cupressaceae have fibers and only primary walls in the sieve cells (Abbe and Crafts, 1939). Large ramified sclereids may develop in the old parts of secondary phloem in Abies (Holdheide, 1951). A characteristic feature of the conifer phloem is the previously mentioned absence of companion cells and the presence of albuminous cells.

The secondary phloem of conifers may contain resin canals. Such canals have been studied in detail in Picea canadensis (Thomson and Sifton, 1925), in which they occur in the rays and are characterized by having series of cyst-like bulbous expansions; they have been interpreted as traumatic structures. With the increase in width of the rays in the outer part of the stem, the resin canals enlarge too, by divisions of epithelial cells. Moreover, the number of layers of the epithelial cells is increased by divisions periclinal with respect to the periphery of the duct. As a result of this activity, the resin duct appears as though surrounded by a cambial zone.

Dicotyledon Phloem

The phloem in dicotyledons shows a wider diversity of patterns of cell arrangement and more variations in the component cells than the phloem in conifers. Storied, intermediate, and nonstoried arrangements of cells are encountered, and the rays may be uniseriate, biseriate, and multiseriate. The elements of the axial system are sieve-tube members usually with companion cells, axial parenchyma cells, and fibers; those of the ray system are parenchyma cells (fig. 12.8). Both systems may contain sclereids, secretory elements of schizogenous and lysigenous origins, laticifers, and various idioblasts with specialized contents. Crystal formation is common and occurs in phloem parenchyma cells, frequently in sclerified parenchyma strands with a crystal in each cell (crystalliferous parenchyma strands, often misinterpreted as septate crystalliferous fibers), and in the rays.

One of the most conspicuous differences in the appearance of the phloem of different species results from the distribution of the fibers (Holdheide, 1951; Möller, 1882; Strasburger, 1891; Zahur, 1959). In certain dicotyledons the fibers occur in tangential bands, more or less regularly alternating with bands containing the sieve tubes and the parenchymatic components of the axial system (figs. 12.8–12.10; pls. 43A; 44A; *Tilia, Vitis, Liriodendron, Magnolia, Corchorus*). Sometimes the fibers are scattered among other cells of the vertical system (*Tecoma, Nicotiana, Cephalanthus, Laurus*), or no fibers are present (*Aristolochia*). The fibers may be very abundant, with sieve tubes and parenchyma cells scattered among them in small strands (*Carya*; Artschwager, 1950). As was mentioned before, in some plants the functioning phloem contains no sclerified elements, but, after the sieve tubes cease to function, fibers and sclereids differentiate (pl. 44B).

The sieve tubes and the parenchyma cells show varied spatial interrelationships. Sometimes the sieve tubes occur in long, continuous radial series (pl. 44B), or, on the contrary, they may form tangential bands alternating with similar bands of parenchyma (pl. 43A). In phloem having tangential bands of fibers alternating with bands of sieve elements and associated parenchyma, the sieve tubes are commonly separated by parenchyma cells from the fibers and the rays.

Many woody dicotyledons have nonstratified phloem, with elongated sieve-tube members bearing mostly compound sieve plates on the inclined end walls (*Betula, Quercus, Populus, Aesculus, Tilia, Liriodendron, Juglans*). In some genera the sieve areas of the sieve plates are distinctly more differentiated than the lateral sieve areas. In others, as in those of Pomoideae (Evert, 1960, 1963a), there is less distinction between the two kinds of sieve areas, and the long, slender sieve elements, with their very much inclined end walls, approach the sieve cells of conifers in their seemingly primitive structure. Slightly inclined (*Fagus, Acer*) and transverse (*Fraxinus, Ulmus, Robinia*) end walls usually bear

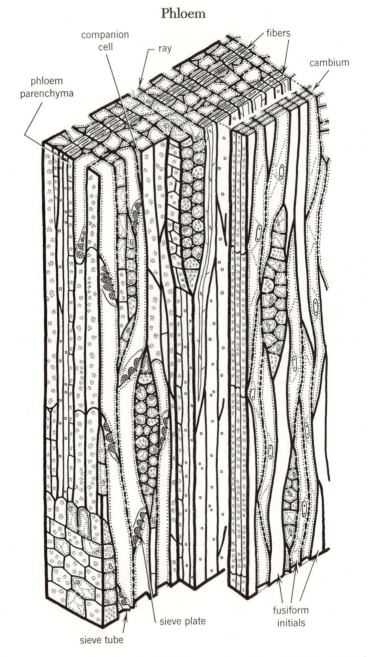

FIG. 12.8. Block diagram of secondary phloem and cambium of *Liriodendron tulipifera* (Tulip Tree), a dicotyledon. (Courtesy of I. W. Bailey. Drawn by Mrs. J. P. Rogerson under the supervision of L. G. Livingston. Redrawn.)

simple sieve plates. The individual sieve-tube members in such plants are relatively short, and if the phloem is derived from a cambium with short initials, the phloem may be more or less distinctly storied (pl. 43B).

If the sieve-tube members possess inclined end walls, the ends of the cells are roughly wedge-shaped and are so oriented that the wide side of the wedge is exposed in the radial section, the narrow in the tangential. The compound sieve plates are borne on the wide sides of the wedge-like ends and are therefore seen in face views in the radial sections (fig. 12.10A, pl. 40A), in sectional views in the tangential (fig. 12.10B, pl. 40B).

As previously mentioned, the secondary phloem rays are comparable to the xylem rays of the same species but may become dilated in the older parts of the tissue. The degree of this dilatation is highly variable. The extreme dilatation of certain of the rays is one of the most conspicuous characteristics of the phloem of Tilia (pl. 28). The wide rays separate the axial system together with the undilated rays into blocks narrowed down toward the periphery of the stem.

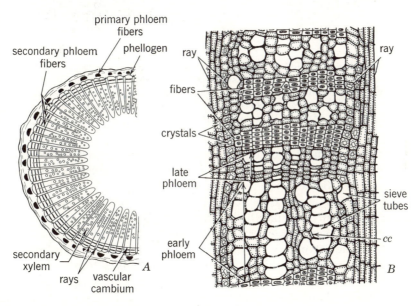

FIG. 12.9. Vitis vinifera (grapevine) phloem in transections. A, one year old branch (cane); B, secondary phloem from a cane. A, epidermis, cortex, and primary phloem were cut off by activity of phellogen which formed cork between primary and secondary phloem. B, sieve elements (unstippled) with sieve areas (breaks in walls) in younger phloem (below), with partly infolded walls in older phloem (above). Companion cell at cc. Fibers in tangential bands. (A, ×4; B, ×100. Esau, Hilgardia 18, 1948.)

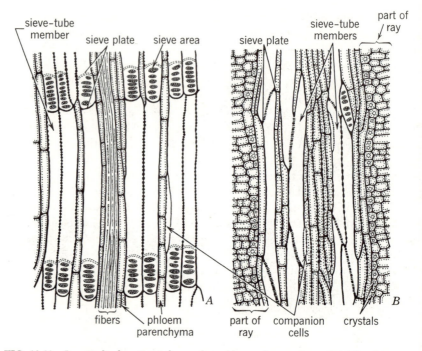

FIG. 12.10. Longitudinal sections of secondary phloem of grapevine, radial (*A*) and tangential (*B*). Compound sieve plates in surface view in *A*, in sectional view in *B*. Beaded effect of lateral walls between adjacent sieve-tube members indicates presence of small sieve areas. Parenchyma walls having similar appearance have primary pit-fields. Crystals in cells along the margins of rays, which are shown only in part. (Both, ×100.)

Herbaceous dicotyledons possessing secondary growth may have secondary phloem resembling that of the woody species (*Nicotiana, Gossypium*). Some herbaceous species, like the vine *Cucurbita*, have a secondary phloem scarcely distinguishable from the primary except in having larger cells (pl. 38*A*). *Cucurbita* has external and internal phloem, and only the external phloem is augmented by secondary growth. The secondary phloem consists of wide sieve tubes, narrow companion cells, and phloem parenchyma cells of intermediate size. There are no fibers and no rays. The sieve plates are simple and have large pores. The lateral walls bear sieve areas that are much less specialized than the single sieve areas of the simple sieve plates. In transections the small companion cells often appear as though cut out of the sides of the sieve tubes. Longitudinally, companion cells usually extend from one end of a sieve-tube member to the other. Sometimes only one companion cell occurs alongside a sieve-tube member, sometimes a vertical row of two or more.

Secondary phloem of relatively simple structure is found in dicotyle-donous storage organs, such as those of the carrot, the dandelion, and the beet (chapter 17). Storage parenchyma predominates in this kind of phloem, and the sieve tubes and companion cells appear as strands anastomosing within the parenchyma.

Differentiation in the Secondary Phloem

The derivatives of the vascular cambium on the phloem side com-monly undergo some divisions before the various phloem elements begin to differentiate. These may simply be tangential divisions in-creasing the number of derivatives, or they may be more specialized divisions. In the conifers the fusiform derivative differentiates into a sieve cell, usually without being subdivided into smaller cells (fig. 12.7). In the dicotyledons there are at least the longitudinal divisions that separate the future companion cells from the associated sieve-tube members (fig. 12.8). But, as was mentioned previously, the fusiform phloem initial may become divided by transverse, oblique, or longitudinal divisions giving rise to assemblages of more than one sieve element with their companion cells or of sieve elements, companion cells, and paren-chyma cells. After all these divisions are completed the sieve-tube members pass through the complex cytologic changes characteristic of these cells, and their primary pit-fields become modified into sieve areas. The differentiating sieve-tube members may be those derived from the cambium during the season of observation or those that overwintered in immature state near the cambium (chapter 6).

The fusiform cells that give rise to axial parenchyma commonly sub-divide into smaller cells by transverse or oblique divisions (parenchyma strand formation), or they differentiate into long, fusiform parenchyma cells. The fibers differentiate from fusiform derivatives by forming secondary walls with or without preceding elongation by intrusive growth.

The phloem cells expand transversely to varied degrees as they diverge from the cambium. Frequently, sieve-tube members show the greatest increase in diameter, whereas fibers may expand only slightly. Ray cells commonly change little during differentiation, except that they expand somewhat. In certain species some of the ray and axial paren-chyma cells develop secondary walls and differentiate into sclereids frequently with intrusive growth preceding the sclerification.

The phloem is considered to be differentiated into a conducting tissue when the sieve elements become enucleate and develop the other associated specialized characteristics, including the conspicuous sieve-area strands between the cells. The width of the yearly increment of active phloem produced in one season varies with the species and seasonal conditions and, as discussed in chapter 6, is considerably narrower than the corresponding increment of xylem. Moreover, in the deciduous species of dicotyledons a given increment of phloem commonly functions in conduction for a single season, in the evergreen dicotyledons and the conifers, probably two seasons (Grillos and Smith, 1959; Huber, 1939). Deviations from these patterns occur. In *Tilia* phloem, for example, sieve elements appear to remain functional for as many as 10 years (Holdheide, 1951). In *Vitis,* the phloem of one season becomes dormant at leaf fall by developing dormancy callose on the sieve areas but becomes reactivated in the following season by removal of most of the callose (pl. 40*D, E;* Esau, 1948; Wilhelm, 1880). At the end of the second season definitive callose is deposited and the protoplast dies.

Because of the relatively narrow width of the yearly increment of phloem and its usually short functioning life, the layer of conducting phloem occupies only a small proportion of the bark. A few examples of the width in millimeters of active phloem in deciduous species are 0.2 for *Fraxinus* and *Tectona* (Zimmermann, 1961); 0.2–0.3 for *Quercus, Fagus, Acer, Betula;* 0.4–0.7 for *Ulmus* and *Juglans;* and 0.8–1.0 for *Salix* and *Populus* (Holdheide, 1951). Sieve elements occupy from 25 to 50 per cent of the area of conducting phloem.

Nonconducting Phloem

The part of the phloem in which the sieve elements have ceased to function may be referred to as *nonconducting phloem.* The formerly widely used term nonfunctioning phloem is ambiguous because the phloem in which the sieve elements are no longer conducting commonly retains living parenchyma cells that continue to store starch and tannins until the tissue is severed from living parts of the plant by the activity of the phellogen.

The various signs of the inactive state of the sieve elements are readily detected. The sieve areas are either covered with a mass of callose (definitive callose) or entirely free of this substance, for callose eventually disappears in old nonfunctioning sieve elements (fig. 12.4*G, H*). The contents of the sieve elements may be completely disorganized, or they may be absent and the cells filled with gas. The determination of the nonconducting state of the phloem is particularly certain if the sieve elements are more or less collapsed or crushed. The companion cells and some parenchyma cells of dicotyledons and the albuminous cells of conifers cease to function and also collapse.

The characteristics of the inactive phloem as a whole are varied in different plants. In certain dicotyledons, like *Liriodendron* (Cheadle and Esau, 1964) *Tilia, Populus,* and *Juglans,* the shape of functionless sieve tubes changes little. In others, like *Aristolochia* and *Robinia*, the sieve elements and associated cells collapse completely, and, since they occur in tangential bands, the crushed cells alternate more or less regularly with the tangential bands of turgid parenchyma cells (pl. 49C, D). In still others the collapse of the sieve elements is accompanied by a conspicuous shrinkage of the tissue and bending of the rays (pl. 44B). In conifers the collapse of the old sieve cells is very marked. The nonconducting phloem of Abietineae shows dense masses of collapsed sieve cells interspersed with intact phloem parenchyma cells, and the rays are bent and folded. In conifers having fibers in the phloem, the sieve cells are crushed between the fibers and the enlarging phloem parenchyma cells (Abbe and Crafts, 1939). In *Vitis vinifera* the nonfunctioning sieve tubes become filled with tyloses-like proliferations from the axial parenchyma cells (Esau, 1948).

The nonconducting phloem frequently undergoes intensive sclerification, particularly by development of fibers or sclereids from axial and ray parenchyma cells. The intrusive growth that may precede sclerification modifies the spatial relations among the cells. The old phloem also accumulates ergastic substances, especially crystals and phenolic compounds. Crystals occur in the conducting phloem as well, but their numbers usually increase concomitantly with the sclerification phenomena. The types and distribution of the crystals are sufficiently characteristic to be useful in comparative studies (Holdheide, 1951; Möller, 1882).

One of the phenomena greatly affecting the appearance of the nonconducting phloem is the dilatation of the parenchymatic components of the tissue. By means of dilatation the phloem is adjusted to the increase in the circumference of the axis resulting from secondary growth. Sometimes the ray cells merely extend tangentially, but more commonly the number of cells is increased in the tangential direction by radial divisions. These divisions may be restricted to the median part of a ray, giving the impression of a localized meristem (Schneider, 1955). Often growth occurs in some rays while others retain their original width. In variable degrees dilatation also involves the axial parenchyma. Some enlargement of parenchyma cells may occur in connection with the collapse of nonfunctioning sieve elements, but these cells may also proliferate to the extent of forming wide wedges of tissue resembling dilated rays (Chattaway, 1955; Whitmore, 1962). Enlargement of parenchyma cells may continue in the rhytidome outside the latest periderm (Chattaway, 1955). The dilatation of the phloem is interrupted when a phellogen arises in the phloem and cuts off the outer part of this tissue by interpolating cork between it and the inner tissue.

The amount of the nonconducting phloem that accumulates in a given plant depends in part on the activity of the phellogen (chapter 14). If the phellogen is superficial and is not replaced by deeper-lying phellogens for many years, the plant may have a broad zone of non-conducting phloem (*Prunus*, Schneider, 1945). If, on the contrary, the phellogen forms year after year in successively deeper layers of the axis, this prevents the accumulation of inactive phloem (*Vitis*, Esau, 1948).

REFERENCES

Abbe, L. B., and A. S. Crafts. Phloem of white pine and other coniferous species. *Bot. Gaz.* 100:695–722. 1939.

Andrews, H. N., Jr. *Studies in Paleobotany.* New York, John Wiley and Sons. 1961.

Artschwager, E. The time factor in the differentiation of the secondary xylem and phloem in pecan. *Amer. Jour. Bot.* 37:15–24. 1950.

Bailey, I. W., and B. G. L. Swamy. The morphology and relationships of *Austrobaileya*. *Arnold Arboretum Jour.* 30:211–226. 1949.

Beck, C. B. *Tetraxylopteris schmidtii* gen. et sp. nov., a probable pteridosperm precursor from the Devonian of New York. *Amer. Jour. Bot.* 44:350–367. 1957.

Biddulph, S., and O. Biddulph. The circulatory system of plants. *Sci. Amer.* 200(2):44–49. 1959.

Blyth, A. Origin of primary extraxylary stem fibers in dicotyledons. *Calif. Univ., Publs., Bot.* 30:145–232. 1958.

Carlquist, S. *Comparative plant anatomy.* New York, Holt, Rinehart and Winston. 1961.

Chattaway, M. M. The anatomy of bark. VI. Peppermints, boxes, ironbarks, and other eucalypts with cracked and furrowed barks. *Austral. Jour. Bot.* 3:170–176. 1955.

Cheadle, V. I. Observations on the phloem in the Monocotyledoneae. II. Additional data on the occurrence and phylogenetic specialization in structure of the sieve tubes in the metaphloem. *Amer. Jour. Bot.* 35:129–131. 1948.

Cheadle, V. I., and K. Esau. Secondary phloem of the Calycanthaceae. *Calif. Univ., Publs., Bot.* 24:397–510. 1958.

Cheadle, V. I., and K. Esau. Secondary phloem of *Liriodendron tulipifera*. *Calif. Univ., Publs., Bot.* 36:143–252. 1964.

Cheadle, V. I., and N. W. Uhl. The relation of metaphloem to the types of vascular bundles in the Monocotyledoneae. *Amer. Jour. Bot.* 35:578–583. 1948.

Cheadle, V. I., and N. B. Whitford. Observations on the phloem in the Monocotyledoneae. I. The occurrence and phylogenetic specialization in structure of the sieve tubes in the metaphloem. *Amer. Jour. Bot.* 28:623–627. 1941.

Committee on Nomenclature, International Association of Wood Anatomists. International glossary of terms used in wood anatomy. *Trop. Woods* 107:1–36. 1957.

Crafts, A. S. *Translocation in plants.* New York, Holt, Rinehart and Winston. 1961.

Currier, H. B. Callose substance in plant cells. *Amer. Jour. Bot.* 44:478–488. 1957.

Currier, H. B., K. Esau, and V. I. Cheadle. Plasmolytic studies of phloem. *Amer. Jour. Bot.* 42:68–81. 1955.

De Bary, A. *Comparative anatomy of the vegetative organs of the phanerogams and ferns.* Oxford, Clarendon Press. 1884.

Duloy, M., F. V. Mercer, and N. Rathgeber. Studies in translocation. II. Submicroscopic anatomy of the phloem. *Austral. Jour. Biol. Sci.* 14:506–518. 1961.

Engleman, E. M. Fine structure of the proteinaceous substance in sieve tubes. *Planta* 59: 420–426. 1963.

Esau, K. Development and structure of the phloem tissue. *Bot. Rev.* 5:373–432. 1939. II. 16:67–114. 1950.

Esau, K. A study of some sieve-tube inclusions. *Amer. Jour. Bot.* 34:224–233. 1947.

Esau, K. Phloem structure in the grapevine, and its seasonal changes. *Hilgardia* 18:217–296. 1948.

Esau, K. *Plants, viruses, and insects.* Cambridge, Mass., Harvard University Press. 1961.

Esau, K., and V. I. Cheadle. Significance of cell divisions in differentiating secondary phloem. *Acta Bot. Neerland.* 4:348–357. 1955.

Esau, K., and V. I. Cheadle. Wall thickening in sieve elements. *Natl. Acad. Sci. Proc.* 44: 546–553. 1958.

Esau, K., and V. I. Cheadle. Size of pores and their contents in sieve elements of dicotyledons. *Natl. Acad. Sci. Proc.* 45:156–162. 1959.

Esau, K., and V. I. Cheadle. An evaluation of studies on ultrastructure of sieve plates. *Natl. Acad. Sci. Proc.* 47:1716–1726. 1961.

Esau, K., and V. I. Cheadle. An evaluation of studies on ultrastructure of tonoplast in sieve elements. *Natl. Acad. Sci. Proc.* 48:1–8. 1962a.

Esau, K., and V. I. Cheadle. Mitochondria in the phloem of *Cucurbita*. *Bot. Gaz.* 124:79–85. 1962b.

Esau, K., V. I. Cheadle, and E. M. Gifford, Jr. Comparative structure and possible trends of specialization of the phloem. *Amer. Jour. Bot.* 40:9–19. 1953.

Esau, K., V. I. Cheadle, and E. B. Risley. Development of sieve-plate pores. *Bot. Gaz.* 123:233–243. 1962.

Esau, K., V. I. Cheadle, and E. B. Risley. A view of ultrastructure of *Cucurbita* xylem. *Bot. Gaz.* 124:311–316. 1963.

Esau, K., H. B. Currier, and V. I. Cheadle. Physiology of phloem. *Ann. Rev. Plant Physiol.* 8:349–374. 1957.

Eschrich, W. Kallose. *Protoplasma* 47:487–530. 1956.

Eschrich, W. Beziehungen zwischen dem Auftreten von Callose und der Feinstruktur des primären Phloems bei *Cucurbita ficifolia*. *Planta* 59:243–261. 1963.

Evert, R. F. Phloem structure in *Pyrus communis* L. and its seasonal changes. *Calif. Univ., Publs., Bot.* 32:127–196. 1960.

Evert, R. F. Ontogeny and structure of the secondary phloem in *Pyrus malus*. *Amer. Jour. Bot.* 50:8–37. 1963a.

Evert, R. F. The cambium and seasonal development of the phloem of *Pyrus malus*. *Amer. Jour. Bot.* 50:149–159. 1963b.

Evert, R. F. Sclerified companion cells in *Tilia americana*. *Bot. Gaz.* 124:262–264. 1963c.

Frey-Wyssling, A., and H. R. Müller. Submicroscopic differentiation of plasmodesmata and sieve plates in *Cucurbita*. *Jour. Ultrastruct. Res.* 1:38–48. 1957.

Grillos, S. J., and F. H. Smith. The secondary phloem of Douglas-fir. *Forest Sci.* 5:377–388. 1959.

Haberlandt, G. *Physiological plant anatomy.* London, Macmillan and Company. 1914.

Hanstein, J. *Die Milchsaftgefässe und die verwandten Organe der Rinde.* Berlin, Wiegandt und Hempel. 1864.

Hartig, T. Vergleichende Untersuchungen über die Organisation des Stammes der einheimischen Waldbäume. *Jahresber. Forsch. Forstwissensch. und Forstl. Naturkunde* 1:125–168. 1837.

Hill, G. P. Exudation from aphid stylets during the period from dormancy to bud break in *Tilia americana* (L.). *Jour. Expt. Bot.* 13:144–151. 1962.

Holdheide, W. Anatomie mitteleuropäischer Gehölzrinden (mit mikrophotographischem

Atlas). In: H. Freund. *Handbuch der Mikroskopie in der Technik.* Band 5. Heft 1. 193–367. Frankfurt am Main, Umschau Verlag. 1951.

Huber, B. Hundert Jahre Siebröhren-Forschung. *Protoplasma* 29:132–148. 1937.

Huber, B. Das Siebröhrensystem unserer Bäume und seine jahreszeitlichen Veränderungen. *Jahrb. f. Wiss. Bot.* 88:176–242. 1939.

Jean, M. Essai sur l'anatomie comparée du liber interne dans quelques familles de Dicotylédones. Étude des plantules. *Botaniste* Ser. 17. fasc. 5–6:225–364. 1926.

Kessler, G. Zur Charakterisierung der Siebröhrenkallose. *Schweiz. Bot. Gesell. Ber.* 68:5–43. 1958.

Kollmann, R. Untersuchungen über das Protoplasma der Siebröhren von *Passiflora coerulea.* I. Lichtoptische Untersuchungen. *Planta* 54:611–640. 1960.

Kollmann, R., and W. Schumacher. Über die Feinstruktur des Phloems von *Metasequoia glyptrostroboides* und seine jahreszeitlichen Veränderungen. II. Vergleichende Untersuchungen der plasmatischen Verbindungsbrücken in Phloemparenchymzellen und Siebzellen. *Planta* 58:366–386. 1962. IV. Weitere Beobachtungen zum Feinbau der Plasmabrücken in den Siebzellen. *Planta* 60:360–389. 1963.

Lecomte, H. Contribution a l'étude du liber des angiospermes. *Ann. des Sci. Nat., Bot.* Ser. 7. 10:193–324. 1889.

Léger, L. J. Recherches sur l'origine et les transformations des éléments libériens. *Soc. Linn. de Normandie, Mém.* 19:49–182. 1897.

Mangin, L. Sur la callose, nouvelle substance fondamentale existant dans la membrane. *Acad. des Sci. Compt. Rend.* 110:644–647. 1890.

Möller, J. *Anatomie der Baumrinden.* Berlin, Julius Springer. 1882.

Nägeli, C. W. Das Wachsthum des Stammes und der Wurzel bei den Gefässpflanzen und die Anordnung der Gefässstränge im Stengel. *Beitr. z. Wiss. Bot.* Heft 1:1–156. 1858.

Peel, A. J., and P. E. Weatherley. Studies in sieve-tube exudation through aphid mouthparts: the effects of light and girdling. *Ann. Bot.* 26:633–646. 1962.

Perrot, E. *Le tissu criblé.* Paris, Librairie Lechevallier. 1899.

Resch, A. Beiträge zur Cytologie des Phloems. Entwicklungsgeschichte der Siebröhrenglieder und Geleitzellen bei *Vicia faba* L. *Planta* 44:75–98. 1954.

Schmidt, E. W. *Bau und Funktion der Siebröhre der Angiospermen.* Jena, Gustav Fischer. 1917.

Schneider, H. The anatomy of peach and cherry phloem. *Torrey Bot. Club Bul.* 72:137–156. 1945.

Schneider, H. Ontogeny of lemon tree bark. *Amer. Jour. Bot.* 42:893–905. 1955.

Smith, F. H. Anatomical development of the hypocotyl of Douglas-fir. *Forest. Sci.* 4:61–70. 1958.

Srivastava, L. M. Cambium and vascular derivatives of *Ginkgo biloba. Arnold Arboretum Jour.* 44:165–192. 1963a.

Srivastava, L. M. Secondary phloem in the Pinaceae. *Calif. Univ., Publs., Bot.* 36:1–142. 1963b.

Srivastava, L. M., and I. W. Bailey. Comparative anatomy of the leaf-bearing Cactaceae. V. The secondary phloem. *Arnold Arboretum Jour.* 43:234–278. 1962.

Strasburger, E. *Über den Bau und die Verrichtungen der Leitungsbahnen in den Pflanzen. Histologische Beiträge.* Band 3. Jena, Gustav Fischer. 1891.

Swanson, C. A. Translocation of organic solutes. In: F. C. Steward. *Plant physiology.* New York, Academic Press. 1959.

Thomson, R. B., and H. B. Sifton. Resin canals in the Canadian spruce (*Picea canadensis* (Mill.) B. S. P.)—an anatomical study, especially in relation to traumatic effects and their bearing on phylogeny. *Roy. Soc. London, Phil. Trans.* Ser. B. 214:63–111. 1925.

Ullrich, W. Beobachtungen über Kalloseablagerungen in transportierenden und nicht-transportierenden Siebröhren. *Planta* 59:239-242. 1962.

Weatherley, P. E. The mechanism of sieve-tube translocation: Observations, experiment and theory. *Adv. of Sci.* 18:571-577. 1962.

Weatherley, P. E., A. J. Peel, and G. P. Hill. The physiology of the sieve tube. Preliminary experiments using aphid mouth parts. *Jour. Expt. Bot.* 10:1-16. 1959.

Whitmore, T. C. Studies in systematic bark morphology. I. Bark morphology in Diptero-carpaceae. II. General features of bark construction in Dipterocarpaceae. *New Phytol.* 61:191-220. 1962.

Wilhelm, K. *Beiträge zur Kenntnis des Siebröhrenapparates dicotyler Pflanzen.* Leipzig, Wilhelm Engelmann. 1880.

Zahur, M. S. Comparative study of secondary phloem of 423 species of woody dicotyledons belonging to 85 families. *Cornell Univ. Agric. Expt. Sta. Mem.* 358. 1959.

Ziegler, H., and T. E. Mittler. Über den Zuckergehalt der Siebröhren- bzw. Siebzellensäfte von *Heracleum Mantegazzianum* und *Picea abies* (L.) Karst. *Ztschr. f. Naturf.* 14b: 278-281. 1959.

Zimmermann, M. H. Movement of organic substances in trees. *Science* 133:73-79. 1961.

13

Secretory Structures

Plant cells produce many substances that appear to be nonutilizable by-products of metabolism and that become more or less isolated from the living protoplasts or removed entirely from the plant body. Examples of such substances are terpenes and related compounds, tannins, and different kinds of crystals (chapter 2). Representatives of the terpenes—hydrocarbons of various degrees of polymerization—are the lower terpenes, as essential oils, and the higher terpenes, as carotenoids, saponins, and rubber (Haagen-Smit, 1958; Moritz, 1958).

Active or passive secretion may be responsible for the removal of the terpenes and other by-products. *Secretion* refers to the act of separation of a substance from the protoplast. In the strict sense, secretion signifies the release of substances that have a special physiologic function (enzymes, hormones). The separation of products eliminated from the metabolism is *excretion* (Kisser, 1958). Commonly, however, no sharp line is drawn between secretion and excretion, partly because the role of many of the by-products of metabolism is not known and partly because the waste products and the physiologically functional secretions may accumulate in the same containers. In this book the term secretion is used to include both secretion in the strict sense and excretion.

Structures concerned with secretion vary widely in degree of specialization and in location in the plant body. Some are external in position, others internal; some are simple glandular hairs, others are many-celled vascularized glands, and still others are intercellular ducts or cavities. The indefinitely elongating cells or the complex cell fusions represented by the laticifers also belong among the secretory structures because they are notable for their content of excretions and secretions.

The secretory structures differ in the relation between the secreted material and the protoplast of the secreting cell (Kisser, 1958). The secretion may remain in the cell producing it or it may leave the cell. Essential oils, balsams, and resins, though true excretions, occur indefinitely as accumulations within cells, some specialized as idioblasts. In many cells these substances appear to be distributed as droplets in the cytoplasm, but they may be walled off from the protoplasts in others. Finally, there are cells that release the excretion either into an intercellular container or to the surface of the plant.

EXTERNAL SECRETORY STRUCTURES

Trichomes and Glands

The surface of the plant bears many forms of secretory structures. Some are epidermal in origin, others include derivatives of the epidermis and of deeper-lying cells (emergences; Kisser, 1958). In some leaves or flowers, more or less large areas of the epidermis are glandular (fig. 13.1C, D); or the glandular epidermis covers emergences such as the shaggy hairs of *Nerium* (fig. 13.1A, B); or the epidermal cells give rise to trichomes of various degrees of complexity. The trichomes may be biseriate hairs; hairs provided with a unicellular or multicellular head (capitate hairs) on a narrow stalk, frequently consisting of one series of cells (fig. 7.10A, B); scales or peltate hairs (figs. 7.10G, H; 13.1E); colleters having a multicellular head on a multicellular stalk. The development of trichomes from the epidermis results from differential enlargement and subsequent division of epidermal cells and their derivatives (Bancher and Hölzl, 1959; Carlquist, 1958).

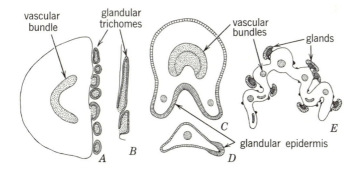

FIG. 13.1. Glandular structures on leaves. *A, B,* multicellular glandular trichomes, with palisade-like secretory layer, of *Nerium oleander. C, D,* glandular epidermis on leaf and stipule of *Salix. E,* leaf from a winter bud of *Betula* with peltate glands bearing a palisade-like glandular epidermis. *A, C–E,* from transverse, *B,* from longitudinal sections of leaves. (*A, B,* ×21; *C, D, E,* ×37.)

The more complex secretory structures may be called glands but no sharp division exists between glandular hairs and glands, and the simple trichomes derived entirely from the epidermis intergrade with emergences. Variations in degree of complexity may be found in closely related species and appear to be of phylogenetic significance (Carlquist, 1959a, b).

Many trichomes and glands excrete the previously mentioned terpenes in various combinations. The floral and extrafloral nectaries produce liquid containing sugar. Plants of saline habitat may excrete salts through their glandular structures. Trichomaceous hydathodes release water, especially in young leaves, and may later absorb water (Kaussmann, 1954). Glands of insectivorous plants excrete nectar, mucilages, or digestive juices.

Active secretory cells have dense protoplasts, rich in proteinaceous substances and with large nuclei that may be polyploid (Stahl, 1957). The density of the protoplasts results from decreasing vacuolation as the active state is reached (Stahl, 1957). In multicellular trichomes and glands the activity concerned with secretion occurs in tissue several cells in depth. Sometimes only the inner cells contain storage products and have dense protoplasts, whereas the epidermis is vacuolated and free of storage products. Phosphatase and hydrogenase have been identified in some glandular structures (fig. 13.5B; Frey-Wyssling and Häusermann, 1960; Stahl, 1957). Ultrastructural studies of the glands of the insectivorous *Drosophyllum* suggests a relation between the vesiculation of dictyosomes and the production of the sticky secretion (Schnepf, 1960, 1963).

The glandular hairs and scales commonly release the secretion between the wall and the cuticle, which extends considerably. Eventually the cuticle bursts. It may become regenerated and the accumulation repeated (Trapp, 1949), or the hair may degenerate after the single act of excretion (Stahl, 1953). The mechanics of the apparent distention of the cuticle is difficult to explain (Kisser, 1958). In the glandular hairs of *Atropa* the essential oil is released without the separation of the cuticle. Moreover, the individual cells become severed successively from the trichome like conidia from the end of a hypha (Hülsbruch, 1961). Apparently the separation of the cells involves a hydration and swelling of the middle lamella, since a pectic ring-like thickening is detectable at the base of the cell shortly before its separation.

The stinging hairs of the nettle (*Urtica*) have a special mechanism for releasing the contents. The hair is like a fine capillary tube, calcified at its lower end and silicified at its upper end. At the base a bladder-like end is imbedded in epidermal cells, somewhat raised above the surface. At the upper end the tube bears a spherical tip which breaks off along a predetermined line when the hair comes in contact with an object. The sharp edge left after the separation of the tip readily penetrates human skin, and the contents of the tube escape into the wound. The poisonous material of the nettle is highly complex and contains a histamine and an acetyl-choline (Feldberg, 1950).

The colleters—the term is derived from the Greek *colla*, glue, referring to the sticky excretion from these structures—are common on bud scales (*Aesculus, Rosa, Carya*). They frequently produce a mixture of terpenes and mucilage. The cuticle is broken during the excretion without being distended. In the colleters of *Azalea* the excretion appears first in the walls between the cells, which become swollen (Kisser, 1958). Colleters develop on young foliar organs and desiccate when the bud opens and the leaves expand.

Nectaries

Nectaries occur on flowers (floral nectaries) and on vegetative parts (extrafloral nectaries). They range in form from glandular surfaces to specialized vascularized glands. Floral nectaries occupy various positions on the flower (Brown, 1938; Fahn, 1952, 1953; Sperlich, 1939). A general trend of phylogenetic migration of the floral nectary from the perianth to the inner floral organs has been deduced from comparative studies (Fahn, 1953). The extrafloral nectaries occur on stems, leaves (fig. 13.5A), stipules, and pedicels of flowers.

In dicotyledonous flowers the nectar may be secreted by the basal parts of stamens (fig. 13.2C) or by a ring-like nectary below the stamens (fig. 13.2E; Caryophyllales, Polygonales, Chenopodiales). The nectary may be a ring or a disc at the base of the ovary (fig. 13.2D, F; Theales, Ericales, Polemoniales, Solanales, Lamiales) or a disc between the stamens and the ovary (fig. 13.2G). Several discrete glands may occur at the base of the stamens (fig. 13.2L). In the Tiliales the nectaries consist of multicellular glandular hairs, usually packed close together to form a cushion-like growth (fig. 13.2I). Such nectaries occur on various floral parts, frequently on sepals. In the perigynous Rosaceae the nectary is located between the ovary and the stamens, lining the interior of the floral cup (fig. 13.2J). In the epigynous flower of the Umbellales the nectary occurs on the top of the ovary (fig. 13.2H). In the Compositae it is a tubular structure at the top of the ovary, encircling the base of the style. In most of the insect-pollinated genera of the Lamiales, Berberidales, and Ranunculales the nectaries are modified stamens, or staminodes (fig. 13.2K). The nectary on the petals of *Frasera* consists of a cup with a glandular floor and a wall provided with numerous hair-like sclerified processes which plug the apical opening

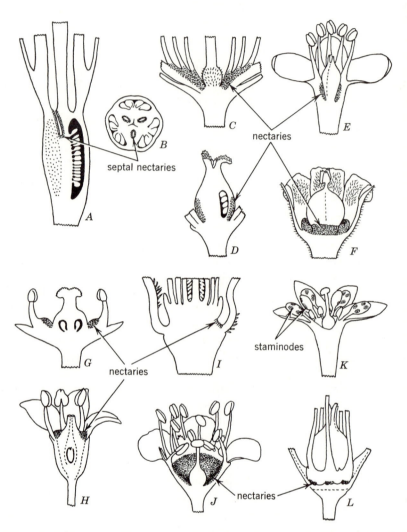

FIG. 13.2. Nectaries. Longitudinal (*A, C–L*) and cross (*B*) sections of flowers. Septal, in Liliales, *Narcissus* (*A*) and *Gladiolus* (*B*). *C*, external, at base of stamens (*Thea*, Theales). *D*, ring at base of ovary (*Euyra*, Theales). *E*, ring below stamens (*Coccoloba*, Polygonales). *F*, disc below ovary (*Jatropha*, Euphorbiales). *G*, disc between ovary and stamens (*Perrottetia*, Celastrales). *H*, disc above inferior ovary (*Mastixia*, Umbellales). *I*, cushion of hairs at base of sepal (*Corchorus*, Tiliales). *J*, lining floral cup (*Prunus*, Rosales). *K*, modified stamens, staminodes (*Cinnamomum*, Laurales). *L*, glands at bases of stamens (*Linum*, Geraniales). (Adapted from Brown, *Amer. Phil. Soc. Proc.* 79, 1938.)

and force the bumble bee to work along the sides of the gland (Davies, 1952).

In the monocotyledons the nectaries frequently occur in the partitions or septae of the ovaries (fig. 13.2*A*, *B*; septal nectaries; Brown, 1938; Okimoto, 1948; Sperlich, 1939). These nectaries have the structure of pockets with a glandular lining and arise in parts of the ovary where the carpel walls are incompletely fused. If they are deeply imbedded in the ovary, they have outlets in the form of canals leading to the surface of the ovary.

The secretory tissue of the nectary may be restricted to the epidermal layer. Usually the secretory epidermal cells have dense cytoplasm and may be papillate or elongated like palisade cells (Agthe, 1951), but in some plants they show no distinguishing cytologic characteristics. In many nectaries the cells beneath the epidermis are also secretory. They are rich in cytoplasm, are closely packed, and have thin walls. Laticifers may be present in the nectary. The nectary is covered by a cuticle.

The sugar of the nectaries, floral and extrafloral, is derived from the phloem. Vascular tissue occurs more or less close to the secretory tissue. In some nectaries the vascular tissue is merely that of the organ bearing the nectary, in others it is part of the nectary. The variations in vascularization of nectaries are related to the type of nectar secreted (Frei, 1955). In nectaries secreting a highly concentrated sugar solution, the ultimate branches of the vascular system terminating below the secretory tissue consist of phloem elements only (*Euphorbia pulcherrima*, *Abutilon striatum*). Such nectaries contrast strikingly with hydathodes in which the ultimate branchings of the vascular system contain only tracheary elements (fig. 13.5*C*). Nectaries and hydathodes may differ in cell arrangement. In the nectary the parenchyma cells are closely packed, whereas in many hydathodes the tissue is permeated with intercellular spaces (pl. 76*A*). Some nectaries (*Ranunculus, Fritillaria*) occupy an intermediate position between the most highly specialized nectaries and the hydathodes. Their ground tissue is moderately compact, xylem and phloem occur in the final branchings of the vascular system, and the nectar shows moderate concentrations of sugar.

The nectar is excreted either through the cell wall and the ruptured cuticle or, in the less highly specialized nectaries, through stomata (Fahn, 1953; Frey-Wyssling and Häusermann, 1960). In some nectaries the stomata are modified in that the guard cells are not able to close the opening. Studies with labeled carbon (C^{14}) have shown that nectaries not only secrete nectar but are also able to absorb it (Shuel, 1961). The absorbed nectar becomes distributed to all parts of the plant (Pedersen et al., 1958), including the stigma (Shuel, 1961).

Osmophors

The fragrance of flowers is commonly produced by volatile substances —mainly essential oils—distributed throughout the epidermis of perianth parts (Weichsel, 1956). In some plants, however, the fragrance originates in special glands named osmophors by Vogel (1962), a term derived from the Greek words scent and bearer. Examples of osmophors are found in Asclepiadaceae, Aristolochiaceae, Araceae, Burmaniaceae, and Orchidaceae. Various floral parts may be differentiated as osmophors and they may assume the form of flaps, cilia, or brushes. The prolongation of the spadix of the Araceae and the insect-attracting tissue in the flowers of Orchidaceae are osmophors. It is possible to identify the osmorphors by staining them with neutral red in whole flowers dipped into a solution of the dye (fig. 13.3).

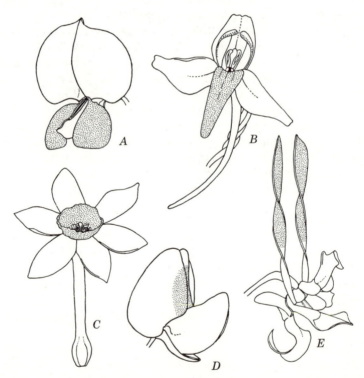

FIG. 13.3. Flowers tested by neutral-red staining for location of osmophors (stippled)— flower parts containing secretory tissue responsible for emission of fragrance. A, *Spartium junceum; B, Platanthera bifolia; C, Narcissus jonquilla; D, Lupinus Cruckshanksii; E, Dendrobium minax.* (After Vogel, *Akad. Wiss. Lit. Mainz, Math.-Nat. Kl. Abh.* 10, 1962.)

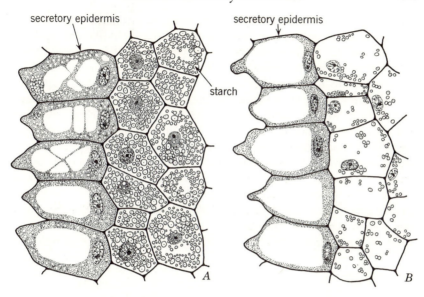

secretory epidermis   secretory epidermis

starch

A   *B*

FIG. 13.4. Sections through secretory tissue of osmophors of a *Ceropegia stapeliaeformis* flower. A, at beginning of secretory activity. B, after emission of fragrance: secretory cells with reduced cytoplasmic density, and starch depleted in tissue below epidermis. (After photographs by Vogel, *Akad, Wiss Lit. Mainz, Math.-Nat. Kl. Abh.* 10, 1962.)

The osmophors have a secretory tissue usually several layers in depth. The emission of the volatile secretions is of short duration and is associated with a utilization of large amounts of storage products (fig. 13.4). The tissue may be compact and vascularized or it may be permeated by intercellular spaces. The oil usually vaporizes immediately, but it also may appear in droplets.

Hydathodes

Hydathodes are structures that discharge water from the interior of the leaf to its surface, a process commonly called guttation (Kramer, 1945). The guttation water contains various salts, sugars and other organic substances, and sometimes glutamine after the plants are fertilized with ammonium sulphite. Guttation may cause injury to plants through accumulation and concentration of guttation products or through interaction of the guttation water with pesticides (Ivanoff, 1962).

Although the hydathodes are usually treated with the secretory structures, one of the common types of hydathode possesses no tissue that is exactly comparable with the glandular tissue of true secretory organs. The hydathodes eliminate water directly from the terminal tracheids of the bundle ends. In many angiosperms the terminal tracheids are in contact with a thin-walled parenchyma (the epithem), deficient in chloroplasts, and provided with intercellular spaces through which the water moves from the tracheids to the epidermis (pl. 76A). The epidermis has openings over the epithem, which often appear as incompletely differentiated stomata lacking the mechanism for opening and closing (Reams, 1953; Stevens, 1956). Each hydathode may have one (fig. 13.5C; *Primula, Aconitum, Delphinium*) or more than one pore (pl. 76A; Umbelliferae, Compositae). In *Equisetum* the epithem occurs along one side of the vascular bundle, rather than at the end, and the number of pores for each hydathode varies from three to fifty (Johnson, 1937). The epithem may be enclosed by suberized cells or by cells having casparian strips (Sperlich, 1939). In some plants the hydathodes are without epithem, and the water moves toward the pore through ordinary mesophyll. In others the hydathodes are rather complex and

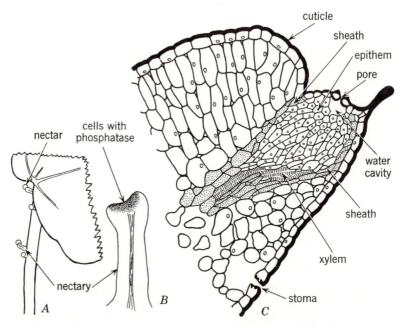

FIG. 13.5. *A,* part of leaf of *Passiflora coerulea* with extrafloral nectaries on petiole and, *B,* section of nectary tested for phosphatase activity, which is concentrated in secretory tissue (stippled). *C,* hydathode of leaf of *Saxifraga lingulata* in longitudinal section. Tannin-containing sheath cells are stippled. (*A, B,* from photographs in Frey-Wyssling and Häusermann, *Schweiz. Bot. Gesell. Ber.* 70, 1960; *C* after Häusermann and Frey-Wyssling, *Protoplasma* 57, 1963.)

appear to be associated with secretory tissue (fig. 13.5C; Sperlich, 1939). Such hydathodes may be interpreted as structures intergrading between nectaries and typical hydathodes. Hydathodes may also be differentiated as secretory trichomes (Kaussmann, 1954).

INTERNAL SECRETORY STRUCTURES

Secretory Cells

The secretory cells are more or less well differentiated from ground-parenchyma cells and contain a variety of substances: balsams, resins, oils, tannins, mucilages, gums, crystals. They are called secretory idio-blasts when they differ conspicuously from neighboring cells among which they are dispersed (pl. 71). The cells may be isodiametric, or more or less elongated into sacs or tubes, or branched (pl. 71C). The secretory cells are commonly classified on the basis of their contents, but such classification is not exact because some of these cells have not been investigated for the chemistry of their contents and others contain mixtures of substances (Kisser, 1958). One of the common types of secretory cell is called an oil cell (pl. 71A). The oily excretion occurs in a spherical intracellular compartment having a distinct limiting membrane—possibly a cellulose wall (Kisser, 1958)—and is attached to the cell wall by a stalk of cellulose. A foamy cytoplasm and lack of nucleus have been noted in such cells (Ziegler, 1960). The cell wall of an oil cell may contain a suberin lamella (Weichsel, 1956). Further examples of secretory cells and lists of taxonomic groups containing them are found in Esau (1960, pp. 163–164) and Metcalfe and Chalk (1950, pp. 1346–1349). Secretory cells occur in all parts of a plant, vegetative and reproductive.

Crystal-containing cells (chapter 2) are often treated as secretory idio-blasts (Foster, 1956). Crystals may occur in parenchyma cells that do not differ from other parenchyma cells in the tissue, but they also may be considerably modified, as for example the lithocysts of *Ficus* (chapter 7) and the mucilage-containing raphid cells (pl. 71B; Kowalewicz, 1956). Crystal-forming cells may die after the deposition of the crystal, or crystals, is completed; or the crystal may be walled off from the living part of the protoplast.

Secretory Spaces

The secretory spaces in the form of cavities or canals are formed by schizogeny or by lysigeny (chapter 3) or sometimes by both phenomena combined. The schizogenous spaces are lined with secretory cells composing the epithelium. The lysigenous spaces are surrounded by more or less disintegrated cells the breakdown of which brought about the formation of the space. The secretory spaces may occur in any part of the plant.

The separation of cells in the formation of a schizogenous secretory space may occur with or without preceding division of cells. Later the cells facing the space divide and thus make possible the enlargement of this space. The spaces may be roundish (Burseraceae, Leguminosae, Myrtaceae) or elongated and canal-like (Coniferae, Anacardiaceae, Araliaceae, Compositae, Umbelliferae). According to Kisser (1958), the excreta are volatile terpenes (Pittosporae, Guttiferae, Myrtaceae, Umbelliferae), viscous balsams (Coniferae, Araliaceae; resin ducts of Coniferae would be more appropriately called balsam ducts, Kisser, 1958), gum-resins (Clusoideae), latex (some Umbelliferae and Cactaceae, *Alisma plantago*), gum or mucilage (Lycopodiaceae, Marattiaceae, Araliaceae, Sterculiaceae). The copal, a resin used in varnishes, is derived from schizogenous ducts of tropical Legminosae (Moens, 1955).

In the epithelial cells of the resin canals of conifers, droplets of the excretion occur in the protoplast next to the wall facing the space (Kisser, 1958). Later, they appear to leave the protoplast and pass through the wall into the space. In some plants (*Lysimachia, Myrsine, Ardisia*) resinous material is excreted into ordinary intercellular spaces and forms a granular layer along the walls.

In the lysigenous spaces the excretions appear to arise in the cells before the latter break down (*Citrus, Eucalyptus*). The dissolution begins in a few cells, then extends to neighboring ones. In *Ruta graveolens* an excretion occurs first from intact cells, then a dissolution of cells sets in (Kisser, 1958). Lysigenous spaces may result as responses to injuries (*Liquidambar orientalis, Styrax benzoin*).

LATICIFERS

Laticifers are cells or series of fused cells containing a fluid called *latex* (plural, *latices*) and forming systems that permeate various tissues of the plant body. The word *laticifer* and its adjectival form *laticiferous* are derived from the word latex, meaning juice in Latin. The latex is often milky or even white in appearance, and therefore the laticifers are sometimes called lactiferous cells or vessels (Jackson, 1953). The term lactiferous is derived from the Latin word for milk, *lac*. Since the latex is highly variable in its physical and chemical composition and is not necessarily milky, the less specific terms laticifer and laticiferous are more appropriate than lactiferous. It is also preferable to employ *laticifer* as a general term (Jackson, 1953) instead of laticiferous ducts or tubes because of the simplicity and wide applicability of this term.

Although the structures bearing the latex may be single cells or series of fused cells, both kinds often produce complex systems of tube-like growth form in which recognition of the limits of individual cells is highly problematical. The term laticifer, therefore, appears most useful if applied to either a single cell or a structure resulting from fusion of cells. A single-cell laticifer can be qualified, on the basis of origin, as a *simple laticifer*, and the structure derived from union of cells as a *compound laticifer*.

The laticifers vary widely in their structure, and the latex in its composition. Latex may be present in ordinary parenchyma cells, as in guayule (*Parthenium argentatum;* Bonner and Galston, 1947), or it may be formed in branching (*Euphorbia*) or anastomosing (*Hevea*) systems of tubes. The ordinary parenchyma cells with latex and the elaborate laticiferous systems intergrade with each other through intermediate types of structures of various degrees of morphologic specialization. Laticifers also intergrade with certain idioblasts that contain tannins (tannin sacs of Leguminosae or *Sambucus*), mucilages, proteinaceous and other compounds. The situation is further complicated by the occurrence of schizogenous canals containing latex (Kisser, 1958). Thus laticifers cannot be delimited precisely.

Latex-containing plants are estimated to include some 12,500 species in 900 genera (Van Die, 1955) of dicotyledons and monocotyledons. Among the lower plants the fern *Regnellidium* has been reported having laticifers (Labouriau, 1952). The plants containing latex range from such small herbaceous annuals as the spurges (*Euphorbia*) to large trees like the rubber-yielding *Hevea*. They occur in all parts of the world, but arborescent types are most common in the tropical floras.

Classification

Laticifers are grouped in two major classes on the basis of their structure: the *articulated* (that is, jointed; pls. 46A, B; 47) and the *nonarticulated* (pl. 46C–E). The former are compound in origin and consist of longitudinal chains of cells in which the walls separating the individual cells either remain intact, become perforated, or are completely removed. The perforation or resorption of the end walls gives rise to laticifers that are tube-like in form and resemble xylem vessels in origin. This type of laticifer has been formerly called laticiferous vessel. The nonarticulated laticifers originate from single cells which through continued growth develop into tube-like structures, often much branched, but typically they undergo no fusions with other similar cells. This type of laticifer is simple in origin and has been formerly called laticiferous cell.

The variations in structure of the two types of laticifers permit the establishment of subdivisions under each. Some of the articulated laticifers consist of long cell chains or compound tubes not connected with each other laterally; others form lateral anastomoses with similar cell chains or tubes, all combined into a net-like structure or reticulum. These two forms of laticifers can be called *articulated nonanastomosing* laticifers (fig. 13.6) and *articulated anastomosing* (pl. 47) laticifers, respectively.

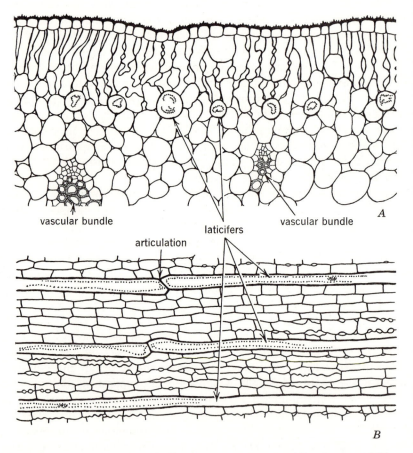

FIG. 13.6. Articulated laticifers of *Allium sativum* in transverse (*A*) and tangential (*B*) sections of leaves. *A*, palisade parenchyma beneath epidermis. Laticifers occur in third layer of mesophyll not in contact with vascular bundles. *B*, laticifers appear like continuous tubes except in places where end wall (articulation) between superposed cells is visible. The end wall is not perforated. (Both, ×79. Drawn from photomicrographs by L. K. Mann.)

The nonarticulated laticifers also vary in degree of complexity in their structure. Some develop into long, more or less straight tubes; others branch repeatedly, each cell thus forming an immense system of tubes. The appropriate names for these two types of structures are *nonarticulated unbranched* laticifers and *nonarticulated branched* (pl. 46*D, E*) laticifers, respectively.

Examples of the various types of laticifers are found in the following families and genera. Articulated anastomosing: Compositae, tribe Cichorieae (*Cichorium, Lactuca, Scorzonera, Sonchus, Taraxacum, Tragopogon*); Campanulaceae, including the Lobelioideae; Caricaceae (*Carica papaya*); Papaveraceae (*Papaver, Argemone*); Euphorbiaceae (*Hevea, Manihot*). Articulated nonanastomosing: Convolvulaceae (*Ipomoea, Convolvulus, Dichondra*); Papaveraceae (*Chelidonium*); Sapotaceae (*Achras sapota*); Liliaceae (*Allium*); Musaceae (*Musa*). Nonarticulated branched: Euphorbiaceae (*Euphorbia*); Asclepiadaceae (*Asclepias, Cryptostegia*); Apocynaceae (*Nerium oleander*); Moraceae (*Ficus, Broussonetia, Maclura*). Nonarticulated unbranched: Apocynaceae (*Vinca*); Urticaceae (*Urtica*); Moraceae (*Cannabis*).

The list given above shows that the type of laticiferous element is not constant in a given family. In the Euphorbiaceae, for example, *Euphorbia* has nonarticulated laticifers, whereas *Hevea* has articulated laticifers. Certain Asclepiadaceae appear to develop two kinds of laticifers, articulated and nonarticulated, in the same plant body and the parenchyma cells, which lie near the articulated elements, assume some of the characteristics of laticiferous cells (Schaffstein, 1932).

Systematic comparative studies on the laticifers are scarce, and the possible phylogenetic significance of the variations in the degree of their specialization has not yet been revealed. Sometimes, however, a comparison of the laticifers in the representatives of the same family or of closely related families suggests possible series of increasing specialization.

In the Aroideae (De Bary, 1884), for example, certain species appear to lack laticifers or any related structures. Others have longitudinal rows of elongated, cylindrical, sac-like cells, with no perforations in their end walls and no lateral anastomoses. Still others contain anastomosing tubes, with open communications established among the individual cells. A similar series is recognizable in the Papaveraceae and the closely allied Fumariaceae (Léger, 1895). Some workers consider that the Fumariaceae have no laticifers (Sperlich, 1939). Its representatives, however, possess certain idioblasts that appear to intergrade with the laticifers of the Papaveraceae. Some of these idioblasts are indistinguishable from other parenchyma cells, except by their peculiarly colored contents, rich in alkaloids; others are larger and occur singly or in chains. In the Papaveraceae similar files of cells are transformed into tubes by perforation of the end walls (*Chelidonium*) or by a partial or complete resorption of the transverse walls and a development of lateral anastomoses joining the tubes with each other (*Papaver*). The contents of these tubes in the Papaveraceae are interpreted as latex. This latex is milky granular in appearance, sometimes highly colored, and rich in alkaloids. The Cruciferae, which are farther removed from the Papaveraceae than the Fumariaceae, also have idioblastic cells resembling laticifers (Sperlich, 1939). These cells contain the enzyme myrosin. They are often long and branched but are not classified as laticifers because their contents cannot be properly called latex.

Composition and Physical State of Latex

Latex is a substance consisting of a liquid matrix with minute organic particles in suspension. The matrix may be regarded as the cell sap of the laticifer (Frey-Wyssling, 1935). Like cell sap it contains various substances in solution and in colloidal suspension: carbohydrates, organic acids, salts, alkaloids, sterols, fats, tannins, and mucilages. The dispersed particles commonly belong to the hydrocarbon family of terpenes, which includes such substances as essential oils, balsams, resins, camphors, carotenoids, and rubber (Bonner and Galston, 1947). Among these substances the resins and particularly rubber, with the empirical formula of $(C_5H_8)_n$, are characteristic components of the latex in many plants. The terpenes occur in different amounts, depending on the kind of plant, and rubber, specifically, is sometimes entirely lacking. Latex may contain a large amount of protein (*Ficus callosa*), sugar (Compositae), or tannins (*Musa*, Aroideae). The latex of some Papaveraceae is well known for its content of alkaloids (*Papaver somniferum*; Fairbairn and Kapoor, 1960), and that of *Carica papaya* for the occurrence of the proteolytic enzyme papain. The latex of *Euphorbia* species was reported to be rich in vitamin B_1 (Urschler, 1956). Crystals of oxalates and malates may be abundant in latex. Certain plants contain starch grains in the laticifers, often together with the enzyme diastase. The starch grains of the genus *Euphorbia* may attain very large size and assume various, sometimes peculiar, shapes—those of spheroids, rods, dumbbells, and bones.

The best-known latex is that of the various rubber-yielding plants (Arreguín, 1958; Whaley, 1948). The rubber content varies widely in different species. Of the approximately 1,800 dicotyledonous species of plants reported to contain rubber, less than one third have been used as rubber producers, and only few of these yield enough pure rubber to make them commercially valuable. In *Hevea*, rubber may constitute

40 to 50 per cent of the latex. The rubber particles suspended in latex vary in size and shape. According to an electron-microscope study (Andrews and Dickenson, 1961) the particles are spherical (pl. 47B) and reach 0.75 micron in diameter. They have a homogeneous internal structure and are bounded by a 100 Å layer, probably a lipo-protein layer responsible for the colloidal stability of the particles. As seen with the light microscope, some particles appear to be composed of smaller particles enclosed in a common membrane (Southorn, 1960). When the latex is released from the plant, the particles clump together; that is, the latex coagulates. This property is utilized in the commercial separation of rubber from latex.

The latex of various plants may be clear (*Morus, Nerium oleander*) or milky (*Asclepias, Euphorbia, Ficus, Lactuca*). It is yellow-brown in *Cannabis* and yellow or orange in the Papaveraceae. The turbidity and milkiness of latex does not depend directly on its composition but results from the difference between the refractive indices of the particles and the dispersion medium.

Students of laticiferous plants have made the curious observation that latex often harbors flagellates. Their presence induces no visible external symptoms in the plants but is suspected of reducing their vigor (Harvey and Lee, 1943).

Laticifers release the latex when they are cut open. The flow of latex is a pressure flow (Bonner and Galston, 1947). In the intact plant the laticifers are under turgor and at the same time are in osmotic equilibrium with the surrounding parenchyma cells. When the laticifer is opened, a turgor gradient is established and the flow occurs toward the cut where the turgor has been reduced to zero (Spencer, 1939c). This flow eventually ceases, and subsequently the turgor is restored (Spencer, 1939a).

Cytology

According to the common concept, the laticifers maintain a living protoplast, the nuclei remain in this protoplast upon maturation of the elements, and the cytoplasm occurs as a parietal layer enclosing the vacuolar sap, or latex. This structure has been recognized with the electron microscope (Andrews and Dickenson, 1961). In nonarticulated laticifers of many plants the nuclei are known to undergo divisions resulting in a multinucleate, coenocytic condition (pl. 46C; Mahlberg, 1959a). Articulated laticifers, in which communications are established between the individual cells, are also multinucleate but apparently only because the protoplasts fuse and not because of a subsequent multiplication of nuclei (Sperlich, 1939). In young laticifers the nuclei are readily distinguished;

later the dense latex obscures their visibility (fig. 13.8*B*, *C*). There are reports that nuclei degenerate in mature laticifers after an extrusion of nucleoli (Milanez, 1946, 1949).

The proof of the presence of parietal cytoplasm is difficult to obtain. As in sieve elements, there is no clear demarcation between the cytoplasm and the vacuole in mature laticifers (Bonner and Galston, 1947; Sperlich, 1939), and in sectioned material the contents suffer considerable displacement. According to Milanez, (1946, 1949), the small vacuoles of the young laticifers of *Hevea* and *Manihot* are absorbed by the cytoplasm instead of fusing into a single large vacuole. Such development implies that, in the mature laticifers, the cytoplasm is highly hydrated and the latex is part of this cytoplasm. Some workers, however, reported having recognized the shrunken cytoplasm in the center of laticifers from which latex had ceased to flow (Frey-Wyssling, 1935; Moyer, 1937). Studies carried out with the articulated laticifers of *Carica papaya* are significant in this connection. In ripe fruits of this plant the loose ground parenchyma was washed away and the laticifers were isolated without too much injury to them. Placed on a 1.5 per cent agar preparation, they remained alive for 3 to 4 days and were subjected to plasmolytic tests. These tests indicated the presence of a protoplasmic sheath lining the wall (Moyer, 1937).

Most evidence suggests that the latex particles are formed in the laticifers themselves, either in the cytoplasm or in plastids (Bonner and Galston, 1947; Frey-Wyssling, 1935; Milanez, 1946, 1949). If the young laticifers have a distinct vacuole, one may assume the subsequent breakdown of the tonoplast and the escape of the latex particles into the vacuolar sap which becomes part of the latex. This interpretation would parallel that given for the relation between the slime and the vacuolar sap in the sieve elements (chapter 12).

Structure of Walls

The walls of the laticifers are nonlignified and apparently plastic (Milanez, 1946; Sperlich, 1939). They may be no thicker than the walls of the adjacent parenchyma cells, or they may be noticeably thicker. The walls often increase in thickness with the age of the element. The thick walls are highly hydrated, and contain cellulose and a high proportion of pectic substances and hemicelluloses (Moor, 1959). The thickening may be uneven, but primary pit-fields are rarely observed. There is a report in the early literature on laticifers that plasmodesmata occur between these structures and the adjacent parenchyma cells (Sperlich, 1939).

Ultrastructural studies of laticifer walls in *Euphorbia splendens* revealed a succession of cellulosic lamellae in three layers with different orientations of microfibrils (Moor, 1959). Growth was interpreted as of the appositional multinet type, with the earlier layers becoming extended and their microfibrils reoriented during the elongation of the cells. The microfibrils of the last-formed layer were distinctly parallel to each other and helical in orientation. This layer began to be formed before the increase of the cell in width was completed and its microfibrils became aggregated into macrofibrils. In agreement with the terminology of most electron microscopists (chapter 3) the wall was interpreted as composed of primary, transitional, and secondary layers, although the three were not sharply separated from one another. In the original terminology of the wood anatomists (chapter 3) all three layers would be primary because of the concomitant growth of the wall in thickness and in surface.

Presence of callose has been recorded in laticifers. In *Hevea*, pluglike masses of callose have been found in the laticifers at the bases of aged leaves (Spencer, 1939*b*). When such leaves were severed from the plant, no latex issued either from the leaf or from the part of the severed petiole remaining on the stem.

Development

Nonarticulated Laticifers. The branched nonarticulated laticifers of Euphorbiaceae, Asclepiadaceae, and Apocynaceae arise during the development of the embryo in the form of relatively few primordia, then grow concomitantly with the plant into branched systems permeating the whole plant body (Cameron, 1936; Mahlberg, 1961, 1963; Schaffstein, 1932; Sperlich, 1939). *Euphorbia* and *Nerium* may be used to exemplify this development. The primordial laticifers appear in the embryo when the cotyledons are initiated. The laticifers become distinct because of their large size and refringent contents. They are located in the plane of the embryo that later represents the cotyledonary node, in the peripheral part of the future vascular cylinder. In some *Euphorbia* species only four primordia have been recognized; in others eight, arranged in four pairs; and in still others, many primordia, distributed in arcs or in a complete circle. In the *Nerium* embryo usually twenty-eight laticifer primordia are present (fig. 13.7A; Mahlberg, 1961). The primordia form protrusions in various directions, and the apices of these protrusions push their way among the surrounding cells by intrusive growth (fig. 13.7B).

When the seed is mature, the embryo has a system of tubes arranged in a characteristic manner. In *Euphorbia* one set of tubes extends from the cotyledonary node downward, following the periphery of the

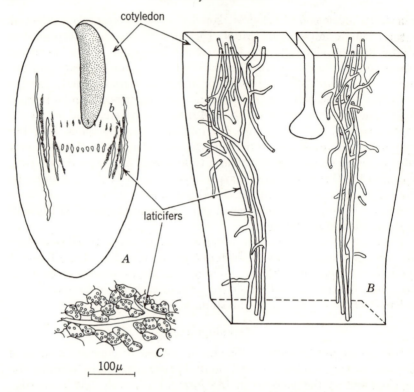

FIG. 13.7. Nonarticulated laticifers of *Nerium oleander*. *A*, immature embryo 550 microns long. Young laticifers at cotyledonary node. They occur along periphery of vascular region. Beginning of laticifer branching at *b*. *B*, 75 microns wide section of mature embryo 5 mm long. Laticifers extend from node into cotyledons and hypocotyl. Short branches extend into mesophyll of cotyledons and cortex of hypocotyl. *C*, branch of laticifer in proliferated mesophyll of cultured embryo. It extends through intercellular spaces. (After Mahlberg, *A*, *B*, *Amer. Jour. Bot.* 48, 1961; *C* from photograph in *Phytomorphology* 9, 1959.)

vascular cylinder of the hypocotyl. Another set passes downward within the cortex, usually near its periphery. The two sets of tubes end near the root meristem at the base of the hypocotylary axis. A third set is prolonged into the cotyledons where the tubes branch, sometimes profusely. A fourth set of tubes extends inward and upward from the nodal primordia toward the shoot apex of the epicotyl where the tubes form a ring-like network. The terminations of this network reach into the third or fourth layers beneath the surface of the apical meristem in both *Euphorbia* and *Nerium*. Thus there are laticifer terminations in the immediate vicinity of both apical meristems, that of the shoot and that of the root. When the seed germinates and the embryo develops into a plant, the laticifers keep pace with this growth by continuously penetrating the meristematic tissues formed by the active apical meristems. When axillary buds or lateral roots arise, they also are penetrated by the intrusively growing tips of the laticifers.

This description of the growth of nonarticulated laticifers does not agree with the concept of Milanez (1959; Milanez and Neto, 1956) that nonarticulated laticifers result from fusion of cells. However, studies of the growth of laticifers in cultured embryos of *Euphorbia marginata* (fig. 13.7C; Mahlberg, 1959b) and of the walls of laticifers in *Euphorbia splendens* (Moor, 1959), as seen under the electron microscope, clearly demonstrate the intrusive type of growth of the nonarticulated laticifers.

During the development of nonarticulated laticifers their nuclei repeatedly divide so that the actively growing distal parts are provided with cytoplasm and nuclei. Since the laticifer tips penetrate the tissues close to the apical meristems, the tube portions below the tips occur for a time in growing tissues and extend in unison with them. Thus the laticifers may be pictured as elongating at their apices by intrusive growth and subsequently extending with the surrounding tissues by symplastic growth (Moor, 1959).

If the plant produces secondary tissues, the nonarticulated laticifers grow into these also. In *Cryptostegia*, for example, the secondary phloem is penetrated by prolongations from the cortical and primary phloem laticifers (Artschwager, 1946). Moreover, the continuity between the laticifer branches in the pith and cortex, established through the interfascicular regions during the primary growth, is seemingly not ruptured by the activity of the vascular cambium during secondary growth. The parts of the laticifer located in the cambium appear to extend by localized growth (intercalary growth) and eventually become imbedded in secondary phloem and xylem (Blaser, 1945).

The question has been raised in the literature whether the laticifers are indefinitely capable of growth and, specifically, whether the tubes in the older portions of the plant retain the ability to invade tissues (Schaffstein, 1932). Laticifer branches penetrating into the pith, the cortex, and the primary phloem of woody species are known to become nonfunctional and to die when the surrounding tissues die. In live tissues, however, they seem to retain the capactiy for further growth. In some experiments, laticifers of *Euphorbia* were observed penetrating from hypocotyl into adventitious shoots that arose on decapitated seedlings. Similarly, laticifers grew into adventitious roots that originated on cuttings. They were also found growing into the dividing tissue beneath a callus formed in grafting. Ultrastructural studies indicate that regions of laticifers with advanced stages of wall development can give rise to new lateral branches (Moor, 1959). All these observations suggest that laticifers of the nonarticulated branched type may be stimulated to resumption of growth, if they are brought in contact with actively growing tissue. In the absence of such tissue in their vicinity they reach a certain maximum of development and then cease to grow. In dormant meristematic tissues laticifers appear to be quiescent (Schaffstein, 1932).

Nonarticulated unbranched laticifers show a simpler pattern of growth than the branched type (Schaffstein, 1932; Sperlich, 1939; Zander, 1928). The primordia of these laticifers have been recognized, not in the embryo, but in the developing shoot (*Vinca, Cannabis*) or in the shoot and root (*Eucommia*). New primordia arise repeatedly beneath the apical meristems, and each elongates into an unbranched tube, apparently by a combination of intrusive and symplastic growth. In the shoot the tubes may extend for some distance in the stem and also diverge into the leaves (*Vinca*). Laticifers may arise in the leaves also, independently of those formed in the stem (*Cannabis, Eucommia*). In some species the unbranched laticifers become multinucleate during development.

Articulated Laticifers. The articulated laticifers develop into extensive tube-like structures, not by growth of individual cells but by constant addition of new primordia to the existing ones. The development of articulated laticifers was studied most intensively in the Cichorieae (Sperlich, 1939), but that of the laticifers in *Hevea* and *Manihot* (Euphorbiaceae) appears to be similar (Scott, 1884, 1886). The primordial laticifers of the Cichorieae are visible in the hypocotyl and the cotyledons of the embryo in the mature seed (Baranova, 1935; Scott, 1882). These primordia are arranged in longitudinal rows, but their end walls are intact. During the first stages of germination the end walls break down and the cell rows are converted into vessels. As the plant develops from the embryo, the vessels are extended by differentiation of further meristematic cells into laticiferous elements. Thus, the laticifers differentiate acropetally in the newly formed plant parts, and they are prolonged not only within the axis but also in the leaves and, later, in the flowers and fruits. The direction of differentiation is similar to that of the nonarticulated branched laticifers, but it occurs by successive conversion of cells into laticiferous elements instead of by apical intrusive growth. Where the vessels lie side by side, parts of the common wall become resorbed (pl. 46*B*). If they are farther apart, the intervening cells may become changed into laticiferous cells with resorption of common walls, or the existing vessels may send out lateral protuberances that fuse with those from another vessel. Thus an anastomosing network of laticifers is formed. Some of the lateral protuberances may end blindly in the tissue.

The Cichorieae produce laticifers also during secondary growth, mainly in the secondary phloem. This development has been followed in some detail in the fleshy roots of *Tragopogon* (Scott, 1882), *Scorzonera* (Baranova, 1935), and *Taraxacum* (Artschwager and McGuire, 1943). Longitudinal rows of derivatives from fusiform cambial initials fuse into tubes through resorption of the end walls. Lateral connections are established—directly or by means of protuberances—among the tubes differentiating in the same tangential plane. The development of articulated laticifers of the nonanastomosing kind is similar to that of the anastomosing laticifers, except that no lateral connections are established among the various tubes (fig. 13.8*B–H*; Karling, 1929).

In grafts performed with *Hevea* (Bonner and Galston, 1947) and *Taraxacum* (Prokofiev, 1945) the establishment of connection between the laticiferous systems of stock and scion was indicated by the evidence of transport of latex from one member of the graft to the other. Both of these genera have anastomosing articulated laticifers, and the interconnection of the laticifers across the graft union is probably a result of their ability to unite with similar elements.

Arrangement in the Plant

Laticifers are frequently distributed rather generally through the plant (fig. 13.9*B*), but sometimes they are more or less restricted to certain tissues (De Bary, 1884; Sperlich, 1939). Most commonly they are associated with the phloem (fig. 13.9*A*; pl. 46*A*). There is much information on the arrangement of the laticifers in the aerial parts of plants, but they occur in roots also (pl. 47*C*).

Nonarticulated Laticifers. In the genus *Euphorbia* the main tubes of the branched nonarticulated laticifers commonly are located in the outer part of the vascular cylinder. From here, branches extend into the cortex and sometimes also into the pith by growing through the interfascicular areas. The cortical branches spread to the epidermis. The minor branches are narrower than the main tubes, and the ultimate ramifications end blindly. In some Apocynaceae, Asclepiadaceae, and Moraceae (*Ficus*, Vreede, 1949) laticifers appear rather generally dispersed in various tissues, including the vascular. In others the main tubes traverse only the pith and form branches at the nodes, some of which penetrate the parenchyma above the leaf insertion (leaf gap) and enter the leaves.

Branched nonarticulated laticifers commonly occur in leaves. Here

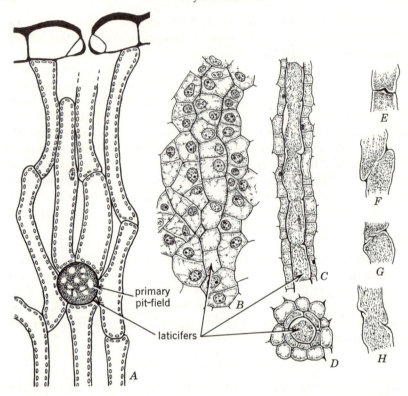

FIG. 13.8. Articulated laticifers. *A*, transection through fleshy scale of *Allium cepa*, showing epidermis with stoma, a few mesophyll cells, and a laticifer with end wall in surface view. *B–H*, laticifer development in *Achras sapota* in longitudinal (*B, C, E–H*) and transverse (*D*) sections. *B*, vertical file of young laticifer cells (from arrow upward) with end walls still intact. *C*, file of cells has been converted into laticiferous vessel by partial dissolution of end walls. Remnants of end walls indicate articulations between members of laticifer. In *D*, flattened cells ensheath laticifer. *E–H*, stages in perforation of end wall. Wall to be perforated first becomes swollen (*E*), then breaks down (*F–H*). (*A*, ×300. *B–H*, adapted from Karling, *Amer. Jour. Bot.* 16, 1929.)

they follow the vascular bundles, ramify in the mesophyll, and often reach the epidermis. In some Euphorbiaceae and in *Ficus* laticifers intrude among the epidermal cells, reach the cuticle, and even continue along the surface of the epidermal cells beneath the cuticle (Sperlich, 1939; Vreede, 1949).

The unbranched nonarticulated laticifers of *Vinca* and *Cannabis* occur in the primary phloem but are apparently absent in the secondary tissues (Schaffstein, 1932; Zander, 1928).

Articulated Laticifers. Articulated laticifers also show various arrangements and a frequent association with the phloem. In the primary body
of the Cichorieae, laticifers appear on the outer periphery of the phloem
(pl. 46*A*) and within the phloem itself. In species with internal phloem,
laticifers are associated with this tissue also (fig. 13.9*A*). The internal
and the external laticifers are interconnected across the interfascicular
areas. The arrangement of laticifers in the secondary body of the
Cichorieae may be exemplified by the use of *Taraxacum kok-saghyz,* a
species utilized commerically for its rubber content (Artschwager and
McGuire, 1943; Kortkov, 1945). The laticifers are within the secondary
phloem. This tissue is formed by the cambium as a series of concentric
layers of parenchyma cells and of layers containing the sieve tubes and
the laticifers. The two kinds of layer alternate with each other radially.
Rays of parenchyma traverse the whole tissue in the radial direction.
Sieve tubes, companion cells, some phloem parenchyma, and the
laticifers are combined into bundles which anastomose and form a network (pl. 47*C*). The laticifers of one growth zone are rarely joined with
the laticifers of another growth zone.

In leaves the articulated laticifers of the Cichorieae accompany the
vascular bundles, ramify more or less profusely in the mesophyll, and
reach the epidermis. The epidermal hairs of the floral involucres of the
Cichorieae become directly connected with laticifers by a breakdown of

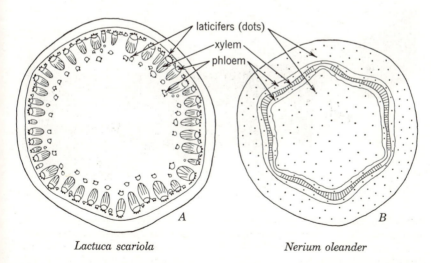

Lactuca scariola   *Nerium oleander*

FIG. 13.9. Distribution of laticifers in transections of stems. *A,* articulated laticifers associated with external and internal phloem. *B,* nonarticulated laticifers dispersed throughout
all tissues, including xylem. (Both, ×13.)

the separating walls and, as a result, the latex readily issues from these hairs when they are broken (Sperlich, 1939).

In several other families the distribution of articulated laticifers is similar to that in Cichorieae. In Caricaceae, however, laticifers apparently occur not only in the phloem but also in the xylem (De Bary, 1884). The laticiferous system that makes *Hevea* (Euphorbiaceae) such an outstanding rubber producer is the secondary system that develops in the secondary phloem (fig. 13.10). The laticifers of *Papaver somniferum* occur in the phloem and become particularly well developed in the mesocarp about two weeks after the petals fall (Fairbairn and Kapoor, 1960). At this time, the capsules are harvested for the commercial extraction of opium.

In the monocotyledons the laticifers of *Musa* are associated with the vascular tissues and also occur in the cortex (Skutch, 1932). In *Allium* the laticifers are entirely separated from the vascular tissue. They lie near the abaxial surface of the leaves or scales (fig. 13.6A), between the second and third layers of parenchyma. The *Allium* laticifers have the

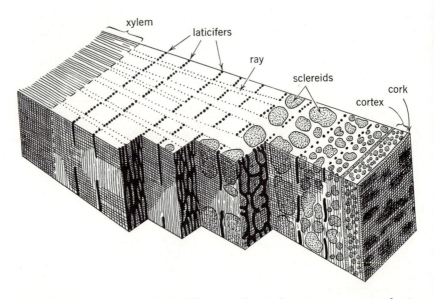

FIG. 13.10. Block diagram of bark of *Hevea brasiliensis*, depicting arrangement of articulated laticifers in secondary phloem. Layers containing sieve tubes and associated parenchyma cells alternate with those in which laticifers (shown in solid black) differentiate. Parenchymatic secondary phloem rays traverse tissue radially. In tangential sections, laticifers of a given growth zone form a reticulum. Sclereids occur in old phloem where sieve tubes and laticifers are nonfunctional. (Adapted from Vischer, *Naturf. Gesell. in Basel, Verhandl.* 35, 1923.)

form of longitudinal chains of cells arranged parallel in the upper parts of the foliar organs and converging toward their bases. The individual cells of the compound laticifers are considerably elongated (fig. 13.6*B*). The end walls are not perforated but have conspicuous primary pit areas (fig. 13.8*A*). Although *Allium* laticifers are classified as nonanastomosing, they form some interconnections at the bases of the leaves or scales.

Possible Function

The laticifers have been the object of intensive study since the early days of plant anatomy (De Bary, 1884; Sperlich, 1939). Because of their distribution in the plant body and their liquid, often milky contents that flow out readily when the plant is cut, the laticiferous system was compared by the early botanists with the circulatory system of animals. The laticifers were called vital sap vessels (Lebenssaftgefässe) and were assumed to have the same function as the blood vessels of animals. Later, they were thought to be related to the vascular elements, particularly the sieve tubes. Still later, the laticifers were interpreted as elements morphologically distinct from the sieve tubes, but related to secretory structures. Views on the function of the laticifers changed in accordance with the various interpretations of their morphological nature. Conclusive information on their role in the life of the plant is still lacking (Bonner and Galston, 1947; Whaley, 1948).

One of the common views advanced was that the laticifers were concerned with food conduction. The evidence for such a role was seen in their high content in food material and in their arrangement in the plant body. However, no actual movement of materials has been observed in the laticifers, only a local and spasmodic one, and no experiments have as yet been designed to test the assumption that food is conducted by the laticifers.

The laticifers have been described also as elements storing food materials. The results of experiments conducted to test this view are contradictory but generally indicate that food materials present in the latex are not readily mobilized when plants are deprived of means to form carbohydrates.

Since latex readily absorbs water from adjacent tissues, it has been thought to be involved with regulation of water balance in the plant. It has also been described as an agency for transport of oxygen, or as material utilized by the plant for protection against animals.

The most widely accepted interpretation of the role of the laticifers is that they form an excretory system. Laticifers accumulate many substances that are commonly recognized as excretory, and such substances are more abundant than food materials in the latex. The terpenes, including rubber and resin, appear to be nonfunctional by-products of cellular metabolism, particularly of that characteristic of young growing tissues. When once deposited in the cells, the terpenes are not known to be utilized again by the plant (Benedict, 1949; Bonner and Galston, 1947).

The highly polymerized terpenes, like rubber, are incapable of passing across cell walls and remain in the cells in which they are formed. It seems significant, therefore, that the formation of high molecular-weight terpenes by plants commonly coincides with the occurence of laticifers, which appear to be adapted as repositories for this type of excreted substance. Resin, on the other hand, is frequently excreted into special-ized intercellular passages, the resin ducts, or to the surface of the plant through excretory trichomes. In the Compositae some groups have laticiferous systems, others have resin ducts, and the two kinds of structures do not occur together (Frey-Wyssling, 1935).

Thus, the laticifers would appear to fit best in the class of excretory structures. At the same time, the variety of substances present in the latex and the variations in its composition in different plants suggest the possibility that the laticifers may have more than one function.

REFERENCES

Agthe, C. Über die physiologische Herkunft des Pflanzennektars. *Schweiz. Bot. Gesell. Ber.* 61:240–274. 1951.

Andrews, E. H., and P. B. Dickenson. Preliminary electron microscope observations on the ultra-structure of the latex vessel and its contents in young tissues of *Hevea brasiliensis. Natl. Rubber Res. Conf. Proc.* 1961:756–765. 1961.

Arreguín, B. Rubber and latex. *Handb. der Pflanzenphysiol.* 10:223–248. 1958.

Artschwager, E. Contribution to the morphology and anatomy of *Cryptostegia* (*Cryptostegia grandiflora*). *U.S. Dept. Agric. Tech. Bul.* 915. 1946.

Artschwager, E., and R. C. McGuire. Contribution to the morphology and anatomy of the Russian dandelion (*Taraxacum kok-saghyz*). *U.S. Dept. Agric. Tech. Bul.* 843. 1943.

Bancher, E., and J. Hölzl. Über die Drüsenhaare von *Solanum tuberosum* Sorte "Sieglinde". *Protoplasma* 50:356–369. 1959.

Baranova, E. A. Ontogenez mlechnoĭ systemy tau-sagyza (*Scorzonera tau-saghyz* Lipsch. et Bosse). [Ontogeny of the laticiferous system of tau-saghyz (*Scorzonera tau-saghyz* Lipsch. et Bosse).] *Bot. Zhur. SSSR* 20:600–616. 1935.

Benedict, H. M. A further study on the nonutilization of rubber as a food reserve by guayule. *Bot. Gaz.* 111:36–43. 1949.

Blaser, H. W. Anatomy of *Cryptostegia grandiflora* with special reference to the latex. *Amer. Jour. Bot.* 32:135–141. 1945.

Bonner, J., and A. W. Galston. The physiology and biochemistry of rubber formation in plants. *Bot. Rev.* 13:543–596. 1947.

Brown, W. H. The bearing of nectaries on the phylogeny of flowering plants. *Amer. Phil. Soc. Proc.* 79:549–595. 1938.

Cameron, D. An investigation of the latex systems in *Euphorbia marginata*, with particular attention to the distribution of latex in the embryo. *Bot. Soc. Edinb. Trans. and Proc.* 32(I):187–194. 1936.

Carlquist, S. Structure and ontogeny of glandular trichomes of Madinae (Compositae). *Amer. Jour. Bot.* 45:675–682. 1958.

Carlquist, S. The leaf of *Calycadenia* and its glandular appendages. *Amer. Jour. Bot.* 46:70–80. 1959*a.*

Carlquist, S. Glandular structures of *Holocarpha* and their ontogeny. *Amer. Jour. Bot.* 46: 300–308. 1959*b.*

Davies, P. A. Structure and function of the mature glands on the petals of *Frasera carolinensis. Kentucky Acad. Sci. Trans.* 13:228–234. 1952.

De Bary, A. *Comparative anatomy of the vegetative organs of the phanerogams and ferns.* Oxford, Clarendon Press. 1884.

Esau, K. *Anatomy of seed plants.* New York, John Wiley and Sons. 1960.

Fahn, A. On the structure of floral nectaries. *Bot. Gaz.* 113:464–470. 1952.

Fahn, A. The topography of the nectary in the flower and its phylogenetic trend. *Phytomorphology* 3:424–426. 1953.

Fairbairn, J. W., and L. D. Kapoor. The laticiferous vessels of *Papaver somniferum* L. *Planta Med.* 8:49–61. 1960.

Feldberg, W. The mechanism of the sting of common nettle. *Brit. Sci. News* 3:75–77. 1950.

Foster, A. S. Plant idioblasts: remarkable examples of cell specialization. *Protoplasma* 46:184–193. 1956.

Frei, E. Die Innervierung der floralen Nektarien dikotyler Pflanzenfamilien. *Schweiz. Bot. Gesell. Ber.* 65:60–114. 1955.

Frey-Wyssling, A. *Die Stoffausscheidung der höheren Pflanzen. Monographien aus dem Gesamtgebiet der Physiologie der Pflanzen und der Tiere.* Band 32. Berlin, Julius Springer. 1935.

Frey-Wyssling, A., and E. Häusermann. Deutung der gestaltlosen Nektarien. *Schweiz. Bot. Gesell. Ber.* 70:150–162. 1960.

Haagen-Smit, A. J. The lower terpenes. *Handb. der Pflanzenphysiol.* 10:52–90. 1958.

Harvey, R. B., and S. B. Lee. Flagellates of laticiferous plants. *Plant Physiol.* 18:633–655. 1943.

Hülsbruch, M. Strobulierende Drüsenköpfchen bei *Atropa Belladonna* L. *Flora* 150:572–599. 1961.

Ivanoff, S. S. Guttation injuries of plants. *Bot. Rev.* 29:202–229. 1962.

Jackson, B. D. *A glossary of botanic terms.* 4th ed. New York, Hafner Publishing Co. 1953.

Johnson, M. A. Hydathodes in the genus *Equisetum. Bot. Gaz.* 98:598–608. 1937.

Karling, J. S. The laticiferous system of *Achras zapota* L. *Amer. Jour. Bot.* 16:803–824. 1929.

Kaussmann, B. Die Trichomhydathoden von *Muehlenbeckia platyclados* Meissn. *Univ. Rostock Wiss. Ztschr.* 3:231–236. 1954.

Kisser, J. Die Ausscheidung von ätherischen Ölen und Harzen. *Handb. der Pflanzenphysiol.* 10:91–131. 1958.

Kowalewicz, R. Zur Kenntnis von *Epilobium und Oenothera.* 1. Über die Raphidenschläuche. *Planta* 47:501–509. 1956.

Kramer, P. J. Absorption of water by plants. *Bot. Rev.* 11:310–355. 1945.

Krotkov, G. A. A review of literature on *Taraxacum kok-saghyz* Rod. *Bot. Rev.* 11:417–461. 1945.

Labouriau, L. G.   On the latex of *Regnellidum* (sic) *diphyllum* Lindm. *Phyton*. 2:57–74. 1952.

Léger, L. J.   Recherches sur l'appareil végétatif des Papavéracées. (Papavéracées et Fumariacées D. C.) *Soc. Linn. de Normandie, Mém.* 18:193–624. 1895.

Mahlberg, P. G.   Karyokinesis in the non-articulated laticifers of *Nerium oleander* L. *Phytomorphology* 9:110–118. 1959*a*.

Mahlberg, P. G.   Development of the non-articulated laticifer in proliferated embryos of *Euphorbia marginata* Pursh. *Phytomorphology* 9:156–162. 1959*b*.

Mahlberg, P. G.   Embryogeny and histogenesis in *Nerium oleander*. II. Origin and development of the non-articulated laticifer. *Amer. Jour. Bot.* 48:90–99. 1961.

Mahlberg, P. G.   Development of non-articulated laticifer in seedling axis of *Nerium oleander*. *Bot. Gaz.* 124:224–231. 1963.

Metcalfe, C. R., and L. Chalk. *Anatomy of the dicotyledons*. Vol. 2. Oxford, Clarendon Press. 1950.

Milanez, F. R.   Nota prévia sôbre os laticíferos de *Hevea brasiliensis*. *Arqu. do Serv. Florestal* 2:39–65. 1946.

Milanez, F. R.   Segunda nota sôbre os laticíferos. *Lilloa* 16:193–211. 1949.

Milanez, F. R.   Contribução ao conhecimento anatômico de *Cryptostegia grandiflora*. I. Embrião. *Rodriguésia* 21–22:347–396. 1959.

Milanez, F. R., and H. M. Neto. Origem dos laticíferos do embrião de *Euphorbia pulcherrima, Willd*. *Rodriguésia* 18–19:351–395. 1956.

Moens, P.   Les formations sécrétrices des copaliers congolais. Étude anatomique, histologique et histogénétique. *Cellule* 57:33–64. 1955.

Moor, H.   Platin Kohle-Abdruck-Technik angewandt auf Feinbau der Milchröhren. *Jour. Ultrastruct. Res.* 2:293–422. 1959.

Moritz, O.   Übersicht über Terpenoide. *Handb. der Pflanzenphysiol.* 10:24–51. 1958.

Moyer, L. S.   Recent advances in the physiology of latex. *Bot. Rev.* 3:522–544. 1937.

Okimoto, M. C.   Anatomy and histology of the pineapple inflorescence and fruit. *Bot. Gaz.* 110:217–231. 1948.

Pedersen, M. W., C. W. LeFevre, and H. H. Wiebe. Absorption of C^{14}-labeled sucrose by alfalfa nectaries. *Science* 127:758–759. 1958.

Prokofiev, A. A.   On the synthesis of rubber in plants—filling of laticiferous vessels with foreign latex. *Acad. Sci. URSS, Compt. Rend. (Doklady)* 48:520–523. 1945.

Reams, W. M., Jr.   The occurrence and ontogeny of hydathodes in *Hygrophila polysperma* T. Anders. *New Phytol.* 52:8–13. 1953.

Schaffstein, G.   Untersuchungen an ungegliederten Milchröhren. *Bot. Centbl. Beihefte* 49:197–220. 1932.

Schnepf, E.   Zur Feinstruktur der Drüsen von *Drosophyllum lusitanicum*. *Planta* 54:641–674. 1960.

Schnepf, E.   Zur Cytologie und Physiologie pflanzlicher Drüsen. 1. Teil. Über den Fangschleim der Insektivoren. *Flora* 153:1–22. 1963.

Scott, D. H.   The development of articulated laticiferous vessels. *Quart. Jour. Micros. Sci.* 22:136–153. 1882.

Scott, D. H.   On the laticiferous tissue of *Manihot Glaziovii* (the Ceàra rubber). *Quart. Jour. Micros. Sci.* 24:194–204. 1884.

Scott, D. H.   On the occurrence of articulated laticiferous vessels in *Hevea*. *Linn. Soc. London, Jour., Bot.* 21:566–573. 1886.

Shuel, R. W.   Influence of reproductive organs on secretion of sugars in flowers of *Streptosolen jamesonii*, Miers. *Plant Physiol.* 36:265–271. 1961.

Skutch, A. F.   Anatomy of the axis of the banana. *Bot. Gaz.* 93:233–258. 1932.

Southorn, W. A. Complex particles in *Hevea* latex. *Nature* 188:165–166. 1960.

Spencer, H. J. The effect of puncturing individual latex tubes of *Euphorbia Wulfenii*. *Ann. Bot.* 3:227–229. 1939a.

Spencer, H. J. On the nature of the blocking of the laticiferous system at the leaf-base of *Hevea brasiliensis*. *Ann. Bot.* 3:231–235. 1939b.

Spencer, H. J. Latex outflow and water uptake in the leaf of *Ficus elastica*. *Ann. Bot.* 3:237–241. 1939c.

Sperlich, A. Das trophische Parenchym. B. Exkretionsgewebe. In: K. Linsbauer. *Handbuch der Pflanzenanatomie*. Band 4. Lief. 38. 1939.

Stahl, E. Untersuchungen an den Drüsenhaaren der Schafgarbe (*Achillea millefolium* L.). *Ztschr. f. Bot.* 41:123–146. 1953.

Stahl, E. Über Vorgänge in den Drüsenhaaren der Schafgarbe. *Ztschr. f. Bot.* 45:297–315. 1957.

Stevens, A. B. P. The structure and development of hydathodes of *Caltha palustris* L. *New Phytol.* 55:339–345. 1956.

Trapp, I. Neuere Untersuchungen über den Bau und Tätigkeit der pflanzlichen Drüsenhaare. *Oberhess. Gesell. f. Nat. u. Heilk. Giessen, Ber. Naturw. Abt.* 24:182–205. 1949.

Urschler, I. Untersuchungen mit dem Phycomyces-Test. *Protoplasma* 46:794–797. 1956.

Van Die, J. A comparative study of the particle fractions from Apocynaceae latices. *Ann. Bogorienses* 2:1–124. 1955.

Vogel, S. Duftdrüsen im Dienste der Bestäubung. Über Bau und Funktion der Osmophoren. *Akad. der Wiss. u. Lit. Mainz, Math.-Nat. Kl. Abhandl.* 10:601–763. 1962.

Vreede, M. C. Topography of the laticiferous system in the genus *Ficus*. *Jard. Bot. Buitenzorg Ann.* 51:125–149. 1949.

Weichsel, G. Natürliche Lagerstätten ätherischer Öle. In: E. Gildermeister and Fr. Hoffmann. *Die ätherischen Öle*. Band 1. 4th ed. Berlin, Akademie-Verlag. 1956.

Whaley, W. G. Rubber—the primary sources for American production. *Econ. Bot.* 2:198–216. 1948.

Zander, A. Über Verlauf und Enstehung der Milchröhren des Hanfes (*Cannabis sativa*). *Flora* 23:191–218. 1928.

Ziegler, A. Zur Anatomie und Protoplasmatik der Ölidioblasten von *Houttuynia cordata*. *Protoplasma* 51:539–562. 1960.

14

The Periderm

CONCEPT

Periderm is a protective tissue of secondary origin. It replaces the epidermis when the axis is increased in girth and the epidermis is destroyed. Periderm formation is a common phenomenon in stems and roots of dicotyledons and gymnosperms that increase in thickness by secondary growth. Structurally, the periderm consists of three parts: the *phellogen*, or cork cambium, the meristem producing the periderm; the *phellem*, commonly called cork, produced by the phellogen toward the outside; and the *phelloderm*, a tissue that resembles cortical parenchyma and consists of the inner derivatives of the phellogen. The term periderm and those referring to its components are derived from Greek words among which *phellem* means cork, *gen*, to produce, *derma*, skin, and *peri*, about.

The term periderm should be distinguished from the nontechnical term *bark* (chapter 12). Bark is applied most commonly to all tissues outside the vascular cambium of the axis, in either a primary or secondary state of growth. It is also used more specifically to designate the tissue that accumulates on the surface of the plant axis as a result of phellogen activity. As the periderm develops, it separates, by means of a nonliving layer of cork cells, variable amounts of primary and secondary tissues of the axis from the subjacent living tissues. The tissue layers thus separated die. The term bark in its restricted meaning refers to these dead tissues together with the layers of periderm. If bark is used with reference to all the tissues outside the vascular cambium, the periderm and the tissues of the axis isolated by it may be combined under the designation of *outer bark*. The technical term for the outer bark is

338

rhytidome (De Bary, 1884). This term is derived from the Greek word meaning wrinkle and refers to the appearance of the outer bark when it consists of layers of cork alternating with layers of tissue cut off by the cork.

The structure and the development of periderm are better known in stems than in roots. Therefore, most of the information on periderm given in this chapter pertains to stems, unless roots are mentioned specifically. Additional data on root periderm appear in chapter 17.

OCCURRENCE

Periderm characteristically appears on the surface of roots and stems and their branches in gymnosperms and woody dicotyledons that undergo a continuous pronounced increase in thickness by secondary growth. Periderm occurs also in herbaceous dicotyledons, in which it is some-times limited to the oldest parts of stem and root. Monocotyledons rarely develop a protective tissue comparable to the periderm of dicotyledons. Foliar organs normally produce no cork, but scales of winter buds in some gymnosperms and dicotyledons are an exception. In stems of extant vascular cryptogams, which as a rule lack secondary growth, no periderm is formed even in species that eventually cast off the epidermis and part of the cortex (Ogura, 1938). In underground stems of some vascular cryptogams the epidermis or the outer cortical layers become suberized.

The formation of periderm in stems of woody plants may be con-siderably delayed, as compared with that of the secondary vascular tissues, or it may never occur, despite the obvious increase in thickness of the stem. In such instances the tissues outside the vascular cambium, including the epidermis, keep pace with the increase in axis circumference by growth and division of cells (species of *Viscum, Menispermum, Ilex, Acer, Citrus, Laurus, Eucalyptus, Acacia*).

Periderm differentiates on surfaces exposed after abscission of plant parts, such as leaves or branches (chapter 16). Cork frequently develops around diseased or dead tissue complexes within the plant body and also beneath the surface of wounds (wound periderm or wound cork; pl. 67).

CHARACTERISTICS OF THE COMPONENTS

In contrast to the vascular cambium, the phellogen is relatively simple in structure, being composed of one type of cell. The phellogen cells are rectangular in transections, and somewhat flattened radially. In longitudinal views they may be rectangular or somewhat irregular in shape. Their protoplasts are vacuolated to varying degrees and may contain tannins and chloroplasts.

The cork cells are approximately prismatic in shape, often somewhat elongated parallel with the long axis of the stem, and usually have the radial diameters shorter than the tangential. Some determinations of their shape show that the cork cells are, like parenchyma cells, basically tetrakaidecahedral in form, with the average of contact faces to one cell approaching fourteen (Lier, 1952). They are usually arranged compactly, without intercellular spaces, and in radial rows, clearly showing their origin from a tangentially dividing meristem (pls. 48C, D; 65).

The phellem owes its protective characteristics to the presence of suberin in the walls. The cork cells may begin to deposit suberin before they attain their full size (Bowen, 1963; De Bary, 1884; Sifton, 1945), and also apparently after they have expanded (Mader, 1954). The suberin occurs as a distinct lamella deposited over the primary cellulose wall as an adcrustation (pl. 48E; Sitte, 1955). It shows layering under the electron microscope probably because of alternation of suberin and waxes (fig. 14.1; Falk and El-Hadidi, 1961). The wax causes the double refraction of the suberin lamella (Mader, 1958). In thick-walled kinds of cork, additional cellulose is added toward the cell lumen, that is, on the inside of the suberin lamella. The cellulosic part of the wall may be lignified. The walls are not pitted but plasmodesmatal pores have been seen under the electron microscope (Sitte, 1955).

Some plants contain within the cork tissue cells that are free of suberin, although otherwise resembling the cork cells. These nonsuberized cells are called *phelloids*, that is, phellem-like cells. They occur in various proportions and various arrangements in the phellem (Mühldorf, 1925; Mylius, 1913; Pfeiffer, 1928). Sclerified phelloids occur in the phellem of some plants. The composition of the phellem may be of diagnostic value (Bamber, 1962). The cork cell walls may be brown or yellow, or they may remain colorless, but these characteristics are not related to suberization. Often the color of the cork cells depends on the presence of colored resinous and tanniniferous compounds in their lumina. Cork cells become devoid of protoplasts after differentiation and are then filled either with air or with the just-mentioned highly colored organic substances. The type of phellem used for bottle cork consists of thin-walled, air-filled cells. The mature cork is a compressible, resilient tissue, highly impervious to water and resistant to oil. The suberin itself contains unsaturated fatty acids and is therefore somewhat permeable; the wax is chiefly responsible for the imperviousness (Sitte,

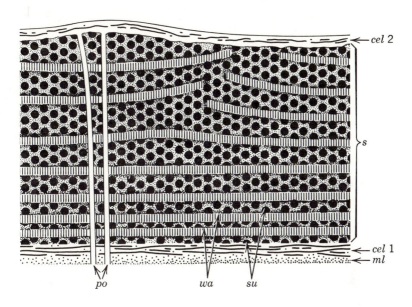

Fig. 14.1.   Interpretation of structure of suberized wall of a cork cell.   Wall consists of (1) middle lamella (*ml*); (2) outer, cellulose-(black lines) containing layer (*cel* 1); (3) suberin layer (*s*) in which suberin lamellae (*su*) alternate with wax (*wa*) lamellae; (4) outer cellulose-containing layer (*cel* 2).   The lines in *wa* indicate orientation of wax molecules.   The presumed plasmodesmatal pores (*po*) are occluded in mature cork.   (Sitte, *Protoplasma* 54, 1962.)

1957).   Since suberin is resistant to enzymes, cork may be present in fossils (Sen, 1961).   Because of air-filled lumina, the cork is light and has thermal insulating qualities (Cooke, 1948).

The phelloderm cells resemble cortical cells in wall structure and contents.   Their shape is similar to that of the phellogen cells.   They may be distinguished from cortical cells by their arrangement in radial series resulting from their origin from the tangentially dividing phellogen.

A periderm of a special kind, the *polyderm*, occurs in roots and underground stems of Hypericaceae, Myrtaceae, Onagraceae, and Rosaceae (Luhan, 1955; Mylius, 1913; Nelson and Wilhelm, 1957).   It contains suberized and nonsuberized cells—the latter are concerned with food storage—in alternating layers.   The suberized layers are one cell deep, the nonsuberized several cells deep.   A polyderm may become twenty and more layers in total thickness.   Only the outermost layers are dead; the others contain living protoplasts, including the suberized cells.

PLACE OF ORIGIN OF PHELLOGEN

With regard to the origin of the phellogen, it is necessary to distinguish between the first periderm (pls. 49C, 65) and the subsequent periderms, which arise beneath the first and replace it as the axis continues to increase in circumference (fig. 14.2, pl. 49D). In the stem the phellogen of the first periderm may be initiated at different depths outside the vascular cambium (Metcalfe and Chalk, 1950). In most stems the first phellogen arises in the subepidermal layer (fig. 14.3C; pl. 48A, B). In a few plants epidermal cells give rise to the phellogen (*Nerium oleander, Pyrus*). Sometimes the phellogen is formed partly from epidermis, partly from subepidermal cells (fig. 14.3A). In some stems the second or third cortical layer initiates the development of periderm (*Robinia pseudoacacia, Gleditschia triacanthos,* and other Leguminosae; species of *Aristolochia, Pinus, and Larix*). In still others the phellogen arises near the vascular region or directly within the phloem (Caryophyllaceae, Cupressoideae, Ericaceae, *Berberis, Camellia, Punica, Vitis;* pls. 49A, B; 54). If the first periderm is followed by others, these are formed repeatedly—but rarely each season—in successively deeper layers of the cortex or phloem. Cork development may occur within the xylem (interxylary cork; Moss and Gorham, 1953) and may be associated with anomalous secondary growth (chapter 15; Metcalfe and Chalk, 1950).

The superficial periderm is commonly initiated parallel to the surface of the stem. If, however, the stem is angled in outline or is ridged, the periderm arises beneath the angles or ridges somewhat deeper than

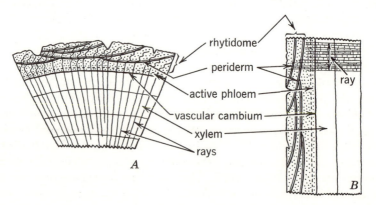

Fig. 14.2. Rhytidome and its location with reference to vascular tissues. *A*, transverse and, *B*, longitudinal sections of part of stem. The rhytidome in this example is composed of periderm and nonliving secondary phloem.

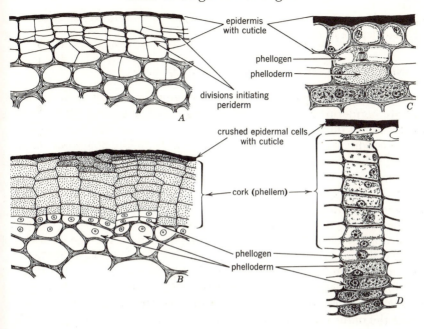

Fig. 14.3. Origin of periderm, partly in epidermis, partly beneath it, in *Pyrus* (*A*, *B*) and beneath epidermis in *Prunus* (*C*, *D*). Nuclei indicate phellogen and phelloderm in *B*. (*A*, *B*, ×350; *C*, *D*, ×490.)

elsewhere. Thus, the prominent parts of the stem are removed, and the outline of the stem becomes less uneven. The deeper-lying initial periderms are also formed completely around the circumference of the axis.

The subsequent periderms show two typical methods of origin. Those that follow deep-seated initial periderms commonly replicate the disposition of such periderms; that is, they encircle the axis like the first periderm (*Vitis*). In contrast, sequent periderms that follow a superficial type of initial periderm usually arise in discontinuous layers located in various parts of the axis circumference. The layers have the shape of shells or scales curved toward the outside, and the successively deeper layers overlap the more peripheral ones (fig. 14.2).

Secondary growth of vascular tissues and formation of periderm are common in roots of dicotyledons and conifers. In most of these roots the first periderm originates deeply in the axis, usually in the pericycle (pls. 85, 88). Some dicotyledonous roots having secondary growth of short duration form only a superficial periderm (chapter 17). Like stems, the roots also may produce periderm layers at successively greater depths in the axis.

INITIATION AND ACTIVITY OF PHELLOGEN

The cells of the epidermis, the collenchyma, or the parenchyma that initiate periderm are living cells, and their change into phellogen is an expression of their ability to resume meristematic activity under appropriate conditions. These cells are usually indistinguishable from neighboring cells. Sometimes, however, the subepidermal layer, where the phellogen arises most frequently in stems, is morphologically distinct from the adjacent cortical cells in that it develops no collenchymatic thickenings and consists of compactly arranged cells of uniform size.

The phellogen is initiated by periclinal divisions (pl. 48A). Usually no obvious cytologic change occurs in preparation for the first divisions. If the cells concerned have starch and tannins, these disappear gradually during the successive divisions. The first periclinal division in a given cell forms two apparently similar cells. Frequently the inner of these two cells divides no further and is then regarded as a phelloderm cell, while the outer functions as the phellogen cell and divides (fig. 14.3C). The outer of the two products of the second division matures into the first cork cell, while the inner remains meristematic and divides again. Sometimes the first division results in the formation of a cork cell and a phellogen cell. Although most of the successive divisions are periclinal, occasional anticlinal divisions in the phellogen increase the number of radial rows of cork cells and enable the periderm to keep pace with the increase in girth of the axis (fig. 14.3B).

The meristematic activity initiating the periderm begins either in localized regions or around the entire circumference (Lier, 1955). When the initial activity is localized the first divisions are frequently those concerned with the formation of lenticels (see below). From the margins of these structures the divisions spread around the stem circumference.

The number of divisions resulting in the formation of cork cells usually exceeds those that give rise to phelloderm cells (fig. 14.3B, D). Some plants have no phelloderm at all; in others this tissue is one to three or more cells in depth. The number of phelloderm cells in the same layer of periderm changes somewhat as the stem ages. In Tilia, for example, the phelloderm may be one cell deep in the first year, two in the second, three or four later. The subsequent periderms formed beneath the first, in later years, contain as much phelloderm as the first or less.

The number of phellem cells in one radial file produced during one year varies from two to twenty, depending on the plant species. If the initial periderm of a stem is retained for many years, the outer cork layers crack and commonly peel off, so that on the stem itself the cork maintains approximately the same thickness. In some stems, however, large amounts of cork accumulate on the surface (*Quercus suber, Aristolochia,* pl. 55C). The initial periderms that are soon replaced by deeper ones, and also the sequent periderms, usually produce only small numbers of cork layers. The phellem is commonly thin in roots, especially in the fleshy kinds. Apparently environmental conditions in the soil promote a rapid decay and sloughing away of the outermost cork layers.

TIME OF ORIGIN OF PHELLOGEN

The time of appearance of the first and subsequent periderms varies in relation to species or larger taxonomic groupings or, sometimes, among individuals of the same species (De Bary, 1884; Douliot, 1889; Möller, 1882; Sanio, 1860). It is affected also by environmental conditions. Thus, for example, in certain northeastern trees insolation was found to be the main factor hastening the appearance of the deep-seated sequent phellogen (Zeeuw, 1941).

Most dicotyledons and gymnosperms develop the initial periderm—whether superficial or deeper lying—during the first year of growth, usually after the primary elongation is completed (De Bary, 1884). Such early periderm frequently arises almost at once around the circumference of the stem. If the periderm appears late in the life of the axis, the divisions leading to its initiation start in localized areas and spread slowly around the circumference. In such instances, several years may pass before the periderm is continuous at a given level of the stem.

The first superficial periderm may be retained for life or for many years (species of *Fagus, Abies, Carpinus, Quercus*). The phellogen cells then undergo periodic anticlinal divisions that increase the circumference of the meristem and of the resulting periderm. An initial periderm formed in deeper parts of the axis may also persist for a long time (*Ribes, Berberis, Punica*). More commonly, however, the first periderms, whether superficial or deep seated, are soon replaced by sequent periderms in successively deeper regions of the axis. Diseases and other external agents may upset the normal pattern of periderm development, either delaying or hastening its appearance, or inducing the formation of deeper periderms when the plant normally develops only a superficial periderm (Kauffert, 1937). The ability of the plant to produce phellogen in deeper layers when the superficial periderm is removed is utilized in the production of commercial cork from the cork oak, *Quercus suber* (Metcalf, 1947). The first superficial cork is removed to the phellogen. The exposed tissue dries out to about ⅛ in. in depth. A new phellogen is established beneath the dry layer and rapidly produces a massive cork of a better quality than the first.

PHYSIOLOGIC ASPECTS OF PERIDERM FORMATION

The physiologic aspects of periderm development have been studied, particularly with reference to wound healing. But the formation of periderm beneath wounds, or in the scar left after leaf fall, or in stems and roots growing in thickness appears to follow the same fundamental sequence (Bloch, 1941; Priestley and Swingle, 1929). The exposed surface is sealed with fatty substances, including suberin. This blocking seems to create internal conditions favorable for meristematic activity that results in cork formation. The process of blocking requires certain external conditions, mainly appropriate amounts of moisture and adequate aeration. Their absence inhibits the blocking and indirectly also the formation of cork. Excessive moisture prevents the maturation of cork, as has been observed in potato tubers grown in too moist soil (Mylius, 1913). Moisture may suppress suberization entirely and induce a development of callus tissue instead of cork (Küster, 1925).

In studies of wound healing in potato tuber and sweet potato root, suberization was found to be preceded by an accumulation of phenolic substances, particularly of chlorogenic acid (Johnson and Schaal, 1957; McClure, 1960), a phenomenon apparently correlated with the suberization. Lignification is also suspected of being associated with wound healing (McClure, 1960). The importance of suberization and periderm development as a protection of wounds from infection by decay organisms has been demonstrated in experiments in which the wound-healing phenomena were retarded or inhibited by chemical treatment (Audia et al., 1962).

MORPHOLOGY OF PERIDERM AND RHYTIDOME

The external appearance of axes bearing periderm or rhytidome is highly variable. This variation depends partly on the characteristics and manner of growth of the periderm itself and partly on the amount and kind of tissue separated by the periderm from the axis.

If the plant has only a superficial periderm, a relatively small amount of primary tissue is cut off, involving either a part of or the entire epidermis or possibly one or two cortical layers. This tissue is eventually sloughed away, and the phellem is exposed. The stem in this instance would be considered to have no rhytidome. If the exposed cork tissue is thin, it commonly has a smooth surface (pl. 28A). If it is thick, the surface is cracked and fissured (pl. 55C). Massive cork usually shows layers that seem to represent annual increments.

In some dicotyledons (*Ulmus* sp.) stems produce a type of winged cork, so called because of a symmetrical longitudinal splitting of the cork into bands projecting like wings from the surface of the stem (Smithson, 1954). Another type of winged cork results from intensive localized activity of phellogen considerably in advance of the formation of the periderm elsewhere (*Euonymus alatus;* Bowen, 1963).

The deeper periderms cut off larger amounts of the original stem tissues and usually form a rhytidome. In some rhytidomes parenchyma and soft cork cells predominate; others contain large amounts of fibers usually derived from the phloem. Fibrous barks form a net-like pattern upon splitting (*Fraxinus, Tilia*); those lacking fibers break up into individual scale-like units (*Acer pseudoplatanus, Pinus;* Holdheide, 1951). The manner of origin of the successive layers of periderm has a characteristic effect upon the appearance of the rhytidome. If the sequent periderms are formed as overlapping scale-like layers (fig. 14.2), the outer tissue breaks up into units related to the layers of the periderm, and the resulting outer bark is referred to as scalebark (*Pinus, Pyrus*). If, on the contrary, the phellogen arises around the entire circumference of the stem, a ringbark is formed, which is characterized by the separation of hollow cylinders (rings) of tissue from the stem. This type of outer bark is common in plants in which the first periderm originates in deep layers of the axis and the subsequent periderms arise more or less concentrically with the first (Cupressaceae, *Lonicera, Clematis, Vitis*). A scalebark with very large individual scales (*Platanus*) may be regarded as being intermediate between the scalebark and the ringbark.

The manner in which the dead tissues separate from the stem is determined also by the nature of the periderm (De Bary, 1884; Mühldorf, 1925; Pfeiffer, 1928). In some plants the separation occurs through thin-walled cork cells. In *Platanus* and *Arbutus*, for example, the dead tissue separates from the periderm in the form of large, thin scales through the outer thin-walled layer of cork, while the subjacent thick-walled cork tissue remains on the stem and has a smooth surface. The thick-walled cork is removed with the new scales during the subsequent period of scaling off. The sloughing of the outer bark sometimes occurs through a break along thin-walled nonsuberized cells in the cork (phelloids) or within the parenchyma of the stem parts that have been isolated by the development of periderm (Chattaway, 1953; Pfeiffer, 1928).

In many plants the periderm cells show considerable cohesion and the succeeding layers of rhytidome adhere to one another. The outer bark then becomes thick, is more or less deeply cracked externally, and gradually wears away. Examples of trees with this kind of outer bark are the redwood, *Sequoia sempervirens* (Isenberg, 1943), and certain species of *Quercus, Betula, Salix, Robinia* (pl. 49C, D). The opposite type of bark, loose and fibrous, occurs in certain *Eucalyptus* species (Chattaway, 1955). This texture results from dilatation of axial phloem parenchyma, the cells of which may enlarge to many times their original size. The parenchyma dilates after it is separated from the underlying tissue by the periderm—presumably before the latter has mature cork—and causes the wide separation of the fiber bundles characteristic of these barks.

PROTECTIVE TISSUE IN MONOCOTYLEDONS

The monocotyledons rarely form a type of periderm resembling that of the dicotyledons (Philipp, 1923; Solereder and Meyer, 1928). In many the epidermis remains intact, sometimes becoming extremely hard (*Calamus*). There may be a modification of the ground parenchyma into a protective tissue by suberization (species of *Livistonia, Typha, Phoenix,* Gramineae) or thickening and sclerification of walls (*Washingtonia filifera*). Such changes occur in spots and spread toward the interior. Some cell division may occur before the suberization.

In monocotyledons with pronounced secondary growth, a special type of protective tissue is formed by repeated divisions of parenchyma cells and subsequent suberization of the products of division. The divisions are periclinal and are repeated several times in the derivatives of the same cell until a linear series of about four to eight cells is formed. The cells then differentiate into cork cells, while other, deeper-lying parenchyma cells undergo similar divisions, and suberization. Thus, the cork arises without the formation of an initial layer, or phellogen, and is referred to as *storied cork* because the linear files of cells form tangential bands as seen in transections. As the formation of cork progresses inwards, nonsuberized cells may become imbedded among the cork cells. Thus a tissue analogous to the rhytidome of the dicotyledons is formed (*Dracaena, Cordyline, Yucca*).

LENTICELS

Lenticels are structurally differentiated portions of the periderm characterized by a relatively loose arrangement of cells. The presence of intercellular spaces in the tissue of the lenticels and the continuity of these spaces with those in the interior of the stem has given rise to the interpretation that lenticels, like the stomata, are concerned with gaseous exchange. Lenticels are generally present on stems and roots. Exceptions are found among stems with a regular formation of complete periderms around the entire circumference (species of *Vitis, Lonicera, Tecoma, Clematis, Rubus*).

Lenticels—from the Latin *lentis*, a lentil—have received their name because of the lenticular shape they commonly assume. As seen from the surface they appear as lenticular masses of loose cells, usually protruding above the surface through a fissure in the periderm. Depending on the orientation of the fissure, transverse and longitudinal lenticels are recognized. The size of lenticels varies from structures that can be scarcely distinguished with the unaided eye to those that are 1 cm and more in length. The large lenticels usually become so with age, since they enlarge in keeping with the increase in the circumference of the stem (*Betula, Abies pectinata, Tamarix indica, Prunus avium*). In some plants, lenticels do not enlarge but become broken up into smaller ones by differentiation of ordinary periderm within the original lenticels (*Pyrus malus, Rhamnus frangula*). Still other lenticels do not perceptibly change in size and form (*Quercus suber, Fraxinus excelsior, Ailanthus*).

In periderms initiated in the subepidermal layer, the first lenticels usually arise beneath stomata. They may appear before the stem ceases its primary growth and before the periderm is initiated, or lenticels and periderm may arise simultaneously at the termination of primary growth. The parenchyma cells about the substomatal chamber divide in various planes, chlorophyll disappears, and a colorless loose tissue is formed. The divisions successively occur deeper and deeper in the cortical parenchyma and become oriented periclinally. Thus a periclinally dividing meristem, the lenticel phellogen, is established. The cells resulting from the initial divisions of the parenchyma beneath the stomata and those produced outwards by the lenticel phellogen constitute the *complementary cells* (cells complementing the periderm), or *filling cells* (Wutz, 1955). As the filling tissue increases in amount, it ruptures the epidermis and protrudes above the surface. The exposed cells die and weather away but are replaced by others developing from the phellogen. By divisions producing cells toward the interior, the phellogen beneath the lenticels forms some phelloderm, usually more than under the cork (Devaux, 1900). The lenticel phellogen is in complete continuity with that formed elsewhere in the stem. Since the number of cells produced in the lenticel region is larger, in both directions, than in regions where cork is formed, the lenticel protrudes above the surface of the periderm and also projects farther inward (pl. 69D). Only in plants with massive cork may the lenticels occur below the surface of the cork (species of *Ulmus, Liquidambar, Quercus*).

Some lenticels arise independently of stomata, either at the same time as the lenticels below the stomata or later. Lenticels may be formed in the part of periderm that produced cork for a while. In such instances the phellogen ceases to produce cork and forms filling cells which break through the layer of cork above them. Lenticels formed in the initial but deeply seated periderms and in all the sequent periderms arise with no reference to the stomata. Lenticels in vertical rows may be confronting the vascular rays, but there is no constant positional relation between stomata and rays (Wutz, 1955). In barks separating in the form of scales, lenticels develop in the newly exposed periderm (*Platanus, Pyrus*). If the bark is adherent and fissured, as in *Robinia* and *Prunus domestica*, the lenticels occur at the bottom of the furrows. If the cork tissue is massive, the lenticels are continued through the whole thickness of the tissue, a feature well illustrated by the commercial cork (*Quercus suber*), in which the lenticels are visible as brown powdery streaks in transverse and radial sections.

The filling tissue varies in degree of contrast with the phellem (Wutz, 1955). In gymnosperms the filling cells are suberized and thus resemble phellem cells except that they may have thinner walls and be radially elongated and that they enclose intercellular spaces among them.

In the dicotyledons three types of lenticels may be distinguished. In the first (*Liriodendron, Magnolia, Populus, Pyrus*) the filling cells are suberized. The tissue, though containing air spaces, is rather compact and may exhibit an annual alternation of thin-walled, looser tissue with thick-walled and more compact tissue. In the second type (*Fraxinus, Quercus, Sambucus, Tilia*) a mass of nonsuberized, loose tissue is succeeded at the end of the season by a compactly arranged closing layer of suberized cells. In the third type (*Betula, Fagus, Prunus, Robinia*) each year several wide, loose, nonsuberized strata regularly alternate with narrow, compact, suberized strata constituting the closing layers, in the sense that they are holding the loose tissue together. The closing layers are successively broken by the new growth from the phellogen.

REFERENCES

Audia, W. V., W. L. Smith, Jr., and C. C. Craft. Effects of isopropyl N-(3-chlorophenyl) carbamate on suberin, periderm, and decay development by Katahdin potato slices. *Bot. Gaz.* 123:255–258. 1962.

Bamber, R. K. The anatomy of the barks of Leptospermoideae. *Austral. Jour. Bot.* 10:25–54. 1962.

Bloch, R. Wound healing in higher plants. *Bot. Rev.* 7:110–146. 1941.

Bowen, W. R. Origin and development of winged cork in *Euonymus alatus*. *Bot. Gaz.* 124:256–261. 1963.

Chattaway, M. M. The anatomy of bark. I. The genus *Eucalyptus*. *Austral. Jour. Bot.* 1: 402–433. 1953. V. *Eucalyptus* species with stringy bark. *Austral. Jour. Bot.* 3:165–169. 1955.

Cooke, G. B. Cork and cork products. *Econ. Bot.* 2:393–402. 1948.

De Bary, A. *Comparative anatomy of the vegetative organs of phanerogams and ferns.* Oxford, Clarendon Press. 1884.

Devaux, H. Recherches sur les lenticelles. *Ann. des Sci. Nat., Bot.* Ser. 8. 12:1–240. 1900.

Douliot, H. Recherches sur le périderme. *Ann. des Sci. Nat., Bot.* Ser. 7. 10:325–395. 1889.

Falk, H., and M. N. El-Hadidi. Der Feinbau der Suberinschichten verkorkter Zellwände. *Ztschr. f. Naturf.* 16b:134–137. 1961.

Holdheide, W. Anatomie mitteleuropäischer Gehölzrinden. In: H. Freund. *Handbuch der Mikroskopie in der Technik.* Vol. 5. Part 1:195–367. 1951.

Isenberg, I. H. The anatomy of redwood bark. *Madroño* 7:85–91. 1943.

Johnson, G., and L. A. Schaal. Accumulation of phenolic substances and ascorbic acid in potato tuber tissue upon injury and their possible role in disease resistance. *Amer. Potato Jour.* 34:200–209. 1957.

Kauffert, F. Factors influencing the formation of periderm in aspen. *Amer. Jour. Bot.* 24: 24–30. 1937.

Küster, E. *Pathologische Pflanzenanatomie.* 3rd ed. Jena, Gustav Fischer. 1925.

Lier, F. G. A comparison of the three-dimensional shapes of cork cambium and cork cells in the stem of *Pelargonium hortorum* Bailey. *Torrey Bot. Club Bul.* 79:312–328; 371–379. 1952.

Lier, F. G. The origin and development of cork cambium cells in the stem of *Pelargonium hortorum*. *Amer. Jour. Bot.* 42:929–936. 1955.

Luhan, M. Das Abschlussgewebe der Wurzeln unserer Alpenpflanzen. *Deut. Bot. Gesell. Ber.* 68:87–92. 1955.

Mader, H. Untersuchungen an Korkmembranen. *Planta* 43:163–181. 1954.

Mader, H. Kork. *Handb. der Pflanzenphysiol.* 10:282–299. 1958.

McClure, T. T. Chlorogenic acid accumulation and wound healing in sweet potato roots. *Amer. Jour. Bot.* 47:277–280. 1960.

Metcalf, W. The cork oak tree in California. *Econ. Bot.* 1:26–46. 1947.

Metcalfe, C. R., and L. Chalk. *Anatomy of the dicotyledons.* 2 vols. Oxford, Clarendon Press 1950.

Möller, J. *Anatomie der Baumrinden.* Berlin, Julius Springer. 1882.

Moss, E. H., and A. L. Gorham. Interxylary cork and fission of stems and roots. *Phytomorphology* 3:285–294. 1953.

Mühldorf, A. Über den Ablösungsmodus der Gallen von ihren Wirtspflanzen nebst einer kritischen Übersicht über die Trennungserscheinungen im Pflanzenreiche. *Bot. Centbl. Beihefte* 42:1–110. 1925.

Mylius, G. Das Polyderm. Eine vergleichende Untersuchung über die physiologischen Scheiden: Polyderm, Periderm und Endodermis. *Biblioth. Bot.* 18(79):1–119. 1913.

Nelson, P. E., and S. Wilhelm. Some aspects of the strawberry root. *Hilgardia* 26:631–642. 1957.

Ogura, Y. Anatomie der Vegetationsorgane der Pteridophyten. In: K. Linsbauer. *Handbuch der Pflanzenanatomie.* Band 7. Lief. 36. 1938.

Pfeiffer, H. Die pflanzlichen Trennungsgewebe. In: K. Linsbauer. *Handbuch der Pflanzenanatomie.* Band 5. Lief. 22. 1928.

Philipp, M. Über die verkorkten Abschlussgewebe der Monokotylen. *Biblioth. Bot.* 23(92): 1–28. 1923.

Priestley, J. H., and C. F. Swingle. Vegetative propagation from the standpoint of plant anatomy. *U.S. Dept. Agric. Tech. Bul.* 151. 1929.

Sanio, C. Vergleichende Untersuchungen über den Bau und die Entwickelung des Korkes. *Jahrb. f. Wiss. Bot.* 2:39–108. 1860.

Sen, J. The nature of cork in ancient buried wood with special reference to normal present-day representatives. *Riv. Ital. Paleontol. e Stratigraf.* 67:77–88. 1961.

Sifton, H. B. Air-space tissue in plants. *Bot. Rev.* 11:108–143. 1945.

Sitte, P. Der Feinbau verkorkter Zellwände. *Mikroskopie* (Wien) 10:178–200. 1955.

Sitte, P. Der Feinbau der Kork-Zellwände. In: E. Treiber. *Die Chemie der Zellwand.* Berlin, Springer-Verlag. 1957.

Smithson, E. Development of winged cork in *Ulmus* x *hollandica* Mill. *Leeds Philos. and Lit. Soc., Sci. Sec., Proc.* 6:211–220. 1954.

Solereder, H., and F. J. Meyer. *Systematische Anatomie der Monokotyledonen.* Heft III. Berlin, Gebrüder Borntraeger. 1928.

Wutz, A. Anatomische Untersuchungen über System und periodische Veränderungen der Lenticellen. *Bot. Studien* No. 4:43–72. 1955.

Zeeuw, C. De. Influence of exposure on the time of deep cork formation in three northeastern trees. *N.Y. State Col. Forestry, Syracuse Univ., Bul.* 56. 1941.

15

The Stem

CONCEPT

According to the formal morphologic concept, the vegetative body of the sporophyte of the vascular plant consists of three organs: stem, leaf, and root. As was discussed in chapter 1, this division is fundamentally theoretical, for the plant is a unit on the basis of its development, evolution, and structure. The boundary between the stem and the leaf is particularly uncertain, and therefore the stem and its appendages are often treated under the broader concept of the shoot (Arber, 1950; Foster, 1949).

The intrinsic unity of the shoot has been recognized since the early days of botany, but the morphologic value of the concepts of leaf and stem and their relation to each other have been interpreted in a variety of ways. The theories that have been advanced to explain the basic structure of the shoot are reviewed in numerous works (Cuénod, 1951; Eames, 1936; Emberger, 1952; Schoute, 1931; and references in Arber, 1950). Briefly, three main concepts are used to interpret the morphologic nature of the shoot: (1) the leaf and the stem are ultimate and discrete units of the plant body; (2) the shoot consists of growth units (phytons, phyllomes, and others), each comprising a leaf and the subjacent part of stem; (3) the axis is a fundamental organ and the leaf is its modification differentiated in the course of phylogeny. Regardless of the merits of the various theories, their discussions have served to emphasize the intimate relation between the stem and the leaf. The proper recognition of this unity is essential for the understanding of the primary structure of the stem.

353

ORIGIN OF THE STEM

The stem, as part of the shoot, is initiated during the development of the embryo (chapter 20). The differentiation of the characteristic organization of the embryo is attained gradually, and its degree varies in different groups of plants. The fully developed embryo commonly consists of an axis, the *hypocotyl-root axis*, bearing, at the upper end, one or more cotyledons and the primordium of the shoot and, at the lower end, the primordium of the root covered with a rootcap (fig. 1.1). The root and shoot primordia may be no more than meristems (apical meristems), or there may be an embryonic root, the *radicle*, at the lower end of the hypocotyl and an embryonic shoot above the insertion of the cotyledons (seemingly lateral to the single cotyledon in monocotyledons; chapter 20). The embryonic shoot consists of an axis with unextended internodes and one or more leaf primordia. This shoot, the first bud, is commonly termed the *plumule* (in Latin, little feather), and its stem part is called the *epicotyl;* or the terms are used synonymously to designate the entire shoot primordium present in the embryo (Darwin, 1892).

The structural relation between the hypocotyl and the cotyledons is comparable to that between the stem and its leaves (chapter 17). Thus, the beginning of shoot organization is found in the hypocotyl-cotyledon system in which the hypocotyl is the first stem unit of the plant and the cotyledons are the first leaves. The hypocotyl could hardly be called an internode. It is located below a node (the cotyledonary node), but not between nodes.

During the germination of the seed, the root meristem forms the first root, while the shoot meristem continues the development of the first shoot by adding new leaves and increments of the axis, which sooner or later differentiate into nodes, and internodes. In plants with branching axes, axillary buds arise on the first shoot and develop into lateral branches.

EXTERNAL MORPHOLOGY OF THE SHOOT

Nodes and Internodes

A common feature of the stem in the primary state of development is its division into nodes and internodes. As may be learned from chapter 5, this division results from the manner of origin of the leaves at the shoot apex and the subsequent growth of the axis bearing these leaves. The shoot apex gives rise to leaf primordia in close succession and consequently the young shoot appears as though composed of a series of superposed shallow discs, each bearing one leaf or more, depending on the arrangement of leaves in a given plant. Later, growth occurs at the bases of these discs, and the leaf insertions become separated from each other. In other words, internodes develop between the nodes by intercalary growth (pls. 14A, 50). The duration of this growth may be shorter or longer, depending on the plant species, the environmental conditions, and the type of stem. Sometimes the internodes are practically undeveloped, and the leaves remain crowded on the axis. For example, no internodes can be distinguished in plants bearing their leaves in a rosette. The rosette stage, however, may be followed by an extension of the internodes in the later-formed part of the axis, usually in preparation for flower development. Bulbs consist of axes with unextended internodes and closely approximated scale leaves. In many rhizomes and in the spurs of fruit trees, and in the needle-bearing short shoots of pines (Sacher, 1955) the internodes remain quite short. Representatives of various plant groups have short and long shoots on the same plant (Troll, 1954). In arborescent plants secondary growth eventually obscures the division of the stem into nodes and internodes, and the external evidence of the relation between this organ and the leaves disappears.

Phyllotaxis

The pattern formed in the stem by the alternation of nodes and internodes is affected by the *phyllotaxis* (or *phyllotaxy;* from the Greek *phyllon,* leaf, and *taxis,* arrangement) and by the manner of attachment of the leaves to the stem. Since these features of the shoot have a bearing upon the structure of the primary vascular system and its development in the stem, they are briefly considered in the following paragraphs.

Some leaves have narrow, others have wide insertions, and still others partly or completely encircle the stem. Each node may bear one, two, or several leaves, the arrangements being called alternate, opposite (or decussate), and whorled, respectively. Students of leaf arrangement attempt to give mathematical expressions to the orderly sequences in which leaves arise at shoot apices and discuss the causal relations that may govern the tendency toward regularity in this process (Dormer, 1955b; Richards, 1951; Snow, 1955; Snow and Snow, 1962; Van Iterson, 1960).

A common method of expressing the phyllotaxis is by reference to the so-called genetic spiral (or helix) and the angular divergence of the leaves succeeding one another along this spiral. The genetic spiral passes through the leaves in their numerical order, that is, the order of their production at the apex. The divergence angle between the leaves is given as a fraction of a circle, which is estimated by finding two superimposed leaves (leaf 1 over leaf 6 in fig. 15.1A) and counting the number of leaves and the number of turns around the axis between the two superimposed leaves. The first value is used as the numerator, the second as the denominator of the fraction (⅖ in fig. 15.1A). The fractions most frequently belong to the so-called Fibonacci summation series, ½, ⅓, ⅖, ⅜, ⁵⁄₁₃, ⁸⁄₂₁, etc., in which each value of the numerator and the denominator is a sum of the two corresponding values that precede it.

The fractional classification of phyllotaxis is criticized because it refers to mature shoots and is not reliable for expressing the arrangement of leaves at their origin. It assumes that certain leaves occur exactly above one another, that is, along orthostichies (singular orthostichy, from the Greek *orthos*, upright, and *stichos*, series), whereas plants with helical leaf arrangements have only parastichies (*para*, beside), that is, helices, which are of different degrees of steepness. Moreover, the divergence angles do not classify the helical systems

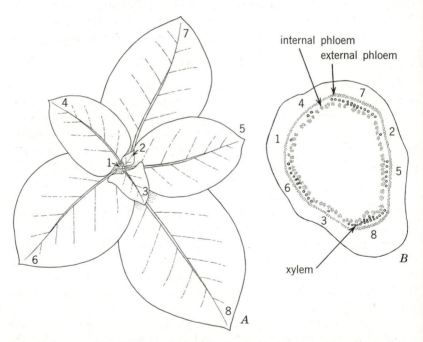

Fig. 15.1. Relation of primary vascular system to leaf arrangement in *Nicotiana tabacum* shoot. *A*, shoot seen from above. *B*, transection of shoot. Numbers 1–8 in *B* indicate leaf traces of the leaves in *A* bearing the same numbers. External and internal phloem tissues are uniformly distributed around circumference of stem; xylem is localized in leaf-trace positions. The plant has a ⅖ phyllotaxy. (*B*, ×12. Esau, *Hilgardia* 13, 1941.)

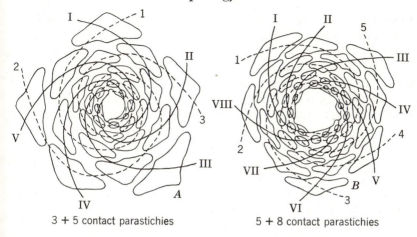

3 + 5 contact parastichies   5 + 8 contact parastichies

Fig. 15.2. Leaf arrangement. Transections of shoots of *Linum perenne.* In each, curved lines connect leaves in series which are called parastichies. These particular series are contact parastichies, for they connect the leaves that were in contact with each other when they emerged at the apex. (Both, ×59.)

because the original angle of divergence at the apex is approximately the same in these systems, close to 137.5°, and the Fibonacci fractions oscillate about this limiting value.

Another method of characterizing a phyllotaxis is by reference to the sets of parastichies that may be recognized when the shoot is viewed from the top. Usually it is possible to identify two sets of parastichies winding in opposite directions. If the parastichies are counted in each direction, certain numbers are more frequent than others. These numbers again pertain to the Fibonacci summation series 1, 1, 2, 3, 5, 8, etc. The leaves arise at the apex in close proximity to one another along some of the parastichies. These are the contact parastichies (Church, 1920). Examples of characteristic numbers of contact parastichies are 2 and 3, 3 and 5, 5 and 8 (fig. 15.2).

According to one of the theories of phyllotaxis (Plantefol, 1946, 1947) certain contact parastichies are fundamental in nature in that they are determined by the activity of the so-called generative centers for leaves, located in the peripheral zone (the initial ring, chapter 5) of the apical meristem. In the dicotyledons, leaf production usually starts along two parastichies beginning with the cotyledons, but the number of helices may increase as the plant grows. This theory is criticized chiefly because of the hypothetical concept of the leaf generative centers (Cutter, 1959).

A method that allows a statistical treatment of values describes the phyllotaxis by giving the divergence angle and the ratio of the radial distances to successive primordia, a ratio called the plastochron ratio (Richards, 1951).

Helical phyllotaxes are rarely constant; the number of parastichies tends to increase during vegetative growth. This increase is associated with the enlargement of the apical meristem (Loiseau, 1959). The mathematical relationships in the leaf arrangement and the shifts in these relationships are part of an overall organization of the plant and have their counterpart in the internal patterns, especially those characterizing the vascular system. As Dormer (1955b) has suggested, the angle of divergence between the successive leaves at the apex is an expression of a mechanism concerned with timing and synchronization of physiological processes that occur in the growing shoot.

TISSUE SYSTEMS

The primary structure of the stem may be conveniently described by reference to the classification introduced in chapter 1. It distinguishes three tissue systems: the dermal, the fundamental (or ground-tissue system), and the vascular. The principal variations in the structure of stems depend on the relative amounts and the spatial arrangement of vascular and ground tissues.

In some of the lower vascular plants (pl. 63A) and in certain aquatics among the angiosperms, the vascular tissue forms a solid cylinder in the center of the axis. Mostly, however, the vascular tissue of stems is variously interpenetrated by ground tissue. The vascular tissue may be arranged, within the ground tissue, as an approximately continuous hollow cylinder (fig. 15.9), or as a cylindrical complex of interconnected strands (fig. 15.3A, pl. 51A), or as anastomosing strands dispersed through all or most of the axis (fig. 15.3B, pl. 58A). In transections of internodes such vascular systems appear, respectively, as a ring of vascular tissue (pl. 62A), as a ring of bundles (pl. 63B), and as bundles scattered individually (pl. 58C). In stems having the vascular system in the form of a solid cylinder, the ground tissue located between the epidermis and the vascular system constitutes the *cortex* (from the Latin word for bark or rind). If the vascular system has the shape of a hollow cylinder, it encloses part of the fundamental tissue, the *pith*. If this cylinder is divided into strands, called the vascular bundles or *fascicles* (from the Latin *fascis*, bundle), the spaces among the strands, the *interfascicular regions*, are occupied by parenchymatic ground tissue. These plates of tissue are sometimes designated as pith rays or medullary rays (from the Latin *medulla*, pith).

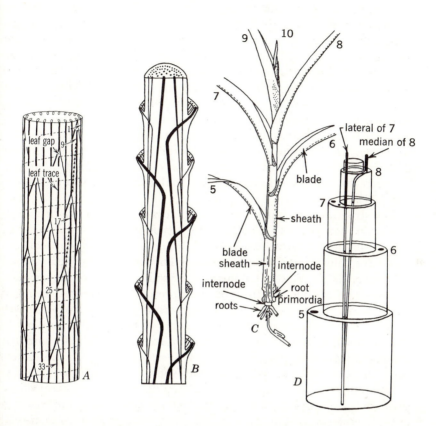

Fig. 15.3. Primary vascular systems in angiosperms. *A*, dicotyledon (*Linum perenne*) with vascular system as a network of leaf traces. In front of each leaf gap a leaf trace diverges toward a leaf. Arrows indicate parastichy of leaves 1–9–17–25–33, etc. The traces of these leaves are connected with each other and with those of leaves in parastichy 6–14–22–30–38, etc. *B*, monocotyledon (palm). One median (thick) and one small lateral (thin) leaf traces are shown for each leaf. Leaves in two-ranked arrangement; therefore, median traces of successive leaves diverge toward leaves at opposite sides of stem. *C*, monocotyledon (*Zea mays*) plant showing two-ranked arrangement of leaves and relation between blade, sheath, internodes, and roots. *D*, part of plant in *C* showing course of median trace of leaf 8 and its connection with lateral trace of leaf 7. Successive units in *D* represent parts of leaf sheaths shown to be completely closed around the stem for sake of simplicity. (Adapted from: *A*, Esau, *Amer. Jour. Bot.* 30, 1943; *B*, Linsbauer, *Schneiders illustriertes Handwörterbuch der Botanik* 1917; *C, D*, Sharman, *Ann. Bot.* 6, 1942.)

The delimitation of the ground tissue into pith and cortex does not exist if the vascular tissue is dispersed as bundles throughout the circumference of the axis (pl. 58C). Still other deviations from a pattern with a distinct separation into cortex, pith, and vascular system occur among plants. The internal phloem of some groups of plants is somewhat removed from the rest of the vascular tissues and may be said to be located within the pith. Certain families have complete vascular bundles scattered within a well-defined pith, the medullary bundles. Vascular bundles may occur on the outer side of the main mass of the vascular system, that is, in the cortex, the cortical bundles. Finally, individual files of vascular elements may differentiate within the ground tissue, as, for example, the sieve tubes that occur throughout the region between the vascular strands and the epidermis in the Cucurbitaceae (fig. 12.1).

The structure and function of the epidermal system have been considered in chapter 7. The cortex of stems typically contains parenchyma with prominent intercellular spaces. Some or all of the cortical cells may have chloroplasts, sometimes in significant numbers (Pearson and Lawrence, 1958). Starch, tannins, and crystals are some of the common inclusions. Collenchyma tissue is often present in the cortex, arranged as a cylinder or as strands near the epidermis or immediately beneath it (fig. 9.1). Sclereids and fibers also occur in the cortex (chapter 10). The cortex of gymnosperms may develop resin ducts (pl. 60), and that of some dicotyledons, lysigenous oil cavities (*Citrus*). Cortical laticifers occur in some of the latex-forming plants (fig. 13.9B).

The pith of stems is largely parenchymatic. It may contain chloroplasts or starch-forming leucoplasts. Intercellular spaces are common in the mature pith. Since the pith is often initiated as a rib meristem, its cells may be arranged in longitudinal files (Rouffa and Gunckel, 1951). This pattern is characteristic of long shoots. In short shoots the arrangement is less orderly (Tolbert, 1961). The pith of many plants is partially destroyed during the growth of the stem. In such instances the internodes are frequently hollow, while the nodes retain their pith (nodal diaphragms). Sometimes, series of horizontal plates of pith are left in the internodes, forming a pattern known as diaphragmed pith (*Juglans, Pterocarya*).

The parenchyma cells of the pith—if the latter is retained at maturity —may show varied degrees of differentiation among themselves (Gris, 1872). Frequently, certain pith cells are specialized as depositories of crystals or tannins. Some may develop rather thick walls or differentiate into sclereids. Either thick- or thin-walled cells may become lignified. Fibers occur rarely (cycads). In many plants some or all of the pith cells become devoid of contents. Specialized structures like laticifers or secretory canals occur in the pith. The outer part of the pith may be somewhat distinct from the bulk of the pith in having, for example, smaller cells and thicker walls. Such a morphologically distinct outer pith region is called the perimedullary zone or medullary sheath. Although the pith is in general less highly differentiated than the vascular tissues and even less so than the cortex, some workers consider this stem region of considerable diagnostic value in systematics (Doyle and Doyle, 1948; Metcalfe and Chalk, 1950).

The nodes of stems differ from the internodes, first of all, in the arrangement of vascular tissues. The nodal vascular system is complicated by the divergence of some vascular tissue into the leaves (fig. 15.4) and the branches. Furthermore, in some herbaceous plants the main interconnections among the vertically oriented bundles occur by means of horizontally oriented strands in the nodal region (pl. 58A). The detailed histology of the vascular bundles may be different at the node (a feature partly determined by the lack of elongation), the cortical and the pith cells may be shorter, and there may be less sclerenchyma and more collenchyma than in the internodes (Prunet, 1891). It seems that the degree of differentiation between the nodes and the internodes is influenced by the relative development of the leaves attached to the nodes (Prunet, 1891). If the leaves are rudimentary, as in undergound stems, nodes and internodes differ little from each other.

In woody plants, the primary stem structure becomes more or less modified by the formation of secondary tissues. The vascular tissues are augmented by the vascular cambium. Frequently the epidermis, or the epidermis and varying amounts of cortex and phloem, become detached from the plant body by the development of periderm (chapter 14). Since eventually the secondary tissues are formed uniformly in the nodal and the internodal regions, the distinctions between the two are not perpetuated in the secondary body.

PRIMARY VASCULAR SYSTEM

Leaf Traces

If the vascular system of a leaf-bearing shoot of a seed plant is considered as a whole, the close connection between the vascular tissues of stem and leaf is clearly apparent. At each node, portions of the vascular system are deflected into the leaf attached at that node (pl. 51A). If the vascular bundles diverging into a leaf are traced backwards into the stem, they may be found to be discrete for variable distances within the stem, and then ultimately to merge with other parts of the vascular system of the stem (fig. 15.3A). A vascular bundle located in the stem but directly related to a leaf, in the sense that it represents the lower part of the vascular supply of this leaf, is termed a *leaf trace* (Hanstein, 1858). A trace may be pictured as extending between the base of a leaf and the point where it is completely merged with other parts of the vascular system in the axis. One or more leaf traces may be associated with each leaf. (In some references all bundles constituting the vascular supply of one leaf are collectively designed as the leaf trace. The single bundle in such a trace is a leaf-trace bundle.)

The concept of leaf traces implies that at least a part of the axial vascular system develops in direct relation to the leaves. It is an old question in botanical literature how much of the vascular system of the stem is foliar by origin (ontogenetic and phylogenetic) and how much of it is *cauline* (from the Latin *caulis*, the stem). This question is related to the broader problem of interpreting the nature of the leaf and of its phylogenetic relation to the stem. In some vascular plants, such as Lycopsida (*Selaginella, Lycopodium*), the leaves are small and simple (microphylls; Foster and Gifford, 1959), and their weak traces are peripherally connected to a prominent cauline vascular cylinder (fig. 15.9A). In Pteropsida (ferns, gymnosperms, and angiosperms), the leaves constitute a prominent part of the shoot (macrophylls or megaphylls; Foster and Gifford, 1959) and their traces are large in relation to the vascular system of the axis. Some workers consider that at least in certain ferns all vascular tissue in the stem is foliar in origin (Verdoorn, 1938); others regard the axial system in this group of plants as a composite structure containing both foliar and cauline vascular components, with the contributions from the leaf traces probably varying in amount in different groups (Wardlaw, 1952). In gymnosperms and angiosperms the primary vascular system of the stem is clearly associated with the leaves and is often described as a system of interconnected leaf traces (Barthelmess, 1935; De Bary, 1884; Esau, 1954), but some authors prefer to treat the leaf-trace complexes as structures distinct from the leaf traces (Dormer, 1954).

If the leaf and the stem are of common phylogenetic origin (chapter 1), the discussions attempting to distinguish between leaf traces and cauline vascular tissue appear to be purely theoretical. The shoot as a whole possesses a vascular system whose form is more or less affected by the development of the leaves. When the leaves are insignificant enations (microphylls) the axial part of the vascular system resembles that of the root in which no appendages arise at the apex; if the leaves are large (megaphylls), most of all of the vascular tissue is directly or indirectly connected with that of the leaves and the vascular system can be described as a system of leaf traces and their complexes. Such a description is not meant to imply that the axis does not have its own vascular system; it states only that the vascular system of the axis has assumed a form reflecting the close relation between the leaf and the axis. In conformity with this concept, the experimental suppression of leaf development modifies the structure of the vascular system in the axis (Wardlaw, 1952).

Since the leaf traces may extend through several internodes between their divergence into a leaf and connection with other leaf traces in the stem, the establishment of the relation between the leaves and the vascular system of the stem must be carried out by the use of serial sections through many internodes and by accounting for every unit of the vascular system seen in the sections. Without such a study the interpretation of bundles as being or not being leaf traces is not reliable.

The arrangement of leaf traces and their complexes varies in different groups of plants and is related to phyllotaxis (Philipson and Balfour, 1963; De Bary, 1884; Ezelarab and Dormer, 1963; Dormer, 1954). In some plants (pl. 51A) the leaf traces form sympodia that are independent of one another; in others (fig. 15.3A) the different trace complexes are interconnected (corresponding to Dormer's, 1954, open and closed systems, respectively). In monocotyledons the common plan is the so-called palm type (fig. 15.3B), which is found not only in the palms but also in many other monocotyledons (De Bary, 1884; Kumazawa, 1961). In this system the numerous leaf traces of a single leaf can be roughly divided into small and large. The small traces have a peripheral course in the stem. The large traces approach the center of the stem in their upper parts but are reoriented toward the periphery in their lower parts. Here they may be united with other peripheral bundles (fig. 15.3D). In monocotyledons with bundles not dispersed but arranged in two or more rings, the trace relationship is similar to that of the palm type but the larger traces do not penetrate the internode as deeply (fig. 15.21). This summary barely indicates the variability of bundle arrangements that are found in the Pteropsida, but the connection between the axis and the leaves plays a dominant role in all of them.

Leaf Gaps

Where the leaf trace diverges into a leaf, in the shoot of a pteropsid plant, it appears as though a section of the vascular cylinder of the stem is deflected to one side. Immediately above such a diverging trace, parenchyma instead of vascular tissue differentiates in the vascular region of the stem. These parenchymatic regions, located adaxially from the diverging leaf traces within the vascular cylinder of the axis, are called *leaf gaps* or *lacunae* (figs. 15.4, 15.5, 15.9). In transections of a stem cut at the level of a leaf gap, the gap resembles an inter-fascicular region.

The gaps are particularly conspicuous in those ferns and angiosperms in which the vascular system in the internodal parts of the axis forms a more or less continuous cylinder. In some ferns the gaps are so high or the leaves so crowded that the gaps formed at the successive nodes overlap one another and the vascular cylinder appears as though dissected into strand-like portions. The transections of such stems show a circle of vascular bundles with parenchymatic regions, the leaf gaps, among them (Ogura, 1938).

In plants with the vascular system composed of anastomosing strands (certain ferns, the gymnosperms, most angiosperms) the recognition of leaf gaps is rather uncertain, because in these plants the parenchyma that occurs above the diverging leaf trace is confluent with the inter-fascicular regions (fig. 15.3A; Bailey and Nast, 1944; Barthelmess, 1935; Nast, 1944). In such stems the gaps become delimited only after the addition of some secondary vascular tissues: in front of the ordinary interfascicular regions the secondary xylem appears closer to the pith than in front of the gaps and, therefore, the gaps project for a greater distance into the secondary xylem cylinder than the interfascicular regions (fig. 15.5C, pl. 61). In plants with the vascular bundles dispersed within the ground tissue the delimitation of gaps is even more problematical.

Despite the difficulties encountered in the application of the concept of leaf gaps to many vascular plants, this concept is commonly utilized in the characterization of the nodes. The trace relationships at the nodes are considered to be of phylogenetic importance, and therefore nodal anatomy receives attention in studies concerned with the systematics and the phylogeny of angiosperms (Bailey, 1956; Canright, 1955; Carlquist, 1961; Sinnott, 1914).

Four main types of nodes are recognized in dicotyledons: two-trace unilacunar, with a single gap and two traces to one leaf connected to the opposite halves of the axial vascular system (fig. 15.5A); one-trace unilacunar, with a single gap and a single trace to a leaf (fig. 15.4A); trilacunar, with three gaps and three traces to a leaf, one median and two lateral (fig. 15.4B, C); and multilacunar, with several to many gaps and traces to a leaf (fig. 15.4D). If the leaves are opposite or whorled, the node is classified on the basis of the number of gaps to each leaf (fig. 15.5B). Such nodal patterns may be called unilacunar opposite, unilacunar whorled, etc. (Carlquist, 1959a).

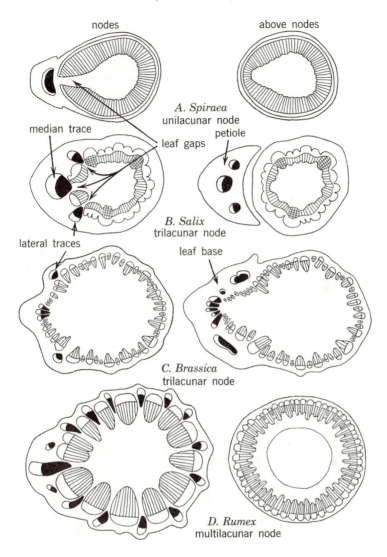

nodes   above nodes

A. Spiraea
unilacunar node
median trace   petiole
                    leaf gaps

                    B. Salix
lateral traces   trilacunar node
                    leaf base

                    C. Brassica
                    trilacunar node

                    D. Rumex
                    multilacunar node

Fig. 15.4.   Nodal anatomy of dicotyledons.   Transections of stems.   Leaf traces (and petiolar bundles in B) are indicated by blackened xylem areas.   In B traces of leaf attached at next higher node are marked by crosshatching.   All plants illustrated have alternate leaf arrangement and each leaf has one (A), three (B, C), or many (D) leaf traces.   Nodes show the same numbers of gaps, or lacunae, as there are leaf traces.   In C the median leaf trace consists of several bundles.   Stem in A has some secondary vascular tissue.   Internode in D is hollow.   (A, B, ×26; C, D, ×6.)

The two-trace unilacunar type is considered to be the most primitive in angiosperms. The one-trace unilacunar and the trilacunar have evolved from the two-trace unilacunar. The trilacunar gave rise to the multilacunar and also to some of the one-trace unilacunar types. Additional modifications, with more than two traces to one gap, appeared. The various phylogenetic changes involved deletions, fusions, and additions of traces. More than one evolutionary type may be present in the same plant, and the cotyledonary node of dicotyledons frequently has the two-trace unilacunar condition. This primitive type is also widespread in vascular plants other than the angiosperms (Bailey, 1956).

Many monocotyledons have leaves with sheathing bases and nodes with a large number of leaf traces separately inserted around the circumference of the stem (fig. 15.21). In ferns the number of traces to a leaf varies from one to many, but regardless of their number they are associated with a single gap (Ogura, 1938). In gymnosperms a unilacunar node is common. In conifers a single gap confronts a single trace (fig. 15.5*C*); in *Ginkgo* (Gunckel and Wetmore, 1946*b*) and *Ephedra* (Marsden and Steeves, 1955) two traces.

Branch Traces and Branch Gaps

Branches developing from axillary buds have vascular connections with the axis of origin. Dicotyledons and gymnosperms commonly have

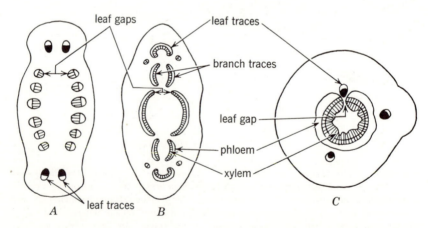

Fig. 15.5. Nodal anatomy. Transections of stems. *A, Clerodendron,* two-trace unilacunar node; two opposite leaves to a node. *B, Veronica,* unilacunar node with two, opposite, leaves; branch traces, two to a branch, in axil of each leaf. *C, Picea* (conifer), alternate leaf arrangement, unilacunar nodes, and some secondary tissues. (*B,* ×14; *C,* ×25.)

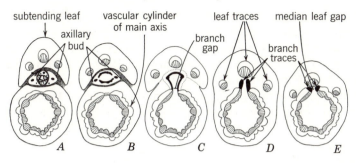

Fig. 15.6. Vascular connection between axillary branch (in bud stage) and main axis in *Salix*. *A–E*, successively lower levels. Vascular system of bud is indicated in black. First two leaves (prophylls) of bud are almost opposite one another. Branch gap and median gap of subtending leaf are confluent. (All, ×9.)

two strands, the two *branch traces* (bud traces in early stages of development), connecting the branch to the main stem (figs. 15.5*B*, 15.6). Some plants have only one branch trace (Murty, 1960; Shah, 1960); others have more than two. The connection of the bud with the main axis appears to be correlated with the extent of tangential continuity or anastomosing within the vascular system of the plant axis. If this system is sympodial or otherwise deficient in tangential interconnection, the bud tends to be connected with a large part of the circumference of the main stem (Dormer, 1955*a*; Ezelarab and Dormer, 1963). Medullary bundles may also be continuous with the bud (Davis, 1961). In monocotyledons the connection of the axillary shoot with the main stem consists of many strands (De Bary, 1884; Kumazawa, 1961).

Like the leaf traces, the branch traces are prolonged within the main axis and merge with the vascular system of the axis. The branch traces form a more or less conspicuous part of the primary vascular cylinder of the main axis, depending on the plant species and the relative time of development of the lateral branch. At the node, branch traces are often closely approximated to the single or the median trace of the leaf subtending the branch, and the two kinds of trace are usually confronted by a common gap in the vascular system of the main axis (fig. 15.6, 15.9*B*).

The occurrence of two branch traces in gymnosperms and many dicotyledons is related to the position of the first two foliar structures, the prophylls (chapter 16), of the axillary shoot (fig. 15.6). The prophylls occur approximately opposite each other, and their median planes cut that of the axillant leaf at right angles (Foster, 1932; Troll, 1937).

The two branch traces are initiated as leaf traces of the two prophylls. They may consist of more than one bundle and may later increase in size because of the development of the vascular supply to one or more of the higher leaves on the branch (Garrison, 1949*a*, *b*). Thus the term branch trace is used somewhat differently than the term leaf trace. Sometimes it refers to a single prophyll trace, sometimes to a complex of traces. In monocotyledons the axillary shoot usually has a single prophyll, interpreted by some authors as a double structure (chapter 16; Arber, 1950; Troll, 1937). It occurs on the adaxial side of the axillary shoot, and two of its veins constitute in their downward prolongation the first two trace bundles of the axillary shoot. Dicotyledons may also have a single prophyll positioned with its back toward the main axis (Fries, 1911; Murty, 1960; Shah, 1960).

Vascular Bundles

A strand-like portion of the primary vascular system of the stem or leaf constitutes a vascular strand or vascular bundle. The vascular bundles merit separate attention because they reflect many details of the histology of the system as a whole and are readily accessible for study.

The phloem and the xylem are associated with each other not only within the system as a whole but usually also within its parts, the vascular bundles. Variations in the arrangement of the vascular tissues in the bundles have led to the establishment of bundle types (De Bary, 1884). One of the most common types in gymnosperms and angiosperms is the *collateral,* in which the phloem occurs on one side of the xylem (figs. 15.7, 15.8). The presence of phloem on both sides of the xylem makes the bundle *bicollateral* (pl. 38A). Such bundles occur in dicotyledons having internal phloem. In some of these plants, however, the internal phloem of stems forms seemingly independent strands in the peripheral part of the pith, and the term bicollateral bundle is not readily applicable, except perhaps with reference to the strands in the leaves where the internal phloem is more closely associated with the other vascular parts (tomato, tobacco).

The third type of vascular bundle is called *concentric* because one kind of vascular tissue completely surrounds the other. Concentric bundles are *amphivasal* (from the Greek words for around and vessel) if the xylem surrounds the phloem (pl. 57D) or *amphicribral* (*cribrum,* Latin for sieve) if the phloem surrounds the xylem (pl. 57C). Examples of amphivasal bundles are found in both monocoytledons and dicotyledons. In the latter, the medullary bundles are commonly amphivasal. In monocotyledons, amphivasal bundles may occur in the internode, or they may be restricted to the nodal regions.

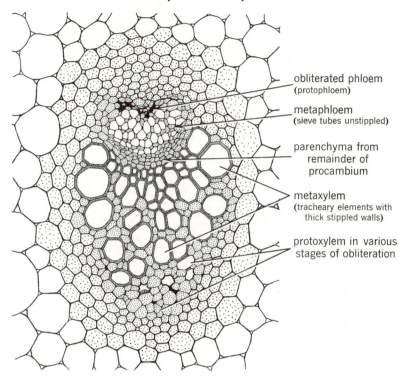

obliterated phloem
(protophloem)

metaphloem
(sieve tubes unstippled)

parenchyma from
 remainder of
 procambium

metaxylem
(tracheary elements with
 thick stippled walls)

protoxylem in various
 stages of obliteration

Fig. 15.7. Transection of vascular bundle of *Ranunculus*. Collateral bundle from an extreme herbaceous dicotyledon lacking secondary growth. (×172.)

Amphicribral vascular bundles are frequently encountered in ferns. In transections these bundles may be circular or oval in outline, or they may be variously curved or lobed (Russow, 1872). In angiosperms the amphicribral condition is apparently rare (De Bary, 1884).

A given bundle may vary in structure in the different parts of its course and may intergrade with another type. Transitional forms between collateral and amphivasal bundles have often been observed, and the amphivasal arrangement is interpreted as more specialized (Cheadle and Uhl, 1948). In some grass bundles the xylem and the phloem meet along a curve, and two large metaxylem vessels appear on the flanks (pl. 57*B*). In another distinctive type the xylem forms, in transections, a V-shaped figure, with the phloem enclosed between the two arms of the V (fig. 15.8).

In most of the lower vascular plants, monocotyledons, and extreme herbaceous dicotyledons, the vascular bundles retain no procambium after the primary vascular tissues mature. They therefore lose their

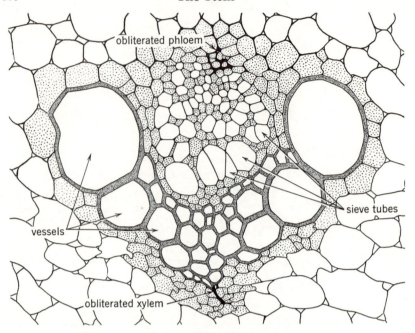

Fig. 15.8. Transection of vascular bundle of *Asparagus.* Collateral bundle from herbaceous monocotyledon lacking secondary growth. Obliterated elements indicate position of protophloem and protoxylem. Intact tissues are metaphloem and metaxylem. (×316.)

potentiality for further growth (figs. 15.7, 15.8; pl. 57C, D). In most dicotyledons and in gymnosperms the vascular bundles, in contrast, have a persisting vascular meristem between the xylem and the phloem. This is the cambium which develops from the procambium at the end of the extension growth of the shoot.

THE CONCEPT OF THE STELE

The early plant anatomists regarded the individual vascular strand as a unit of primary vascular construction (De Bary, 1884). Later, the continuity of the vascular system in the plant body came to be emphasized. This attitude was reflected in Sachs' (1875) classification of the plant tissues into three systems, the dermal, the fundamental, and the vascular, but it was formulated most emphatically by Van Tieghem and Douliot (1886) when they interpreted the primary vascular system, whether compact and simple or loose and complex, as a unit combining the vascular tissues and the associated fundamental tissue. This unit was named the *stele,* a term derived from the Greek word meaning column. The stelar concept became the basis of the stelar theory, which postulated that the primary bodies of the stem and the root were basically alike because each consisted of a central core, the stele, enclosed within the cortex. The core was interpreted as including the vascular system and the so-called conjunctive tissue, that is, the interfascicular regions, the gaps, the pith (if present), and some fundamental tissue on the outer periphery of the vascular system, the pericycle. In relation to the structural variations of the primary vascular system, different types of steles were recognized.

The concept of stele was accepted by many morphologists and was extended to apply to all vascular plants. However, the interpretation of the phylogeny of the stele and the classification of its types have subsequently undergone many changes, and there is still no uniform treatment of the subject (Bower, 1930; Campbell, 1921; Jeffrey, 1898–99, 1903; Nast, 1944; Ogura, 1938; Schoute, 1903). Some authors even have expressed doubt of the usefulness of the concept (Brebner, 1902; Bugnon, 1924; Hasselberg, 1937; Meyer, 1916). Contributors to the physiological anatomy have made little or no use of the stelar concept (Haberlandt, 1914; contributors to K. Linsbauer, *Handbuch der Pflanzenanatomie,* cited in chapter 1); and in much of the later literature the stele is used as a convenient abbreviation for the vascular system. Nevertheless, the stelar theory has been of unmistakable value in emphasizing the unity of structure of the vascular system and in stimulating extensive comparative research. As a result of this research the literature on the stele has become voluminous and rich in terms. In the following discussion a few of the terms regarding the stele are reviewed (see also Foster and Gifford, 1959).

The simplest type of stele, and also the most primitive phylogenetically, contains a solid column of vascular tissue enclosing no pith. This is the *protostele (protos,* first, in Greek). In the simplest protostele the xylem forms the core and the phloem surrounds it as a relatively uniform layer (pl. 63A). In more complex types the xylem and the phloem intermingle as strands or plates of variable shape (species of *Lycopodium* and *Selaginella*). Protosteles are most common in the lower vascular plants, but they occur also in the earliest part of the shoot of ferns and in the stems of some angiospermous water plants. The pithless central core characteristic of many angiospermous roots is commonly interpreted as protostele.

Presence of pith differentiates the second form of stele, the *siphonostele,* that is, tubular stele (fig. 15.9). The siphonostele and its variations are characteristic of most Pteropsida. The phloem and the xylem show two main distributional patterns in the siphonostele. In the *ectophloic siphonostele* the phloem occurs only on the outer side of the xylem cylinder; in the *amphiphloic siphonostele* or *solenostele* (*solen* and *siphon* from Greek words, both meaning tube) it also differentiates on the inner side of the xylem (internal phloem). In its simplest form the siphonostele has no leaf gaps (fig. 15.9A). Relatively small leaf gaps, not overlapping each other in the internodes, are found in some other siphonosteles (fig. 15.9B, C). The transections of such steles in the internodes show a continuous ring of vascular tissue. In many ferns the leaf gaps are large and overlap to the extent that the vascular system appears dissected into a net-like structure, with each segment constituting a concentric vascular bundle. Such vascular structure distinguishes the siphonostelic type called the *dictyostele* (*dictyo*, net, in Greek).

Another modification of the siphonostele is the *eustele* (true stele, from the Greek), in which the vascular system consists of collateral or bicollateral strands, with the leaf gaps and the interfascicular regions not clearly delimited from one another (fig. 15.3A, pl. 51A). The designation of such a stele as eustele was originally chosen because it is the stele type of the highly evolved vascular plants, the gymnosperms and dicotyledons (Brebner, 1902). The most complex stele containing a system of dispersed strands, as in monocotyledons, is called *atactostele* (pl. 58; from the Greek *atactos*, without order).

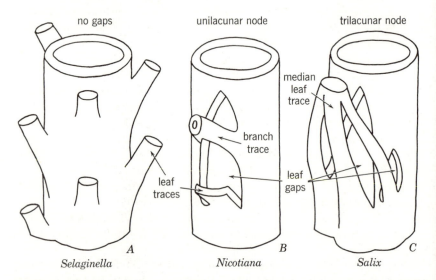

Fig. 15.9. Examples of steles. A, siphonostele without leaf gaps. B, C, siphonosteles with leaf gaps. B, connection of branch stele to main stele, with branch traces and single leaf trace having a common gap. A, based on Ogura, *Handb. d. Pflanzenanat.* 7(36), 1938.)

The stelar concept and the classification of stelar types were conceived and developed with regard to axes in the primary state of growth. The cambial secondary growth that occurs in the axes of gymnosperms and dicotyledons obscures the original structure of the stele. The interfascicular regions and later the leaf gaps are commonly discontinued as such because of the formation of a continuous cylinder of secondary vascular tissues, the primary phloem is removed from the primary xylem by this growth, and the eustelic condition ceases to be recognizable except in the pattern formed by the primary xylem remaining next to the pith.

The stelar theory, with its emphasis on the unity of the vascular system, appears to be in conflict with the previously reviewed concept that the vascular system of many plants, especially the higher, is essentially a system of leaf traces. If the leaf traces are assumed to be units of structure, then, obviously, the vascular system of the stem would be interpreted as a composite structure. But, with reference to the phylogeny and the ontogeny of the shoot, it is more appropriate to regard the leaf traces as subordinate parts of a larger unit, the vascular system of the plant. Then the different forms of steles could be looked upon as expressions of different degrees of relation between the leaves and the axis—or of the absence of such relation if the leaves are absent—in conformity with the concept of the lack of discontinuity between the leaf and the stem. One extreme is the protostelic form, which is least influenced by leaf development; the other is the eustele, in which the primary vascular system differentiates largely or entirely in relation to the leaves.

The essential unity of leaf and stem is well expressed in the simple concept of a connected system of vascular tissues, on the one hand, and of nonvascular tissues, on the other (Brebner, 1902). From the standpoint of descriptive anatomy and physiology, such a concept may be used successfully in the place of that of the stele, especially with reference to the seed plants. If it is necessary to refer to the vascular region, as distinguished from cortex and pith, the term *vascular cylinder* (Foster, 1949) may be employed.

DELIMITATION OF THE VASCULAR REGION

The three primary tissue systems forming the stem—the epidermal, the fundamental, and the vascular—are variously delimited from each other. Commonly the epidermis is clearly set off from the subjacent ground tissue. The demarcation between the ground tissue and the vascular is sometimes plainly indicated, but sometimes it is indefinite. This boundary is more certain in the axes of the lower vascular plants and in the roots of seed plants than in the stems of the latter. The development of the views on the morphologic nature of the limiting layers between the vascular and the fundamental systems has been profoundly affected by the stelar theory because a morphologic delimitation of the stele was considered an important evidence for the support of the stelar theory (Schoute, 1903).

Endodermis

According to the stelar concept, two limiting layers occur at the perivascular boundary: the pericycle, located outside the vascular tissues, and the endodermis surrounding the pericycle. In the original concept of the stele, the term endodermis was applied to the inner layer of the cortex. The seed plants, especially their roots, demonstrate a clear ontogenetic relation between the endodermis and the cortex, as far as the two can be traced distally into the meristematic region (chapters 5, 17). In the lower vascular plants the origin of the endodermis appears to be variable (Demalsy, 1958; Ogura, 1938); it may arise from the same meristem as the vascular tissue.

The morphologically specialized endodermis forms a single layer of compactly arranged cells, parenchymatic in appearance but with distinctive wall characteristics. The most outstanding of these is the band of wall material in the radial and transverse walls, which is chemically different from the rest of the wall (fig. 17.1A, pl. 37D). This band was first recognized as a wall structure by Caspary (1865–66) and is therefore known as the *casparian strip* or *band*. The casparian strip contains lignin and suberin. In older axes the endodermal cells may become modified by a deposition of a suberin lamella over the entire inner surface of the wall. Still later, a secondary layer of cellulose, sometimes lignified, may cover the suberin lamella, and finally the cellulose wall may become incrusted with oxidized products of phenolic and other substances (Van Fleet, 1961). The cellulose layer is often most massive on the inner tangential wall (fig. 17.3).

The endodermis is commonly clearly differentiated topographically and morphologically in the stems of the lower vascular plants and is found here with casparian strips and with the additional suberin lamella (Guttenberg, 1943; Ogura, 1938). It occurs in these stems around the periphery of the vascular cylinder, sometimes also between the pith and the vascular tissues. In some ferns it encloses individual vascular bundles. In seed plants the endodermis is best known in the roots, but there are a number of angiosperms, mostly herbaceous, the stems of which develop an endodermis with casparian strips (Carlquist, 1959b; Courtot and Baillaud, 1960; Guttenberg, 1943; Van Fleet, 1961).

Underground rhizomes have an anatomically identifiable endodermis more frequently than the aerial axes. Sometimes the endodermis develops casparian strips in a herbaceous stem when the plant attains the flowering state (Datta, 1945; Warden, 1935). The aerial axes of woody dicotyledons and gymnosperms typically are devoid of an endodermis with distinctive wall characteristics.

In young stems of angiosperms the endodermis often assumes the form of a *starch sheath*, a layer with more abundant starch deposition than the adjacent cortical cells. The starch sheath usually extends from a few millimeters to a few centimeters below the apical meristem (Fischer, 1900). It is more conspicuous during the summer than at other times of the year (Schoute, 1903). In older stem regions this layer ceases to accumulate starch differentially and either undergoes no further differentiation or, in some plants, develops casparian strips (Bond, 1931; Datta, 1945; Warden, 1935). The starch sheath sometimes appears as a continuous layer (pl. 46A); sometimes as interrupted arcs outside the individual vascular strands; and sometimes it is more than one cell layer in depth and has a diffuse outer boundary. No starch sheath commonly occurs in the stems of gymnosperms, although the innermost cortical layers may have somewhat more starch than the outer ones.

Studies on the histochemical aspects of endodermal differentiation suggest that the endodermis has no special morphologic significance but arises as a result of reaction between substances originating in the vascular system and in the cortex. They suggest, further, that the endodermis with specialized walls, the starch sheath, and the layer detectable only histochemically are different manifestations of the chemical reactions that occur at the perivascular boundary. The chemical systems characterizing this layer—various enzymes and substances acted upon by the enzymes—may be recognized while the layer is still meristematic (Van Fleet, 1961). As is typical of all differentiating tissues, the endodermis undergoes a continual change in its chemical structure and, depending on environmental conditions, assumes one or another of its forms. The influence of the environment on the morphologic and cytologic differentiation of the endodermis is illustrated by the development of an endodermis with casparian strips in the place of a starch sheath in stems of etiolated plants (Van Fleet, 1961). Studies on experimentally induced alterations in the configuration of the vascular system indicate, furthermore, that the differentiation of the endodermis is not restricted to one certain tissue region but appears at the boundary of the vascular system and the ground parenchyma regardless of the origin of cells at this boundary (Wardlaw, 1947).

The variability in the morphologic differentiation of the cells located at the boundary between the vascular and nonvascular regions and the consistent histochemical specificity of these cells require a broad definition of the endodermis. The selection of one or another of the manifestations of chemical reactions at the perivascular boundary for the definition does not give a proper conception of this boundary (Van Fleet, 1961). Broadly defined, the endodermis is a layer (sometimes layers) of cells located between the vascular region and the ground-tissue region and characterized by specific enzymic systems whose activities may result in morphologic differentiation of the cells. Thus defined, the term endodermis is applicable to layers that may or may not have casparian strips or any other specialized wall characteristic; and it is physiologically significant since it emphasizes the specificity of reactions at the perivascular boundary.

Pericycle

The pericycle was early defined as part of the fundamental tissue of the stele (Van Tieghem, 1882; Van Tieghem and Douliot, 1886). As has been stated previously in this chapter, the stems and roots of many lower vascular plants and the roots of higher vascular plants typically show an anatomically distinct endodermis and a layer or more of paren-chyma—the pericycle—between the vascular tissues and the endodermis. In the stems of gymnosperms and angiosperms, the delimitation of the vascular region is variable. In many stems there is no layer separating the cortex from the vascular tissues, for the protophloem differentiates next to the innermost cortical layer. The sieve elements of the protophloem are soon obliterated, and the remaining cells often differentiate into fibers (chapters 10, 12). In some stems of dicotyledons a continuous or nearly continuous cylinder of fibers occurs on the periphery of the vascular cylinder. The fibers may originate from the same meristem as the phloem (*Pelargonium*) or from tissue outside the phloem but inside the starch sheath (*Aristolochia, Cucurbita;* Blyth, 1958; Carothers, 1959; Mourré, 1958). Thus, in some stems nonphloic tissue occurs between the cortex and the phloem. This kind of tissue was used by Van Tieghem (1882) when he introduced the concept of pericycle, but later the concept came to be applied to all stems and roots. In many, probably most, seed plants the term pericycle refers to the outermost part of the phloem (Metcalfe and Chalk, 1950).

The recognition of the phloic origin of the tissue occurring at the periphery of the vascular system in the stems of many higher vascular plants dates back to the days when the concept of pericycle was being developed (Léger, 1897; see also Esau, 1943*b*, 1950), but in view of the popularity of the stelar theory the idea of the existence of a definite limiting layer of the stele seemed attractive and was accepted even by those who recognized the inconstancy in the origin and nature of the layer called pericycle (Brebner, 1902). This book emphasizes the evidence that the segregation of the different tissues in the plant body varies in distinctness; and that the presence or absence of an anatomic delimitation of cortex, endodermis, pericycle, medullary rays, leaf gaps, and pith constitutes a variation in the relative distribution of the vascular and ground tissues. On the one hand are the plant axes with an almost diagrammatic division into cortex, vascular cylinder, and pith (if present), and with distinctly differentiated endodermis and pericycle; on the other, are the axes having no sharp boundary between the vascular and fundamental tissues, and lacking a pericycle. In the extreme condition, the vascular system is dispersed to such an extent that no cortex or pith can be delimited (stems of many monocotyledons).

PRIMARY VASCULAR DIFFERENTIATION

As the procambium differentiates among the derivatives of the apical meristem, it assumes the outlines of the future vascular system that will develop from it. Thus, one may find a solid cylinder of procambium in some plants, a hollow cylinder in certain others, and a system of procambial strands in still others. The differentiation of the primary vascular tissues from the procambium follows various developmental patterns. The maturation of the first vascular elements in the procambial strand or cylinder may occur while the procambium is still actively dividing, or it may happen after most of the divisions have been completed and the procambium clearly shows the outline and the internal pattern of the future vascular system (Wetmore, 1943). The former relation is commonly found in the aerial parts of the seed plants, in which the separation between vascular and ground tissues is not precise. The relatively early delimitation of the procambial system, on the other hand, is characteristic of stems of many lower vascular plants and of most roots, that is, axes in which the different tissue systems are rather clearly delimited at maturity.

Transverse Course of Differentiation

To characterize the course of vascular differentiation as seen in transverse sections of an axis, the position of the successively appearing elements is referred to the center of the axis or, in some vascular systems, to the center of individual vascular strands. The xylem shows three fundamental patterns of differentiation. In the first, the initial mature xylem elements are located farthest from the center of the axis. In other words, if the differentiation is visualized in time, the progress of maturation of the xylem elements occurs in a centripetal order (pl. 86, chapter 17). Such xylem is called *exarch* (in Greek, beginning outside). In the second, the initial xylem elements occur nearest, and the latest occur farthest, from the center of the axis (pls. 56; 57A, B; figs. 15.15, 15.16); that is, the differentiation is centrifugal, and the xylem is called *endarch* (in Greek, beginning within). In the third, the differentiation progresses in two or more directions from the first mature xylem elements (Foster and Gifford, 1959). The resulting primary xylem is called *mesarch* (in Greek, beginning in the middle). The exarch and mesarch types of primary xylem appear to be more primitive than the endarch and are commonly associated with procambial systems that are completely demarcated before vascular differentiation occurs.

The phloem associated with all three types of xylem shows a centripetal direction of differentiation, unless it is located inside the xylem, as in stems with internal phloem. Such phloem differentiates centrifugally. The terms exarch and endarch are not applied to the phloem, probably because they were coined with reference to the xylem before the development of the phloem was properly understood.

Chapters 11 and 12 consider the classification of primary xylem and phloem into protoxylem and metaxylem, protophloem and metaphloem, respectively. The tissues distinguished by the prefix proto are the first to differentiate and are followed by the metaxylem and the metaphloem. If the xylem is exarch, the protoxylem appears on the outer margin of the xylem system or strand, the metaxylem at or near the center. In the endarch xylem the relative positions of the two parts of the xylem are reversed. In the mesarch xylem the protoxylem is flanked on two sides or is surrounded by metaxylem. The protophloem usually appears farthest from the xylem, the metaphloem nearest to it (fig. 15.7). As was previously stressed (chapters 11, 12), the protophloem and the protoxylem mature early and become more or less modified in structure before the primary body completes its development. These changes often make it difficult to recognize the position of the first vascular elements, particularly those of the protophloem, in the fully developed primary vascular system.

The beginning of delimitation of the procambium below the shoot apices of seed plants is recognized by the differential staining—probably mainly a result of differential vacuolation—among the derivatives of the apical meristem and by the distinctive patterns of growth. The ground meristem cells early show increased vacuolation, whereas the procambial cells longer remain densely cytoplasmic (pl. 56A). The procambial cells undergo repeated longitudinal divisions but expand transversely to a limited degree (fig. 15.10). Thus, eventually the procambium becomes distinguishable by its dense narrow cells, elongated parallel with the longitudinal axis of the organ (fig. 15.11). In older parts of the axis the procambial cells become more vacuolated but retain their elongated shape and rather short transverse diameters.

The longitudinal divisions in the differentiating procambium may occur in various planes or may early become oriented in the tangential plane. Because of this difference in growth, the procambial cells may show, in transections of stems, either a random arrangement or a radial seriation resembling that in the cambial zone. The occurrence of radial alignment of cells in the procambium has led to many erroneous assumptions concerning occurrence and time of inception of secondary growth in various groups of plants (see Esau, 1943b). The primary xylem differentiating from procambial cells sometimes retains the arrangement in radial files exhibited by the meristem (fig. 15.15A; Esau, 1942). Sometimes subsequent divisions and changes in cell shape in the differentiating tissue obscure the initial radial seriation (pl. 64A; Esau, 1945). In the primary phloem, radial seriation is less commonly found than in the primary xylem.

The part of the procambium that eventually gives rise to the phloem

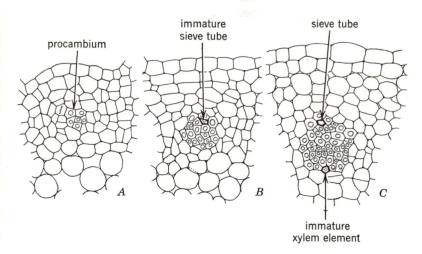

Fig. 15.10. Successive stages in development of procambium (cells with nuclei) in transections of *Linum perenne* stem. First phloem and xylem elements begin to differentiate before procambial strand completes increase in diameter. This increase occurs by divisions within strand and by addition of cells from adjacent ground meristem. (All, ×430. Esau, *Amer. Jour. Bot.* 29, 1942.)

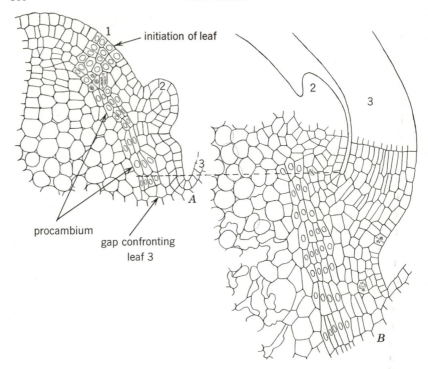

Fig. 15.11. Longitudinal sections through shoot tip of *Linum perenne.* Early stage in differentiation of leaf-trace procambium pertaining to leaf primordium 1, just initiated by periclinal divisions in second layer of tunica. Nuclei indicate cells pertaining to leaf primordium and procambium. *A, B,* from two sections of same shoot, 14 microns apart. Leaves 2 and 3 were numbered arbitrarily, not in ontogenetic sequence. *A,* procambial strand deviates from vertical path near gap associated with leaf 3. Lower end of this strand appears in *B.* Broken line connecting *A* and *B* shows level at which procambial strand changed from one section to the other. Discontinuities of procambium at lower end in *A* and at upper end in *B* are thus only apparent. Strand is continuous throughout and shows an intensification of procambial characteristics in downward direction. (Both, ×365. Esau, *Amer. Jour. Bot.* 29, 1942.)

is frequently distinct in its morphology from that forming the xylem. It may show denser staining and different planes of division than the xylary part. The terms *phloic procambium* and *xylary procambium* may be used to stress the early differentiation of the meristem into the two parts. The occurrence of this kind of differentiation clearly indicates that the procambium is intrinsically a vascular tissue in early stages of differentiation. In this sense, it is a meristematic tissue, a provascular tissue, although in plants with secondary growth a part of it maintains meristematic characteristics and changes into the vascular cambium.

Longitudinal Course of Differentiation

Procambium. Although in a general way it is possible to picture the procambium as forming a system similar to that later presented by the mature primary vascular system, the developmental relation between the meristem and the final product is highly complex. The differentiation of the vascular elements occurs simultaneously in more than one direction, both transversely and longitudinally, and the various steps in tissue formation overlap at the same levels of the axis. Each developmental stage—the formation of procambium, the differentiation of phloem, and the differentiation of xylem—presents special aspects (Esau, 1943b; Philipson, 1949; Sifton, 1944; Wetmore, 1943).

The differentiation of procambium has been given most attention in plants in which the primary vascular system can be described as a system of leaf traces. In these plants the vascularization at the shoot apex is closely associated with the development of leaves. In fact, initiation of leaves and initiation of the vascular tissue connected with these leaves commonly appear as parts of the same growth process. As was mentioned, the delimitation of the future vascular tissue becomes evident when the ground-meristem cells stain lighter than the prospective vascular region (pl. 52). This differentiation in the ground tissue characterized by an increasing vacuolation and enlargement of cells is so closely correlated in the stem and the developing leaf primordium that there is, from the beginning, a complete unity of the prospective vascular system of stem and leaf as it is demarcated by the changes in the ground tissue (pl. 52, 53).

Such initial unity has been described in shoot development in many pteropsid plants of different degrees of specialization (reviews by Esau, 1943b, 1954; Gustin and De Sloover, 1955). However, the interpretation of the meristematic tissue that constitutes the precursor of the vascular region is not settled. The first question is whether this tissue consists wholly or in part of procambium or whether it is a precursor of the procambium. The best-supported view is that a part of this tissue is procambial and the rest is less determined meristematic tissue. Some of the latter subsequently changes into additional procambium, and the remainder differentiates as parenchyma of the interfascicular regions and of the leaf gaps. In the seed plants the initial procambium at a given level is that of the leaf traces of the nearest leaves above. The procambium differentiating later pertains to leaves arising at higher levels of the shoot.

The second question—and it is related to the first—is whether the less determined part of the potential vascular system is a meristematic tissue that has been delayed in its differentiation, a *residual meristem,* or whether it also is partly differentiated vascular tissue. The tissue is continuous with the eumeristematic peripheral zone of the apical meristem where the leaf primordia originate. A possible cytologic or histochemical differentiation between the two tissues has not been established. Residual meristem, therefore, is the least objectionable descriptive term for the precursor of the vascular region. Moreover, the term is applicable not only to the future vascular region but also to other tissue regions that show a less pronounced degree of vacuolation than the other tissues at the same level of the shoot (pls. 52, 53).

The prospective vascular system of the shoot, as initially blocked out by the differentiation phenomena in the pith and the cortex of the stem and in the abaxial and adaxial parts of the leaf primordium, may be pictured as a cylinder of tissue with prolongations into the leaf primordia. The procambial strands constitute part of this system. As was described before, the differentiation of the procambium results from a special character of division and elongation of cells. Students of initial vascularization are concerned with the question whether the divisions initiating the procambium progress from the leaf primordium downward toward a connection with the more mature part of the vascular system in the axis, or from the axis upward toward the leaf primordium; that is, whether the procambium differentiates basipetally (in Greek, seeking the base) or acropetally (in Greek, seeking the apex) within the youngest part of the shoot. In differentiating basipetally the procambium of a leaf trace would be initially discontinuous. The acropetal differentiation could be continuous or discontinuous.

The determination of the longitudinal course of procambial differentiation is technically difficult. The change of the derivatives of the apical meristem into procambial cells is gradual, and, therefore, observers vary in their interpretation of when procambium is actually present. Procambial cells are easily missed in sections in which they are cut on the bias and in which they deviate from the vertical path in relation to leaf gaps or other interfascicular regions. If a leaf gap occurs below a trace bundle, and if the section passes through both the trace and the gap, the trace appears as though it were interrupted at the lower end (fig. 15.11A). Adjacent sections are needed to reveal its connection with the older traces below (fig. 15.11B). The study of procambial differentiation must be based on complete information on phyllotaxis, nodal anatomy, and trace connections in a given plant, and it must employ serial transverse and longitudinal sections. Furthermore, since the procambium is initiated close to the apical meristem, the activity of this meristem and the phenomena involved in leaf formation should be correlated with vascularization.

Among the numerous investigations dealing with the longitudinal course of procambial differentiation, relatively few are complete enough to be reliable. These few studies indicate certain important variations in procambial development (Esau, 1943b, 1954; Gustin and De Sloover, 1955). Several conifers and dicotyledons with the vascular tissues organized into leaf-trace systems had procambium differentiating acropetally and continuously from existing vascular tissue in the stem toward the apex and in most the procambium was identifiable beneath the youngest leaf primordia (Esau, 1942; Lawalrée, 1948; McGahan, 1955). In some species, the procambium of one or more traces was detected in the axis before the pertinent primordium was initiated at the apex (De Sloover, 1958; Gunckel and Wetmore, 1946a; Sterling, 1945, 1947). In dormant buds of *Abies*, on the other hand, several primordia were devoid of trace procambium (Parke, 1963).

The study of procambial differentiation in monocotyledons is especially difficult because of the numerous traces to a leaf and the complex course of the bundles in the stem. Several studies on Gramineae (*Oryza, Zea*) indicate that one or more of the leaf traces, usually the larger and earlier, differentiate acropetally, whereas the smaller traces differentiate from the node downward in the axis and upward in the leaf itself; furthermore, a given strand may have more than one locus of initiation (Inosaka, 1962; Kumazawa, 1961; Maeda, 1962). Possibly other monocotyledons and some dicotyledons with similarly complex vascular systems have some basipetally differentiating leaf traces, but for palms an entirely acropetal differentiation has been described (Tomlinson, 1961).

The establishment of connection between the axillary or the adventitious bud and the main axis has received some attention (De Sloover, 1958; Fukumoto, 1960; Gulline, 1960; references in Esau, 1954). The procambium of the prophyll traces connecting the axillary bud with the main axis may be identifiable as early as the bud itself and be continuous with the procambium of the main axis from the beginning. On the other hand, parenchyma may differentiate between the bud and the vascular cylinder of the axis. Then the procambium usually differentiates from the bud toward the axis, that is, basipetally. Adventitious buds commonly establish their vascular connection with the parent structure (stem, leaf, or root) by basipetal differentiation of procambium. If the adventitious buds arise on an older stem with secondary growth, they become directly connected with the secondary vascular tissues without the formation of bud traces in the axis (Dermen, 1959).

The course of procambial initiation in the shoot apex is of considerable interest in connection with the search for causes underlying the establishment of phyllotactic patterns in plants. Many hypotheses have been formulated to explain the occurrence of phyllotaxes and the mechanics of leaf formation at the apex (Cutter, 1959; Esau, 1954; Snow, 1955; Wardlaw, 1952). Some of these seek the causes of the specific leaf arrangements in the apex itself (chapter 5); others suggest that the procambium, in developing acropetally, plays a role in determining the leaf patterns at the apex. Surgical treatments of shoot apices resulting in the development of new shoot apices with procambium initially discontinuous from that of the older shoot parts graphically demonstrate the ability of shoots to organize their vascular system. These results parallel the events commonly observed in adventitious shoots produced with or without experimental stimulation. Studies on tissue cultures, on the other hand, indicate that a growing parenchyma tissue is capable of initiating the differentiation of vascular tissue without an apical meristem, but that such tissue becomes organized into a system characteristic of shoots and roots only after the corresponding apical meristems develop (Gautheret, 1959; Steward et al., 1958).

Doubtless, both of the two opposite views, one, that the apex determines the position of the leaves and thereby also that of their vascular traces and, the other, that the vascular system of the mature shoot parts determines the position of the leaves through the acropetally developing leaf traces, oversimplify the relationships in the growing plant. It seems more likely that the phyllotaxis and the vascular organization are determined by a common mechanism concerned, first, with the establishment of polarity in the embryo, or the adventitious shoot, or the plant arising in a tissue culture, and, second, with the subsequent timing and synchronization of the various processes occurring in the growing plant (Dormer, 1955b; Philipson, 1949; Richards, 1948).

Xylem and Phloem. The establishment of procambium is followed by the differentiation of certain of its cells into phloem and xylem elements. The longitudinal course of this differentiation has been studied most comprehensively in conifers and dicotyledons (De Sloover, 1958; Esau, 1943b, 1954; Girolami, 1953; Gustin and De Sloover, 1955; McGahan, 1955). The information on vascular differentiation in monocotyledons and taxa below those of the seed plants is meager (Esau, 1954). The first phloem elements (sieve tubes in the angiosperms, sieve cells or related elements in the gymnosperms) usually differentiate acropetally along the outer periphery of the procambium from their connection with the phloem of older leaf traces into the leaf primordium. In its initial stages this differentiation may be continuous or discontinuous. The development of the phloem elements begins before there is any xylem in the trace. Therefore, if the procambium is studied in transverse sections, the first phloem elements may be found in it before the first xylem elements (figs. 15.10, 15.12, pl. 56).

Xylem differentiation in seed plants is initiated on the inner side of the trace procambium, commonly near the base of the leaf or at its node, and from there it progresses acropetally into the leaf and basipetally

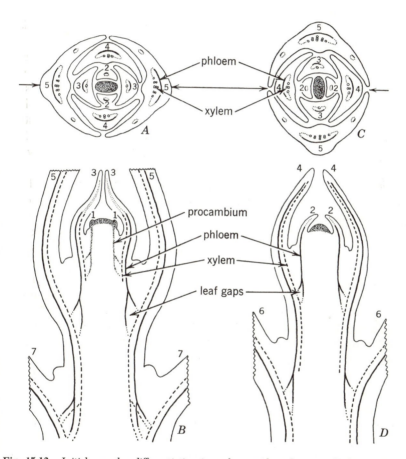

Fig. 15.12. Initial vascular differentiation in a shoot with a decussate leaf arrangement. Longitudinal (B, D) and transverse (A, C) sections. Leaves are numbered in pairs. The two longitudinal views represent same shoot seen in two median planes oriented at right angles to each other. Planes of sectioning of B, D, are indicated by means of arrows in A, C. Dense stippling, shoot apex and youngest leaf primordia. Details: leaf pair 1, procambium only; pair 2, some mature phloem, continuous with phloem of older parts of stem; pair 3, some phloem and xylem, latter in discontinuous strands; pair 4, some phloem and xylem, latter in discontinuous strands; pair 5, phloem and xylem, latter connected with older xylem. (Actual connection of young xylem and phloem with older tissues below is not shown.) Transections show lateral spread of vascular differentiation in vascular bundles.

into the stem. Within the stem the new xylem unites with that of the older traces or with the secondary xylem if cambial activity occurs in the lower parts of the stem before the trace xylem reaches those levels (O'Neill, 1961). Several vertical files of tracheary elements may be initiated successively in the isolated locus before the first file becomes connected with the xylem below. In other words, isolated xylem bundles with mature elements may be present in the upper levels of the shoot. The number of leaves with the isolated xylem is variable in different species and may change in the same plant during development (Esau, 1954; O'Neill, 1961). In the young leaf primordium itself, a rather extensive xylem system may develop before its prolongation in the axis unites it with the system below (Esau, 1945). The establishment of connection between the isolated xylem and the mature xylem in the axis appears to speed up the acropetal course of xylem differentiation in the leaf (Jacobs and Morrow, 1957).

Variations have been observed in the pattern of vascularization just described. In addition to the basipetally differentiating xylem from the leaf base, some xylem of the same trace may be differentiating acropetally within the stem (De Sloover, 1958; Esau, 1943b, 1954), and the first xylem of a trace may be initiated in the stem in an isolated locus and appear here before its appearance at the leaf base (Jacobs and Morrow, 1957).

Several files of tracheary elements, possibly all of protoxylem, may have the same course of differentiation (De Sloover, 1958; Esau, 1943a; Jacobs and Morrow, 1957). Bidirectional differentiation has been described for the metaxylem of some plants (De Sloover, 1958).

With regard to the procambium of the axillary bud traces, the few available studies indicate an acropetal differentiation of phloem and an initiation of xylem differentiation at the bases of prophylls.

As was mentioned before, the trace interconnections in a shoot are related to the phyllotaxis of that shoot. Similarly, the timing in the development of phloem and xylem occurs according to the phyllotactic pattern (fig. 15.13). The timing of the various events of vascularization is related to the length of leaf primordia and this length is proportional in leaves of different ages. Thus, if no major shifts occur in the correlations, it is possible, by measuring an older leaf, to determine the size of a younger leaf and its stage of vascularization (Jacobs and Morrow, 1957, 1958). The quantitative relation between leaf size and vascular differentiation may shift during the transition from vegetative to reproductive state: there may be a speeding up of vascularization in that leaves containing the first mature phloem and xylem elements are smaller than during the vegetative stage (Jacobs and Raghavan, 1962).

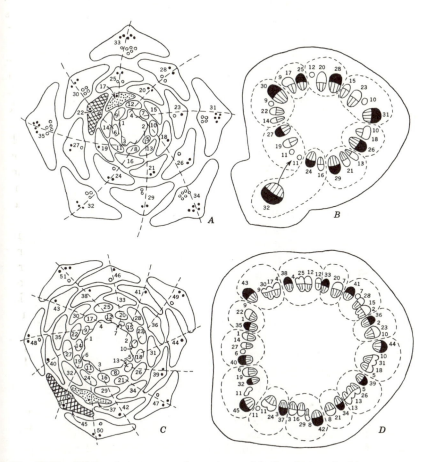

Fig. 15.13. Relation between vascular systems of leaf and stem in *Linum perenne*. Transections. Shoot tips with the uppermost leaves (*A*, *C*) and stems (*B*, *D*). Section in *B*, 5.3 mm below *A*; *D*, 8.8 mm below *C*. Curved lines in *A*, *C*, parastichies of leaves whose traces are most directly connected with each other. Broken lines in *B*, *D*, parts of vascular system each composed of traces pertaining to one parastichy in *A*, *C*. Numbers assigned to leaves and their traces follow age sequence of leaves, beginning with youngest. The two shoots show different leaf arrangements. *C*, *D*, as contrasted with *A*, *B*, show (1) closer sequence of leaves, (2) larger number of leaves without mature vascular elements (youngest leaves with mature sieve tubes are stippled, and those with mature sieve tubes and xylem elements are crosshatched), (3) thicker stem, and (4) larger number of vascular bundles in transection of stem. Details: dots, sieve tubes; circles, tracheary elements; bundles with blackened phloem, leaf traces; bundles with phloem left blank, composites of leaf traces. (*A*, *C*, *D*, ×66; *B*, ×79. After Esau, *Amer. Jour. Bot.* 30, 1943.)

Causal Relations in Vascularization

The possible role of the apical meristem in inducing and synchronizing the events of vascularization has been briefly discussed above. Experimental work has revealed some of the more direct factors involved in vascular differentiation. Two variables have been singled out in particular, auxins and sugar. The effect of auxin on xylem differentiation has been effectively demonstrated by studies of regeneration of severed xylem in an internode of *Coleus* (Jacobs, 1954). This regeneration occurred through the pith tissue in a basipetal direction and could be inhibited by removal of the leaf and bud—the sources of auxin—above the wound. If the stump of that leaf was treated with lanolin containing auxin, regeneration proceeded normally. An indirect limiting factor was found to be the ability of the internode to transport the available auxin. The relation of auxin to xylem differentiation can be used to explain the relation between leaf size and the degree of xylem differentiation in the normal development of a shoot (Jacobs and Morrow, 1957). Tissue-culture studies also demonstrate the requirement for auxin in xylem differentiation. This auxin can be provided either by grafting a shoot into the callus or by placing it in agar into an incision in the callus (Wetmore and Sorokin, 1955). Such treatment induces differentiation of nodules and strands of xylem in an originally homogeneous callus and they are positioned in relation to the graft or to the insert with auxin.

Application of sugar and auxin in agar to the surface of callus revealed the importance of sugar for the differentiation of phloem (Wetmore and Rier, 1963). Varying concentrations of sugar alter the proportions of xylem to phloem: low concentrations are favorable for xylem differentiation, high concentrations for phloem differentiation. Middle concentrations—probably those prevailing in normally growing plants—induce differentiation of both tissues, usually with a cambium between them, in the dicotyledon material used.

Vascularization and Growth of the Axis

Primary Growth of the Axis. As was explained earlier, the increments of the shoot produced by the apical meristem in connection with the initiation of leaf primordia become articulated into nodes and internodes chiefly by the growth of the internodes. The internodal elongation is a typical example of intercalary growth and varies not only in degree but also in timing and in distribution within the internode. The variation in the extent of this growth determines the differentiation into short shoots and long shoots; and typically the lower internodes of the first axis of the plant or of a branch are shorter than the subsequent ones.

The growth of an internode, including both cell division—often of the rib-meristem type—and cell enlargement, may progress acropetally (*Helianthus, Syringa*) or basipetally (Gramineae, Liliaceae, *Equisetum*). The elongation of successive internodes may occur step by step (*Helianthus*) or may overlap (Gramineae, *Syringa*). In some internodes the chief element of elongation is cell enlargement, in others cell division. Auxin and other growth-regulating substances are known to be involved in internodal elongation (Sachs et al., 1960; Wetmore and Garrison, 1961).

The primary increase in diameter of the axis also occurs through cell division and cell enlargement. In various degrees it is characteristic of seed plants and more primitive taxa (Rauh and Falk, 1959; Troll and Rauh, 1950; Wetter and Wetter, 1954). In dicotyledons and gymnosperms this growth may be rather diffuse, or more or less restricted to the pith or cortex. In many monocotyledons cell division is largely localized in a peripheral mantle-like zone, the *primary thickening meristem*. This meristem resembles a cambium in that it forms cells in radial series (fig. 15.14, pl. 58B; Eckardt, 1941). If there is intense thickening directly beneath the apical meristem, the leaf insertions are raised to the level of the apex or above it (Rauh and Rappert, 1954). If this growth is located mainly in the pith, the procambial strands assume a strongly curved or even a horizontal position at the uppermost levels of the shoot (Weber, 1956). Usually, there is an acceleration in the growth in width combined with an increase in the size of the apical meristem. Thus, the plant assumes an obconical form if in the meantime secondary growth does not cause an additional increase in the thickness of the axis below (chapter 1; Troll and Rauh, 1950).

The use of the term primary with reference to the growth phenomena just described needs some discussion in addition to that given in chapter 4. The classification into primary growth, that is, growth that occurs among the more or less direct derivatives of the apical meristem, and secondary growth, that is, growth resulting from the activity of the vascular cambium, is not sufficiently comprehensive and is not uniformly treated in the literature. The type of growth responsible for the initial widening of the shoot may not be restricted to the upper levels of the axis. Fleshy plant parts or monocotyledonous stems, for example, may undergo a similar type of growth at some distance from the apical meristem. This later parenchymatic thickening growth is called by some authors secondary growth (Hagemann, 1959; Troll and Rauh, 1950) or diffuse secondary growth (Tomlinson, 1961). It intergrades with the so-called anomalous secondary growth observed in some fleshy plant structures (Orsós, 1941). One should refer here also to the dilata-

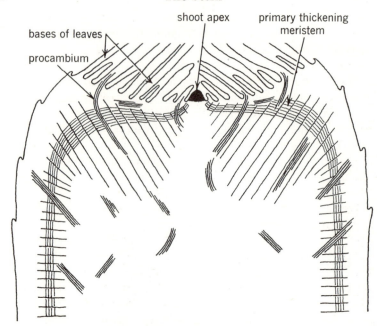

Fig. 15.14. Upper part of shoot of a monocotyledon illustrating meristems concerned with its growth. Apical meristem produces axial tissue downward and leaf primordia laterally. Beneath the primordia, derivatives of apical meristem divide periclinally and form anticlinal rows (indicated by widely spaced parallel lines). Increase in axis thickness results. Periclinal divisions may be localized in a mantle-like tissue region, the primary thickening meristem. This meristem may be prolonged in the peripheral part of the axis and may be continuous with cambium producing secondary tissues. Primary thickening meristem forms ground parenchyma and procambial strands. (Based on Eckardt, *Bot. Arch.* 42, 1941.)

tion growth in the bark (chapters 12, 14), which is far removed from an apical meristem and can hardly be thought of as primary growth.

Thus, to be useful, the terms primary and secondary with regard to growth (and the resulting tissues) must be conceived broadly, with the time element as the chief criterion. On this basis, some secondary growth results from the activity of spatially restricted meristems (the cambia), and some comes about by cell division and cell enlargement dispersed in parenchyma. Thus, we can classify secondary growth into cambial and diffuse (chapter 4). Descriptive terms, such as elongation growth, thickening growth, and dilatation growth, frequently suffice for designating the phenomena concerned.

Growth of the Primary Vascular System. The complexities of development and of mature structure of the primary vascular system of the shoot result, in part, from the circumstance that this system is initiated before the shoot undergoes its primary growth in width and length. The vascular system, delimited at the apex in its meristematic state, expands and elongates with the axis, and this growth overlaps with the differentiation and maturation of procambial cells into vascular elements. In plants with prominent leaf traces (ferns and seed plants) there is the added complication that the vascular system is initiated, not uniformly within a given level of the axis but in relation to the leaves, and therefore some parts of it develop conspicuously in advance of others.

The cell divisions occurring during the primary thickening of stems of seed plants are not immediately distinguishable from those that bring about the differentiation of the procambium, a feature that makes the recognition of the procambium in its early stages frequently quite uncertain (fig. 15.10). During the expansion of the vascular system the first procambial bundles become farther removed from each other and new ones differentiate among them from the residual meristem (fig. 15.15A, B). The successive origin of the vascular bundles at a given level of the stem and the varied relations of the bundles to each other (some are leaf or branch traces, others are complexes of traces) cause the commonly observed variation in size and structure of the parts of the vascular system in a given transection of a stem of a higher vascular plant (figs. 15.13, 15.15, 15.16). Some strands are large, others small, and the composition of their vascular tissues varies conspicuously. Since the strands in a given internode are initiated at various times, some are affected more, others less, by the elongation of the internode. The early-formed bundles develop more of the kind of xylem that has extensible types of secondary walls (annular and helical) and show more destruction of xylem than those that arise later. Moreover, a single strand shows structural differences at different levels. Because of the characteristically downward differentiation of the xylem in the traces, the amount of stretching and destruction of the first xylem, as well as the number of elements with extensible secondary walls, is greater at the higher levels. The phloem, too, shows more obliteration in older bundles of a given internode.

In chapters 11 and 12 the parts of the primary phloem and xylem which differentiate first were termed protophloem and protoxylem. These parts of the primary vascular tissues may now be defined more exactly with regard to the shoot of a higher vascular plant: they are the first vascular elements of a system of strands and not of each individual bundle. In a given internode, for example, protoxylem occurs only in the older, larger bundles. When younger bundles appear among the older, the latter are forming metaxylem; that is, the internode is at the stage of metaxylem differentiation, so the first xylem of the younger bundles is also metaxylem. Since the phloem differentiates before the xylem and acropetally, more bundles have protophloem than protoxylem. If a given leaf trace is followed through its entire length, it is usually found to have more protoxylem at the base of the leaf where xylary differentiation began, than farther down; and if the trace is long in terms of internodes traversed, it may have no protoxylem and even no metaxylem at its lower end. The primary xylem of such a trace usually becomes continuous with the secondary xylem in a dicotyledon but apparently may end blindly in palms (Tomlinson, 1961) and probably other monocotyledons.

The distinction between bundles that differentiate early and those that differentiate late during the development of a given internode may be illustrated most conveniently by a monocotyledon having no secondary growth. In *Zea*, for example, the bundles near the center of the axis have the full complement of protoxylem and protophloem. In mature internodes the protoxylem of such bundles contains a lacuna formed in connection with the destruction of the tracheary elements during the extension of the axis, and the protophloem is entirely crushed (pl. 57*B*). (Some investigators think that the lacuna is not present in nonprocessed, living material; Ricardi and Torres, 1956.) Bundles located closer to the stem periphery have smaller protoxylem lacunae and a smaller amount of crushed protophloem. The outermost and smallest bundles have only metaphloem and metaxylem tissues and show no evidence of destruction of vascular elements.

The relation of the structure of the vascular system to the elongation of the axis is of particular interest in plants with prolonged intercalary growth in the internodes (many monocotyledons). Vascular connections are early established through the intercalary meristem and all elements of the xylem have extensible types of secondary walls. While growth occurs in this meristem, the first mature vascular elements are destroyed, but others differentiate in the meantime. The question has been raised in the literature whether the formation of new elements through the intercalary meristem keeps pace with the destruction. In some plants, intact xylem elements are constantly present—at least one file in a bundle—in the active intercalary meristem (Golub and Wetmore, 1948; Stafford, 1948). In others, no intact elements appear to be present after the first ones are destroyed at the beginning of the intercalary growth and until new elements differentiate at the end of this growth (Buchholz, 1920). The protoxylem lacunae are suggested as the possible conduits of water during the intercalary elongation. Phloem differentiation has not been investigated with special reference to the intercalary meristems.

The primary vascular system shows, at various levels of the same plant, certain structural differences which are related to the changes in the thickness of the axis from lower to higher levels. The thickening of the axis is commonly accompanied by an increase in the number of strands visible in a transection of the stem. In plants whose vascular system appears as a system of leaf traces, an increase in the number of bundles may be brought about by an increase in the number of traces to a leaf, or an extension of the traces through a larger number of internodes, or both. The phyllotaxis may change concomitantly.

SECONDARY GROWTH OF THE VASCULAR SYSTEM

The increase in the amount of vascular tissues by means of secondary growth from a vascular cambium is characteristic of dicotyledons and gymnosperms. A few monocotyledons also enlarge their vascular system, by a special form of secondary growth. Among the lower vascular plants, secondary growth appears to have been rather common in extinct forms but is rare in living representatives (Eames, 1936). Examples of living vascular cryptogams interpreted as having a vascular cambium are *Isoetes* (Lycopsida) and *Botrychium* (eusporangiate fern).

Origin of the Vascular Cambium

If all the procambial cells differentiate into primary vascular tissues, no cambium is formed (pls. 56, 57A, B). If some procambium remains in a meristematic state, after the completion of primary growth, it becomes the cambium of the secondary body (pl. 64). This cambium is called *fascicular*, since it originates within the bundles or larger segments of the primary vascular system. Commonly the bands of fascicular cambium become interconnected by additional bands of meristem, the *interfascicular* cambium, which originates from the interfascicular parenchyma (pl. 64C, D). The completely formed cambium of the stem has the shape of a hollow cylinder extending through the nodes and internodes. If the axis is branched, the cambium of the main axis is continuous with that of the branches, and it may extend some distance into the leaves.

The procambium and the cambium may be looked upon as two developmental stages of the same meristem; they intergrade with regard to their morphologic and physiologic characteristics. The typical features of the cambium of arborescent dicotyledons and gymnosperms —the segregation of its initials into fusiform and ray initials, the occurrence of apical intrusive growth, the precise method of division in a tangential plane during the formation of xylem and phloem (chapter 6)

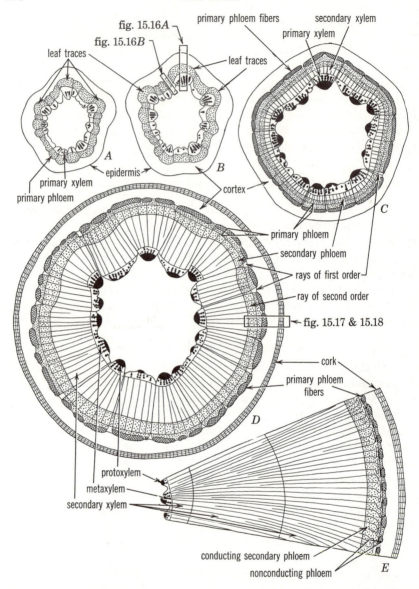

Fig. 15.15. Primary (*A, B*) and secondary (*C–E*) structure of *Prunus* stem in transverse sections. *A–D*, developmental stages beginning with differentiation of primary vascular tissues and ending with first secondary increment of xylem and phloem. *E*, segment of stem with three secondary increments. The largest vascular bundles in *A, B*, are leaf traces (3 to one leaf); the others are composites of leaf traces. Narrow interfasicular regions between bundles, black lines. *C–E*, protoxylem, solid black; primary phloem fibers, crosshatched.

—are acquired gradually, and some of these characteristics appear before the primary growth is completed, that is, while the meristem is still in the stage of procambium. For instance, the procambial cells gradually become as vacuolate as the cambium, and in many plants the primary vascular tissues, or at least the xylem, are formed by repeated tangential divisions. So far, the only definite character separating primary and secondary growth has been recorded in the xylem of arborescent species. As was discussed in chapter 11, the first secondary tracheary elements in such species are shorter than the last primary elements of the same kind.

The origin of the interfascicular cambium in the more or less vacuolated interfascicular parenchyma results from a resumption of meristematic activity by a potentially meristematic tissue. Usually, no cytologic changes are noticeable in connection with this return to meristematic activity (pl. 64C, D). If the interfascicular regions are relatively wide, the first divisions initiating a cambium in these regions occur next to the bundles in continuity with the fascicular cambium.

Common Form of Secondary Growth

In the stems of gymnosperms and dicotyledons most commonly studied for secondary growth, the cambium arises in the form of a cylinder between the primary xylem and phloem and remains in the same relative position indefinitely, producing secondary xylem toward the inside of the axis, secondary phloem toward the outside (figs. 15.15–15.18, pl. 65). The details of origin and activity are somewhat varied. The following three patterns may be encountered. (1) The primary vascular tissues form an almost continuous vascular cylinder in the internodes (the interfascicular regions are very narrow), and the secondary vascular tissues do the same (pl. 62; *Tilia, Nicotiana, Veronica, Syringa*). (2) The primary vascular tissues form a system of strands, but the secondary vascular tissues arise as a continuous cylinder (fig. 15.15, pls. 60, 61; conifers, *Sambucus, Salix, Prunus*, and many other herbaceous and woody dicotyledons). (3) The primary vascular tissues form a system of strands, the interfascicular cambium produces only ray parenchyma, and, therefore, the secondary vascular tissues also appear as strands (pl. 55; vine types of stems, as *Aristolochia* and *Vitis*).

Pith and primary vascular system expand in width while primary vascular tissues are differentiating and at beginning of secondary growth (A–C). Rays of first order originate in interfasicular regions; of second order, in vascular bundles. Epidermis replaced by cork in D, E. C–D, strands of primary phloem fibers are pulled apart into smaller strands; parenchyma fills resulting spaces. (All, ×22.)

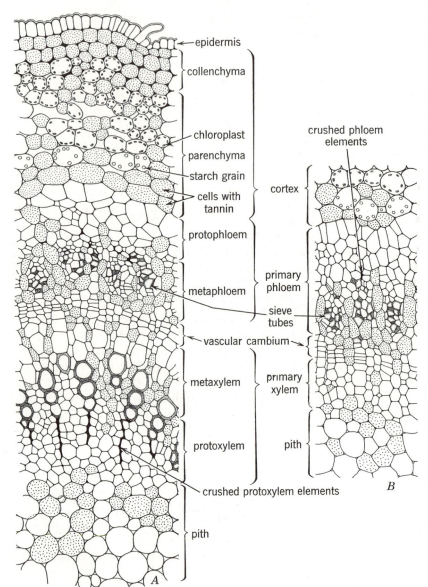

Fig. 15.16. Details of *Prunus* stem structure from fig. 15.15*B*. End of primary growth before maturation of last metaphloem and metaxylem cells but after initiation of first cambial divisions. In *A*, leaf-trace region with full complement of primary vascular tissues. In *B*, protophloem and metaphloem and small amount of immature metaxylem. Tannin-containing cells are stippled. Large cells in outer primary phloem are immature fibers. (Both, ×350.)

Further minor deviations of quantitative nature occur in relation to the phylogenetic reduction of secondary activity. In some herbaceous dicotyledons with secondary growth, the interfascicular cambium may produce only fibers or only sclerified parenchyma on the xylem side (*Medicago, Salvia*), or the secondary growth may be so small in amount that it remains limited to the vascular bundles (pl. 63C; *Trifolium, Cucurbita*).

Anomalous Secondary Growth

The secondary growth of some dicotyledons and gymnosperms deviates considerably from the form of growth just described. The deviating methods of secondary thickening are called atypical or anomalous, although the typical and atypical forms of growth are not sharply separated from one another. Moreover, the anomalous type of growth may be more common than is known at present: the tropical flora in which it is frequently found (De Bary, 1884; Obaton, 1960; Pfeiffer, 1926) has not been adequately studied anatomically. The developmental details of anomalous secondary growth vary considerably. In some plants the cambium occurs in normal position, but the resulting tissue has an unusual relative distribution of xylem and phloem. Some of the Bignoniaceae have an uneven growth of xylem and phloem so that the xylem becomes lobed, the lobes alternating with bands of phloem. In such genera as *Strychnos* (Loganiaceae), *Leptadenia* (Asclepiadaceae, fig. 15.19A), and *Thunbergia* (Acanthaceae; Mullenders, 1947) strands of phloem are included in the xylem (included phloem). In other plants, part of the cambium originates in an abnormal position. For example, in Chenopodiaceae, Amaranthaceae, Nyctaginaceae, Menispermaceae, *Cycas* (Pant and Mehra, 1962), and *Gnetum* secondary growth begins from a vascular cambium in the normal position; then another vascular cambium arises in the phloem or outside it and produces xylem toward the inside and phloem toward the outside. Still another supernumerary cambium arises outside the first supernumerary layer and also forms xylem toward the inside and phloem toward the outside. In this sequence many cambia and many alternating layers of xylem and phloem may be formed (fig. 15.19B). Frequently, the successive cambia are ontogenetically interrelated, in that sister cells of one cambium layer become the cambial cells of another layer. The cambia in abnormal position may be restricted in their extent and form separate units of secondary tissues. Anomalous growth sometimes results from intensified growth of parenchyma distant from the cambium. In *Bauhinia* and in many Bignoniaceae, for example, the originally regularly formed continuous xylem is split into irregular units by the growth of pith and xylem parenchyma.

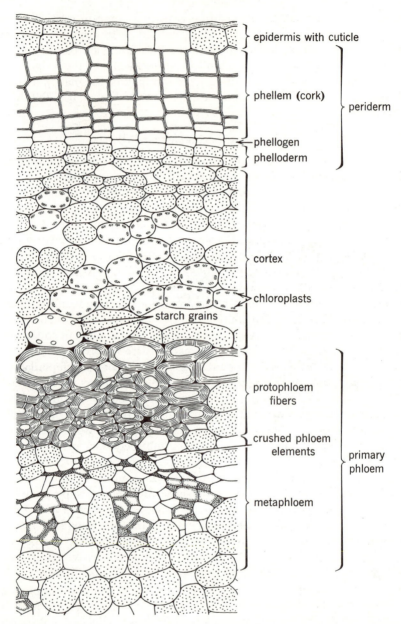

Fig. 15.17. Details of *Prunus* stem structure from the outer part of the rectangle in fig. 15.15*D*. Tannin-containing cells (at bottom of figure) separate primary from secondary phloem shown in fig. 15.18. (×445.)

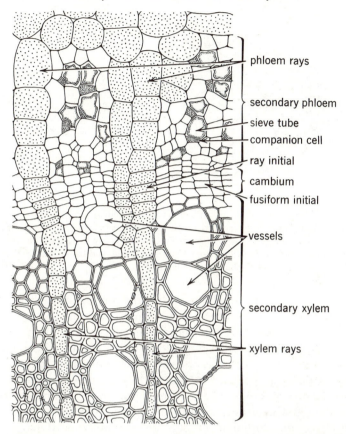

Fig. 15.18. Details of *Prunus* stem structure from the inner part of the rectangle in fig. 15.15D. Outward from cambium, sieve tubes become wider and their walls (left white) thicker. Vessels are in different stages of differentiation; one nearest cambium is without secondary walls. (×445.)

In view of the variability of the so-called anomalous structure, which may be primary and secondary, its precise definition is difficult and depends on how narrowly the normal type is circumscribed. Medullary bundles, for example, are often regarded as anomalous formations, although they may occur in otherwise typically formed stems. Stems of vines with discrete vascular strands in the secondary body are sometimes discussed with the ordinary dicotyledonous stem types, sometimes with the anomalous. Obviously, atypical growth represents no distinct class of phenomena. The designation anomalous serves simply to assemble growth patterns that appear to be less common, at least among plants investigated thus far.

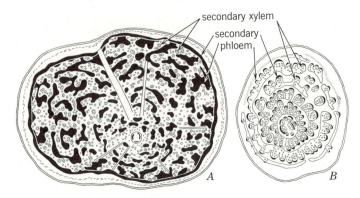

Fig. 15.19. Transections of stems with anomalous secondary growth. *A, Leptadenia spartium,* Asclepiadaceae, with secondary phloem strands imbedded in secondary xylem (included phloem). *B, Boerhaavia diffusa,* Nyctaginaceae, with successive increments of secondary vascular tissues, each composed of xylem, phloem, and parenchyma. Each increment arises from a separate cambial layer. (Both, ×9.)

Anomalous types of growth are widely distributed among the various taxonomic groups. Sometimes a whole family shows atypical secondary thickening, sometimes only one genus or even a smaller group. It may be associated with specific physiologic adaptations. For instance, anomalous secondary growth is often found in climbing plants and lianas (Obaton, 1960), and primary and secondary anomalies occur in stems modified as storage organs in the form of rhizomes, tubers, and corms. In such storage structures there is usually a shortening of internodes and an extensive development of storage parenchyma (Orsós, 1941). Anomalous growth is not restricted to stems but is equally common in roots (chapter 17).

Secondary Growth in Monocotyledons

Although most monocotyledons lack secondary growth from a vascular cambium, by an intense and protracted thickening growth they may produce such large bodies as those of the palms. As was mentioned earlier, the monocotyledons may have a well-defined primary thickening meristem. The activity of this meristem resembles that concerned with secondary growth found in certain monocotyledons. Furthermore, there may be a developmental continuity between the two meristems when both are present in a given plant. It will be useful, therefore, to relate the primary thickening to the secondary (Ball, 1941; Eckardt, 1941). The apical meristem produces directly only a small part of the primary body. Most of it is formed by the thickening meristem. This meristem is located beneath the leaf primordia and produces anticlinal rows of cells by periclinal divisions (fig. 15.14, pl. 58B). The derivatives of the meristem differentiate into a tissue consisting of ground parenchyma traversed by procambial strands, which eventually mature into vascular bundles. The internodes elongate after the axis becomes rather wide. Upon completion of elongation, there is a further increase in thickness by division and enlargement of ground parenchyma cells (Solereder and Meyer, 1928). In palms such thickening growth may be considerable. It is appropriately called diffuse secondary growth (Tomlinson, 1961), diffuse because it does not result from meristematic activity in a restricted region and secondary because it occurs far from the apical meristem.

Secondary growth from a restricted meristem occurs in herbaceous and woody Liliflorae (*Aloe, Sansevieria, Yucca, Agave, Dracaena*) and other groups of monocotyledons (Cheadle, 1937). The meristem concerned with this growth is commonly called cambium and appears to be a direct continuation of the primary thickening meristem (Chouard, 1937; Eckardt, 1941). Unlike the latter, however, the cambium functions in the part of the axis that has completed its elongation. The cambium originates in the parenchyma outside the vascular bundles. This part of the axis is sometimes identified as cortex, sometimes as pericycle, but the difficulty of delimiting a pericycle in the stems of seed plants has been previously emphasized.

The cambial cells vary in shape. As seen in longitudinal sections, they may be fusiform or rectangular, sometimes truncated at one end, tapering at the other (Cheadle, 1937). At first, cells are produced toward the interior of the stem; later a small amount of tissue is formed also toward the periphery. The cells given off inwards differentiate into vascular strands and parenchyma (pl. 68A). The outer derivatives all become parenchyma. In the development of the vascular bundles, individual derivatives of the cambium divide longitudinally; then two or three of the resulting cells form bundles by further longitudinal divisions. In a vertical direction many tiers of cells combine to make a bundle.

The mature bundles are oval in transections. In different species they are either predominantly collateral, or amphivasal. Their phloem consists of short sieve-tube members, with transverse end walls and simple sieve plates, companion cells, and phloem parenchyma. The tracheary elements are tracheids. These are very long since they undergo intensive apical intrusive growth. The tracheids are associated with a small amount of xylem parenchyma which appears to be lignified. The parenchyma in which the bundles are imbedded is either thin walled or thick walled and is lignified. The small amount of parenchyma formed outwards usually remains thin walled and contains crystals. Sometimes these parenchyma cells divide transversely and become shorter than the meristem cells.

The secondary vascular bundles and the associated parenchyma tend to be seriated radially (pl. 68A). In contrast, the primary strands are arranged with no detectable order, and the ground parenchyma shows no radial seriation of cells. In general, however, the basic structure of the primary and the secondary bodies is quite similar, for both consist of a ground tissue traversed by vascular strands. The primary and the secondary bodies are also physically continuous in that the secondary bundles are connected with the peripheral prolongations of the leaf traces.

Effect of Cambial Activity on the Primary Body

In dicotyledons and gymnosperms with prolonged secondary growth the primary body is modified to various degrees. Commonly the primary xylem and the pith are simply covered by the secondary tissues without much change (fig. 15.15), except that sooner or later the protoplasts of the living cells in these tissues die. Crushing of the pith and the interfascicular areas occurs in certain vine types of stems (pl. 55). The primary phloem is pushed toward the outside and is more or less compressed. (Loss of function by the protoxylem and the primary phloem, the frequent development of fibers in the protophloem, and the sclerification of the interfascicular parenchyma in the phloem region are phenomena that appear to be independent of cambial activity.) The effect of secondary growth on the cortex and the epidermis varies with plant species. In some, these parts of the axis, by growth, keep pace with the increase in circumference of the inner tissues; in others they are removed, sooner or later, by periderm formation (chapter 14).

The features characterizing the nodal structure are not perpetuated in the secondary body. A cambium develops in the parenchyma of the leaf gap and forms vascular tissues in continuity with those bordering the gap. This phenomenon is referred to as the closing of the gap (fig. 15.20). The parenchyma cells near the margin of the gap are the first to change into cambium; those in the inner portion change later. This process occurs gradually, and the gap parenchyma is propagated as such within the secondary body until the cambium differentiates throughout the entire tangential width of the gap. Wide gaps extend farther into the secondary body than the narrow ones.

In the leaf trace itself complicated changes occur during secondary growth. The lower end of the trace is affected like the other segments of the primary vascular system. The primary xylem is buried by the

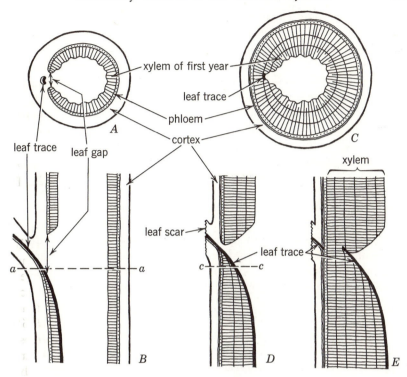

Fig. 15.20. Closing of leaf gaps by secondary growth. A, B, transverse and longitudinal sections through nodal region of stems in first year of growth. Single leaf trace (xylem shown in black) confronts the single gap. B, leaf trace diverges toward base of petiole. C, transverse and, D, E, longitudinal sections through stems several years old. C, secondary xylem has differentiated on flanks and outside xylem of leaf trace; gap no longer continuous with cortex as in A. Two stages in closing of gaps D, E, and rupture of leaf trace, E. In both, the leaf had abscised. A, C, correspond with levels aa and cc in figs. B, D, respectively.

secondary tissues while the phloem is pushed outward. The upper part of the trace, however, diverges outward and crosses the plane of the cambium. The part of the cambium that differentiates above the trace, in the gap region, produces vascular tissue between the trace and the vascular cylinder. This tissue, continuously increasing in amount, exerts a pressure on the trace and eventually causes its rupture (fig. 15.20E). The break is filled with parenchyma which is changed into cambium and connects the cambium of the lower part of the trace with that formed in the gap. After this cambium has formed some secondary tissues, the end of the trace below the break becomes imbedded in secondary xylem (fig. 15.20E). The upper severed end is carried outward, and in time it may be thrown off, together with the cortex, by the activity of the periderm. Since the cambium within the trace itself pushes the trace phloem outward, the buried part of the trace consists of xylem only. In rare instances, the trace passes through the cortex almost horizontally and is not broken. Such a trace is buried only after the secondary body extends beyond the former position of the leaf scar.

The breaking and the burial of the leaf trace occur regularly and early in deciduous species in which the trace constitutes a connection between the vascular systems of leaf and stem for one year only. In evergreen species the connection is maintained for a longer time. In the conifers (but not in the dicotyledonous evergreens) the leaf trace is said to be ruptured every year and a new connection established between the xylem of the stem and the trace part above the break (Tison, 1903).

Rupture of xylem of traces by secondary growth was reported for dormant buds of woody species (Braun, 1960). When such buds grow out in the second year or later they may have no xylem connection with the main axis until a continuity of secondary vascular tissues is established between the bud and the axis.

Grafting and Wound Healing

In successful grafting, a complete union between the stock and the scion is formed. The differentiation of the connecting vascular tissues is preceded by a proliferation of parenchyma tissue—the callus—from both components of the graft. This parenchyma fills the space between the stock and the scion where their surfaces are not in complete contact with each other (pl. 68B). The callus is commonly produced by recent cambial derivatives (Barker, 1954; Buck, 1954) and also by parenchyma of the phloem rays and of the immature xylem rays (Sharples and Gunnery, 1933). The contributions from the stock and the scion in a graft union may be approximately equal. In a pine graft, however, the contribution from the stock was found to predominate (Mergen, 1955).

The initial phenomena involved in the formation of callus are those often considered in pathological anatomy (Krenke, 1933; Küster, 1925). On the cut surfaces of the stock and the scion some of the living parenchyma cells are destroyed in cutting. The breakdown products form a necrotic layer, the isolation layer. It corresponds to the cicatrice that appears on the surface of open wounds. Intact cells near the cut surfaces enlarge until their dimensions surpass considerably those of other similar cells. Such enlargement is called *hypertrophy.* The increase in size may occur for several cells in depth. Subsequently, the large cells divide abundantly, producing callus. Such multiplication of cells above the normal growth is called *hyperplasia.* The isolation layer is broken and subsequently absorbed. The calluses of the stock and the scion intermingle, and eventually vascular cambium is formed through the mixed callus in line with the cambium in the stock and the scion. The cambium appears first where the cambia of the stock and the scion are in contact with the callus cells. Then divisions forming the cambium in the callus progress toward each other until they meet. The tissues resulting from the activity of this cambium are in continuity with the xylem and the phloem of both members of the graft. Sieve elements and tracheary cells may differentiate from callus cells before the cambium appears (Crafts, 1934). Then the cambium arises between these phloem and xylem elements, which occur in longitudinal files extending between the stock and the scion.

A formation of vascular cambium across a callus occurs also in connection with the process of healing of a deep wound such as may be inflicted by removal of a strip of bark (Sharples and Gunnery, 1933). The callus develops from all the exposed surfaces and partly fills the cavity (pl. 66). The cambium begins to develop in this callus—by a change of callus cells into cambial cells—wherever the intact vascular cambium impinges upon it (pl. 67A). Thus, the callus cambium differentiates from all margins of the wound toward the center, the process being comparable to the closing of a diaphragm. The new vascular cambium forms xylem and phloem in continuity with the same tissues in the uninjured portion of the stem (pl. 67B). A periderm develops in the peripheral region of the callus in line with the original stem periderm, if the latter is present (pl. 67). In shallow wounds the periderm develops under the cicatrice without callus formation. Callus may also be absent in the healing of slit-like wounds (Zasche, 1960).

The close developmental relation between the cambia in the callus and in the components of the graft explains why a proper matching of the cambia of the stock and the scion speeds up the formation of cambial connection (Bradford and Sitton, 1929). Improper matching does not necessarily prevent union, but it usually delays it. The establishment of graft union involves many problems, some of which cannot be explained as resulting from faulty technique. Plants may fail to unite readily because of their peculiar structure—monocotyledons, for example, are notably difficult to join by grafting (Muzik, 1958; Muzik and La Rue, 1954)—or inherent incompatibility between the stock and the scion may be the chief obstacle to successful union (Roberts, 1949).

TYPES OF STEM

It is customary to distinguish woody stems from herbaceous, ordinary dicotyledonous stems from vine types, dicotyledonous from monocotyledonous, and stems with normal structure from those with anomalous structure. These groupings, however, are not necessarily based on sharp distinctions. In some instances the differences are mainly quantitative; in others stem types placed into different groups intergrade through transitional types.

Of particular interest is the separation into woody and herbaceous stem types. The prevalent view is that in angiosperms the woody type of plant is more ancient than the herbaceous (Bailey, 1944; Cheadle, 1942; Takhtajan, 1959). The more primitive angiosperms are composed overwhelmingly of woody plants. In orders and families having both woody and herbaceous representatives, the primitive types are more woody than the advanced ones. More than half the families of dicotyledons have no herbaceous species, and the few families that are entirely herbaceous are highly specialized, for example, insectivorous plants, water plants, and parasites. Herbaceous plants also usually possess highly advanced type of xylem throughout the stems and roots.

The evolution of herbaceous dicotyledons from the woody involved a decrease in the activity of the vascular cambium often supplemented by a widening of the interfascicular regions, a change resulting in the formation of a vascular system composed of strands. Sometimes, instead of the individual interfascicular regions becoming high and wide, whole segments of vascular tissue became changed into fibrous or parenchymatic tissue, making the individual vascular bundles of the system particularly discrete. Still other histologic changes may be associated with the evolution of the herbaceous habit (Cumbie and Mertz, 1962). The distinctness of the vascular bundles is common for herbaceous stems, but it does not represent a typical state, since many herbaceous genera, and often whole families, have a primary vascular cylinder that is conspicuously interrupted by parenchyma only at the leaf gaps (Caryophyllaceae, Hypericaceae, Onagraceae, Solanaceae, Polemoniaceae, Ericaceae). Among the monocotyledons, too, the herbaceous stem appears to have arisen primarily through the loss of secondary thickening (Bailey, 1944; Cheadle, 1942). Some herbaceous stems are modified through close association with leaves or by assuming foliar characteristics (cladode). Critical studies are necessary to reveal the nature of such stems and the degree of participation of leaves in their structure (James and Kyhos, 1961; Kaussmann, 1955; Schlittler, 1960).

In the following pages several specific examples of stems of the higher Tracheophyta are described with the aid of illustrations. (Further details on most of these stems may be found in Foster, 1949, Exercise 13.)

Conifer

Pinus. The primary structure of the stem is revealed close to the apex. Here the helically arranged foliar structures (scales) are crowded on the axis, whose internodes are still unextended (pl. 51B). The axils of the scales bear the buds of the short shoots which later produce the needle-like leaves (Sacher, 1955). Because of the crowding of the scales on the young stem, its peripheral part (cortex and epidermis) is confluent with the bases of the scales (pl. 60). The outer boundary of the cortex becomes clearly delimited after the internodal elongation. The primary vascular system consists of collateral strands separated from one another, in transections, by interfascicular regions. The strands are leaf traces connected with each other in a sympodial manner (pl. 51A). Traces to the buds are also present, two to each bud. The nodes are unilacunar, with the gaps of the different nodes overlapping and with more than one visible in each transection (*Picea*, fig. 15.5C). The secondary growth produces a continuous cylinder of xylem and phloem (pl. 61). Opposite the gaps the cambium becomes continuous only gradually so that the gap parenchyma projects into the secondary wood. After the secondary growth has been occurring for some time, the primary xylem of the original bundles may be recognized next to the pith, but the primary phloem is completely obliterated. In the secondary vascular cylinder the amount of phloem is considerably smaller than that of xylem. The demarcation between the cortex and the vascular cylinder is obscure. The endodermis is not identifiable morphologically and the primary phloem forms no peripheral fibers. During secondary growth the outer limit of the phloem may be determined by following the phloem rays to their outermost limits. Sometimes there is a concentration of tannin-containing cells outside the phloem. The cortex is typically paren-chymatic, with many cells containing tannins. Early in the development of the stem, resin ducts appear in the cortex (pl. 60). As the stem increases in circumference, the resin ducts become wider, especially in the tangential direction (pl. 61). The initial periderm arises beneath the epidermis and is not replaced by deeper periderms for several years.

Woody Dicotyledon

Tilia. The primary vascular system consists of closely approximated segments so that in transections the vascular ring appears continuous (pl. 62A; Smith, 1937). The leaves are two-ranked, and the nodes are trilacunar. Beneath the epidermis is a single layer of parenchyma cells; then follows a multiseriate layer of collenchyma. The rest of the cortex is parenchymatic and contains chlorophyll. The endodermis appears as a starch sheath. Fibers are formed in the protophloem, and when mature they constitute a clearly defined but discontinuous outer boundary of the vascular system. The pith is parenchymatic but early shows the development of mucilage canals. Similar canals are formed in the cortex (Strasburger, 1891).

During primary thickening the pith and the vascular cylinder increase in diameter in the presence of considerable amounts of mature xylem and phloem. (Compare young and old stems in pl. 62.) The growth of the vascular cylinder probably results from a lateral expansion of the narrow interfascicular regions and the enlargement of the xylem-parenchyma cells which occur in radial rows. When the pith attains its mature size the peripheral cells have smaller lumina, thicker walls, and larger amounts of deeply colored tanniniferous inclusions than the interior pith cells; they constitute the medullary sheath (pl. 62). Its cells remain alive and store starch while the interior cells become devoid of protoplasts relatively early. The morphologic differentiation of the peripheral pith helps to delimit the inner boundary of the xylem. Otherwise this boundary is somewhat difficult to detect because the tracheary elements of the protoxylem are destroyed during internodal elongation and the parenchyma cells long remain unlignified (Raimann, 1890). *Tilia* exemplifies a plant in which the primary xylem shows radial seriation. Its delimitation from the secondary is somewhat facilitated by the greater density of the secondary xylem as contrasted with the metaxylem (pl. 62B).

The secondary tissues form a continuous cylinder. The primary xylem constitutes an insignificant part of the vascular cylinder after a few years of secondary growth (pl. 28A). The secondary phloem has a distinctive appearance because of the alternation of fiber bands and bands containing sieve tubes and parenchyma cells and because of the lateral expansion of many of the rays (pl. 28A; chapter 12). The initial periderm arises in the layer of parenchyma located between the epidermis and the collenchyma and is not replaced by deeper periderms for many years (Strasburger, 1891).

Dicotyledonous Vine

In *Aristolochia* (Blyth, 1958; Schellenberg, 1899; Strasburger, 1891) the primary vascular system consists of collateral strands separated from each other by wide and high interfascicular regions (pl. 55A). In transections of stems the strands form a discontinuous oval about a parenchymatic pith. The leaves have a two-ranked arrangement, and their bases encircle the stem halfway. The nodes are trilacunar. The median trace consists, in part of its course, of three strands (*mt*, pl. 55A). These three strands and the two laterals of the same set of traces are the smallest bundles in a given transection of stem. The primary tissues outside the vascular system are the following: an epidermis; parenchyma and collenchyma of the cortex, both with chlorophyll; a cylinder of perivascular sclerenchyma (chapter 10) composed of fibrous cells with

blunt ends and concerned with starch storage; and parenchyma interpolated between the sclerenchyma and the vascular strands. A starch
sheath occurs outside the sclerenchyma. According to the stelar
terminology, the sclerenchyma and the subjacent parenchyma would be
called the pericycle.

During the secondary growth, vascular tissues are formed only within
the strands. The interfascicular part of the cambium, which is not
sharply delimited, forms parenchyma similar to that in the primary interfascicular regions and, therefore, the strands remain discrete (pl. 55B–C).
Growth increments are visible in the secondary xylem and also in the
part of the rays associated with this xylem. In both tissues relatively
small cells are formed at the end of a season's growth. The phloem
contains no fibers. In the secondary phloem tangential bands of parenchyma alternate with bands containing sieve tubes and associated parenchyma cells. When the sieve tubes cease to function and are crushed,
a conspicuous banding appears in the phloem, the compressed cells
alternating with noncompressed parenchyma. Concomitantly with the
increase in the circumference of the stem, the individual vascular strands
widen out toward the periphery. From time to time new rays are
interpolated in these widening vascular wedges (pl. 55C). The primary
interfascicular regions and their secondary continuations extend mostly
from node to node, whereas the rays interpolated later within the
vascular strands are successively lower.

The pith and the rays that are continuous with the pith are partly
crushed during secondary growth. This crushing is probably a result
of the resistance offered to the expanding vascular system by the
continuous perivascular sclerenchyma cylinder. Eventually this cylinder
is ruptured, mostly in front of the rays (pl. 55B), with the adjacent
parenchyma invading the breaks. In some species the first of these
parenchyma cells differentiate into sclereids.

The periderm develops in the subepidermal collenchyma or somewhat deeper. The development begins in a few places, and it takes
several years until the periderm spreads over the entire surface of the
stem. Longitudinally the isolated periderms appear as vertical strips
extending from node to node (Czaja, 1934). The cork shows layering
because of an alternation of radially unextended cells with cells that
are larger in radial direction (pl. 55C). A considerable amount of phelloderm is formed by the phellogen.

Cucurbita (Blyth, 1958; Zimmermann, 1922) has bicollateral vascular
bundles appearing in two series, the outer composed of leaf traces, the
inner of trace complexes (pl. 63C). The node is trilacunar, and three of
the five bundles in the outer series belong to the leaf at the node nearest

to the given transection. (Further details on vascular structure in chapter 12.) The ground-tissue system resembles that of *Aristolochia*. Beneath the uniseriate epidermis is the collenchyma, which forms wide bands and alternates with bands of chlorenchyma. The chlorenchyma bands occur beneath the parts of the epidermis bearing the stomata. The deeper lying cortical parenchyma has few chloroplasts. The endodermis is differentiated as a starch sheath. Inside the starch sheath is a cylinder of perivascular sclerenchyma. Some parenchyma intervenes between the sclerenchyma and the vascular bundles. (This tissue region of *Cucurbita*, consisting of sclerenchyma and parenchyma, was used by Van Tieghem, 1882, when he formulated the concept of the pericycle.)

Cambial activity and associated phenomena in Cucurbitaceae are similar to those in *Aristolochia*. However, in the less woody species, secondary growth is often restricted to the vascular strands, and the sclerenchyma cylinder is not ruptured. In *Cucurbita* the pith breaks down during primary growth.

The presence of a continuous cylinder of sclerenchyma outside the vascular system is not a constant characteristic of the vine types of stem. There may be protophloem fibers associated with the individual strands, as in *Vitis* (Esau, 1948). The vine type of structure in this genus is expressed in the presence of relatively wide and high rays (chapter 12). In *Vitis* the vascular strands are not displaced toward the pith during secondary growth, but the pith dies early.

Herbaceous Dicotyledon

Various transitional stem structures may be found between the woody type illustrated by *Tilia* and the extreme dicotyledonous herbs with no secondary growth in the stem (pl. 63*F*). In *Pelargonium* (pl. 63*E*; Blyth, 1958; Carothers, 1959) the primary vascular system consists of closely approximated strands varying in size. During secondary growth a continuous vascular cylinder is produced. The vascular cylinder is clearly set off from the cortex since fibers differentiate on its periphery. In the stem of *Medicago* (alfalfa), Leguminosae, the vascular bundles, as seen in transections, are not too varied in size and are clearly separated from each other (pl. 63*B*). Some secondary growth occurs at the base of the stem, but the interfascicular cambium produces mainly sclerenchyma on the xylem side. The stems of Leguminosae have various patterns of secondary growth and rarely lack cambial activity (Cumbie, 1960). The extremely herbaceous stem of *Ranunculus* resembles those of some monocotyledons in that it has a somewhat dispersed arrangement of the vascular bundles and lacks vascular cambium (pl. 63*F*).

The stems described above may all be considered typical of the groups of plants to which they belong. In some plants, however, parts of stems assume more or less modified aspects, often in relation to specialization as storage organs. One of the best examples of a stem that serves mainly for storage is the potato tuber (Artschwager, 1924), whose anatomy contrasts rather strikingly with that of the aerial vegetative stem (Artschwager, 1918). In the latter, normal elongation occurs during development; in the tuber, the internodes remain short, but there is lateral expansion increasing the amount of storage parenchyma. The leaf traces constitute a prominent part of the vascular system of the aerial stem. This system, therefore, shows variable structure related to the position of the leaves. In the tuber the vascular system is morphologically more homogeneous because the traces of the scale-like leaves subtending the axillary buds are very small. Both the aerial stem and the tuber have internal and external phloem, but in the tuber the internal phloem is dispersed in the wide pith, so that only a narrow parenchymatic zone in the interior is free of phloem elements. The internal phloem is rich in parenchyma and appears to be the principal storage tissue of the tuber.

Herbaceous Monocotyledon

Vascular systems composed of widely spaced strands and not restricted to one ring in transections are relatively uncommon in the dicotyledons (Ranunculaceae, Nymphaeaceae, Piperaceae), but in the monocotyledons similar and more complex systems are usual (Metcalfe, 1946). Most monocotyledons have leaf sheaths protecting the internodes, which for a relatively long time continue intercalary growth. In *Musa* (banana) the leaf sheaths are combined into a stem-like structure. Stems of monocotyledons are often modified into rhizomes (*Iris, Gladiolus*) or shoots into bulbs (*Allium*). The stem part of a bulb is reduced to a plate with short internodes, no pith, and congested traces of leaves and adventitious roots (Mann, 1952).

Grass Stem. In transections of the internodes of most grasses the usual three tissue systems, the epidermal, the fundamental, and the vascular, are visible. The vascular bundles are commonly distributed according to two basic plans. Either they are in two circles, one of smaller bundles nearer the periphery, the other of larger bundles somewhat deeper within the stem (pl. 63*D*; *Triticum, Avena, Hordeum, Secale, Oryza*); or they are scattered throughout the transection, with smaller bundles densely arranged near the periphery and larger bundles more widely spaced in the center (pl. 58*C*; *Zea, Saccharum, Sorghum, Bambusa*). The vascular bundles are collateral, each enclosed in a sheath of sclerenchyma (pl. 57*B*).

In grasses in which the vascular bundles are arranged in two circles, there is commonly a continuous cylinder of sclerenchyma close to the epidermis, with the outer smaller bundles imbedded in it (pl. 63D). On the outer sides of these bundles occur fiber strands that reach to the epidermis. Bands of parenchyma with chloroplasts alternate with the fiber bands. The chlorenchyma bands extend parallel to each other through the internode and terminate at the nodes. In some places the bands coalesce. The chlorenchyma is located beneath parts of the epidermis where stomata are concentrated. Inside the ring of sclerenchyma is the fundamental parenchyma in which the vascular bundles are imbedded. The central part of this parenchyma, which is free of vascular tissue in the internodes, may be spoken of as pith. In many grasses, especially in Festucoideae, the pith breaks down in the internodes, but not in the nodes (fig. 15.21, pls. 59C, 63D). In others it is retained throughout the stem (Brown et al., 1959a). The internodal lacuna develops during the elongation of the stem (Kaufman, 1959). In grasses with the vascular bundles in a scattered arrangement, no sclerenchyma cylinder is formed (pl. 58C), but the subepidermal parenchyma may be strongly sclerified (chapter 10). Deviations from the patterns just described occur in some grasses (Metcalfe, 1960).

The vascular system of grasses apparently consists of leaf traces and axillary bud traces and their combinations (Percival, 1921; Inosaka, 1962; Kumazawa, 1961). Because of the complex arrangement of the traces and the presence of transversely oriented nodal networks of bundles (pl. 58A) the relation between the bundles in the axis and those of the lateral organs is not revealed unless an exceptionally detailed study is carried out (Kumazawa, 1961). Some inadequately documented papers report the presence of bundles not related to lateral appendages.

The leaves of grasses are typically two-ranked in arrangement and their sheaths completely encircle the stem (fig. 15.3C). Most grasses have pulvini—local enlargements—above the junction of the sheaths and the stem (culm). The pulvinus is particularly well developed in the Festucoideae in which it is located in the sheath (pl. 59C, D); in the Panicoideae the pulvinus occurs in the internode but is poorly developed or absent in the sheath (Brown et al., 1959b).

Each leaf has numerous traces, some large, others smaller and alternating with the larger. If the traces of a given leaf of Zea are studied in successive sections downward from the node, their course is found to be as follows (Kumazawa, 1961). Within the node the large bundles bend inward, whereas the small ones remain near the periphery (fig. 15.3D). The median of the large bundles may reach the center of the stem. The other large bundles occupy positions intermediate between central and peripheral. With slight alterations in their positions, the traces extend downward through one or more internodes. Farther below, the larger traces become reoriented to a peripheral position, often appearing on the side opposite the part of the leaf to which they are attached above (fig. 15.3*D*). This reorientation is accompanied by a diminution in size and by fusions with other small bundles in the peripheral portions of the stem. The outermost small bundles, however, form an independent system connected to axillary shoots and small leaf bundles.

In the wheat stem (Percival, 1921) the course of the vascular bundles through the internode and the leaf sheath is practically parallel (fig. 15.21*A*). Near the node the leaf sheath is considerably thickened, attaining its maximum thickness just above its union with the stem, that is, in the pulvinus (fig. 15.21*B–D*). The stem, on the other hand, decreases in thickness in the same direction and has the smallest diameter above the junction with the leaf sheath. The stem is hollow in the internode, solid at the node. The sheath is open on one side at higher levels, closed near the node (fig. 15.21*C, D*). In the pulvinus region the intercalary growth continues the longest, and the tissues remain capable of further elongation after this activity ceases (shown for *Hordeum* in pl. 59*C, D*). In this part of the shoot sclerenchyma does not form, and lignification is at minimum. Massive collenchymatic bundle caps differentiate in connection with the leaf sheath bundles (fig. 15.21*D*).

Below the junction of leaf sheath and stem the smaller of the leaf traces are prolonged in the peripheral part of the axis. The larger leaf traces become part of the inner cylinder of strands. The bundles of the internode located above the leaf insertion assume, just above the node, a horizontal and oblique course (fig. 15.21*E, F*) and are reoriented toward a more peripheral position in the node and below it (fig. 15.21*G, H*). In these horizontal and oblique positions the bundles variously branch and coalesce, and their total number is reduced. Through a mutual reorientation, the large leaf traces and the bundles from the internode above the insertion of the leaf together form the inner cylinder of bundles of the next lower internode (fig. 15.21*H*). In this cylinder about half of the bundles are leaf traces from the nearest leaf above, and the other half are bundles from the internode above the insertion of this leaf. The peripheral bundles are mostly leaf traces from the nearest leaf above.

The transverse bundles in the nodal regions are a conspicuous feature of grass stems (pl. 58*A*). These bundles interconnect the leaf traces of the main shoot among themselves. The transverse bundles appear somewhat later in the ontogeny of the stem, and some workers interpret them as prolongations of the small peripheral traces (Bugnon, 1924;

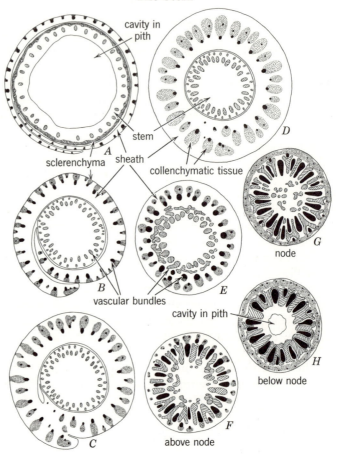

Fig. 15.21. Nodal anatomy of *Triticum* stem. Transections at various levels of stem beginning with middle of internode (*A*), through lower part of internode (*B–F*), node (*G*), and ending just below node (*H*). Bundles of sheath and their prolongations as traces in stem are shown in black; vascular tissue of internode and its continuation through node is hatched. Fine stippling indicates sclerenchyma, coarse stippling in *D*, collenchymatic tissue replacing sclerenchyma in the pulvinus region. Toward node, sheath increases and stem decreases in thickness. (All, ×7.6.)

Sharman, 1942). The transverse bundles are not specifically associated with the axillary buds or the adventitious roots attached at the nodes. The bud traces are prolonged vertically in the main axis, and the roots are peripherally connected to the vascular system of the main stem (Bugnon, 1924).

REFERENCES

Arber, A. *The natural philosophy of plant form.* Cambridge, Cambridge University Press. 1950.

Artschwager, E. F. Anatomy of the potato plant, with special reference to the ontogeny of the vascular system. *Jour. Agr. Res.* 14:221–252. 1918.

Artschwager, E. F. Studies on the potato tuber. *Jour. Agr. Res.* 27:809–835. 1924.

Bailey, I. W. The development of vessels in angiosperms and its significance in morphological research. *Amer. Jour. Bot.* 31:421–428. 1944.

Bailey, I. W. Nodal anatomy in retrospect. *Arnold Arboretum Jour.* 37:269–287. 1956.

Bailey, I. W., and C. G. Nast. The comparative anatomy of the Winteraceae. IV. Anatomy of the node and vascularization of the leaf. *Arnold Arboretum Jour.* 25:215–221. 1944.

Ball, E. The development of the shoot apex and the primary thickening meristem in *Phoenix canariensis* Chaub., with comparisons to *Washingtonia filifera* Wats. and *Trachycarpus excelsa* Wendl. *Amer. Jour. Bot.* 28:820–832. 1941.

Barker, W. G. A contribution to the concept of wound repair in woody stems. *Canad. Jour. Bot.* 32:486–490. 1954.

Barthelmess, A. Über den Zusammenhang zwischen Blattstellung und Stelenbau unter besonderer Berücksichtigung der Koniferen. *Bot. Arch.* 37:207–260. 1935.

Blyth, A. Origin of primary extraxylary stem fibers in dicotyledons. *Calif. Univ., Publs., Bot.* 30:145–232. 1958.

Bond, G. The stem endodermis of the genus *Piper*. *Roy. Soc. Edinb., Trans.* 56:695–724. 1931.

Bower, F. O. *Size and form in plants.* London, Macmillan and Company. 1930.

Bradford, F. C., and B. G. Sitton. Defective graft unions in the apple and the pear. *Mich. Agr. Expt. Sta. Tech. Bul.* 99. 1929.

Braun, H. J. Der Anschluss von Laubbaumknospen an das Holz der Tragachsen. *Deut. Bot. Gesell. Ber.* 73:258–264. 1960.

Brebner, G. On the anatomy of *Danaea* and other Marattiaceae. *Ann. Bot.* 16:517–552. 1902.

Brown, W. V., W. F. Harris, and J. D. Graham. Grass morphology and systematics. I. The internode. *Southwest. Nat.* 4:115–125. 1959*a*.

Brown, W. V., G. A. Pratt, and H. M. Mobley. Grass morphology and systematics. II. The nodal pulvinus. *Southwest. Nat.* 4:126–130. 1959*b*.

Buchholz, M. Über die Wasserleitungsbahnen in den interkalaren Wachstumszonen monokotyler Sprosse. *Flora* 14:119–186. 1920.

Buck, G. J. The histology of the bud graft union in roses. *Iowa State Coll. Jour. Sci.* 28:587–602. 1954.

Bugnon, P. Contribution à la connaissance de l'appareil conducteur chez les Graminées. *Soc. Linn. de Normandie, Mém.* 26:21–40. 1924.

Campbell, D. H. The eusporangiate ferns and the stelar theory. *Amer. Jour. Bot.* 8:303–314. 1921.

Canright, J. E. The comparative morphology and relationships of the Magnoliaceae. IV. Wood and nodal anatomy. *Arnold Arboretum Jour.* 36:119–140. 1955.

Carlquist, S. Studies on Madinae: anatomy, cytology, and evolutionary relationships. *Aliso* 4:171–236. 1959*a*.

Carlquist, S. Vegetative anatomy of *Dubautia, Argyroxiphium,* and *Wilkesia* (Compositae). *Pacific Sci.* 13:195–210. 1959*b*.

Carlquist, S. *Comparative plant anatomy.* New York, Holt, Rinehart and Winston. 1961.

Carothers, Z. B. Observation on the procambium and primary phloem of *Pelargonium domesticum*. *Amer. Jour. Bot.* 46:397–404. 1959.

Caspary, R. Bemerkungen über die Schutzscheide und die Bildung des Stammes und der Wurzel. *Jahrb. f. Wiss. Bot.* 4:101–124. 1865–66.

Cheadle, V. I. Secondary growth by means of a thickening ring in certain monocotyledons. *Bot. Gaz.* 98:535–555. 1937.

Cheadle, V. I. The role of anatomy in the phylogenetic studies of the Monocotyledoneae. *Chron. Bot.* 7:253–254. 1942.

Cheadle, V. I., and N. W. Uhl. Types of vascular bundles in the Monocotyledoneae and their relation to the late metaxylem conducting elements. *Amer. Jour. Bot.* 35:486–496. 1948.

Chouard, P. La nature et le rôle des formations dites "secondaires" dans l'édification de la tige des Monocotylédones. *Soc. Bot. de France Bul.* 83:819–836. 1937.

Church, A. H. *On the interpretation of phenomena of phyllotaxis.* London, Oxford University Press. 1920.

Courtot, Y., and L. Baillaud. A propos de la gaine casparyenne de la tige de certaines Labiées. *Ann. Sci. Univ. Besançon, Bot.* 15:47–52. 1960.

Crafts, A. S. Phloem anatomy in two species of *Nicotiana*, with notes on the interspecific graft union. *Bot. Gaz.* 95:592–608. 1934.

Cuénod, A. Du rôle de la feuille dans l'édification de la tige. *Soc. Sci. Nat. Tunis. Bul.* 4:3–15. 1951.

Cumbie, B. G. Anatomical studies in the Leguminosae. *Trop. Woods* 113:1–47. 1960.

Cumbie, B. G., and D. Mertz. Xylem anatomy of *Sophora* (Leguminosae) in relation to habit. *Amer. Jour. Bot.* 49:33–40. 1962.

Cutter, E. G. On a theory of phyllotaxis and histogenesis. *Cambridge Phil. Soc. Biol. Rev.* 34:243–263. 1959.

Czaja, A. T. Zur Entwicklungsphysiologie des Periderms und die Entstehung der Kork-krusten. *Planta* 23:105–145. 1934.

Darwin, C. *The power of movement in plants.* New York, D. Appleton and Company. 1892.

Datta, A. Comparative study of the vegetative and flowering axes of *Leonurus sibiricus* L. *Ind. Acad. Sci. Proc.* 22:10–17. 1945.

Davis, E. L. Medullary bundles in the genus *Dahlia* and their possible origin. *Amer. Jour. Bot.* 48:108–113. 1961.

De Bary, A. *Comparative anatomy of the vegetative organs of phanerogams and ferns.* Oxford, Clarendon Press. 1884.

Demalsy, P. Nouvelles recherches sur le sporophyte d'*Azolla*. *Cellule* 59:233–268. 1958.

Dermen, H. Adventitious bud and stem relationship in apple. *Jour. Washington Acad. Sci.* 49:261–268. 1959.

De Sloover, J. Recherches sur l'histogénèse des tissus conducteurs. II. Le sens longitudinal de la différenciation du procambium, du xylème et du phloème chez *Coleus, Ligustrum, Anagallis* et *Taxus*. *Cellule* 59:55–202. 1958.

Dormer, K. J. The acacian type of vascular system and some of its derivatives. I. Intro-duction, Menispermaceae, Lardizabalaceae, Berberidaceae. *New Phytol.* 53:301–311. 1954.

Dormer, K. J. *Azarum europaeum*—a critical case in vascular morphology. *New Phytol.* 54:338–342. 1955a.

Dormer, K. J. Mathematical aspects of plant development. *Discovery* 16:59–64. 1955b.

Doyle, M. H., and J. Doyle. Pith structure in conifers. I. Taxodiaceae. *Roy. Irish Acad., Proc. Sect. B.* 52:15–39. 1948.

Eames, A. J. *Morphology of vascular plants. Lower groups.* New York, McGraw-Hill Book Company. 1936.

Eckardt, T. Kritische Untersuchungen über das primäre Dickenwachstum bei Monokotylen, mit Ausblick auf dessen Verhältnis zur sekundären Verdickung. *Bot. Arch.* 42:289–334. 1941.

Emberger, L. Tige, racine, feuille. *Ann. Biol.* 28:109–125. 1952.

Esau, K. Vascular differentiation in the vegetative shoot of *Linum*. I. The procambium. *Amer. Jour. Bot.* 29:738–747. 1942. II. The first phloem and xylem. *Amer. Jour. Bot.* 30:248–255. 1943*a*.

Esau, K. Origin and development of the primary vascular tissues in seed plants. *Bot. Rev.* 9:125–206. 1943*b*.

Esau, K. Vascularization of the vegetative shoots of *Helianthus* and *Sambucus*. *Amer. Jour. Bot.* 32:18–29. 1945.

Esau, K. Phloem structure in the grapevine, and its seasonal changes. *Hilgardia* 18:217–296. 1948.

Esau, K. Development and structure of the phloem tissue. II. *Bot. Rev.* 16:67–114. 1950.

Esau, K. Primary vascular differentiation in plants. *Cambridge Phil. Soc. Biol. Rev.* 29: 46–86. 1954.

Ezelarab, G. E., and K. J. Dormer. The organization of the primary vascular system in Ranunculaceae. *Ann. Bot.* 27:21–38. 1963.

Fischer, H. Der Pericykel in den freien Stengelorganen. *Jahrb. f. Wiss. Bot.* 35:1–27. 1900.

Foster, A. S. Investigations on the morphology and comparative history of development of foliar organs. IV. The prophyll of *Carya Buckleyi* var. *arkansana*. *Amer. Jour. Bot.* 19:710–728. 1932.

Foster, A. S. *Practical plant anatomy.* 2nd ed. New York, D. Van Nostrand Company. 1949.

Foster, A. S., and E. M. Gifford, Jr. *Comparative morphology of vascular plants.* San Francisco, W. H. Freeman and Company. 1959.

Fries, R. E. Ein unbeachtet gebliebenes Monokotyledonenmerkmal bei einigen Polycarpicae. *Deut. Bot. Gesell. Ber.* 29:292–301. 1911.

Fukumoto, K. Studies on adventitious bud formation. I. Morphological and histological observations on the adventitious buds on tomato leaves. *Bot. Mag. Tokyo* 73:348–354. 1960.

Garrison, R. Origin and development of axillary buds: *Syringa vulgaris* L. *Amer. Jour. Bot.* 36:205–213. 1949*a*. *Betula papyrifera* March. and *Euptelea polyandra* Sieb. et Zucc. *Amer. Jour. Bot.* 36:379–389. 1949*b*.

Gautheret, R. J. *La culture des tissus végétaux. Techniques et réalisations.* Paris, Masson and Cie. 1959.

Girolami, G. Relation between phyllotaxis and primary vascular organization in *Linum*. *Amer. Jour. Bot.* 40:618–625. 1953.

Golub, S. J., and R. H. Wetmore. Studies of development in the vegetative shoot of *Equisetum arvense* L. II. The mature shoot. *Amer. Jour. Bot.* 35:767–781. 1948.

Gris, A. Extrait d'un mémoire sur la moelle des plantes ligneuses. *Ann. des Sci. Nat., Bot.* Ser. 5. 14:34–79. 1872.

Gulline, H. F. Experimental morphogenesis in adventitious buds of flax. *Austral. Jour. Bot.* 8:1–10. 1960.

Gunckel, J. E., and R. H. Wetmore. Studies of development in long shoots and short shoots of *Ginkgo biloba* L. I. The origin and pattern of development of the cortex, pith and procambium. *Amer. Jour. Bot.* 33:285–295. 1946*a*. II. Phyllotaxis and the organization of the primary vascular system; primary phloem and primary xylem. *Amer. Jour. Bot.* 33:532–543. 1946*b*.

Gustin, R., and J. De Sloover. Recherches sur l'histogénèse des tissus conducteurs. I. Problèmes posés et données acquises. *Cellule* 57:97–128. 1955.

Guttenberg, H. von. Die physiologischen Scheiden. In: K. Linsbauer. *Handbuch der Pflanzenanatomie.* Band 5. Lief. 42. 1943.

Haberlandt, G. *Physiological plant anatomy.* London, Macmillan and Company. 1914.

Hagemann, W. Vergleichende morphologische, anatomische und entwicklungsgeschichtliche Studien an *Cyclamen persicum* Mill. sowie einigen weiteren *Cyclamen*-Arten. *Bot. Stud.* Heft 9. 1959.

Hanstein, J. Über den Zusammenhang der Blattstellung mit dem Bau des dicotylen Holzringes. *Jahrb. f. Wiss. Bot.* 1:233–283. 1858.

Hasselberg, G. B. E. Zur Morphologie des vegetativen Sprosses der Loganiaceen. *Symb. Bot. Upsaliensis* II:3. 1937.

Inosaka, M. [Studies on the development of vascular system in rice plant and the growth of each organ viewed from the vascular connection between them.] *Bul. Fac. Agr. Miyazaki Univ.* 7:15–116. 1962.

Jacobs, W. P. Acropetal auxin transport and xylem regeneration—a quantitative study. *Amer. Nat.* 88:327–337. 1954.

Jacobs, W. P., and I. B. Morrow. A quantitative study of xylem development in the vegetative shoot apex of *Coleus*. *Amer. Jour. Bot.* 44:823–842. 1957.

Jacobs, W. P., and I. B. Morrow. Quantitative relations between stages of leaf development and differentiation of sieve tubes. *Science* 128:1084–1085. 1958.

Jacobs, W. P., and V. Raghavan. Studies in the histogenesis and physiology of *Perilla*—I. Quantitative analysis of flowering in *P. frutescens* (L.) Britt. *Phytomorphology* 12:144–167. 1962.

James, L. E., and D. W. Kyhos. The nature of the fleshy shoot of *Allenrolfea* and allied genera. *Amer. Jour. Bot.* 48:101–108. 1961.

Jeffrey, E. C. The morphology of the central cylinder in the angiosperms. *Canad. Inst. Toronto, Trans.* 6:599–636. 1898–99.

Jeffrey, E. C. The structure and development of the stem in the Pteridophyta and gymnosperms. *Roy. Soc. London, Phil. Trans.* Ser. B. 195:119–146. 1903.

Kaufman, P. B. Development of the shoot of *Oryza sativa* L.—III. Early stages in histogenesis of the stem and ontogeny of the adventitious root. *Phytomorphology* 9:382–404. 1959.

Kaussmann, B. Histogenetische Untersuchungen zum Flachssprossproblem. *Bot. Stud.* Heft 3. 1955.

Krenke, N. P. *Wundkompensation, Transplantation und Chimären bei Pflanzen.* Berlin, Julius Springer. 1933.

Kumazawa, M. Studies on the vascular course in maize plant. *Phytomorphology* 11:128–139. 1961.

Küster, E. *Pathologische Pflanzenanatomie.* 3rd ed. Jena, Gustav Fischer. 1925.

Lawalrée, A. Histogénèse florale et végétative chez quelques Composées. *Cellule* 52:215–294. 1948.

Léger, L. J. Recherches sur l'origine et les transformations des éléments libériens. *Soc. Linn. de Normandie, Mém.* 19:49–182. 1897.

Loiseau, J. E. Observations et expérimentation sur la phyllotaxie et le functionnement du sommet végétatif chez quelques Balsaminacées. *Ann. des Sci. Nat., Bot.* Ser. 11. 20:1–214. 1959.

Maeda, E. Structure and development of the vegetative shoot in rice plant. *Proc. Crop Plant Devlpmt., Fac. Agr. Nagoya Univ.* 1:1–28. 1962.

Mann, L. K. Anatomy of the garlic bulb and factors affecting bulb development. *Hilgardia* 21:195–251. 1952.

Marsden, M. P. F., and T. A. Steeves. On the primary vascular system and the nodal anatomy of *Ephedra*. *Arnold Arboretum Jour.* 36:241–258. 1955.

McGahan, M. W. Vascular differentiation in the vegetative shoot of *Xanthium chinense*. *Amer. Jour. Bot.* 42:132–140. 1955.

Mergen, F. Anatomical study of slash pine graft unions. *Fla. Acad. Sci. Quart. Jour.* 17: 237–245. 1955.

Metcalfe, C. R. The systematic anatomy of the vegetative organs of the angiosperms. *Cambridge Phil. Soc. Biol. Rev.* 21:159–172. 1946.

Metcalfe, C. R. *Anatomy of the monocotyledons.* I. Gramineae. Oxford, Clarendon Press. 1960.

Metcalfe, C. R., and L. Chalk. *Anatomy of the dicotyledons.* 2 vols. Oxford, Clarendon Press. 1950.

Meyer, F. J. Die Stelärtheorie und die neuere Nomenklatur zur Beschreibung der Wasser-leitungsbahnen der Pflanzen. *Bot. Centbl. Beihefte* 33:129–168. 1916.

Mourré, J. L'ontogenie du sclérenchyme chez *Cucurbita pepo* L. *Rev. de Cytol. et de Biol. Vég.* 19:99–149. 1958.

Mullenders, W. L'origine du phloème interxylémien chez *Stylidium et Thunbergia*. Étude anatomique. *Cellule* 51:5–48. 1947.

Murty, Y. S. Studies in the order Piperales—I. A contribution to the study of vegetative anatomy of some species of *Peperomia*. *Phytomorphology* 10:50–59. 1960.

Muzik, T. J. Role of parenchyma cells in graft union in vanilla orchid. *Science* 127:82. 1958.

Muzik, T. J., and C. D. La Rue. Further studies on the grafting of monocotyledonous plants. *Amer. Jour. Bot.* 41:448–445. 1954.

Nast, C. G. The comparative morphology of the Winteraceae. VI. Vascular anatomy of the flowering shoot. *Arnold Arboretum Jour.* 25:454–466. 1944.

Obaton, M. Les lianes ligneuses a structure anomale des forêts denses d'Afrique occidentale. *Ann. des Sci. Nat., Bot.* Ser. 12. 1:1–220. 1960.

Ogura, Y. Anatomie der Vegetationsorgane der Pteridophyten. In: K. Linsbauer. *Hand-buch der Pflanzenanatomie.* Band 7. Lief. 36. 1938.

O'Neill, T. B. Primary vascular organization of *Lupinus* shoot. *Bot. Gaz.* 123:1–9. 1961.

Orsós, O. Die Gewebeentwicklung bei der Kohlrabiknolle. *Flora* 35:6–20. 1941.

Pant, D. D., and B. Mehra. *Studies in gymnospermous plants.* Cycas. Allahabad, India, Central Book Depot. 1962.

Parke, R. V. Initial vascularization of the vegetative shoot of *Abies concolor*. *Amer. Jour. Bot.* 50:464–469. 1963.

Pearson, L. C., and D. B. Lawrence. Photosynthesis in aspen bark. *Amer. Jour. Bot.* 45: 383–387. 1958.

Percival, J. *The wheat plant.* New York, E. P. Dutton and Company. 1921.

Pfeiffer, H. Das abnorme Dickenwachstum. In: K. Linsbauer. *Handbuch der Pflanzen-anatomie.* Band 9. Lief. 15. 1926.

Philipson, W. R. The ontogeny of the shoot apex in dicotyledons. *Cambridge Phil. Soc. Biol. Rev.* 24:21–50. 1949.

Philipson, W. R., and E. E. Balfour. Vascular patterns in dicotyledons. *Bot. Rev.* 29:382–404. 1963.

Plantefol, L. Fondements d'une théorie phyllotaxique nouvelle: la théorie des hélices foliaires multiples. *Ann. des Sci. Nat., Bot.* Ser. 11. 7:153–299. 1946. 8:1–71. 1947.

Prunet, A. Recherches sur les noeuds et les entre-noeuds de la tige de Dicotylédones. *Ann. des Sci. Nat., Bot.* Ser. 7. 13:297–373. 1891.

Raimann, R. Über unverholzte Elemente in der innersten Xylemzone der Dicotyledonen. *Akad. der Wiss. Wien, Math.-Nat. Cl. Sitzber.* Abt. 1. 98:40–75. 1890.

Rauh, W., and H. Falk. *Stylites* E. Amstutz, eine neue Isoëtaceae aus den Hochanden Perus. 2. Teil: Zur Anatomie des Stammes mit besonderer Berücksichtigung der Verdickungsprozesse. *Heidelberg. Akad. der Wiss., Math.-Nat. Kl. Sitzber.* 1959:87–160. 1959.

Rauh, W., and F. Rappert. Über das Vorkommen und die Histogenese von Scheitelgruben bei krautigen Dikotylen, mit besonderer Berücksichtigung der Ganz- und Halbrosettenpflanzen. *Planta* 43:325–360. 1954.

Ricardi, M., and F. Torres. Estudio crítico de la anatomía de los haces de *Zea mays* L. *Soc. Biol. Conception* [Chile] *Bol.* 31:141–144. 1956.

Richards, F. J. The geometry of phyllotaxis and its origin. *Symp. Soc. Expt. Biol.* 2:217–245. 1948.

Richards, F. J. Phyllotaxis: its quantitative expression and relation to growth in the apex. *Roy. Soc. London, Phil. Trans.* Ser. B. 235:509–564. 1951.

Roberts, R. H. Theoretical aspects of graftage. *Bot. Rev.* 15:423–463. 1949.

Rouffa, A. S., and J. E. Gunckel. Leaf initiation, origin, and pattern of pith development in the Rosaceae. *Amer. Jour. Bot.* 38:301–307. 1951.

Russow, E. Vergleichende Untersuchungen der Leitbündel-Kryptogamen. *Mém. Acad. Imp. Sci. St. Petérsbourg.* Ser. VII. 19:1–207. 1872.

Sacher, J. A. Dwarf shoot ontogeny in *Pinus lambertiana. Amer. Jour. Bot.* 42:784–792. 1955.

Sachs, J. *Textbook of botany.* Oxford, Clarendon Press. 1875.

Sachs, R. M., A. Lang, C. F. Bretz, and J. Roach. Shoot histogenesis: subapical meristematic activity in a caulescent plant and the action of gibberellic acid and Amo-1618. *Amer. Jour. Bot.* 47:260–266. 1960.

Schellenberg, H. C. Zur Entwicklungsgeschichte des Stammes von *Aristolochia sipho* L'Herit. *Botan. Untersuchungen. Festschrift für Schwendener* 1899:301–320. Berlin, Gebrüder Borntraeger. 1899.

Schlittler, J. Die Asparageenphyllokladien erweisen sich auch ontogenetisch als Blätter. *Bot. Jahrb.* 79:428–446. 1960.

Schoute, J. C. *Die Stelär-Theorie.* Jena, Gustav Fischer. 1903.

Schoute, J. C. On phytonism. *Rec. des Trav. Bot. Néerland.* 28:82–96. 1931.

Shah, J. J. Morpho-histogenetic studies in Vitaceae—I. Origin and development of the axillary buds in *Cayratia carnosa* Gagnep. *Phytomorphology* 10:157–174. 1960.

Sharman, B. C. Developmental anatomy of the shoot of *Zea mays* L. *Ann. Bot.* 6:245–282. 1942.

Sharples, A., and H. Gunnery. Callus formation in *Hibiscus Rosa-sinensis* L. and *Hevea brasiliensis* Müll. Arg. *Ann. Bot.* 47:827–840. 1933.

Sifton, H. B. Developmental morphology of vascular plants. *New Phytol.* 43:87–129. 1944.

Sinnott, E. W. Investigations on the phylogeny of the angiosperms. I. The anatomy of the node as an aid in the classification of angiosperms. *Amer. Jour. Bot.* 1:303–322. 1914.

Smith, E. P. Nodal anatomy of some common trees. *Bot. Soc. Edinb., Trans. and Proc.* 32:260–277. 1937.

Snow, M., and R. Snow. A theory of the regulation of phyllotaxis based on *Lupinus albus. Roy. Soc. London, Phil. Trans.* 244:483–514. 1962.

Snow, R. Problems of phyllotaxis and leaf determination. *Endeavour* 14:190–199. 1955.

Solereder, H., and F. J. Meyer. *Systematische Anatomie der Monokotyledonen.* Heft III. Berlin, Gebrüder Borntraeger. 1928.

Stafford, H. A. Studies on the growth and xylary development of *Phleum pratense* seedlings in darkness and in light. *Amer. Jour. Bot.* 35:706–715. 1948.

Sterling, C. Growth and vascular development in the shoot apex of *Sequoia sempervirens* (Lamb.) Endl. II. Vascular development in relation to phyllotaxis. *Amer. Jour. Bot.* 32:380–386. 1945.

References 421

Sterling, C. Organization of the shoot of *Pseudotsuga taxifolia* (Lamb.) Britt. II. Vascularization. *Amer. Jour. Bot.* 34:272–280. 1947.

Steward, F. C., M. O. Mapes, and K. Mears. Growth and organized development in cultured cells. II. Organization in cultures grown from freely suspended cells. *Amer. Jour. Bot.* 45:705–708. 1958.

Strasburger, E. *Über den Bau und die Verrichtungen der Leitungsbahnen in den Pflanzen. Histologische Beiträge.* Band 3. Jena, Gustav Fischer. 1891.

Takhtajan, A. *Die Evolution der Angiospermen.* Jena, Gustav Fischer. 1959.

Tison, A. Les traces foliaires des Conifères dans leur rapport avec l'épaissisement de la tige. *Soc. Linn. de Normandie, Mém.* 21:57–82. 1903.

Tolbert, R. J. A seasonal study of the vegetative shoot apex and the pattern of pith development in *Hibiscus syriacus. Amer. Jour. Bot.* 48:249–255. 1961.

Tomlinson, P. B. *Anatomy of the monocotyledons. II. Palmae.* Oxford, Clarendon Press. 1961.

Troll, W. *Vergleichende Morphologie der höheren Pflanzen.* Band I. *Vegetationsorgane.* Heft I. Berlin, Gebrüder Borntraeger. 1937.

Troll, W. *Praktische Einführung in die Pflanzenmorphologie. Erster Teil. Der vegetative Aufbau.* Jena, Gustav Fischer. 1954.

Troll, W., and W. Rauh. Das Erstarkungswachstum krautiger Dikotylen, mit besonderer Berücksichtigung der primären Verdickungsvorgänge. *Heidelberg. Akad. der Wiss., Math.-Nat. Kl. Sitzber.* 1. Abh. 1950.

Van Fleet, D. S. Histochemistry and function of the endodermis. *Bot. Rev.* 27:165–220. 1961.

Van Iterson, G., Jr. New studies on phyllotaxis. *K. Nederland. Akad. van Wetensch.* 63:137–150. 1960.

Van Tieghem, P. Sur quelques points de l'anatomie des Cucurbitacées. *Soc. Bot. de France Bul.* 29:277–283. 1882.

Van Tieghem, P., and H. Douliot. Sur la polystélie. *Ann. des Sci. Nat., Bot. Ser.* 7. 3:275–322. 1886.

Verdoorn, F., ed. *Manual of pteridology.* The Hague, Martinius Nijhoff. 1938.

Warden, W. M. On the structure, development and distribution of the endodermis and its associated ducts in *Senecio vulgaris. New Phytol.* 34:361–385. 1935.

Wardlaw, C. W. Experimental investigations of the shoot apex of *Dryopteris aristata* Druce. *Roy. Soc. London, Phil. Trans. Ser. B.* 232:343–384. 1947.

Wardlaw, C. W. *Phylogeny and morphogenesis.* London, Macmillan and Company. 1952.

Weber, H. Histogenetische Untersuchungen am Sprossscheitel von *Espeletia. Mainz. Akad. der Wiss. u. der Lit., Math.-Nat. Kl. Abhandl.* 1956:567–618. 1956.

Wetmore, R. H. Leaf-stem relationships in the vascular plants. *Torreya* 43:16–28. 1943.

Wetmore, R. H., and R. Garrison. The growth and organization of internodes. *Recent Adv. in Bot.* 1961:827–832. 1961.

Wetmore, R. H., and J. P. Rier. Experimental induction of vascular tissues in callus of angiosperms. *Amer. Jour. Bot.* 50:418–430. 1963.

Wetmore, R. H., and S. Sorokin. On the differentiation of xylem. *Arnold Arboretum Jour.* 36:305–317. 1955.

Wetter, R., and C. Wetter. Studien über das Erstarkungwachstum und das primäre Dickenwachstum bei leptosporangiaten Farnen. *Flora* 141:598–631. 1954.

Zasche, H. Untersuchungen über den Wundverschluss von Längswunden bei *Corylus colurna. Ztschr. f. Bot.* 48:304–329. 1960.

Zimmermann, A. *Die Cucurbitaceen.* Heft 1. *Beiträge zur Anatomie und Physiologie.* Jena, Gustav Fischer. 1922.

16

The Leaf

CONCEPT

The leaf is the principal appendage, or lateral organ, borne by the stem. As was emphasized in chapter 15, the term organ is applied to the leaf in a purely descriptive sense; the leaf and the stem are parts of one unit, the shoot. But the concept of the leaf as a morphologically distinct entity has not been completely abandoned (Troll, 1939); one school of thought even divides the leaf in two basic elements, the lower leaf (Unterblatt) and the upper leaf (Oberblatt; Weberling, 1955). Leaves are commonly classified into microphylls and macrophylls (or megaphylls) according to their presumed phylogenetic origin (Eames, 1936). A microphyll, as found, for example, in *Lycopodium, Selaginella, Isoetes*, and *Psilotum*, is interpreted as a lateral outgrowth of the stem (enation theory of leaf origin). The macrophyll, characteristic of Pteropsida, is regarded as having been derived from a branch that became limited in growth and assumed a leaf-like form. Because of the determinate growth of such a branch, the main branch overtops it and the leaf appears as a lateral appendage (overtopping theory of leaf origin). The concept of microphylls and macrophylls as two radically distinct formations is not universally accepted. It has been suggested that the leafless condition of the Rhyniales, presumed to typify precursors of leafy plants, is not truly primitive; that microphylls are reduced macrophylls; and that the organizational and histological aspects of development of the lateral appendages are similar in all vascular plants (Wardlaw, 1957).

The leaf usually contains the same tissue systems as the stem—the dermal, the vascular, and the fundamental. (The microphyll has little or no vascular tissue.) The epidermis forms the outermost layer, and the vascular tissue is variously distributed in the ground tissue. Authors who use the concept of the stele regard the vascular system of the leaf as a prolongation of the stelar tissue of the stem and homologize the ground tissue of the leaf with the cortex.

Although fundamentally alike in structure, the stem and the leaf differ from each other in details of growth and in relative arrangement of tissues. The leaf shows determinate apical growth as contrasted with the open, indeterminate type of growth exhibited by the stem in its apical meristem. The structural differences of the two organs appear to be related to their principal functions. In the stem the columnar shape, the vertical orientation of the vascular system, and the abundance of mechanical elements and of storage parenchyma suggest efficiency in longitudinal conduction of materials, support of the aerial body, and storage of food. In the ordinary foliage leaf the relatively large external surface, the extensive air space system, the abundance of chloroplasts in the ground tissue, and the close spatial relation between the vascular and the ground tissues suggest a specialization related to photosynthesis. These characteristics favor exposure of the chloroplasts to the light and give ready access of water and gases to the cells concerned with photosynthesis.

The structural distinction between the stem and the leaf is enhanced by certain concomitants of the specialization related to photosynthesis (Wylie, 1947). In contrast to the stem, the foliage leaf commonly lacks storage tissues, develops no periderm, and consists mainly of primary tissues. In the absence of any substantial amount of secondary growth, the leaf is restricted in its capacity to restore its tissues, which are constantly exposed to weathering and other injurious outside influences. In perennial plants new leaves are repeatedly formed and the old ones shed. Thus, the leaves are usually limited in growth and longevity and are restricted in mass.

Some investigators emphasize the similarity between the stem and the leaf by pointing out that detached and cultured leaves may be induced to produce adventitious roots and secondary tissues from a vascular cambium (Gupta, 1960; Samantarai and Kabi, 1954); and that even reproductive axes may arise as outgrowths from true leaves (Stork, 1956). These phenomena remind one of Arber's (1950) interpretation of the leaf as a partial shoot.

The concept of leaf is applied in the seed plants to many forms of lateral appendages of the axis varying in structure and function. This variation requires a segregation of the foliar organs into different types. A common classification distinguishes among foliage leaves, cataphylls, hypsophylls, and cotyledons. Foliage leaves are the principal photosynthetic organs. Cataphylls (from the Greek words *cata*, down, and *phyllon*, leaf, meaning leaves inserted at low levels of plant or shoot) are exemplified by scales occurring on buds and underground stems; they are concerned with protection or storage or both. Hypsophylls (from the Greek words *hypso*, high, and *phyllon*, meaning leaves inserted at high levels of the plant) are represented by the various floral bracts, some possibly protective in function. The first cataphylls on a lateral branch are called *prophylls* (from the Greek *pro*, before, and *phyllon*). Most monocotyledons have one prophyll (fig. 16.18D), some have two; the dicotyledons usually have two (chapter 15) but may have one (Bugnon, 1952–53; Rüter, 1918). The cotyledons are the first leaves of the plant. If the flower is interpreted as a modified shoot, the floral organs also constitute a type of foliar structure. The various kinds of leaf-like organs just enumerated may intergrade with each other by transitional forms, and each, especially the foliage leaf, varies widely in external form and anatomy.

A generalized term for the foliar members of the plant in the morphological literature is *phyllome* (Arber, 1950). Phyllomes include foliage leaves (or simply leaves), scales, bracts, and floral appendages. The differences in structure and form of phyllomes result from early divergencies in method of growth, distribution of meristems, and rates of maturation (Cross, 1938; Foster, 1928, 1931, 1936). If phyllomes have a common origin phylogenetically, their divergences have arisen as modifications of their ontogenies.

MORPHOLOGY OF THE FOLIAGE LEAF

This chapter deals almost entirely with the foliage leaf. The variations in the structure of a foliage leaf are manifold. In angiosperms the main part of the photosynthetic tissue is typically expanded into a flattened structure, the *blade*, or *limb*, or *lamina* (from the Latin, thin leaf). In sessile (in Latin, sitting) leaves, this blade is directly attached to the stem, in others it is attached by means of a stalk, the *petiole* (in Latin, foot). In most monocotyledons and certain dicotyledons (Polygonaceae, Umbelliferae) the base of the leaf is expanded into a *sheath* around the stem. Leaves may be simple or compound. A simple leaf has one blade. In a compound leaf two or more blades, the *leaflets*, are attached to a common axis or *rachis* (in Greek, backbone). The lamina of a leaf, or of a leaflet in the compound leaf, varies greatly in shape and size. There are lancet-shaped leaves and various broad forms. Others are cylindrical or somewhat flattened, as the needles of the conifers. Leaves may lack a blade and have the petiole simulating the blade (a *phyllode*). In some plants the leaves are fleshy and contain relatively large amounts of nonphotosynthetic tissue; in others the foliar structures are mere scales, and the main photosynthetic activity occurs in the chlorenchyma of the stem. Sometimes stems specialized with relation to photosynthetic activity are flattened like leaves and are then called *cladodes* (from the Greek *clados*, branch).

Leaves may have basal appendages, the *stipules* (from the Latin, stubble), or they may be exstipulate. A relationship exists between the type of nodal anatomy and the occurrence of stipules and leaf sheaths in dicotyledons (Sinnott and Bailey, 1914). Most plants with trilacunar nodes have stipules, whereas most of those with unilacunar nodes lack stipules, and plants with multilacunar nodes have leaves with sheathing bases.

The foliage leaves of extant gymnosperms are represented by widely divergent forms (Foster and Gifford, 1959). Cycadales have large pinnate leaves and *Ginkgo* has the well-known petiolate fan-shaped leaf. In the Coniferales the leaves are always simple and most commonly have the form of needles or scales (Laubenfels, 1953). In the Gnetales, the leaves of *Ephedra* are scale-like and inconspicuous; those of *Gnetum* are petiolate and have a lamina resembling that of dicotyledons; and *Welwitchia* is unique with its two huge strap-shaped leaves that continue to elongate for years through the activity of the meristematic base of the blade (Rodin, 1958).

In discussions of the form and anatomy of the leaf, it is customary to designate the leaf surface that is continuous with the surface of the part of the stem located above the leaf insertion as the upper, ventral, or adaxial side; the opposite side as the lower, dorsal, or abaxial.

HISTOLOGY OF ANGIOSPERM LEAF

The foliage leaves vary greatly in their internal structure and the differences are related to taxonomic groupings and to evolutionary adaptations of plants to different habitats. In the following references, the basic features of angiosperm and gymnosperm leaves are reviewed. More or less extensive treatments of leaf structure with reference to ecology are available in reviews and research papers (Grieve, 1955; Hasman and Inanç, 1957; Jones, 1955; Morretes and Ferri, 1959; Philpott, 1956; Shields, 1950; Stålfelt, 1956; Vasilevskaĭa, 1954; see also Esau, 1960). Broad surveys of leaf structure from the taxonomic aspect are given in the series from Kew (Metcalfe, 1960; Metcalfe and Chalk, 1950; Tomlinson, 1961).

The Leaf

Epidermis

The complex morphologic and physiologic organization of the leaf epidermis makes the term and the concept of epidermal tissue system highly appropriate for this part of the plant body. The epidermis of leaves is composed of the various types of epidermal cells described for the aerial parts of plants in chapter 7: epidermal cells, composing the main mass of the epidermal tissue; guard cells of the stomata, commonly accompanied by subsidiary cells; various trichomes; silica and cork cells in the Gramineae; bulliform cells in various monocotyledons (pl. 70A), fiber-like cells in various groups of plants (chapter 10). Stomata are particularly characteristic of leaves and occur either on one, or the other, or both surfaces, but they are most common on the abaxial surface (fig. 16.2, pl. 72A). The multiple epidermis described at some length in chapter 7 is most commonly encountered in leaves (fig. 16.2A). The subsurface cells of such an epidermis often are large, thin walled, and colorless and are interpreted as water storage cells.

In the terrestrial higher vascular plants, the leaf epidermis is a living

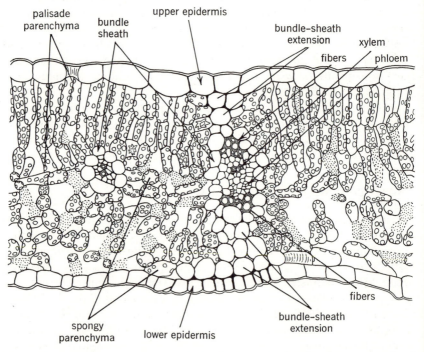

Fig. 16.1. Transverse section of pear leaf. Both vascular bundles shown are enclosed in bundle sheaths, but only the larger has bundle-sheath extensions reaching to epidermis on both sides of leaf. Mesophyll cells with chloroplasts, except those having crystals. Bundle-sheath cells with few chloroplasts (not shown in drawing). (×285.)

tissue and typically contains no well-differentiated chloroplasts. Certain plants, however, contain abundant chlorophyll in the epidermis. Water plants may show more abundant chloroplasts in the epidermis than in the parenchyma beneath it (Sauvageau, 1891). Small amounts of chlorophyll are found in plastids of the leaf epidermis of angiosperms with and without relation to any special environmental conditions (chapter 7).

Except for the presence of intercellular spaces between the guard cells of the stomata and those associated with hydathodes, the epidermis of foliage leaves usually shows compact organization. The continuity of the epidermis is one of the features that contributes toward its effectiveness in protecting the leaf tissues from excessive loss of water and in offering mechanical support. The anticlinal walls of epidermal cells that are located in the interveinal areas may be undulate (pl. 72A).

The wall structure of the leaf epidermis varies widely, the most constant characteristics being the presence of cutin in the walls, notably the outer, and of cuticular layers on the surface. The epidermal walls may be thin in plants requiring moderately moist, or mesophytic, habitats (mesomorphic plants) and in water plants (hydromorphic plants). In xeromorphic plants, that is, plants that can subsist in dry, or xerophytic, environments, the epidermis may have thick, lignified walls. The deposition of silica in the epidermal walls, sometimes in the form of opals completely filling the cell lumen (Parry and Smithson, 1958), is characteristic of grasses and related plants.

Mesophyll

The ground tissue of the leaf which is enclosed within the epidermis is called the mesophyll (from the Greek words *mesos*, in the middle, and *phyllon*, leaf). The mesophyll is usually specialized as a photosynthetic tissue. It is living, lacunose (that is, with many intercellular spaces) parenchyma containing chloroplasts. In many plants, particularly in dicotyledons of the mesomorphic type, the mesophyll is commonly differentiated into palisade and spongy parenchyma (figs. 16.1, 16.2, pls. 73A, 77A). The palisade tissue consists of cells elongated at right angles to the epidermis and arranged like a row of stakes (hence the use of the word *palus*, from the Latin stake). The spongy parenchyma appears less regular, and its name has reference to the conspicuous intercellular-space system permeating it.

The individual cells of the palisade parenchyma are commonly elongate-prismatic in shape, but may range from nearly isodiametric to several times longer than wide (Meyer, 1962). In some plants the palisade cells are irregular in shape, having either relatively small lateral protuberances or long arm-like processes that make the whole cell appear branched (fig. 16.2B).

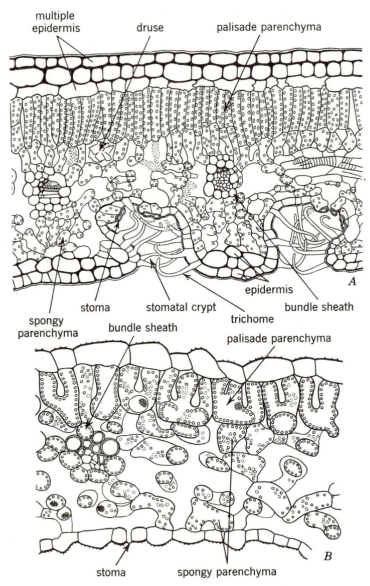

multiple
epidermis   druse   palisade parenchyma

                                                           A

                                               epidermis
        stoma   stomatal crypt   bundle sheath
  spongy   trichome
parenchyma   bundle sheath
                                      palisade parenchyma

                                                           B
      stoma   spongy parenchyma

Fig. 16.2. Transections of leaves, *A*, *Nerium oleander* (dicotyledon) and, *B*, *Lilium* (monocotyledon). In *A*, multiple epidermis; stomata in stomatal crypts. In *B*, arm-palisade cells. (Both, ×300.)

Palisade cells occur beneath the epidermal surface layer (pl. 73A), un-less there is a multiple epidermis (fig. 16.2A) or a specialized hypo-dermis. There may be more than one layer of palisade cells (fig. 16.1), with a uniform or variable length of the cells in the different layers. Frequently, in such a multiseriate palisade tissue, the outermost cells are the longest, the innermost the shortest.

In mesomorphic plants of temperate regions, the palisade mesophyll is commonly restricted to the adaxial side of the leaf. In xerophytes the palisade tissue often occurs on both sides of the leaf, with the spongy tissue much reduced or absent. The increase in proportion of palisade tissue in relation to xeromorphy occurs in both dicotyledons and monocotyledons (Kasapligil, 1961; Shields, 1950). Leaves with relatively undifferentiated mesophyll, as found in many hydrophytes, have no palisade tissue.

If the palisade tissue occurs on one side of the leaf blade, and the spongy tissue on the other, the leaf is called *dorsiventral,* that is, having distinct dorsal and ventral sides. If the palisale parenchyma is present on both sides, the leaf is *isolateral,* that is, has equal sides. A modification of the isolateral leaf is the *centric,* found in narrow cylin-drical leaves, with the adaxial and abaxial mesophyll continuous (Met-calfe and Chalk, 1950).

The spongy parenchyma shows many forms of cells, either nearly iso-diametric, or elongated in the same direction as the palisade cells and connected with each other by lateral extensions of various lengths, or, most commonly, elongated parallel with the surface of the leaf (pl. 72B). The spongy mesophyll shows a wider variety in organization than does the palisade mesophyll.

The distinction between palisade and spongy tissues varies in degree of sharpness. If the palisade mesophyll is multiseriate, there is com-monly a transition between the two kinds of mesophyll, the innermost layer of palisade parenchyma approaching the spongy parenchyma in shape, size, and arrangement of cells.

The degree of differentiation of the mesophyll and the proportion of palisade and spongy parenchyma vary in relation to plant species and habitat. A well-known phenomenon is the stronger development of the palisade tissue in leaves exposed to light during differentiation (sun leaves) as contrasted with leaves differentiating in the shade (shade leaves). Differences in mesophyll structure occur in leaves developing at different levels of the same plant, a phenomenon ultimately related to light conditions during the growth of the various leaves. Xeromor-phic leaves have a relatively more strongly developed palisade tissue than mesomorphic leaves.

In most grasses of the temperate regions the mesophyll is not differentiated into palisade and spongy parenchyma (pl. 70A, B) and generally appears rather homogeneous. In many tropical grasses the mesophyll cells surrounding the vascular bundles are oriented with their long diameters at right angles to the bundle. Thus, in transections, the mesophyll cells appear to radiate from the vascular bundles (Metcalfe, 1960).

The basic contrasts in the organization of palisade and spongy parenchyma suggest different functional specialization of the two tissues. The palisade parenchyma appears to be the most highly specialized type of photosynthetic tissue (Meyer, 1962). In a leaf with a mesophyll differentiated into palisade and spongy parenchyma, most of the chloroplasts occur in the palisade parenchyma. Examples of comparative percentages of chloroplasts in the palisade and spongy tissues, respectively, are *Fragaria elatior*, 86 and 14; *Ricinus communis*, 82 and 18; *Brassica rapa*, 80 and 20; *Helianthus annuus*, 73 and 27; *Phaseolus multiflorus*, 69 and 31 (Schürhoff, 1924). It is commonly assumed that, because of the shape and arrangement of the palisade cells, the chloroplasts are brought into a most favorable position with reference to the light. During active photosynthesis the chloroplasts line the walls, one layer in thickness (pls. 72C, 73A). The considerable wall surface in the narrow cells of the most common type of palisade tissue accommodates numerous chloroplasts in a single layer. In the wider cells arm-like protrusions or ridges increase the wall surface (fig. 16.2B).

Another well-known concept is that the lacunose structure of the mesophyll makes possible a thorough gaseous exchange between the outside air and the photosynthetic tissue. Because of the large intercellular system in the mesophyll, an extended cell-wall surface is exposed to the intercellular air; that is, the mesophyll has a large surface. This surface is called the internal surface of the leaf as contrasted with the external surface, that is, the surface of the epidermis exposed to the external air.

The extent of the development of the internal aerating system may be illustrated by means of figures. The proportion of air by volume in normal leaves varies between 77 parts per 1,000 in *Camphora officinalis* and 713 parts per 1,000 in *Pistia texensis* (Sifton, 1945). Data comparing the internal with the external surface of leaves are also instructive. In a study of the entire foliage of a 21-year-old *Catalpa* tree, the internal surface of 5,100 m^2 was found to be associated with 390 m^2 of external surface (Turrell, 1934). The relative extent of the two surfaces varies in different ecologic types of leaves. In certain dicotyledonous leaves the ratios of internal to external surface were found to be relatively low in shade leaves (6.8 to 9.9), intermediate in mesomorphic leaves (11.6 to 19.2), and high in xeromorphic sun leaves (17.2 to 31.3;

Turrell, 1936). The size of leaves also affects this ratio. Large alfalfa leaves were found to have a larger volume of intercellular spaces and a higher ratio of internal to external surface than small leaves (Turrell, 1942). The intercellular space system may be continuous throughout the leaf (Williams, 1948) or restricted to isolated blocks (Meidner, 1955).

As in the numbers of chloroplasts, the palisade tissue demonstrates its high degree of specialization with regard to photosynthetic activity also in its relation to the intercellular space system. Although the spongy parenchyma has much larger intercellular spaces than the palisade tissue (pls. 72B, C; 73A), the palisade tissue has a larger free surface. An analysis of leaves of several species of dicotyledons showed that per unit volume of leaf tissue the palisade tissue exposes to the intercellular air 1.6 to 3.5 times the surface exposed by the spongy parenchyma (Turrell, 1936). In view of these data the high ratio of internal to external surface in sun leaves is explained by the higher proportion of palisade cells in such leaves. The high ratio of internal to external surface may be accompanied by a high concentration of chlorophyll (Turrell, 1939). The ratio of internal to external surface is strongly and positively correlated with the rate of transpiration (Turrell, 1944). Thus, the structure favorable for photosynthesis induces at the same time a high loss of water (Stålfelt, 1956). The compactly arranged, cutinized, and cuticularized epidermis and the presence of a fine film of fatty material on the walls of the mesophyll cells exposed to the intercellular spaces (chapter 3) apparently reduce but do not completely control the high transpiration rate which is a concomitant of the structural specialization for photosynthesis (Wylie, 1947).

In some plants, particularly in those of aquatic or marshy habitat, the mesophyll assumes the character of an aerenchyma (chapter 8). Aerenchymatic and ordinary mesophyll develop the intercellular spaces mainly by schizogeny (chapter 3). But in some species the intercellular spaces result from a breakdown of parenchyma cells, probably by tearing, or rexigeny (*Oryza*, Kaufman, 1959; *Typha, Juncus*, Sifton, 1945; *Musa*, Skutch, 1927).

Vascular System

The arrangement of the vascular bundles, that is, the *venation*, imparts a characteristic appearance to leaves. The word venation is derived from the term *vein* (from the Latin *vena*, vein), which, in botany, is applied sometimes to a vascular bundle or a group of closely approximated bundles, sometimes to bundles together with the spatially associated nonvascular tissues. In this chapter the term vein denotes a vascular bundle or a group of closely spaced bundles.

A leaf may have a single vein or two or more. Examples of single-veined leaves are found among the conifers and in *Equisetum*, whereas multiveined leaves are common in the higher ferns and angiosperms. The two prevalent types of venation patterns in the angiosperms are the reticulate, or netted, and the parallel. In reticulate venation, which is widespread among the dicotyledons, vascular bundles of many sizes form an anastomosing network (fig. 16.4, 16.6, pl. 80A), with the smaller bundles diverging from the larger. In parallel-veined leaves, which are characteristic of monocotyledons, bundles of relatively uniform size are arranged longitudinally but converge toward each other and merge at the apex or at both ends of the blade or at its margins (fig. 16.3B-D). The longitudinal veins are laterally interconnected by small bundles throughout the blade. These connections are often arranged in a ladder-like manner (fig. 16.3D), but they may also form various other patterns (Schuster, 1910). Some monocotyledons have a modified parallel venation with the veins arranged longitudinally for a distance, then diverging laterally in a pinnate manner (fig. 16.10E; Troll, 1939). Parallel venation is found in some dicotyledons also (*Plantago, Trago-*

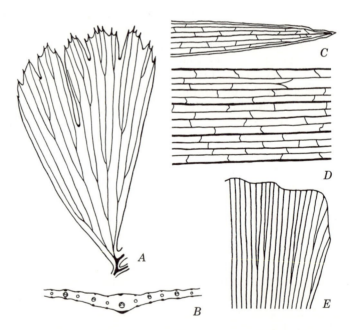

Fig. 16.3. Leaf-venation patterns. Cleared leaves (*A, C–E*) and transection (*B*). *A*, open dichotomous in a dicotyledon, *Kingdonia uniflora; B–D*, parallel in a grass, *Avena.* Cross anastomoses between longitudinal bundles in *C, D. E*, open dichotomous in *Ginkgo.* (*A*, from photograph in Foster, *Amer. Jour. Bot.* 47, 1960.)

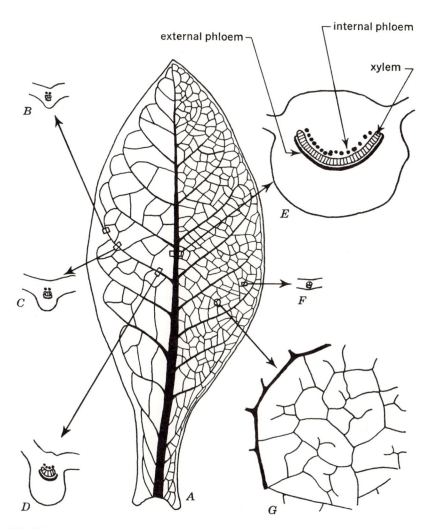

Fig. 16.4. Venation of *Nicotiana tabacum* leaf. *A*, mature leaf showing midvein, principal lateral veins, and coarse vascular network. In small transections of veins (*B–F*), xylem is hatched, phloem (external and internal) is black. Smaller veins (*F*) lack internal phloem. *G*, portion of blade showing smallest veins and their free endings in mesophyll. Leaf depicted was obtained from midway on the stalk and had 543 mm of veins to a square centimeter of lamina. (*A*, ×⅓; *G*, ×20. After Avery, *Amer. Jour. Bot.* 20, 1933.)

pogon), and, reciprocally, some monocotyledons have netted venation (Araceae, Smilacoideae, Taccaceae, Orchidaceae; Schuster, 1910).

Both systems, the reticulate and the parallel, are termed *closed* because the veins anastomose with one another. Two relic genera of dicotyledons, *Kingdonia* and *Circaeaster*, have been found to have an *open dichotomous* (repeatedly forked) venation (fig. 16.3*A*) resembling that of *Ginkgo* (fig. 16.3*E*) and of some ferns (Foster, 1963; Foster and Arnott, 1960). The adjective open refers here to the characteristic that large subdivisions of the system end freely in the interior of the leaf or at its margins.

In the anastomosing vascular system of dicotyledonous leaves the veins are of many sizes. The largest vein often occurs in a median position and forms the midvein, and the somewhat smaller veins diverge from it laterally (fig. 16.4; pinnately veined leaf). In other leaves there may be several large veins, comparable in size, spreading out from the base of the blade toward the margins (palmately veined leaf). The large veins commonly occur in enlarged portions of the blade which appear like ridges (ribs) on the abaxial surface of the leaf (figs. 16.4, 16.5). These ridges consist of parenchyma with a relatively small amount of chlorophyll and some supporting tissue, usually collenchyma. The vascular bundles of the large veins are imbedded in the parenchyma and are thus somewhat separated from the mesophyll proper (fig. 16.5). The small veins (minor venation), in contrast, form a network among the large veins within the mesophyll. They occur in the median part of the mesophyll, usually beneath the palisade cells, that is, in the uppermost layer of the spongy parenchyma (figs. 16.1, 16.2).

The minor venation of dicotyledonous leaves exhibits a wide range of intergrading patterns (fig. 16.6). The branchings of these veins subdivide the mesophyll into a series of successively smaller polygons, with the ultimate branches, the bundle or vein endings, extending into the smallest subdivisions of the mesophyll, the areole, and terminating there freely. The areoles may lack free vein endings (fig. 16.6*A*). Lineolate venation patterns, with parallel-oriented minor veins, occur in the Quiinaceae (pl. 80*B, C*; Foster, 1952) and Rubiaceae (Pray, 1959).

In monocotyledons the longitudinal bundles may be of almost equal thickness, or they may vary in size, the larger veins alternating with the smaller. The median bundle may be larger than the others and associated with a prominent rib (fig. 16.3*B*). The lateral veins may or may not form ribs. In some large grasses the median part of the blade is thickened into a midrib by the differentiation of massive colorless parenchyma on the adaxial side (fig. 16.10*H*). Numerous vascular bundles occur in such a midrib. In many monocotyledons the smallest

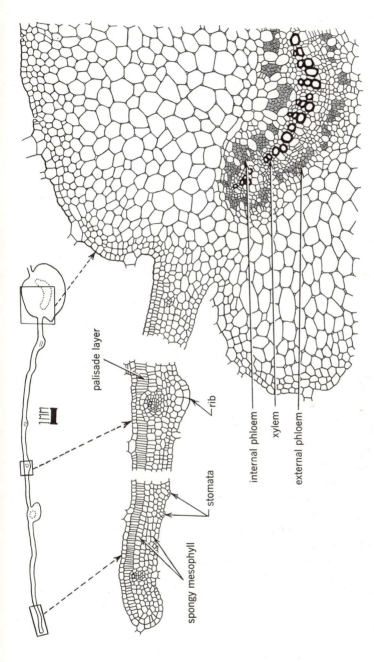

Fig. 16.5. Structure of leaf of *Nicotiana tabacum*. Transections including midvein, smaller vein, and some mesophyll. From lamina midway between base and tip of a leaf 135 mm long. Leaf was not fully differentiated as evidenced by densely arranged mesophyll cells. (After Avery, *Amer. Jour. Bot.* 20, 1933.)

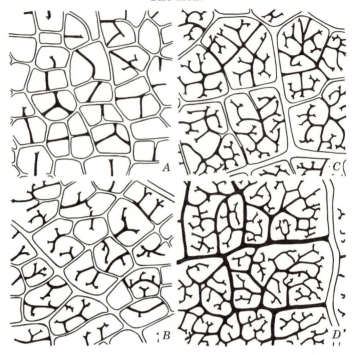

Fig. 16.6.   Veins and bundle-sheath extensions.   Double lines, veins with sheath extensions; solid black, no sheath extensions.   A to D, density of veins with sheath extensions decreases whereas complexity of minor venation increases.   Negative correlation between smallest distance from vein to vein (vein spacing) and spacing of sheath extensions.   Spacing in microns are A to D: 124, 103, 89, 85 for veins; 199, 225, 378, 1581 for sheath extensions.   A, Tilia americana; B, Quercus marcocarpa; C, Morus alba; D, Ricinus communis.   (After Wylie, Iowa Acad Sci. Proc. 53, 1947.)

bundles extend from one large vein to another, but in some, free vein endings occur in the mesophyll (Pray, 1955b; Schuster, 1910).

The venation of cataphylls, hypsophylls, and cotyledons is similar to but simpler than that of the foliage leaves of the same plant; it appears as though it were ontogenetically underdeveloped (Höster and Zimmermann, 1961; Müller, 1944).

The histologic composition of vascular bundles of various sizes shows quantitative and qualitative differences.   The largest bundles contain xylem and phloem in amounts comparable to those of the bundles in the petiole or the leaf trace.   In the collateral bundles the xylem occurs on the adaxial, the phloem on the abaxial, side (fig. 16.1).   If the leaf traces are bicollateral, the adaxial phloem occurs also in the leaf, but may lack in small veins (figs. 16.4, 16.5).   The vascular tissue of the largest veins in the dicotyledonous leaves forms either one bundle or several (fig. 16.9; Plymale and Wylie, 1944). As seen in transections of veins, the vascular bundles may be arranged in a circle (fig. 16.9E; *Liriodendron, Vitis*) or a semicircle (*Ambrosia*), or they may be distributed irregularly (*Silphium, Helianthus*). When the vascular bundle is single, it is crescent-shaped in some plants (fig. 16.9C; *Cercis, Ulmus, Tilia, Abutilon*), circular in others (fig. 16.9G; *Catalpa, Acer, Quercus*). The larger veins in dicotyledonous leaves may have primary and

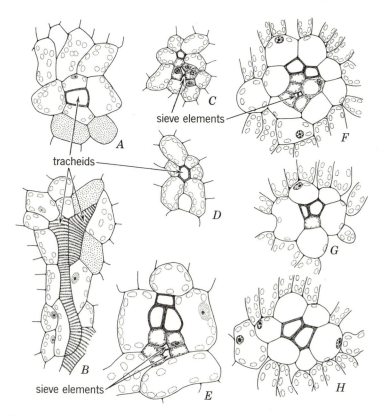

Fig. 16.7. Structure of small leaf veins in dicotyledons. *B*, tangential section, all others transverse. *A, B, Vitis vinifera,* bundle ends consist of tracheids enclosed by bundle-sheath cells. (Stippling indicates tannins.) *C, D, Humulus,* one bundle with tracheids, sieve elements, and some parenchyma cells, the other (bundle end) with single tracheid. *E, Nicotiana tabacum,* small bundle with tracheary elements, sieve elements, and parenchyma. *F–H, Prunus* (peach). *F,* two tracheary elements, two sieve elements, and some parenchyma; *G,* two tracheary elements and a parenchyma cell occupying the position of phloem; *H,* two tracheary elements (bundle end). Bundle-sheath cells have relatively numerous chloroplasts in *A–E,* few or none in *F–H.* (*A, B,* ×470; *C–H,* ×600.)

secondary tissues; the smaller are usually entirely primary. Cambial activity is more pronounced in leaves of evergreen species than in those of the deciduous (Shtromberg, 1959). Veins of various sizes, but not the smallest, have vessels in the xylem and sieve tubes in the phloem. In the small veins the tracheary elements are represented by tracheids. Near the ends of the ultimate vein branchings the phloem part may contain only parenchyma (fig. 16.7G), and the vein endings in the dicotyledons frequently consist of tracheids only (fig. 16.7A, B, D, H). But in some species sieve elements accompany the tracheids to the ends, or even extend further than the tracheids (Morretes, 1962; Pray, 1955b). The sieve elements of the bundle ends are often associated with exceptionally large companion cells. The tracheids usually have helical and sometimes annular thickenings. At the very end of the bundle there may be a single tracheid, a pair of tracheids lying parallel to each other, or an irregular group of them (Strain, 1933). Sclereids may differentiate beyond the last tracheids of the vein endings, in contact with these elements (pl. 73B; Foster, 1946, 1947). In some genera large ovoid or irregularly branched tracheids, frequently with pitted walls, terminate the veinlets. These cells are sometimes interpreted as water reservoirs and are called storage tracheids (Pirwitz, 1931).

The small veinlets of monocotyledons also have few conducting elements. The transverse anastomoses in grass leaves may contain a single file of tracheary elements and a single file of sieve-tube elements (fig. 16.8B–E). In the small vascular bundles of monocotyledons, as well as dicotyledons, the sieve elements may occur next to the tracheary elements (fig. 16.8; Morretes, 1962; Pray, 1955b).

The especially significant characteristic of the vascular system of the leaf is its close spatial relation to the mesophyll. Measurements performed on six species of dicotyledons, herbaceous, shrubby, and arborescent, have shown that the total length of the veins averaged 102 cm per sq cm of blade (Plymale and Wylie, 1944). The thorough distribution of the vascular tissue within the mesophyll is illustrated by the small size of the areas free of veins. According to some measurements, the interveinal spacings in dicotyledonous leaves average about 130 microns (Wylie, 1939).

A significant correlation has been found to exist between vein distribution and those structural features of the nonvascular tissues of the leaf that may have an influence upon conduction. Thus, the larger the volume of tissue having relatively little lateral contact among its cells (pl. 72–C, palisade tissue)—an arrangement resulting in a comparatively low efficiency in lateral conduction—the closer together are the vascular bundles. On the contrary, the greater the amount of tissues with extensive lateral contacts among the component cells (such as

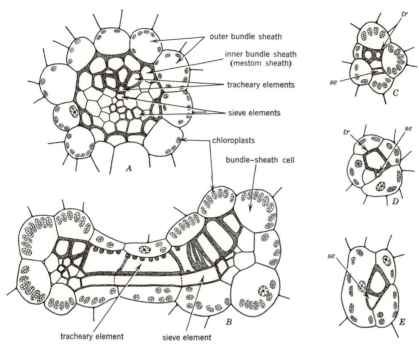

Fig. 16.8. Small vascular bundles of grass leaves. *A*, longitudinal bundle from a transection of *Triticum* leaf. *B*, two longitudinal bundles connected by a transverse bundle as seen in transection of *Zea* leaf. *C*, one of the smallest longitudinal bundles of *Zea*. *D*, *E*, transverse anastomoses in *Zea* as exposed in sections cut parallel with long axis of leaf and perpendicular to epidermal layers. Details: *se*, sieve element; *tr*, tracheary element. (All, ×540.)

epidermis, pl. 72*A*, and spongy parenchyma, pl. 72*B*), the larger are the interveinal distances (Philpott, 1953; Wylie, 1939). These data agree with the observation that sun leaves, in which the palisade tissue usually shows particularly strong development, contain a greater length of veins than shade leaves (Schuster, 1908).

Bundle Sheaths

As has been described above, the large vascular bundles of dicotyledonous leaves are surrounded by parenchyma with small numbers of chloroplasts, whereas the small bundles occur in the mesophyll. The small bundles are not in contact with intercellular spaces but are commonly enclosed within a layer of compactly arranged parenchyma, the bundle sheath (fig. 16.7). In dicotyledons the bundle-sheath parenchyma is also called border parenchyma.

The bundle sheaths of dicotyledonous leaves usually consist of cells elongated parallel with the course of the bundle and having walls as thin as those of the adjacent mesophyll. In some plants these cells have chloroplast complements similar to those of the mesophyll cells (fig. 16.7C–E); in others they have few or no chloroplasts (fig. 16.7F–H). Individual sheath cells may contain crystals. The bundle sheaths extend to the ends of the bundles and completely enclose the terminal tracheids (fig. 16.7B).

In many dicotyledons, plates of cells similar to those in the bundle sheath extend from the bundle sheath toward one or both epidermises, some terminating in the mesophyll, others reaching the epidermis (fig. 16.1, pl. 77A). These *bundle-sheath extensions* have received careful attention in the literature (Wylie, 1952). Measurements carried out in leaves of certain dicotyledons have shown that 99 percent of the total vein length is invested in sheath parenchyma (Armacost, 1945). In 10 species of mesomorphic species, bundle-sheath extensions were found along 58 per cent of the total vein length (Wylie, 1943). If the bundle sheaths and their extensions are concerned with conduction, their presence materially increases the contact between the mesophyll and the conducting cells.

Certain observations do suggest that the sheaths and their extensions take part in conduction. Potassium ferrocyanide solution introduced into leaves was found to move out through the veins into the bundle sheaths and through the sheath extensions into the epidermis where it spread throughout. Then a correlation was established between the distribution of veins and the presence of sheath extensions. If bundle-sheath extensions are numerous and spaced close together, the vascular net is less dense and less elaborate than if sheath extensions are less numerous (fig. 16.6; Wylie, 1947).

The parenchymatic bundle sheaths are the most common, but in certain dicotyledons bundles of various sizes are enclosed in sclerenchyma (Winteraceae, Melastomaceae; Bailey and Nast, 1944; Foster, 1947). In some of the Winteraceae even the terminal veinlets are ensheathed by sclerenchyma.

Bundle sheaths occur in monocotyledonous leaves also. The best known are those of the Gramineae (Schwendener, 1890). Grass leaves have two kinds of sheaths: entirely parenchymatic, with chloroplasts, and relatively thick-walled sheaths, without chloroplasts. The thick-walled sheath was termed *mestom sheath* by Schwendener, because mestom was previously used to designate the conducting elements of a vascular bundle. If present, the mestom sheath occurs next to the vascular tissue, and outside it is a second thin-walled sheath with chloroplasts.

Among Gramineae (Brown, 1958; Metcalfe, 1960) many representatives of Panicoideae have single thin-walled sheaths around the small vascular bundles (fig. 16.8B–E, pl. 70B)—the sheaths around the larger bundles may be relatively thick walled—whereas those of Festucoideae frequently have two sheaths (fig. 16.8A, pl. 70A). The inner, or mestom, sheath consists of elongated living cells with blunt or pointed ends. The wall thickening is variable, even in different parts of the same sheath, and is often pitted. Sometimes the inner walls are thicker than the outer. In small bundles the inner sheath may be restricted to the phloem side.

The bundle sheath of angiospermous leaves is an endodermis. Although the casparian strip is mostly indistinguishable, the walls and the contents of the sheath cells may react to various dyes and indicators like those of the typical endodermis of other plant parts (chapter 15). Furthermore, casparian strips have been detected in mestom sheaths in young leaves of certain Gramineae and Cyperaceae (Van Fleet, 1950).

A bundle sheath may be also a starch sheath. In some dicotyledons and in the grasses that have single-layered bundle sheaths the parenchymatic sheath forms starch (Rhoades and Carvalho, 1944). In genera like *Zea* and *Sorghum* the chloroplasts in the sheath cells are particularly large (fig. 16.8B–E) and appear to be the only plastids in the leaf concerned with forming starch during active photosynthesis. Ultrastructurally, these chloroplasts were seen to be devoid of grana in *Zea* (pl. 3B; Brown, 1960). In Festucoideae having double sheaths about the bundles (fig. 16.8A), the inner sheath contains no chloroplasts, and those in the outer are somewhat smaller than in the rest of the mesophyll. Starch is produced in all green cells so that the sheath is not visibly differentiated with regard to starch formation.

Developmentally, the bundle sheaths of the dicotyledons, the single parenchymatic sheaths in Panicoideae, and the outer of the two sheaths in Festucoideae appear to be part of the ground tissue. The inner, mestom sheath is probably of procambial origin.

Supporting Structures

In many leaves, supporting structures are not so prominently developed as in the stem, and much of the strength of such leaves depends on the arrangement of cells and tissues. In leaves with flat blades the soft mesophyll is partly supported by the vascular system which so completely permeates it. In dicotyledonous leaves the bundle sheaths, with their extensions reaching to the compactly arranged epidermis, probably also contribute to the support of the blade (Wylie, 1943). Typically, dicotyledonous leaves develop collenchyma beneath the epidermis over the large veins and often along the margin of the blade.

Some of the bundle-sheath extensions may be collenchymatically thickened. Many dicotyledonous leaves have sclereids within the mesophyll. Abundant development of sclerenchyma is common among xerophytes in which this tissue is thought to reduce the injurious effects of wilting (Stålfelt, 1956). Monocotyledonous leaves develop relatively large amounts of sclerenchyma, in the form of fibers, in association with the vascular bundles (pl. 70C) or in separate strands, a feature especially common in palms (Tomlinson, 1961). In grasses, strands of fibers occur on one or both sides of the vascular bundles and are connected to the bundle sheaths and also to the epidermis. The epidermis may have thick-walled long cells over the sclerenchyma strands so that all the sclerenchyma and the vascular bundles together form girder structures traversing the entire thickness of the blade.

The epidermis offers considerable support by its compact arrangement and the relatively strong walls impregnated with cutin and bearing a tough cuticle on the outer surface. In some plants, notably grasses, the epidermis is lignified and silicified in varying degrees.

Secretory Structures

Leaves bear diverse secretory structures concerned with a discharge of water from the interior, with or without appreciable amounts of dissolved materials. Among these materials are salts and complex organic substances, such as resins, mucilages, gums, oils, and nectar. (The secretory structures are described in chapter 13.)

Petiole

The tissues of the petiole are comparable to primary tissues of the stem. There is a close similarity between petiole and stem with regard to the structure of epidermis. The ground parenchyma of the petiole is like the stem cortex in arrangement of cells and in number of chloroplasts, which are fewer than in the mesophyll of the leaf blade. The supporting tissue of the petiole is collenchyma or sclerenchyma. These may have both disposition and structure similar to those in the stem. Sometimes, however, the petiole has one or the other kind of supporting tissue that may be absent in the stem. In relation to the arrangement of the vascular tissues in the stem, the vascular bundles of the petiole may be collateral, bicollateral, or concentric. Primary phloem fibers may differentiate in both the stem and the petiole, or the corresponding phloem cells may develop only thickened primary walls in the petiole (chapter 10).

The petioles of different plants show considerable variation in the distribution of the vascular tissues within the body of the petiole (figs.

16.9, 16.10; Bouygues, 1902; Petit, 1887). As seen in transections, the vascular tissues frequently form a continuous or a multistranded arc open toward the adaxial side of the petiole (fig. 16.9*B*, *D*, *L*; *Olea*, *Euonymus*, *Stellaria*, *Nicotiana*). The bundles may form a circle (*Ricinus*, *Paeonia*, *Aquilegia*, *Hedera*, *Geranium*, *Smilax*), sometimes with additional bundles within the circle and outside it (fig. 16.9*F*; *Tilia*, *Robinia*, *Juglans*, *Wistaria*, *Rhododendron*). The bundles may be numerous and arranged in several superposed arcs (fig. 16.10*G*; *Canna*, *Eryngium*, *Petasites*), or they may be scattered (fig. 16.10*D*; many monocotyledons, *Rumex*). The petiolar bundles are variously interconnected among themselves, so that their numbers and the patterns of their arrangement may vary from level to level (Gerresheim, 1913; Rippel, 1913).

If the petiole has only one collateral bundle, the phloem is on the abaxial side, the xylem on the adaxial (fig. 16.9*B*). In bicollateral bundles the phloem occurs on both sides of the xylem (fig. 16.9*D*). If the vascular tissues occur in arcs or circles, as seen in transections, the phloem is usually oriented toward the periphery of the petiole (fig.

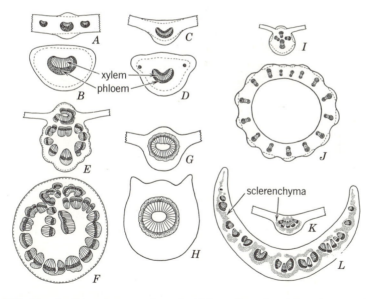

Fig. 16.9. Vascular tissues in transections of midribs and petioles of dicotyledonous leaves. *A, B, Euonymus; C, D, Nerium; E, F, Platanus; G, H, Citrus; I, J, Cucurbita; K, L, Mahonia.* In each pair of drawings the midrib appears above the petiole, and both are oriented with the adaxial side upward. (*A, B, ×9; C, D, ×4.7; E, F, K, L, ×8; G, H, ×6; I, J, ×3.*)

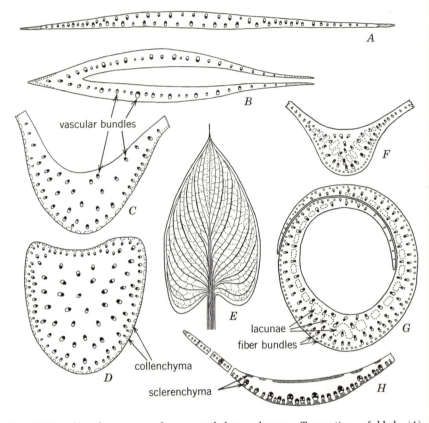

Fig. 16.10. Vascular system of monocotyledonous leaves. Transections of blade (*A*) and sheath (*B*) of *Iris* leaf. Transections of midrib (*C*) and petiole (*D*) and surface view (*E*) of *Zantedeschia* leaf. Transections of midrib (*F*) and sheath (*G*) of *Canna* leaf. *H*, transection of midrib and parts of blade of *Zea*. In vascular bundles xylem is black and phloem white. (*A–D, F, G,* ×4; *H,* ×6; *E,* approx. ½ size.)

16.9*B*, *H*). Other arrangements are found in petioles with numerous vascular strands (fig. 16.9*F*, *L*). If an endodermis is distinguishable, it may encircle individual bundles or an entire complex of bundles.

The rachis and the pedicels of leaflets in compound leaves are comparable in structure to the petioles of simple leaves, but the amounts of tissues in the pedicels are relatively small.

The petioles of some plants (Leguminosae, Oxalidaceae, Marantaceae, Aroideae) have pad-like swellings, the pulvini (chapter 4), that are able to curve and thus change the position of the leaf or leaflet (Weintraub, 1952). The movements of the leaves may be stimulated by environmental factors (light, gravity) or may be autonomous. Pulvini differ anatomically from the other parts of the petiole (Arslan, 1954; Brauner and Brauner, 1947). The vascular tissue is grouped in the center and the periphery is occupied by parenchyma (fig. 16.11). The pulvinus appears swollen because of the large volume of parenchyma, and its surface is often wrinkled. The changes in the curvature of the pulvinus depend on a differential contraction and expansion of cells, whereas the flexibility of the whole structure is assured by the anatomic peculiarities. Various explanations have been advanced to account for the changes in the volume of cells (Weintraub, 1952). One of these relates the changes to the activity of specialized contractile vacuoles (Datta, 1959–60). In *Mimosa* a thread-like structure, which is constantly displaced by cytoplasmic streaming, is attached to the vacuole (Toriyama, 1960, 1962).

HISTOLOGY OF GYMNOSPERM LEAF

Considerable work has been done on leaves of gymnosperms, some of comparative and systematic nature (Feustel, 1921; Florin, 1931; Fulling,

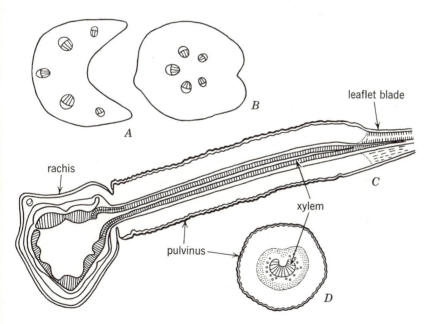

Fig. 16.11. Structure of pulvini. *A*, petiole and, *B*, pulvinus in transections of leaf of peanut (*Arachis hypogaea*). *C*, *D*, pulvinus of *Robinia pseudoacacia*. *C*, longitudinal section of pulvinus of leaflet attached to rachis in transection. *D*, transection. Compact arrangement of vascular tissue in pulvinus in *B–D*. Wrinkled surface in pulvinus in *C* and *D*. (*A*, *B*, ×20; *C*, ×20; *D*, ×25. *A*, *B*, based on photographs in Yarbrough, *Amer. Jour. Bot.* 44, 1957; *C*, *D*, after Brauner and Brauner, *Rev. Fac. Sci. Univ. Istanbul,* 12, 1947.)

1934; Gathy, 1954; Orr, 1944; Sprecher, 1907), others more limited in scope. Among the coniferous leaves the pine needle has been studied in most detail (Huber, 1947; Strasburger, 1891; Sutherland, 1933).

The needle-like leaves of conifers have a low ratio of surface to volume, a typical xeromorphic character. The pine needle, as seen in transection, is semicircular (pl. 78A), or triangular, or terete. The shape depends on the number of needles in the fascicle borne on the short shoot (Dolivo, 1948). The center of the needle is traversed by one or two vascular bundles surrounded by a special vascular tissue, called *transfusion tissue,* and a thick-walled endodermis. Outside the endodermis is the mesophyll. The periperal layers are the hypodermis and the epidermis.

As in other conifers the leaf epidermis in the pine is heavily cuticularized and has such thick walls that the cell lumina are almost obliterated (chapter 7, pl. 79). The fiber-like cells of the hypodermis also have thick walls and form a compact layer interrupted only beneath the stomata (pl. 79). The occurrence and the arrangement of the hypodermal sclerenchyma vary in different conifers, and some lack this tissue entirely. The epidermis bears numerous stomata on one or all sides in the different conifers. In many genera, including *Pinus,* the stomata occur in longitudinal rows parallel with the vascular bundles. The front cavity of the stoma is typically filled with a whitish or dark, granular or alveolate occluding material. Since this material is porous and the pores are filled with air, the stomata appear white superficially, a circumstance that facilitates their recognition from the surface. The guard cells are sunken and are overtopped by the subsidiary cells (chapter 7, pls. 77D, 79A).

The mesophyll cells have internal ridges on the walls projecting into the cell lumina (pl. 79B; Reinhardt, 1905). In the pine and some other conifers the mesophyll is not differentiated into palisade and spongy parenchyma (pl. 78A). Certain conifers (*Abies, Cunninghamia, Dacrydium, Sequoia, Taxus, Torreya*) and other gymnosperms (*Cycas, Ginkgo*) show such differentiation, and some have palisade parenchyma on both sides (*Araucaria, Podocarpus*). The mesophyll cells of *Pinus* and other conifers are arranged in horizontal layers separated from one another by intercellular spaces (pl. 77C, D). The horizontal strata are not completely detached from each other. Interconnecting files of cells make the whole tissue appear as an anastomosing system with a prevailingly horizontal (anticlinal) orientation of the spaces (Cross, 1940).

Gymnosperm leaves have resin ducts in the mesophyll. Their number varies even in individual genera, although there appears to be a constant minimum. In *Pinus* two lateral ducts occur rather consistently (pl. 78A); others may be present, variable in number and position. The resin ducts of *Pinus* are lined with thin-walled secretory epithelial cells. Outside these cells is a sheath of fibers with thickened lignified walls (pl. 79B). This sclerenchyma is in contact with the hypodermis. The resin ducts of conifer leaves vary in length. Some are continuous from the leaf into the cortex of the stem (*Cryptomeria, Cunninghamia;* Cross, 1941, 1942); others are restricted to the leaf, sometimes in the form of elongated cysts or sacs (*Picea;* Marco, 1939).

The vascular system of gymnosperm leaves ranges from a single vein in median position, as is common in conifers, to complex branched venations, open dichotomous in *Ginkgo* and most cycads, reticulate in *Gnetum.* In transverse sections of pine needles the vascular bundles are oriented somewhat obliquely, with the xylem pointing toward the adaxial side, the phloem toward the abaxial side (pl. 78A). The xylem is endarch. The protoxylem is partly crushed in mature needles. Outward from the crushed elements are some helically thickened tracheids —probably also part of the protoxylem—then some metaxylem tracheids with bordered pits. The primary xylem elements are in radial rows, and the rows of tracheary elements are interspersed with rows of parenchyma cells oriented like the rays in a secondary tissue. The individual parenchyma cells are vertically elongated and have transverse end walls. The sieve cells are also in radial rows alternating with rows of parenchyma cells. The parenchyma of the phloem is more abundant than that of the xylem. In the xylem the parenchyma forms starch. In the phloem some parenchyma cells form starch, whereas others appear to be albuminous cells (chapter 12) which lack starch but have dense cytoplasm. Some parenchyma cells have crystals. As a rule pine needles are shed in the third year, sometimes in the fourth. The vascular bundles increase somewhat in thickness after the first year through the activity of a vascular cambium (Strasburger, 1891).

The transfusion tissue surrounding the vascular bundles in a pine needle consists mainly of two kinds of cells, living parenchyma cells with nonlignified walls, and thin-walled but lignified tracheids with bordered pits. The parenchyma cells contain deeply staining tanniniferous and resinous substances, and also starch during part of the year. Next to the xylem, the transfusion tracheids are somewhat elongated; farther away from the bundles they are shorter and more like the parenchyma cells in shape. The tracheids appear to be nonliving cells. Their thin walls seem unable to offer sufficient resistance to the turgid living cells adjacent to them, and their lumina are somewhat compressed (pl. 78B). Next to the phloem the transfusion tissue contains cells similar to the albuminous cells in having dense cytoplasm and prominent nuclei (pl. 78B).

The transfusion tracheids and the transfusion parenchyma cells form continuous systems, and the two systems interpenetrate each other (Huber, 1947). Parenchyma cells are more abundant near the endodermis; tracheids are abundant closer to the vascular bundles. The vascular bundles appear to be separated from the transfusion cells by sclerenchyma except on their flanks where the transfusion tracheids and the marginal albuminous cells are concentrated (Strasburger, 1891).

The transfusion tissue is of universal occurrence in gymnosperms, but it shows various spatial relations with respect to the vascular bundles (Gathy, 1954; Lederer, 1955). It curves about the xylem in *Araucaria, Dammara, Sciadopitys;* occurs on two sides of the vascular bundle in *Cunninghamia, Cupressus, Juniperus, Thuja, Torreya, Sequoia,* and *Taxus;* and is present most abundantly on two sides of the phloem in *Larix.* In addition to the transfusion tissue associated with the vascular tissue, a so-called accessory transfusion tissue is identified in *Podocarpus* (Griffith, 1957), *Dacrydium* (Lee, 1952), and *Cycas* (Lederer, 1955). It is composed of elongated cells, some considered to be tracheids, extending outward from near the vein into the mesophyll. The origin and function of the transfusion tissue have not been satisfactorily determined, but the tissue is commonly assumed to be concerned with translocation between the vascular bundle and the mesophyll.

The endodermis surrounding the transfusion tissue in the pine needle consists of relatively thick-walled cells, sometimes containing starch. This cell layer is clearly differentiated in the Pinaceae and in certain other conifers, but it is ill-defined in others. It is sometimes described as having casparian bands in younger stages of development, a secondary wall with suberin or lignin or both in older stages. Intercellular spaces are lacking between the endodermal cells and throughout most of the vascular region (Strasburger, 1891).

DEVELOPMENT OF THE LEAF

Origin from Apical Meristem

The morphologic and cyto-histologic aspects of leaf initiation at the shoot apex were considered in chapter 5. A brief summary of the events will suffice here. Divisions in the peripheral zone of the apical meristem initiate a lateral protrusion, the leaf buttress, upon which the erect portion of the leaf later develops. In many plants, the leaf primordium arises so close to the shoot apex that the latter changes its shape and size periodically in relation to the lateral extension of the buttress. In some others, the primordium is initiated relatively low on the apical cone, which remains unaffected in its appearance in the part above the leaf primordium. In still other plants, the leaves, besides being inserted low on the apical cone, are also so small that they form no protrusion meriting the name of a leaf buttress (*Hippuris, Elodea*). The initial lateral protrusion of the axis formed during the growth of a leaf primordium commonly results from periclinal divisions below the distal zone of the apical meristem. In a wide variety of angiosperms these divisions occur in one or more of the layers near the surface, but not in the surface layer itself. The surface layer undergoes anticlinal divisions as the subsurface divisions produce a bulge. In some angiosperms, however, the surface layer is concerned with the initiation of the first protrusion by dividing periclinally. In such instances the outer covering is formed by the anticlinally dividing outer derivatives of the superficial layer.

The two growth zones of angiospermous shoot apices, the tunica and the corpus, variously participate in the formation of the leaf primordium (Foster, 1936). The degree of their participation is determined by the quantitative relationship between the tunica and the corpus and by the depth at which the periclinal divisions initiating the leaf are located. In *Scrophularia nodosa*, for example, the apical meristem has a single tunica layer, and the first divisions to form a leaf occur in the corpus. In *Vinca minor*, with a three-layered tunica, the leaf is initiated in the innermost layer of the tunica (Schmidt, 1924). The *Acacia* phyllodes involve both the tunica and the corpus in their initiation, although the tunica is three-layered (fig. 16.12, pl. 75; Boke, 1940). In Gramineae, some of which have one layer of tunica, others two, the leaf primordia arise through periclinal divisions in the first two layers of the apex, regardless of the number of tunica layers (figs. 16.19; 16.20; Kaufman, 1959; Sharman, 1945; Thielke, 1951). The gymnosperms, with a less precise apical organization than that expressed in the tunica-corpus complex, appear to show variations in the initiation of the leaf primordia similar to those in the angiosperms. In *Taxodium distichum*, for example, leaf growth starts with periclinal divisions in the subsurface layer, accompanied by anticlinal divisions in the surface layer (Cross, 1940), whereas in many other conifers (Korody, 1937; Sacher, 1955) and in *Zamia* (Johnson, 1943) periclinal divisions occur in the surface and the subsurface layers.

Whereas the location and the orientation of the divisions initiating leaf primordia may be readily observed in sections, the degree of participation of the various layers of the shoot apex in the final constitution of a leaf is difficult to judge (figs. 16.19, 16.20). Periclinal cytochimeras (chapter 5) have proved to be useful for the analysis of the composition of leaves in relation to the initial layers of the shoot apex. Such analysis of the cranberry leaf may be cited as an example (Dermen, 1947).

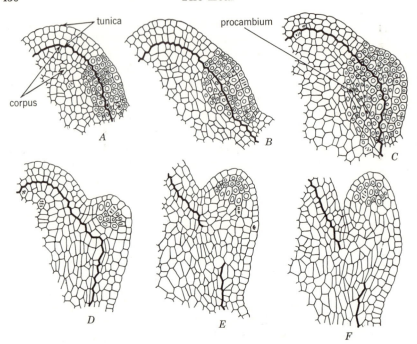

Fig. 16.12. Development of foliar organ (phyllode) in *Acacia*. Longitudinal sections of shoot apices. Heavy lines delimit tunica and derivatives from corpus and derivatives. Nuclei indicate cells most directly concerned with growth of primordium. *A*, periclinal divisions have occurred in outer layer of corpus and third layer of tunica. *B*, periclinal divisions are spreading into second layer of tunica. *C*, leaf buttress and, beneath it, procambium of leaf trace. *D*, meristematic activity in part of buttress initiates upward growth of primordium. *E*, *F*, upward growth is continued; periclinal and other divisions in subapical initials of primordium and growth in surface of protoderm. (All, ×175. After Boke, *Amer. Jour. Bot.* 27, 1940.)

Three layers of the apical meristem take part in the formation of this leaf, the two layers of the biseriate tunica and the outermost layer of the corpus. The leaf epidermis is derived from the outer layer of the tunica entirely by anticlinal divisions. The derivatives of the second layer of the tunica and those of the corpus together contribute toward the formation of the mesophyll and the vascular tissues. The derivatives of the tunica are represented at the leaf tip and the margins, those of the corpus in its central part.

Early Growth and Histogenesis

After the leaf is initiated at the apex of the shoot, its further growth depends on cell division and cell enlargement. The timing and distribution of these processes determine the size and the shape of the leaf, as well as the internal structure. In the herbaceous and some woody plants growth occurs uninterruptedly until the full size is attained. In many trees, the leaves originate during one season, interrupt their growth during the winter—they remain in the bud—and resume growth the following spring. The winter buds contain a part of or the full complement of foliar structures to be borne by the mature shoot, and the primordia of the foliage leaves are considerably advanced in their development as regards the delimitation of the various meristems. The following spring the leaves unfold as cells divide and expand (Artiushenko and Sokolov, 1952). In some woody species part of the leaves borne by a new shoot are present in the bud; the others are initiated during the season when they also become mature (*Syringa vulgaris, Ligustrum vulgare, Tilia vulgaris, Ulmus campestris, Ulmus montana*). The two kinds of leaves may differ in morphology (*Populus trichocarpa;* Critchfield, 1960).

Dicotyledonous Foliage Leaf. The direction and the amount of growth in a leaf, following its initiation by the apical meristem, vary in relation to the eventual form assumed and the size attained at maturity. In dicotyledons with ordinary leaves having an expanded blade and a relatively narrow base, with or without a petiole, the development of the leaf may be divided in three stages: (1) formation of the foliar buttress (fig. 16.12A–C), (2) formation of the leaf axis (fig. 16.12D–F), and (3) formation of the lamina (figs. 16.13, 16.14). This division is somewhat artificial because the successive stages overlap.

As was mentioned before, the leaf buttress is formed by meristematic activity below the distal apical region. The position of the initial divisions depends on the phyllotaxis of the shoot and on the circumferential spread of the future leaf. If the leaf has a narrow insertion, the divisions remain localized; if the leaf has a large base or completely ensheaths the stem, the divisions are propagated circumferentially in both directions from the point of their initiation (more common in monocotyledons than in dicotyledons; Tucker, 1962).

Through a change in the direction of growth an erect peg-like protuberance (fig. 16.12D–F), often somewhat flattened on the adaxial side (fig. 16.13A), arises on the buttress. This protuberance is the axis of the young leaf. It may be regarded as consisting of the midrib-petiole part of the primordium (only midrib part in sessile leaves) bearing the meristematic precursors of the future lamina. The meristematic activity of the leaf axis is at first concentrated at the apex (fig. 16.12D–F). Later it occurs throughout; that is, apical growth is followed by intercalary growth. Apical growth is generally of short duration, and the

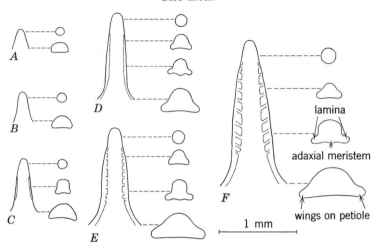

FIG. 16.13. Growth of *Nicotiana tabacum* leaf. Longitudinal and transverse sections. In transections, adaxial sides of primordia are turned downward. Details: growth of primordium in height; it is first an axis without blade (*A, B*), later, marginal-meristem activity on two flanks begins to form blade (*C–F*); marginal activity of limited duration in petiole forms wings. Dotted lines, external boundaries of midvein and lateral veins. (After Avery, *Amer. Jour. Bot.* 20, 1933.)

distinctness of the cells concerned with this growth varies in different plants. In some leaves, apical growth is interpreted as resulting from the activity of a *subapical initial* which gives rise to the internal tissue of the leaf axis, while the protoderm divides anticlinally concomitantly with the increase in length of the primordium (fig. 16.15*D*); in others a single subapical initial has not been recognized (fig. 16.12*F*).

As the leaf axis is elevated above the buttress, procambium is differentiated in its median part in continuity with the procambium in the buttress and the internode below it (pl. 75*C, D*). The leaf axis also increases in thickness, often through the activity of a strip of cells beneath the adaxial protoderm, the *adaxial meristem* (pls. 75*D*, 76*C*; Foster, 1936; Troll, 1939). The divisions in this meristem may be so orderly that the resulting tissue appears like a cambial derivative.

The lamina is initiated in the early stages of elongation of the leaf axis from two bands of meristematic cells located along two margins of the leaf axis (figs. 16.13, 16.14, 16.16*C*, pl. 76*C*). These bands of cells are called the *marginal meristems* (Foster, 1936). The panels of tissue developing from the marginal meristems may spread sideways from the leaf axis or they may turn toward the shoot axis. The leaf primordia attain somewhat variable lengths before the activity of the marginal meristems begins, but in general they are less than 1 mm long (fig. 16.17*B*), and may not have completed their apical growth (Avery, 1933; Foster, 1936; MacDaniels and Cowart, 1944).

The outermost cell layer—the protoderm—of the marginal meristem typically divides by anticlinal walls in dicotyledons. Thus the protoderm is continuous from its origin in the outermost layer of tunica, through the stage of initiation of the leaf primordium, and during the growth of the leaf. But the surface cells may divide periclinally and contribute cells to the interior of the lamina (*Daphne,* Hara, 1957; plants with variegated leaves, Renner and Voss, 1942).

The origin of the interior cell layers of the lamina from the subsurface cells of the marginal meristem varies in different species, but usually a more or less regular pattern is established close to the margin. Investigators give much weight to these patterns and use them, together with the occasionally observed mitoses, for identifying the presumed initials in the marginal meristems. As a result, the concept has developed that the marginal meristem is composed of a file of superficial initials, the

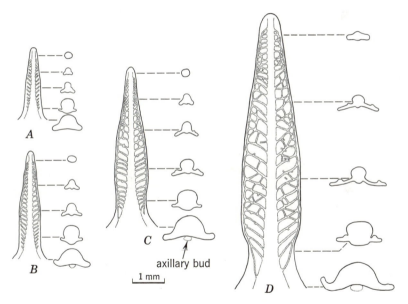

FIG. 16.14.   Growth of *Nicotiana tabacum* leaf.   Longitudinal and transverse sections. Stages succeeding those illustrated in fig. 16.13.   Dotted lines, external boundaries of veins. Details: continued increase in height of primordium; further growth of lamina and appearance on it of ridges associated with some veins; increase in thickness of parenchyma associated with midvein, on both abaxial and adaxial sides; lack of adaxial thickening in petiole region; development of network of veins in basipetal direction.   (After Avery, *Amer. Jour. Bot.* 20, 1933.)

marginal initials, which extend the protoderm of the lamina by anticlinal divisions, and a file of subsurface initials, the *submarginal initials,* located beneath the marginal initials and forming the interior tissue of the lamina by various combinations of periclinal, anticlinal, and oblique divisions (figs. 16.15A–C; 16.16A, B; Foster, 1936).

Many studies indicate that the presumed initials in the marginal meristems may be indistinct, or that the relation between them and the cell layers in the blade is variable in leaves of the same plant and even in one leaf, or that a group of cells in the submarginal position (as seen in a transection of a leaf) contribute cells to the interior tissues of the blade (Girolami, 1954; Hara, 1957; Roth, 1960, 1961; Schneider, 1952). In a study of marginal growth of *Xanthium* leaf (fig. 16.17A), including counts of mitoses in several hundred sections, only two divisions were found in the submarginal position at the edge of the lamina (Maksymowych and Erickson, 1960). It is conceivable that the organization of the marginal meristem is no more precise with regard to initials than that of the shoot apex. As reviewed in chapter 5, the similarity between the initials and their derivatives in shoot apices of seed plants prevents the positive identification of the initials. Possibly, plants with more precise apical organization have a more regular marginal growth in leaves (see ferns: Pray, 1960, 1962; Saha, 1963).

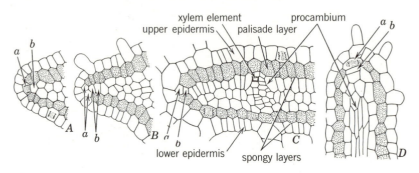

FIG. 16.15. Leaf development in *Nicotiana tabacum.* A–C, transections through edges of young laminae in three successive stages of development serving to interpret activity of marginal meristem. Submarginal initial may divide periclinally, producing cells *a* and *b* in A and C. Cell *a* is now the initial and may divide next by an anticlinal wall (between cells *a* in B). Stippled cells, anticlinal derivatives of initial *a*. Periclinal derivatives are unstippled. Vascular bundles arise among these derivatives. Protoderm increases in surface by anticlinal divisions. C shows, to the right, plate-meristem structure. D, median longitudinal section of primordium illustrating meristematic activity at its apex during initial growth in length. Subapical initial (*a*) adds to interior tissue of primordium by alternating periclinal (wall between *a* and *b*) and anticlinal (division figure in *a*) divisions. Anticlinal derivatives are stippled. Periclinal derivatives are unstippled. Procambium differentiates among them. (Adapted from Avery, *Amer. Jour. Bot.* 20, 1933.)

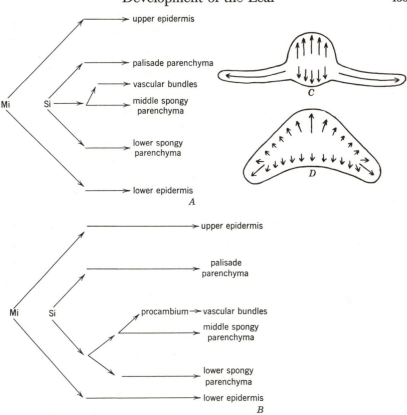

FIG. 16.16. A, B, diagrams illustrating common interpretation of growth of leaf lamina. Based on assumption of presence of marginal and submarginal initials and orderly sequences of divisions among the derivatives of these initials. A, *Nicotiana tabacum;* B, *Carya Buckleyi.* C, D, contrasting patterns of growth in leaf with thin lamina (C, *Oenothera*) and one of relatively uniform thickness (D, *Honkenya*). (Adapted from: A, B, Foster, *Bot. Rev.* 2, 1936; C, D, Roth, *Flora* 150, 1961.)

Like the apical growth of the leaf axis, the marginal growth of the lamina varies in duration. During marginal growth and after its cessation the lamina expands also by intercalary growth throughout its extent. The cells produced by the marginal meristem divide for a more or less extended time in various planes until a characteristic number of layers is established. This number remains constant during the further intercalary expansion of the lamina, except in the regions of procambial differentiation where additional divisions occur in various planes. The relative constancy in the number of layers and the consequent stratified appearance of the young lamina result from the restriction of divisions

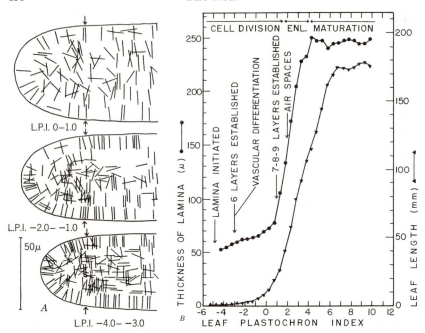

FIG. 16.17. Growth of leaf in *Xanthium italicum*. *A*, composite diagrams showing orientation of cell plates observed in numerous transverse sections of margins of leaves of three ages indicated by leaf plastochron indices (L.P.I.). Negative indices denote leaves below 10 mm in length. Adaxial surface is below in each diagram. In protoderm, cell plates perpendicular to blade surface (anticlinal divisions) predominate. In deeper tissue and left from arrows, divisions are irregular and reveal no exact cell lineage pattern. To the right of arrows plate-meristem activity is indicated by predominance of divisions perpendicular to blade surface. *B*, summary graph relating various processes in growth and differentiation of leaf to leaf plastochron index. (From Maksymowych and Erickson, *Amer. Jour. Bot.* 47, 1960.)

to anticlinal planes, that is, planes oriented at right angles to the surface of the leaf. Thus each layer increases in area but not in thickness. As was mentioned in chapter 4, a meristem composed of parallel cell layers growing in one plane is termed plate meristem. The establishment of the characteristic number of layers of the plate meristem occurs more or less close to the margin of the leaf, depending on the sequence of divisions in the marginal meristem and its derivatives. This sequence may vary markedly in different genera. In some leaves the divisions in the submarginal position produce only one layer of cells, in others two layers, in still others more than two layers (Foster, 1936; Hara, 1957; Roth, 1960, 1961; Schneider, 1952). Leaves that are thick and have no expanded blade—centric leaves, for example—do not have a plate-meristem type of growth (fig. 16.16*D*). Periclinal divisions predominate and cells may be arranged in anticlinal rows instead of periclinal layers as is typical of the plate meristem (Roth, 1960, 1961). The subdivision of the submarginal derivatives is usually illustrated by means of precise diagrams largely based on the interpretation of cell patterns seen in leaf sections (fig. 16.16A, B). These diagrams are approximations and do not indicate the possible variability in the ontogenetic relations between the cell layers. The interpretation of the ontogenetic patterns preceding the plate meristem activity is as problematical as that concerning the putative marginal and submarginal initials (Hara, 1957; Maksymowych and Erickson, 1960).

The above description considers the early developmental phenomena in a simple dicotyledonous leaf. A compound leaf also is initiated as a leaf axis upon a buttress. This leaf axis is a primordium of the petiolerachis part of the leaf and bears the meristems that produce the leaflets. These arise at the margins of the leaf axis as protuberances; in its initial stages, the origin of leaflets resembles marginal growth, but one that is restricted to portions of the leaf axis. If the leaf has a terminal leaflet, this leaflet is formed at the apex of the axis. Each leaflet resembles a simple leaf in its development and histogenesis. It appears first as a leaflet axis, which shows apical and later intercalary growth, and eventually develops a lamina from two bands of marginal meristem (Foster, 1936; Tepfer, 1960). The direction of appearance of the leaflet primordia on the leaf axis may be basipetal, acropetal, or divergent (starting in the middle and progressing in two directions; Foster, 1936; Troll, 1939).

Dicotyledonous Cataphyll. As was mentioned at the beginning of this chapter, the cataphylls show early growth peculiarities that determine their development as cataphylls rather than as foliage leaves. When compared with foliage leaves, the cataphylls of deciduous species may exhibit the following deviating anatomic characteristics (Foster, 1928): poorly differentiated mesophyll, usually without palisade tissue; scanty vascular system, often of an open dichotomous type, as though vascular anastomoses were arrested in their development; stomata few or absent. In some cataphylls sclerenchyma is small in amount or absent, in others there may be fibers or sclereids (*Camellia, Fagus, Quercus, Populus*). The outer bud scales may produce a periderm beneath the abaxial epidermis (*Aesculus*). The bud scales of evergreen species differ less from the foliage leaves than those of deciduous species (Vasilevskaĭa and Shilova, 1960).

Like the foliage leaf, the cataphyll originates by periclinal and anticlinal divisions in the peripheral zone of the apical meristem and forms a leaf axis as the first structure distinct from the stem. Sooner or later, the development of this primordium begins to deviate from that of the foliage leaf of the same plant. The following are some of the common developmental differences between the scales and the leaves. Whereas the axis of a foliage leaf increases in thickness by the activity of the adaxial meristem, a cataphyll shows little or no adaxial growth. The marginal activity, however, is accelerated in the scale and is also directed more definitely laterally, rather than adaxially, as in many foliage leaves. The rapid marginal growth, combined with the lack of thickening of the midrib, gives the scale its characteristic vaginant (sheathing) form. In the bud scales of *Rhododendron* the marginal initials divide periclinally and contribute cells to the ground meristem (Foster, 1937). The cataphyll tissues mature rapidly, usually without such a high degree of differentiation as the foliage leaf tissues.

Monocotyledonous Leaf. Among the leaves of monocotyledons the grass leaf has been studied most intensively from the developmental aspect (Abbe et al., 1941; Bugnon, 1921; Kaufman, 1959; Sharman, 1942, 1945). The grass leaf has a narrow blade and a base sheathing the stem (fig. 16.18A–C). The periclinal divisions initiating the leaf appear at one side of the apical cone and are then propagated to both

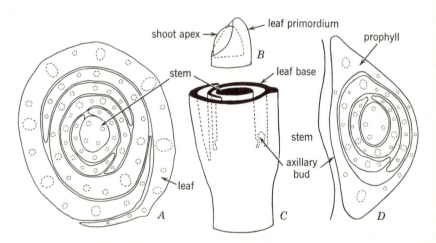

FIG. 16.18.   Leaf-stem relation in grasses. *A–C, Zea mays.* A, transection of young shoot with stem surrounded by successively older leaf primordia in two-ranked arrangement. *B,* shoot apex partly enclosed by the youngest leaf primordium. *C,* part of shoot, including a node bearing base of leaf ensheathing stem. To the left, margins of leaf overlap each other. *D,* axillary bud of *Avena* (oat) in transection. Prophyll with two prominent vascular bundles; it is flattened on the stem side. (*A,* ×40; *B, C,* after Sharman, *Ann. Bot.* 6, 1942; *D,* from photograph in Bonnett, *Univ. Illinois Agr. Expt. Sta. Bul.* 672, 1961, ×30.)

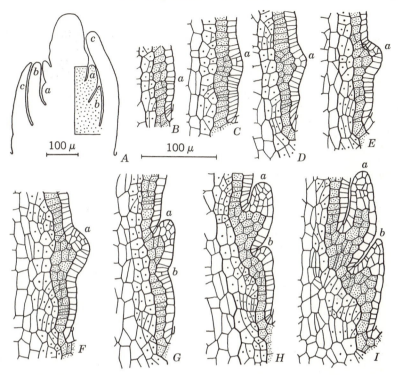

FIG. 16.19. Development of grass leaf, *Agropyron repens*. Median longitudinal sections through leaves. Stippling in *A*, area represented in *I*. *B–I*, origin and early growth of median part of leaf primordium (*a*) and of base of older primordium located one node lower (*b*). *B–F*, emergence of leaf buttress after periclinal divisions in two outer cell layers. *G–I*, upward growth of primordium. Stipples, derivatives of second tunica layer; single dots, derivatives of outermost layer of corpus. (Adapted from Sharman, *Bot. Gaz.* 106, 1945.)

sides of this initiation center until they encircle the stem (figs. 16.19*B*, 16.20*B–E*), a feature related to the sheathing nature of the leaf. The center of initiation of divisions occurs above and opposite the median part of the next lower leaf in conformity with the two-ranked arrangement of grass leaves (fig. 16.20*A*). The lateral spread of divisions results in the formation of a crescentic protrusion and then, with further growth, of a collar-like structure encircling the stem (fig. 16.18*B*, pl. 92*B*). If the sheath is open, the edges of this protrusion meet at the side of the stem opposite the point of initiation of the divisions and one of the edges is prolonged above the other so that the edges overlap (fig. 16.18*C*). If the sheath is closed (rare in Gramineae, typical in Cyperaceae), the encircling growth forms a complete ring.

If the concept of the leaf buttress is to be applied to the grass shoot,

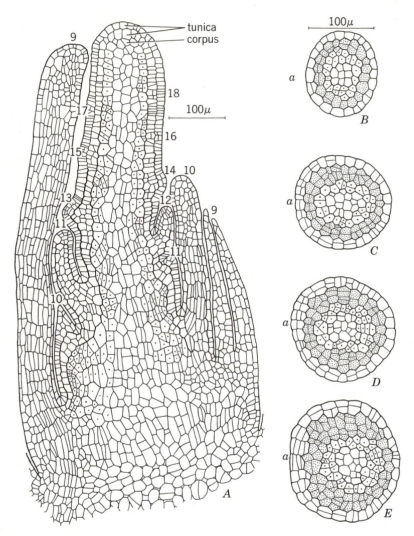

FIG. 16.20. Development of grass leaf, *Agropyron repens*. *A*, median longitudinal section through shoot tip and leaf primordia 9–18. Leaves 12–18 had not completed encircling the stem. Leaves 9–11 had completed the encirclement; parts of them appear on both sides of stem. In leaf 9, overlapping edges (right) appear as a double structure. *B–E*, transections of shoot at origin of leaf primordium. Periclinal divisions (at *a* in *B*), spread around circumference of shoot during formation of sheathing base of leaf. Letter *a* indicates location of apex of primordium. Stipples, derivatives of tunica; single dots, outer corpus derivatives. (Adapted from Sharman, *Bot. Gaz.* 106, 1945.)

this structure must be identified as the protrusion encircling the stem (figs. 16.19*B–F*, 16.20*B–E*). The upward growth of the grass leaf from its buttress begins at the point where the initial divisions occur (fig. 16.19*G*). This growth, which may be termed apical growth, starts before the encirclement of the axis is completed, and throughout its development the leaf remains highest at the point of its origin and slopes down along the margins (fig. 16.18*B*). The upward growth of the margins is similar to the marginal growth described for dicotyledonous leaves. But in the grass leaf the apical and marginal growth are less distinct from the growth of the leaf axis than are similar processes in a dicotyledonous leaf with a narrow base.

In the early stages of development the grass leaf assumes a hood-like shape (fig. 16.18*B*) and has no boundary between the leaf blade and the leaf sheath, although the part that encircles the apical meristem may be regarded as the primordium of the sheath. The boundary between blade and sheath becomes established when the ligule (a thin projection from the top of the leaf sheath) develops from the adaxial protoderm (Kaufman, 1959; Thielke, 1951, 1957). The auricles, if present in the species, appear at the same time. The blade continues to elongate by intercalary growth which lasts longest at the base of the blade, that is, above the ligule. The intercalary meristematic activity forming the sheath occurs below the ligule. Since the sheath begins to grow relatively late, it lags behind the blade in development. The leaf completes its elongation when the blade fully emerges from the enclosing sheaths (Begg and Wright, 1962). But at this time the sheath continues to be potentially meristematic at its base and can be stimulated to elongate by defoliation or lodging (chapter 4). The internode below the leaf still elongates when the leaf has ceased to grow. In *Zea*, specifically, blade elongation is completed before that of the internode below. In the lower parts of the plant, sheath elongation is also completed before that of the corresponding internode, but in the upper parts, sheath and internode elongate contemporaneously (Heimsch and Stafford, 1952). The elongation of the successive leaves of grasses is more or less closely integrated. Many grass leaves show transverse bands resulting from the pressure exercised by the collars of the older leaf sheaths upon the younger leaves. A comparison of distances between these marks in successive leaves of the same plant indicates that the growth of the different parts of one blade is correlated with that of the different sheaths enclosing it in the bud and with that of the next higher parts of the leaf next in succession (Panje, 1961).

Marginal growth with marginal and submarginal initials has been described for leaves of *Zea* (Mericle, 1950), *Oryza* (Kaufman, 1959), and two broad-leaved monocotyledons (Pray, 1957). In *Hosta*, a third broad-leaved monocotyledon, no definite submarginal initials were detected (Pray, 1957).

The derivatives of the marginal meristem in a grass leaf may become oriented in parallel layers and divide anticlinally (plate meristem) during the increase in the surface area of the leaf (Mericle, 1950). As in dicotyledonous leaves the procambial strands originate in a middle layer through divisions in various planes and thus interrupt the original parallel stratification. The leaf sheath in the rice leaf, in contrast to the blade, does not show plate meristem growth (Kaufman, 1959). During the apical and the marginal growth, periclinal divisions often occur in the protoderm in monocotyledonous leaves so that part of the inner tissue is of protodermal origin. In the sheath of many grasses two-layered margins develop from the protoderm and distinguish this part of the leaf from the blade (Kaufman, 1959; Thielke, 1951).

The development of the leaves in monocotyledons varies in complexity. In Gramineae, Amaryllidaceae, some Liliaceae, and others, the leaf primordium has distinct abaxial and adaxial surfaces, and its initial development results from the activity of one continuous meristematic layer extending from the apex down along the entire free margin. In certain other monocotyledons, however, apical growth is discontinued in its original position and a growth center is established abaxially from the adaxial margin (fig. 16.21). The structure developing from the abaxial growth center may be cylindrical in shape (*Allium cepa; Juncus glaucus*) or flattened, either along the median plane of the leaf (*Iris*, fig. 16.10*A*, *B*) or at right angles to this plane (*Allium lineare*). Anatomically such leaves appear as though the blade were rolled into a tube or folded. Some authors refer to such leaves as unifacial and interpret them as derived from only the abaxial side of the leaf (Roth, 1949; Thielke, 1948). The unifacial part of the leaf may be rather short as, for example, in the Araceae.

As in the dicotyledons, the cataphyll and the foliage leaf in the mono-cotyledon diverge from one another at an early stage of development (Chang and Sun, 1948; Sun, 1948). As exemplified by *Narcissus* (Denne, 1960), the distinction between the scale and the leaf is determined by the distribution of intercalary growth. Until they are 1 mm long, the scale and the leaf are similar. Later, active cell division may become restricted to the base of the primordium and a scale develops. If an intercalary growth region appears somewhat above the base, a foliage leaf is formed.

Among the developmental phenomena in monocotyledonous leaves the segmenting of palm leaves attracts attention. The segmenting is a

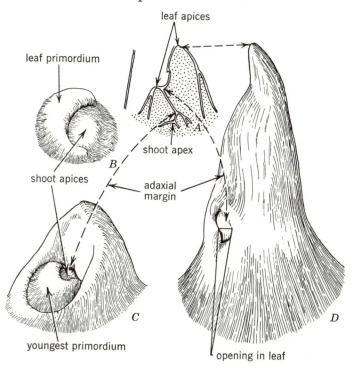

leaf apices

leaf primordium

shoot apex

shoot apices

adaxial margin

youngest primordium

opening in leaf

FIG. 16.21. Early development of *Allium cepa* leaf. *A*, median section through shoot apex with associated primordia. *B–D*, three-dimensional aspects of shoot apex with primordia in three stages of development. Leaf primordium arises on one side of shoot apex (*B*) and encircles it completely (*C*). The leaf sheath is closed (*C*, *D*). Adaxial margin ceases growth and is supplanted by apex located somewhat abaxially (leaf apices in *A*, *C*, *D*). Blade is tubular. (*B–D* drawn by Alva D. H. Grant.)

remarkably complicated process and, consequently, the interpretation of the mechanism involved is controversial (Tomlinson, 1961). According to one view differential growth in the lamina meristem is combined with splitting of cell walls and dissociation of tissues (Eames, 1953; Venkatana-rayana, 1957); according to another, differential growth alone accounts for the segmented form of the leaf (Periasamy, 1962). The evidence for the occurrence of splitting appears to be rather strong.

Gymnosperm Leaf. The investigated gymnosperm leaves show fundamental similarity with angiosperm leaves regarding their development (Cross, 1940–42; Johnson, 1943). As exemplified by Taxodiaceae, periclinal divisions near the surface at the side of the apical cone initiate a buttress. Apical growth of short duration and a longer-lasting inter-calary growth form the leaf axis. Marginal activity of limited duration initiates the narrow blade. The intercalary growth concerned with the further development of the blade is also small in amount. Although based on anticlinal divisions, it contributes chiefly to the length of the leaf and thus resembles rib-meristem activity rather than that of a plate meristem (pl. 77B). This kind of growth, combined with the limited amount of marginal growth, results in the development of an elongated narrow leaf.

Cataphyll development was studied in *Pinus* (Sacher, 1955). In the primordium, cells at the apex divide anticlinally and periclinally, and marginal growth is contemporaneous with the apical. The method of cell division is continuously changing in the marginal meristem and is completed with the formation of biseriate and finally uniseriate wings.

Differentiation of Mesophyll

The mesophyll differentiates from the derivatives of the marginal meristem after these derivatives have undergone the intercalary growth (of the plate-meristem type in flat laminae). The establishment of the characteristic differences between the palisade and the spongy parenchyma results from an unequal growth in the various layers of the leaf: there are differences in duration of cell division and cell enlargement and in the direction of cell expansion in the epidermis and in the various layers of the mesophyll. In dorsiventral dicotyledonous leaves, cell division commonly ceases first in the upper epidermis and continues longest in the future palisade tissue (Avery, 1933; Heslop-Harrison, 1962; MacDaniels and Cowart, 1944). The areas where the procambial strands occur must be excluded from consideration in this connection, since new procambium may be initiated by cell division after the meristematic activity has ceased elsewhere in the mesophyll (Avery, 1933). On the other hand, in areas where vascular bundles have been formed, the associated mesophyll may cease dividing early (MacDaniels and Cowart, 1944).

The difference in amount of cell division between the upper epidermis and the palisade parenchyma may be illustrated by the numerical relationships between cells of the two tissues in young and old apple leaves (MacDaniels and Cowart, 1944). The ratios between the numbers and diameters of the epidermal and the palisade cells were 1:1 in a young leaf. In the mature leaf the diameters of the epidermal cells were 3 to 4 times larger than those of the palisade cells, and there were 8 to 10 palisade cells to one epidermal cell. As exemplified by *Xanthium* leaf (Maksymowych, 1963), the upper epidermis and the palisade parenchyma differ in duration of cellular enlargement and rates of expansion. In the epidermis this rate is high in the horizontal plane but low in the vertical. The opposite relation characterizes the palisade tissue. While the division and the enlargement of the palisade cells remain correlated with the growth of the epidermal cells, the palisade cells are closely packed (pl. 74A, B). When the growth of the two kinds of cells becomes differential, intercellular spaces develop in the palisade tissue (pl. 74C). In *Xanthium* leaf the appearance of intercellular spaces was observed at cessation of cell division and beginning of cell expansion (fig. 16.17B). The palisade cells divide mainly by anticlinal walls except in leaves without an expanded blade; there, periclinal divisions may precede palisade differentiation (Roth, 1960, 1961). The palisade cells also elongate at right angles to the surface, and during the formation of intercellular spaces they separate from each other along the anticlinal walls. All these phenomena are responsible for the characteristic appearance of the palisade, that is, of a tissue composed of orderly rows of elongated cells largely separated from each other along their anticlinal walls.

The developmental relation between the lower epidermis and the spongy mesophyll in bifacial leaves appears to be somewhat variable. This epidermis may cease cell division before the spongy mesophyll but continue longer with cell enlargement (Avery, 1933), or it may show cell division after this activity has stopped in the spongy tissue (MacDaniels and Cowart, 1944). In both situations the epidermis grows in surface without formation of intercellular spaces, while the spongy parenchyma grows in a tangential plane by some enlargement of cells and by a loosening of contacts (pl. 74).

Among the various tissue elements of the leaf, the epidermal hairs, the stomata, and the large veins complete their differentiation before the mesophyll (MacDaniels and Cowart, 1944). The stomata develop concomitantly with or after the development of the intercellular spaces in the mesophyll.

Development of Vascular Tissues

The development of the vascular system of a foliage leaf is an integral part of leaf growth and overlaps with the various growth phenomena described above. The procambium of the midvein in a dicotyledon is differentiated in the leaf axis in early stages of lamina development. This is an acropetal process in the sense that it progresses in the upward direction as the primordium elongates above the buttress. As the lamina is formed, procambium differentiates in its middle layers, giving rise first to the largest lateral veins, then to the successively smaller veins of various orders, until a reticulate venation is formed (figs. 16.13, 16.14). This differentiation occurs throughout the intercalary growth stage of the leaf (Schneider, 1952) in successively more highly vacuolated ground tissue. The larger veins are initiated through a greater depth of tissue than the smaller. The smallest veins may be uniseriate in origin, that is, they may arise from cell series one cell in diameter (Pray, 1955a). The differentiation of procambium appears to be typically a continuous process since the successively formed procambial strands arise in continuity with those formed earlier (Pray, 1955a, c). The smaller veins probably develop as a unit between previously differentiated procambial strands, but the vein endings differentiate progressively from strands delimiting the areoles (Pray, 1963), and they may branch. (The concept that the bundle ends arise through breaking of previously established connections between veins is not supported by critical studies; cf. Pray, 1963.)

The pattern formed by the intercostal venation (that is, venation between the larger veins) of angiosperms is related to the growth pattern of the plate meristem (Pray, 1959). In the polygonal areoles of *Liriodendron* leaf the plate meristem is composed of isodiametric cells undergoing anticlinal divisions in random planes; the veins also have no preferred orientation. In *Hosta,* the plate-meristem cells are elongated at right angles to the primary veins, and the minor veins are roughly at right angles to the larger veins. In certain ferns the ground tissue is established by the marginal meristem in radial files branching toward the periphery as the leaflets increase in surface area. This pattern foreshadows the dichotomous branching of the lateral veins (Pray, 1960, 1962).

The longitudinal initiation of the venation in a dicotyledon follows a complicated sequence. The procambium of the midvein differentiates acropetally. The lateral veins of the first order develop from the midrib toward the margins (figs. 16.13, 16.14; Pray, 1955a). In a broad-leaved monocotyledon also, the larger veins develop acropetally. The small veins in both dicotyledons and monocotyledons differentiate basipetally so that the leaf apex is the first to complete the development of the procambial system (fig. 16.14). In the leaf of *Zea* (Sharman, 1942) the median and the main lateral procambial strands differentiate in the developing leaf in an acropetal direction. The small lateral strands alternating with the larger differentiate from the tip of the leaf downward after some protophloem appears in the larger strands. The transverse anastomoses are the last strands to appear and they, too, follow the basipetal course.

As in the stem, the vascular elements mature in the leaf before its procambial system is completely differentiated. The phloem, as far as it has been studied, precedes the xylem in maturation. In its initial stages of differentiation it appears to follow an acropetal course in continuity with the earlier-formed phloem (Esau, 1943; Pray, 1955a, c); critical information on this topic is still meager. In *Zea* (Sharman, 1942), the protophloem differentiates acropetally, first in the median vein, then in the large lateral veins before the basipetally differentiating procambium is initiated. The protoxylem follows the protophloem and differentiates in the same direction. The differentiation of the protophloem and the protoxylem coincides with the period of elongation of the leaf. After this extension is completed, metaphloem and metaxylem differentiate basipetally in the larger strands, which first developed protophloem and protoxylem, and in the smaller basipetally differentiating strands, which have no protophloem and no protoxylem. The protoxylem and protophloem are destroyed during the elongation, especially in the intercalary regions. There is some question whether the obliterated xylem is immediately replaced by new xylem or whether the intercalary region is for a time devoid of intact conducting elements (Sharman, 1942).

A remarkable instance of destruction of xylem during intercalary growth has been observed in the gymnosperm *Welwitschia* (Rodin, 1958). As was mentioned before, the leaf of this plant elongates by means of a basal meristem for many years. The xylem maturing across this meristem is continually being destroyed and replaced by new tracheary elements. The phloem has not been investigated.

The double wave of differentiation of the xylem, first in acropetal and then in basipetal direction, is common in monocotyledonous leaves. In dicotyledons the initial differentiation of xylem is also characteristically acropetal, but the subsequent development of this tissue follows a less orderly sequence than in the monocotyledons, probably in conformity with the less strictly basipetal course of differentiation of dicotyledonous leaves (De Sloover, 1958; Esau, 1943).

Growth and Form

Although the growth of the leaf is strongly influenced by the environment, the basic form of this growth is genetically controlled (Humphries and Wheeler, 1963). The main intrinsic factors determining the final shape of the leaf are (1) shape of the leaf primordium; (2) number, distribution, and orientation of cell divisions; (3) amount and distribution of cell enlargement not associated with cell division (Ashby, 1948a). In this listing of factors, cell division is assumed to be accompanied by cell enlargement between divisions. Cell division alone would not contribute to the enlargement and the change in shape of the leaf (Haber and Foard, 1963). Figure 16.17B, illustrates the relative unimportance of the cell-division stage in increasing the thickness and the length of the leaf as compared with the cell-expansion stage.

A comparison between the prevailing type of monocotyledonous leaf with its stem-ensheathing base and a dicotyledonous leaf having a narrow base illustrates the influence of the shape of primordium on the final leaf form. On the other hand, a comparison of needle-shaped leaves with those having an expanded blade shows that similar peg-like primordia may develop into leaves unlike in shape. Results of surgical experiments on ferns involving isolation, by means of incisions, of sites of future primordia or of incipient primordia from the apical meristem and existing leaf primordia are interpreted as indicating that the primordium is at first undetermined—a bud may develop in its place after the operation. It becomes determined during development apparently by the apex as a whole (Wardlaw, 1956). In dicotyledons a change in the shape of the leaf was obtained by similar operations (Sussex, 1955), but whether this change resulted from a release from the influence of the apex or from a response to reduction of the area for growth is not certain (Snow and Snow, 1959). It seems that in the ferns the degree of determination of leaf primordia and their differentiation from buds are especially small (Gregory, 1956).

The determination of leaf shape by cell division and cell enlargement finds various expressions (Foster, 1936; Papen, 1935). Fern leaves, for example, characteristically show a prolonged apical activity and an acropetal progression of intercalary growth and maturation of tissues. In contrast, the leaves of seed plants have a short period of apical growth and a prolonged one of intercalary growth. In narrow leaves (Gramineae, *Tragopogon, Linum, Plantago*) cessation of intercalary activity and the following maturation of tissues occur in a more or less strict basipetal direction. In broad leaves basipetal maturation is combined with lateral expansion. The pattern of leaf development may be recognized by the differentiation of the stomata (Ziegenspeck, 1944). In leaves maturing strictly basipetally, the stomata differentiate in the same direction. In leaves combining basipetal maturation and a lateral growth the different developmental stages of the stomata are mixed in a mosaic fashion.

The stages of cell division with a small amount of cell expansion and cell enlargement without cell division may be clearly definable in a growing leaf (fig. 16.17B), but they also may overlap to a considerable degree. In certain leaves of *Lupinus* and *Helianthus*, cell division was observed till the leaves reached one-half to three-fourths of their maximal area, whereas cell expansion commenced soon after the leaf was initiated and continued after cell division stopped (Sunderland, 1960). The duration of cell division varies in different leaves on the same plant.

Cell division controls form through its rate, duration, and distribution in the growing leaf. In leaf development in an aquatic heterophyllous species of *Ranunculus* the difference between the expanded and the narrowly dissected leaf was related to a different form of cell division in the latter: the lobes were produced at a greater rate and intercalary cell division was prolonged in these lobes but marginal growth was inhibited (Bostrack and Millington, 1962). Avery's (1933) study of a developing tobacco leaf (fig. 16.22) graphically demonstrated the determination of leaf form by the differential distribution of growth in different leaf areas (localized growth) and by the greater growth in one dimension than in another (polarized growth). The growth of such a leaf may be characterized as being anisotropic (Ashby, 1948a). This anisotropy is expressed in both differential cell division and differential cell enlargement. An analysis of growth of *Xanthium* leaf (Maksymowych, 1959) has related the distribution of growth to a time scale, the leaf plastochron index (chapter 5). Many physiologic changes show a straight-line relationship to the stage of growth expressed by means of the plastochron index (Michelini, 1958). The distribution of growth in a leaf, resulting in a particular form, is part of this series of coordinated events.

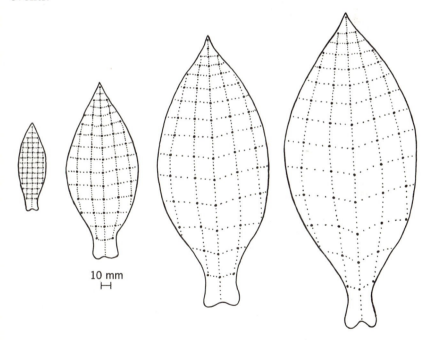

FIG. 16.22. Development of *Nicotiana tabacum* leaf. Four stages of development as seen in surface view. When leaf was ¼ final size (left), its entire surface was marked into 5-mm squares. Their unequal change in shape and size shows that growth is variable in different parts of leaf. (After Avery, *Amer. Jour. Bot.* 20, 1933.)

Growth phenomena are coordinated not only within the leaf itself, but also between the leaf and the plant as a whole. The well-known phenomenon of heteroblastic (from the Greek words other, or different, and shoot) development of leaves, that is, the change in form and size of successive leaves on a plant or a shoot, illustrates such coordination. In many grasses, for example, the blades of successive leaves are progressively longer and reach a maximum length before heading occurs. Concomitantly, the blade and sheath relationship changes. Both cell division and cell expansion appear to be involved in determining the changes, as judged by the studies of epidermal cells (Borrill, 1959, 1961; Maeda, 1959). The smaller size of older leaves of *Fragaria* was ascribed to a curtailment of the cell-division period (Arney, 1954). With regard to *Ipomoea*, studies directed toward determining the causal relations in the heteroblastic leaf development (Ashby, 1948*b*; Ashby and Wangermann, 1950) suggested that, despite their sensitivity to treatments, the gradients in number and size of epidermal cells from leaf to leaf up the shoot occur primarily as a response to the position of the leaves on the shoot; and the difference in position may be related to physiologic changes associated with the increasing age of the plant and its apical meristem (Allsopp, 1954; Ashby, 1950; Crotty, 1955).

ABSCISSION OF LEAVES

The periodic defoliation of perennial plants is a complex phenomenon involving the development of features bringing about the separation of the leaf from the stem, without injury to the living tissues in the stem, and giving protection to the newly exposed surface from desiccation and invasion by microorganisms. This development occurs in a region commonly called the *abscission region* or *zone* (from the Latin *abscissus*, cut off). Within this zone it is commonly possible to distinguish between the *separation layer* through which the actual break occurs, and the *protective layer* (pl. 69A–C). The separating leaf may be said to abscise.

The characteristics of the abscission region vary widely in different plants, as may be learned from the voluminous literature on the subject (Pfeiffer, 1928). Most studies on leaf abscission pertain to dicotyledons, but monocotyledons, conifers, and ferns also have received attention.

In simple dicotyledonous leaves the abscission zone occurs within the petiole or at its base. In compound leaves abscission zones develop in the petiole of the leaf as a whole and also at the base of the individual leaflets. The various abscission zones of compound leaves are similar in structure, although those associated with the leaflet separation may be somewhat simpler.

The features facilitating the separation are of two kinds: (1) peculiarities of histologic structure of the part of the petiole where the abscission zone is located and (2) the presence of a separation layer directly concerned with the severing of the connection between the leaf and the stem. The abscission zone differs from the adjacent parts of the petiole in having a minimum of strengthening tissues. Except in the vascular tissues, the cells are mainly parenchymatic; in the vascular tissues the lignified cells may be represented by tracheary elements only. Moreover, these elements may be exceptionally short (Scott et al., 1948). Thus, the abscission region is structurally weak.

In many herbaceous species, shrubs, and trees, branches become abscised also. This phenomenon occurs at various stages of development of the branch, and branches may be shed in living condition with all leaves attached (Eames and MacDaniels, 1947). The abscission of branches in many species occurs through a swollen part resembling a pulvinus, the abscission joint (Pijl, 1952). Leaves also may have abscission joints. Such joints differ from pulvini in that the vascular tissues are not contracted into a central strand, but there is a strong development of ground parenchyma. The xylem is weakly lignified and lacks sclerenchyma. The abscission joints may also have an annular groove with a strongly lignified disk below it. True pulvini may also be involved in abscission. In *Phaseolus*, leaflet abscission occurs at the abrupt transition from the pulvinus to the lower part of the rachis (Brown and Addicott, 1950).

Commonly, chemical changes in cell walls bring about the separation of the leaf. Three types of dissolution phenomena are listed (Addicott and Lynch, 1955): (1) removal of middle lamella, (2) removal of middle lamella and part of the primary wall, (3) dissolution of entire cells. In the removal of middle lamella enzymic conversions of calcium pectate into pectic acid and of the latter into water-soluble pectin appear to take place (Facey, 1950; Yager, 1960). The remaining cellulosic wall assumes a gelatinous consistency. Dissolution may fail to occur in the abscission of leaves in herbaceous dicotyledons and many monocotyledons. Under such conditions physical stresses alone appear to be effecting the abscission. Mechanical breakage without chemical changes was observed in the abscission of *Picea* needles (Facey, 1956).

The separation layer consists of at least two superposed rows of cells in which chemical changes in the cell walls take place. The morphologic distinctness of these cell layers varies. In many woody plants the separation layer is prepared by divisions in the ground tissue, which range in number from one or two to several in each cell (pl. 69A; Pfeiffer, 1928).

The process of separation commonly starts from the periphery of the petiole and progresses toward the interior (pl. 69C). In the vascular bundles the separation layer is continued through the living cells, but the sieve elements and the tracheary and other nonliving cells that may be present are broken mechanically. The tracheary cells may be occluded by tyloses before the leaf abscises, and then the living tylose cells complement the separation layer.

The protection of the surface exposed after leaf fall occurs in several ways. Two main phenomena may be distinguished: (1) the formation of a scar, or cicatrice, and (2) the development of periderm beneath the scar. The basic feature of cicatrization is the deposition of substances that are usually said to protect the new surface from outside injuries and loss of water. These substances are localized beneath the separation layer in a region several cell layers in depth. Together these layers constitute the protective layer of the abscission zone. Sometimes modifications similar to those of the protective layer occur above the separation layer on the leaf side (Pfeiffer, 1928). The materials deposited in the protective layer are variously identified as suberin, wound gum, and lignin. Suberin gives the usual reactions of fatty substances and is laid down, as in cork cells, in the form of a lamella on the inside of the cellulose wall. The presence of lignin is inferred from the positive staining reaction with phloroglucinol and hydrochloric acid. The wound gum shows many of the same microchemical reactions as lignin and therefore the distinction between the two is not always certain. The wound gum appears in the walls, the intercellular spaces, and frequently also in the tracheary elements.

The cicatrization may affect ground-tissue cells without any previous changes in the tissue. In other instances, divisions occur in preparation for the development of the protective layer. The periderm which develops beneath the protective layer is continuous with the periderm of the stem. In some plants periderm develops immediately as part of the abscission phenomenon (Salix, Aesculus). The timing of the various changes connected with leaf fall varies widely. The separation layer may be prepared early, during leaf differentiation, or it may become discernible only shortly before abscission (Leinweber and Hall, 1959). Similarly, cicatrization may occur before or after leaf fall. The development of the cicatrice is affected by environmental conditions.

Among the internal factors determining the characteristics and the timing of development of the abscission zone in plants—those concerned with leaf fall and also with the abscission of other organs—auxin is clearly involved (Jacobs, 1962). Numerous chemical substances affect abscission, and the regulation of abscission of leaves, flowers, fruits, and bark of trees has become a common agricultural practice.

References 473

REFERENCES

Abbe, E. C., L. F. Randolph, and J. Einset. The developmental relationship between shoot apex and growth pattern of leaf blade in diploid maize. *Amer. Jour. Bot.* 28:778–784. 1941.

Addicott, F. T., and R. S. Lynch. Physiology of abscission. *Ann. Rev. Plant Physiol.* 6:211–238. 1955.

Allsopp, A. Juvenile stages of plants and the nutritional status of shoot apex. *Nature* 173: 1032–1035. 1954.

Arber, A. *The natural philosophy of plant form.* Cambridge, Cambridge University Press. 1950.

Armacost, R. R. The structure and function of the border parenchyma and vein-ribs of certain dicotyledon leaves. *Iowa Acad. Sci. Proc.* 51:157–169. 1945.

Arney, S. E. Studies of growth and development in the genus *Fragaria.* III. The growth of leaves and shoot. *Ann. Bot.* 18:349–365. 1954.

Arslan, N. *Phaseolus multiflorus* pulvinuslarinda turgor reaksiyonlarinin gerileme kabiliyeti üzerine incelemeler. [Studies on the reversibility of the turgor reactions in the pulvini of *Phaseolus multiflorus.*] *Istanbul Univ. Rev. Fac. Sci. Ser. B.* 19:131–167. 1954.

Artiushenko, Z. T., and S. IA. Sokolov. O roste plastinki lista u nekotorykh drevesnykh porod. [Growth of leaf blade in some tree genera.] *Bot. Zhur. SSSR* 37:610–628. 1952.

Ashby, E. Studies in morphogenesis of leaves. I. An essay on leaf shape. *New Phytol.* 47: 153–176. 1948a. II. The area, cell size and cell number of leaves of *Ipomoea* in relation to their position on the shoot. *New Phytol.* 47:177–195. 1948b. VI. Some effects of length of day upon leaf shape in *Ipomoea caerulea. New Phytol.* 49:375–387. 1950.

Ashby, E., and E. Wangermann. Studies in morphogenesis of leaves. IV. Further observations on area, cell size and cell number of leaves of *Ipomoea* in relation to their position on the shoot. *New Phytol.* 49:23–35. 1950.

Avery, G. S., Jr. Structure and development of the tobacco leaf. *Amer. Jour. Bot.* 20:565–592. 1933.

Bailey, I. W., and C. G. Nast. The comparative morphology of the Winteraceae. V. Foliar epidermis and sclerenchyma. *Arnold Arboretum Jour.* 25:342–348. 1944.

Begg, J. E., and M. J. Wright. Growth and development of leaves from intercalary meristems in *Phalaris arundinacea* L. *Nature* 194:1097–1098. 1962.

Boke, N. H. Histogenesis and morphology of the phyllode in certain species of *Acacia. Amer. Jour. Bot.* 27:73–90. 1940.

Borrill, M. Inflorescence initiation and leaf size in some Gramineae. *Ann. Bot.* 23:217–227. 1959.

Borrill, M. The developmental anatomy of leaves in *Lolium temulentum. Ann. Bot.* 25:1–11. 1961.

Bostrack, J. M., and W. F. Millington. On the determination of leaf form in an aquatic heterophyllous species of *Ranunculus. Torrey Bot. Club Bul.* 89:1–20. 1962.

Bouygues, H. Structure, origine et développement de certaines formes vasculaires anormales du pétiole des Dicotylédones. *Soc. Linn. de Bordeaux, Actes* 57:41–176. 1902.

Brauner, L., and M. Brauner. Untersuchungen über den Mechanismus der phototropishen Reaktion der Blattfiedern von *Robinia Pseudacacia. Istanbul Univ. Rev. Fac. Sci. Ser. B.* 12:35–79. 1947.

Brown, H. S., and F. T. Addicott. The anatomy of experimental leaflet abscission in *Phaseolus vulgaris. Amer. Jour. Bot.* 37:650–656. 1950.

Brown, W. V. Leaf anatomy in grass systematics. *Bot. Gaz.* 119:170–178. 1958.

Brown, W. V. A cytological difference between the Eupanicoideae and the Chloridoideae (Gramineae). *Southwest. Nat.* 5:7–11. 1960.

Bugnon, F. Valeur morphologique des préfeuilles adossées chez les Dicotylédones. II. *Beupleurum rotundifolium et Alnus glutinosa. Bul. Sci. de Bourgogne* 14:129–134. 1952–53.

Bugnon, P. La feuille chez les Graminées. *Soc. Linn. de Normandie, Mém.* 21:1–108. 1921.

Chang, C. Y., and C.-N. Sun. Morphology and development of the vegetative shoot of *Arisaema consanguineum* Schott with special reference to the development of the cataphyll and the foliage leaf. *Natl. Peking Univ. Seme-Cent. Papers, Coll. Sci.* 1948:169–182. 1948.

Critchfield, W. B. Leaf dimorphism in *Populus trichocarpa. Amer. Jour. Bot.* 47:699–711. 1960.

Cross, G. L. A comparative histogenetic study of the bud scales and foliage leaves of *Viburnum opulus. Amer. Jour. Bot.* 25:246–258. 1938.

Cross, G. L. Development of the foliage leaves of *Taxodium distichum. Amer. Jour. Bot.* 27:471–482. 1940.

Cross, G. L. Some histogenetic features of the shoot of *Cryptomeria japonica. Amer. Jour. Bot.* 28:573–582. 1941.

Cross, G. L. Structure of the apical meristem and development of the foliage leaves of *Cunninghamia lanceolata. Amer. Jour. Bot.* 29:288–301. 1942.

Crotty, W. J. Trends in the pattern of primordial development with age in the fern *Acrostichum daneaefolium. Amer. Jour. Bot.* 42:627–636. 1955.

Datta, M. A new interpretation of the structure of the contractile vacuole in sensitive pulvini. *Bose Res. Inst. Calcutta, Trans. Biol. and Phys. Res.* 23:1–20. 1959–60.

Denne, M. P. Leaf development in *Narcissus pseudonarcissus* L. II. The comparative development of scale and foliage leaves. *Ann. Bot.* 24:32–47. 1960.

Dermen, H. Periclinal cytochimeras and histogenesis in cranberry. *Amer. Jour. Bot.* 34:32–43. 1947.

De Sloover, J. Le sens longitudinal de la différenciation du procambium, du xylème et du phloème chez *Coleus, Ligustrum, Anagallis* et *Taxus. Cellule* 59:55–202. 1958.

Dolivo, A. Anatomie comparée des aiguilles de douze espèces de pins. *Soc. Bot. de Genève Bul.* 39:8–33. 1948.

Eames, A. J. *Morphology of vascular plants. Lower groups.* New York, McGraw-Hill Book Company. 1936.

Eames, A. J. Neglected morphology of the palm leaf. *Phytomorphology* 3:172–189. 1953.

Eames, A. J., and L. H. MacDaniels. *Introduction to plant anatomy.* 2nd ed. New York, McGraw-Hill Book Company. 1947.

Esau, K. Origin and development of primary vascular tissues in seed plants. *Bot. Rev.* 9:125–206. 1943.

Esau, K. *Anatomy of seed plants.* New York, John Wiley and Sons. 1960.

Facey, V. Abscission of leaves in *Fraxinus americana* L. *New Phytol.* 49:103–116. 1950.

Facey, V. Abscission of leaves in *Picea glauca* (Moench.) Voss and *Abies balsamea* L. *North Dakota Acad. Sci. Proc.* 10:38–43. 1956.

Feustel, H. Anatomie und Biologie der Gymnospermenblätter. *Bot. Centbl. Beihefte* 38:177–257. 1921.

Florin, R. Untersuchungen zur Stammesgeschichte der Coniferales und Cordaitales. *Svenska Vetensk. Acad. Handl.* Ser. 5. 10:1–588. 1931.

Foster, A. S. Salient features of the problem of bud-scale morphology. *Cambridge Phil. Soc. Biol. Rev.* 3:123–164. 1928.

Foster, A. S. Phylogenetic and ontogenetic interpretations of the cataphyll. *Amer. Jour. Bot.* 18:243–249. 1931.

References 475

Foster, A. S.   Leaf differentiation in angiosperms. *Bot. Rev.* 2:349–372. 1936.

Foster, A. S.   Structure and behavior of the marginal meristem in the bud scales of *Rhododendron*. *Amer. Jour. Bot.* 24:304–316. 1937.

Foster, A. S.   Comparative morphology of the foliar sclereids in the genus *Mouriria* Aubl. *Arnold Arboretum Jour.* 27:253–271. 1946.

Foster, A. S.   Structure and ontogeny of the terminal sclereids in the leaf of *Mouriria Huberi* Cogn. *Amer. Jour. Bot.* 34:501–514. 1947.

Foster, A. S.   Foliar venation in angiosperms from an ontogenetic standpoint. *Amer. Jour. Bot.* 39:752–766. 1952.

Foster, A. S.   The morphology and relationships of *Circaeaster*. *Arnold Arboretum Jour.* 44:299–327. 1963.

Foster, A. S., and H. J. Arnott.   Morphology and dichotomous vasculature of the leaf of *Kingdonia uniflora*. *Amer. Jour. Bot.* 47:684–698. 1960.

Foster, A. S., and E. M. Gifford, Jr.   *Comparative morphology of vascular plants.* San Francisco, W. H. Freeman and Company. 1959.

Fulling, E. H.   Identification, by leaf structure, of the species of *Abies* cultivated in the United States. *Torrey Bot. Club Bul.* 61:497–524. 1934.

Gathy, P.   Les feuilles de *Larix*. Étude anatomique. *Cellule* 56:331–353. 1954.

Gerresheim, E.   Über den anatomischen Bau und die damit zusammenhängende Wirkungsweise der Wasserbahnen in Fiederblättern der Dicotyledonen. *Biblioth. Bot.* 19(81): 1–67. 1913.

Girolami, G.   Leaf histogenesis in *Linum usitatissimum*. *Amer. Jour. Bot.* 41:264–273. 1954.

Gregory, F. G.   General aspects of leaf growth. In: F. L. Milthorpe. *The growth of leaves.* London, Butterworths. 1956.

Grieve, B. J.   The physiology of sclerophyll plants. *Roy. Soc. West. Austral. Jour.* 39:31–45. 1955.

Griffith, M. M.   Foliar ontogeny in *Podocarpus macrophyllus*, with special reference to the transfusion tissue. *Amer. Jour. Bot.* 44:705–715. 1957.

Gupta, S. C.   Secondary growth in *Urtica dioica* L. *Agra Univ. Jour. Res. Sci.* 9:93–102. 1960.

Haber, A. H., and D. E. Foard.   Nonessentiality of concurrent cell divisions for degree of polarization of leaf growth. II. *Amer. Jour. Bot.* 50:937–944. 1963.

Hara, N.   On the types of the marginal growth in dicotyledonous foliage leaves. *Bot. Mag. Tokyo* 70:108–114. 1957.

Hasman, M., and N. Inanç.   Investigations on the anatomical structure of certain submerged, floating and amphibious hydrophytes. *Istanbul Univ. Rev. Facul. Sci. Ser. B.* 22:137–153. 1957.

Heimsch, C., and H. Stafford.   Developmental relationships of the internodes of maize. *Torrey Bot. Club Bul.* 79:52–58. 1952.

Heslop-Harrison, J.   Effect of 2-thiouracil on cell differentiation and leaf morphogenesis in *Cannabis sativa*. *Ann. Bot.* 26:375–387. 1962.

Höster, H. R., and W. Zimmermann.   Die Entwicklung des Leitbündelsystems im Keimblatt von *Pulsatilla vulgaris* mit besonderer Berücksichtigung der Protoxylem-Differenzierung. *Planta* 56:71–96. 1961.

Huber, B.   Zur Mikrotopographie der Saftströme im Transfusionsgewebe der Koniferennadel. *Planta* 35:331–351. 1947.

Humphries, E. C., and A. W. Wheeler.   The physiology of leaf growth. *Ann. Rev. Plant Physiol.* 14:385–410. 1963.

Jacobs, W. P.   Longevity of plant organs: internal factors controlling abscission. *Ann. Rev. Plant Physiol.* 13:403–436. 1962.

Johnson, M. A.   Foliar development in *Zamia*. *Amer. Jour. Bot.* 30:366–378. 1943.

Jones, H. Further studies on heterophylly in *Callitriche intermedia:* Leaf development and experimental induction of ovate leaves. *Ann. Bot.* 19:369–388. 1955.

Kasapligil, B. Foliar xeromorphy of certain geophytic monocotyledons. *Madroño* 16:43–70. 1961.

Kaufman, P. B. Development of the shoot of *Oryza sativa* L.—II. Leaf histogenesis. *Phytomorphology* 9:277–311. 1959.

Korody, E. Studien am Spross-Vegetationspunkt von *Abies concolor, Picea excelsa* und *Pinus montana. Beitr. z. Biol. der Pflanz.* 25:23–59. 1937.

Laubenfels, D. J. de. The external morphology of coniferous leaves. *Phytomorphology* 3:1–20. 1953.

Lederer, B. Vergleichende Untersuchungen über das Transfusionsgewebe einiger rezenter Gymnospermen. *Bot. Studien* Heft 4:1–42. 1955.

Lee, C. L. The anatomy and ontogeny of the leaf of *Dacrydium taxoides. Amer. Jour. Bot.* 39:393–398. 1952.

Leinweber, C. L., and W. C. Hall. Foliar abscission in cotton. III. Macroscopic and microscopic changes associated with natural and chemically induced leaf-fall. *Bot. Gaz.* 121:9–16. 1959.

MacDaniels, L. H., and F. F. Cowart. The development and structure of the apple leaf. *N.Y. (Cornell) Agr. Expt. Sta. Mem.* 258. 1944.

Maeda, E. Studies on the mechanism of leaf formation in the crop plants. I. Changes of structure of successive leaves of wheat. *Crop Sci. Soc. Japan Proc.* 27:451–457. 1959.

Maksymowych, R. Quantitative analysis of leaf development in *Xanthium pensylvanicum. Amer. Jour. Bot.* 46:635–644. 1959.

Maksymowych, R. Cell division and cell elongation in leaf development of *Xanthium pensylvanicum. Amer. Jour. Bot.* 50:891–901. 1963.

Maksymowych, R., and R. O. Erickson. Development of the lamina in *Xanthium italicum* represented by the plastochron index. *Amer. Jour. Bot.* 47:451–459. 1960.

Marco, H. F. The anatomy of spruce needles. *Jour. Agr. Res.* 58:357–368. 1939.

Meidner, H. The determination of paths of air movement in leaves. *Physiol. Plant.* 8:930–935. 1955.

Mericle, L. W. The developmental genetics of the Rg mutant in maize. *Amer. Jour. Bot.* 37:100–116. 1950.

Metcalfe, C. R. *Anatomy of the monocotyledons.* I. *Gramineae.* Oxford, Clarendon Press. 1960.

Metcalfe, C. R., and L. Chalk. *Anatomy of the dicotyledons.* 2 vols. Oxford, Clarendon Press. 1950.

Meyer, F. J. Das trophische Parenchym. A. Assimilationsgewebe. In: *Handbuch der Pflanzenanatomie.* Band 4. Teil. 7A. 1962.

Michelini, F. J. The plastochron index in developmental studies of *Xanthium italicum* Moretti. *Amer. Jour. Bot.* 45:525–533. 1958.

Morretes, B. L. de. Terminal phloem in vascular bundles of leaves of *Capsicum annuum* and *Phaseolus vulgaris. Amer. Jour. Bot.* 49:560–567. 1962.

Morretes, B. L. de, and M. G. Ferri. Contribuição ao estudo da anatomia das fôlhas de plantas do cerrado. *São Paulo Univ. Facul. de Filos., Ciên. e Let., Bol.* 243 *Bot.* 16:1–70. 1959.

Müller, E. Die Nervatur der Nieder- und Hochblätter. *Bot. Arch.* 45:1–92. 1944.

Orr, M. Y. The leaf anatomy of *Podocarpus. Bot. Soc. Edinb. Trans. and Proc.* 34:1–54. 1944.

Panje, R. R. On constriction bands, and the system of integrated growth-zones in the leaf-blades of grasses. *Phytomorphology* 11:257–262. 1961.

Papen, R. von. Beiträge zur Kenntnis des Wachstums der Blattspreite. *Bot. Arch.* 37:159–206. 1935.

Parry, D. W., and F. Smithson. Silification of branched cells in the leaves of *Nardus stricta* L. *Nature* 182:1460–1461. 1958.

Periasamy, K. Morphological and ontogenetic studies in palms—I. Development of the plicate condition in the palm-leaf. *Phytomorphology* 12:54–64. 1962.

Petit, L. Le pétiole des Dicotylédones au point de vue de l'anatomie comparée et de la taxinomie. *Soc. des Sci. Phys. et Nat. Bordeaux, Mém.* Ser. 3. 3:217–404. 1887.

Pfeiffer, H. Die pflanzlichen Trennungsgewebe. In: K. Linsbauer. *Handbuch der Pflanzenanatomie.* Band 5. Lief. 22. 1928.

Philpott, J. A blade tissue study of leaves of forty-seven species of *Ficus*. *Bot. Gaz.* 115:15–35. 1953.

Philpott, J. Blade tissue organization of foliage leaves of some Carolina shrub-bog species as compared with their Appalachian mountain affinities. *Bot. Gaz.* 118:88–105. 1956.

Pijl, L. van der. Absciss-joints in the stems and leaves of tropical plants. *Nederland. Akad. van Wetensch. Proc. Ser. C., Biol. and Med. Sci.* 55:574–586. 1952.

Pirwitz, K. Physiologische und anatomische Untersuchungen an Speichertracheiden und Velamina. *Planta* 14:19–76. 1931.

Plymale, E. L., and R. B. Wylie. The major veins of mesomorphic leaves. *Amer. Jour. Bot.* 31:99–106. 1944.

Pray, T. R. Foliar venation of angiosperms. II. Histogenesis of the venation in *Liriodendron*. *Amer. Jour. Bot.* 42:18–27. 1955*a*. III. Pattern and histology of the venation of *Hosta*. *Amer. Jour. Bot.* 42:611–618. 1955*b*. IV. Histogenesis of the venation of *Hosta*. *Amer. Jour. Bot.* 42:698–706. 1955*c*.

Pray, T. R. Marginal growth of leaves of monocotyledons: *Hosta, Maranta* and *Philodendron. Phytomorphology* 7:381–387. 1957.

Pray, T. R. Pattern and ontogeny of the foliar venation of *Bobea elatior* (Rubiaceae). *Pacific Sci.* 13:3–13. 1959.

Pray, T. R. Ontogeny of the open dichotomous venation in the pinna of the fern *Nephrolepis*. *Amer. Jour. Bot.* 47:319–328. 1960.

Pray, T. R. Ontogeny of the closed dichotomous venation of *Regnelidium*. *Amer. Jour. Bot.* 49:464–472. 1962.

Pray, T. R. Origin of vein endings in angiosperm leaves. *Phytomorphology* 13:60–81. 1963.

Reinhardt, M. O. Die Membranfalten in den *Pinus*-Nadeln. *Bot. Ztg.* 63:29–50. 1905.

Renner, O., and M. Voss. Zur Entwicklungsgeschichte randpanaschierter Formen von *Prunus, Pelargonium, Veronica, Dracaena. Flora* 35:356–376. 1942.

Rhoades, M. M., and A. Carvalho. The function and structure of the parenchyma sheath plastids of the maize leaf. *Torrey Bot. Club Bul.* 71:335–346. 1944.

Rippel, A. Anatomische und physiologische Untersuchungen über die Wasserbahnen der Dicotylen-Laubblätter mit besonderer Berücksichtigung der handnervigen Blätter. *Biblioth. Bot.* 19(82):1–74. 1913.

Rodin, R. J. Leaf anatomy of *Welwitschia*. I. Early development of the leaf. *Amer. Jour. Bot.* 45:90–95. 1958.

Roth, I. Zur Entwicklungsgeschichte des Blattes, mit besonderer Berücksichtigung von Stipular- und Ligularbildungen. *Planta* 37:299–336. 1949.

Roth, I. Histogenese der äquifazialen Blätter einiger Strand und Dünenpflanzen. I. *Flora* 149:604–636. 1960. II. *Flora* 150:95–116. 1961.

Rüter, E. Über Vorblattbildung bei Monokotylen. *Flora* 10:193–261. 1918.

Sacher, J. A. Cataphyll ontogeny in *Pinus lambertiana*. *Amer. Jour. Bot.* 42:82–91. 1955.

Saha, B.   Morphogenetic studies on distribution and activities of leaf meristems in ferns. *Ann. Bot.* 27:269–279. 1963.

Samantarai, B., and T. Kabi.   Secondary growth in the leaves of *Chenopodium album* L. and *Amaranthus gangeticus* L. and the partial shoot theory of the leaf. *Phytomorphology* 4:446–452. 1954.

Sauvageau, C.   Sur les feuilles de quelques Monocotylédones aquatiques. *Ann. des Sci. Nat., Bot.* Ser. 7. 13:103–296. 1891.

Schmidt, A.   Histologische Studien an phanerogamen Vegetationspunkten. *Bot. Arch.* 8: 345–404. 1924.

Schneider, R.   Histogenetische Untersuchungen über den Bau der Laubblätter, insbesondere ihres Mesophylls. *Österr. Bot. Ztschr.* 99:253–285. 1952.

Schürhoff, P. N.   Die Plastiden. In: K. Linsbauer. *Handbuch der Pflanzenanatomie.* Band 1. Lief. 10. 1924.

Schuster, W.   Die Blattaderung des Dicotylenblattes und ihre Abhängigkeit von äusseren Einflüssen. *Deut. Bot. Gesell. Ber.* 26:194–237. 1908.

Schuster, W.   Zur Kenntnis der Aderung des Monocotylenblattes. *Deut. Bot. Gesell. Ber.* 28:268–278. 1910.

Schwendener, S.   Die Mestomscheiden der Gramineenblätter. *Preuss. Akad. der Wiss., Phys.-Math. Kl., Sitzber.* 22:405–426. 1890.

Scott, F. M., M. R. Schroeder, and F. M. Turrell.   Development, cell shape, suberization of internal surface, and abscission in the leaf of the Valencia orange, *Citrus sinensis.* *Bot. Gaz.* 109:381–411. 1948.

Sharman, B. C.   Developmental anatomy of the shoot of *Zea mays* L.   *Ann. Bot.* 6:245–282. 1942.

Sharman, B. C.   Leaf and bud initiation in the Gramineae. *Bot. Gaz.* 106:269–289. 1945.

Shields, L. M.   Leaf xeromorphy as related to physiological and structural influences. *Bot. Rev.* 16:399–447. 1950.

Shtromberg, A. Ĭ.   Deĭatel'nost' kambiĭa v listĭakh nekotorykh drevesnykh dvudol'nykh rasteniĭ. [Cambial activity in leaves of some woody dicotyledons.] *Akad. Nauk SSSR Dok.* 124:699–702. 1959.

Sifton, H. B.   Air-space tissue in plants. *Bot. Rev.* 11:108–143. 1945.

Sinnott, E. W., and I. W. Bailey.   Investigations on the phylogeny of the angiosperms. 3. Nodal anatomy and the morphology of stipules. *Amer. Jour. Bot.* 1:441–453. 1914.

Skutch, A. F.   Anatomy of leaf of banana, *Musa sapientum* L. var. Hort. Gros Michel. *Bot. Gaz.* 84:337–391. 1927.

Snow, M., and R. Snow.   The dorsiventrality of leaf primordia. *New Phytol.* 58:188–207. 1959.

Sprecher, A.   *Le Ginkgo biloba L.* Diss. Genf. Genève, Imprimerie Atar. 1907.

Stålfelt, M. G.   Morphologie und Anatomie des Blattes als Transpirationsorgan. *Handb. der Pflanzenphysiol.* 3:324–341. 1956.

Stork, H. E.   Epiphyllous flowers. *Torrey Bot. Club Bul.* 83:338–341. 1956.

Strain, R. W.   A study of vein endings in leaves. *Amer. Midland Nat.* 14:367–375. 1933.

Strasburger, E.   *Über den Bau und die Verrichtungen der Leitungsbahnen in den Pflanzen. Histologische Beiträge.* Band 3. Jena, Gustav Fischer. 1891.

Sun, C.-N.   Morphology and development of the vegetative shoot of *Amorphophallus rivieri* Dur. with special reference to the ontogeny of the cataphyll and the foliage-leaf. *Natl. Peking Univ. Semi-Cent. Papers, Coll. Sci.* 1948: 183–193. 1948.

Sunderland, N.   Cell division and expansion in the growth of the leaf. *Jour. Expt. Bot.* 11: 68–80. 1960.

Sussex, I. M.   Morphogenesis in *Solanum tuberosum* L.: Experimental investigation of leaf dorsiventrality and orientation in the juvenile shoot. *Phytomorphology* 5:286–300. 1955.

Sutherland, M.   A microscopical study of the structure of leaves on the genus *Pinus*. *New Zeal. Inst. Trans. and Proc.* 63:517–568. 1933.

Tepfer, S. S.   The shoot apex and early leaf development in *Clematis*. *Amer. Jour. Bot.* 47:655–664. 1960.

Thielke, C.   Beiträge zur Entwicklungsgeschichte unifazialer Blätter. *Planta* 36:154–177. 1948.

Thielke, C.   Über die Möglichkeiten der Periklinalchimärenbildung bei Gräsern. *Planta* 39:402–430. 1951.

Thielke, C.   Über die Differenzierungsvorgänge bei Cyperaceen. I. Der Bau des vegetativen Vegetationskegels und die Anfangsstadien der Blattentwicklung. *Planta* 48:564–577. 1957.

Tomlinson, P. B.   *Anatomy of the monocotyledons*. II. *Palmae*. Oxford, Clarendon Press. 1961.

Toriyama, H.   Observational and experimental studies of sensitive plants. XI. On the thread-like apparatus and the chloroplasts in the parenchymatous cells of the petiole of *Mimosa pudica*. *Cytologia* 25:267–279. 1960. XIV. On the changes of a new cellular element of *Mimosa pudica* in diurnal and nocturnal conditions. *Cytologia* 27: 276–284. 1962.

Troll, W.   *Vergleichende Morphologie der höheren Pflanzen*. Band 1. *Vegetationsorgane*. Heft. 2. Berlin, Gebrüder Borntraeger. 1939.

Tucker, S. C.   Ontogeny and phyllotaxis of the terminal vegetative shoots of *Michelia fuscata*. *Amer. Jour. Bot.* 49:722–737. 1962.

Turrell, F. M.   Leaf surface of a twenty-one-year-old catalpa tree. *Iowa Acad. Sci. Proc.* 41:79–84. 1934.

Turrell, F. M.   The area of the internal exposed surface of dicotyledon leaves. *Amer. Jour. Bot.* 23:255–264. 1936.

Turrell, F. M.   The relation between chlorophyll concentration and the internal surface of mesomorphic and xeromorphic leaves grown under artificial light. *Iowa Acad. Sci. Proc.* 46:107–117. 1939.

Turrell, F. M.   A quantitative morphological analysis of large and small leaves of alfalfa with special reference to internal surface. *Amer. Jour. Bot.* 29:400–415. 1942.

Turrell, F. M.   Correlation between internal surface and transpiration rate in mesomorphic and xeromorphic leaves grown under artificial light. *Bot. Gaz.* 105:413–425. 1944.

Van Fleet, D. S.   The cell forms, and their common substance reactions, in the parenchyma-vascular boundary. *Torrey Bot. Club Bul.* 77:340–353. 1950.

Vasilevskaĭa, V. K.   *Formirovanie lista zasukhoustoĭchivykh rasteniĭ*. [Formation of leaf of drought resistant plants.] Akad. Nauk Turkmen. SSR. 1954.

Vasilevskaĭa, V. K. and N. V. Shilova.   Osobennosti stroeniĭa listovykh organov Pyrolaceae Lindl. i ikh znachenie dlĭa pobegoobrazovaniĭa. [Characteristics of structure of foliar organs of Pyrolaceae Lindl. and their significance in formation of lateral shoots.] *Leningrad Univ. Vest.* 1960:5–13. 1960.

Venkatanarayana, G.   On certain aspects of the development of the leaf of *Cocos nucifera* L. *Phytomorphology* 7:297–305. 1957.

Wardlaw, C. W.   Inception of leaf primordia. In: F. L. Milthorpe. *The growth of leaves*. London, Butterworths. 1956.

Wardlaw, C. W.   Experimental and analytical studies of pteridophytes. XXXVII. A note on the inception of microphylls and megaphylls. *Ann. Bot.* 21:427–437. 1957.

Weberling, F.   Morphologische und entwicklungsgeschichtliche Untersuchungen über die Ausbildung des Unterblattes bei dikotylen Gewächsen.   *Beitr. z. Biol. der Pflanz.* 32: 27–105.   1955.

Weintraub, M.   Leaf movements in *Mimosa pudica* L.   *New Phytol.* 50:357–382.   1952.

Williams, W. T.   The continuity of intercellular spaces in the leaf of *Pelargonium zonale*, and its bearing on recent stomatal investigations.   *Ann. Bot.* 12:411–420.   1948.

Wylie, R. B.   Relations between tissue organization and vein distribution in dicotyledon leaves.   *Amer. Jour. Bot.* 26:219–225.   1939.

Wylie, R. B.   The role of the epidermis in foliar organization and its relations to the minor venation.   *Amer. Jour. Bot.* 30:273–280.   1943.

Wylie, R. B.   Conduction in dicotyledon leaves.   *Iowa Acad. Sci. Proc.* 53:195–202.   1947.

Wylie, R. B.   The bundle sheath extension in leaves of dicotyledons.   *Amer. Jour. Bot.* 39: 645–651.   1952.

Yager, R. E.   Possible role of pectic enzymes in abscission.   *Plant Physiol.* 35:157–162. 1960.

Ziegenspeck, H.   Vergleichende Untersuchungen der Entwicklung der Spaltöffnungen von Monokotyledonen und Dikotyledonen im Lichte der Polariskopie und Dichroskopie. *Protoplasma* 38:197–224.   1944.

17

The Root

The root constitutes the underground part of the plant axis, specialized as an absorbing and anchoring organ. It occurs in the sporophytes of the vascular plants. As a group among the vascular plants, only the Psilotales lack such an organ. The sporophytes of these primitive tracheophytes are attached to the ground by means of rhizomes bearing hair-like absorbing structures, the rhizoids (Eames, 1936).

The morphologic relation between the root and the stem is variously interpreted. Since the two organs have many similarities in structure and show physical continuity, they are commonly treated as two parts of the same unit axis, and similar terms are applied to their tissue systems. In such treatment the designation of the stem and the root as organs serves to bring out their morphological and physiological specialization. Certain concepts deny the existence of complete homology between the two organs. According to one view, only part of the shoot, namely its inner region, is represented in the root (Arber, 1950). Another entirely opposite suggestion is that the vascular cylinder of the root, the stele, may be homologous with the entire shoot axis, the peripheral root tissues having no counterparts in the shoot (Allen, 1947).

It is particularly common to question the morphologic equivalence of the epidermis in the two organs because of the developmental and structural differences of this tissue system in the stem and in the root (chapter 7). There is also some difficulty in correlating the morphology of the primary vascular cylinder in root and stem. The old view that the entire primary vascular cylinder (or central cylinder) of the root is a single bundle was superseded by the interpretation of this cylinder

481

as a system of units corresponding to the system of bundles in the shoot; and when the concept of stele was introduced, the vascular cylinder of the root was interpreted as the stele of this organ.

No complete agreement exists regarding the interpretation of the parenchymatic region that occurs in the center of the vascular cylinder of many roots, particularly in the monocotyledons. Commonly this region is referred to as pith or pith-like, but there is some question whether it should not rather be regarded as potential vascular tissue that fails to differentiate as such. The introduction of the stelar concept did not resolve the problem because of the different views on the phylogenetic origin of the pith in the stele. According to one interpretation, the pith is stelar in origin (Schoute, 1903). According to another, the pith was derived from the cortex and appeared in the evolution of the siphonostele with the development of leaves and leaf gaps (Jeffrey, 1898–99). Such pith would not be present in roots because they lack leaves.

In this book a topographic-morphologic definition of the tissue systems of the root is employed, conforming with that used in the treatment of the stem. The superficial layer of the root in the primary state of growth is the epidermis. Beneath this layer is the fundamental tissue system in the form of a cortex, which surrounds the vascular system. If a clearly defined parenchymatic core appears in the center of the root, it is considered a part of the fundamental tissue system and is called pith.

ORIGIN

As was mentioned in chapter 1, the root and the stem appear to be closely related phylogenetically. The primitive axis-like plant body is considered to have differentiated into a shoot and a root with reference to different habitats and functions of the aerial and underground parts. The greater uniformity of the underground habitat, as contrasted with the aerial, is thought to be one of the factors causally connected with the relative simplicity of the root and its retention of some of the primitive structural features which eventually disappeared in the stem.

Ontogenetically the root origin is somewhat variable (Troll, 1949). The seed plants possess a radicle or only a root meristem at the root end (root pole) of the embryo from which the first root of the plant develops upon germination. In gymnosperms and dicotyledons this root (the tap-root, or primary root) commonly produces, by elongation and branching, the root system of the plant. In monocotyledons the first root, derived from the root meristem of the embryo, usually dies early in the growth of the plant, and the root system of the mature plant develops as a composite structure from numerous roots borne on the stem above the place of origin of the first root. Some of these stem-borne roots may be initiated in the embryo; others arise later. In the vascular cryptogams also, the main root system consists of roots that arise on the stem (Troll, 1949).

The first apical meristem of the root of seed plants arises not superficially like that of the epicotyl, but more or less deeply in the tissue of the root end of the embryo (chapter 20; Troll, 1949, however, regards the embryo root as exogenous). The deep-seated origin of the lateral roots (pl. 15B) and of those occurring adventitiously on the stem is even more clearly expressed. Thus, typically, the roots originate endogenously, the shoots exogenously; but adventitious buds may be initiated as deeply as the lateral roots (Torrey, 1958).

The roots originating at the root pole of the embryo and all their branches formed in normal sequence are usually distinguished from roots originating in various other manners by the designation of the latter as *adventitious roots*. This broad usage of the term is employed in the present book. It refers to roots that arise on aerial plant parts, on undergound stems, and on relatively old roots. Some workers prefer to restrict the appellation adventitious to roots arising from mature tissues or from parts of the plant where roots would not arise under ordinary conditions of growth. In this strict usage the roots arising from young stem tissues in dicotyledons, monocotyledons, and the lower vascular plants would not be called adventitious (*cladogenous roots;* Troll, 1949). In the German literature, the root system based on stem-borne adventitious roots is called homorhizic (implying that all roots are equivalent), as contrasted to an allorhizic root system composed of two kinds of roots, the taproot and the lateral root. The adventitious origin of the root is regarded as an ancient character, since it is widely distributed in extant ferns and has been found in fossil ferns (Baranova, 1951).

MORPHOLOGY

Roots vary widely in their morphology (Weaver, 1926) and exhibit structural and developmental differences correlated with more or less pronounced physiological specializations (Guttenberg, 1940). Most dicotyledons and gymnosperms possess a root system based on the taproot and its branches. The taproot produces the lateral or branch roots in an acropetal sequence, that is, with the youngest laterals located nearest the apical meristem, the oldest nearest the base (end of taproot merging with the hypocotyl). The taproot is often called the primary root; branches of the first order, the secondary roots; and branches of the secondary roots, the tertiary roots. Some plants may have root branches of fourth and even fifth orders (Dittmer, 1948). In perennial species the taproots and the older laterals undergo secondary growth. At this stage of development they serve as conductors of food and water, and as storage and anchorage organs. Absorption, on the other hand, is carried on mainly by the ultimate branchings which are in a primary state of growth. The fine absorptive branches—the feeder roots—remain short and are often fragile and short-lived (Jones, 1943; Preston, 1943; Wilcox, 1954; Zgurovskaĩa, 1958). Adventitious roots may also constitute normal complements of the root system in these groups of plants. Many gymnosperms develop such roots from the hypocotyl (Guttenberg, 1941). Some dicotyledons, usually rhizome-bearing plants, resemble the monocotyledons in having mainly adventitious roots.

The root systems of monocotyledons are commonly composed of stem-borne adventitious roots (Guttenberg, 1940; Tomlinson, 1961). There may be several orders of branches in the individual roots, or branching may be lacking. The roots are devoid of secondary growth and are relatively homogeneous in size and form. They frequently constitute the so-called fibrous root systems and are found in the grasses and on bulbs and rhizomes in Liliaceae, Iridaceae, and other families. In the grasses some of the adventitious roots may start developing in the embryo so that the embryo possesses two or more root primordia in the hypocotyl, in addition to the terminal radicle. All these primordia together are commonly spoken of as seminal roots. Some or all of these die after other adventitious roots develop. The formation of numerous adventitious roots in the Gramineae is associated with the important phenomenon of tillering characteristic of many grasses. It consists in the production of numerous shoots with unelongated internodes by axillary buds and the development of adventitious roots in connection with these shoots.

The above description characterizes the most common and widespread types of root systems concerned with absorption, conduction, storage, and anchorage of the plant in the soil. Some roots are more definitely specialized with reference to one particular function and correspondingly show morphologic peculiarities. Many roots develop as fleshy storage organs, with or without anomalous secondary growth. Others serve mainly as supporting organs, such as the prop roots in the mangrove plants and, on a smaller scale, in the grasses and sedges. Roots may be specialized as aerating organs (pneumatophores, Tomlinson, 1961) or modified into thorns. Certain vines (*Ficus pumila*) and epiphytes develop aerial roots that attach the shoots to the surface upon which the plant may be growing.

Reference to morphologic forms of roots is incomplete without mention of mycorrhizae and root nodules. Mycorrhizae are associations of roots and fungi, usually interpreted as symbiotic. They occur widely among woody and herbaceous angiosperms and in gymnosperms (Guttenberg, 1940, 1941; Kelley, 1950). The mycorrhizal roots are often short, and their internal structure deviates somewhat from that of noninvaded roots; their rootcap cells may become decomposed within the fungal mantle (Clowes, 1951, 1954; Morrison, 1956). The development of root nodules is caused ordinarily by bacteria entering through the root hairs and stimulating a proliferation of cortical cells (Allen and Allen, 1954). In some plants root nodules have been interpreted as modified lateral roots (Pommer, 1956). Root nodules are particularly characteristic of the Leguminosae (Arora, 1956; Bond, 1948; Guttenberg, 1940) but are found also in some other families (Guttenberg, 1941; Pommer, 1956).

PRIMARY STRUCTURE

Epidermis

A comprehensive discussion of the epidermis appears in chapter 7. The root epidermis consists of closely packed elongated cells with thin walls. According to some not generally accepted reports (chapter 7), this wall bears a cuticle. If the epidermis persists it may become conspicuously cutinized or suberized (Guttenberg, 1940; Kroemer, 1903). Thickened outer walls occur in root parts growing in air and also in roots that retain their epidermis for a long time (many monocotyledons and some dicotyledons). The walls of a long-persisting epidermis may also show lignification or may be impregnated with dark-colored substances.

The root epidermis is typically uniseriate. A well-known example of a multiseriate epidermis is the velamen (chapter 7) of air roots of tropical Orchidaceae and epiphytic Araceae and of some terrestrial monocotyledons (Gessner, 1956; Mulay and Deshpande, 1959). The velamen is a one to many layered parchment-like sheath consisting of compactly arranged nonliving cells with thickened walls. The thickenings may be densely spiral, reticulate, or pitted. An exodermis occurs beneath the velamen. During dry weather the cells are filled with air; during rain they become filled with water. The velamen is commonly interpreted as absorptive tissue, but this view is questioned because tests with radioactive phosphorus have failed to show passage of water from the velamen into the cortex in air roots of some orchids (Dycus and Knudson, 1957).

A typical characteristic of the root epidermis is the development of root hairs. Ordinarily, the root hairs are confined to a region between one and several centimeters in length near the tip (Farr, 1928). They are absent in the nearest proximity of the apical meristem, and they die off in the older root parts. Exceptional longevity and persistence of the root hairs—probably with a loss of the function of absorption—has been observed in a number of plants (chapter 7). It has also been observed that relatively long, filamentous soil-grown roots of dicotyledons and monocotyledons bear live root hairs over their entire extent (Scott, 1963). Root hairs vary in width and length (Dittmer, 1949). In some plants all epidermal cells are capable of initiating root hairs, in others, only certain ones. The formation of root hairs from a subepidermal layer has been recorded in *Citrus* (Hayward and Long, 1942).

Rootcap

The rootcap (pl. 15A) is commonly regarded as a structure that protects the root meristem and assists the root in the penetration of the soil during its growth. The latter function is suggested by the mucilaginous consistency of the walls of the outermost rootcap cells, a characteristic that presumably reduces the friction between the growing root tip and the soil. In some plants the rootcap cells are mechanically strong and possibly could serve to force soil particles apart (Guttenberg, 1940).

The cells of the rootcap are living parenchyma cells, often containing starch. The starch grains are commonly localized at the transverse wall nearer the ground, a phenomenon that has led to the interpretation of the starch grains as statoliths involved in the geotropic reaction of the root. The starch is rather persistent, in the sense that it is not readily utilized, except under conditions of extreme starvation (Netolitzky, 1935). Mucilaginous walls occur between the rootcap and the protoderm (pl. 82B) and also in the peripheral cells of the cap. In air roots of many tropical plants the rootcap may be covered with a mucilage layer several millimeters thick, which can dry into a hard crust (Weber, 1953). The mucilaginous condition of the walls in soil roots is assumed to facilitate the separation of the rootcap from the flanks of the growing root and the sloughing of the cells on the outer surface of the rootcap. During the sloughing process the separating cells have turgid protoplasts surrounded by a continuous wall layer, even after they are obviously disconnected from the rootcap. Environmental conditions affect the structure of the rootcap. For example, rootcaps of roots ordinarily growing in the soil undergo a reduction in size and a structural loosening when the plants are transferred to a water culture (Richardson, 1955).

Cortex

The root cortex may be homogeneous and simple in structure, or it may contain a variety of cell types. The degree of differentiation is apparently related to the longevity of the cortex. In the roots of gymnosperms and dicotyledons, which possess secondary growth and shed their cortex early, the cortex consists mainly of parenchyma. In roots retaining their cortex, as in many of the monocotyledons, abundant sclerenchyma may develop in addition to parenchyma. The innermost cortical layer of roots of seed plants growing in soil differentiates as an *endodermis* (fig. 17.2*A*, pl. 81*B*). Roots often develop a specialized layer—the *exodermis*—beneath the epidermis (pl. 81*B*) or beneath the velamen.

As seen in transverse sections, the cortical cells may be arranged in orderly radial rows or they may alternate with one another in the successive concentric layers. Sometimes the radial alignment is combined with a pronounced concentric layering, a pattern especially common in water plants and often associated with the presence of large intercellular spaces. In many roots a radially seriated inner cortex is combined with a less orderly appearing outer cortex. The presence of schizogenous intercellular spaces is typical of the root cortex. These spaces arise in the early ontogeny of the root, usually before the divisions forming the cortex are completed and before any vascular elements mature in the vascular cylinder. In wheat roots, air spaces were detected 50 to 100 microns from the boundary between the meristem and the rootcap (Burström, 1959). They contained pure CO_2 at this level.

The schizogenous intercellular spaces may become large. Large spaces may also result from a more or less extensive breakdown of cells by processes of lysigeny or rhexigeny. Thus the cortical tissue may assume the aspect of aerenchyma. The schizogenous lacunae have smooth outlines and are sometimes symmetrically arranged; the lysigenous or rhexigenous ones are bounded by broken walls and are rather irregular in shape and distribution. Cortical lacunae are found in Gramineae, Cyperaceae, various palms, and other monocotyledons (Guttenberg, 1940; Pillai and Pillai, 1962; Tomlinson, 1961). Aerenchymatic root cortex is common in plants of aquatic and moist habitats (Hasman and Inanç, 1957; Kacperska-Palacz, 1962; Katayama, 1961), but may occur also in grasses of relatively dry habitats (Beckel, 1956).

The orderly arrangement of cortical cells frequently observed in roots results from the method of cell division during the origin of this tissue region. As was shown in chapter 5, the root cortex is often one or two layers of cells in thickness at its origin from the apical meristem (pl. 82B). Repeated periclinal divisions augment the number of layers in radial extent, while anticlinal divisions increase the circumference and the length of the cortex. In most roots the sequence of periclinal divisions is centripetal; that is, of the two cells formed by a given periclinal division only the inner cell repeats the periclinal division. Thus a series of cells is cut off toward the periphery of the root, and, developmentally, the outer cortex is older than the inner (Guttenberg, 1940, 1943; Kroemer, 1903; Williams, 1947). After the periclinal divisions are completed, the innermost layer, the endodermis, develops casparian strips; but some endodermal characteristics detectable histochemically may appear during the meristematic activity at the boundary of the vascular cylinder (Van Fleet, 1961).

The centripetal sequence of divisions in the root cortex is common in the dicotyledons but it also occurs in many monocotyledons (Typhaceae, Pontederiaceae, Alismaceae, Cannaceae). In some dicotyledons (Ranunculaceae) and in many monocotyledons (Gramineae, Cyperaceae, Juncaceae, Commelinaceae, Aroideae) the internal part of the cortex shows centripetal growth; the external, centrifugal or irregular growth (Flahault, 1878; Janczewski, 1874a). Apparently, in the lower vascular plants also, part or all of the root cortex grows centripetally (Janczewski, 1874a; Williams, 1947).

The cortical parenchyma of roots is usually devoid of chlorophyll but is capable of developing chloroplasts, as demonstrated, for example, by their differentiation in intact and excised wheat roots grown in light (Burström and Hejnowicz, 1958). Roots of some water plants and aerial roots of many epiphytes normally have chloroplasts. Starch is often present, and various idioblasts and secretory structures may occur. Sclerification is common in the monocotyledons, including grasses (Soper, 1959), but is rare in dicotyledons. If sclerenchyma is present, it assumes a cylindrical arrangement, several cell layers in depth, either beneath the epidermis directly, or beneath an exodermis, or next to the endodermis. The sclerenchyma cells may be elongated like fibers, or they may be short. Some palms contain cortical fibers scattered individually or grouped into strands (Tomlinson, 1961). The cortical cells in roots of many gymnosperms have band-like or reticulate thickenings which may be lignified (Guttenberg, 1941; Wilcox, 1962a). Some dicotyledons (Cruciferae, Pomoideae, Prunoideae, Spiraeoideae, Caprifoliaceae) also develop prominent reticulate or band-like thickenings in cortical cells outside the endodermis (fig. 17.11; Guttenberg, 1940). Because of the transectional shape of the walls bearing the bands the structures are called phi thickenings. The cortex of lower vascular plants consists of thin-walled parenchymatic and variously sclerified cell complexes (Ogura, 1938). Sometimes a collenchymatic differentiation occurs in roots (Guttenberg, 1940; Van Fleet, 1950).

Endodermis. An endodermis characterized by casparian strips on its anticlinal walls (figs. 17.1, 17.2) is almost universally present in roots. The strip is formed during the early ontogeny of the cell and is a part of the primary wall. It varies in width and is often much narrower than the wall in which it occurs. It is typically located close to the inner tangential wall.

The chemistry of the casparian strip is a matter of controversy. It has been variously described as composed of lignin or suberin or both. According to some studies (Van Fleet, 1961), the casparian strip is initiated as a localized deposition of phenolic and unsaturated fatty substances between the radial walls—that is, in the middle lamella—where they form partly oxidized films. The primary wall becomes incrusted with and later thickened by deposits of similar substances on the inside of the wall. The incrustation of the cell wall by the material constituting the casparian strip presumably blocks the submicroscopic capillaries in the wall (Frey-Wyssling, 1959). Moreover, the cytoplasm of the endodermal cell is relatively firmly attached to the casparian strip, so that it does not readily separate from the strip when the tissue is subjected to the effects of plasmolytic or other agents normally

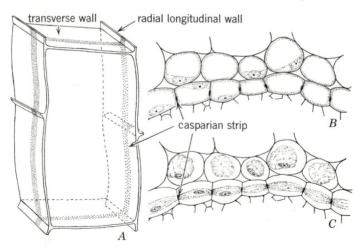

FIG. 17.1. Endodermal cells. *A,* entire cell showing location of casparian strip. *B, C,* effect of treatment with alcohol on cells of endodermis and of ordinary parenchyma. *B,* cells before treatment, *C,* after. Casparian strip is seen only in sectional views in *B, C.*

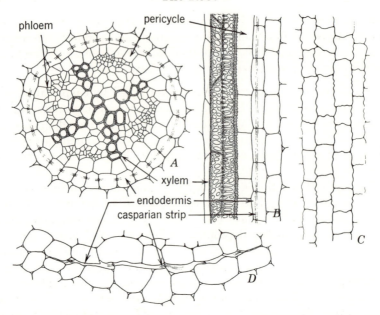

FIG. 17.2. *Convolvulus arvensis* (morning-glory) root. Endodermis in relation to other tissues. A, transection of vascular cylinder and part of cortex. Details: tetrarch xylem, uniseriate pericycle, uniseriate endodermis with casparian strips, intercellular spaces outside the endodermis. B, radial section through xylem, pericycle, and endodermis. C, tangential section through endodermis in casparian strip region, showing the characteristic waviness of walls. D, transection from an older root showing crushing of endodermis during secondary growth of vascular cylinder. (A–C, ×225; D, ×135. After Kennedy and Crafts, *Hilgardia* 5, 1931.)

causing a contraction of protoplasts (fig. 17.1B, C). Thus the casparian strip appears to form a barrier at which the soil solution is forced to pass through the selectively permeable cytoplasm rather than through the cell wall.

The casparian strips differentiate after the centripetal growth of the cortex is completed. At this level of the root, primary xylem development in the vascular cylinder may be more or less advanced. In gymnosperms and dicotyledons having secondary growth, the roots commonly develop no other kind of endodermis than that with casparian strips. In many of these plants the endodermis is discarded, together with the cortex, when the periderm develops in the pericycle. If the periderm is superficial and the cortex is retained, either the endodermis is stretched and crushed (fig. 17.2D) or it keeps pace with the expansion of the vascular cylinder by radial anticlinal divisions, the new walls developing casparian strips in continuity with the old ones (Guttenberg, 1943).

In the absence of secondary growth (most monocotyledons and a few dicotyledons) the endodermis commonly undergoes certain wall modifications. Workers distinguish two developmental stages in addition to the first stage when only the casparian strip is present. In the second stage a suberin (or endodermin, Frey-Wyssling, 1959) lamella covers the entire wall on the inside of the cell, so that the casparian strip is separated from the cytoplasm and the connection between the two ceases to be evident. In the third stage a thick cellulose layer is deposited over the suberin lamella, sometimes mainly on the inner tangential wall (figs. 17.3, 17.4). The thick wall, as well as the original wall in which the casparian strip is located, may become lignified. The casparian strip may or may not be identifiable after the thickening of the endodermal wall has occurred. The thick endodermal wall, here classified as secondary, may have pits. The successive development of endodermal walls is clearly expressed in monocotyledons. In dicotyledons the distinction between the second and third stages of endodermal development may not be sharp (Guttenberg, 1943), and in the lower vascular plants the differentiation is terminated with the

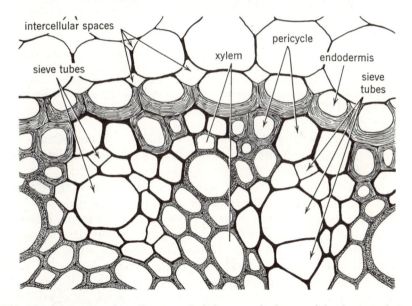

FIG. 17.3. Transection from *Zea* root. Endodermis in third stage of development characterized by presence of thick walls. Thickening is confined to radial and inner tangential walls. Pericycle consists in part of sclerenchyma. Part of a xylem strand and two phloem strands flanking the xylem are shown. Narrow sieve tubes, located next to pericycle, are associated with two companion cells each. Inward from these are one or two wide sieve tubes. Parenchyma occurs between phloem and xylem. That associated with xylem is sclerified. (×690.)

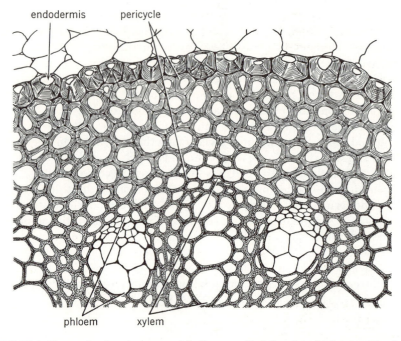

FIG. 17.4. Transection from inner part of *Smilax* root. Endodermis in third stage of development characterized by thick walls. The thickening is greatest on the radial and inner tangential walls. Pericycle, many-layered and sclerenchymatic. Part of a xylem strand is flanked by two phloem strands, each with many sieve tubes. The parenchyma between xylem and phloem is sclerified. (×257.)

deposition of the suberin lamella (Ogura, 1938). An endodermis with casparian strips and later wall modifications occurs in aerial roots (Napp-Zinn, 1953).

The development of the wall structures distinguishing the different stages of endodermal differentiation does not occur simultaneously throughout the entire endodermis at a given level. There are, therefore, more or less extended parts of the root where the endodermis is partly in one state, partly in another, and often cells in all three stages of development are found at the same level. The change from one stage to another usually follows a pattern suggesting a relation of this change to the proximity of the phloem. The casparian strips and the subsequent wall modifications appear first on the face of the phloem strands and then spread toward the parts of the endodermis opposite the xylem (Clowes, 1951; Guttenberg, 1943). Such unequal development of the endodermis opposite the xylem and the phloem often results in the occurrence of a thick-walled endodermis facing the phloem and cells with only casparian strips—the passage cells—facing the xylem. The name passage cells is based on the assumption that the cells allow a limited transfer of material between the cortex and the vascular cylinder. The passage cells either remain unmodified as long as the root lives or develop thick walls like the rest of the endodermis.

Exodermis. The subepidermal cortical layers of the root are often differentiated as a protective tissue containing suberin in its walls. Some workers apply the general term of hypodermis to the morphologically specialized subepidermal layers in both root and shoot; others distinguish the root hypodermis under a special designation of exodermis because of its distinctive histologic characteristics (Guttenberg, 1943).

The exodermis resembles the endodermis histochemically and structurally, and the causal factors of development of these tissues appear to be similar (Van Fleet, 1950). Exodermal cells may have casparian strips, but more commonly they are described as having a suberin lamella on the inside of the primary wall (Guttenberg, 1940, 1941; Kroemer, 1903). Usually the suberin lamella is covered by centripetally developing cellulose layers, which may attain considerable thickness (fig. 17.5) and be lignified. Sometimes the suberin lamella is not distinct, although fatty materials and lignin may be identifiable. The protoplasts appear to be retained in these cells.

The exodermis varies in thickness from one to several layers and is sometimes accompanied by sclerenchyma in the subjacent parts of the cortex (pineapple root; Krauss, 1949). The exodermis contains either one kind of cells, all elongated and suberized (some Gramineae, *Linum usitatissimum, Lactuca sativa*), or some of the cells are short and unsuberized (*Allium cepa; Asparagus officinalis;* Guttenberg, 1943).

Vascular Cylinder

The central part of the root is occupied by the vascular cylinder composed of the vascular system and the associated parenchyma. The

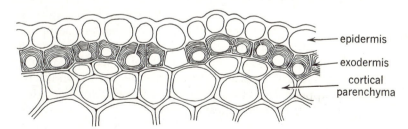

FIG. 17.5. Transection from outer part of *Smilax* root. Thick-walled exodermis beneath epidermis. One exodermal cell is not thickened. (×400.)

vascular system of the root is more clearly delimited from the cortex than that of the shoot because of several distinctive anatomic features of the root. First, the vascular tissue is compactly arranged and is not interrupted by leaf gaps; second, this tissue is surrounded by a commonly distinct uniseriate or multiseriate tissue zone, the pericycle (pericambium in German literature); and third, a morphologically differentiated endodermis (the innermost layer of the cortex in the seed plants) typically surrounds the pericycle (fig. 17.2; pl. 81A, B).

Pericycle. The pericycle of relatively young roots consists of thin-walled parenchyma (fig. 17.2). In angiosperms and gymnosperms it is concerned with meristematic activities. The lateral roots in these taxa arise in the pericycle; the phellogen originates here in most roots having secondary growth; and part of the vascular cambium is formed from pericyclic cells. (These activities, however, are not characteristic of the root pericycle of the vascular cryptogams; Guttenberg, 1943; Ogura, 1938.) In the monocotyledons, which usually lack secondary growth, the pericycle often undergoes sclerification in older roots, partly (fig. 17.3) or entirely (fig. 17.4).

In the angiosperms the pericycle is commonly uniseriate, but in many monocotyledons (some Gramineae, *Smilax, Agave, Dracaena,* palms) and a few dicotyledons (*Celtis, Morus, Salix, Castanea, Calycanthus*) it consists of several layers (fig. 17.4). The gymnosperms typically have a multiseriate pericycle. Sometimes the pericycle is uniseriate opposite the phloem and wider opposite the xylem. Roots without pericycle are rare but may be found among water plants and parasites. The pericycle may be interrupted by the differentiation of xylem (many Gramineae and Cyperaceae) or phloem elements (Potamogetonaceae) next to the endodermis (Guttenberg, 1943). The pericycle may contain laticifers and secretory ducts (Bruch, 1955; Williams, 1954).

Vascular System. The phloem of the root occurs in the form of strands distributed near the periphery of the vascular cylinder, beneath the pericycle (figs. 17.2–17.4). The xylem either forms discrete strands, alternating with the phloem strands (fig. 17.6D, pl. 81B) or occupies the center as well, with the strand-like parts projecting from the central core like ridges (figs. 17.2, 17.6A–C, 17.9, pl. 81A). If no xylem differentiates in the center, the center is occupied by a pith (pl. 81B). Plants with internal phloem in the stem may have such phloem in the root also (Obaton, 1949, Van Tieghem, 1891b).

As was discussed in chapter 15, the root typically has an exarch xylem; that is, its elements mature in the centripetal direction (fig. 17.9). Since the earliest xylem in a given plant organ is commonly called

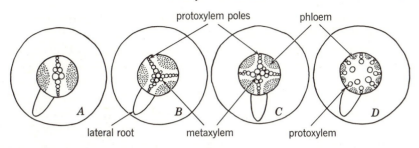

FIG. 17.6. Arrangement of primary vascular tissues and orientation of lateral root. In relation to number of radially arranged ridges in the xylem system, the roots are diarch (A), triarch (B), tetrarch (C), and polyarch (D). A–C illustrate common patterns in dicotyledonous roots, D in a monocotyledonous root. Lateral roots are shown as having originated opposite a protoxylem pole (B, C), between the xylem and the phloem poles (A), and opposite the phloem (D).

protoxylem, the root may be said to have the protoxylem located near the periphery of the vascular cylinder, the metaxylem farther inward (fig. 17.9). In the phloem, too, differentiation is centripetal, the protophloem occurring closer to the periphery than the metaphloem. Since the protophloem and the protoxylem mark, by their appearance, the beginning of vascular differentiation and thus may be used later as points of reference for the determination of the direction of vascular differentiation in the transverse plane, the location of these first vascular cells may be referred to as poles, *protophloem* and *protoxylem poles*, or simply phloem and xylem poles. Commonly there are equal numbers of protophloem and protoxylem poles.

Depending on the number of protoxylem poles, one, two, three, or more, the roots are called monarch, diarch, triarch, and so on (fig. 17.6). The term polyarch may be used when the number is high. In these designations the last part of the word, arch, stems from the Greek word meaning beginning. Monarch, diarch, and the other words thus indicate the number of loci where xylary differentiation begins, whereas exarch signifies that this beginning is peripheral with reference to the later xylem.

The number of protoxylem poles is in general characteristic in the different large groups of plants, but is not invariable. Like the presence or absence of pith, it may be related to the diameter of the vascular cylinder. In wider cylinders the number of poles may be larger, and pith is more likely to be present than in narrow vascular cores. Such variations sometimes occur in the same plant and in the same root (Cheadle, 1944; Guttenberg, 1940; Preston, 1943). Frequently the number of xylem strands is higher in the proximal (basal) end of a given root than in its distal (apical) end (Wilcox, 1954), but the change may occur in the opposite direction as well.

Some studies indicate that the diameter of the root and of the central cylinder and, concomitantly, the number of protoxylem strands increase when the rate of growth of the root increases (Wilcox, 1962*b*). There is reason to assume that the size of the vascular cylinder is determined by auxin-controlled changes in the size of the apical meristem (Torrey, 1957). In excised roots or in those with incised apices the vascular pattern may be different from normal in the beginning of the experiment, for example, diarch instead of triarch or, conversely, triarch instead of diarch, but it tends to revert back to normal in time. These results are interpreted as suggesting a control of the vascular pattern through the activity of the apical meristem (Reinhard, 1956; Torrey, 1955).

In dicotyledons the taproot is frequently di-, tri-, or tetrarch, but it may have five to eight and even more poles (many Amentiferae, *Castanea*). Only one xylem strand occurs in the slender root of the hydrophyte *Trapa natans*. The seedlings taproots of monocotyledons show numbers of protoxylem strands similar to those of dicotyledons, but the adventitious roots often have considerably higher numbers, as many as 100 and more in Palmae and Pandanaceae. High numbers of xylem strands in adventitious roots are associated with long transverse diameters and presence of pith. The roots of many gymnosperms may be diarch or polyarch (Wilcox, 1962*a*). A monarch condition has been observed in the smallest roots of the Araucariaceae (Guttenberg, 1941). The roots of the vascular cryptogams have one to many protoxylem and protophloem strands (Ogura, 1938).

Monocotyledons show a varied spatial relation between the peripheral xylem strands (usually composed partly of protoxylem, partly of metaxylem) and the more interior wide metaxylem vessels (Cheadle, 1944; Guttenberg, 1940). In some roots a single vessel occupies the center and is separated by nontracheary elements from the peripheral strands (pl. 84A). In others variable numbers of large metaxylem vessels are arranged in a circle around the pith (pl. 83). The number of these large vessels is not necessarily correlated with that of the peripheral strands. In some roots each strand terminates toward the center with a large vessel; in others two strands converge toward one large vessel. In the woody monocotyledons the inner metaxylem elements may form two to three circles (*Latania*), or they may be rather widely separated from each other (*Phoenix dactylifera*), or even scattered throughout the center (*Raphia Hookeri*). In some monocotyledons (*Cordyline, Musa,* Pandanaceae) phloem strands are scattered among the tracheary elements in the center of the root (Esau, 1960).

Although in transections the different strands and individual vessels might seem isolated from one another, they are commonly interconnected by lateral anastomoses (Guttenberg, 1940; Meyer, 1925). Where the xylem has the form of a star or of a diarch plate in transections, the tracheary elements are extensively interconnected. If there are peripheral radiating strands terminating at the central pith, the xylem and phloem strands have relatively few lateral connections with strands of their own kind. In some roots, however, the strands are apparently isolated from one another and, in some monocotyledons, also from the large central metaxylem vessel.

As seen in transections, the xylem elements at the poles are smaller in diameter than the more central ones (fig. 17.9) but the transition between narrow and wide elements is usually gradual, and therefore it is difficult to draw a line of demarcation between protoxylem and metaxylem. On the basis of size differences, few elements at each pole could be called protoxylem, sometimes only one element. Similarly, one to few sieve elements may constitute one protophloem pole. As was discussed in chapter 11, the wall sculpture is not a reliable criterion for distinguishing between protoxylem and metaxylem, especially in roots. The elongation of the axial part containing protoxylem is much more limited in the root than in the shoot, and therefore the extensible types of primary xylem elements may be few or absent in roots. The exact morphology of the tracheary elements of the protoxylem—whether they are tracheids or vessel members—has not been definitely established. The metaxylem contains tracheids and vessel members in the angiosperms.

In angiosperm roots the first mature sieve elements are easily recognized, because in contrast to the surrounding, still meristematic cells they contain only small amounts of stainable material in their lumina (pl. 83A). These sieve tubes differentiate on the outer periphery of the phloem strands and may or may not have companion cells (Resch, 1961). Their number in each strand is somewhat variable. In the more mature portions of the root other sieve tubes appear centripetally from the first (fig. 17.9). The later sieve tubes are part of the metaphloem. They are commonly associated with companion cells and parenchyma cells. In very small roots the metaphloem may be absent. Fibers occur in the primary phloem of some plants (Papilionaceae, Annonaceae, Malvaceae; Guttenberg, 1943). The first phloem elements of gymnosperm roots appear to be at a low level of specialization: they do not have typical sieve areas (Wilcox, 1954, 1962a). They are called precursory phloem instead of protophloem (chapter 12). The subsequently formed phloem contains elements with the usual characteristics of the primary sieve cells of gymnosperms.

Parenchyma cells are associated with the conducting cells in the xylem and the phloem. In older roots of species having no secondary growth this parenchyma often becomes sclerified (fig. 17.4, pl. 81*B*). The pith of roots consists of parenchyma essentially similar to that located among the vascular elements, but sometimes having thinner walls (pl. 81*B*, 83*B*).

Certain Coniferae show a characteristic distribution of resin ducts in the primary vascular region of the seedling taproot (Guttenberg, 1943). The Araucariaceae have resin ducts in the primary phloem, four to five in each phloem strand in the larger roots, fewer in the smaller. In the Pinaceae there is either a single central resin duct (*Abies, Cedrus, Tsuga*) or one duct at each protoxylem pole (*Picea, Larix, Pseudotsuga*). Taxaceae, Taxodiaceae, and Cupressaceae lack resin ducts in the primary vascular cylinder.

DEVELOPMENT

Histogenesis and Initial Vascularization

Before any specific tissue elements differentiate from the derivatives of the apical meristem, the derivatives pass through a period of division and elongation, and these growth phenomena overlap with maturation stages of the first vascular (phloem) elements (figs. 17.7, 17.8; Jensen and Kavaljian, 1958). From photographic records obtained microscopically from live growing roots, relative elemental growth rates (growth of infinitely small portions of the root) were determined for different regions of the root. In *Phleum* root, the maximum of such growth occurred 600 to 650 microns from the apex (Goodwin and Avers, 1956). In *Zea* root, the elemental growth rate was small near the apex, rose to a maximum at 4 mm from the rootcap tip, and fell to zero at 10 mm (Erickson and Sax, 1956*a*). Efforts were made to distinguish growth by cell division from growth by cell elongation. According to a study of *Triticum* root, the duration of mitotic stages is constant in the different parts of the meristematic root and, therefore, the distribution of mitoses is a measure of frequency of cell division (Hejnowicz, 1959). In *Zea* root (Erickson and Sax, 1956*b*), the relative elemental rate of cell formation rises to a maximum 1.25 mm from the rootcap tip and declines to zero at about 2.5 mm. Beyond this level, cell elongation alone is responsible for the further increase in length of the root.

According to studies on *Allium cepa* roots (Jensen and Kavaljian, 1958), the first stage of development of cells basal to the apex is a radial enlargement without increase in cell length except that associated with cell division (fig. 17.8). Shortly before the final diameter is reached, cells

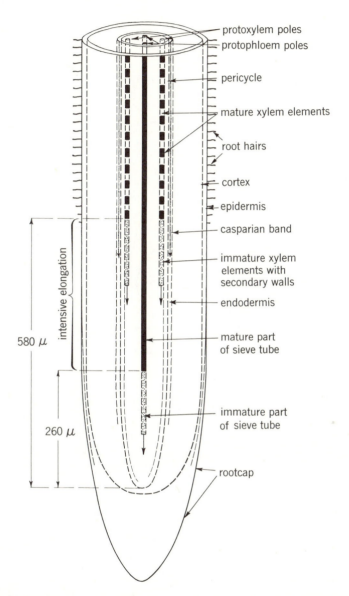

FIG. 17.7. Vascular differentiation in a root tip of *Nicotiana* (tobacco). Longitudinal section. Rootcap and epidermis have common origin. Cortex and vascular cylinder have separate initials at apex. Pericycle is delimited close to the apex. In vascular cylinder, sieve tubes mature first. (After Esau, *Hilgardia* 13, 1941.)

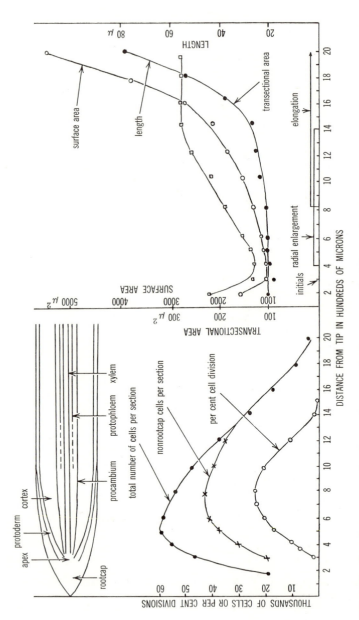

FIG. 17.8. Growth processes in first 2 mm of root tip of *Allium cepa*. The following determinations are used to characterize growth at levels indicated in the diagram of the root tip. Left: total number of cells in a section 100 microns thick; number of cells exclusive of root-cap cells; per cent of nuclei in division. Right: transectional area, length, and surface area of cells in each 100 micron section. (After Jensen and Ashton, *Plant Physiol.* 35, 1960.)

begin to elongate, slowly at first, then more rapidly. Cell division is the property of cells undergoing radial enlargement and early elongation. The cells in the two stages of enlargement differ in wall structure (Jensen and Ashton, 1960). In the region of radial enlargement the wall is still low in all components. During the transition to the elongation stage all components increase in amount per cell and pectin increases also per unit wall area. After the radial elongation stage, cellulose, pectin, and noncellulosic polysaccharides markedly increase per unit wall area. During elongation the increase in cell components is directly proportional to the increase in surface area of the cell.

The architecture of the apical meristem of the root and its developmental relation to the primary tissue systems of this organ have been discussed in chapter 5. At somewhat variable distances from the apical initials the meristems of the epidermis, the cortex, and the vascular cylinder become delimited from each other (pl. 82B). In the root this delimitation is usually more precise and occurs closer to the apex than in the shoot. It results initially from differential cell division and cell enlargement (Jensen and Kavaljian, 1958). In angiosperms the boundary between the primordial cortex and the future vascular region is particularly definite because the innermost layer of the cortex repeatedly divides by periclinal walls and contributes cells toward the outside (pl. 83A), while independent cell divisions in the central part of the root are forming the vascular cylinder. The pericycle becomes distinct from the central part of the vascular cylinder close to the apical meristem (fig. 17.7).

The histogenetic terminology regarding the vascular cylinder in the roots is problematical. The central cylinder may be treated as composed of xylem and phloem units imbedded in ground tissue. But the two conducting tissues are spatially so closely associated that it is also appropriate to treat the entire primary vascular region as derived from one procambial entity. If such treatment is adopted, the presence of the pith in some roots could be interpreted as a differentiation of a potentially vascular meristem into ground tissue or as evidence that the procambium of such roots has the shape of a hollow cylinder enclosing some ground meristem. The position of the pericycle in the histogenetic pattern of the root also requires scrutiny. The question to be answered is whether the precursor of the pericycle in roots should be regarded as procambium or as ground tissue. These terminological difficulties are clearly traceable to the circumstance that vascular and nonvascular tissues are not definitely separated from each other in their origin and ontogeny.

The phloic and xylary parts of the procambium of the root are morphologically differentiated close to the apical meristem. Commonly the prospective xylem is early distinguished from the prospective phloem by the increasing size and vacuolation of cells (Torrey, 1953; Wilcox, 1962a). At this time the cells in the phloic procambium are still dividing. In conifers the precursory phloem may vacuolate before the metaxylem (Wilcox, 1954). The vacuolation, the concomitant reduction in stainability, and the enlargement of cells in the xylary procambium usually occur in a reverse order from that followed by the maturation of the xylem cells (Popham, 1955). That is, the future metaxylem cells enlarge and vacuolate before the protoxylem cells (pl. 83A), but the latter are the first to develop secondary walls and to reach functioning state (pl. 86A). Because of this developmental pattern the metaxylem cells attain a larger ultimate size than the protoxylem cells. This size contrast is particularly conspicuous in the monocotyledons and in those dicotyledons whose roots lack secondary growth. In such plants the primordia of the largest metaxylem elements may be recognized among the most recent derivatives of the apical initials (pl. 82; Heimsch, 1951; Young, 1933). After these primordia are individualized, they cease dividing longitudinally but undergo some transverse divisions. They increase in width and length, while the surrounding cells are still dividing.

The early delimitation of the phloic and xylary parts of the procambium makes vascular differentiation in the root appear simpler than in the shoot, in which, at a given level, the maturation of elements overlaps with active procambial divisions that enlarge the earlier procambial strands and add new ones. As was stressed in chapter 15, the vascular ontogeny of the shoot is complex because the vascular system of this organ differentiates largely or entirely in relation to leaf primordia. In contrast, the vascular system of the root differentiates independently of the lateral organs, and as a result, the recognition of the direction of differentiation of the primary vascular meristem in the root constitutes no problem. As the apical meristem adds new cells to the root, the delimitation of the procambial tissue follows into the new portions of the root. In other words, the procambium differentiates acropetally.

The differentiation and maturation of the xylem and the phloem follow the procambium in its acropetal course (fig. 17.7; Esau, 1943a; Torrey, 1953) and the first phloem elements typically mature closer to the apex than the first xylem elements. The root exhibits a simpler pattern of maturation of the first vascular elements than the shoot, in which the xylem is initiated discontinuously and subsequently develops bidirectionally in relation to the leaf primordia.

The timing of the maturation of the first vascular elements is related to the growth of the root as a whole (fig. 17.7). Periclinal divisions in the cortex cease near the level where sieve tubes mature. The maximum elongation occurs below this region. But the protoxylem typically matures after most of the elongation is completed. At the same level or slightly farther from the apex, casparian strips develop in the endodermis and the epidermis forms root hairs.

There seems to be a causal relation between the rate of growth of the root and the proximity of mature elements to the apical meristem, and both are affected by environmental conditions, by the type of root, and by the developmental stage of the root (Heimsch, 1951; Wilcox, 1954, 1962a). In barley, for example, the main roots of the adventitious system which were rapidly elongating had mature tracheary elements farther from the apices than either the main roots which were approaching their maximum length or the small lateral roots. Considering all these roots, the various vascular elements have been found maturing the following distances from the apical meristem: sieve tubes, 250 to 750 microns; protoxylem elements, 400 to 8,500 or more microns; the early metaxylem, 550 microns to 1.5 or more cm; and the large central metaxylem vessels at still greater distances. In *Abies*, mature protoxylem occurred 7 mm from the apex in fast growing roots, 500 microns at the end of growth, and 50 microns during dormancy. In isolated pea roots treated with indoleacetic acid, xylem differentiated close to the apical meristem not only because root elongation was inhibited, but also because the treatment accelerated the maturation of xylem (Torrey, 1953).

The apical growth of the root is not a uniformly continuous process. In *Abies procera* (Wilcox, 1954), for example, root growth slows down periodically; cell maturation approaches the apex; fatty materials, probably suberin, are deposited in the cortex and the rootcap; and the root becomes dormant. The deposition of the fatty materials occurs in a layer of cells continuous with the endodermis and covering the protomeristem on its sides and toward the rootcap. Externally, such root tips are brown. When growth is resumed the brown cap is broken and the root tip pushes beyond it. Roots of dicotyledons may show similar alternation of periods of growth and rest (Zgurovskaĭa, 1958). Apparently, internal factors determine the growth changes, rather than seasonal phenomena (Wilcox, 1954).

Primary and Secondary Growth

Roots, like stems, show a wide variation in the amount and characteristics of secondary growth. Some herbaceous dicotyledons lack secondary growth, others have mere vestiges of such growth, and still others produce considerable amounts of secondary tissue (pl. 81C). The taproot and the main branch roots of gymnosperms and aborescent dicotyledons typically have secondary growth (pls. 28*B*, 81*D*), but the small branch roots are devoid of it. With some exceptions, the roots of monocotyledons contain only primary tissues (pl. 81*B*). *Dracaena*, a monocotyledon, has been mentioned most often as having secondary growth in stem and root (Cheadle, 1937; De Silva, 1936). The secondary tissues in roots of dicotyledons and gymnosperms are basically similar to those in the stems of the same plants, but the initiation of cambial activity has its distinctive features in the two organs in relation to the differences in arrangement of the primary vascular tissues.

Roots without Secondary Growth. Absence of secondary growth is characteristic of monocotyledonous roots. Several figures illustrate various stages of development of monocotyledonous roots: the delimitation of the tissue regions subjacent to the apical meristem (pl. 82*B*); the maturation of the first sieve tubes, one at each phloem pole, and the enlargement of the metaxylem elements (pl. 83*A*; compare with pl. 82*A*); the maturation of the protoxylem and the development of an endodermis with casparian strips (pl. 84). The completion of primary growth is attained with the maturation of early and late metaxylem and of metaphloem and the sclerification of parenchyma cells associated with the vascular elements (pls. 81*B*, 83*B*), the development of thick secondary walls in the endodermis (figs. 17.3, 17.4), and the differentiation of an exodermis (fig. 17.5, pl. 81*B*). Since no secondary growth occurs, the cortex is retained and no periderm develops. The peripheral protective tissues are the epidermis and the exodermis; or the epidermis is destroyed and the exodermis replaces it as the surface layer.

A dicotyledonous root with small amount of secondary growth may be illustrated by the tetrarch root of *Ranunculus* (pl. 81*A*). Figure 17.9*B* shows the central part of such a root at the level where the first sieve tubes and xylem elements are mature. The outermost sieve tube at each phloem pole is the protophloem sieve tube; the others are part of the metaphloem. In the xylem region the smallest outermost elements constitute the protoxylem. A uniseriate pericycle appears outside the vascular elements and is surrounded by the similarly uniseriate endodermis with casparian strips. The central metaxylem elements have no secondary walls but are conspicuously enlarged. In figure 17.9*C* all the primary phloem is present. This phloem is composed of sieve tubes and companion cells (the sieve tubes are stippled). Some cambial divisions initiating secondary growth have occurred on the inner edges of the phloem strands. The metaxylem elements have thickened their walls. Figure 17.9*D*, depicts the central part of the root with the

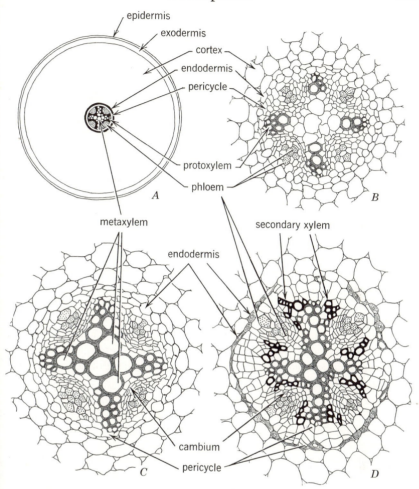

FIG. 17.9. Differentiation of vascular tissues in a tetrarch root of *Ranunculus*. Transections. *A*, entire root in mature state. *B–D*, details of vascular cylinder and adjacent cortical layers in three stages of development. (*A*, ×27; *B–D*, ×150.)

primary vascular cells all mature. The cambium has produced a few cells between the phloem and the xylem. Some of these cells have differentiated into secondary tracheary elements (with walls shown in solid black); the others have remained parenchymatic. Pericyclic cells outside the xylem poles have divided. The endodermis has developed secondary walls, mainly opposite the phloem; elsewhere it has been crushed. In some *Ranunculus* roots all endodermal cells develop secondary walls. Because the secondary growth is so small in amount,

the cortex is retained in the mature state (fig. 17.9*A*, pl. 81*A*). A multiseriate, relatively thin-walled exodermis differentiates in the cortex.

Roots with Secondary Growth. Secondary growth of a woody
dicotyledon root is exemplified by the roots of *Pyrus*, pear (Esau, 1943*b*)
and *Salix*, willow. Figure 17.10 illustrates the growth of the pear root
diagrammatically, and plates 85–87 give some of the histologic details
of the willow root. The xylary procambium becomes delimited from
the phloic procambium by a decrease in density of staining. Differentiation of the first sieve tubes at each phloem pole is followed by that of
the protoxylem elements at the xylem poles (fig. 17.11, pls. 86*A*, 87*A*).
Centripetal differentiation of further xylem and phloem cells in successively older root parts completes the primary differentiation of the
vascular tissues (fig. 17.10*B*, pl. 86*B*). In the pear, the late metaxylem
elements in the center of the root enlarge relatively little and mature
after the secondary growth is initiated. In the willow, the center is
occupied by sclerenchyma (pls. 86*B*, 87*C*). The pericycle is parenchymatic (pl. 86*A*). The endodermis has casparian strips (fig. 17.11),
and in the pear it accumulates considerable amounts of tanniniferous
compounds. Tannin deposition is frequently first restricted to cells
facing the phloem. Later it spreads to all endodermal cells and also
to some cortical cells farther out and to pericyclic cells. In *Pyrus* the
cortical cells outside the endodermis develop thickenings resembling
those of a collenchyma (fig. 17.11).

The vascular cambium appears first on the inner edges of the phloem
strands (fig. 17.10*C*, pl. 86*B*). While these cambial cells form some
secondary elements, the pericyclic cells outside the protoxylem poles
divide in a manner similar to that previously shown in the root of
Ranunculus (fig. 17.9*D*). The inner derivatives of these divisions
complete the cylinder of cambium by joining the strips located on the
inner faces of the phloem strands. The vascular cambium assumes a
circular outline in transections because on the inner boundary of the
phloem the secondary xylem is deposited earlier than outside the protoxylem (fig. 17.10*C*, *D*). The secondary vascular tissues assume the form
of a continuous cylinder and completely imbed the primary xylem (fig.
17.10*E*; pls. 85*A*, 87*C*). The sieve elements of the primary phloem are
crushed, and in the pear some of the remaining cells differentiate into
fibers. The secondary phloem contains fibers also (pl. 87*B*). In *Pyrus*
the cambium which arises in the pericycle outside the xylem poles forms
wide vascular rays (fig. 17.10*E*).

The periclinal divisions in the pericycle which are not involved in the
formation of the vascular cambium occur not only outside the xylem

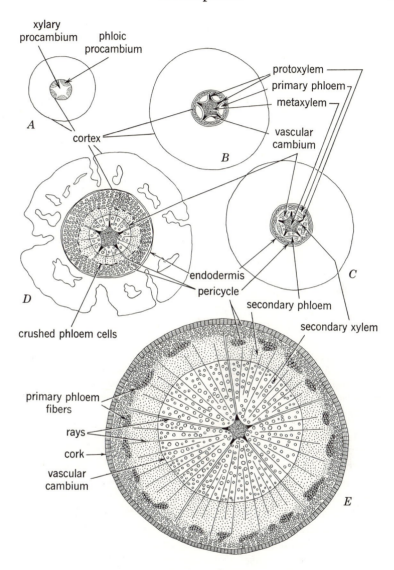

FIG. 17.10. Development of *Pyrus* (pear) root. Transections. *A*, vascular cylinder in procambial state. *B*, primary growth completed. *C*, strips of vascular cambium between phloem
and xylem have produced some secondary vascular tissues. *D*, vascular cambium, now a
cylinder, has produced additional secondary tissues; pericycle has undergone periclinal divisions; endodermis partly crushed; cortex breaking down. *E*, secondary growth has progressed
further; periderm has appeared; cortex has been shed. Cambium opposite protoxylem poles
has formed wide rays (*D, E*). (All, ×29.)

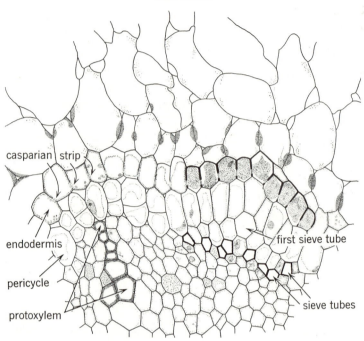

FIG. 17.11. Transection through part of pear root with one phloem and one xylem pole. Pericyclic cells in front of protoxylem pole have divided tangentially. Faint areas in the radial walls of endodermis are sectional views of casparian strips. Endodermis outside phloem has accumulated tannins. Cortical cells outside endodermis have collenchymatic wall thickenings. (×540. Esau, *Hilgardia* 15, 1943.)

poles but spread around the circumference of the root. These are divisions preparatory for the formation of the periderm. Their number varies with growth conditions (Mylius, 1913). A phellogen arises among the outer cells of the proliferated pericycle. Outwards this phellogen forms cork tissue (fig. 17.10*E*, pl. 85); inwards it may produce phelloderm. It is difficult to distinguish between the phelloderm and the parenchyma derived from the growth of the pericycle preceding the initiation of phellogen activity. The expansion of the vascular cylinder by secondary growth causes the rupture and sloughing of the cortex together with the endodermis (fig. 17.10*D*, pl. 85*B*).

Dicotyledonous roots, especially those with a limited amount of secondary growth, sometimes retain their cortex (*Actaea, Convolvulus arvensis*). Such roots may develop an exodermis or a superficial periderm. In *Citrus sinensis* there first appears a periderm of subepidermal origin; later a deeper periderm arises in the pericycle (Hayward and Long, 1942). A variety of protective tissues were observed in roots of plants growing in the Alps (Luhan, 1955): persisting thick-walled epidermis (*Gentiana, Ranunculus*); exodermis (*Primula*); exogenously originating periderm (*Artemisia* and other Compositae); dead and collapsed but persisting cortex (*Linaria, Myosotis, Polygonum*); subdivided and suberized endodermis (*Gentiana*); polyderm (*Geum, Potentilla*); periderm of deep seated origin (Saxifragaceae). The polyderm (chapter 14) consists of rows of parenchyma interspersed with uniseriate rows of cells showing casparian strips or more extensive suberization; that is, cells of endodermal type. The tissue arises in the pericycle by tangential divisions and may consist of many layers, parenchyma cells alternating with endodermoid cells. With the death of the cortex, the polyderm becomes exposed to the surface. Its outermost cells die but the inner, including the suberized cells, remain alive. Polyderm is characteristic of certain Rosaceae, Hypericaceae, Onagraceae, and Myrtaceae (Guttenberg, 1943; Mylius, 1913). Periderm derived from a storied cambium (chapter 14) is rather frequent in massive aerial roots of palms (Tomlinson, 1961).

An aspect of secondary growth of considerable interest to horticulturists and ecologists is the natural grafting of roots. Where growing roots come in contact with one another they become united by secondary growth. Often sizeable groups of trees become interconnected, as may be demonstrated by translocation of water, minerals, poisons, dyes, and isotopes (Borman and Graham, 1959). Some of this movement was observed to a distance of over 40 feet (Borman, 1962). Root grafting maintains life in old stumps. Possibly the stimulus inducing cambial activity is transmitted from stronger to weaker trees and to stumps. Most frequently root grafting was observed within species, but apparently it may also occur between species (Beskaravaĭnyĭ, 1955; Lotova and Liarskaĭa, 1959).

Development of Lateral Roots

The lateral roots arise at some distance from the apical meristem deep in the tissue (endogenous origin). In both gymnosperms and angiosperms the branch roots commonly are initiated in the pericycle of the parent root and subsequently grow through the cortex of the latter (pl. 15*B*). In the lower vascular plants the branch roots originate, as a rule, in the endodermis, although exceptions may occur (Ogura, 1938).

During the initiation of a lateral root in an angiosperm, a group of pericyclic cells undergoes periclinal and anticlinal divisions (fig. 17.12*A*, *B*), which eventually result in the formation of a protrusion, the lateral root primordium (fig. 17.12*C*). By continued growth, the primordium gradually penetrates the cortex (pl. 15*B*). Before it emerges on the surface

pericycle   endodermis   phloem of parent root   cortex

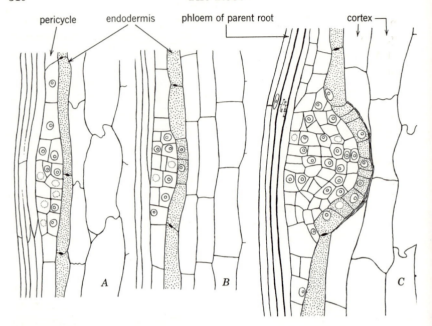

A   B   C

FIG. 17.12. Lateral root development. Longitudinal sections through young taproots of carrot. Divisions in pericycle initiate root primordium (*A*). Endodermis divides anticlinally and keeps pace with growth of primordium (*B*, *C*). Cortical parenchyma cells are compressed in front of primordium in *C*. (All, ×350. Esau, *Hilgardia* 13, 1940.)

of the parent root, the apical meristem, the primary tissue regions of the young root axis, and the rootcap become delimited by oriented cell divisions (fig. 17.13*A*). The apical meristem has not necessarily the same architecture as that of the parent root, but it may develop such with further growth. Regarding the mechanism of growth of the lateral root through the cortex of the parent root, some workers assume that the lateral root partly digests the cortical tissue as it advances, others consider that the penetration is entirely mechanical (Guttenberg, 1940). There is common agreement, however, that the advancing lateral root forms no connection with the tissues it penetrates.

In many plants the endodermis of the parent root participates in the initial growth of the branch root (Janczewski, 1874*b*). Sometimes it undergoes only anticlinal divisions and forms a single layer on the surface of the primordium (figs. 17.12, 17.13); sometimes it divides periclinally also and forms more than one layer. Before or soon after the lateral root emerges on the surface, the tissue derived from the endodermis dies and eventually is shed. The derivatives of the endodermis, sometimes combined with those of other cortical layers, may

form a rootcap-like structure called *pocket* (Tasche in German, Guttenberg, 1960). The pocket is especially large in water plants, in which
the rootcap may be completely absent (*Hydrocharis, Lemna, Eich-
hornia*). In *Pistia*, the epidermis is derived from the innermost layer of
the pocket. Whether or not the endodermis takes part in the formation
of the lateral primordium apparently depends on the proximity of the
lateral root origin to the apical meristem. If the lateral roots arise
rather far from the apical meristem, at the level where some xylem is
mature and the endodermis has casparian strips, the endodermis is little
or not at all concerned with the phenomenon. If the new primordium
is initiated while the endodermis is still essentially meristematic
(Cucurbitaceae, Papilionaceae, many water plants), the endodermis may
contribute tissue to the root primordium (Berthon, 1943). The view

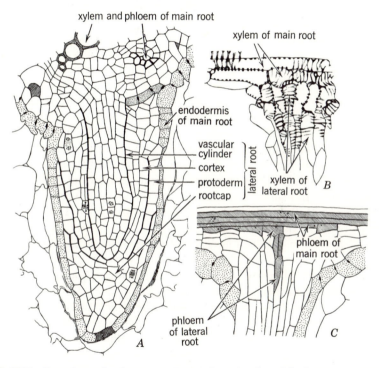

FIG. 17.13. Lateral root development in *Daucus* (carrot) in longitudinal sections. *A*, entire
section of root that had not completely traversed the cortex of main root. Layer of endodermis
enclosing root primordium is beginning to break down. At base of lateral root some cells have
developed casparian strips in connection with endodermis of main root. *B*, *C*, sections through
bases of lateral roots, illustrating elements connecting vascular tissues of main and lateral roots.
Xylem and phloem elements at base of lateral root are derived from pericyclic cells. (All, × 196.
Esau, *Hilgardia* 13, 1940.)

has been expressed that the presence of a large pocket is a primitive evolutionary character (Voronin, 1957).

There is a certain regularity in the spacing of lateral roots with reference to the xylem and phloem poles of the parent root (fig. 17.6; Guttenberg, 1940). If the parent root has more than two xylem poles, the branch roots arise either opposite these poles (common in dicotyledons), or opposite the phloem poles (Gramineae, Cyperaceae, Juncaceae). In diarch roots the lateral primordia are formed between the phloem and the xylem or opposite the xylem (Knobloch, 1954). In roots of this kind, there may be a row of primordia on both sides of each xylem pole. Thus the number of rows of branch roots may equal the number of xylem poles or may be double that number. In the fleshy root of the carrot, additional branch roots arise at the bases of the earlier roots as these die. They are formed in cushions of tissue of pericyclic origin (Esau, 1940; Thibault, 1946).

The vascular systems of the main and the lateral roots are delimited independently of each other, and the connection between the two is established through the intervening cells. Since the lateral root originates in the pericycle, the distance between its vascular region and that of the parent root is short. The intervening cells are derivatives of the pericycle. They differentiate into tracheary and sieve elements in continuity with similar elements in the main and the lateral roots (fig. 17.13B, C). The timing of this differentiation, that is, whether vascular elements mature first nearest the vascular tissues of the main root and then successively farther (acropetally) into the lateral root (Torrey, 1951), or whether some elements mature in the lateral root before the connection with the main root is established basipetally (Thibault, 1946) has not been adequately investigated.

The connection between lateral and parent roots varies in degree of complexity. In the seed plants, when the lateral root is diarch, the longer transverse diameter of its xylem plate (the diameter connecting the two xylem poles) is oriented parallel with the long axis of the parent root, and when, at the same time, the lateral root faces a protoxylem pole of the main root, there is a most direct connection between the xylem systems of the two roots. The two phloem strands of such a branch root are united with two phloem poles of the parent root. If the lateral root is formed between a xylem and a phloem pole of the main root, it is connected to these two poles. In the monocotyledons the xylem of the lateral root is often joined to two or more xylem strands of the main root (*Monstera*; Guttenberg, 1940). Moreover, the connection may occur not only with the peripheral xylem strands but also with the late metaxylem vessels through the modification into tracheary elements of the vascular parenchyma cells intervening between the peripheral strands and the late metaxylem vessels (Rywosch, 1909). Such differentiation may extend considerably to the sides of the actual insertion of the lateral root. Thus, sometimes the insertion of a lateral root has a marked although localized influence on the structure of the vascular cylinder of the parent root (Fourcroy, 1942). In plants with secondary growth, the secondary tissues of the main and the lateral roots differentiate in continuity with each other, and the xylem of the bases of the lateral roots is imbedded in the xylem of the main root.

Experimental studies on initiation of lateral roots indicate that complex factors determine the formation of a lateral root meristem (Torrey, 1959). There is a requirement for substances not synthesized in the root itself and apparently derived from the cotyledons or shoots; and the absence of lateral roots within a certain distance from the apex suggests that the terminal meristem produces substances inhibiting the appearance of lateral roots. The distance between the youngest lateral roots and the apex is not always constant, however. In rapidly growing roots of *Libocedrus* the lateral roots were found to be more widely spaced and originating at a greater distance from the apical meristem than in slowly growing roots (Wilcox, 1962b).

Development of Adventitious Roots

Adventitious roots, in the wide sense of the term reviewed previously, may occur on the hypocotyl of a seedling, at nodes and internodes of stems, and in roots. They may arise in connection with buds or independently (Bannan, 1942; Guttenberg, 1940). They may be formed in young organs (embryos and intercalary meristems in Gramineae) or in older tissues that have not lost their meristematic potentialities. Most adventitious roots arise endogenously, although examples of exogenous origin are also known (Guttenberg, 1940). Adventitious roots may arise from primordia laid down previously and remaining dormant until stimulated to growth, or they may be new formations (Baranova, 1951; Carlson, 1950; Siegler and Bowman, 1939).

Stem-borne adventitious roots constitute the main vascular system in lower vascular plants, in most monocotyledons, in dicotyledons propagating by means of rhizomes or runners, in water plants, in saprophytes, and in parasites. Roots that form on cuttings, directly from the stem or from the callus tissue, are also adventitious. The subject of adventitious roots in cuttings has been considerably explored, particularly in connection with the research on growth-promoting substances (Carlson, 1950; Swingle, 1940).

The principal histologic aspects of the origin of adventitious roots may be summed up in a statement that such roots usually are initiated in the vicinity of differentiating vascular tissues of the organ which gives rise to them (Priestley and Swingle, 1929; Swingle, 1940). If the organ is young, the adventitious primordium is initiated by a group of cells near the periphery of the vascular system. If it is older, the seat of this origin is located deeper, near the vascular cambium. In young stems, the cells forming the root primordium are commonly derived from the interfascicular parenchyma; in older stems, from a vascular ray. Sometimes the adventitious roots appear to be initiated by divisions in the cambial zone (Smith, 1936). Often the seat of the first divisions forming a root primordium in stems is identified in the literature as pericycle (Stangler, 1956). As was discussed in chapter 15, in many stems the region formerly defined as pericycle is, by origin, partly primary phloem, partly interfascicular parenchyma between two strands of primary phloem. Some authors specifically mention that the adventitious roots arise in the phloem region (Petri el al., 1960; Satoo, 1955; Satoo, and Fukuhara, 1955). The origin of adventitious roots in the interfascicular region, in the vascular ray, or in the cambium places the young root close to both xylem and phloem of the parent axis and facilitates the establishment of vascular connection between the two organs.

The primordia of adventitious roots are initiated by divisions of parenchyma cells. In dicotyledons and gymnosperms these may be parenchyma cells of the phloem region, as detailed above, or they may be callus cells. In cuttings, the origin of roots from callus is a familiar phenomenon. Variations in the origin of adventitious roots may be found in the same plant (Wilcox, 1955). In the stems of monocotyledons the adventitious roots arise in the parenchyma in perivascular position.

Before the adventitious root emerges from the stem it differentiates a rootcap and the usual tissue systems in the body of the root. This differentiation is similar to that observed in lateral roots; and in both the formation of the vascular connection with the parent axis has not been critically studied. When vascular elements differentiate in the adventitious root, the parenchyma cells—callus cells or others—located at the proximal end of the primordium differentiate into vascular elements and provide the vascular connection with the initiating organ. Adventitious roots originating in relatively old stems may grow obliquely through the outer tissues of the stem probably because of the resistance offered by sclerenchyma in the phloem or outside it (Satoo, 1955; Stangler, 1956; Tomlinson, 1961). Poor rooting capacity of stems may be related to their high degree of sclerification (Beakbane, 1961).

Development of Buds on Roots

The formation of buds on roots makes possible the propagation of plants by root cuttings and is an important means of spread of noxious weeds. Buds develop on roots of various ages and structure. They frequently arise endogenously like lateral or adventitious roots. Endogenous origin was observed in naturally growing roots (Kondrat'eva-Mel'vil', 1957; Vasilevskaĩa, 1957) and in isolated roots (Seeliger, 1959; Torrey, 1958). The bud may originate in the pericycle of a younger root and be at first deceptively similar to a root primordium (Bakshi and Coupland, 1959). In an older root it may be found in a callus-like proliferation of ray tissue giving rise to more than one bud (Vasilevskaĩa, 1957), or it may be initiated exogenously in the callus-like growth derived from the phellogen (Murray, 1957). Buds often arise near lateral roots and, if these or their traces are still alive, the buds may become connected to the lateral root trace. If a bud arises in the pericycle, the vascular connection with the initiating root is formed by acropetal differentiation; if the bud is initiated near the surface, vascular differentiation is basipetal (Kondrat'eva-Mel'vil', 1957).

STRUCTURE IN RELATION TO FUNCTION

Absorbing Root

Many studies have been carried out to determine the part of the root concerned with the intake of water and salts (Kramer, 1959; Steward and Sutcliffe, 1959), but only a few include attempts to consider accurately the structure of the absorbing zone. Absorption of water and salts occurs mainly in the young growing part of the root. Thus the region engaged in absorption is structurally heterogeneous, and is constantly changing in its anatomic and physiologic characteristics. The assumptions are justified, and are supported by experimental evidence, that at least certain phenomena of absorption are dependent on metabolic activity associated with growth and that the factors determining salt intake are not necessarily the same as those responsible for the entry of water (Hoagland, 1937).

Apparently little water enters through the rootcap and the apical meristem. In plants grown in culture solutions the maximum rates of absorption of water are usually observed several centimeters from the apical meristem where some or most of the primary xylem is mature and the endodermis has casparian strips but no other wall formation which would decrease permeability (Kramer, 1959).

The distribution of the rates of water absorption along the root is known to vary in relation to the length of the root, its age, rate of growth, and other internal conditions. Some of these variations are probably associated with structural differences. Thus, when growth slows down at the approach of winter, the absorbing zone at the root

tip may be eliminated completely by the formation of various impermeable layers. A suberized exodermis and an endodermis develop to
within a short distance of the apical meristem, and fatty substances
appear in the superficial rootcap cells and in the epidermal cells intervening between the suberized exodermis and the rootcap (Guttenberg,
1943; Hayward and Long, 1942; Wilcox, 1954). The difference between
slowly and rapidly growing roots with regard to proximity of mature
xylem to the apical meristem might be related to variable rates of
absorption.

With regard to absorption of salts, most studies deal with the
accumulation of substances, but it is not certain whether the most
active absorption occurs at the level of most intensive accumulation.
Generally, maximum accumulation of substances that were tested
occurred close to the apical meristem (Canning and Kramer, 1958;
Steward and Sutcliffe, 1959), at levels where cells are actively concerned
with division and enlargement.

Absorption of water and salts is not limited to the young, largely
unsuberized root parts. Roots with secondary growth and a periderm
have been shown to be capable of absorbing considerable amounts of
water (Kramer, 1959), but the main centers of absorption and accumulation are the young roots (Steward and Sutcliffe, 1959).

The root hairs are usually interpreted as structures substantially
increasing the absorbing surface of the roots (Kramer, 1959). In agreement with this concept, the root hairs come to their full development
in the root zone where the most active absorption of water takes place.
The ability of these structures to absorb water has been demonstrated
experimentally, but hairless epidermal cells are also capable of absorbing
water (Rosene, 1943). In many woody plants root hairs are not common (Kramer, 1959). Mycorrhizae are sometimes thought to be
compensating for the absence of root hairs, but their efficiency as
nutrient-absorbing structures has not been adequately tested. Possibly
some fungi are more effective than others, and environmental conditions
also appear to affect their activity (Melin, 1953). In experiments with
poplar seedlings, faster growth was observed in those infected with
endotrophic mycorrhizae than in the noninfected ones (Clark, 1963).

The structure of the root is of particular interest with regard to the
movement of water and salts from the absorbing cells to the conducting
tissues, and their release from the living cells of the vascular cylinder
into the nonliving tracheary elements. Figure 17.14 illustrates the
pathway of the soil solution in the root by means of a drawing of a part
of a wheat root in transection. The arrows indicate the direction of
movement in certain selected cells. Among these the living cells are
stippled. The most notable features of this pathway are: (1) the presence

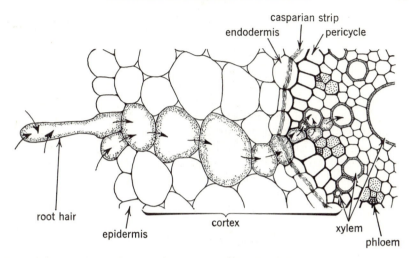

FIG. 17.14. Part of transection of wheat root, illustrating the kinds of cell that may be traversed by water and salts absorbed from the soil before they reach the tracheary elements of the xylem. Arrows indicate direction of movement through a selected series of cells. Among these, the living cells are partly stippled. Casparian strip in endodermis is shown as though exposed in surface views of end walls. ($\times$330.)

of abundant intercellular spaces in the cortex, (2) the lack of such spaces in the vascular cylinder, and (3) the presence of a specialized endodermis between the two systems. Some investigators stress the contrasting environments of the cortex and of the vascular cylinder. The well-aerated cortex is high in metabolic activity and capable of accumulating salts; the poorly aerated vascular cylinder is low in activity and unable to hold salts (Crafts and Broyer, 1938). The endodermis located between the two distinct systems acts as a barrier that facilitates the development of hydrostatic pressure in the vascular cylinder by preventing a leakage of solutes from the vascular cylinder into the cortex and thereby also has to do with the entry of the solutes into the nonliving tracheary cells. The incrustation of the wall with fatty and other materials in the casparian-strip region is presumed to hinder the movement of substances through the walls, whereas the connection of the cytoplasm to the strip prevents the passage between the protoplast and the wall. Hence, all materials crossing the endodermis would be forced to pass the living protoplasm and be subjected to its regulatory activity. (See also Arnold, 1952; Steward and Sutcliffe, 1959.)

Storage Root

Roots of ordinary structure are important storage organs of the plant; in addition, roots may become specifically adapted for this function by

distinct developmental peculiarities.   Primary roots store food, notably
starch, in the cortex, which is often wide.   In roots having a limited
amount of secondary growth, the cortex may remain as a storage tissue.
The secondary tissues of the root accumulate starch in the same kind of
cells as those of the stem, that is, in various parenchymatic and some
sclerenchymatic cells of the xylem and the phloem.   In general, roots
possess a higher proportion of parenchyma cells than do stems.

The special adaptations for storage are commonly expressed in the
development of fleshy bodies in parts of the root system.   Frequently,
the hypocotyl and the base of the taproot jointly form one fleshy
structure (*Daucus, Pastinaca, Beta*).   Some fleshy organs have large
amounts of storage parenchyma associated with an otherwise ordinary
arrangement of tissues.   This type of development is exemplified by the
carrot and fennel, in which the hypocotyl and the upper part of the taproot, after sloughing the cortex in a normal manner (pl. 88), become
fleshy through a massive development of parenchyma in the phloem and
the xylem (Bruch, 1955; Esau, 1940).

In contrast, the sugar beet forms its fleshy hypocotyl-root organ by
anomalous growth (Artschwager, 1926; Seeliger, 1919).   Both the table
beet and the sugar beet show a usual type of primary and early secondary
development.   Later, however, a series of supernumerary cambia arise
outside the normal vascular core and produce several increments of
vascular tissue, each consisting of a layer of parenchyma, and collateral
strands of xylem and phloem imbedded in parenchyma (pl. 89).   Sugar
occurs as a storage product in parenchyma cells, particularly those
closely associated with the vascular strands.

Anomalous growth of a different kind is found in the sweet potato,
Ipomoea batatas (Esau, 1960).   In the normally developed but highly
parenchymatic primary and secondary xylem, anomalous cambia arise
around individual vessels or vessel groups.   These cambia produce
phloem rich in parenchyma and with some laticifers away from the
vessels, and tracheary elements toward them.   The fleshy roots, the
rhizomes, and the stems of many Cruciferae (turnip, radish, kohlrabi,
rutabaga, and others) show a diffuse secondary growth of parenchyma
in the pith (if pith is present) and in the secondary xylem, and a differentiation of concentric vascular bundles within this parenchyma (Lund
and Kiaerskou, 1885; Soeding, 1924).   Monocotyledons may also form
fleshy roots by diffuse secondary growth (Weber, 1958).

Despite their variations in structure, the storage organs all have in
common an abundance of parenchyma and a thorough permeation of
this parenchyma with vascular elements.   The close association between
the two kinds of tissues may be brought about by (1) a proliferation of

the parenchyma among the normally located vascular elements, (2) a massive development of parenchyma followed by a differentiation of additional vascular elements in this parenchyma, or (3) a development of whole new systems of parenchyma-vascular tissues outside the normally placed system of the same kind.

Anchorage Root

The well-known function of the root as an organ fastening the plant to the soil need not be emphasized here. The development of sclerenchyma in the old root leads to the formation of a strong rigid anchorage organ, but the firm attachment to the soil is also dependent on the development of the many branches in the branched type of a root system and of the many adventitious roots in the fibrous type. Of the two types of systems the fibrous generally penetrates the soil less deeply but binds the surface soil more tightly than the branched type. The root hairs, too, play a part in binding the soil (Dittmer, 1948). They are particularly efficient in anchoring young plants and preventing their being pushed upward by the growth of the root apex (Farr, 1928).

One aspect of the anchorage of the plant to the soil deserving the anatomist's attention is the contraction of roots that during a certain stage of development of the plant draws the shoot apex near or below the ground level and places it in an optimal environment for growth and for development of adventitious roots. Root contraction is a common phenomenon and is widely distributed among herbaceous perennial dicotyledons and monocotyledons (Arber, 1925; Gravis, 1926; Rimbach, 1929). In one study, the contraction of roots was recorded, by actual testing, in 450 species of 315 genera in 82 families (1 gymnosperm, 15 monocotyledon, 66 dicotyledon families; Rimbach, 1929). The Gramineae appear to lack contractile roots (Arber, 1934). Examples of well-known economic plants showing root contraction are alfalfa (fig. 17.15), sugar beet, carrot, and sweet clover (Bottum, 1941). Contraction of roots is also observed in the bulbous monocotyledons which draw the bulbs to considerable depths into the soil (Chan, 1952). In *Rubus* the terminal bud may become rooted when it is brought into contact with the soil and is subsequently pulled into the ground by a shortening of the roots (Rimbach, 1898).

The contraction occurs in taproots, in lateral roots, and in adventitious roots. In root parts showing maximum contraction, 10 to 70 per cent shortening has been reported (Rimbach, 1898). The contraction begins soon after the elongation of the root is completed and continues for variable lengths of time. In some plants it continues from 1 to 5 months, and in *Taraxacum* and *Panax Ginseng* it is said to occur for years in the same root (Grushvitskiĭ, 1952; Rimbach, 1898). In some plants only certain of the roots undergo contraction, and these are rather specialized morphologically; in others, no morphologic differentiation of contractile roots occurs. The highly specialized contractile roots, or contractile root parts, exhibit histologic peculiarities. They show relatively little lignification, have a high proportion of parenchyma, and, in general, appear little differentiated. Contractile roots thicken during the contraction. In some species this thickening is associated with the development of the root as a storage organ (fig. 17.15A, B; *Melilotus, Asparagus officinalis;* Bottum, 1941; Rimbach, 1899); in others the parenchyma collapses with age and the root appears wrinkled.

Apparently the histologic details of root contraction vary considerably in different plants. In some (*Medicago, Melilotus,* sugar beet) a radial extension of parenchyma cells and an assumption of a sinuous course by the lignified tissues (particularly of the central xylem core; fig. 17.15C) have been observed in connection with root shortening (Bottum, 1941; Jones, 1928; Rimbach, 1929). One must postulate that in these plants

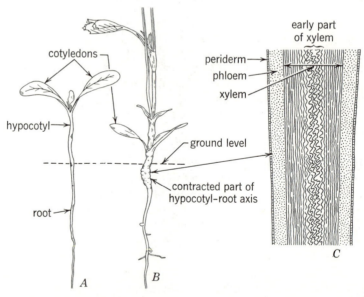

FIG. 17.15. Illustration of root contraction in alfalfa (*Medicago sativa*) seedlings. The younger in *A* bears its cotyledons high above the ground level; the older in *B* has pulled the cotyledons close to the ground by contracting the hypocotyl and the upper part of the root. The contracted part is considerably thickened. *C,* from a cleared longitudinal section of root. Central xylem system (consisting of primary and some secondary xylem) became undulate after the root contracted. (*A, B,* drawn by R. H. Miller.)

the radial extension of the parenchyma cells is combined with their vertical contraction. In some monocotyledons the contraction has been found restricted to the inner cortex, the outer cortex dying and becoming wrinkled (Rimbach, 1929). In the Umbelliferae certain cell groups, located among those that are expanding radially, die and collapse. This change in the volume of the tissue apparently permits the mutual adjustment between the expanding cells and those elements that are oriented longitudinally and are undergoing a bending (Berckemeyer, 1929). In certain species of *Oxalis* the collapsing cells occur in transverse zones, and the reduction in length of the root is thought to result from a reduction in volume rather than from growth phenomena (Davey, 1946).

COMPARATIVE STRUCTURE OF SHOOT AND ROOT

Primary Body

The preceding pages of this chapter give ample evidence that the root has many distinctive characteristics differentiating it from the shoot, particularly in the seed plants. It will be useful to assemble the data comparing the two parts of the plant. The dissimilarities between the root and the shoot are evident in their earliest stages of development. The apical meristem of the shoot is truly apical, for it occupies a superficial position; that of the root is subterminal, for it is covered by the rootcap. The architecture of the two meristems differs also in that the relation between the regions of the primary body and the apical initials is often more precise in the root than in the shoot. It is not uncommon, for example, that in the root the vascular cylinder and the cortex have separate initials, whereas in the stem these two tissue regions are closely related in their early ontogeny. The epidermis of the root has a more varied ontogenetic origin than that of the shoot. The leaf primordia arise directly from the apical meristem of the shoot, and the branches more or less directly; and both are exogenous. The lateral roots arise independently of the apical meristem and are endogenous.

In the higher vascular plants the vascular system of the shoot differentiates largely or entirely in relation to the leaves. The root develops its vascular system as an axial structure independent of the lateral organs. The lack of influence of the lateral organs on the organization of the root is also reflected in the absence of a segmentation into nodes and internodes in the root. Leaf gaps and pith are characteristic of the vascular cylinder in the stems, except in certain lower vascular plants. There are no gaps in the root and frequently no pith.

The primary vascular tissues of the shoot are commonly arranged in the form of more or less discrete bundles, each containing both xylem and phloem, in collateral or in bicollateral combinations. The root lacks vascular bundles, in the sense of units combining xylem and phloem, but develops radially alternate phloem and xylem strands, the latter being discrete or united in the center into one continuous body. In the seed plants the root and the stem contrast strikingly with regard to the direction of differentiation of primary xylem in the horizontal plane. This direction is centrifugal in the shoot (endarch xylem) and centripetal in the root (exarch xylem). In lower vascular plants (Psilopsida and Lycopsida) the primary xylem is exarch in both root and stem, in ferns, commonly mesarch in the stem.

The boundaries between the tissue systems are quite precise in the root. The vascular cylinder forms a compact core delimited from the cortex by an endodermis and surrounded by a distinct nonvascular tissue region, the pericycle. In the stem of the higher vascular plants, the vascular tissues are not compactly arranged, a morphologically specialized endodermis is rare, and commonly no distinct region meriting the special name of pericycle occurs between the cortex and the vascular tissues. The difference in the precision of the tissue delimitation is evident in the two organs of the same plant, and, therefore, if the tissue regions are followed upward from the root, their limits will appear more diffuse as the level of the shoot is approached.

The root and the stem show some differences in the manner of primary growth. The root has a shorter elongation region than the shoot, and a sharper transition between the region of small, actively dividing cells and that composed of large, expanding cells (Sinnott and Bloch, 1941). Concomitantly with the small amount of elongation, the root frequently develops no extensible types of protoxylem elements (with annular and helical secondary walls), whereas in the shoot such elements are common.

Secondary Body

Whereas the primary bodies of the shoot and the root show fundamental differences, which can be traced directly into the meristems, the secondary bodies of the two organs are much alike in both origin and structure, and the existing differences are of a quantitative rather than a qualitative kind (pl. 28). The secondary vascular tissues of the root commonly have a greater proportion of living to nonliving cells than similar tissues in the stem (Riedl, 1937). This difference appears to be related to the different environments under which the root and the stem develop, for the underground stems (rhizomes) have more similarities with roots than with stems in the structure of their secondary tissues.

Furthermore, root and stem can be made to produce tissues resembling those of the opposite organ by reversing their environmental conditions, that is, by exposing the root to an aerial environment and, contrarily, burying the stems in the soil (Bannan, 1934; Beakbane, 1941; Miyawaki, 1957). The quantitative nature of the difference is also suggested by its variability in roots of the same plants.

In detail, the differences between the structure of secondary vascular tissues of stem and root may be enumerated as follows. Compared with the stem, the root may have: a higher bark-to-wood ratio (assuming that the bark includes all extracambial tissues); a lower percentage of area of bark occupied by fibers; a smaller number of fibers in the xylem; larger vessels of more uniform size, although sometimes fewer in numbers; a poor differentiation of growth increments; a larger volume of ray tissue in the gymnosperms, wider and longer tracheids with multiseriate arrangement of pits and frequent occurrence of pits on tangential walls; a larger ratio of area of living cells to area of nonliving cells in both the phloem and the xylem; more starch; and less tannic substances. The first periderm of the roots commonly arises in the pericycle, that of the stem in the peripheral layers of the axis.

VASCULAR CONNECTION BETWEEN THE SHOOT AND THE ROOT

Concept of the Transition Region

The connection between the morphologically distinct primary vascular systems of the shoot and the root in the seed plants is of interest from the developmental as well as from the phylogenetic viewpoint and is, therefore, extensively treated in the botanical literature. Since this connection involves spatial adjustments between systems with differently oriented parts and with different directions of differentiation in the horizontal plane, it shows some features that are intermediate or transitional between those of the shoot and the root. The change from one type of structure to the other, as viewed in successive levels of the root-shoot connection, is commonly called vascular transition, and the region of the plant axis where it occurs is called the transition region.

As was shown in chapters 1 and 15, the shoot of a higher vascular plant arises at one end of the embryo axis (the hypocotyl), the root at the other. Accordingly, the connection between the two is established through the hypocotyl. The basic features of this connection are delimited in the form of a procambial system during the development of the embryo (Miller and Wetmore, 1945; Spurr, 1950). The differentiation of the vascular elements from the procambial cells follows the delimitation of the procambium (it may begin during the development of the embryo or after germination). Its sequence and direction are determined not only by the form of the initial procambial pattern but also by the distribution of growth in the different parts of the seedling. Therefore, a proper understanding of the transition region may be gained only if this plant part is studied throughout its development. Most of the literature on vascular transition deals with partly differentiated seedlings, so that despite its volume it gives only a partial picture of the phenomenon.

Although the structure of the transition region is variable in the different groups of plants and is generally complex, an understanding of this structure has been unnecessarily obscured by the interpretation that there is a transition between the root and the stem rather than between the root on the one hand, and the cotyledons and the epicotylary shoot on the other. The transition region represents a connection, not between two axial organs with somewhat different arrangements of tissues, but between an organ with an axial vascular system and one whose vascular system develops in relation to leaves. A study of the transition region must, therefore, explain the relation between the vascular system of the root and the traces of the first foliar organs of the plant.

Structure of the Transition Region

In most dicotyledons and gymnosperms, the characteristics intermediate between those of the vascular systems of root and shoot are present within the system connecting the root and the cotyledons (Chauveaud, 1911; Guttenberg, 1941; Hill and de Fraine, 1908–10; Thomas, 1914). In other words, the transition in these plants occurs between the root and the cotyledons. Whereas the root has a more or less compact core of vascular tissue, the levels intermediate between the root and the cotyledonary node show strands diverging above into the cotyledons. Using the concept of leaf traces, one can say that in the transition region the cotyledonary traces diverge from the vascular system of the hypocotyl-root axis. This divergence differs from that of the leaf traces in the shoot in that the cotyledonary traces are connected with a system with exarch xylem and an alternate arrangement of xylem and phloem. The cotyledonary traces are more or less affected in their structure by this association. The matter may be explained best by a simple example.

The seedling in fig. 17.16 has two cotyledons, a small epicotylary shoot between the cotyledons, a hypocotyl, and a root with a diarch xylem plate flanked on both sides by phloem strands. Each cotyledon has, in median position, a double vascular bundle composed of two partially

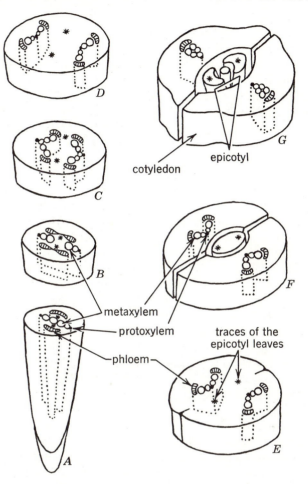

cotyledon

epicotyl

metaxylem

protoxylem

traces of the
epicotyl leaves

phloem

FIG. 17.16. Connection between root and cotyledons (the transition region) in a dicotyledonous seedling (*Beta vulgaris*). The root is diarch (*A*). The primary vascular system of the root diverges, above, into two cotyledons.

merged strands. Such structure of the median cotyledonary strands is common in various groups of plants (Hill and de Fraine, 1913; Thomas, 1914). In some seedlings the double nature of the median strand may be less pronounced than in fig. 17.16, in others more pronounced, and in still others there may be two separate strands in median position. The double nature of the median cotyledonary strands is considered to have phylogenetic significance (Bailey, 1956).

The median parts of the cotyledons are located in direct line above the protoxylem poles. In the root the xylem is strictly centripetal in its differentiation, the metaxylem occupying the center. In the lower parts of the hypocotyl the protoxylem maintains its peripheral position, but the metaxylem, instead of differentiating toward the center, diverges laterally from the protoxylem. Such order of differentiation leaves the center of the axis unoccupied by vascular elements. In other words, a pith differentiates in this part of the seedling. At successively higher levels the distance between the protoxylem poles increases, for the hypocotyl axis widens toward the cotyledonary node. Concomitantly, the plates of metaxylem associated with each protoxylem pole do not join in the intercotyledonary plane. Thus, instead of one xylem plate as in the root, there are, higher up, two distinct xylem complexes, each pertaining to one double cotyledonary trace. In the upper hypocotyl and in the bases of the cotyledons the direction of xylem differentiation is such that in each cotyledonary trace the protoxylem pole occupies a deeper position in the axis than the metaxylem. This orientation signifies that the xylem approaches the endarch condition. It is entirely endarch still higher in each cotyledon, where the metaxylem differentiates, not in the form of two diverging plates, but as one double plate directed outwards from the protoxylem pole. Thus fig. 17.16 illustrates the transition from exarch to endarch xylem.

The differences in the orientation of the phloem at various levels of the seedling are less pronounced than those of the xylem. Instead of the two phloem strands appearing in the root, there are four in the hypocotyl. Considering the structure from the base upward, one could say that the phloem branches, each phloem strand of the root giving two branches in the hypocotyl. Each of the four hypocotylary phloem strands is associated with one metaxylem plate (fig. 17.16C). In the part of each cotyledon where the xylem is endarch, the phloem differentiates as one mass on the abaxial side of the double cotyledonary bundle. This bundle is, therefore, collateral. Thus fig. 17.16 shows the transition from the radially alternate arrangement in the root to the collateral in the cotyledons.

The epicotylary shoot of seedlings having such a transition region as that depicted in fig. 17.16 develops after the primary vascular system of the root-hypocotyl-cotyledon unit is delimited and partly differentiated. The traces of the first two leaf primordia of the epicotyl (these usually appear almost at the same time, opposite each other and alternating with the cotyledons) alternate with the cotyledonary traces in the hypocotyl, and all these traces together encircle a central pith. In the root, where the xylem of the cotyledonary traces is merged with the diarch primary xylem plate, the vascular tissues of the epicotylary traces are prolonged directly without change in orientation along the flanks of the diarch plate and along the inner margins of the primary phloem strands. In other words, the epicotylary traces are connected with the part of the xylem that occurs on the flanks of the diarch plate and with the part of the phloem developing centripetally from the initial phloem. These tissues may be entirely secondary or partly secondary and partly primary. The traces of the first leaves on the epicotyl are collateral and have endarch xylem. Since they are connected in the root with similarly oriented tissues, there is no transition between the root and the epicotyl in the type of seedling depicted in fig. 17.16, but rather a simple direct connection between similarly oriented tissues. The epicotyl seems to be superimposed over the initially complete root-hypocotyl-cotyledon unit.

The type of transition region just described is common among the dicotyledons. However, there are many deviations from this type. Seedlings may have several traces to each cotyledon, one double median and two or more lateral. Frequently, the lateral traces are relatively small and are connected with the root in a manner described for the epicotylary traces in the preceding paragraphs, that is, without any change in the orientation of tissues. The cotyledon-root connection varies also in relation to the structure of the vascular system of the root. If, for example, this vascular system has tetrarch xylem, two of the xylem poles may be continuous with two median cotyledonary traces, the other two with the two pairs of the lateral cotyledonary traces. In some plants, like the Cucurbitaceae, each cotyledon has many vascular bundles and a highly complex vascular system in the transition region (Hayward, 1938).

The epicotyl-root relation also varies in the dicotyledons, and apparently the closeness of connection between the two parts depends on the timing in the development of the epicotyl. If it initiates leaf primordia relatively early, the first traces may be connected with the primary tissues of the root; if the foliar organs appear later, the connection is formed with the secondary tissues (Compton, 1912a). In some plants the epicotyl is connected with the root apparently only indirectly through the cotyledonary traces (certain Cucurbitaceae, Hayward, 1938; *Cynara*, Phillips, 1937). A notable deviation in the vascular transition is found in the dicotyledons with hypogeous cotyledons, that is, cotyledons remaining below the surface of the ground after germination (*Pisum sativum, Vicia sativa, Vicia faba, Lens esculenta, Cicer arietinum*). In the representatives of this group of Leguminosae the traces of the first foliage leaves may be connected with the primary vascular tissues of the root. Depending on the closeness of connection between the root and the epicotyl, the transitional characteristics of the vascular

system are extended more or less far into the epicotylary shoot, sometimes through more than one internode (Compton, 1912a; Muller,
1937).

Evidently the extent of the dicotyledonous seedling axis which shows
the features of transition is variable. The transition region, in other
words, may be short or long; or, with reference to the position of the
root, it may be high or low. In some plants transitional characteristics
are evident throughout the hypocotyl; in others they are restricted to
the upper hypocotyl and part of the cotyledons. In the latter types of
seedlings the hypocotyl is said to have root structure. In the seedlings
with hypogeous cotyledons the transition region is particularly long, for
it extends into one or more of the internodes above the cotyledons.

The specific characteristics of the vascular transition in the monocotyledons are related to the presence of a single cotyledon and the
shortness of the lower internodes. The latter feature is probably the
main cause of the frequently close connection between the epicotyl and
the root in this group of plants (Arber, 1925). In many monocotyledons
one part of the root system is connected with the cotyledon, the other
with the first leaf of the epicotyl, and both connections exhibit transitional
features (*Allium cepa*, Hayward, 1938; *Asparagus officinalis*, Mullendore, 1935; palms, Drabble, 1905; *Yucca*, Arnott, 1962). However, in
some of the monocotyledons the transition occurs, like in so many
dicotyledons, only between the root and the cotyledon, with the whole
primary vascular system of the root prolonged into the single cotyledon
(*Anemarrhena*, Arber, 1925).

The transition region of the Gramineae is particularly complex (Avery,
1930; Boyd and Avery, 1936; McCall, 1934), because the vascular
system of the root is connected with more than one leaf above the
scutellum, which many workers consider to be the single cotyledon in
this group of plants. The vascular transition of *Triticum* may be used
as an example (fig. 17.17; Boyd and Avery, 1936). The polyarch
vascular cylinder of the root is connected with the vascular system of
the foliar organs through the plate-like vascular system located below
the insertion of the scutellum (nodal plate of the scutellum according
to some authors). The vascular tissue prolonged upward from the nodal
plate is separated into strands which, at lower levels, show irregular
arrangement and transitional features and, at higher levels, form a hollow cylinder, all parts of which have endarch xylem and collateral
arrangement of xylem and phloem. This system consists of traces and
trace complexes of the scutellum, the coleoptile, and the first and second
foliage leaves. Thus, there is a relatively abrupt transition from the
vascular cylinder of the root with exarch xylem and alternate arrange-

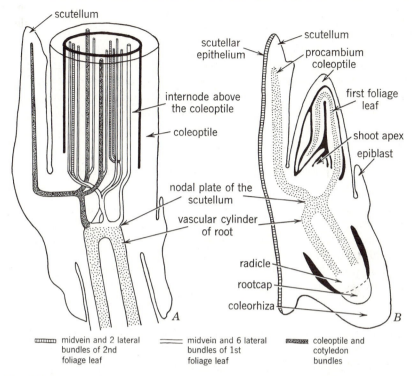

scutellum

scutellum

scutellar epithelium

scutellum

procambium

coleoptile

internode above the coleoptile

first foliage leaf

coleoptile

shoot apex

epiblast

nodal plate of the scutellum

vascular cylinder of root

radicle

rootcap

coleorhiza

A

B

| ▭▭▭ midvein and 2 lateral bundles of 2nd foliage leaf | ▭▭▭ midvein and 6 lateral bundles of 1st foliage leaf | ▰▰▰ coleoptile and cotyledon bundles |

FIG. 17.17. Longitudinal views of transition region of wheat seedling (*A*) and embryo (*B*). Epithelial surface of scutellum is appressed to endosperm in the seed (see fig. 19.2). The epicotylary shoot is enclosed by the coleoptile, the radicle by the coleorhiza. (*B*, ×25. *A*, after Boyd and Avery, *Bot. Gaz.* 97, 1936; *B*, adapted from McCall, *Jour. Agr. Res.* 48, 1934.)

ment of xylem and phloem, to a system of leaf traces with endarch xylem and collateral arrangement of xylem and phloem.

The transition region of the gymnosperms resembles that of many dicotyledons in that it represents primarily a connection between the root and the cotyledons (Guttenberg, 1941; Hill and de Fraine, 1908–10). The variations in the structure of the transition region in these plants result, in part, from the variable numbers of cotyledons and of traces to each cotyledon. A similar direct continuity between the vascular tissue of the root and that of the first leaves exists also in the sporelings of the lower vascular plants (Campbell, 1921; Hill and de Fraine, 1913).

The vascular system showing transitional characteristics is entirely primary. When cambial activity occurs in plants having secondary growth, the secondary tissues are formed in complete continuity between the stem and the root (fig. 1.2). The vascular cambium arises in the same position, between metaxylem and metaphloem, in the root, the hypocotyl, and the epicotyl and produces derivatives in the same direction, phloem toward the outside, xylem toward the inside, in all three parts of the plant. Thus, the secondary growth obscures the initial differences in the structure of root, hypocotyl, and epicotyl. Moreover, it separates the primary phloem from the primary xylem by carrying the primary phloem outward and leaves only the xylem pattern of the transition region buried in the center of the axis.

Morphologic Significance of the Transition Region

The peculiar structure of the transition region makes it difficult to classify this part in relation to the other organs of the plant. As a result, more than one theory have been formulated with reference to the structural and evolutionary significance of this region of the plant (Compton, 1912b; Duchaigne, 1951). One common concept is that the seedling plant has a unit vascular system, morphologically equivalent in all its parts, and that the difference in orientation of its parts at the various levels may be described, figuratively, as branching, twisting, rotation, and inversion (Eames and MacDaniels, 1947; Lenoir, 1920; Van Tieghem, 1891a). The proponents of this concept recognize that the elements differentiate in the same positions where they occur in the mature state and use the expressions implying motion of parts merely to emphasize the unity of the system. The opposite view is that the seedling system is initially discontinuous, consisting of a radicular-hypocotylary part on the one hand, and a cotyledonary part on the other, and that the two are joined in the upper hypocotyl (theory of accord; Dangeard, 1913).

A double origin of the vascular system of the seedling is also postulated in a physiologic interpretation of the transition region (Thoday, 1939). The seedling is pictured as having a unique structure in that it consists of a short axis bearing at opposite ends, but at close proximity to each other, two self-determining centers of different kind. Each of these two opposite poles, the shoot pole and the root pole, is capable of impressing its own inherent pattern on the meristematic tissues to which it gives rise. The cotyledons, and, if the epicotyl is precocious, the first foliar primordia also, influence the structure of the upper part of the seedling axis, and the root leaves its impression at the base of it. In the intervening region the two patterns are mutually accommodated.

One of the most elaborately developed theories—the theory of basifugal acceleration of Chauveaud (1911; see Duchaigne, 1951)—postulates that, from the evolutionary standpoint, the different arrangements of the vascular system in the different plant parts are not equivalent, but that the alternate arrangement in the root is primitive, the collateral or superposed arrangement in the shoot, advanced. The different structures at the successive levels of the transition region result from an acropetal acceleration in the development of the different evolutionary types in the transition region; that is, at higher levels the more advanced evolutionary stages appear earlier than at the lower levels, and at the highest position the primitive stage is completely omitted. The understanding of the structure and the evolution of the transition region would seem to be of much importance for the interpretation of homologies between the root and the shoot. Although the vascular system of the root and the cotyledons is a unit from the early stages of embryogeny, the epicotyl often appears to be a separate structure attached to the root-hypocotyl-cotyledon unit. The study of the relation between the various tissue systems of the epicotyl and the axis below it might prove highly significant in the interpretation of such matters as the comparative nature of the epidermis and the cortex of root and shoot, the morphologic value of the region called pericycle, the meaning of protoxylem and metaxylem, protophloem and metaphloem, and the developmental relation between the primary and secondary tissues.

REFERENCES

Allen, G. S. Embryogeny and the development of the apical meristems of *Pseudotsuga*. III. Development of the apical meristems. *Amer. Jour. Bot.* 34:204–211. 1947.

Allen, O. N., and E. K. Allen. Morphogenesis in the leguminous root nodule. *Brookhaven Symp. in Biol.* 6:209–234. 1954.

Arber, A. *Monocotyledons: A morphological study*. Cambridge, Cambridge University Press. 1925.

Arber, A. *The Gramineae*. Cambridge, Cambridge University Press. 1934.

Arber, A. *The natural philosophy of plant form*. Cambridge, Cambridge University Press. 1950.

Arnold, A. Über den Funktionsmechanismus der Endodermiszellen der Wurzeln. *Protoplasma* 41:189–211. 1952.

Arnott, H. J. The seed, germination, and seedling of *Yucca*. *Calif. Univ., Publs., Bot.* 35:1–164. 1962.

Arora, N. Histology of the root nodules on *Cicer arietinum* L. *Phytomorphology* 6:367–378. 1956.

Artschwager, E. Anatomy of the vegetative organs of the sugar beet. *Jour. Agr. Res.* 33:143–176. 1926.

Avery, G. S., Jr. Comparative anatomy and morphology of embryos and seedlings of maize, oats, and wheat. *Bot. Gaz.* 89:1–39. 1930.

Bailey, I. W. Nodal anatomy in retrospect. *Arnold Arboretum Jour.* 37:269–287. 1956.

Bakshi, T. S., and R. T. Coupland. An anatomical study of the subterranean organs of *Euphorbia esula* in relation to its control. *Canad. Jour. Bot.* 37:613–620. 1959.

Bannan, M. W.   Origin and cellular character of xylem rays in gymnosperms. *Bot. Gaz.* 96:260–281. 1934.

Bannan, M. W.   Notes on the origin of adventitious roots in the native Ontario conifers. *Amer. Jour. Bot.* 29:593–598. 1942.

Baranova, E. A.   Zakonomernosti obrazovaniĭa pridatochnykh korneĭ u rasteniĭ. [Laws of formation of adventitious roots in plants.] *Trudy Glav. Bot. Sada* 2:168–193. 1951.

Beakbane, A. B.   Anatomical studies of stems and roots of hardy fruit trees. III. The anatomical structure of some clonal and seedling apple rootstocks stem- and root-grafted with a scion variety. *Jour. Pomol. and Hort. Sci.* 18:344–367. 1941.

Beakbane, A. B.   Structure of the plant stem in relation to adventitious rooting. *Nature* 192:954–955. 1961.

Beckel, D. K. B.   Cortical disintegration in the roots of *Bouteloua gracilis*. *New Phytol.* 55:183–190. 1956.

Berckemeyer, W.   Über kontraktile Umbelliferenwurzeln. *Bot. Arch.* 24:273–318. 1929.

Berthon, R.   Sur l'origine des radicelles chez les Angiospermes. *Acad. des Sci. Compt. Rend.* 216:308–309. 1943.

Beskaravaĭnyĭ, M. M.   Srastanie korneĭ drevesnykh porod v raĭone g. Kamyshina. [Concrescence of roots of woody genera in the region of Kamyshin.] *Agrobiol.* 3:78–89. 1955.

Bond, L.   Origin and developmental morphology of root nodules of *Pisum sativum*. *Bot. Gaz.* 109:411–434. 1948.

Borman, F. H.   Root grafting and non-competitive relationships between trees. In: T. T. Kozlowski. *Tree growth*. New York, Ronald Press Company. 1962.

Borman, F. H., and B. F. Graham, Jr.   The occurrence of natural root grafting in eastern white pine, *Pinus strobus* L., and its ecological implications. *Ecol.* 40:677–691. 1959.

Bottum, F. R.   Histological studies on the root of *Melilotus alba*. *Bot. Gaz.* 103:132–145. 1941.

Boyd, L., and G. S. Avery, Jr.   Grass seedling anatomy: the first internode of *Avena* and *Triticum*. *Bot. Gaz.* 97:765–779. 1936.

Bruch, H.   Beiträge zur Morphologie und Entwicklungsgeschichte der Fenchelwurzel (*Foeniculum vulgare* Mill.). *Beitr. z. Biol. der Pflanz.* 32:1–26. 1955.

Burström, H.   Growth and formation of intercellularies in root meristems. *Physiol. Plant.* 12:371–385. 1959.

Burström, H., and Z. Hejnowicz.   The formation of chlorophyll in isolated roots. *Kungl. Fysiogr. Sallsk. Lund Förhandl.* 28:65–69. 1958.

Campbell, D. H.   The eusporangiate ferns and the stelar theory. *Amer. Jour. Bot.* 8:303–314. 1921.

Canning, R. E., and P. J. Kramer.   Salt absorption and accumulation in various regions of roots. *Amer. Jour. Bot.* 45:378–382. 1958.

Carlson, M. C.   Nodal adventitious roots in willow stems of different ages. *Amer. Jour. Bot.* 37:555–561. 1950.

Chan, T.-T.   The development of the narcissus plant. *Daffodil and Tulip Yr. Bk.* 17:72–100. 1952.

Chauveaud, G.   L'appareil conducteur des plantes vasculaires et les phases principales de son évolution. *Ann. des Sci. Nat., Bot.* Ser. 9. 13:113–438. 1911.

Cheadle, V. I.   Secondary growth by means of a thickening ring in certain monocotyledons. *Bot. Gaz.* 98:535–555. 1937.

Cheadle, V. I.   Specialization of vessels within the xylem of each organ in the Monocotyledoneae. *Amer. Jour. Bot.* 31:81–92. 1944.

Clark, F. B.   Endotrophic mycorrhizae influence yellow poplar seedling growth. *Science* 140:1220–1221. 1963.

Clowes, F. A. L.   The structure of mycorrhizal roots of *Fagus sylvatica*.   *New Phytol.* 50: 1–16. 1951.

Clowes, F. A. L.   The root cap of ectotrophic mycorrhizas.   *New Phytol.* 53:525–529. 1954.

Compton, R. H.   An investigation of the seedling structure in the Leguminosae.   *Linn. Soc. London, Jour., Bot.* 41:1–122. 1912*a*.

Compton, R. H.   Theories of the anatomical transition from root to stem.   *New Phytol.* 11: 13–25. 1912*b*.

Crafts, A. S., and T. C. Broyer.   Migration of salts and water into the xylem of the roots of higher plants.   *Amer. Jour. Bot.* 25:529–535. 1938.

Dangeard, P. A.   Observations sur la structure des plantules chez les Phanérogames dans ses rapports avec l'évolution vasculaire.   *Soc. Bot. de France Bul.* 60:73–80, 113–120. 1913.

Davey, A. J.   On the seedling of *Oxalis hirta* L.   *Ann. Bot.* 10:237–256. 1946.

De Silva, B. L. T.   Secondary thickening in the roots of *Dracaena*.   *Ceylon Jour. Sci. Sec. A, Bot.* 12:127–135. 1936.

Dittmer, H. J.   A comparative study of the number and length of roots produced in nineteen angiosperm species.   *Bot. Gaz.* 109:354–358. 1948.

Dittmer, H. J.   Root hair variations in plant species.   *Amer. Jour. Bot.* 36:152–155. 1949.

Drabble, E.   The transition from stem to root in some palm seedlings.   *New Phytol.* 5:56– 66. 1905.

Duchaigne, A. L.   *Le passage de la racine à la tige*.   Thesis.   Univ. Poitiers. 1951.

Dycus, A. M., and L. Knudson.   The role of the velamen of the aerial roots of orchids.   *Bot. Gaz.* 119:78–87. 1957.

Eames, A. J.   *Morphology of vascular plants. Lower groups.*   New York, McGraw-Hill Book Company. 1936.

Eames, A. J., and L. H. MacDaniels.   *An introduction to plant anatomy.*   2nd ed.   New York, McGraw-Hill Book Company. 1947.

Erickson, R. O., and K. B. Sax.   Elemental growth rate of the primary root of *Zea mays*.   *Amer. Phil. Soc. Proc.* 100:487–498. 1956*a*.

Erickson, R. O., and K. B. Sax.   Rates of cell division and cell elongation in the growth of the primary root of *Zea mays*.   *Amer. Phil. Soc. Proc.* 100:499–514. 1956*b*.

Esau, K.   Developmental anatomy of the fleshy storage organ of *Daucus carota*.   *Hilgardia* 13:175–226. 1940.

Esau, K.   Origin and development of primary vascular tissues in seed plants.   *Bot. Rev.* 9: 125–206. 1943*a*.

Esau, K.   Vascular differentiation in the pear root.   *Hilgardia* 15:299–324. 1943*b*.

Esau, K.   *Anatomy of seed plants.*   New York, John Wiley and Sons. 1960.

Farr, C. H.   Root hairs and growth.   *Quart. Rev. Biol.* 3:343–376. 1928.

Flahault, C.   Recherches sur l'accroissement terminal de la racine chez les Phanérogames.   *Ann. des Sci. Nat., Bot.* Ser. 6. 6:1–168. 1878.

Fourcroy, M.   Perturbations anatomiques intéressant le faisceau vasculaire de la racine au voisinage des radicelles.   *Ann. des Sci. Nat., Bot.* Ser. 11. 3:177–198. 1942.

Frey-Wyssling, A.   *Die Pflanzenzellwand.*   Berlin, Springer-Verlag. 1959.

Gessner, F.   Der Wasserhaushalt der Epiphyten und Lianen.   *Handb. der Pflanzenphysiol.* 3:915–950. 1956.

Goodwin, R. H., and C. J. Avers.   Studies on roots. III. An analysis of root growth in *Phleum pratense* using photomicrographic records.   *Amer. Jour. Bot.* 43:479–487. 1956.

Gravis, A.   Contribution à l'étude anatomique du raccourcissement des racines.   *Acad. Roy. de Belg., Bul. de Cl. des Sci.* Ser. 5. 12:48–69. 1926.

Grushvitskiĭ, I. V. "Vtĩagivaĩushchie korni"—vazhnaĩa biologicheskaĩa osobennost' zhen'-shenĩa (*Panax Ginseng* S.A.M.). ["Contractile roots"—important biologic peculiarity of ginseng (*Panax Ginseng* Mey)]. *Bot. Zhur.* SSSR 37:682–685. 1952.

Guttenberg, H. von. Der primäre Bau der Angiospermenwurzel. In: K. Linsbauer. *Handbuch der Pflanzenanatomie*. Band 8. Lief. 39. 1940.

Guttenberg, H. von. Der primäre Bau der Gymnospermenwurzel. In: K. Linsbauer. *Handbuch der Pflanzenanatomie*. Band 8. Lief. 41. 1941.

Guttenberg, H. von. Die physiologischen Scheiden. In: K. Linsbauer. *Handbuch der Pflanzenanatomie*. Band 5. Lief. 42. 1943.

Guttenberg, H. von. Grundzüge der Histogenese höherer Pflanzen. I. Angiospermen. *Handbuch der Pflanzenanatomie*. Band 8. Teil 3. 1960.

Hasman, M., and N. Inanç. Investigations on the anatomical structure of certain submerged, floating and amphibious hydrophytes. *Istanbul Univ. Rev. Facul. Sci. Ser. B*. 22:137–153. 1957.

Hayward, H. E. *The structure of economic plants*. New York, The Macmillan Company. 1938.

Hayward, H. E., and E. M. Long. The anatomy of the seedling and roots of the Valencia orange. *U.S. Dept. Agr. Tech. Bul.* 786. 1942.

Heimsch, C. Development of vascular tissues in barley roots. *Amer. Jour. Bot.* 38:523–537. 1951.

Hejnowicz, Z. Growth and division in the apical meristem of wheat roots. *Physiol. Plant.* 12:124–138. 1959.

Hill, T. G., and E. de Fraine. On the seedling structure of gymnosperms. Part I. Taxaceae, Podocarpaceae, Cupressaceae, Abietineae. *Ann. Bot.* 22:689–712. 1908. Part II. Abietineae and Araucarieae. *Ann. Bot.* 23:189–227. 1909a. Part III. Ginkgoaceae and Cycadaceae. *Ann. Bot.* 23:433–458. 1909b. Part IV. Gnetales. *Ann. Bot.* 24:319–353. 1910.

Hill, T. G., and E. de Fraine. A consideration of facts relating to structure of seedlings. *Ann. Bot.* 27:257–272. 1913.

Hoagland, D. R. Some aspects of the salt nutrition of higher plants. *Bot. Rev.* 3:307–334. 1937.

Janczewski, E. de. Recherches sur l'accroissement terminal des racines dans les Phanérogames. *Ann. des Sci. Nat., Bot.* Ser. 5. 20:162–201. 1874a.

Janczewski, E. de. Recherches sur le développement des radicelles dans les Phanérogames. *Ann. des Sci. Nat., Bot.* Ser. 5. 20:208–233. 1874b.

Jeffrey, E. C. The morphology of the central cylinder in the angiosperms. *Canad. Inst. Toronto, Trans.* 6:599–636. 1898–99.

Jensen, W. A., and M. Ashton. Composition of developing primary wall in onion root tip cells. I. Quantitative analyses. *Plant Physiol.* 35:313–323. 1960.

Jensen, W. A., and L. G. Kavaljian. An analysis of cell morphology and the periodicity of division in the root tip of *Allium cepa*. *Amer. Jour. Bot.* 45:365–372. 1958.

Jones, F. R. Winter injury to alfalfa. *Jour. Agr. Res.* 37:189–211. 1928.

Jones, F. R. Growth and decay of the transient (noncambial) roots of alfalfa. *Amer. Soc. Agron. Jour.* 35:625–634. 1943.

Kacperska-Palacz, A. The aeration system in some grasses appearing chiefly on lowland meadows. *Rocz. Nauk Rolnicz.* 75:295–318. 1962.

Katayama, T. Studies on the intercellular spaces in rice. I. *Crop Sci. Soc. Japan Proc.* 29:229–233. 1961.

Kelley, A. P. *Mycotrophy in plants*. Waltham, Mass., Chronica Botanica Company. 1950.

Knobloch, I. W. Developmental anatomy of chicory.—The root. *Phytomorphology* 4:47–54. 1954.

Kondrateva-Mel'vil', E. A. Obrazovanie kornevykh otpryskov u nekotorykh travianistykh dvudol'nykh. [Shoot formation on roots in some herbaceous dicotyledons.] *Univ. Leningrad Vest. Ser. Biol.* 12:22–37. 1957.

Kramer, P. J. Transpiration and the water economy of plants. In: F. C. Steward. *Plant physiology* Vol. 2. New York, Academic Press. 1959.

Krauss, B. H. Anatomy of the vegetative organs of the pineapple *Ananas comosus* (L.) Merr. III. The root and the cork. *Bot. Gaz.* 110:550–587. 1949.

Kroemer, K. Wurzelhaut, Hypodermis und Endodermis der Angiospermenwurzel. *Biblioth. Bot.* 12(59):1–159. 1903.

Lenoir, M. Évolution du tissu vasculaire chez quelques plantules des Dicotylédones. *Ann. des Sci. Nat., Bot.* Ser. 10. 2:1–123. 1920.

Lotova, L. I., and R. P. Liarskaia. Nekotorye anatomicheskie osobennosti srastaniia kornei gimalaiskogo i atlasskogo kedrov. [Some anatomic features of the concrescence of roots of *Cedrus deodara* and *C. atlantica.*] *Nauch. Dok. Vys. Shk. Biol. Nauk* 4:99–104. 1959.

Luhan, M. Das Abschlussgewebe der Wurzeln unserer Alpenpflanzen. *Deut. Bot. Gesell. Ber.* 68:87–92. 1955.

Lund, S., and H. Kiaerskou. Morphologisk-anatomisk Beskrivelse af *Brassica oleracea* L., *B. campestris* (L.) og *B. Napus* (L.) (Havekaal, Rybs og Raps), samt Redegjörelse for Bestövnings-og Dyrkningsforsög med disse Arter. *Bot. Tidsskr.* 15:1–150. 1885. (Review in *Bot. Centbl.* 27:326–331. 1886.)

McCall, M. A. Developmental anatomy and homologies in wheat. *Jour. Agr. Res.* 48:283–321. 1934.

Melin, E. Physiology of mycorrhizal relations in plants. *Ann. Rev. Plant Physiol.* 4:325–346. 1953.

Meyer, F. J. Untersuchungen über den Strangverlauf in den radialen Leitbündeln der Wurzeln. *Jahrb. f. Wiss. Bot.* 65:88–97. 1925.

Miller, H. A., and R. H. Wetmore. Studies in the developmental anatomy of *Phlox drummondii* Hook. II. The seedling. *Amer. Jour. Bot.* 32:628–634. 1945.

Miyawaki, A. Quantitative und morphologische Studien über die ober- und unterirdischen Stämme von einigen Krautarten. *Bot. Mag. Tokyo* 69:481–488. 1957.

Morrison, T. M. Mycorrhiza of silver beech. *New Zeal. Jour. Forest.* 7:47–60. 1956.

Mulay, B. N., and B. D. Deshpande. Velamen in terrestrial monocots. I. Ontogeny and morphology of velamen in the Liliaceae. *Indian Bot. Soc. Jour.* 38:383–390. 1959.

Mullendore, N. Anatomy of the seedling of *Asparagus officinalis*. *Bot. Gaz.* 97:356–375. 1935.

Muller, C. La tige feuillée et les cotylédons des Viciées a germination hypogée. *Cellule* 46:195–354. 1937.

Murray, B. E. The ontogeny of adventitious stems and roots of creeping-rooted alfalfa. *Canad. Jour. Bot.* 35:463–475. 1957.

Mylius, G. Das Polyderm. Eine vergleichende Untersuchung über die physiologischen Scheiden: Polyderm, Periderm und Endodermis. *Biblioth. Bot.* 18(79):1–119. 1913.

Napp-Zinn, K. Studien zur Anatomie einiger Luftwurzeln. *Österr. Bot. Ztschr.* 100:322–330. 1953.

Netolitzky, F. Das trophische Parenchym. C. Speichergewebe. In: K. Linsbauer. *Handbuch der Pflanzenanatomie*. Band 4. Lief. 31. 1935.

Obaton, M. Organisation vasculaire dès germination chez les *Datura* (*D. sanguinea* Ruiz et Pav., *D. metel* L., *D. stramonium* L.) *Ann. des Sci. Nat., Bot.* Ser. 11. 10:179–200. 1949.

Ogura, Y. Anatomie der Vegetationsorgane der Pteridophyten. In: K. Linsbauer. *Handbuch der Pflanzenanatomie*. Band 7. Lief. 36. 1938.

Petri, P. S., S. Mazzi, and P. Strigoli. Considerazioni sulla formazione delle radici avventizie con particolare riguardo a: *Cucurbita pepo, Nerium oleander, Menyanthes trifoliata* e *Solanum lycopersicum. Nuovo Gior. Bot. Ital.* 67:131–175. 1960.

Phillips, W. S. Seedling anatomy of *Cynara scolymus. Bot. Gaz.* 98:711–724. 1937.

Pillai, A., and S. K. Pillai. Air spaces in the roots of some monocotyledons. *Indian Acad. Sci. Proc. B.* 55:296–301. 1962.

Pommer, E.-H. Beiträge zur Anatomie und Biologie der Wurzelknöllchen von *Alnus glutinosa* Gaertn. *Flora* 143:603–634. 1956.

Popham, R. A. Levels of tissue differentiation in primary roots of *Pisum sativum. Amer. Jour. Bot.* 42:529–540. 1955.

Preston, R. J., Jr. Anatomical studies of the roots of juvenile lodgepole pine. *Bot. Gaz.* 104:443–448. 1943.

Priestley, J. H., and C. F. Swingle. Vegetative propagation from the standpoint of plant anatomy. *U.S. Dept. Agr. Tech. Bul.* 151. 1929.

Reinhard, E. Ein Vergleich zwischen diarchen und triarchen Wurzeln von *Sinapis alba. Ztschr. f. Bot.* 44:505–514. 1956.

Resch, A. Zur Frage nach den Geleitzellen im Protophloem der Wurzel. *Ztschr. f. Bot.* 49:82–95. 1961.

Richardson, S. D. The influence of rooting medium on the structure and development of the root-cap in seedlings of *Acer saccharinum* L. *New Phytol.* 54:336–337. 1955.

Riedl, H. Bau und Leistungen des Wurzelholzes. *Jahrb. f. Wiss. Bot.* 85:1–75. 1937.

Rimbach, A. Die kontraktilen Wurzeln und ihre Thätigkeit. *Beitr. z. Wiss. Bot.* 2:1–28. 1898.

Rimbach, A. Beiträge zur Physiologie der Wurzeln. *Deut. Bot. Gesell. Ber.* 17:18–35. 1899.

Rimbach, A. Die Verbreitung der Wurzelverkürzung im Pflanzenreich. *Deut. Bot. Gesell. Ber.* 47:22–31. 1929.

Rosene, H. F. Quantitative measurement of the velocity of water absorption in individual root hairs by a microtechnique. *Plant Physiol.* 18:588–607. 1943.

Rywosch, S. Untersuchungen über die Entwicklungsgeschichte der Seitenwurzeln der Monokotylen. *Ztschr. f. Bot.* 1:253–283. 1909.

Satoo, S. Origin and development of adventitious roots in layered branches of 4 species of conifers. *Jap. Forest. Soc. Jour.* 37:314–316. 1955.

Satoo, S., and M. Fukuhara. Some experiments on vegetative propagation of *Paulownia tomentosa* and anatomical observations on the origin and development of adventitious shoots and roots. *Jap. Forest. Soc. Jour.* 37:317–320. 1955.

Schoute, J. C. *Die Stelär-Theorie.* Jena, Gustav Fischer. 1903.

Scott, F. M. Root hair zone of soil-grown roots. *Nature* 199:1009–1010. 1963.

Seeliger, I. Zur Entwicklungsgeschichte und Anatomie wurzelbuertiger Sprosse an isolierten Robinienwurzeln. *Flora* 148:119–124. 1959.

Seeliger, R. Untersuchungen über das Dickenwachstum der Zuckerrübe (*Beta vulgaria* L. var. *rapa* Dum.). *Biol. Reichsanst. f. Land u. Forstw. Arb.* 10:149–194. 1919.

Siegler, E. A., and J. J. Bowman. Anatomical studies of root and shoot primordia in 1-year apple roots. *Jour. Agr. Res.* 58:795–803. 1939.

Sinnott, E. W., and R. Bloch. Division in vacuolate plant cells. *Amer. Jour. Bot.* 28:225–232. 1941.

Smith, A. I. Adventitious roots in stem cuttings of *Begonia maculata* and *B. semperflorens. Amer. Jour. Bot.* 23:511–515. 1936.

Soeding, H. Anatomie der Wurzel-, Stengel- und Rübenbildung von Oelraps und Steckrübe (*Brassica Napus* L. var. *oleifera* und var. *napobrassica*). *Bot. Arch.* 7:41–69. 1924.

Soper, K. Root anatomy of grasses and clovers. *New Zeal. Jour. Agr. Res.* 2:329–341. 1959.

Spurr, A. R. Organization of the procambium and development of the secretory cells in the embryo of *Pinus strobus* L. *Amer. Jour. Bot.* 37:185–197. 1950.

Stangler, B. B. Origin and development of adventitious roots in stem cuttings of chrysanthemum, carnation, and rose. *N.Y. (Cornell) Agr. Expt. Sta. Mem.* 342. 1956.

Steward, F. C., and J. F. Sutcliffe. Plants in relation to inorganic salts. In: F. C. Steward. *Plant physiology.* Vol. 2. New York, Academic Press. 1959.

Swingle, C. F. Regeneration and vegetative propagation. *Bot. Rev.* 6:301–355. 1940.

Thibault, M. Contribution à l'étude des radicelles de carotte. *Rev. Gén. de Bot.* 53:434–460. 1946.

Thoday, D. The interpretation of plant structure. *Nature* 144:571–575. 1939.

Thomas, E. N. Seedling anatomy of Ranales, Rhoedales, and Rosales. *Ann. Bot.* 28:695–733. 1914.

Tomlinson, P. B. *Anatomy of the monocotyledons.* II. *Palmae.* Oxford, Clarendon Press. 1961.

Torrey, J. G. Cambial formation in isolated pea roots following decapitation. *Amer. Jour. Bot.* 38:596–604. 1951.

Torrey, J. G. The effect of certain metabolic inhibitors on vascular tissue differentiation in isolated pea roots. *Amer. Jour. Bot.* 40:525–533. 1953.

Torrey, J. G. On the determination of vascular patterns during tissue differentiation in excised pea roots. *Amer. Jour. Bot.* 42:183–198. 1955.

Torrey, J. G. Auxin control of vascular pattern formation in regenerating pea root meristems grown in vitro. *Amer. Jour. Bot.* 44:859–870. 1957.

Torrey, J. G. Endogenous bud and root formation by isolated roots of *Convolvulus* grown in vitro. *Plant Physiol.* 33:258–263. 1958.

Torrey, J. G. Experimental modification of development in the root. In: D. Rudnick. *Cell, Organism and Milieu.* New York, Ronald Press Company. 1959.

Troll, W. Über die Grundbegriffe der Wurzelmorphologie *Österr. Bot. Ztschr.* 96:444–452. 1949.

Van Fleet, D. S. A comparison of histochemical and anatomical characteristics of the hypodermis with the endodermis in vascular plants. *Amer. Jour. Bot.* 37:721–725. 1950.

Van Fleet, D. S. Histochemistry and function of the endodermis. *Bot. Rev.* 27:165–220. 1961.

Van Tieghem, P. *Traité de Botanique.* 2nd ed. Paris, Librairie F. Savy. 1891*a*.

Van Tieghem, P. Sur les tubes criblés extralibériens et les vaisseaux extraligneux. *Jour. de Bot.* 5:117–128. 1891*b*.

Vasilevskaĭa, V. K. Anatomiĭa obrazovaniĭa pochek na korniakh nekotorykh drevesnykh rasteniĭ. [Anatomy of bud formation on roots of some woody plants.] *Univ. Leningrad Vest. Ser. Biol.* 12:5–21. 1957.

Voronin, N. S. Ob evoliŭtsii korneĭ u rasteniĭ. 2. Evoliŭtsiĭa kornerozhdeniĭa. [Evolution of plant roots. 2. Evolution of root origin.] *Moskov. Obshch. Isp. Prirody, Otd. Biol. Biul.* 62:35–49. 1957.

Weaver, J. E. *Root development of field crops.* New York, McGraw-Hill Book Company. 1926.

Weber, H. Die Bewurzelungsverhältnisse der Pflanzen. Freiburg, Verlag Herder. 1953.

Weber, H. Die Wurzelverdickungen von *Calathea macrosepala* Schum. und einigen anderen monokotylen Pflanzen. *Beitr. z. Biol. der Pflanz.* 34:177–193. 1958.

Wilcox, H. Primary organization of active and dormant roots of noble fir, *Abies procera.* *Amer. Jour. Bot.* 41:812–821. 1954.

Wilcox, H.   Regeneration of injured root systems in noble fir.   *Bot. Gaz.* 116:221–234. 1955.

Wilcox, H.   Growth studies of the root of incense cedar, *Libocedrus decurrens.*   I. The origin and development of primary tissues.   *Amer. Jour. Bot.* 49:221–236.   1962*a.*

Wilcox, H.   Growth studies of the root of incense cedar, *Libocedrus decurrens.*   II. Morphological features of the root system and growth behavior.   *Amer. Jour. Bot.* 49: 237–245.   1962*b.*

Williams, B. C.   The structure of the meristematic root tip and origin of primary tissues in the roots of vascular plants.   *Amer. Jour. Bot.* 34:455–462.   1947.

Williams, B. C.   Observations on intercellular canals in root tips with reference to the Compositae.   *Amer. Jour. Bot.* 41:104–106.   1954.

Young, P. T.   *Histogenesis and morphogenesis in primary root of Zea Mays.*   Thesis. New York, Columbia University.   1933.

Zgurovskaĩa, L. N.   Anatomo-fiziologicheskoe issledovanie vsasyvaĩushchikh, rostovykh i provodĩashchikh korneĩ drevesnykh porod.   [Anatomic-physiologic investigation of absorbing, growing and conducting roots of woody genera.]   *Akad. Nauk Inst. Lesa Trudy* 41:5–32.   1958.

18

The Flower

The present and the following two chapters deal with the angiospermous flower and the structures derived from it, the fruit and the seed. The phylogeny and morphologic nature of the flower and its parts are subjects of much discussion in the literature. The old classical theory (Eames, 1961) homologizes the flower with a shoot, that is, it regards the flower as an entity consisting of an axis (*receptacle*) and foliar appendages (floral parts, or organs). The axis is relatively short and has determinate growth. The floral parts are divided into sterile and fertile, or reproductive (fig. 18.1). Megasporogenesis and microsporogenesis are carried out on separate floral organs, which may occur on the same or different flowers.

The floral parts concerned with megasporogenesis constitute, collectively, the *gynoecium* (from the Greek words meaning woman and house). The basic unit of the gynoecium is the *carpel* (in Greek, fruit), which is commonly regarded as a megasporophyll. One or more carpels may enter into the composition of a gynoecium. *Pistil* (in Latin, pestle) is another term referring to the megasporangial part of the flower. The pistil may consist of one carpel (simple pistil) or of several (compound pistil). If the gynoecium is composed of a single carpel or of several united carpels, the pistil and the gynoecium refer to the same entity. If the gynoecium consists of more than one separate carpel, it also consists of more than one separate pistil. The abandonment of the term pistil has been advocated (Parkin, 1955) but it continues to be useful. Some authors substitute *ovary* for pistil, but this word denotes only the lower part of the pistil. The other parts are the *style* and the *stigma*.

The carpels enclose the *ovule* or *ovules* (in Greek, egg) borne on the *placenta* (in Latin, cake, or flat plate). The *nucellus*, which is the central part of the ovule, is usually interpreted as the megasporangium. The functioning megaspore germinates within the megasporangium and gives rise to the female gametophyte, the *embryo sac*. Because of this developmental sequence, the gynoecium is commonly referred to as the female part of the flower.

The floral parts forming the microspores are called, collectively, the *androecium* (from the Greek words meaning man and house). The individual units of the androecium are the *stamens* (in Latin, filament). Classically, the stamen is interpreted as a microsporophyll, and the part of the stamen called *pollen sac*, as the microsporangium. The pollen sacs are contained within the *anther* (based on the Greek word for flowering). A microspore develops into the male gametophyte, the *pollen grain* (pollen from the Latin, fine flour). Since gametogenesis occurs in the anther the androecium is referred to as the male part of the flower.

The sterile parts of the flower are the *petals* (in Greek, flower leaves), collectively called the *corolla* (in Latin, small crown), and the *sepals* (in Greek, a covering) composing the *calyx* (in Greek, a cup). The calyx and the corolla constitute the *perianth* (from Greek words, about and flower). If the perianth is not differentiated into sepals and petals (fig. 18.10A), the individual members of the perianth are called *tepals* (from the Latin *tepalum*, an anagram of *petalum*). Flowers commonly have nectaries borne on their various parts (chapter 13). Some of these are modified stamens, or staminodes.

The literature dealing with the question of morphologic nature of the flower is extensive and has been more or less comprehensively reviewed (Andrews, 1963; Arber, 1950; Barnard, 1961; Eames, 1961; Kaussmann, 1963; Melville, 1962, 1963; Pervukhina, 1957a, b; Rao, 1961; Takhtajan, 1959; Wilson and Just, 1939). Most of the proponents of the concept interpreting the flower as a modified shoot assume that the floral organs are appendicular structures in the same sense as leaves, both kinds of appendages possibly having undergone parallel evolutionary development. Emphasis is thus placed upon unity of types of structures; that is, foliage leaves and floral organs are both regarded as leaf-like appendages or phyllomes. Discussions on the nature of the flower frequently refer to the concept of leaves as derivatives of branches (Emberger, 1951). The floral organs, though resembling leaves in extant angiosperms, evolved from cauline assemblages similar to those that gave rise to the foliage leaves. From such an aspect, the question to be asked is not how the leaf became a floral organ but how this organ evolved from a branch system.   Apparently plants bore ovules before the leaves—as we know them now—were in existence (Camp and Hubbard, 1963).

Sepals and petals are basically leaf-like in external form.   They may intergrade with one another and with the small bracts (bracteoles) subtending the flower.   In some flowers, however, the petals intergrade with the stamens through structures bearing characters of both (Moseley, 1958).   Moreover, frequently stamens and petals differ from other floral parts in having a single vascular trace.   These features are used to suggest that in some taxa the petals have evolved from the stamens.

The specialized types of stamens, characterized by a distinct differentiation into a filament and an anther, appear rather unlike the leaves, but in many Ranales the stamens are wide, leaf-like structures with no differentiation of a filament (Bailey and Smith, 1942; Canright, 1952). This form is the basis for the view that primitively the stamen may have been leaf-like.   The fascicled types of stamens (Malvaceae, Guttiferae, and other dicotyledons), on the other hand, are thought to indicate an origin from primitive dichotomous branch systems—systems of telomes —bearing terminal sporangia (Wilson, 1942).   Still another theory proposes that stamens originated from the *gonophyll,* a leaf-like structure bearing fertile branches (Melville, 1963).   Through condensations and deletions the gonophyll gave rise to the modern stamen either through a line with fascicled stamens or through one in which the stamen is laminar.

The classical concept of the carpel interprets it as a leaf-like appendage (Arber, 1937; Bailey and Swamy, 1951; Savchenko, 1957; Troll, 1939). By folding and fusion of margins and by unions with one another the carpels are assumed to have evolved into pistils.

The German literature deals extensively with the question regarding the type of a leaf to which the carpel may be compared.   The postulate has been advanced that many carpels have the same growth form as a peltate leaf, that is, a leaf in which the stalk is attached to the lower surface of the blade (Baum, 1952; Baum-Leinfellner, 1953; Leinfellner, 1950; Schaeppi and Frank, 1962).   The degree of peltation is considered to be variable and absent in some forms.   Peltation is recognized in stamens and perianth parts as well (Jäger, 1961; Leinfellner, 1955, 1956a).

The concept of the carpel as a sporophyll bearing sporangia is frequently criticized because not all gynoecia of angiosperms can be interpreted by reference to it.   Some authors consider that the ontogeny of the carpel makes it quite distinct from leaves (Grégoire, 1938; Plantefol, 1948); others find that the vascular anatomy of many flowers suggests an independence between carpellary and placental vascular traces

(Melville, 1962; Sterling, 1963). The inconsistencies in the sporophyll concept of the carpel are proposed to be resolved by the gonophyll theory (Melville, 1962). According to this theory, the basic component of the gynoecium is a leaf with an epiphyllous fertile branch, the two together comprising the gonophyll. Evolutionary modifications have resulted in a close association of the fertile branch—the placental axis bearing the ovules—with the laminar part of the gonophyll.

If the floral parts are ultimately derived from branch systems, the flower is a condensed and highly modified inflorescence and the term flower covers reproductive structures of angiosperms in various stages of condensation (Melville, 1962, 1963; Nozeran, 1955). This interpretation modifies the term flower to one referring to a biological unit rather than to a morphological one and makes it applicable not only to single flowers but also to more or less condensed inflorescences (Melville, 1963).

STRUCTURE

Arrangement of Flower Parts

The apical meristem of the flower usually ceases its activity after the reproductive structures have been initiated, an expression of the determinate type of growth. In certain groups of angiosperms considered to be primitive the determinate growth is less pronounced than in the more advanced families. In the primitive groups, the activity of the apical meristem is prolonged and therefore the number of floral parts is relatively large and indefinite. Moreover, these parts occur on a rather elongated axis, with sepals, petals, stamens, and carpels succeeding each other acropetally in the order named. The similarity between such a flower and a vegetative shoot is not difficult to visualize, especially if the flower parts are arranged helically.

In the more highly specialized flower types the growth period is shorter and the number of floral parts is smaller and more definite. Moreover, the shortening of the period of activity of the apical meristem is associated with a development of distinguishing characteristics that obscure or even efface the evidences of similarity between a flower and a vegetative shoot. Such characteristics are: whorled (or cyclic) instead of helical arrangement of parts; cohesion of parts within one whorl; adnation of parts of two or more different whorls; loss of parts; zygomorphy (bilateral symmetry) instead of actinomorphy (radial symmetry); and epigyny (inferior ovary) instead of hypogyny (superior ovary). The words synsepalous, sympetalous, and syncarpous are used to characterize flowers with united sepals, petals, and carpels, respectively. If the gynoecium occupies a position similar to that in an epigynous flower but is not adnate to the noncarpellary tissue, the flower is called perigynous and the ovary superior. Epigynous flowers (those with inferior ovaries) are especially difficult to interpret morphologically because the gynoecium is imbedded in noncarpellary tissue and appears to be inserted below the other floral parts.

The flowers of different degrees of specialization form an intergrading series of morphologic types. The degree of fusion of sepals, petals, stamens, and carpels varies widely, and the union is not necessarily equally pronounced in the different whorls of the same flower. The perianth may not be differentiated into calyx and corolla, or the sepals and petals may intergrade with each other. Transitional forms may also occur between the petals and the stamens. The flower may lack certain parts. If it lacks either the gynoecium or the androecium, it is called unisexual.

Vascular System

Investigations on the vascular system of the flower occupy a prominent place in the literature on the anatomy of the flower. A commonly accepted postulate is that the vascular system is conservative and, therefore, might be expected to reveal at least some of the evolutionary changes that have been obliterated in the external form (Puri, 1951). Thus, the vascular anatomy of the flower has been frequently studied to find explanations of the morphology of flowers (Douglas, 1944; Eames, 1961; Leroy, 1955; Smith and Smith, 1942a; Moseley, 1961; Ozenda, 1949; Wilson and Just, 1939), to obtain additional data for establishing taxonomic relations (Nast, 1944; Palser, 1961; Paterson, 1961; Rao et al., 1958; Wilkinson, 1949), and to construct evolutionary schemes (Melville, 1962, 1963).

The vascular system of relatively unspecialized flowers with superior ovaries is comparable to that of a vegetative shoot in which strands diverge into the lateral organs from an axial system of bundles. Many authors draw a complete parallel between the patterns of vascularization in the shoot and the flower, and apply the concepts of stele, traces, and gaps with reference to both structures (Eames, 1961). If the receptacle is elongated, the floral parts may be arranged according to a phyllotactic pattern correlated with an orderly arrangement and interconnection of vascular traces (Tucker, 1961). But the shortness of internodes characteristic of so many flowers, the union of parts, the epigynous condition, and various other modifications in the interrelations of floral parts make the vascular system of flowers less regular than that of vegetative shoots and obscure the relation between the vascular system of the axis and that of the floral organs (Moseley, 1961; Nast, 1944; Sporne, 1958).

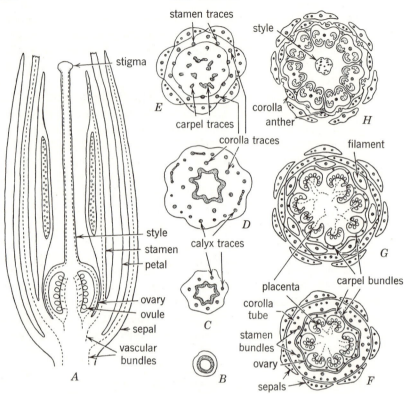

FIG. 18.1. Cultivated tomato flower in longitudinal (A) and transverse (B–H) sections. Broken lines in A represent vascular tissue. At right, vascular bundles are shown diverging from the floral axis toward sepal and stamen, at left, toward petal. Vascular bundles also diverge toward ovary wall and toward central part of ovary and the ovules. Bundles traversing ovary wall are continuous throughout style. Transections were taken, B–H, at successively higher levels beginning with the pedicel (B). Vascular tissue is indicated by stippling and broken lines. Stamen bases are adnate to corolla tube. H, blackened areas in style represent stigmatoid tissue; circles near periphery, vascular bundles. (×8.)

In a hypogynous flower with relatively little fusion of parts, the vascular system may be readily depicted in terms of traces to the various floral appendages (fig. 18.1A). The pedicel shows a cylindrical vascular region enclosing a pith and delimited on the outside by the cortex (fig. 18.1B). In the receptacle or torus (the part of the axis bearing the floral parts), at the level of attachment of sepals, traces diverge into these appendages (fig. 18.1C, D). Each sepal frequently has as many traces as a foliage leaf of the same plant. Above this level, traces diverge into the corolla, one or more to each petal in dicotyledons (fig. 18.1D, E), one to many to each tepal in the monocotyledons (Kaussmann, 1941).

Still higher, the traces to the stamens become discernible, predominantly one to each stamen (fig. 18.1E–G), and finally is found the carpellary supply (fig. 18.1E–G). The frequent number is three traces to each carpel, one median and two lateral (fig. 18.7), but more than three traces have been recorded (for example, in Gentianaceae; Krishna and Puri, 1962). Small branches of carpellary vascular bundles, often derived from the laterals, connect the carpellary system with the ovules (fig. 18.1A, F, G). The placental bundles may also be branches from the dorsal bundles, as in some Ranales (Canright, 1960; Periasamy and Swamy, 1956), or be independent from the carpellary traces (Melville, 1962; Ozenda, 1949). The vascular system is prolonged into the style (fig. 18.1A, H).

Some of the common modifications in the arrangement of the vascular system are associated with the fusion of floral parts. In many flowers the lateral bundles of the adjacent carpels are fused with each other. Similar fusions occur in other floral organs. The reduction in the numbers of traces and bundles may also occur if some of them do not develop.

The vascular system of epigynous flowers shows additional complications related to the apparently basal position of the gynoecium. The vascularization of such flowers has been frequently studied (Bersillon, 1956; Douglas, 1944, 1957; Gauthier, 1950; Pervukhina, 1962; Puri, 1952b; Smith and Smith, 1942b) with the result that some authors have developed rather definite ideas on the nature of the noncarpellary tissue enclosing the gynoecium. In most epigynous flowers this tissue is interpreted as appendicular in origin, composed of the bases of sepals, petals, and stamens that underwent a concrescence during the evolution of the flower. The vascular system is thought to reflect this structure in that the bundles pertaining to members of different whorls are variously fused but all show the usual orientation of xylem and phloem (fig. 18.2A). In some epigynous flowers (Calycanthaceae, Santalaceae, and probably Juglandaceae), however, the ovary is said to be partially enclosed in receptacular tissue. The vascular bundles are prolonged from the axis to the level below the insertion of floral parts, other than the carpels, where traces to these parts diverge. The main bundles, instead of ending here, continue farther from the periphery in a downward direction—with a corresponding inverse position of the xylem and the phloem—and at lower levels give branches to the carpels (fig. 18.2B). This orientation of the vascular system is interpreted as a result of an invagination of the axis (actually intercalary growth of the tissue enclosing the gynoecium).

In general, the vascular elements in the bundles of the flower are

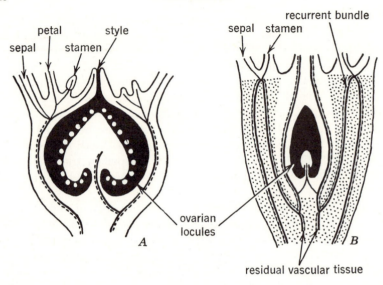

FIG. 18.2. An interpretation of flowers with inferior ovaries. Longitudinal sections. Xylem is shown by solid lines; phloem by broken lines. A, *Samolus floribundus*, with ovary imbedded in floral tube. B, *Darbya*, with ovary imbedded in invaginated receptacle (stippled part). Receptacular nature of outer tissue in B is indicated by presence of recurrent bundles with inverted orientation of xylem and phloem, and of residual vascular tissue at base of ovary. (Adapted from Douglas, *Bot. Rev.* 10, 1944.)

comparable to those in foliage leaves. The tissues are mostly primary, although some secondary growth may occur later, during fruit development, particularly in the pedicel. The vascular system of the sepals, the petals, and the carpels is more or less elaborately ramified (Jäger, 1961; Sprotte, 1940; Tepfer, 1953; Unruh, 1941). Stamens rarely show a branched vascular system (Moseley, 1958; Puri, 1951). In general character the venation of the perianth parts of monocotyledons and dicotyledons shows distinctive characteristics similar to those in the foliage leaves of these two groups of plants (Kaussmann, 1941). Perianth parts of many flowers exhibit an open venation (Kaussmann, 1963).

Sepal and Petal

The sepal and the petal are essentially leaf-like in form and anatomy but generally simpler in detailed structure than a foliage leaf. They consist of ground parenchyma, often called mesophyll, a vascular system permeating the ground tissue, and epidermal layers on the abaxial and adaxial sides (pl. 91*D*, *E*). Crystal-containing cells, idioblasts, and laticifers may occur in the ground tissue or in association with the vascular elements. The sepals of Geraniaceae have a thick-walled hypodermis with a druse in each cell (Kenda, 1956). The sepals are commonly green. The chloroplast distribution in the sepals depends on their position. If the sepals are upright and are closely applied to the petals, most chloroplasts are on the abaxial side; if the sepals are recurved, the chloroplasts are most abundant on the adaxial side. The mesophyll is rarely differentiated into palisade and spongy parenchyma. Commonly it is simple in structure and consists of approximately isodiametric cells loosely arranged into a lacunose tissue. The epidermis of the sepals shows a deposition of cutin and a development of stomata and trichomes similar to those on the foliage leaves. The vascular system resembles that in the leaves but is less elaborate.

Petals show a wider variety of shapes than the sepals and are usually distinguished from the sepals by their color. The vascular system may consist of one or several large veins and a system of small veinlets. The patterns formed by these veinlets vary greatly (Glück, 1919; Gumppenberg, 1924). Commonly the veinlets are dichotomously branched. The mesophyll is few cells in thickness, except in flowers with fleshy corollas. The tissue is parenchymatic with the cells either closely packed or loosely arranged.

The epidermis of petals shows certain peculiarities in the shape of cells (Hiller, 1884) and in the structure of cuticle (Martens, 1934). The anticlinal cell walls may be straight or wavy or may bear internal ridges. The undulation and ridging vary widely in degree of expression in different plants. In some the anticlinal walls are only slightly wavy; in others the undulations are so deep that the cells are star-like in shape as seen from the surface. The ridges, which arise through a localized centripetal growth of cell walls, may appear as small buttons in sectional views, or as long bars, straight or bent, solid or hollow. The degree of waviness or ridging may vary in the same petal. For instance, the anticlinal walls are usually straight at the base of the petal and along the veins, even if they are wavy elsewhere. Frequently, the undulate walls are restricted to or are more pronounced on the lower side.

Intercellular spaces may develop in the epidermis in connection with the differentiation of ridges. In some species the two wall layers composing a ridge split apart and the space between the two layers becomes filled with air. These spaces are open toward the interior of the petal but appear to be closed with the cuticle on the exterior (Hiller, 1884). Ridged walls occur mainly in the dicotyledons, although they have been found in some members of the Liliaceae also.

The tangential walls of the epidermis may be horizontal or convex to various degrees. The inner tangential wall is commonly slightly convex over the entire extent. The outer wall is often strongly convex, or it may bear one or more capitate or cone-shaped papillae (*Viola, Nasturtium*). The papillose structure is more common in the adaxial epidermis than in the abaxial and does not develop at the base of the petals. Various trichomes may occur on the petals, usually similar to those found on the leaves of the same plants. The stomata which occur on the petals either resemble those on the foliage leaves or are incompletely differentiated (Watson, 1962).

The cuticle of the corolla is rarely smooth. Commonly it is striated, and the lines form various patterns in different plants (pl. 24A). The development of these patterns has been suggested as resulting from two phenomena: first, a temporarily excessive production of cutin and the consequent increase in surface and folding of the cuticle; second, a stretching of the cuticle and a reorientation of the initial folds by cell extension (Martens, 1934). Cuticular patterns formed by folds were seen also at the ultrastructural level (Bringmann and Kühn, 1955).

The color of petals is caused by the presence of chromoplasts or pigments in the cell sap (Paech, 1955). The pigment color is usually modified by acids and other components of the cell sap. Starch is often formed in young petals. Volatile oils imparting the characteristic fragrance to the flowers commonly occur in the epidermal cells of the petals, sometimes in parts of flowers differentiated as osmophors (chapter 13).

Stamen

The well-known type of stamen, with a single-veined filament bearing at the upper end a two-lobed, four-loculed anther, is phylogenetically an advanced structure (figs. 18.3A, 18.13A). As was mentioned before, among the Ranales leaf-like stamens are found. In the least modified form, such stamens have three veins and bear the microsporangia on the abaxial surface between the midvein and the lateral veins (Bailey and Smith, 1942; Bailey and Swamy, 1949; Canright, 1952; Melville, 1963). The reduction of the three veins to one is apparently a concomitant of the reduction in width of the sporophyll, and particularly of the modification of the base of the sporophyll into a filament. The presence of a single vascular bundle is the prevailing condition in angiosperms. An extensive survey (Wilson, 1942) has shown that 95 per cent of angiosperms have a single vascular bundle in the stamen. This strand traverses the filament and may end at the base of the anther or may be prolonged into the tissue between the anther lobes, the so-called connective, terminating blindly near the apex. The vascular bundle is not connected by any vascular elements with the sporogenous tissue, but if the ground parenchyma of the anther develops secondary thickenings

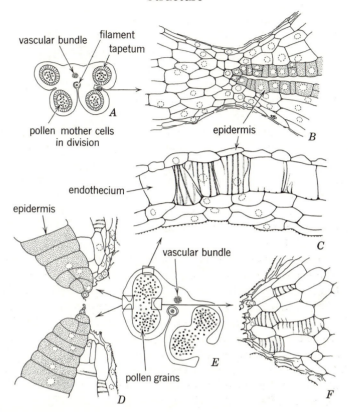

vascular bundle / filament / tapetum

pollen mother cells in division

epidermis   B

A

endothecium

epidermis

vascular bundle   C

vascular bundle

pollen grains   E

D   F

FIG. 18.3. Structure of *Lilium* anther. *A*, transection of stamen taken during division of pollen mother cells into tetrads. *B*, detail of wall part between members of a pair of locules of one anther lobe. *E*, dehisced anther containing mature pollen grains. During dehiscence a break occurred between epidermis (stippled in *D*) and subjacent cells (partly collapsed in *F*). A break also occurred between certain epidermal cells (small epidermal cells in *D*) and caused opening of anther locules. *C*, details of wall structure somewhat removed from dehiscence region. An endothecium with secondary wall thickenings is present. Similar thickenings occur elsewhere in the parenchyma of anther (*F*). (*A, E,* ×9; *B–D, F,* ×120.)

the cells in the vicinity of the sporogenous tissue remain thin walled and there are also vertical bands of similar thin-walled cells interpolated between the vascular strand and the anther lobes. The vascular bundle of the anther may be amphicribral in dicotyledons, but is reported to be collateral in monocotyledons (Leinfellner, 1956*b*). The anthers vary in shape and number of locules (Trapp, 1956).

The ground tissue of the filament is vacuolated parenchyma without a prominent intercellular-space system. It often contains pigments in the vacuoles. The epidermis is cutinized, bears trichomes in some species, and may have stomata (Kenda, 1952), possibly permanently open as in hydathodes (Aleksandrov and Dobrotvorskaĩa, 1960). The ground tissue of the anther and the connective is also parenchymatic but is highly specialized in the vicinity of the sporogenous cells (pl. 91*B–E*). This specialized tissue forms the wall layers or the parietal layers of the microsporangia (anther locules, or pollen sacs).

Anther Wall. The wall layers vary in number and are established through a series of divisions parallel to the periphery of the anther locule (fig. 18.4*A*; pl. 91*B*, *C*). The parietal layers facing the epidermis are ontogenetically related to the sporogenous tissue. Both the parietal cells and the pollen mother cells arise from the same initial cells, the archesporial cells. The wall layers occurring internally to the pollen sacs, however, arise from the ground tissue in contact with the archesporial cells (pl. 91*B*).

The outermost wall layer, the *endothecium* (from the Greek words for inner and case), is located beneath the epidermis. In anthers that open at maturity by longitudinal slits, the endothecium commonly develops secondary thickenings as the stamen approaches maturity (fig. 18.3*C*). These thickenings occur on the anticlinal and the inner tangential cell walls. In the anticlinal walls the secondary thickenings frequently have the form of strips or ridges oriented perpendicularly to the epidermal layer. The cell walls facing the sporogenous tissue may have uniform or irregular thickenings. Because of these thickenings the endothecium is often called the fibrous layer. The pattern of the thickening is variable and may be useful in taxonomic studies (Dormer, 1962). The endothecium also may have uniformly thick walls (Venkatesh, 1957). The protoplasts either disappear as the cell layer completes its development, or they remain alive until the pollen is shed. Wall thickenings similar to those in the endothecium may develop rather generally throughout the ground parenchyma of the anther.

The innermost of the parietal layers is the *tapetum* (from the Greek, carpet) (fig. 18.4*A*, *B*; pl. 91*E*). The tapetal cells are characterized by densely staining protoplasts and prominent nuclei. The nuclei show various behavior in different plants (Cooper, 1933; Wunderlich, 1954). In some, they do not divide after all the tapetal cells have been formed; in others, one or more nuclear divisions occur without being followed by cytokineses so that the cells become bi- or multinucleate (fig. 18.4*B*; *Lactuca, Taraxacum*). Sometimes the nuclear divisions are not carried to completion: the chromosomes divide but do not form separate nuclei. Such behavior results in polyploidy of the tapetal nuclei (Cooper, 1933; Witkus, 1945). In general, tapetal cells become richer in chromatic

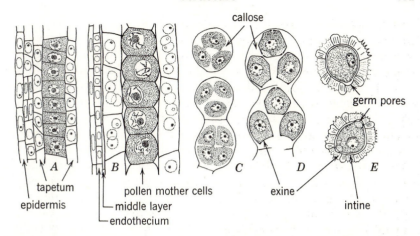

FIG. 18.4. Pollen differentiation in *Cichorium endivia* (endive). Longitudinal sections of anthers. *A*, pollen mother cells compactly arranged, tapetum present, wall layer between tapetum and epidermis in division. *B*, pollen mother cells rounding off, all wall layers present, tapetum multinucleate. *C*, protoplasts of tetrads of microspores imbedded within pollen mother cell wall of callose. Some microspores still connected with one another by cytoplasmic bridges. *D*, microspores show beginning of development of exine. *E*, pollen grains with exine and intine. (×470.)

material. This occurs through either multiplication of nuclei, restitution of nuclei during various stages of mitosis, or endopolyploidy (Carniel, 1963). The tapetal layer attains its maximal development at the tetrad stage in microspore formation. In some angiosperms the tapetum remains as a discrete layer—apparently functioning as a secretory tissue —until the pollen is mature. In many others, however, the cell walls disintegrate and the cells assume the appearance of plasmodial masses (Iijima, 1962). The latter gradually disintegrate as the pollen develops (Schnarf, 1927).

The tapetum is apparently concerned with the nutrition of the pollen mother cells and young microspores. Ultrastructural studies suggest that the material of the outer pollen wall (exine) is synthesized in the tapetum (Heslop-Harrison, 1962). But autoradiographic methods have failed to show any relation between the DNA of the tapetal nuclei and that of the microspores (Takats, 1962).

The parietal layers intervening between the endothecium and the tapetum frequently are crushed and destroyed so that, after the maturation of the pollen and the disintegration of the tapetum, the anther locule is bordered on the outside only by the epidermis and the endothecium.

In many plants the release of the pollen occurs through dehiscence (from the Greek, to yawn), that is, spontaneous opening of the anther. The opening, or *stomium*, may be a longitudinal slit located between the two pollen locules of each half of the anther. Before the dehiscence, the partition between the two locules of the same anther lobe may break down (fig. 18.3*D–F*). After this event, only one cell layer, the epidermis, separates the locule from the outside in the region of dehiscence. This part of the epidermis consists of particularly small cells and is easily broken when the pollen is mature (fig. 18.3*D*). Another common type of stomium is oriented transversely near the apex of the anther lobe. When such a stomium is formed, the apex of each anther lobe separates like a cap and leaves a pore (poricidal dehiscence; many Ericaceae, *Solanum*). Pores may also be formed laterally. It has been suggested that the long, slit-like stomium is a more primitive character than one shaped like a pore (Venkatesh, 1955, 1957). In species of *Senna*, the anther is provided with lateral sutures that do not serve as stomia (Venkatesh, 1957). The epidermal cells along these sutures divide and apparently serve as plugs. Dehiscence occurs at the sterile tip of the anther where short linear stomia are present. The tissue located between these stomia and the pollen sacs breaks down and the pollen emerges through the stomium. In some plants, anthers do not dehisce but open by an irregular breaking and exfoliating of tissue fragments (Coulter and Chamberlain, 1912).

Pollen

The development of the sporogenous tissue in the anther involves certain characteristic phenomena of wall formation. The cells that eventually undergo meiosis, the pollen mother cells, are closely packed in their early stages of development (fig. 18.4*A*). During meiosis these cells usually separate from one another, and the protoplasts round off and become enclosed in a thick gelatinous wall that has been identified as callose (Waterkeyn, 1961). This wall is designated as the pollen mother cell wall or special wall. The microspore mother cells may assume a distinctive arrangement in the pollen sac. In Gramineae and Cyperaceae, for example, they appear, in transections of the anthers, like sectors of a circle (Carniel, 1961). As is well known, normal meiosis results in the formation of four nuclei, the microspore nuclei (fig. 18.4*C*). Each nuclear division may be followed immediately by cytokinesis (successive formation of walls), or the four protoplasts may be walled off simultaneously at the end of meiosis (simultaneous formation of walls; Maheshwari, 1950; Schnarf, 1927). The first type of division is particularly common in monocotyledons, the second in dicotyledons. The simultaneous wall formation may occur by development of cell plates or by furrowing.

The first wall delimiting the microspore protoplasts from each other is of the same material, callose, as the special wall around the entire tetrad of microspores (fig. 18.4C; Reeves, 1928; Waterkeyn, 1961). Later, each microspore forms its own wall, the *sporoderm* (figs. 18.4D, E, 18.5).

According to a submicroscopic study of *Tradescantia* anthers (Bal and De, 1961), mature pollen grains have abundant mitochondria, dictyosomes, and endoplasmic reticulum; in the earlier stages these entities are not fully differentiated. In younger cells leucoplasts with starch are present; later the plastids become scarce. The number of nuclei in the mature pollen grains is of taxonomic significance and is also associated with certain physiological characteristics of the grains (Brewbaker, 1959).

The sporoderm is usually described as consisting of two layers (figs. 18.4E, 18.5C, D), the *exine* (outer wall) and the *intine* (inner wall). The exine is differentiated into a sculptured *ektexine*, or *sexine*, and a nonsculptured *endexine*, or *nexine* (Erdtman and Vishnu-Mittre, 1958; Faegri, 1956). Some workers recognize a third layer, the *medine*, located between the exine and the intine (Saad, 1963). The exine consists mainly of a lipoidal substance *sporopollenin*, which is less soluble than cutin or suberin (Frey-Wyssling, 1959). The research on the structure of walls of pollen grains and spores is highly technical and is designated by a special term, *palynology* (Aleshina, 1962; Faegri, 1956).

Most pollen grains are *aperturate*, that is, provided with pores or

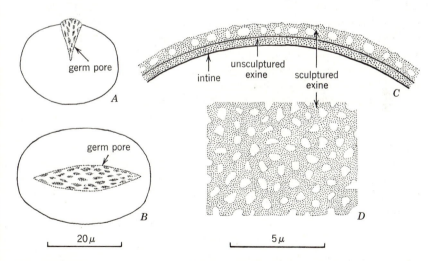

FIG. 18.5. Pollen structure in *Saxofridericia compressa*. A, equatorial view, narrow axis. B, equatorial view, long axis. C, section of pollen grain wall showing the layers. D, outer layer of exine in section parallel with the surface. Germ pore is also called aperture. (After Carlquist, *Aliso* 5, 1961.)

furrows (colpi, sing. colpa; fig. 18.5A, B). These apertures are not actual openings but places where the exine is very thin (Faegri, 1956) and the intine well developed. The pollen tube emerges through the aperture during the germination of the pollen grain apparently by pushing aside the intine (Bailey, 1960). The apertures are also regarded as the flexible parts of the sporoderm that permit the change in shape and size of the pollen grain caused by varying water content (Faegri, 1956). The number of apertures varies from one to many.

As seen from the surface, the exine of many species has spines, depressions, areolations (division into distinct spaces), and other types of ornamentations (Bradley, 1960). These external markings and the shape of the pollen grains are characteristics that may be utilized in taxonomic studies (Wodehouse, 1935, 1936). Ultrastructurally the ektexine often shows a porous structure (Larson and Lewis, 1962).

The intine varies in thickness and in a given species is more or less thickened in the aperture region (fig. 18.4E). It has no ornamentations. The intine consists mainly of polyuronides or a mixture of polyuronides and polysaccharides but its inner part contains also cellulose (Bailey, 1960). In conifers the outer intine is reported to contain callose (Martens and Waterkeyn, 1962).

When the pollen tube emerges from the pollen grain it grows by addition of wall material at its apex (Schoch-Bodmer, 1945). The pollen tube wall contains cellulose and is cutinized (Frey-Wyssling, 1959). It has also been described as having an outer lamella of pectin and an inner lamella of a mixture of callose and cellulose (Müller-Stoll and Lerch, 1957). The cytoplasm accumulates at the tip of the tube and may completely disappear from its basal part. In such instances, the older parts of the elongating pollen tube are successively sealed off by plugs of callose (fig. 18.8H; Brink, 1924; Schoch-Bodmer, 1945). Accumulation of callose is intensified under conditions of incompatibility, possibly in relation to the reduced growth rate of the pollen tube (Tupý, 1959). In plants forming no plugs of callose (*Fagopyrum esculentum*) the whole tube probably has a thin layer of cytoplasm in addition to the accumulation at the apex (Schoch-Bodmer, 1945). Cytoplasmic streaming has been observed in pollen tubes, even in parts sealed off by the callose plugs (Iwanami, 1956).

Carpel

Relation to the Gynoecium. Whatever may be the phylogenetic origin of the carpel, in many extant angiosperms with superior ovaries it resembles a leaf. As mentioned before, the carpels may or may not be united with other carpels. If the carpels are free, the gynoecium is apocarpous (fig. 19.1A, C), if they are united the gynoecium is syncarpous (fig. 19.1B). An apocarpous gynoecium may have a single carpel (*Prunus*, Leguminosae).

The carpel of an apocarpous gynoecium appears as a leaf-like folded structure, differentiated, in the specialized condition, into a basal fertile part, the ovary, and an upper sterile part, the style (fig. 19.1A, C). According to an older concept, the folded carpel has infolded or involuted margins, that is, margins turned toward the interior of the folded carpel, and these margins bear the placentae that give rise to the ovules. A later view, based on studies of the woody Ranales, states that in the primitive form the carpel is a conduplicately (from the Latin, doubling) folded structure, that is, a structure folded lengthwise without involution of margins (fig. 18.6H). Such a carpel shows laminar placentation; the ovules are borne not on the margins but on the inner (ventral) surface, more or less distant from the margins (fig. 18.6B), and may be vascularized by connection with the dorsal bundle rather than the ventral (Bailey and Swamy, 1951; Canright, 1960; Periasamy and Swamy, 1956). The apparent involution and marginal placentation (fig. 18.6D, pl. 91A) are thought to have resulted from phylogenetic change in the ontogeny of the carpel, a decrease in the extension of its folded adaxial part (the unstippled areas in fig. 18.6C, D). An argument offered in opposition to the concept of conduplicate carpel states that the surfaces coming in contact in the folded carpel are not ventral but marginal (Puri, 1961). The evidence on the phylogenetic reduction of the adaxial margins (fig. 18.6) does not support this argument.

The interpretation of the phylogenetic differentiation of the dicotyledonous carpel into ovary, style, and stigma has been consequentially developed with reference to the carpel of the woody Ranales (Bailey, 1954; Canright, 1960). The unspecialized carpel is a styleless, unsealed, conduplicate leaf-like structure with laminar placentation. The stigmatic tissue occurs on the free margins of the carpel (fig. 18.6H, I), on its inner surface, and at times also on the outer surface. Successive phylogenetic stages involve closure of the carpel, reduction in the number of ovules and their restriction to the lower part of the carpel (the ovary), and differentiation of the upper part into the style with a stigma localized on its apex. The closure of the carpel occurs through a growing together (concrescence) of the ventral surfaces along the margins that are in contact with each other. The concrescence is ontogenetic and may leave a conspicuous suture; or the union may be so complete that the evidence of a suture is partly or entirely obliterated.

The evolutionary changes in the structure of the gynoecium of the angiospermous flower also involve various manners of union of carpels

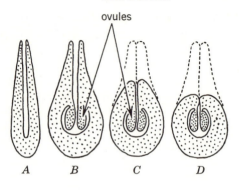

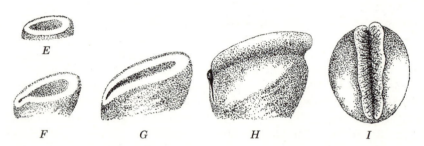

FIG. 18.6. *A–D*, carpels of Ranales in transections. *A*, open folded sterile carpel. *B*, open folded fertile carpel with locule enclosing ovules. *C, D*, stages in phylogenetic closure of folded carpels. Folded ventral part (delimited by broken lines) is retracted phylogenetically: becomes less and less extended during ontogeny. *E–I*, carpel of *Degeneria vitiensis* in several stages of development. *E*, carpel primordium as a shallow cup. Cessation of divisions on ventral side results in formation of notch (*F*). Uneven growth of the rim transforms carpel into a conduplicate structure (*G*). Free edges grow out into flanges with flaring margins (*H*, side view, and *I*, surface view). Internal surfaces of flanges become stigmatic in mature carpel. (After, *A–D*, Bailey and Swamy, *Amer. Jour. Bot.* 38, 1951; *E–I*, Swamy, *Arnold Arboretum Jour.* 30, 1949.)

of the same flower (Bailey, 1954). The carpels may become joined by their margins to the receptacle (fig. 18.7*B*), or they may grow together laterally in a folded closed condition (fig. 18.7*C*), or they may become laterally united in a folded open condition (fig. 18.7*A*). The union of carpels may occur during their ontogeny (figs. 18.10*F, G*, 18.11) or they may grow as a unit structure (pl. 93) and are then interpreted as congenitally fused, that is, fused from their inception (Baum, 1949).

The manner of union of carpels is related to differences in internal structure, such as number of locules in the ovary and the arrangement of placentae, the placentation (Puri, 1952*a*). Each carpel typically has two placentae (figs. 18.6*B–D*, 18.7). If the carpel has a congenitally united lower part (fig. 18.6H), the placentae may fuse in this part and the placental region then has the shape of a U (Leinfellner, 1951a; Schaeppi and Frank, 1962). In syncarpous gynoecia, the junction of carpels in a folded condition (figs. 18.1, 18.7B, C) may result in an ovary with as many locules as there are carpels and with the placentae arranged around a central column of tissue (*axile placentation*). If the carpels are united with each other in an open condition, the ovary is usually not divided into locules and the ovules are borne on the ovary wall or on extensions from it (fig. 18.7A; *parietal placentation*). The parietal placentation is considered to have evolved from the axile (Takhtajan, 1959).

Various deviations from the basic structures of the ovary just described are encountered in different angiosperms. Division of the ovary into compartments may occur in other ways than by the folding of carpels. The placentae may be borne upon a central column of tissue not connected by partitions with the ovary wall (*free central placentation*), or may occur at the very base of a unilocular ovary (*basal placentation;* pl. 94A, B). The free central placentation apparently results or has resulted from disappearance of partitions in terms of either ontogeny or phylogeny (Hartl, 1956b). Syncarpy and apocarpy may be present in the same pistil if the carpels are joined only at the base. The type of syncarpy may also vary in different parts of the pistil since the individual carpels may have a congenitally united lower part and an open upper part; the type of concrescence of carpels may be different in the two parts (Morf, 1950).

The controversial views on the nature of the carpel are reflected in the interpretations of the placenta. According to one of the common

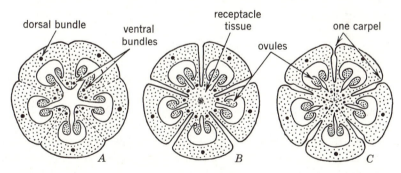

FIG. 18.7. Transections of gynoecia of Ranales illustrating syncarpous tendencies. *A*, whorl of five open folded carpels laterally concrescent. *B*, whorl of folded carpels adnate to receptacle with their free margins. *C*, whorl of folded carpels concrescent in their ventral parts. (After Bailey and Swamy, *Amer. Jour. Bot.* 38, 1951.)

concepts, the column of tissue bearing the ovules in ovaries with axile or free central placentation may be entirely carpellary (fig. 18.7C) or partly axial and partly carpellary (fig. 18.7B). Presence of vascular tissue other than that of the carpels in the central column is one of the evidences used to identify the axial nature of the central tissue. In both situations the placentae would be part of the carpels. When the ovules are borne on the carpels the species is said to be *phyllosporous* (Lam, 1961). The opposite view, chiefly concerning species with central and basal placentations, considers that the placentae and ovules may be cauline structures (Pankow, 1962). When the ovules are borne on cauline tissue the species is designated as *stachyosporous* (Lam, 1961).

The structure of inferior ovaries also presents problems of interpretation, especially with regard to the questions whether any carpellary tissue lines the lower part of the ovarian cavity (Guenot, 1954) and whether the extracarpellary tissue is axial (receptacular) or appendicular (floral tube). As was mentioned before, the use of vascular anatomy has led to the concept that the cup (hypanthium) of extracarpellary tissue is appendicular in some plants, receptacular in others. Some authors, however, see no distinction between the inferior ovaries and prefer to consider the cup as uniformly receptacular (Leinfellner, 1954; Puri, 1952b).

The ovary wall is not highly differentiated before and during anthesis (time when fertilization takes place in the flower). It consists largely of parenchyma and vascular tissue and bears a cuticularized epidermis on the outer surface. In the Compositae, calcium oxalate crystals occurring in the cells of the ovary wall were found to differ according to species (Dormer, 1961). The ovary wall undergoes more or less profound changes during the development of the fruit and then may show marked specializations (chapter 19).

Style and Stigma. The development of the style occurred as a concomitant of sterilization of the apical part of the carpel (Bailey and Swamy, 1951). In an apocarpous gynoecium each carpel usually has one simple style. In syncarpous gynoecia the styles of the component carpels may be variously united with each other (Baum, 1948d; Parkin, 1955). The carpels may be united only at their bases, leaving the styles free, or partly so (fig. 18.8; Theaceae, Hypericaceae). In highly modified flowers the carpels are united from base to apex and form a gynoecium with a single ovary, style, and stigma (fig. 18.1; Solanaceae, Oleaceae). If the styles are free, the stylar portions derived from the individual carpels are often called style branches, a designation giving an erroneous concept of the structure of the compound style; the branches are morphologically entire styles (Baum, 1948d). The term *stylode* has been proposed as a replacement for stylar branch (Parkin, 1955).

The style and the stigma have structural and physiological peculiarities that make possible the germination of the pollen and the growth of the pollen tube from the stigma to the ovules. On the stigma the protoderm differentiates into a glandular epidermis with cells rich in cytoplasm, often papillate in shape, and covered with a cuticle (Schnarf, 1928). This epidermis excretes a sugary liquid. Thus, the stigma resembles a nectary (chapter 13) in structure and function. The cells beneath the epidermis may be as rich in cytoplasm as the epidermis, and then they constitute a part of the glandular tissue. In many plants, the stigmatic epidermal cells develop into short, densely crowded hairs (cherry, bean) or into long, branched hairs (grasses and other wind-pollinated plants; pl. 93F).

An outstanding feature of the organization of the carpel is that the stigma is connected with the interior of the ovary by a tissue cytologically similar to the glandular stigmatic tissue (Coulter and Chamberlain, 1912; Schnarf, 1928). This tissue is interpreted as a medium facilitating the progress of the pollen tube through the style and supplying the developing pollen tube with food. It is commonly called conducting tissue, a term easily confused with that referring to the vascular tissue. The terms transmitting tissue and pollen-transmitting tracts serve as substitutes (Arber, 1937). In the following discussion this tissue is referred to as *stigmatoid tissue* on the basis of its apparent cytologic and physiologic similarity to the tissue of the stigma.

The carpels of the more primitive dicotyledons (fig. 18.6) do not show a differentiation into stigmatic and stigmatoid tissues, for, as was stated previously, the surfaces of the flaring margins and the inner surface of the open carpel are lined with stigmatic glandular hairs. With the increase in specialization of the carpels, characterized by their gradual closure and the development of the style, the stigma proper became restricted to a part of the style, but the continuity of the stigmatic tissue with the placentae was maintained. The internal glandular surfaces became modified into pollen-transmitting (Bailey and Swamy, 1951), or stigmatoid, tissue.

In relation to the variation in degree of concrescence of carpels and in methods of growth of the styles, the styles may be open or solid in both the apocarpous and the syncarpous gynoecia. The open styles are described as having a canal. In a syncarpous gynoecium the compound style may have one common canal (*Viola, Erythronium*), or each component style may have its own canal (*Lilium, Citrus*). The stigmatoid tissue lining the stylar canal resembles the glandular tissue of the stigma and may be papillose. In some plants starch has been observed in this tissue, and a cuticle has been identified on the surface exposed to the canal. The stigmatoid tissue may line the entire canal, or it may be restricted to localized parts in the form of one or more longitudinal bands. In many plants the stigmatoid tissue is several cells in thickness and if, at the same time, it is distributed in longitudinal bands, one can speak of strands of stigmatoid tissue. Stigmatoid tissue occurs on the placenta within the ovary and in some species on the funiculus of the ovule as well. In certain plants the stigmatoid tissue is brought close to the micropyle by a placental proliferation in the form of a small protuberance, the *obturator* (Schnarf, 1928). Developmental studies on the styles of *Datura* and *Cucurbita* have shown that the multilayered stigmatoid tissue lining the stylar canals and placentae in these plants originates from the epidermis by periclinal divisions (Kirkwood, 1906; Satina, 1944).

In most angiosperms the styles are solid; that is, they have no canals (fig. 18.1). The stigmatoid tissue is present, nevertheless, usually in the form of strands of considerably elongated cells staining deeply with cytoplasmic stains. If the gynoecium with a single solid style is syncarpous, the stigmatoid tissue of the style forms several strands. Commonly the stigmatoid tissue has a course independent from that of the vascular bundles, but it may be associated with the bundles (*Zea*, Kiesselbach, 1949).

Syncarpous gynoecia may have openings that enable a pollen grain germinating on a stigma of one of the styloids or any part of the stigma of a single style to reach any part of the ovary rather than only the one to which a given stigma or part of stigma is related. The opening (*compitum*, Carr and Carr, 1961) may consist of a canal (fig. 18.8C), pore, or split in the septum between locules. In unilocular ovaries with parietal placentation the crossing over of the pollen tube may occur in the style itself. Some syncarpous gynoecia have no compital structure and, therefore, function like apocarpous gynoecia with regard to pollination.

With regard to the possible factors that might direct the growth of pollen toward the ovule, some workers stress the evidence that there is a chemotactic attraction between the pollen tube and the tissues of the stigma and of the ovule; others consider that the structure of the stigmatoid tissue and its distribution in the pistil are sufficient to account for the direction of growth of the pollen tube (Brink, 1924; Renner and Preuss-Herzog, 1943; Schnarf, 1928). The presence of pollen tubes in or on the stigmatoid tissue has been repeatedly ascertained in various plants (fig. 18.8; Borthwick, 1931; Doak, 1937; Maheshwari, 1950; Pope, 1946; Schnarf, 1928).

The relation of the pollen tube to the stigmatoid tissue is somewhat different in styles with and without open canals. In the former, the pollen tubes may have an entirely superficial course. After the germination of the pollen grain on the stigma the pollen tube grows among the papillae or hairs or on the surface of the nonpapillate cells. The course in the stylar canal is essentially the same as on the stigma. Frequently the cuticle disappears in the stylar canal before pollination, and the walls of the glandular tissue become swollen and soft (Schnarf,

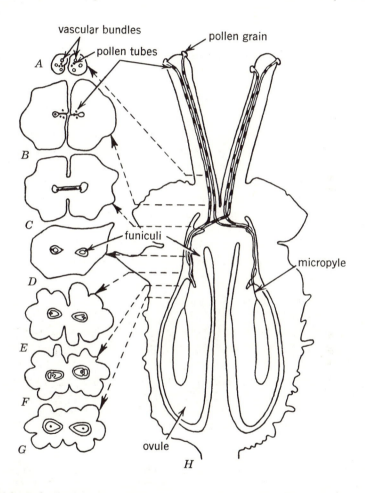

FIG. 18.8. Path of pollen tubes within a flower. Transverse (*A–G*) and longitudinal (*H*) sections of *Daucus carota* (carrot) flowers. The blackened parts of pollen tubes represent plugs of callose. Pollen tube passes through tissue of style (*A, B*), then emerges into stylar canal (*C*). Farther down it follows the funiculus (*D–E*) and finally enters the micropyle (*F–G*). At level where the stylar canals are interconnected, the pollen tubes may cross over from one carpel to the other (*C, H*). (*A–G*, ×13; *H*, ×24. After Borthwick, Bot. Gaz. 92, 1931.)

1928). The pollen tube may also penetrate the lining of the stylar canal to somewhat deeper layers and proceed there by growing between cells. If the style is solid, the pollen tube usually passes through the stigmatoid tissues by intercellular growth. Reports that pollen tubes penetrate the cells themselves are not well substantiated (Schnarf, 1928). In grasses, the pollen tube may take an intercellular course on the stigma itself. As was mentioned previously, the grass stigma commonly bears long hairs. These may be multicellular columns, both vertically and horizontally (Kiesselbach, 1949; Pope, 1946). The pollen tube penetrates into the interior of the column of cells and proceeds from there into the stigmatoid tissue of the style. After the pollen tube reaches the ovarian cavity, it follows the stigmatoid tissue lining the ovary wall and the placenta and eventually comes in contact with the ovule.

The intercellular growth of the pollen tube appears to involve a digestion of the intercellular substance (Schoch-Bodmer and Huber, 1947). In agreement with this assumption pollen tubes give a positive reaction for an enzyme capable of digesting pectic substances (Paton, 1921). The stigmatoid tissue, however, appears to undergo a partial weakening in its structure before the pollen tube passes through it. Its walls assume a swollen aspect (the tissue resembles collenchyma in this state, fig. 18.9A), and the connection between cells is loosened, as demonstrated by the ease with which the tissue may be macerated. In fact, the walls appear as though they have been converted into a

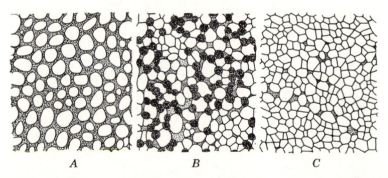

| A | B | C |

FIG. 18.9. Relation between pollen tube and stigmatoid tissue. Transections of stigmatoid tissue of *Lythrum salicaria* without pollen tubes (A), with young, densely cytoplasmic pollen tubes (B), and with old pollen tubes having scanty cytoplasm (C). Undisturbed mature stigmatoid tissue has thick collenchymatous walls (A). Pollen tubes remove these wall thickenings (B). Cell lumina of stigmatoid tissue also shrink. In the exhausted stigmatoid tissue old pollen tubes are almost indistinguishable (C). (All, ×400. After Schoch-Bodmer and Huber, *Naturf. Gesell., Zürich, Vrtljschr. 92*, 1947.)

mucilage (Schnarf, 1928). When the pollen tube passes through the stigmatoid tissue, it occupies the space formerly filled with cell wall material (fig. 18.9B). The protoplasts of the stigmatoid tissue may also become exhausted and sometimes even shrivel and die. Because of these relationships the entry of pollen tubes, even if these are very numerous, does not cause the expansion of the stigmatoid tissue (Schoch-Bodmer and Huber, 1947). The pollen tubes may be said to replace some of the stigmatoid tissue (fig. 18.9C).

The exhaustion of the protoplasts of the stigmatoid tissue by the pollen tube indicates an effect of chemical nature. In studies on Gramineae the pollen was found to have an effect on the stigmatic tissue after a short period of contact, that is, even before germination: the cells of the stigma showed increased stainability of nuclei (Kato and Watanabe, 1957).

The stigmatoid tissue and the vascular bundles constitute the most specialized parts of the style. The ground tissue is parenchymatic, and the outer epidermis shows no peculiar features. It bears a cuticle and may have stomata.

Ovule

The ovule developing from the placenta of the ovary is the seat of formation of the megaspores (or macrospores) and of the development of the embryo sac (female gametophyte) from a megaspore. Sporogenesis, the development of the embryo sac, and the many variations in the details of these phenomena, have been the subject of numerous investigations (Coulter and Chamberlain, 1912; Gerasimova-Navashina, 1954; Schnarf, 1927, 1928, 1931; Maheshwari, 1950) and are not reviewed here. Concomitant with the development of the embryo from the fertilized egg, and of the endosperm from the product of the triple fusion (two polar nuclei and one sperm nucleus), the ovule develops into a seed. Histologically, the ovule is rather simple as compared with the resulting seed.

Commonly the ovule is differentiated into the following morphologic parts (pl. 94C): the *nucellus* (from the Latin, small kernel), a central body of tissue containing some vegetative and some sporogenous cells; one or two *integuments* (from the Latin, covering) enclosing the nucellus; the *funiculus* (from the Latin, rope), the stalk by means of which the ovule is attached to the placenta. The size of the nucellus, the number of integuments, and the shape of the ovule are important distinguishing characteristics of ovules in different groups of angiosperms. If the nucellar apex points away from the funiculus, the ovule is termed *atropous* (synonym of *orthotropous; a*, not; *tropos*, turned, in Greek), that is, not turned. If the ovule is completely inverted so that the nucellar apex is turned toward the funiculus, it is called *anatropous* (*ana*, up, in Greek; pl. 94C). Between these two extreme forms of ovules, there are several variously named intermediate ones with various degrees of curvature (Bocquet, 1959; Maheshwari, 1950; Schnarf, 1927).

The ovule primordium arises from the placenta as a conical protuberance with a rounded apex. The first sporogeneous cell (archesporial cell) becomes evident, in the still undifferentiated protuberance, by its size and often also by denser appearance of its cytoplasm. This cell occurs beneath the protoderm at the apex of the primordium. Slightly below the apex the inner integument (or the single integument) is initiated by perclinal divisions in the protoderm. It arises as a ring-like welt and grows upward (fig. 18.11B). With the appearance of the integument, the nucellus of the primordium becomes delimited as the part enveloped by the integument (pl. 94A). The latter grows faster than the nucellus and encloses it partially or completely. Usually a narrow, canal-like opening remains at the top of the integument. This is the *micropyle* (fig. 18.11C; pl. 94C; from the Greek, *micros*, small, and *pyle*, gate-like). The outer integument, if such develops at all, arises in the protoderm slightly below the inner integument and develops in a manner similar to that of the inner (fig. 18.11C, pl. 94B). It frequently does not reach the apex of the ovule in its upward growth. In the anatropous and other curved ovules the growth of the integuments is asymmetrical, being more pronounced on the side of the ovule which eventually becomes convex (pl. 94).

There is no agreement on the morphologic nature of the ovule and its parts. Some workers consider the ovule a foliar structure, others an axial. The nucellus is commonly regarded as the megasporangium, but the interpretation of the homology of the integuments constitutes a major morphologic problem (Meeuse, 1963; Roth, 1957).

The nucellus, the integuments, and the funiculus cannot be sharply delimited from one another either morphologically or cytologically. The nucellus is usually clearly outlined above the level where the integuments originate (pl. 94). From this level upward the nucellus and the integument (or integuments) have each their distinct epidermal layers (pl. 94B). Below this level, that is, at the base of the nucellus, the nucellus and the integuments are confluent with the funiculus. The region of the ovule where all its parts merge with one another is called the *chalaza* (pl. 94C; in Greek, small tubercle).

The ovules of certain plants show considerable deviations from the structure just outlined (Maheshwari, 1950; Schnarf, 1927). Some have no integuments, and others have more than two. The nucellus may be entirely confluent with the integuments, a condition supposedly different from that interpreted as absence of integument. Ovules may have other outgrowths than the integuments, such as the *aril* (*Euonymus europaeus*) derived from the funiculus, and the *caruncle* (*Ricinus*), an integumentary protuberance near the micropyle. In some plants the integument so completely overgrows the nucellus that no micropyle remains; in others the integuments do not reach the apex of the nucellus.

The nucellus varies in size in different groups of plants. It may be so small that it comprises little more than an epidermis and the sporogenous tissue enclosed by it (pl. 94*A, C*). In other plants a more or less massive vegetative tissue envelops the sporogenous tissue (pl. 94*B*). The integuments also show variations in thickness. The thinnest integument is two cells thick; that is, it consists only of two epidermal layers (pl. 94*B*). Sometimes the micropylar end is somewhat thicker in the two-layered integuments. Most angiosperms have two-layered integuments, although some dicotyledonous families have integuments of three and more layers (Netolitzky, 1926). In relation to the size of nucelli the ovules are classified into *crassinucellate* (*crassus*, thick in Latin) and *tenuinucellate* (*tenuis*, slender in Latin). Crassinucellate ovules with two integuments are considered to be more primitive than the tenuinucellate with a single integument.

The ovules have a vascular system connected with that of the placenta. The presence of integumentary bundles is sometimes considered a primitive characteristic (Walton, 1952), but such bundles occur in the more specialized as well as in the less specialized angiosperms, and therefore their phylogenetic significance is uncertain (Kühn, 1928). Most commonly there is a single strand ending in the chalaza with no prolongations into the integuments. In some species the bundle extends beyond the chalaza as a single strand or is variously branched. Such an intraovular system occurs in the integument. If two integuments are present, vascular tissue may be found in both integuments or only in the outer. Rarely does vascular tissue occur in the nucellus (Kühn, 1928; Maheshwari, 1950; Schnarf, 1931). The vascular tissue is primary and appears to be in a functioning state during the maturation of the seed.

The distribution of cuticles in the ovules deserves special mention because of their prominence and physiologic importance in the seed that develops from the ovule. The cuticles of the ovules and seeds are called by various names: cuticles, suberized membranes, semipermeable membranes, and fatty membranes. They are here referred to as cuticles in keeping with the most prevalent designation (Schnarf, 1927). Cuticles are reported to be present in ovules in relatively early stages of development.   The entire surface of the ovule primordium bears a cuticle. After the development of the integuments three cuticular layers may be distinguished: the outer, on the outside of the outer integument and the funiculus; the median, double in nature, between the two integuments; the inner, also double in nature, between the inner integument and the nucellus.   In ovules with a single integument the median cuticle is absent.   If the nucellus is small and its vegetative tissue is disorganized during the development of the embryo sac, the cuticle of the micropylar part of the nucellus may also be dissolved (Scrophulariaceae, Labiatae, Campanulaceae).

Parts of the ovule are disorganized during the development of the embryo sac, and the resulting materials are presumably utilized by the growing female gametophyte.   The vegetative tissue of the nucellus is partly or entirely resorbed.   In the latter instance the embryo sac comes in contact with the inner epidermis of the integument.   Large nucelli may be partially retained, and in some plant groups they form a storage tissue (*perisperm*) in the seed (Centrospermae; fig. 20.4A).   The nucellar epidermis is sometimes highly resistant and may proliferate into a nucellar cap with relatively thick walls (*Allium*).

The integuments undergo certain histologic changes or are disorganized to varying degrees.   Particularly common is the differentiation of the inner epidermis of the integument into the so-called nutritive jacket or *integumentary tapetum* consisting of deeply staining cells elongated perpendicularly with reference to the surface of the embryo sac (pl. 94C).   Such differentiation is characteristic of families in which the nucellus is early disorganized and the integument comes in contact with the embryo sac (Sympetalae).   The physiologic significance of the integumentary tapetum is not agreed upon, and it might be variable (Schnarf, 1927).   Some connection with the nutrition of the embryo is suggested by the disintegration of the ovule tissue located next to the tapetum (fig. 19.4) and the persistence of the tapetum until the contents of the embryo sac complete their development (fig. 19.5).

ORIGIN AND DEVELOPMENT

The change from vegetative to reproductive activity in the apical meristem follows a sequence that is determined by the nature of the plant (Hillman, 1962).   Herbaceous annuals pass, during one season, through an uninterrupted sequence of vegetative growth, floral initiation, and floral development.   Woody species, at least in the North Temperate zone, commonly initiate the flowers in one season and complete their development during the next.   The degree of differentiation that the flowers attain before the end of the first season is highly variable (Roberts, 1937). Floral initiation is affected by external factors, but only within the limits of reactivity of the plant to a given environment. Plants show, for example, characteristic responses to length of day and to temperature and produce flowers under specific combinations of these two factors (Hillman, 1962).

Flowers arise at the apex of the main shoot, or on lateral branches, or on both. The lateral branches may form further branches of various orders before producing flowers. In different angiosperms the grouping of the flowers, called inflorescences, are highly variable and bear special names (Rickett, 1944). The formation of all types of inflorescences involves, in the activity of a given apical meristem, a cessation of the vegetative stage and the initiation of the reproductive stage. Frequently, the first visible sign of determination of the flowering stage is the enhanced development of axillary buds (Hagemann, 1963; Haupt, 1952; Rohweder, 1963). In species with cymose inflorescences a change from alternate, five-ranked arrangement of leaves to the single-ranked arrangement of floral primordia occurs at the beginning of the reproductive stage (Prior, 1960).

Questions pertaining to the developmental relation between the apical meristems in the vegetative and reproductive states and to the significance of the structural differences of the meristem in the two states have been considered in chapter 5.

Organogenesis

Much can be learned about floral development by comparing flowers in different stages of development in material dissected under magnifications of moderate degrees. Payer (1857) employed this method in his classic comparative study of organ development in flowers, and in modern times it has been applied with particular success to investigations of floral differentiation in the Gramineae (Barnard, 1957a, b; Bonnett, 1936, 1937, 1940, 1948; Evans and Grower, 1940; Sharman, 1947; 1960a, b). A correlation of the observations on dissected material with those on flowers sectioned with a microtome gives a rather comprehensive picture of the main phenomena in the development of the specific form of flowers and their parts.

Depending on the structure of the flower, the parts may appear in acropetal order at successively higher levels like the leaves on a vegetative shoot (*Ranunculus*), or the parts of a given kind may arise at the same level or nearly so (*Capsella*). In the former instance the floral parts are arranged helically; in the latter they are in whorls (cycles). If the parts arise in a helical sequence, the helices of the various parts are usually not continuous with one another. The calyx members, however, may appear along helices that are continuations of those of the foliage leaves (Plantefol, 1948). The flower parts either arise in a continuous acropetal sequence of sepals, petals, stamens, and carpels, or else this sequence is more or less modified. In *Capsella*, for example, the stamen and carpel primordia appear before those of the petals. There may be a difference in the rhythm of development of floral parts. In Papaverceae, for example, the sepals arise considerably in advance of the other parts (Bersillon, 1956). Petals, stamens, and carpels appear in rapid succession and overlap one another in timing of their origin.

The successive formation of the different floral parts—as contrasted with that of similar parts during vegetative growth—apparently is governed by complex determination phenomena involving, among other mechanisms, those of hormonal balances (Gavaudan and Debraux, 1951; Heslop-Harrison, 1959). Surgical experimentation with developing flowers of *Primula* indicates that the flower passes through a succession of physiological states which permit and regulate the formation of each organ in turn (Cusick, 1956).

As was mentioned previously, the floral parts may remain discrete at maturity, or they may become variously united within the whorls and between whorls. Three developmental patterns may bring about the union of parts: (1) the whorl arises as a unit structure; that is, the parts of the whorl show congenital unity; (2) the parts of a whorl or of adjacent whorls become joined during ontogeny; (3) the union of parts results from a combination of the two phenomena, the ontogenetic and congenital unions. The calyx and corolla tubes in *Datura*, for example, arise by ontogenetic fusion (Satina, 1944). Those of *Frasera* are congenitally united, for they are formed by intercalary growth of a ring of tissue at the base of the primordia of calyx and corolla (McCoy, 1940). In *Vinca*, however, the corolla tube consists of two parts, one formed by intercalary growth of receptacular tissue at the base of the petals, the other resulting from the union of the bases of the initially free petals (Boke, 1948).

The development of an initially open carpel into a closed structure involves a clearly expressed ontogenetic union of carpel margins (Baum, 1948*a*, *b*, *d*). The lower part of the carpel, however, may have a sac-like seamless form from the inception of the primordium (fig. 18.6). The formation of syncarpous gynoecia is associated with congenital and ontogenetic union in varying proportions (Baum, 1948*a*, *b*; Boke, 1949; Leinfellner, 1950, 1951*b*). There may be also an ontogenetic union between the carpels and the stamens (Baum, 1948*c*). On the other hand, the perianth parts and the stamens may originate together from unit primordia and become distinct during later growth (Ehrenberg, 1945; Jones and Emsweller, 1936; Roth, 1959b; Sattler, 1962).

The features discussed above may be elucidated by means of specific examples of floral development. The flower of *Allium cepa* (onion) is relatively unspecialized in having an undifferentiated perianth of free parts and a superior ovary (fig. 18.10A). Its carpels are united, however. The six-parted perianth consists of two whorls of tepals, an outer and an inner. The six stamens occur in the axils of the six perianth members. The three carpels are united into a gynoecium with a three-loculed ovary and an axile placentation. The style is thin and has a slightly three-lobed stigma. An individual flower is a globose protuberance before the flower parts appear. The outer three tepals arise first. The stamens in the axils of these tepals arise simultaneously with the tepals

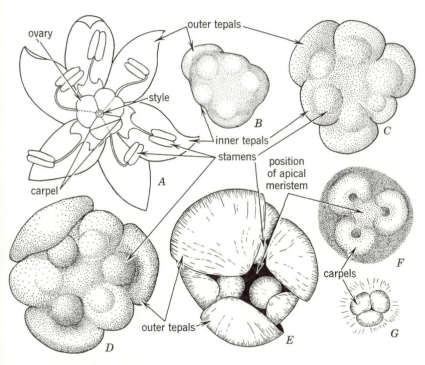

FIG. 18.10. Development of *Allium cepa* (onion) flower. All views from above. *A*, opened flower. Three outer tepals and three inner tepals constitute the perianth. Each tepal subtends a stamen. Those in the axils of the inner tepals have wide bases. The syncarpous gynoecium consists of three carpels (separated by solid lines). The lines of dehiscence (broken lines) alternate with lines of union (loculicidal dehiscence). *B–E*, four stages in development of tepals and stamens. *F, G*, two stages in development of carpels. (*A*, ×9; *B–F*, ×70; *G*, ×28. *B–G*, after Jones and Emsweller, *Hilgardia* 10, 1936.)

and from the same primordia (fig. 18.10B). The outer tepals and the associated stamens arise in a clockwise direction. The inner tepals and the stamens subtended by them also arise together, but in a counterclockwise direction (fig. 18.10B, C). With further growth, the tepals overarch the stamens (fig. 18.10D, E). When this stage is reached, the carpels are initiated. They occur within the inner staminal whorl in alternation with its members. At first they project over the surface of the receptacle in the form of three horseshoe-shaped welts of meristematic tissue (fig. 18.10F). Then they grow upward and toward the center where their margins meet and fuse (fig. 18.10G). The compound style is formed by apical growth of the three carpels, the three parts uniting completely (fig. 18.11A–D). The base of the style eventually appears deeply imbedded in the center of the ovary because the carpels bulge upward during the differentiation of the ovules (fig. 18.11D). The ovules are initiated before the carpel margins fuse. They are anatropous and have two integuments (fig. 18.11).

The flower of *Lactuca sativa* (lettuce) may be used to illustrate the growth of a highly specialized flower, one with an inferior ovary (epigy-

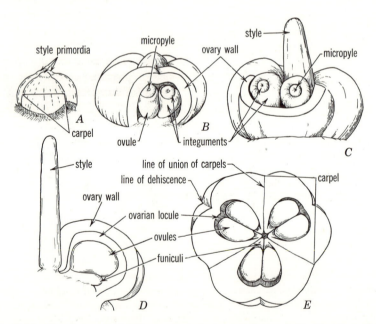

FIG. 18.11. Development of *Allium cepa* (onion) gynoecium. *A–D*, side views of three developmental stages; *B–D*, partly cut open to expose ovules. The apices of the three carpels extend and form jointly a compound style. *E*, ovary opened by a transverse cut and seen from above. (All, ×28. After Jones and Emsweller, *Hilgardia* 10, 1936.)

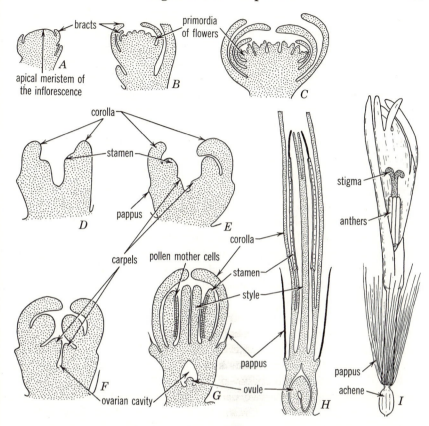

FIG. 18.12. Development of *Lactuca sativa* (lettuce) flower. *A–H*, longitudinal sections of young inflorescences (*A–C*) and flowers (*D–H*). *I*, entire flower. It has an inferior ovary and a sympetalous zygomorphic corolla. Stamens are adnate to corolla (epipetalous), and their anthers are joined into a column. Style is compound (composed of two styles, one from each of two carpels). Two strap-shaped ends of style bear stigmatic hairs. (*A–C*, ×29; *D–F*, ×153; *G, H*, ×27; *I*, ×7. After Jones, *Hilgardia* 2, 1927.)

nous flower) and a zygomorphic sympetalous corolla (Jones, 1927). Lettuce belongs to the Compositae in which the flowers occur in capitate (head-like) inflorescences. The individual flowers arise acropetally on the flattened receptacle, so that the outermost flowers of the head are the oldest, the innermost the youngest (fig. 18.12*A–C*). In an individual flower, the petal lobes appear first, as five protuberances on the margin of the floral primordium. However, immediately after their appearance they are thrust upward by intercalary growth of a ring of tissue upon which the corolla is inserted. As a result of this growth the central part of the flower primordium becomes cup-shaped (fig. 18.12*C*, primordium in center). The stamens, which are initiated after the corolla, seem to be inserted below the corolla, but actually they occur closer to the center or apex of the flower than the other floral parts (fig. 18.12*D*). The pappus, which is interpreted sometimes as a set of epidermal trichomes (Puri, 1951), sometimes as the calyx, appears almost at the same time as the stamens. It arises below and opposite the stamens on the outer surface of the rim of the cup-like primordium which higher up bears the corolla and the stamens (fig. 18.12*D*, *E*).

In its further growth the corolla develops as a tubular structure with a unilateral strap-shaped prolongation (zygomorphic ligulate corolla). Two stages may be distinguished in the growth of the corolla tube. First, intercalary growth above the insertion of the stamens forms the upper part of the corolla tube (fig. 18.12*G*). Second, intercalary growth below the insertion of the stamens forms the lower part of the tube (fig. 18.12*H*) in which the bases of the corolla and of the stamens are congenitally fused (epipetalous stamens). The second stage occurs comparatively late in the development of the flower. In the Compositae with actinomorphic tubular corollas the growth of the upper part of the corolla is uniform throughout. In zygomorphic corollas, as in lettuce, the upper part grows asymmetrically (fig. 18.12*I*). The free parts of the stamens elongate also, and each becomes differentiated into a filament and an anther (fig. 18.12*H*).

The carpels develop at the morphologically highest position of the flower, that is, within the cavity of the cup-like primordium. The two carpels become visible as two protuberances located seemingly below the stamens (fig. 18.12*E*, *F*). These two carpel units overarch the ovarian cavity (fig. 18.12*F*) and become prolonged above into a solid compound style with a two-parted stigma (fig. 18.12*G*, *H*). In Compositae the cup enclosing the ovary is commonly interpreted as consisting of adnate bases of the floral whorls joined to the carpel bases; in other words, the ovary is enclosed by the floral tube.

The development of an inflorescence and flower of a representative of the Gramineae may be illustrated by reference to the study on *Triticum* and *Avena* (Barnard, 1955; Bonnett, 1936, 1937). The wheat inflorescence is a spike and consists of several groups of flowers, each referred to as a spikelet. The spikelets are attached directly to the main axis, the rachis (pl. 92*A*). A spikelet of a grass (fig. 18.13*D*, pl. 92*I*) consists of a short axis, the rachilla, bearing several chaff-like, two-ranked (distichous), overlapping bracts (commonly called glumes). The two lowermost bracts bear no flowers in their axils and are called empty glumes. Above the empty glumes are others that subtend flowers, usually referred to as florets (fig. 18.13*A*–*C*). The wheat spikelet has four to six florets, each subtended by two bracts: the lower or abaxial,

called the lemma, and the upper or adaxial, called the palea. The
reproductive parts of a grass floret consist of three stamens with thread-
like filaments and rather large anthers, and a single, unilocular pistil with
a short style and two feathery stigmas (fig. 18.13A, pl. 93F). At the
base of the ovary and opposite the palea are two lodicules (fig. 18.13A,
pl. 93E), small scales involved in the opening of the bracts during
anthesis.

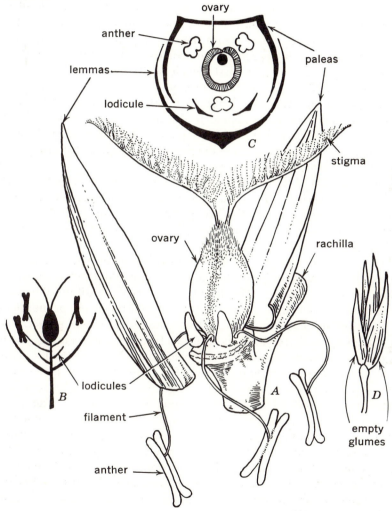

FIG. 18.13. The grass flower. A, partly dissected grass flower at anthesis. B, longitudinal
and, C, transverse diagrams of flower. D, spikelet. Lodicules are small scales outside the
stamens. (From A. M. Johnson, *Taxonomy of the Flowering Plants*, Appleton-Century-
Crofts, Inc., 1931.)

The reproductive stage of a wheat plant begins while the plant is still in the rosette stage. The initiation of reproductive stage is quickly followed by a sudden and vigorous elongation of the shoot, the subsequent culm. The addition of leaf primordia ceases, and even the further development of the existing leaf buttresses is stopped. Some of the younger buttresses may be obliterated as the apex expands in length and width. Whereas the foliage leaf primordia arise as single ridges gradually encircling the shoot axis (pl. 92B–D; chapter 16), the spikelet development is initiated by the appearance of double ridges (pl. 92E, F). The spikelet proper differentiates from the upper of the paired ridges (pl. 92G). A spikelet is interpreted as an axillary bud, and the lower ridge as the subtending leaf. The first spikelets differentiate in the middle of the spike (pl. 92E, F), and differentiation then progresses acropetally and basipetally (pl. 92G, H). Within the individual spikelet differentiation is acropetal, the parts appear in the sequence of: empty glumes, first flower, second flower, and so forth (pl. 92G–I). Within an individual floret the parts arise in a close overlapping sequence: lemma, palea, lodicules, stamens, and gynoecium.

The primordia of the lemma, palea, and lodicules are ridge-like; that is, they resemble leaf primordia. The stamen primordia, on the other hand, are rounded (pl. 92H) like bud primordia, one of the features that is used to interpret the stamen as a cauline structure (Sharman, 1960b; Surkov, 1961).

The gynoecium occupies the apex of the floral meristem. A crescent-shaped ridge, which is highest on the side toward the lemma, arises just below the apex (pl. 93A). The apical mound itself constitutes the ovule primordium. The ridge grows entirely around the ovule primordium (pl. 93B) and initiates two styles on two sides of its margin (pl. 93C, D). Continued upward growth of the margins below the styles brings about the closure of the ovarian cavity. The stigmatic hairs are the last parts of the gynoecium to develop (pl. 93E, F). Thus, the grass gynoecium arises as a unit and does not reveal, ontogenetically, the three-carpellate origin usually ascribed to the gynoecium of Gramineae. The same method of origin and growth of the gynoecium has been observed in various other Gramineae (Barnard, 1957a) and in Cyperaceae (Barnard, 1957b), except that in the latter some species have three styles. The apical position of the ovule is used for interpreting it as an axial structure, but the opinions on the number of carpels are divided (Barnard, 1957a). The carpellary part, or ovary wall, of the gynoecium is considered to be leaf-like in its method of origin and growth.

The rhythm of development of a flower as a whole has certain distinguishing characteristics that are closely correlated with the important phenomena of mitosis and meiosis occurring during the formation of spores and gametes (Erickson, 1948). The sequence of formation of floral parts is more rapid than that of the foliage leaves so that the ontogeny of the flower may have an explosive character (Bersillon, 1956). Morphologic observations and studies on comparative weights of developing flowers and their parts show that these parts may have divergent rates of growth after they are initiated (Sosa-Bourdouil, 1945). The petals, for example, may appear before the stamens but may develop more slowly. Sometimes the principal period of growth of the petals occurs only after the stamens complete their growth. Both the petals and the stamens may accelerate their growth rate shortly before anthesis (Pearson, 1933). The remarkable speed with which the stamens may attain their final length is well illustrated by the rate of elongation of 2.5 mm per minute observed in the growing anther filaments of rye (Schoch-Bodmer, 1939). The stamens may lag behind the gynoecium in development at first, then rapidly attain the final length which brings the anther into a most favorable position for release of pollen (fig. 18.13A). The ovary usually enlarges uniformly like a vegetative organ. Sometimes, however, the enlargement slows down before fertilization, and if fertilization fails to take place, the gynoecium dies. Comparative studies on floral parts show that the reproductive parts constitute a relatively large mass of the flower as a whole (Sosa-Bourdouil, 1945).

Histogenesis

Research on histogenesis of floral parts is used extensively for the interpretation of the morphologic nature of the flower and in comparative taxonomic studies. The sepals and petals originate, like the foliage leaves, from periclinal divisions in one or more subsurface layers of the apical meristem. Such origin of the perianth parts is apparently common in both the dicotyledons and the monocotyledons (Barnard, 1960; Kaussmann, 1941; Rohweder, 1963; Tepfer, 1953; Tucker, 1959). In their upward growth (pl. 90B–D), the perianth parts show apical activity of short duration followed by some intercalary growth. Marginal activity followed by intercalary growth is responsible for the increase in width of the perianth primordia. In *Vinca* the marginal meristem of the petals is more active than that of the sepals and is involved in the formation of the upper part of the floral tube which arises through the ontogenetic fusion of the corolla lobes (Boke, 1948).

Some workers find that the stamens are initiated just like the members of the perianth (Boke, 1948, 1949; Holt, 1954; Kaussmann, 1941; Lawalrée, 1948; Rohweder, 1963; Tepfer, 1953; Tucker, 1959; Wilson and Just, 1939). Others report that the stamens have a deeper origin than the perianth parts and are, therefore, axial structures (Barnard, 1960; Satina and Blakeslee, 1941; Sharman, 1960a).

After their initiation the stamens (pl. 90D) undergo apical growth of short duration, followed by intercalary growth. If the stamen filament is flattened, it shows marginal growth; otherwise such growth is suppressed (Kaussmann, 1941; Tepfer, 1953). The anthers have a special form of marginal activity which produces the characteristic two-lobed, four-loculed structure, rather than a flat blade (pl. 91D; Boke, 1949).

With regard to the gynoecium, frequently the origin of the placentae and ovules are considered to be distinctive from that of the carpel. Some authors find that the carpel resembles a leaf in the manner it originates from the apical meristem, whereas the placenta or the single, basally attached ovule is initiated like an axial structure (Barnard, 1957a, b; Pankow, 1962; Roth, 1959a). Others postulate that the primary ontogenetic relation of the ovule is with the carpel (Eckardt, 1957) and that the method of growth, which is obviously correlated with the future form of an entity, is hardly a safe criterion of homology (Sattler, 1962). Still another view has been advanced on the basis of developmental relations in cytochimeras of *Datura* (Satina and Blakeslee, 1943). All parts of the gynoecium, carpels, placentae, and ovules are cauline in nature because they arise in the third layer of the apical meristem, whereas the foliage leaves are initiated in the second. In their future growth, the carpels (pls. 90E, F, 91A) undergo apical and marginal growth (Boke, 1949; Tepfer, 1953; Tucker, 1959; Sprotte, 1940).

Histologic studies have revealed the manner of ontogenetic union of flower parts. As was mentioned previously, the union of perianth parts or of the carpels may be congenital, or it may occur, partly or entirely, during ontogeny. The ontogenetic union is brought about by fusion of the margins of parts that come in contact with each other during growth. In the petals of *Vinca* such union occurs through apposition of two epidermal layers, with the line of union eventually becoming obscured (pl. 91D, E; Boke, 1948). Evidence of the fusion of perianth parts is thoroughly obliterated if divisions, periclinal and others, occur in the apposed epidermal layers (*Datura;* Satina, 1944). The degree of union of carpels also varies from a rather loose one to a thorough interlocking of the epidermal cells, accompanied by divisions in these cells and a complete effacement of the suture (Baum, 1948a, b, c; Hartl, 1956a). In dicotyledons the carpels of syncarpous gynoecia are generally more firmly joined than in monocotyledons (Baum, 1948c).

Vascular Development

Information on vascular development in the flower is meager. Some consideration has been given the question of the direction of differentiation of procambium. The assumption that there is an acropetal differentiation of procambium in the flower and a basipetal differentiation in the vegetative shoot has been used in support of the concept that the flower is a unique structure and not comparable to the shoot (Grégoire, 1938). Later research has shown that there is no such simple and straightforward difference between the flower and the shoot. Acropetal differentiation of procambium is common in the vegetative shoot in a wide variety of plants (chapter 15). In the flowers, both acropetal and basipetal differentiation of procambium have been reported (Boke, 1949; Lawalrée, 1948; Paterson, 1961; Tucker, 1959).

According to the classic study of Trécul (1881), the xylem in the flower follows a pattern of differentiation similar to that in the shoot; that is, it appears in one or more loci and then progresses bidirectionally toward the distal and the proximal parts of the flower. In *Perilla* vascular differentiation appears to be speeded up when the reproductive stage is induced (Jacobs and Raghavan, 1962).

ABSCISSION

The abscission of floral parts has been less intensively investigated than that of the leaves (chapter 16), but the basic phenomena appear to be similar in the separation of all these structures (Pfeiffer, 1928). Abscission of parts or of entire structures occurs at various stages in the reproductive process. The completion of flowering may be followed by the shedding of parts of flowers, of entire flowers, or of inflorescences. Particularly common is the shedding of petals. The petals may fall without previous wilting (*Canna, Aquilegia, Cydonia, Rosa, Geranium, Linum*). They also abscise in a wilted or dried state, either close to the level of their insertion (*Lilium, Tulipa,* most Cruciferae, *Cucurbita*), or a short distance above the insertion, with the basal part remaining attached to the flower (*Althaea, Datura, Nicotiana*). If the petals are not shed at the end of flowering, they remain temporarily or permanently attached to the fruit in the dry state (*Agapanthus, Hypericum, Convallaria*). In some monocotyledons the perianth becomes green and persists in the fruit (*Veratrum, Eucomis, Paris*).

Petals are often constricted in the abscission zone. Usually no cell division precedes abscission, and the separation layer is poorly differentiated. The cells in this layer remain small, little vacuolated, and closely packed. They may contain chloroplasts or chromoplasts, and also raphides. The cells are roundish or polygonal in outline, occasionally tabular, with their long diameters oriented transversely with reference to the long axis of the petal. If the petal is much constricted, collenchyma may be present beneath the epidermis. Apparently the separation results from a softening of the middle lamella. Cell division may occur in the separation layer (Griesel, 1954). The protection of the scar seems to involve an impregnation of the walls with fatty substances without the deposition of a suberin lamella or formation of cork. Sepals, staminal filaments, and styles may abscise after flowering in essentially the same manner as the petals (Kendall, 1918; Pfeiffer, 1928). The abscission of entire flowers is characteristic of plants with unisexual flowers. The staminal flowers are regularly abscised after the pollen is shed (Yampolsky, 1934). These flowers may fall singly (Cucurbitaceae) or as entire inflorescences (catkins of the Amentiferae). If fertilization does not take place, carpellate and bisexual flowers may drop also (*Solanum tuberosum, Nicotiana tabacum, Lycopersicon esculentum*). Floral abscission can be induced artificially by various treatments (Kendall, 1918; Laurie and Duffy, 1948). The separation layer in pedicels of flowers is, in some species, preformed during development. Surface grooves are sometimes present in pedicels but do not necessarily coincide with the abscission zone.

REFERENCES

Aleksandrov, V. G., and A. V. Dobrotvorskaî̈a. Obrazovanie tychinok i formirovanie fibroznogo sloîà v pyl'nikakh tsvetkov nekotorykh rastenii. [Formation of stamens and of fibrous layer in anthers of flowers of some plants.] *Bot. Zhur. S.S.S.R.* 45:823–831. 1960.

Aleshina, L. A. Kriticheskiĭ obzor noveĭshikh rabot po stroeniîù obolochki pyl'tsevykh zeren pokrytosemennykh rastenii. [Critical review of the newest works on structure of pollen-grain wall of angiosperms.] *Bot. Zhur. S.S.S.R.* 47:1210–1213. 1962.

Andrews, H. N. Early seed plants. *Science* 142:925–931. 1963.

Arber, A. The interpretation of the flower: a study of some aspects of morphological thought. *Cambridge Phil. Soc. Biol. Rev.* 12:157–184. 1937.

Arber, A. *The natural philosophy of plant form.* Cambridge, Cambridge University Press. 1950.

Bailey, I. W. *Contributions to plant anatomy.* Waltham, Mass., Chronica Botanica Company. 1954.

Bailey, I. W. Some useful techniques in the study and interpretation of pollen morphology. *Arnold Arboretum Jour.* 41:141–151. 1960.

Bailey, I. W., and A. C. Smith. Degeneriaceae, a new family of flowering plants from Fiji. *Arnold Arboretum Jour.* 23:356–365. 1942.

Bailey, I. W., and B. G. L. Swamy. The morphology and relationships of *Austrobaileya*. *Arnold Arboretum Jour.* 30:211–226. 1949.

Bailey, I. W., and B. G. L. Swamy. The conduplicate carpel of dicotyledons and its initial trends of specialization. *Amer. Jour. Bot.* 38:373–379. 1951.

Bal, A. K., and D. N. De. Developmental changes in the submicroscopic morphology of the cytoplasmic components during microsporogenesis in *Tradescantia*. *Devlpmt. Biol.* 3:241–254. 1961.

Barnard, C. Histogenesis of the inflorescence and flower of *Triticum aestivum* L. *Austral. Jour. Bot.* 3:1–20. 1955.

References 579

Barnard, C. Floral histogenesis in the monocotyledons. I. The Gramineae. *Austral. Jour. Bot.* 5:1–20. 1957a. II. The Cyperaceae. *Austral. Jour. Bot.* 5:115–128. 1957b. IV. The Liliaceae. *Austral. Jour. Bot.* 8:213–225. 1960.

Barnard, C. The interpretation of the angiosperm flower. *Austral. Jour. Sci.* 24:64–72. 1961.

Baum, H. Über die postgenitale Verwachsung in Karpellen. *Österr. Bot. Ztschr.* 95:86–94. 1948a.

Baum, H. Die Verbreitung der postgenitalen Verwachsung im Gynözeum und ihre Bedeutung für die typologische Betrachtung des coenokarpen Gynözeums. *Österr. Bot. Ztschr.* 95:124–128. 1948b.

Baum, H. Postgenitale Verwachsung in und zwischen Karpell- und Staubblattkreisen. *Akad. der Wiss. Wien, Math.-Nat. Kl. Sitzber.* Abt. 1. 157:17–38. 1948c.

Baum, H. Ontogenetische Beobachtungen an einkarpelligen Griffeln und Griffelenden. *Österr. Bot. Ztschr.* 95:362–372. 1948d.

Baum, H. Der einheitliche Bauplan der Angiospermengynözeen und die Homologie ihrer fertilen Abschnitte. *Österr. Bot. Ztschr.* 96:64–82. 1949.

Baum, H. Über die "primitivste" Karpellform. *Österr. Bot. Ztschr.* 99:632–634. 1952.

Baum-Leinfellner, H. Die Peltationsnomenklatur der Karpelle. *Österr. Bot. Ztschr.* 100:424–426. 1953.

Bersillon, G. Recherches sur les Papaveracées. Contribution à l'étude du développement des Dicotylédones herbaceés. *Ann. des Sci. Nat., Bot.* Ser. 11. 16:225–447. 1956.

Bocquet, G. The campylotropous ovule. *Phytomorphology* 9:222–227. 1959.

Boke, N. H. Development of the perianth in *Vinca rosea* L. *Amer. Jour. Bot.* 35:413–423. 1948.

Boke, N. H. Development of the stamens and carpels in *Vinca rosea* L. *Amer. Jour. Bot.* 36–535:547. 1949.

Bonnett, O. T. The development of the wheat spike. *Jour. Agr. Res.* 53:445–451. 1936.

Bonnett, O. T. The development of the oat panicle. *Jour. Agr. Res.* 54:927–931. 1937.

Bonnett, O. T. Development of the staminate and pistillate inflorescences of sweet corn. *Jour. Agr. Res.* 60:25–37. 1940.

Bonnett, O. T. Ear and tassel development in maize. *Mo. Bot. Gard. Ann.* 35:269–287. 1948.

Borthwick, H. A. Development of the macrogametophyte and embryo of *Daucus carota*. *Bot. Gaz.* 92:23–44. 1931.

Bradley, D. E. The electron microscopy of pollen and spore surfaces. *Grana Palynol.* 2:3–8. 1960.

Brewbaker, J. L. Biology of the angiosperm pollen grain. *Indian Jour. Genet. and Plant Breed.* 19:121–133. 1959.

Bringman, G., and R. Kühn. Elektronenmikroskopische Befunde zur Morphologie der Cuticula von Blüten gärtnerischen und landwirtschaftlichen Nutzpflanzen. *Ztschr. f. Naturforsch.* 10b:47–58. 1955.

Brink, R. A. The physiology of pollen. *Amer. Jour. Bot.* 11:218–228, 283–294, 351–364, 417–436. 1924.

Camp, W. H., and M. M. Hubbard. On the origins of the ovule and cupule in Lyginopterid pteridosperms. *Amer. Jour. Bot.* 50:235–243. 1963.

Canright, J. E. The comparative morphology and relationships of the Magnoliaceae. I. Trends of specialization in the stamens. *Amer. Jour. Bot.* 39:484–497. 1952. III. Carpels. *Amer. Jour. Bot.* 47:145–155. 1960.

Carniel, K. Beiträge zur Entwicklungsgeschichte des sporogenen Gewebes der Gramineen und Cyperaceen. I. *Zea mays*. *Österr. Bot. Ztschr.* 108:228–237. 1961.

Carniel, K. Das Antherentapetum. Ein kritischer Überblick. Österr. Bot. Ztschr. 110: 145–176. 1963.

Carr, S. G. M., and D. J. Carr. The functional significance of syncarpy. Phytomorphology 11:249–256. 1961.

Cooper, D. C. Nuclear divisions in the tapetal cells of certain angiosperms. Amer. Jour. Bot. 20:358–364. 1933.

Coulter, J. M., and C. J. Chamberlain. Morphology of angiosperms. New York, D. Appleton and Company. 1912.

Cusick, F. Studies of floral morphogenesis. I. Median bisections of flower primordia in "Primula bulleyana" Forrest. Roy. Soc. Edinb. Trans. 63:153–166. 1956.

Doak, C. C. The pistil anatomy of cotton as related to environmental control of fertilization under varied conditions of pollination. Amer. Jour. Bot. 24:187–194. 1937.

Dormer, K. J. The crystals in the ovaries of certain Compositae. Ann. Bot. 25:241–254. 1961.

Dormer, K. J. The fibrous layer in the anthers of Compositae. New Phytol. 61:150–153. 1962.

Douglas, G. E. The inferior ovary. Bot. Rev. 10:125–186. 1944. II. Bot. Rev. 23:1–46. 1957.

Eames, A. J. Morphology of the angiosperms. New York, McGraw-Hill Book Company. 1961.

Eckardt, T. Vergleichende Studie über die morphologischen Beziehungen zwischen Fruchtblatt, Samenanlage und Blütenachse bei einigen Angiospermen. Neue Hefte z. Morph. No. 3:1–91. 1957.

Ehrenberg, L. Zur Kenntnis der Homologieverhältnisse in der angiospermen Blüte. Bot. Notiser 1945:438–444. 1945.

Emberger, L. La valeur morphologique et l'origine de la fleur. In: Morphogénèse. Colloque Internat. du Centre Nat. de la Rech. Sci. 28:279–295. 1951.

Erdtman, G., and Vishnu-Mittre. On terminology in pollen and spore morphology. Grana Palynol. 1:6–9. 1958.

Erickson, R. O. Cytological and growth correlations in the flower bud and anther of Lilium longiflorum. Amer. Jour. Bot. 35:729–739. 1948.

Evans, M. W., and F. O. Grover. Developmental morphology of the growing point of the shoot and the inflorescence in grasses. Jour. Agr. Res. 61:481–520. 1940.

Faegri, K. Recent trends in palynology. Bot. Rev. 22:639–664. 1956.

Frey-Wyssling, A. Die pflanzliche Zellwand. Berlin, Springer-Verlag. 1959.

Gauthier, R. The nature of the inferior ovary in the genus Begonia. Inst. Bot. Univ. Montreal Contr. 66:1–91. 1950.

Gavaudan, P., and G. Debraux. L'unité du plan de composition foliaire et floral chez les Rosa et les théories de la fleur. Acad. des Sci. Compt. Rend. 232:430–431. 1951.

Gerasimova-Navashina, E. N. Razvitie zarodyshevogo meshka, dvoĭnoe oplodotvorenie i vopros o proiskhozhdenii pokrytosemennykh. [Development of the embryo sac, double fertilization, and the problem of origin of angiosperms.] Bot. Zhur. S.S.S.R. 39:655–680. 1954.

Glück, H. Blatt- und blütenmorphologische Studien. Jena, Gustav Fischer. 1919.

Grégoire, V. La morphogénèse et l'autonomie morphologique de l'appareil floral. I. Le carpelle. Cellule 17:287–452. 1938.

Griesel, W. O. Cytological changes accompanying abscission of perianth segments of Magnolia grandiflora. Phytomorphology 4:123–132. 1954.

Guenot, T. Recherches sur la valeur morphologique de la paroi de l'ovaire infère chez Samolus Valerandi L. Bul. Sci. de Bourgogne 15:85–120. 1954.

Gumppenberg, O. von. Beiträge zur Entwicklungsgeschichte der Blumenblätter mit besonderer Berücksichtigung der Nervatur. *Bot. Arch.* 7:448–490. 1924.

Hagemann, W. Weitere Untersuchungen zur Organisation des Sprossscheitelmeristems; der Vegetationspunkt traubiger Floreszenzen. *Bot. Jahrb.* 82:273–315. 1963.

Hartl, D. Morphologische Studien am Pistill der Scrophulariaceen. *Österr. Bot. Ztschr.* 103:185–242. 1956a.

Hartl, D. Die Beziehungen zwischen den Planzenten der Lentibulariaceen und Scrophulariaceen nebst einem Excurs über Spezialisationsrichtungen der Plazentation. *Beitr. z. Biol. der Pflanz.* 32:471–490. 1956b.

Haupt, W. Untersuchungen über den Determinationsvorgang der Blütenbildung bei *Pisum sativum. Ztschr. f. Bot.* 40:1–32. 1952.

Heslop-Harrison, J. Growth substances and flower morphogenesis. *Linn. Soc. London, Jour., Bot.* 56:269–281. 1959.

Heslop-Harrison, J. Origin of exine. *Nature* 195:1069–1071. 1962.

Hiller, G. H. Untersuchungen über die Epidermis der Blühtenblätter. *Jahrb. f. Wiss. Bot.* 15:411–451. 1884.

Hillman, W. S. *The physiology of flowering.* New York, Holt, Rinehart and Winston. 1962.

Holt, I. V. Initiation and development of the inflorescences of *Phalaris arundiacea* L. and *Dactylis glomerata* L. *Iowa State Coll. Jour. Sci.* 28:603–621. 1954.

Iijima, M. A. A comparative study on tapetum. *Cytologia* 27:375–385. 1962.

Iwanami, Y. Protoplasmic movement in pollen grains and tubes. *Phytomorphology* 6:288–295. 1956.

Jacobs, W. P., and V. Raghavan. Studies in the histogenesis and physiology of *Perilla*—I. Quantitative analysis of flowering in *P. frutescens* (L.) Britt. *Phytomorphology* 12:144–167. 1962.

Jäger, I. Vergleichend-morphologische Untersuchungen des Gefässbündelsystems peltater Nektar- und Kronblätter sowie verbildeter Staubblätter. *Österr. Bot. Ztschr.* 108:433–504. 1961.

Jones, H. A. Pollination and life history studies of lettuce (*Lactuca sativa* L.). *Hilgardia* 2:452–479. 1927.

Jones, H. A., and S. L. Emsweller. Development of the flower and macrogametophyte of *Allium cepa. Hilgardia* 10:415–428. 1936.

Kato, K., and K. Watanabe. The stigma reaction. II. The presence of the stigma reaction in intra-specific and inter-generic pollinations in the Gramineae. *Bot. Mag.* (Tokyo) 70:96–101. 1957.

Kaussmann, B. Vergleichende Untersuchungen über die Blattnatur der Kelch-, Blumen-, und Staubblätter. *Bot. Arch.* 42:503–572. 1941.

Kaussmann, B. *Pflanzenanatomie.* Jena, Gustav Fischer. 1963.

Kenda, G. Stomata an Antheren. I. Anatomischer Teil. *Phyton* (*Ann. Rei. Bot.*) 4:83–96. 1952.

Kenda, G. Das Hypoderm der Geraniaceen-Kelchblätter. *Phyton* (*Ann. Rei. Bot.*) 6:207–210. 1956.

Kendall, J. N. Abscission of flowers and fruits in the Solanaceae, with special reference to *Nicotiana. Calif. Univ., Publs., Bot.* 5:347–428. 1918.

Kiesselbach, T. A. The structure and reproduction of corn. *Nebr. Agr. Expt. Sta. Res. Bul.* 161. 1949.

Kirkwood, J. E. The pollen tube in some of the Cucurbitaceae. *Torrey Bot. Club Bul.* 33:327–341. 1906.

Krishna, G. G., and V. Puri. Morphology of the flower of some Gentianaceae with special reference to placentation. *Bot. Gaz.* 124:42–57. 1962.

Kühn, G. Beiträge zur Kenntnis der intraseminalen Leitbündel bei den Angiospermen. *Bot. Jahrb.* 61:325–379. 1928.

Lam, H. J. Reflections on angiosperm phylogeny. I and II. Facts and theories. *K. Akad. van Wetensch. te Amsterdam, Afd. Natuurk., Proc.* 64:251–276. 1961.

Larson, D. A., and C. W. Lewis. Pollen wall development in *Parkinsonia aculeata. Grana Palynol.* 3:21–27. 1962.

Laurie, A., and J. Duffy. Anatomical studies of the abscission of *Gardenia* buds. *Amer. Soc. Hort. Sci. Proc.* 51:575–580. 1948.

Lawalrée, A. Histogénèse florale et végétative chez quelques Composées. *Cellule* 52:215–294. 1948.

Leinfellner, W. Der Bauplan des synkarpen Gynözeums. *Österr. Bot. Ztschr.* 97:403–436. 1950.

Leinfellner, W. Die U-förmige Plazenta als der Plazententypus der Angiospermen. *Österr. Bot. Ztschr.* 98:338–358. 1951a.

Leinfellner, W. Die Nachahmung der durch kongenitale Verwachsung entstandenen Formen des Gynözeums durch postgenitale Verschmelzungsvorgänge. *Österr. Bot. Ztschr.* 98: 403–411. 1951b.

Leinfellner, W. Die Kelchblätter auf unterständigen Fruchtknoten und Achsenbechern. *Österr. Bot. Ztschr.* 101:315–327. 1954.

Leinfellner, W. Beiträge zur Kronblattmorphologie. V. Über den homologen Bau der Kronblattspreite und der Staubblattanthere bei *Koelreuteria paniculata. Österr. Bot. Ztschr.* 102:89–98. 1955.

Leinfellner, W. Inwieweit kommt der peltat-diplophylle Bau des Angiospermen-Staubblattes in dessen Leitbündelanordnung zum Ausdruck? *Österr. Bot. Ztschr.* 103:381–399. 1956a.

Leinfellner, W. Die Gefässbündelversorgung des *Lilium*-Staubblattes. *Österr. Bot. Ztschr.* 103:346–352. 1956b.

Leroy, J. F. Étude sur les Juglandaceae. A la recherche d'une conception morphologique de la fleur femelle et du fruit. *Mém. du Muséum Nat. D'Hist. Naturelle, Ser. B., Bot.* 6:1–246. 1955.

Maheshwari, P. *An introduction to the embryology of angiosperms.* New York, McGraw-Hill Book Company. 1950.

Martens, P. Recherches sur la cuticule. IV. Le relief cuticulaire et la différenciation épidermique des organes floraux. *Cellule* 43:289–320. 1934.

Martens, P., and L. Waterkeyn. Structure du pollen "ailé" chez les Conifères. *Cellule* 62: 171–222. 1962.

McCoy, R. W. Floral organogenesis in *Frasera carolinensis. Amer. Jour. Bot.* 27:600–609. 1940.

Meeuse, A. D. J. From ovule to ovary: a contribution to the phylogeny of the megasporangium. *Acta Biotheor.* 16:127–182. 1963.

Melville, R. A new theory of the angiosperm flower: I. The gynoecium. *Kew Bul.* 16: 1–50. 1962. II. The androecium. *Kew Bul.* 17:1–66. 1963.

Morf, E. Vergleichend-morphologische Untersuchungen am Gynoeceum der Saxifragaceen. *Schweiz. Bot. Gesell. Ber.* 60:516–590. 1950.

Moseley, M. F. Morphological studies on the Nymphaeaceae. I. The nature of the stamens. *Phytomorphology* 8:1–29. 1958. II. The flower of *Nymphaea. Bot. Gaz.* 122:233–259. 1961.

Müller-Stoll, W. R., and G. Lerch. Über Nachweis, Enstehung und Eigenschaften der Kallosebildungen in Pollenschläuchen. *Flora* 144:297–334. 1957.

Nast, C. G. The comparative morphology of the Winteraceae. VI. Vascular anatomy of the flowering shoot. *Arnold Arboretum Jour.* 25:456–466. 1944.

Netolitzky, F. Anatomie der Angiospermen-Samen. In: K. Linsbauer. *Handbuch der Pflanzenanatomie.* Band 10. Lief. 14. 1926.

Nozeran, R. Contribution a l'étude de quelques structures florales. (Essai de morphologie comparée.) *Ann. des Sci. Nat., Bot.* Ser. 11. 16:1–227. 1955.

Ozenda, P. *Recherches sur les Dicotylédones apocarpiques. Contribution a l'étude des Angiospermes dites primitives.* Thesis. Paris, École Normale Supérieure. Publ. Ser. Biol. Fasc. II. 1949.

Paech, K. Colour development in flowers. *Ann. Rev. Plant Physiol.* 6:273–298. 1955.

Palser, B. F. Studies of floral morphology in the Ericales. V. Organography and vascular anatomy in several United States species of the Vacciniaceae. *Bot. Gaz.* 123:79–111. 1961.

Pankow, H. Histogenetische Studien an den Blüten einiger Phanerogamen. *Bot. Stud.* Heft 13:1–106. 1962.

Parkin, J. A plea for a simpler gynoecium. *Phytomorphology* 5:46–57. 1955.

Paterson, B. R. Studies of floral morphology in the Epacridaceae. *Bot. Gaz.* 122:259–279. 1961.

Paton, J. V. Pollen and pollen enzymes. *Amer. Jour. Bot.* 8:471–501. 1921.

Payer, J. B. *Traité d'organogénie comparée de la fleur.* Texte, 748 pp. Atlas, 154 plates. Paris, Librarie de Victor Masson. 1857.

Pearson, O. H. Study of the life history of *Brassica oleracea. Bot. Gaz.* 94:534–550. 1933.

Periasamy, K., and B. G. L. Swamy. The conduplicate carpel of *Cananga odorata. Arnold Arboretum Jour.* 37:366–372. 1956.

Pervukhina, N. V. Strobil′naĭa teoriĭa proiskhozhdeniĭa tsvetka i ee kritika. [Strobiloid theory of origin of flower and its critique.] In: *Morfologiĭa i anatomiĭa rastenii.* Vol. IV. Leningrad, Izdatel′stvo Akademii Nauk SSSR. 1957*a.*

Pervukhina, N. V. Rol′ telomnoĭ teorii v razvitii vzglĭadov na tsvetok pokrytosemennykh. [Role of telome theory in the development of views on the flower of angiosperms.] In: *Morfologiĭa i anatomiĭa rastenii.* Vol. IV. Leningrad, Izdatel′stvo Akademii Nauk SSSR. 1957*b.*

Pervukhina, N. V. Priroda nizhneĭ zaviĭazi zontichnykh i nekotorye voprosy "teorii tsvetka." [Nature of the inferior ovary in Umbelliferae and some questions on the "theory of the flower."] In: *Morfologiĭa i anatomiĭa rastenii.* Vol. V. Leningrad, Izdatel′stvo Akademii Nauk SSSR. 1962.

Pfeiffer, H. Die pflanzlichen Trennungsgewebe. In: K. Linsbauer. *Handbuch der Pflanzenanatomie.* Band 5. Lief. 22. 1928.

Plantefol, L. L'ontogénie de la fleur. *Ann. des Sci. Nat., Bot.* Ser. 11. 9:35–186. 1948.

Pope, M. N. The course of the pollen tube in cultivated barley. *Amer. Soc. Agron. Jour.* 38:432–440. 1946.

Prior, P. V. Development of the helicoid and scorpioid cymes of *Myosotis laxa* Lehm. and *Mertensia virginica* L. *Iowa Acad. Sci. Proc.* 67:76–81. 1960.

Puri, V. The role of floral anatomy in the solution of morphological problems. *Bot. Rev.* 17:471–553. 1951.

Puri, V. Placentation in angiosperms. *Bot. Rev.* 18:603–651. 1952*a.*

Puri, V. Foral morphology and inferior ovary. *Phytomorphology* 2:122–129. 1952*b.*

Puri, V. The classical concept of angiosperm carpel: a reassessment. *Indian Bot. Soc. Jour.* 40:511–524. 1961.

Rao, V. S. *Floral anatomy.* New York, Scholars Library. 1961.

Rao, V. S., K. Sirdeshmukh, and M. G. Sardar. The floral anatomy of the Leguminosae. *Jour. Univ. Bombay Sect. B.* 26:65–138. 1958.

Reeves, R. G. Partition wall formation in the pollen mother cells of *Zea Mays. Amer. Jour. Bot.* 15:144–122. 1928.

Renner, O., and G. Preuss-Herzog. Der Weg der Pollenschläuche im Fruchtknoten der Oenotheren. *Flora* 36:215–222. 1943.

Rickett, H. W. The classification of inflorescences. *Bot. Rev.* 10:187–231. 1944.

Roberts, R. H. Blossom bud development and winter hardiness. *Amer. Jour. Bot.* 24:683–685. 1937.

Rohweder, O. Anatomische und histogenetische Untersuchungen an Laubsprossen und Blüten der Commelinaceen. *Bot. Jahrb.* 82:1–99. 1963.

Roth, I. Die Histogenese der Integumente von *Capsella bursa-pastoris* und ihre morphologische Deutung. *Flora* 145:212–235. 1957.

Roth, I. Histogenese und morphologische Deutung der Plazenta von *Primula*. *Flora* 148:129–152. 1959a.

Roth, I. Histogenese und morphologische Deutung der Kronblätter von *Primula*. *Bot. Jahrb.* 79:1–16. 1959b.

Saad, S. I. Sporoderm stratification: the "medine," a distinct layer in pollen wall. *Pollen et Spores* 5:17–39. 1963.

Satina, S. Periclinal chimeras in *Datura* in relation to development and structure (A) of the style and stigma (B) of calyx and corolla. *Amer. Jour. Bot.* 31:493–502. 1944.

Satina, S., and A. F. Blakeslee. Periclinal chimeras in *Datura stramonium* in relation to development of leaf and flower. *Amer. Jour. Bot.* 28:862–871. 1941.

Satina, S., and A. F. Blakeslee. Periclinal chimeras in *Datura* in relation to the development of the carpel. *Amer. Jour. Bot.* 30:453–462. 1943.

Sattler, R. Zur frühen Inflorescenz- und Blütenentwicklung der Primulales sensu lato mit besonderer Berücksichtigung der Stamen-Petalum-Entwicklung. *Bot. Jahrb.* 81:358–396. 1962.

Savchenko, M. I. O prirode plodolistika pokrytosemennykh rasteniĭ. [On the nature of the carpel in angiosperms.] In: *Morfologiíà i anatomiíà rastenii.* Vol. IV. Leningrad, Izdatel'stvo Akademii Nauk SSSR. 1957.

Schaeppi, H., and K. Frank. Vergleichend-morphologische Untersuchungen über die Karpellgestaltung, insbesondere der Plazentation bei Anemonen. *Bot. Jahrb.* 81:337–357. 1962.

Schnarf, K. Embryologie der Angiospermen. In: K. Linsbauer. *Handbuch der Pflanzenanatomie.* Band 10. Lief. 21. 1927; Lief. 23. 1928.

Schnarf, K. *Vergleichende Embryologie der Angiospermen.* Berlin, Gebrüder Borntraeger. 1931.

Schoch-Bodmer, H. Beiträge zur Kenntnis des Streckungswachstums der Gramineen-Filamente. *Planta* 30:168–204. 1939.

Schoch-Bodmer, H. Über das Spitzenwachstum der Pollenschläuche. *Schweiz. Bot. Gesell. Ber.* 55:154–168. 1945.

Schoch-Bodmer, H., and P. Huber. Die Ernährung der Pollenschläuche durch das Leitgewebe. (Untersuchungen an *Lythrum Salicaria* L.) *Naturf. Gesell. in Zürich, Vrtljschr.* 92:43–48. 1947.

Sharman, B. C. The biology and developmental morphology of the shoot in the Gramineae. *New Phytol.* 46:20–34. 1947.

Sharman, B. C. Developmental anatomy of the stamen and carpel primordia in *Anthoxanthum odoratum*. *Bot. Gaz.* 121:192–197. 1960a.

Sharman, B. C. Development of the inflorescence and spikelets of *Anthoxanthum odoratum* L. *New Phytol.* 59:60–64. 1960b.

Smith, F. H., and E. C. Smith. Floral anatomy of the Santalaceae and some related forms. *Oreg. State Monographs, Stud. in Bot.* 5. 1942a.

Smith, F. H., and E. C. Smith. Anatomy of the inferior ovary of *Darbya*. *Amer. Jour. Bot.* 29:464–471. 1942b.

Sosa-Bourdouil, C. Sur le développement comparé des organes floraux. *Soc. Bot. de France Bul.* 92:154–158. 1945.

Sporne, K. R. Some aspects of floral vascular systems. *Linn. Soc. London, Proc.* 169:75–84. 1958.

Sprotte, K. Untersuchungen über Wachstum und Nervatur der Fruchtblätter. *Bot. Arch.* 40:463–506. 1940.

Sterling, C. The affinities of *Prinsepia* (Rosaceae). *Amer. Jour. Bot.* 50:693–699. 1963.

Surkov, V. A. Ontogenez i morfologicheskaĩa priroda chlenov tsvetka u zlakov. *Bot. Zhur. S.S.S.R.* 46:1134–1143. 1961.

Takats, S. T. An attempt to detect utilization of DNA breakdown products from the tapetum for DNA synthesis in the microspores of *Lilium longiflorum*. *Amer. Jour. Bot.* 49:748–758. 1962.

Takhtajan, A. *Die Evolution der Angiospermen*. Jena, Gustav Fischer. 1959.

Tepfer, S. S. Floral anatomy and ontogeny in *Aquilegia formosa* var. *truncata* and *Ranunculus repens*. *Calif. Univ., Publs., Bot.* 25:513–648. 1953.

Trapp, A. Zur Morphologie und Entwicklungsgeschichte der Staubblätter sympetaler Blüten. *Bot. Stud.* Heft 5:1–93. 1956.

Trécul, A. Recherches sur l'ordre d'apparition des premiers vaisseaux dans les organes aériens. *Ann. des Sci. Nat., Bot.* Ser. 6. 12:251–381. 1881.

Troll, W. Die morphologische Natur der Karpelle. *Chron. Bot.* 5:38–41. 1939.

Tucker, S. C. Ontogeny of the inflorescence and the flower of *Drimys winteri* var. *chilensis*. *Calif. Univ., Publs., Bot.* 30:257–336. 1959.

Tucker, S. C. Phyllotaxis and vascular organization of the carpels in *Michelia fuscata*. *Amer. Jour. Bot.* 48:60–71. 1961.

Tupý, J. Callose formation in pollen tubes and incompatibility. *Biol. Plant.* (Praha) 1:192–198. 1959.

Unruh, M. Blattnervatur und Karpellennervatur. Kleiner Beitrag zur morphologischen Deutung des Karpels. *Beitr. z. Biol. der Pflanz.* 27:232–241. 1941.

Venkatesh, C. S. The structure and dehiscence of the anther in *Memecylon* and *Mouriria*. *Phytomorphology* 5:435–440. 1955.

Venkatesh, C. S. The form, structure and special walls of dehiscence of anthers of *Cassia*—III. Subgenus *Senna*. *Phytomorphology* 7:253–273. 1957.

Walton, J. L'évolution des téguments et de la protection du sporange. *Ann. Biol.* 28:C.129–C.133. 1952.

Waterkeyn, L. Étude des dépôts de callose au niveau des parois sporocytaires au moyen de la microscopie de fluorescence. *Acad. des Sci. Compt. Rend.* 252:4025–4027. 1961.

Watson, L. The taxonomic significance of stomatal distribution and morphology in Epacridaceae. *New Phytol.* 61:36–40. 1962.

Wilkinson, A. M. Floral anatomy and morphology of *Triosetum* and of the Carprifoliaceae in general. *Amer. Jour. Bot.* 36:481–489. 1949.

Wilson, C. L. The telome theory and the origin of the stamen. *Amer. Jour. Bot.* 29:759–764. 1942.

Wilson, C. L., and T. Just. The morphology of the flower. *Bot. Rev.* 5:97–131. 1939.

Witkus, E. R. Endomitotic tapetal cell divisions in *Spinacia*. *Amer. Jour. Bot.* 32:326–330. 1945.

Wodehouse, R. P. *Pollen grains*. New York, McGraw-Hill Book Company. 1935.

Wodehouse, R. P. Evolution of pollen grains. *Bot. Rev.* 2:67–84. 1936.

Wunderlich, R. Über das Antherentapetum mit besonderer Berücksichtigung seiner Kernzahl. *Österr. Bot. Ztschr.* 101:1–63. 1954.

Yampolsky, C. The cytology of the abscission zone in *Mercurialis annua*. *Torrey Bot. Club Bul.* 61:279–289. 1934.

19

The Fruit

Fertilization of the egg commonly induces the development of a seed from the ovule and of a fruit from the ovary. (The style and stigma usually wither after pollination.) Formation of a fruit may also occur without seed development and without fertilization, a phenomenon known as parthenocarpy (from the Greek *parthenos*, virgin, and *carpos*, fruit).

In view of the variation in the structure of flowers, the fruits are highly diversified in their morphology. Furthermore, fruits derived from the same type of flower may follow distinctive ontogenies. Then, changes leading to fruit development are not restricted to the ovary but often involve noncarpellary (accessory) parts of the flower, such as the receptacle in the strawberry, the calyx in the mulberry, the bracts in the pineapple, and the floral tube or the receptacle in epigynous flowers. Another complication in fruit development results from the various aggregations of separate carpels into unit structures. These carpels may be derived from one flower (aggregate fruit from an apocarpous gynoecium) or from several flowers (multiple fruit). A concomitant of the complexity of fruit structure is the lack of general agreement on the definition and classification of fruits.

Botanically, the classification of fruits should reflect the fundamental structure of the flowers from which the fruits are derived. An example of such classification is that of Winkler (1939). It treats the fruit as the product of the entire gynoecium and any floral parts that may be associated with the gynoecium in the fruiting stage. Winkler's classification of the fruits is based primarily on four features: (1) choricarpelly

586

(carpels free, Sammelfrucht or *aggregate fruit*); (2) syncarpelly (carpels united, Einheitsfrucht or *unit fruit*); (3) epichlamydy (hypogynous flower, Freifrucht or *free fruit*); (4) hypochlamydy (perigynous and epigynous flower, Becherfrucht or *cup fruit*). An individual carpel in an aggregate fruit forms the *fruitlet* (Winkler, 1940). Features 1 and 2 may be combined with either 3 or 4. Examples of some of the combinations are the choricarpellous epichlamydous fruit of *Ranunculus* (fig. 19.1*A*); the syncarpellous epichlamydous fruit of *Solanum* (fig. 19.1*B*); the choricarpellous hypochlamydous fruit of *Rosa* (fig. 19.1*C*); and the

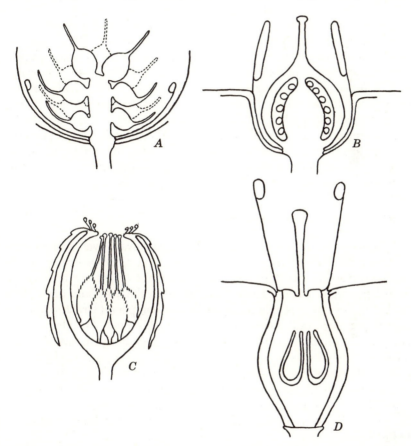

FIG. 19.1. Longitudinal sections of flowers from which the following types of fruits, according to Winkler (1939), are derived: *A, Ranunculus,* aggregate free fruit, choricarpellous, epichlamydous; *B, Solanum,* unit free fruit, syncarpellous, epichlamydous; *C, Rosa,* aggregate cup fruit, choricarpellous, hypochlamydous; *D, Cornus,* unit cup fruit, syncarpellous, hypochlamydous.

syncarpellous hypochlamydous fruit of *Cornus* (fig. 19.1*D*). In this scheme the choricarpellous epichlamydous gynoecium is considered the most primitive, the syncarpellous hypochlamydous the most advanced; and the follicle is regarded as the most primitive type of gynoecial unit (Juhnke and Winkler, 1938; Winkler, 1939). Combinations of other characters, especially the arrangement and manner of union of carpels and the nature of the fruit wall and its dehiscence, may be used for further subdivisions of the larger groupings (Baumann-Bodenheim, 1954).

A classification based on presumed evolution of fruit types (Levina, 1961) recognizes four basic types of fruits: apocarpous, syncarpous in the narrow sense (axile placentation), paracarpous (parietal placentation), and lysicarpous (free central placentation). These four types are subdivided according to their evolutionary modifications. The apocarpous type shows two trends of such modification: polycarpy to monocarpy and many-seeded to single-seeded condition. The trends for the syncarpous (wide sense) types are hypogyny to epigyny and many-seeded to single-seeded condition. The reduction in the number of seeds is regarded as particularly significant because the development of the one-seeded condition has resulted in many morphologic specializations related to the protection of seeds, their dissemination, and other functions of the fruit.

FRUIT WALL AND PERICARP

When an ovary develops into a fruit, the ovary wall (carpel wall) becomes the *pericarp* (in Greek *peri*, around, and *carpos*, fruit). In the unit cup fruits (fruits derived from syncarpous epigynous flowers) the pericarp merges more or less completely with the accessory parts of the fruit. No term is available to designate the compound structure consisting of pericarp and accessory parts. In this book Winkler's (1939) definition of the fruit (product of gynoecium together with any accessory parts that may be associated with it in fruit) is adopted, and the term *fruit wall* is applied to the pericarp of fruits derived from superior ovaries and to the combination of pericarp and noncarpellary parts found in fruits originating from inferior ovaries. Some authors widen the term pericarp to include the noncarpellary tissue (Baumann-Bodenheim, 1954).

In the flower, the ovary wall consists of little-differentiated parenchyma cells, vascular tissues, and outer and inner epidermal layers. During maturation the pericarp frequently undergoes an increase in the number of cells. Its ground tissue either remains relatively homogeneous and parenchymatic or differentiates into parenchyma and sclerenchyma. The pericarp may become differentiated into three parts, more or less distinct morphologically: the exocarp or epicarp, the mesocarp, and the endocarp; that is, the outer, median, and inner layers, respectively. Sometimes only an exocarp and an endocarp may be distinguished, or the exocarp and the endocarp may be limited to the outer and inner epidermal layers of the ovary wall. The terms applied to the different layers of the pericarp have little value for showing the origin of the various tissues of the fruit wall, but they are useful for the description of mature fruits. A stricter, but no less artificial, definition employs the terms epicarp and endocarp to the outer and inner epidermal layers exclusively (Sterling, 1953). The terms exocarp, mesocarp, and endocarp are sometimes used for describing a fruit wall in which accessory and carpellary tissues are not distinguishable (pepo of Cucurbitaceae).

The fruit wall encloses the ovarian locule in which the seed or seeds are borne (fig. 19.1*B*). A vascular system, with characteristic variations in the different types of fruits, is present in the pericarp and the other parts of the fruit (fig. 18.1*A*). The basic arrangement of the vascular system is considered in chapter 18. During the development of fruits the vascular tissues are more or less increased in amount through a differentiation of additional vascular bundles within the ground parenchyma.

HISTOLOGY OF THE FRUIT WALL

Two structural types of fruit walls are recognized, the parenchymatic fleshy, often succulent type and the sclerenchymatic dry type. With reference to the structure of the fruit wall, fruits are referred to as fleshy or dry. They may be dry dehiscent fruits if the fruit wall splits open at maturity, or dry indehiscent if the fruit wall remains closed. Dry or fleshy, dehiscent or indehiscent fruit walls occur in fruits derived from both superior and inferior ovaries.

Dry Fruit Wall

Dehiscent Fruit Wall. If the ovary differentiating into a dry fruit contains several ovules, it commonly dehisces at maturity. Such a fruit may develop from a single carpel (follicle, legume) or from several united carpels (capsule). The pericarp of follicles usually has a relatively simple structure. There may be a narrow exocarp of thick-walled cells and a thin-walled parenchymatic mesocarp and endocarp. The three main longitudinal vascular bundles (one median and two lateral) and the transversely oriented branches from the main bundles become enclosed in sclerenchymatic sheaths. As the fruit approaches maturity, the pericarp dries. Apparently the differential drying of the parenchymatic and sclerenchymatic tissues of the pericarp creates tensions that cause the splitting of the follicle along the line where the margins of the carpel became fused during the ontogeny of the flower.

The legume commonly has a more complicated structure than the follicle (Fahn and Zohary, 1955; Monsi, 1943). In some Leguminosae, for example, the ovary wall shows a considerable increase in the number of cells after fertilization and then matures into a pericarp with a thick-walled exocarp, a thin-walled parenchymatic mesocarp, and a highly sclerified endocarp. The exocarp may be represented by the epidermis *Pisum, Vicia*), or it may include a subepidermal layer of elongated cells with thick walls (*Phaseolus, Glycine*). The sclerenchymatic endocarp consists of several rows of thick-walled cells oriented at an angle to the long axis of the fruit and is covered, internally, by a thin-walled epidermis. The thick-walled part of the endocarp may be differentiated into two distinct layers. In one of these layers, located next to the mesocarp, the cellulose microfibrils in the cell walls are oriented in helices of low pitch; in the other the helices are steeply pitched. This cell-wall structure in the endocarp is interpreted as a feature facilitating the dehiscence of the fruit. As a result of the different orientation of the microfibrils in the two wall layers, the latter undergo their strongest contraction in different planes. The two lines of dehiscence, one following the line of union of carpel margins and the other located in the region of the median bundle, may consist of conspicuously thin-walled parenchyma cells.

An ovary wall maturing into the pericarp of a capsule may have but little increase in the number of cells, as in tobacco; or numerous cell divisions may occur before the pericarp matures, as in certain lilies. The pericarps of capsules have both sclerenchymatic and parenchymatic tissues in variable distributions. The pericarp of *Linum usitatissimum*, for example, has an exocarp of sclerified and lignified cells and a mesocarp and an endocarp of parenchymatic cells. That of *Nicotiana tabacum* shows, in contrast, a thick-walled endocarp, two or three cells in thickness, and parenchymatic lacunose exocarp and mesocarp. The dehiscence of capsules may be longitudinal (Kaden, 1962) and occurs either along the lines of juncture of the carpels (*Convolvulus;* septicidal dehiscence, really separation into the individual carpels; Stopp, 1950) or along the plane of the median bundle of each carpel (loculicidal dehiscence; *Allium*, fig. 18.11E). In the two examples just cited, the longitudinal split extends the whole length of the pericarp. In some capsules, as in tobacco, dehiscence is confined to the terminal part of the fruit. A few plants have circumscissile dehiscence (from the Latin *circumscissus,* cut around), that is, dehiscence by a transverse lid (*Portulaca, Plantago;* Subramanyam and Raju, 1953). Such dehiscence is made possible through the development of a zone of mechanical weakness between the lid and the base (Rethke, 1946). This zone may differ from the adjacent fruit parts in cell number, cell size, density of protoplasts, wall thickness, and various combinations of these features. A softening of middle lamella and cell wall before dehiscence has also been reported (Holden, 1956). The capsule of *Trematolobelia* does not dehisce but develops pores (Carlquist, 1962). Parenchyma tissue disintegrates and the pores—the meshes within the coarse network of highly sclerenchymatic vascular tissue—are exposed. The seeds fall out through the pores.

Indehiscent Fruit Wall. When the ovary contains a single ovule, it usually develops into an indehiscent fruit. The pericarp of many indehiscent one-seeded fruits resembles a seed coat in structure. In fact, commonly the seed coat of such fruits acquires no mechanical character- istics or is more or less obliterated during fruit development. If the pericarp and the testa (seed coat) are adherent in the fruit, the fruit is a grain or caryopsis (most Gramineae). If the seed is attached to the peri- carp at one point only, the fruit is an achene. Examples of achenes derived from hypogynous flowers occur in Ranunculaceae. The term achene is also used with reference to the bicarpellate fruit of Compositae in which the ovary is inferior and, therefore, the pericarp is confluent with the floral tube.

The caryopses of Gramineae show certain conspicuous differences in the development of their fruit coats. Most commonly, as in *Triticum* and *Hordeum* (Krauss, 1933), the protective layer is developed in the pericarp. The two parts that compose the grain coats, the pericarp and the seed coat, are distinct in the ovary before fertilization. The ovary wall of wheat consists of the following cell layers, beginning from the outside: the outer epidermis, one cell layer in depth; many layers of colorless parenchyma cells; chlorophyll-containing parenchyma tissue consisting of one or two layers of cells over most of the grain and of several layers in the region where the grain is grooved; one layer of small cells of the inner epidermis. At this time both integuments are intact, each consisting of two layers of cells. The nucellus also is present and consists of several layers of thin-walled cells bounded by a distinct nucellar epidermis.

The changes in the ovary wall begin with the inner epidermis which partly disintegrates. The remaining cells elongate parallel with the long axis of the grain, and their walls lignify (fig. 19.2, tube cell). The

chlorenchyma cells elongate transversely with respect to the long axis
of the grain; their chlorophyll disappears and their walls thicken and
lignify (fig. 19.2, cross cells). The parenchyma outside the chlorenchyma
is partly resorbed, and the remaining spaces are filled with air (fig. 19.2,
crushed parenchyma). One to four layers of this parenchyma persist
in the mature grain but become compressed (fig. 19.2, subepidermal
layer). The outer epidermis is compressed also and is covered with a
cuticle.

The nucellus and the integuments in wheat undergo even more
profound changes than the ovary wall. The nucellar tissue, with the
exception of the epidermis, is absorbed by the enlarging endosperm and
embryo. The nucellar epidermis is eventually compressed into a hyaline
layer covered with a cuticle (fig. 19.2, crushed nucellar cells). The inner
layer of the inner integument becomes compressed (fig. 19.2, inner layer
of ii). The outer layer of this integument is crushed into a hyaline
membrane covered with a cuticle (fig. 19.2, cuticular layer). The outer
integument disintegrates. Within and next to the grain coats lies
the proteinaceous endosperm layer, the aleuron layer, which encloses
the starchy endosperm (fig. 19.2A, C). The bran of wheat includes the

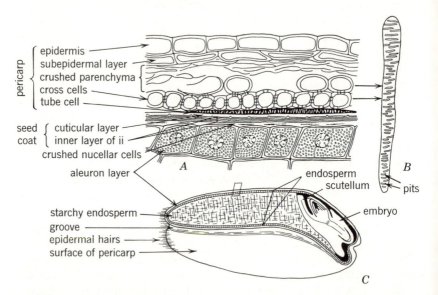

FIG. 19.2. The caryopsis (C) of *Triticum* (wheat) and its pericarp (A, B). The caryopsis was
cut longitudinally parallel with the groove. Small rectangle in C, above, indicates location of
section in A. Cross cells are elongated perpendicular to long axis of grain. B, one cross cell
from a transection of grain. Tube cell is part of inner epidermis of pericarp. Letters ii signify
inner integument. (A, B, ×300; C, ×7.)

pericarp, remnants of the integuments and the nucellus, and the aleuron layer (Bradbury et al., 1956a).

The degree of developmental modification of the seed coats and the pericarp varies in different cereals (Narayanaswami, 1955). In the grain of *Zea* (Kiesselbach and Walker, 1952), the outer pericarp is much compressed and consists of cells with thick, pitted walls. The inner pericarp remains thin-walled and is much distorted. The central part of the pericarp disintegrates. The integuments also disintegrate completely, but the nucellar epidermis is retained as a thick-walled layer showing fatty properties and covered with a cuticle (Randolph, 1936).

The cuticular layers of the caryopsis located outside the nucellus are of importance with regard to the water absorption by the grain. The cuticles are derived from the integuments and the nucellar epidermis, possibly also from the inner epidermis of the pericarp. The remnants of the seed coats combined with the cuticles are sometimes referred to as the semipermeable layer (Bradbury et al., 1956b). Tests with fluorochromes indicate reduced penetrability of this layer (Ziegenspeck, 1952).

The achene type of fruit may be illustrated by the fruit of lettuce, *Lactuca sativa*, a composite. The lettuce achene is derived from an inferior ovary (figs. 18.12, 19.3). The postulated adnation between the carpels and the floral tube is so complete that throughout the development of the fruit wall no distinction can be made between the pericarp and the floral tube (Borthwick and Robbins, 1928).

In an ovule taken before anthesis the integument consists of many layers of cells. The innermost, next to the embryo sac, constitutes the integumentary tapetum. (The nucellus is largely absorbed during the development of the gametophyte.) The fruit wall, composed of rather small parenchyma cells, is in contact with the ovule. At this early stage some of the inner fruit-wall cells are already disorganized and have left cavities (fig. 19.4A, B). After anthesis, while the achene enlarges the integument is increasing in thickness but is also disorganizing next to the integumentary tapetum (fig. 19.4C, D). Eventually this tapetum and all the parenchyma of the integument are destroyed (figs. 19.3A–C; 19.5B, C). Only the outer epidermis of the integument persists and develops thick walls (fig. 19.5D). The vascular bundle located in the integument is also identifiable in the mature fruit. The outer layer of the endosperm develops into a compact layer. This layer and another beneath it are retained in the mature fruit and develop thick walls (fig. 19.5). A cuticle becomes conspicuous between the endosperm and all that remains of the integument (fig. 19.5D). It may be a combination of the nucellar and the integumentary cuticles (Schnarf, 1927). The

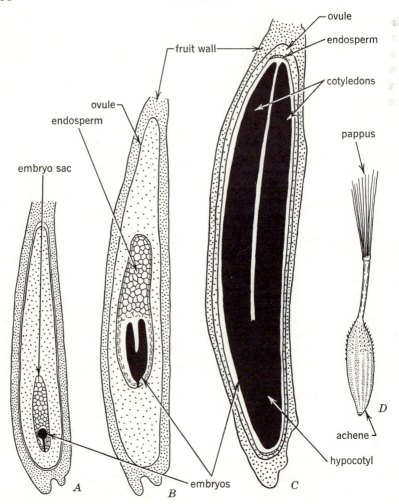

FIG. 19.3. Development of embryo and fruit (achene) in *Lactuca sativa* (lettuce). *A–C*, longitudinal sections of achenes with embryos before (*A*) and after (*B, C*) emergence of cotyledons. Details: increase in size of the embryo sac, its encroachment upon ovule, development of endosperm in embryo sac, and replacement of endosperm by embryo. *D*, mature achene with pappus. (*A–C*, ×33; *D*, ×6. After Jones, *Hilgardia* 2, 1927.)

inner layers of the fruit wall become completely disorganized, but the outer layers persist. Certain parts of the remaining layers project in the form of ribs and develop into sclerenchyma (fig. 19.5). The fruit-wall cells between the ribs are large and have thin, slightly lignified walls. In the mature achene all the persisting layers are compressed close together and their identity becomes obscured (fig. 19.5*D*).

Fleshy Fruit Wall

Many ovaries, monocarpellate or multicarpellate, develop into indehiscent fruits with fleshy fruit walls. The fleshy fruit character is considered to be relatively new from the evolutionary aspect (Pijl, 1955). As with the dry fruit wall, the fleshy fruit wall may consist either of the ovary wall (a pericarp) or of such a wall fused with the noncarpellary tissue in which it is imbedded (cup fruits of Winkler, 1939). According

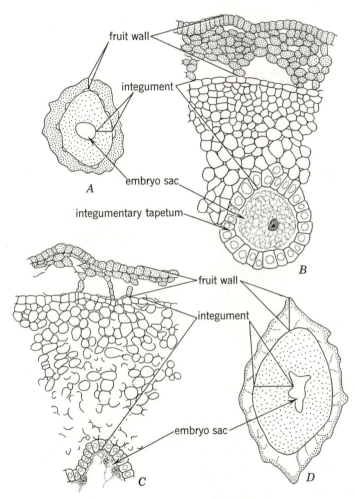

FIG. 19.4. Development of *Lactuca sativa* (lettuce) achene. A, D, entire transections of ovaries; B, C, details of transections. A, B, sampled 2 hours before anthesis; C, D, 3 days after anthesis. (A, D, ×45; B, C, ×215. After Borthwick and Robbins, *Hilgardia* 3, 1928.)

to the type of fleshy fruit, the entire ovary wall or the external part of it differentiates into a parenchymatic tissue whose cells retain their protoplasts in the mature fruit. An immature fleshy fruit wall has a firm texture, but it becomes softer as the fruit ripens. Chemical changes in the cell contents and in the structure of the walls are responsible for the softening (Reeve, 1959). The cells may even become dissociated from each other.

The ripening of the fruit wall is generally accompanied by color

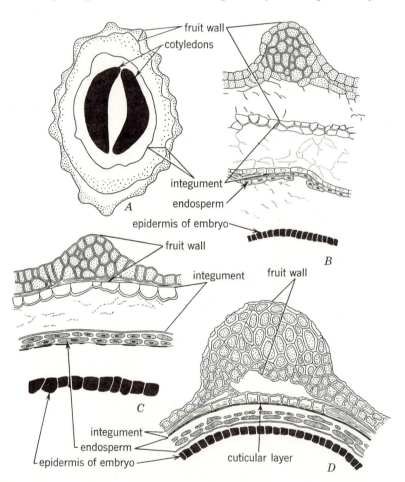

FIG. 19.5. Development of *Lactuca sativa* (lettuce) achene. *A*, transection, 7 days after anthesis; *B–D*, details of transections taken the following numbers of days after anthesis: 7 (*B*), 10 (*C*), and 19 (*D*). *D* is from a mature achene. (*A*, ×45; *B–D*, ×215. After Borthwick and Robbins, *Hilgardia* 3, 1928.)

changes. Immature fruits have numerous chloroplasts in the outermost cells and are consequently green. The disappearance of chlorophyll and the development of carotenoid pigments induces a change to a yellow, orange, or red color (tomato, *Pyracantha*). Ripening fruits may form anthocyanins which give the tissue a red, purple, or blue color. These pigments may be distributed in the entire fruit wall, as in some cherries, or they may be restricted to peripheral parts of the fruit wall, as in the plum or Concord grape. The outer epidermis frequently accumulates tannins.

The ripening of the fruit is associated with changes in the composition of carbohydrates (Miller, 1958). In some fruits (apple, pear, banana), starch accumulates during ripening but later disappears, whereas sucrose increases in amount. In fruits without reserve starch (plum, peach, citrus), ripening is characterized by a decrease in acid content and an increase in sugars. In the ripening avocado, the sugar content decreases and fat content increases.

If the entire ground tissue develops into a fleshy tissue, the fruit is a berry. All the fleshy tissue of the berry may originate from the ovary wall, as in the grape; or the main body of the mature fruit may consist of the placenta, as in the tomato. In the development of the tomato berry there are but few cell divisions in the ovary wall and in the septa dividing the ovary into locules. In contrast, each placenta shows active multiplication of cells and an increase in volume, so that the locule is filled with fleshy tissue and the seed is completely imbedded. The placental tissue constitutes the pulp of the mature fruit. It undergoes a mucilaginous degeneration during the maturation of the fruit (Czaja, 1963). The pericarp has a cutinized epidermis and subepidermal collenchyma. The inner tissue is parenchymatic, and the inner epidermis is thin-walled. When berries have definite locules, the inner epidermis of the fruit wall may have thick walls and sometimes a cuticle (Kraus, 1949). In some berries the locules are filled by proliferations of pericarp wall as well as of placentae (*Physalis Alkekengi*); in others, by the growth of partition walls (*Bryonia dioica;* Kraus, 1949).

The citrus fruit, the hesperidium, is closely related to a berry. It develops from a multicarpellate ovary with axile placentation. As the fruit develops, cell multiplication occurs throughout the ovary, and eventually the pericarp becomes differentiated into three layers (Scott and Baker, 1947). The outer, the exocarp or flavedo, is compact, collenchymatic, and contains oil glands. The mesocarp, the albedo, is spongy because of the loose connection among cells. The endocarp is compact and gives rise to the juice sacs which at maturity fill the locules.

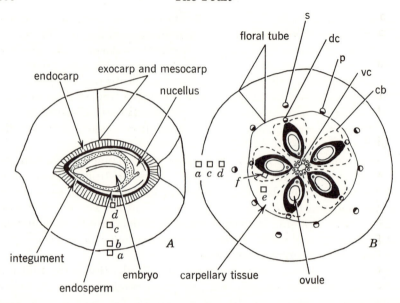

FIG. 19.6. Longitudinal section of peach fruit (*A*) and transection of apple fruit (*B*). Details: *s*, sepallary bundles; *p*, petallary bundles; *dc*, dorsal carpellary bundles; and *vc*, ventral carpellary bundles; *cb* are carpellary bundles connecting dorsal with ventral carpellary bundles. The rectangles accompanied by small letters indicate positions of sections shown in figs. 19.7 and 19.8. (*A*, after Lee and Tuckey, *Bot. Gaz.* 104, 1942; *B*, after MacDaniels *N.Y. Agr. Expt. Sta. Mem.* 230, 1939.)

The juice sacs develop as multicellular hairs (Hartl, 1957). The distal part of each hair becomes enlarged, then the interior cells break down, and the cavity becomes filled with juice. The basal part of the hair develops into a stalk supporting the juice sac.

The pepo of the Cucurbitaceae is a berry-like fruit derived from an inferior ovary. The fruit wall has a massive mesocarp, heterogeneous in structure (Matienko, 1957). The inner and the outer epidermal layers constitute the exocarp and the endocarp. The mesocarp consists of the following tissues: collenchyma; parenchyma which may contain chloroplasts; sclerenchyma in some genera (watermelon, gourd); fleshy parenchyma; juicy parenchyma in juicy species. Carotenoid pigments may be present in the juicy layer. The vascular bundles occur in the fleshy mesocarp. In some genera (watermelon, gourd), the inner epidermis clings to the seed as a transparent membrane. Some Cucurbitaceae fruits develop periderm. In *Cucumis*, for example, the periderm forms a corky net. This development appears to be a response to the cracking of the fruit surface (Meissner, 1952).

If the ovary wall matures into a pericarp with a conspicucus stony endocarp and a fleshy mesocarp, the fruit is called a drupe. In one example of drupe, the peach (*Prunus persica*), the pericarp of the mature fruit is composed of three distinct parts (fig. 19.6A): the thin exocarp or skin, the thick fleshy mesocarp, and the stony endocarp. The exocarp includes the epidermis and several layers of collenchyma beneath it. The epidermis bears a cuticle and numerous unicellular hairs (fig. 19.7A). The fleshy mesocarp consists of loosely packed parenchyma cells which increase in size from the periphery toward the interior (fig. 19.7B, C). In the same direction the cells change in shape from ovoid, with the longest diameter parallel to the surface of the fruit, to cylindrical, with the longest diameter in the radial direction. The smaller cells near the periphery contain most of the chloroplasts in the immature fruit (fig. 19.7B). Chemical and histologic differences in the mesocarp differentiate the "melting-fleshed" types of peaches from the canning "cling" types. The former show a decrease of wall thickness and eventual disorganization of cells as the fruit ripens. The endocarp is composed of tightly packed sclereids and forms the pit, or stone, of the fruit (fig. 19.7D). The outer surface of the pit is very rough and pitted. Vascular bundles occur within channels in the endocarp. Branch bundles diverge from this system into the mesocarp. The endocarp is the first part of the fruit that reaches its maximum size (Ragland, 1934).

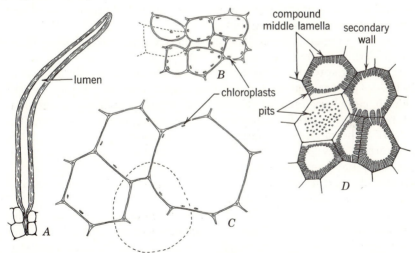

FIG. 19.7. Tissue elements of *Prunus* (peach) fruit. Drawings from a longitudinal section of a fruit about 3 cm in diameter. (See fig. 19.6A.) A, epidermal hair. B, C, parenchyma of mesocarp taken closer to (B) and farther from (C) the surface of fruit. D, group of sclereids from endocarp. (All, ×300. Slide by R. M. Brooks.)

A stony endocarp composed of elongated curved sclereids, varying in orientation in different layers, occurs in the drupelets of raspberry (*Rubus;* Reeve, 1954*b*). The succulent pulp constitutes the mesocarp. The exocarp, represented by the epidermis, forms hairs that hold the drupelets together at maturity (Reeve, 1954*a*).

A fleshy fruit derived from an inferior ovary is here illustrated by the apple fruit or pome (*Pyrus malus*), whose development has been investigated (MacArthur and Wetmore, 1939, 1941; Smith, 1940, 1950). Most workers accept the appendicular interpretation of the extracarpellary part of the pome fruit and describe the flesh of the fruit as composed of the floral tube and carpellary tissue. As seen in transections of the fruit, the floral tube region consists of fleshy parenchyma with a ring of vascular bundles (fig. 19.6*B*). There are five petallary and five sepallary bundles alternating with one another. Branches from these bundles permeate the parenchyma as an anastomosing system. The subepidermal parenchyma of the floral tube region consists of several layers of tangentially elongated cells with thick walls (fig. 19.8*A*, *B*). No intercellular spaces occur here until the late developmental stages. The ground parenchyma located somewhat deeper has abundant intercellular spaces (fig. 19.8*C*). The still deeper-lying ground parenchyma consists of cells roughly elliptical in outline and having an approximately radial orientation (fig. 19.8*D*). This part of the fruit shows particularly intensive growth during development, first by cell division and cell enlargement, then by cell enlargement only.

The ovary region (the core) consists of five carpels (fig. 19.6*B*). These are folded, but their margins are not fused. In some varieties the margins later spread and curve away from the center of the fruit (Bell, 1940). The vascular system of this region consists of five median (dorsal) carpellary bundles outside and opposite each locule and ten lateral (ventral) carpellary bundles forming a ring inside the locules (fig. 19.6*B*). The median and lateral bundles anastomose with each other and form a network, chiefly following the outline of the locules. The boundary between the ovary and the floral tube may or may not be discernible, and it occurs between the median carpellary bundles and the ten main bundles of the floral tube.

The ovary wall is considered to be differentiated into the fleshy parenchymatic exocarp and the cartilaginous endocarp lining the locules (MacDaniels, 1940). The exocarp consists of parenchyma cells (fig. 19.8*E*). The cartilaginous endocarp consists of sclereids with such thick walls that the cell lumina are almost occluded (fig. 19.8*G*). In the region of the median carpellary bundles, sclerenchymatic cells are absent so that the hard endocarp of each carpel forms two disconnected sheets of tissue, one on either side of the locule. The cartilaginous endocarp is the first tissue of the apple that attains its maximum development. The fleshy exocarp follows next, then the extracarpellary tissue. The latter continues to grow up to the time of ripening of the fruit.

The epidermis of the apple fruit consists of radially elongated cells in young stages (fig. 19.8A), but toward maturity the tangential diameter

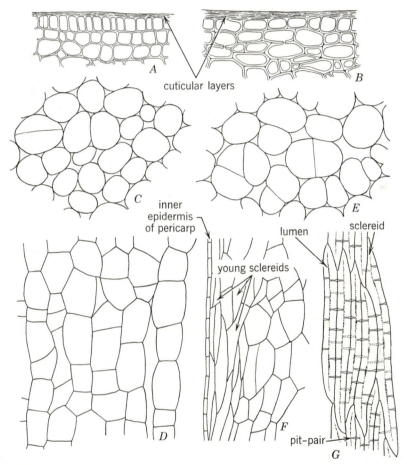

FIG. 19.8. Tissue elements of *Malus* (apple) fruit. (See fig. 19.6B.) A, B, epidermis and subjacent collenchymatic tissue from young (A) and mature (B) fruits. C, D, parenchyma of the floral-tube part of the flesh. C was taken closer to the surface, D farther away. E, parenchyma from exocarp. F, G, endocarp from young (F) and mature (G) fruits. A, C–E, from transections of a fruit 1 cm in diameter; F, from radial longitudinal section of similar fruit. B, transverse and, G, tangential longitudinal sections from mature fruit. (A–E, ×178; F, G, ×310. Slide by R. M. Brooks.)

surpasses the radial (fig. 19.8*B*). Throughout the growth of the apple the cuticle on the epidermis increases in thickness (Tetley, 1930, 1931). Granules of wax may accumulate on the cuticle (Mazliak and Chaperot, 1959). Stomata are present in the young epidermis. Later they cease to function and are replaced by lenticels consisting mostly of patches of suberized cells (Clements, 1935). Such lenticels arise also beneath scars left by fallen trichomes and breaks in the skin (Krapf, 1961). Unicellular epidermal hairs occur on young fruits but fall off later. Russeting of apples results from the replacement by cork of the outer layers of fruit over parts of its surface (Tetley, 1930).

In another pome fruit, *Pyrus communis* (pear), sclereids constitute a characteristic tissue element of the fleshy part of the fruit wall. Usually, concentric divisions occur in parenchyma cells surrounding the initial small clusters. Thus, radiating files of cells are formed from which additional sclereids differentiate. After the sclereids reach maturity they no longer change (Sterling, 1954). The softening of the mature fruit is caused by degradation of cell walls and collapse of parenchyma cells.

Development. Fleshy fruits are often used for developmental studies of the general problems of growth and morphogenesis (Luckwill, 1959; Nitsch, 1953). Experimental evidence indicates that auxins are involved in inducing the initial development of the fruit and also its subsequent growth. The initial hormone synthesis resulting from pollen-tube growth is augmented and then replaced by that occurring in the growing seeds (Gustafson, 1961).

Of the two usual processes that regulate growth, cell division and cell enlargement, the latter may be particularly pronounced in fruit development (Nitsch, 1952). The cells in the watermelon, for example, may become so large that they are discernible with the unaided eye. The ovary commonly passes through a period of cell division with a small amount of cell enlargement succeeded by one of cell enlargement without cell devision, but the two stages together give a sigmoid curve and are not sharply distinguishable. Commonly, cell division gradually ceases after anthesis and cell enlargement occupies the longest period of growth (Bain and Robertson, 1951; Nitsch, 1952; Sinnott, 1939). The size of the fruit may depend mainly on cell multiplication or cell enlargement or both and the form is determined by polarization of both elements of growth (Kano et al., 1957; Sinnott, 1944).

In the orange, rapid morphologic and physiologic changes occur during the main cell enlargement period, and the fruit continues to increase in size at a reduced rate after it reaches maturity (Bain, 1958).

The avocado fruit has an initial period of cell division and cell enlargement and, in contrast to most other fleshy fruits, continues with cell
multiplication as long as it remains on the tree (Schroeder, 1953).

ABSCISSION

In the abscission of fruits the separation layer may be prepared by
cell division or differentiated without divisions. In fruit clusters there
are often two to three separation layers. First the fruit abscises, then
the axial parts (Fehér, 1925). Some fruits separate together with their
stalks (*Carpinus, Ulmus, Salix, Populus, Pyrus, Tilia, Robinia*). In
certain *Prunus* species the first abscission occurs at the base of the fruit,
the second at the base of the pedicel, the third at the base of the spur.
The scar left by the spur is healed over by a periderm soon after
abscission. In *Castanea, Quercus,* and *Fagus* the fruit separates from
the involucre without preceding cell division. The cells at the base of
the fruit die after a sclerification of their walls and break away from the
still living, relatively thin-walled cells of the involucre. The fruit of
Umbelliferae has special separation layers along which the two halves
of the fruit (the two mericarps) break away at maturity. The separation
layer consists of parenchyma tissue with many intercellular spaces. The
tissue collapses at maturity. In many Compositae the abscission region
of the achenes is constricted, and its ground tissue consists of small-
celled parenchyma. At maturity the cells separate from one another or
shrink and thus bring about the loosening of connection between the
fruit and the receptacle (John, 1921). In the grass *Aegilops triaristata*
the fertile part of the spike abscises by a break through thin-walled dead
cells (Markgraf, 1925).

In the apple the abscission phenomenon appears to vary, depending
on the stage of development of the fruit (McCown, 1943). If a flower
or an immature fruit becomes separated, abscission is preceded by cell
enlargement and division. The separation of mature fruits, on the other
hand, occurs without any cell division. In the abscission of some tropical
fruits the process of fruit separation from the stalk is interpreted as a
concomitant of the progressive softening and disintegration of tissues
during the late stages of ripening (Barnell, 1939). The shedding of the
fruit stalk occurs after the fruit falls and is associated with a development of a distinct separation layer. Abscission cork is formed in the
fruit-stalk scar.

Fruits may abscise with the seeds still enclosed. The subsequent
separation of seeds may be entirely passive without formation of an
abscission zone, or it may involve the development of a relatively poorly

differentiated separation layer between the funiculus and the placenta (Pfeiffer, 1928). The cells of this layer are commonly thin-walled and give the cellulose reaction. In Leguminosae, the separation layer between the seeds and the placenta shows a combination of thicker-walled sclerified and thinner-walled nonsclerified elements. Absence of a separation layer is exemplified by berries in which the seeds are severed from the placenta after the placenta breaks down.

REFERENCES

Bain, J. M. Morphological, anatomical, and physiological changes in the developing fruit of the Valencia orange, *Citrus sinensis* (L.) Osbeck. *Austral. Jour. Bot.* 6:1–24. 1958.

Bain, J. M., and R. N. Robertson. The physiology of growth of apple fruits. **I**. Cell size, cell number and fruit development. *Austral. Jour. Sci. Res. Ser. B. Biol. Sci.* 4:75–91. 1951.

Barnell, E. Studies in tropical fruits. V. Some anatomical aspects of fruit-fall in two tropical arboreal plants. *Ann. Bot.* 3:77–89. 1939.

Baumann-Bodenheim, M. G. Prinzipien eines Fruchtsystems der Angiospermen. *Schweiz. Bot. Gesell. Ber.* 64:94–112. 1954.

Bell, H. P. Calyx end structure in Gravenstein apple. *Canad. Jour. Res. Sect. C, Bot. Sci.* 18:69–75. 1940.

Borthwick, H. A., and W. W. Robbins. Lettuce seed and its germination. *Hilgardia* 3:275–305. 1928.

Bradbury, D., I. M. Cull, and M. M. MacMasters. Structure of the mature wheat kernel. I. Gross anatomy and relationships of parts. *Cereal Chem.* 33:329–342. 1956*a*.

Bradbury, D., M. M. MacMasters, and I. M. Cull. Structure of the mature wheat kernel. II. Microscopic structure of pericarp, seed coat, and other coverings of the endosperm and germ of hard red winter wheat. *Cereal Chem.* 33:342–360. 1956*b*.

Carlquist, S. *Trematolobelia:* Seed dispersal; anatomy of fruit and seeds. *Pacific Sci.* 16:126–134. 1962.

Clements, H. Morphology and physiology of the pome lenticels of *Pyrus malus. Bot. Gaz.* 97:101–117. 1935.

Czaja, A. T. Neue Untersuchungen an der Testa der Tomatensamen. *Planta* 59:262–279. 1963.

Fahn, A., and M. Zohary. On the pericarpical structure of the legumen, its evolution and relation to dehiscence. *Phytomorphology* 5:99–111. 1955.

Fehér, D. Untersuchungen über den Abfall der Früchte einiger Holzpflanzen. *Deut. Bot. Gesell. Ber.* 43:52–61. 1925.

Gustafson, F. G. Development of fruits. *Handb. der Pflanzenphysiol.* 14:951–956. 1961.

Hartl, D. Struktur und Herkunft des Endokarps der Rutaceen. *Beitr. z. Biol. der Pflanz.* 34:35–49. 1957.

Holden, D. J. Factors in dehiscence of the flax fruit. *Bot. Gaz.* 117:294–309. 1956.

John, A. Beiträge zur Kenntnis der Ablösungseinrichtungen der Kompositenfrüchte. *Bot. Centbl. Beihefte* 38:182–203. 1921.

Juhnke, G., and H. Winkler. Der Balg als Grundelement des Angiospermengynaceums. *Beitr. z. Biol. der Pflanz.* 25:290–324. 1938.

Kaden, N. N. Tipy prodol'nogo vskryvaniíā plodov. [Types of longitudinal dehiscence of fruits.] *Bot. Zhur. S.S.S.R.* 47:495–505. 1962.

References   605

Kano, K., T. Fujimura, T. Hirose, and Y. Tsukamoto. Studies in the thickening growth of garden fruits. I. On the cushaw, egg-plant and pepper. *Kyoto Univ. Res. Inst. Food Sci. Mem.* 12:45–90. 1957.

Kiesselbach, T. A., and E. R. Walker. Structure of certain specialized tissues in the kernel of corn. *Amer. Jour. Bot.* 39:561–569. 1952.

Krapf, B. Entwicklung und Bau der Lentizellen des Apfels und ihre Bedeutung für die Lagerung. *Landw. Jahrb. der Schweiz* 10:387–440. 1961.

Kraus, G. Morphologisch-anatomische Untersuchungen der entwicklungsbedingten Veränderungen an Achse, Blatt und Fruchtknoten bei einigen Beerenfrüchten. *Österr. Bot. Ztschr.* 96:325–360. 1949.

Krauss, L. Entwicklungsgeschichte der Früchte von *Hordeum, Triticum, Bromus* und *Poa* mit besonderer Berücksichtigung ihrer Samenschalen. *Jahrb. f. Wiss. Bot.* 77:733–808. 1933.

Levina, R. E. O klassifikatsii i nomenklature plodov. [On classification and nomenclature of fruits.] *Bot. Zhur. S.S.S.R.* 46:488–495. 1961.

Luckwill, L. C. Fruit growth in relation to internal and external chemical stimuli. In: D. Rudnick. *Cell, organism and milieu.* New York, Ronald Press Company. 1959.

MacArthur, M., and R. H. Wetmore. Developmental studies in the apple fruit in the varieties McIntosh Red and Wagener. I. Vascular anatomy. *Jour. Pomol. and Hort. Sci.* 17:218–232. 1939. II. An analysis of development. *Canad. Jour. Res. Sect. C, Bot. Sci.* 19:371–382. 1941.

MacDaniels, L. H. The morphology of the apple and other pome fruits. *N.Y. (Cornell) Agr. Expt. Sta. Mem.* 230. 1940.

Markgarf, F. Das Abbruchgewebe der Frucht von *Aegilops triaristata* Willd. *Deut. Bot. Gesell. Ber.* 43:117–120. 1925.

Matienko, B. T. Ob anatomo-morfologicheskoï prirode tsvetka i ploda tykvennykh. [On the anatomico-morphological nature of the flower and fruit of cucurbits.] In: *Morfologiâ i Anatomiâ Rasteniĭ.* IV. Leningrad, Izdatel'stvo Akademii Nauk SSSR. 1957.

Mazliak, P., and D. Chaperot. Recherches morphologiques sur le revêtement cireux de la pomme Calville Blanc. *Rev. Gén. de Bot.* 66:645–653. 1959.

McCown, M. Anatomical and chemical aspects of abscission of fruits of the apple. *Bot. Gaz.* 105:212–220. 1943.

Meissner, F. Die Korkbildung der Früchte von *Aesculus-* und *Cucumis*-Arten. *Österr. Bot. Ztschr.* 99:606–624. 1952.

Miller, E. V. The accumulation of carbohydrates by seeds and fruits. *Handb. der Pflanzenphysiol.* 6:871–882. 1958.

Monsi, M. Untersuchungen über den Mechanismus der Schleuderbewegung der Sojabohnen-Hülse. *Jap. Jour. Bot.* 12:437–474. 1943.

Narayanaswami, S. The structure and development of the caryopsis in some Indian millets. V. *Eleusine coracana* Gaertn. *Papers Michigan Acad. Sci., Arts and Letters* 40:33–46. 1955.

Nitsch, J. P. Plant hormones in the development of fruits. *Quart. Rev. Biol.* 27:33–57. 1952.

Nitsch, J. P. The physiology of fruit growth. *Ann. Rev. Plant Physiol.* 4:199–236. 1953.

Pfeiffer, H. Die pflanzlichen Trennungsgewebe. In: K. Linsbauer. *Handbuch der Pflanzenanatomie.* Band 5. Lief. 22. 1928.

Pijl, L. van der. Sarcotesta, aril, pulpa and the evolution of the angiosperm fruit. I and II. *Nederland. Akad. van Wetensch. Proc. Ser. C., Biol. and Med. Sci.* 58:154–161, 307–312. 1955.

Ragland, C. H. The development of the peach fruit, with special reference to split-pit and gumming. *Amer. Soc. Hort. Sci. Proc.* 31:1–21. 1934.

Randolph, L. F.   Developmental morphology of the caryopsis in maize.   *Jour. Agr. Res.* 53: 881–916. 1936.

Reeve, R. M.   Fruit histogenesis in *Rubus strigosus*.   I. Outer epidermis, parenchyma, and receptacle.   *Amer. Jour. Bot.* 41:152–160. 1954*a*.   II. Endocarp tissues.   *Amer. Jour. Bot.* 41:173–181. 1954*b*.

Reeve, R. M.   Histological and histochemical changes in developing and ripening peaches. II. The cell walls and pectins.   *Amer. Jour. Bot.* 46:241–248. 1959.

Rethke, R. V.   The anatomy of circumscissile dehiscence.   *Amer. Jour. Bot.* 33:677–683. 1946.

Schnarf, K.   Embryologie der Angiospermen.   In: K. Linsbauer.   *Handbuch der Pflanzenanatomie.* Band 10. Lief. 21. 1927.

Schroeder, C. A.   Growth and development of the Fuerte avocado fruit.   *Amer. Soc. Hort. Sci.* 61:103–109. 1953.

Scott, F. M., and K. C. Baker.   Anatomy of Washington navel orange rind in relation to water spot.   *Bot. Gaz.* 108:459–475. 1947.

Sinnott, E. W.   A developmental analysis of the relation between cell size and fruit size in cucurbits.   *Amer. Jour. Bot.* 26:179–189. 1939.

Sinnott, E. W.   Cell polarity and the development of form in curcurbit fruits.   *Amer. Jour. Bot.* 31:388–391. 1944.

Smith, W. H.   The histological structure of the flesh of the apple in relation to growth and senescence.   *Jour. Pomol. and Hort. Sci.* 18:249–260. 1940.

Smith, W. H.   Cell-multiplication and cell-enlargement in the development of the flesh of the apple fruit.   *Ann. Bot.* 14:23–38. 1950.

Sterling, C.   Developmental anatomy of the fruit of *Prunus domestica* L.   *Torrey Bot. Club Bul.* 80:457–477. 1953.

Sterling, C.   Sclereid development and the texture of Bartlett pears.   *Food Res.* 19:433–443. 1954.

Stopp, K.   Karpologische Studien.   I. Vergleichend-morphologische Untersuchungen über die Dehiszenzformen der Kapselfrüchte.   *Abhandl. Akad. Wiss. Lit. Mainz Math.-Nat. Kl.* 1950 (7):165–210. 1950.

Subramanyam, K., and M. V. S. Raju.   Circumscissile dehiscence in some angiosperms. *Amer. Jour. Bot.* 40:571–574. 1953.

Tetley, U.   A study of the anatomical development of the apple and some observations on the "pectic constituents" of the cell walls.   *Jour. Pomol. and Hort. Sci.* 8:153–172. 1930.

Tetley, U.   The morphology and cytology of the apple fruit with special reference to the Bramley's seedling variety.   *Jour. Pomol. and Hort. Sci.* 9:278–297. 1931.

Winkler, H.   Versuch eines "natürlichen" Systems der Früchte.   *Beitr. z. Biol. der Pflanz.* 26:201–220. 1939.

Winkler, H.   Zur Einigung und Weiterführung in der Frage des Fruchtsystems.   *Beitr. z. Biol. der Pflanz.* 27:92–130. 1940.

Ziegenspeck, H.   Die Wegsamkeit der Pigmentschicht der Getreidekörner (Endoderminschicht) für Fluorochrome.   *Protoplasma* 41:425–431. 1952.

20

The Seed

This chapter deals with the seed of angiosperms. A seed develops from an ovule and, at maturity, consists of the following parts (figs. 20.3C, 20.5B): the young, partially developed sporophyte called the *embryo;* variable amounts of *endosperm* (from the Greek words, within and seed), sometimes none; and the protective layers on the surface, the seed coat or *testa* (from the Latin, brick or tile), which is derived from the integument or integuments. Various external markings on the seed are traceable to certain structural details of the ovule. The micropyle may be completely obliterated, or it may remain in the form of an occluded pore. A scar, the *hilum* (from the Latin, a trifle), which is considered to be relatively permeable to water, occurs where a seed abscises from the funiculus. In anatropous ovules the funiculus is adnate to the ovule, and the abscission of the seed occurs at the lower level of the funiculus, that is, near the placenta. The funiculus is recognizable in such a seed as a longitudinal ridge, the *raphe* (from the Greek, seam). A *caruncle* and an *aril* mentioned in connection with the ovule are present in some seeds. In some plants the raphe produces an exceptionally large, oil-containing appendage that attracts ants and thus ensures seed dispersal (Berg, 1958). The seeds of angiosperms vary widely in structure but are relatively constant in narrower groups and, therefore, may be used in connection with taxonomic studies (McClure, 1957).

EMBRYO

The embryo shows a variety of developmental patterns and attains different sizes and degrees of differentiation. Commonly, the future

607

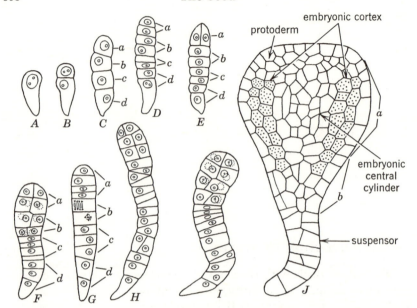

FIG. 20.1. Embryo development in *Daucus carota* (carrot). Longitudinal sections. The lower end of embryo in each drawing is the end directed toward micropyle. A–C, stages in development of linear four-celled embryo. D, E, two common variations in eight-celled embryos: difference in division of cell *a* of the four-celled embryo. F–I, older embryos varying in cell arrangement. J, embryo differentiated into main body and suspensor. Initial organization of tissue regions is present in J. Relation of parts of certain embryos to cells of the four-celled embryo (C) is indicated by the letters *a–d*. (All, ×500. After Borthwick, *Bot. Gaz.* 92, 1931.)

vegetative organs of the plant are initiated during the development of the embryo, at least in the form of their apical meristems. As was reviewed in chapters 1 and 15, the embryo consists of an axis, the hypocotyl-root axis, bearing, at one end, the root meristem and, at the other, the cotyledon or cotyledons and the meristem of the first shoot. Sometimes a shoot bud, the epicotyl, and a primordial root, the radicle, are present in the embryo. Usually a rootcap develops at the root end of the embryo. In the peanut (*Arachis hypogaea*), the embryo has not only a leafy epicotyl but also two lateral shoot primordia arising in the cotyledonary axils (Yarbrough, 1957). The grass family has a highly differentiated embryo with many parts, including adventitious root primordia. Some dicotyledonous embryos also have such primordia (Steffen, 1952). On the other hand, an embryo may be poorly differentiated, even lack cotyledons (*Cistanche*, a parasitic herb; Kadry, 1955). A procambial system, continuous throughout the hypocotyl and the cotyledons, is commonly differentiated in the embryo. Some vascular elements may mature in an embryo before the seed germinates.

When the embryo is initiated by the division of the zygote, most frequently this division is transverse (figs. 20.1*B*, 20.2*B*). When each of the two resulting cells divides, the orientation of the two new walls may vary. Commonly the cell oriented toward the micropyle, the proximal cell, divides transversely. The distal cell may divide transversely, vertically, or obliquely. As a result, the four-celled embryo appears either as a single file of cells (fig. 20.1*C*) or as a three-tiered structure with the distal tier composed of two cells (fig. 20.2*D*). The distribution of the subsequent divisions is usually unequal in the various tiers of the

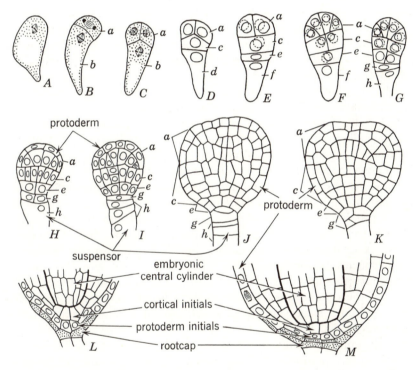

FIG. 20.2. Embryo development in *Lactuca sativa* (lettuce). Longitudinal sections. Lower end of embryo in each drawing is the end directed toward micropyle. *A*, zygote in division. *B–G*, embryos in successive stages of development, showing establishment of several horizontal tiers of cells. In *G*, cell *h* later gives rise to all of the suspensor cells, and the tiers above *h* develop into the main body of embryo. *H–M*, further developmental stages resulting in the initial organization of tissue regions. In *L*, *M*, only lower parts of embryos. *K*, flattened apex characteristic of stage preceding emergence of cotyledons. (All, ×400. After Jones, *Hilgardia* 2, 1927.)

four-celled embryo. Furthermore, the divisions become specifically oriented in the different tiers so that the embryo differentiates into the main body or embryo proper and the suspensor (figs. 20.1*J*, 20.2*H*, pl. 96*A–D*). The embryo proper is at this stage a relatively massive body, whereas the suspensor has the form of a stalk, variable in length, uniseriate, or more massive, and is attached to the wall of the embryo sac at the micropylar end. Before it differentiates into the main body and the suspensor the young embryo is sometimes called proembryo.

The development of an embryo follows an orderly pattern characteristic of a given plant group from the outset. The differentiation into a root pole and a shoot pole indicates an early establishment of polarity (Wardlaw, 1955), and the difference between the two poles increases through differential divisions and cell enlargement in the successive stages of embryogeny (Meyer, 1958). The embryo illustrates the inception of the organization characteristic of the adult plant and is being used for studies of causal relations in organized growth. The behavior of embryos cultured *in vitro* suggests that the embryonic form results from an interaction between the embryo and environmental factors within the ovule, such as nutrition, space, and others (Norstog, 1961). The successful raising of plants from cultured dissociated parenchyma cells indicates that any nonspecialized cell has the potentiality to produce organized growth (chapter 4). The embryonic development from a zygote is one of the manifestations of this potentiality.

Students of embryo development treat the sequence of divisions and the fate of the resulting cells of the four-celled embryo as one of the main problems of embryology. The parts of the mature embryo derived from each tier of the four-celled embryo may differ from genus to genus, or from species to species, but they may also vary within the same species (Bhadurim, 1936; Borthwick, 1931); and they do not determine the later histogenesis in the embryo (Guttenberg, 1960). Despite these limitations, the characteristics of embryos in their early developmental stages are given much weight in comparative studies (Iakovlev, 1958; Johansen, 1950; Souèges, 1938–1951). Several types of early embryo ontogenies have been conceived, chiefly by Souèges (summary of types in Guttenberg, 1960). These types differ in the mode of the first divisions of zygote, in the sequence of divisions in the products of the terminal cell of the two-celled embryo, and in the amount of contribution to the embryo by the basal cell of the two-celled embryo. A unique type of embryogeny occurs in *Paeonia:* a massive proembryo, at first coenocytic, then cellular, forms numerous embryo primordia one of which becomes dominant (Cave et al., 1961). The evolutionary significance of these types has not been fully established (Takhtajan,

1959) but they have been utilized for taxonomic purposes (Crété, 1955; Lebègue, 1952; Souèges, 1956). The structure of the mature embryo and its position in the seed and relation to the endosperm also are distinctive in different groups of plants and thus may serve for identifying seeds (Martin, 1946). Mature embryo characteristics play an important role in the classification of Gramineae (Reeder, 1957).

Dicotyledon and Monocotyledon Embryos. The number of cotyledons —one in monocotyledons, two in dicotyledons—is considered to be the primary distinction between the two groups of angiosperms, but certain dicotyledons normally develop a single cotyledon (Haccius, 1952*b*; Takhtajan, 1959). There are also dicotyledons with more than two cotyledons (Haskell, 1954). The ontogenetic fusion of cotyledonary sheaths is another striking deviation found in some dicotyledons (Haccius, 1953).

Embryos of dicotyledons and monocotyledons may be similar in form up to the stage when the main body of the embryo is globose in shape. Subsequently, the dicotyledonous embryo assumes a bilobed shape, because of the appearance of the two cotyledons (fig. 20.2*K*; pl. 95*B*, *C*), whereas the monocotyledonous embryo is changed into a more or less cylindrical structure by the extension of the single cotyledon (pl. 96*E*, *F*). The cotyledons of the dicotyledonous embryo arise as two meristematic protrusions on the apical end of the embryo. The emergence of the cotyledons is preceded by a lateral expansion of the apical end of the embryo (fig. 20.1*J*, pl. 95*B*). This change in shape results from localized growth based on periclinal divisions in the two opposite positions in the embryo where the cotyledons later appear. The periclinal divisions occur in several superficial layers which may include the outermost layer (Miller and Wetmore, 1945; Nast, 1941). The initiation of the cotyledons brings about a change in embryo symmetry from radial to bilateral. The widening of the apical end of the embryo in preparation for the emergence of the cotyledons is comparable to the formation of leaf buttresses in vegetative shoots. Subsequently, the cotyledons grow upward upon their buttresses (fig. 19.3, pl. 95) and expand laterally. As in foliage leaves apical and marginal growth may be recognized in the development of the cotyledons (Miller and Wetmore, 1945; Nast, 1941; Steffen, 1952).

The part of the apex that is left in the notch between the two cotyledons constitutes the apical meristem of the epicotyl (pl. 95*D*). In monocotyledonous embryos the epicotylary meristem occurs in a depression which is discernible at the base of the cotyledon early during the increase in thickness of the embryo. This depression is at first quite shallow (pl. 96E) but increases in depth as the tissue on its lower margin grows upward (pl. 96G). In the final stage of embryo development the apical meristem appears on one side of the cotyledon and is completely surrounded by a sheath-like lateral extension from the base of the cotyledon. Thus, the first epicotyl apex of the monocotyledon bears the same relation to the cotyledon as the vegetative shoot apex to a foliage leaf. As was shown earlier, the relation between the cotyledons and the embryo apex on the one hand, and that between the foliage leaves and the vegetative shoot apex on the other are comparable in the dicotyledons also. Some investigators, therefore, interpret the embryo apex before it forms the cotyledons as the first shoot apex of the plant (Mahlberg, 1960; Spurr, 1950).

The relation of the single cotyledon of monocotyledons to the apex of the embryo is a matter of controversy. According to one view, the cotyledon is terminal in origin, the shoot apex is lateral, and the whole plant is a sympodium of lateral shoots (Souèges, 1954). Other authors consider the terminal position of the cotyledon to be only apparent: the lateral position of the apical meristem results from its displacement by the cotyledon (Baude, 1956; Haccius, 1952a; Swamy and Lakshmanan, 1962). In the dicotyledons having a single cotyledon the embryogenetic events resemble those in monocotyledons and support the idea of the initially lateral origin of the cotyledon (Haccius, 1954).

The differentiation of the root pole involves the organization of the protomeristem and of the rootcap (fig. 20.2). The meristem may resemble that of the growing root in its relation to the primary tissue regions or it may attain that pattern only after the seed germinates. In some embryos a cell, the hypophysis (from the Greek, under and growth), which is concerned with the organization of the root pole, is definable in early embryogeny (Guttenberg, 1960).

The early embryogeny is treated comprehensively in the botanical literature (Johansen, 1950; Schnarf, 1929, 1931; Swamy and Padmanabhan, 1962). Studies on the later embryogeny, particularly those concerned with the organization of the system of primary meristems (protoderm, procambium, and ground meristem) and of the apical meristem are, on the contrary, rather limited in number (Arnott, 1962; Buell, 1952a; Miller and Wetmore, 1945; Nast, 1941; Reeve, 1948; for gymnosperms, Allen, 1947a, b; Spurr, 1949, 1950).

The delimitation of the three meristems, the protoderm, the procambium, and the ground meristem, begins in the embryo long before it reaches its final size. The protoderm is initiated by periclinal divisions commonly starting in the distal tier ond progressing toward the proximal end (fig. 20.2H–K). The protoderm does not extend to the suspensor but merges with the apical meristem of the root (fig. 20.2L, M). The procambial system is first delimited by the vacuolation and decrease in stainability of the ground tissue (pl. 95B, C). Later the procambial cells assume their characteristic narrow elongated shape. The procambium may become discernible before the cotyledons emerge and, as the latter begin to grow, the procambium is organized in them in continuity with that in the embryo axis (pl. 95). The procambial system in the hypocotyl-root axis varies in shape, depending on how much of this axis is organized as a root. In the Gramineae, for example, most of this axis is root-like in structure, in *Juglans*, less than one-sixth of it is (Nast, 1941). The timing in the histogenetic differentiation of embryos varies in different groups of plants. Slender embryos with a regular and constant sequence of cell division have a relatively earlier differentiation of histogens than massive embryos with less regular early meristematic activity.

Embryos in various stages of differentiation may contain chloroplasts (Poddubnaïa-Arnol'di, 1952). Chemical analyses and experimental studies with light indicate that the chloroplasts are photosynthetically active and that the pigment may be chlorophyll (Kantor, 1955; Meeuse and Ott, 1962). A well-developed cuticle and stomata have been recorded in the embryos of cycads (Pant and Nautiyal, 1962).

Grass Embryo. The embryo of Gramineae, especially that of cultivated cereals, has received much attention. As exemplified by the wheat embryo, it has the following structure. An axis bears a scutellum (in Latin, tray) on one side, a radicle with a rootcap covered by the coleorhiza (from Greek, sheath and root) at the root pole, and a plumule covered by the coleoptile (in Greek, sheath) at the shoot pole. In young embryos the coleorhiza is continuous with the suspensor. Above, the coleorhiza bears a flap-like projection, the epiblast (from Greek, upon and shoot), inserted opposite the scutellum. Some Gramineae (*Zea*) have no epiblast. The plumule has several leaf primordia. The procambial system extends throughout the embryo and permits the recognition of the node of the scutellum. Since the radicle starts below this node, and the scutellum is considered to be the cotyledon, no hypocotyl is identifiable unless the term is applied to the scutellar node. Seminal adventitious roots occur above this node.

The scutellum is a shield-like structure. Its expanded abaxial surface bears a secretory epithelial epidermis, which is in contact with the endosperm. The cone-shaped coleoptile has a pore at the apex through which the first foliage leaf emerges. The coleoptile has stomata on both surfaces. In *Oryza* all of the stomata on the adaxial surface and in *Avena* some of them serve as hydathodal pores (Butterfass, 1956).

The morphologic nature of the parts of the grass embryo is a subject of much speculation. According to a common view, the scutellum is a cotyledon, the coleoptile the first leaf, and the axial interval between the two is the first internode (Guignard, 1961). The epiblast is often interpreted as the second, rudimentary cotyledon (Negbi and Koller, 1962; Roth, 1955). As a cotyledon the epiblast would fit into the two-ranked leaf sequence: scutellum, epiblast, coleoptile, first foliage leaf. Some authors, however, consider the epiblast to be part of the coleorhiza because of the similarity between the two, including the occurrence of root hairs on both (Brown, 1960; Foard and Haber, 1962; Guignard, 1961).

The foliar interpretation of the coleoptile is not universal. Some regard it as an outgrowth of the scutellum, the scutellar sheath, rather than a product of the apical meristem (Pankow and Guttenberg, 1957). The term mesocotyl (meso from Greek in the middle, cotyl refers to cotyledon) as applied to the axis between the scutellum and coleoptile refers to the unity of the two structures. Another assumption is that the coleoptile and mesocotyl are new acquisitions without homologues in other embryos (Brown, 1960). There is also a concept that the scutellum is the embryonic axis, the epiblast the rudiment of the cotyledon bearing in its axil a bud, the plumule, covered by a prophyll, the coleoptile (Jacques-Felix, 1958). The foliar nature of the coleoptile, however, is clearly indicated by the form of this structure in *Streptochaeta*, a primitive grass: the coleoptile of this plant is an open leaf with a median bundle located opposite the free margins (Reeder, 1953).

The coleorhiza is identified as part of the suspensor or of the hypocotyl (Roth, 1957); or it is regarded as the suppressed primary root, the radicle as an adventitious root (Guignard, 1961; Negbi and Koller, 1962; Pankow and Guttenberg, 1957).

STORAGE TISSUE

The endosperm usually develops after fertilization from the product of triple fusion, that is, the fusion of the two polar nuclei with one male gamete (Brink and Cooper, 1947). The fusion nucleus is commonly termed the primary endosperm nucleus. The timing of the initial divisions of this nucleus and of the zygote is variable, but commonly the endosperm starts to develop first. A lack of correlation between the development of embryo and endosperm has been observed in connection with apomictic phenomena (that is, development of embryo without gametic union; Cooper and Brink, 1949; Esau, 1946). In some taxa the storage tissue is derived from the nucellus, part of which may be retained in the seed and accumulate storage materials (fig. 20.4A). This type of storage tissue is called *perisperm* (from the Greek, about and seed). Thus storage in angiosperm seeds may occur in triploid (endosperm) and diploid (perisperm) tissues; in gymnosperms, the storage tissue is of megagametophyte origin and is, therefore, haploid.

Seeds lacking endosperm or perisperm in the mature state are called *exalbuminous* (from the Latin *albumen,* the white of an egg). In such seeds the embryo is large in relation to the seed as a whole. It fills the seed almost completely, and its body parts, particularly the cotyledons, store the food reserves (Leguminosae, Cucurbitaceae, Compositae; fig. 19.3C). Seeds with endosperm or perisperm are called *albuminous.* In such seeds the embryo varies in size in relation to the amount of endosperm left at maturity. The monocotyledons commonly have albuminous seeds (fig. 19.2, pl. 96G).

Three principal methods of endosperm formation are recognized: (1) many nuclei are formed by free nuclear division which may or may not be followed by formation of cell walls; (2) cell walls are formed immediately after the first nuclear division; (3) after the first mitosis the embryo sac is divided in two unequal chambers of which the larger (chalazal) usually develops noncellular endosperm and the smaller (micropylar) shows somewhat variable behavior. The endosperms resulting from the three methods of development are called (1) nuclear or, more appropriately (Rao, 1959), noncellular; (2) cellular, and (3) helobial (after the Helobiae, a monocotyledonous taxon). The helobial type is found only in monocotyledons; the presumed helobial endosperm described in dicotyledons results from aberrant ontogenies (Swamy and Parameswaran, 1963). The recognition of this aspect has a bearing on the question of phylogeny of endosperm types (Swamy and Parameswaran, 1963; Wunderlich, 1959).

In the noncellular type of endosperm the free nuclei commonly occur in the parietal layer of cytoplasm enclosing a central vacuole. The nuclear divisions are synchronized: at each division mitoses proceed across the embryo sac, beginning at the micropylar end. The formation of walls starts relatively late. Two methods of wall formation have been reported: (1) by phragmoplasts and cell plates and (2) by furrowing (Schnarf, 1928). After the cell walls have been formed, cell division continues, with every mitosis followed by cytokinesis, until the embryo sac is filled with cellular endosperm. The divisions occurring after the free nuclear stage may show no particular orientation, or they may be as orderly as those in a vascular cambium. In the cellular type of endosperm the first mitosis is followed by cytokinesis, and the formation of cell walls usually continues throughout the growth of the endosperm.

Deviations from these ontogenetic patterns are encountered (Swamy and Parameswaran, 1963). In cocos, the free nuclei suspended in a clear fluid later become associated with cytoplasm into free spherical cells. These cells and the remaining free nuclei migrate toward the periphery where eventually a cellular endosperm is initiated (Cutter et al., 1955). The central cavity remains filled with sap, the coconut milk.

The structure of the fully developed endosperm varies widely. It may be a thin-walled, highly vacuolated tissue, storing no food. Such endosperm is usually used up, partly or completely, by the developing embryo (fig. 19.3). In many plants the endosperm is differentiated as a storage tissue (fig. 19.2). As such it may have thin or thick walls, sometimes very thick and horny in appearance (*Asparagus*, Robbins and Borthwick, 1925). As a rule endosperm has no intercellular spaces. In some grasses the mature endosperm lacks cell walls and has a mushy, oily consistency (Dore, 1956; Matlakówna, 1912; Müller, 1943). The endosperm may invade the ovular tissue in the form of haustoria (Chopra, 1955; Sarfatti, 1960). The ruminate (in Latin, chewed) endosperm results from growth of the seed coat into the embryo-sac (Periasamy, 1962).

Diverse substances are stored in the endosperm (Crocker and Barton, 1953; Miller, 1958; chapters 2, 8). The principal storage carbohydrate is starch in the form of starch grains. Starch is combined in various proportions with proteins, oils, and fats. The starch grains arise in plastids, singly or in multiples (Buttrose, 1960). Cereals often combine small and large starch grains. The small grains appear to have a distinct origin, a phenomenon which has led to various interpretations regarding their formation (nonplastid starch, Aleksandrov and Aleksandrova, 1954; starch arising in vesicles that bud off from the amyloplasts, Buttrose, 1960; mitochondrial starch, Îakovlev, 1950).

Hemicelluloses, giving mannose and other monosaccharides on hydrolysis (Meier, 1958), constitute as cell wall components the carbohydrate reserves of some seeds (endosperm of *Diospyros, Phoenix, Strychnos, Coffea, Iris, Asparagus;* cotyledons of *Tropaeolum, Primula, Impatiens, Lupinus*). The most notable example of a seed with hemicellulose storage is the ivory nut (*Phytelephas macrocarpa*). A hemicellulose called amyloid resembles starch in that it stains blue with iodine. Amyloid has been found in the walls of endosperm and cotyledons of many species (Kooiman, 1960).

Proteins found in seeds occur in two principal forms: (1) the glutens, amorphous in structure; (2) the aleuron grains, composed of a proteinaceous substrate with a crystalloid body (protein crystal) and a globoid body (double phosphate of calcium and magnesium with an organic radical). Glutens are common in the starch-containing cells of cereal grains. Aleuron grains occur in all endosperm cells of *Ricinus* (castor bean) and in the peripheral endosperm layer (aleuron layer) of Polygonaceae and Gramineae. Leguminous seeds are almost the only ones regularly accumulating large quantities of proteins (Miller, 1958). The protein bodies of certain legumes and their degradation during germination have been studied with the electron microscope (Bagley et al., 1963; Varner and Schidlovsky, 1963). Subcellular particles high in protein and lipids have been isolated from cotton seed (Yatsu and Altschul, 1963).

In many families (Juncaceae, Gyperaceae, most Gramineae, Commelinaceae, Cannaceae, Polygonaceae, Caryophyllaceae, and others) the starch-containing endosperm or perisperm cells are nonliving (Müller, 1943). The aleuron layer in some of these families was found to be living. The small amount of endosperm in Chenopodiaceae is living, whereas the perisperm is nonliving. In some plants the endosperm, as well as the embryo, contains chloroplasts (Ioffe, 1957). Starch grains and proteins may be sufficiently characteristic to serve in taxonomic studies (Avdulov, 1931; Blagoveschchenskiĭ, 1958; Tateoka, 1962).

SEED COAT

The young testa or seed coat developing from the integument or integuments consists of more or less vacuolate thin-walled cells. During the maturation of the seed, the testa undergoes varied degrees of structural alteration. There may be a change in contents and wall structure as well as destruction of some or all of the original integumentary layers (Netolitzky, 1926).

Some seed-coat differences among plants may be traced to differences in the structure of the ovule, such as the number and thickness of the integuments and the arrangement of the vascular tissues. Similar ovules, however, may become highly dissimilar during development. There may be variations in the intensity of cellular destruction; in the degree of sclerification and the distribution of mechanical cells; in the deposition of coloring and other organic substances; and in the differentiation of specialized trichomes, such as hairs, papillae, and hooks (Esau, 1960). The epidermis of the seed frequently develops very thick walls and is filled with coloring matter (pl. 96*F*, *G*). In some Cucurbitaceae the hard outer part of the seed coat separates from the inner papery layers, which remain attached to the embryo together with the remains of the nucellus and endosperm (Singh, 1953). In the Leguminosae the protodermal cells elongate at right angles to the surface and differentiate into macrosclereids (chapter 10; Corner, 1951; Reeve, 1946a, b). In *Gossypium* the epidermal cells elongate into hairs, the commercially used cotton fibers. In many seeds (*Linum, Plantago psyllium;* some Cruciferae and Compositae) the epidermal walls are highly hygroscopic and become mucilaginous on contact with moisture. The mucilage may or may not contain cellulose (Freytag, 1958; Frey-Wyssling, 1959). The potentially mucilaginous wall sometimes more or less replaces the contents of the cell.

The mechanically protective tissue may differentiate in the outer or the inner integument. In *Asparagus* such tissue is represented by the outer epidermis of the outer integument, in *Capsella* by the second layer of the outer integument, in *Reseda* by the first layer of the inner integument. The seeds are also protected by cuticles that originate in the ovule. The seed cuticles usually combine into a continuous membrane (probably interrupted in the hilar region) which encloses the embryo and the associated endosperm (if the latter is present).

The seeds of Leguminosae attract considerable attention because of the peculiar wall formation, the light line, in the outer epidermis of the testa. The epidermis forms the palisade layer consisting of sclereids (chapter 10; Steiner and Janke, 1955). The light line, which occurs somewhat above the middle of the cells, is a particularly compact wall region, more transparent and more highly doubly refractive than the rest of the wall (Frey-Wyssling, 1959). In *Cercidium* the light line was found to have a transverse orientation of microfibrils in contrast with the dominantly parallel, longitudinal orientation elsewhere in the wall (Scott et al., 1962). The light line is commonly interpreted as playing a role in the impermeability of the testa, especially in the hard legume seeds, but the exact nature of impermeability of this line is not known (Frey-Wyssling, 1959).

The hard legume seeds attain and maintain a high degree of desiccation, possibly because of the combination of an impermeable testa and the valvular action of the hilum (Hyde, 1954). The fissure, which occurs along the groove of the hilum, opens when the seed is surrounded by dry air and closes when the outside air is moist.

The structure of the seed coat is best understood if studied developmentally. Below, the development of the testa is described for the following kinds of seed: (1) seed derived from an ovule with two integuments and having a mechanically strong seed coat (*Asparagus officinalis*); (2) seed derived from an ovule with two integuments and having a mechanically weak seed coat (*Beta vulgaris*); (3) seed derived from an ovule with a single integument and having a mechanically weak seed coat (*Lycopersicon esculentum*).

Asparagus Seed (Robbins and Borthwick, 1925). The anatropous ovule of asparagus has two integuments and a relatively large nucellus (fig. 20.3*A*, *B*). In the mature seed the integuments are transformed into a black, finely rugose, somewhat brittle seed coat. The nucellus is completely absorbed during the enlargement of the embryo sac. The mature embryo is a slender cylindrical structure (fig. 20.3*C*) completely imbedded in a massive horny endosperm with walls of hemicellulose. At the time of pollination the outer integument consists of from five to ten layers of cells, the inner of only two (fig. 20.3*B*). The cells are small and closely packed. During the first 16 days after pollination the seed coat reaches its maximum thickness as a result of cell enlargement (fig. 20.3*D*, *E*). In addition, the outer wall in the outer integument develops a pronounced thickening, and some yellowish granular substance is deposited in the inner layer of the inner integument. When the seed coat is of maximum thickness, two cuticles are discernible, one located between the two integuments, the other, a thicker one, between the inner integument and the nucellus (fig. 20.3*E*).

In subsequent developmental stages the seed coat progressively desiccates and shrinks and is gradually compressed by the enlarging endosperm (fig. 20.3*F*, *G*). About 30 days after pollination the cells of the inner integument are disintegrated and compressed so that the two fatty membranes closely approach each other (fig. 20.3*G*). In the mature seed they become indistinguishable from one another though they can be separated by treatment with alkali (fig. 20.3*H*). The walls of the outer epidermal cells continue to thicken until, at maturity, the cell lumina are completely filled with dark-brown wall material (fig. 20.3*H*). The outer surface is covered with a thin transparent membrane which appears to be pectic in nature and is hydrophilous. Thus, the principal structural features of the mature seed coat are: a thick-walled epidermis offering mechanical protection and bearing a surface membrane which readily absorbs water and a thick cuticular membrane enclosing the endosperm and embryo.

Beta Seed (Artschwager, 1927; Bennett and Esau, 1936). The campylotropous ovule has two integuments, each two cells in thickness (pl. 94*B*). The nucellus is relatively large. During the development of the seed, a curved sac (the caecum) is formed by breakdown of nucellar cells in continuity with the embryo sac at its chalazal end. The embryo, which eventually fills the embryo sac and the caecum, curves around the remaining part of the nucellus which becomes a storage tissue, the perisperm (fig. 20.4*A*). The endosperm is reduced to a single layer at the micropylar end of the embryo sac. The mature seed is a shiny lenticular structure with a thin seed coat.

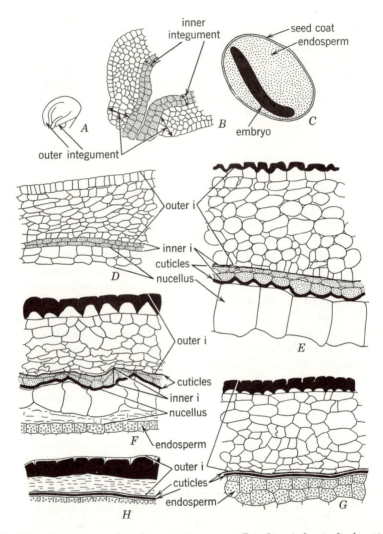

FIG. 20.3. Development of seed coat in *Asparagus officinalis*. *A*, longitudinal section of ovule. *B*, integuments at pollination time. *C*, entire seed sampled 44 days after pollination. *D–H*, seed coat in different stages of development. These sections were made the following numbers of days after pollination: 8 (*D*), 16 (*E*), 20 (*F*), 29 (*G*). *H* is from mature seed. Letter i indicates integument. (*B*, *D–H*, ×140; *C*, ×7. After Robbins and Borthwick, *Bot. Gaz.* 80, 1925.)

The seed coat develops from the two integuments (fig. 20.4B). The protoplasts of the outer layer of the outer integument die, and the cells become filled with brown resinous material (fig. 20.4C, D). The inner layer of the outer integument may increase in thickness by cell division, but it remains thin-walled and parenchymatic. The outer layer of the inner integument disintegrates. The inner layer of the inner integument develops somewhat thickened, delicately sculptured walls (fig. 20.4D). The outer surface of the seed is covered with a cuticle. No cuticle has been identified between the two integuments, but there is a conspicuous cuticle on the inner side of the inner integument (fig. 20.4). This cuticle stops abruptly in the chalazal region where the vascular tissue approaches the perisperm. A tightly packed layer of cells, rich in tannin, intervenes between the vascular tissues and the perisperm. When the seed is mature, the walls of these tannin cells give a positive fat reaction. Although the mature beet seed coat is mechanically weak, the seed is well protected because it is retained within the fruit, which

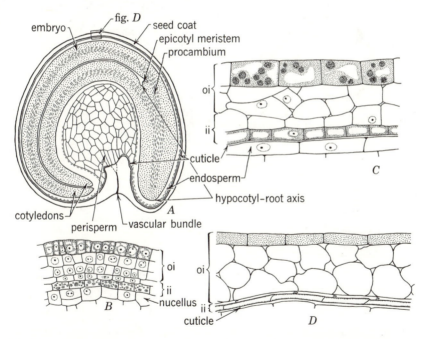

FIG. 20.4. Seed of *Beta vulgaris* (sugar beet) in longitudinal section (A), and its seed coat in three developmental stages (B–D). The letters ii signify inner integument, oi, outer integument. The rootcap is not visible because the root end of the hypocotyl lies in a different plane than does the rest of the hypocotyl. (A, ×20; B–D, ×310; A adapted from Bennett and Esau, *Jour. Agr. Res.* 53, 1936.)

develops an extremely hard wall. Usually several such fruits remain united in a structure known as the seed ball which, if planted, permits the emergence of the seedlings after the effect of moisture loosens the upper lid-like parts of the fruits along a predetermined line of dehiscence.

Lycopersicon Seed (Netolitzky, 1926; Smith, 1935). The seed is derived from an anatropous ovule, and the seed coat from a thick single integument. The small nucellus and the large integument are largely digested during the development of the seed. The integumentary tapetum investing the embryo sac after the nucellar epidermis breaks down is conspicuously differentiated. All the tissue outside it, except the outer epidermis of the integument, is digested (fig. 20.5A, B). The epidermis develops thickenings on the inner tangential walls and the innermost parts of the anticlinal walls (fig. 20.5C, D). The walls of the epidermal cells become mucilaginous and disintegrate except for the

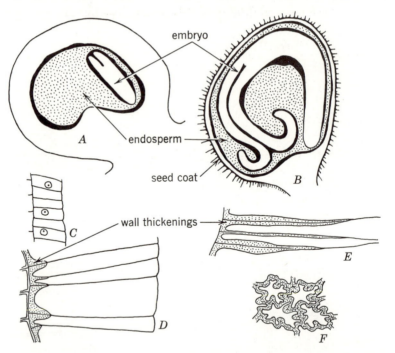

FIG. 20.5. Development of seed coat in *Lycopersicon esculentum* (tomato). Longitudinal sections of ovules sampled 25 (A) and 40 (B) days after pollination. In B, hair-like extensions on surface are wall thickenings of mucilaginous epidermal cells remaining after the cells break down. C–E, developmental stages of epidermis in longitudinal section. F, transverse section of epidermis through wall thickenings. (C–F, ×170. Adapted from Smith, N.Y. Agr. Expt. Sta. Mem. 184, 1935.)

thickened radial walls (Czaja, 1963). These remain and appear like hairs (fig. 20.5E, F). The mucilage readily separates from the seed. In the mature seed the testa includes the integumentary tapetum, remains of the epidermis, and remains of the digested integumentary parenchyma. This seed coat encloses a curved filiform embryo and an endosperm which practically fills the part of the seed not occupied by the embryo (fig. 20.5B). A cuticle occurs between the seed coat and the endosperm.

NUTRITIONAL ASPECTS IN SEED DEVELOPMENT

The characteristic feature of the seed plants is that not only does the megagametophyte develop in the tissues of the sporophyte but the new sporophyte is also supported by the old sporophyte during its early growth. The development of these gametophytic and sporophytic bodies obviously involves an active transfer of food from the old sporophyte to the new structures. This transfer occurs not merely by translocation of food through the vascular tissues to a close proximity of reproductive structures but also by an extensive digestion of tissues. Microsporogenesis and microgametogenesis, too, are associated with destruction of tissues, but on a much smaller scale than is the formation of the female reproductive structures and of the new sporophyte.

The digestive phenomena encountered during the development of the seed occur in the following chronological order. In normal sporogenesis the early growth of one of the megaspores (the functioning megaspore) into an embryo sac involves the destruction of the three nonfunctioning megaspores. Subsequently, the embryo sac increases in size by encroaching upon the nucellus, which is digested either partially or entirely (Coe, 1954). In the latter instance, there is frequently a differentiation of the integumentary tapetum, or endothelium, next to the embryo sac. During the development of the embryo several phenomena may occur: formation of endosperm; partial or complete digestion of the endosperm by the embryo; digestion of the parenchyma of the nucellus (if the nucellus is still present at this stage) and of the integuments.

A simple enumeration of phenomena fails to bring out the complexity of the relationship between the tissues being digested and those that apparently utilize the products of breakdown. The endosperm itself, for example, utilizes such products and is, at the same time, absorbed by the embryo. The relation between the embryo and the endosperm is not entirely clear. Commonly embryo development depends on the presence of the endosperm, even in apomictic species (Rutishauser, 1954), but under certain conditions the embryo appears to be able to draw directly on the food supply in the integument (Cooper and Brink, 1949). The role of the integumentary tapetum is not known definitely either. The breakdown of the parenchyma outside the tapetum suggests that the latter might be the source of digestive enzymes. With reference to the transfer of food from the tapetum to the embryo sac, it is of interest that a cuticle is originally present between the nucellus and the inner integumentary epidermis. There is, however, a cuticle-free avenue to the embryo sac in the chalazal region. Some plants develop highly specialized mechanisms of food absorption. Various cells of the embryo sac, and also the endosperm and the suspensor, may develop haustoria that penetrate into adjacent tissues (Maheshwari, 1950).

The accumulation of starch in the developing ovule, as exemplified by *Dianthus* (Buell, 1952b), is related to the stages in embryogenesis. At the time of fertilization a maximum supply of reserve food accumulates in the placenta, ovary wall, and ovule. After fertilization the starch content of the placenta rapidly declines. A large quantity of starch is present in the mature embryo sac which is used in early embryo development. The nucellus has two periods of accumulation. One starts in early stages of ovule development and reaches a maximum at maturity of the embryo sac. The other is the accumulation of the definitive deposit in the perisperm. A contrasting situation is encountered in the spider lily (*Hymenocallis;* Flint and Moreland, 1943) in which no concentration of organic reserve takes place. But the integuments develop into chlorenchyma with stomata and, as experiments indicate, the development of the embryo depends on the photosynthetic activity of this tissue.

The embryo is not known to have any vascular connection with the old sporophyte. In fact, its cellular connection with the next-lying tissues is usually ephemeral. In the early stages of its development the embryo is attached to the embryo-sac wall by means of the supensor, but the suspensor frequently appears shrivelled before the embryo is full grown. The exact nature of the connection between the suspensor and the embryo-sac wall, particularly whether there are any plasmodesmata in this connection, has not been determined. In the transfer of food from the endosperm to the embryo during the germination of albuminous seeds parts of the embryo may be differentiated as absorbing organs. In the Gramineae, for example, the cotyledon (the scutellum) has a glandular epidermis which is in contact with the endosperm, and, in the onion, the tip of the cotyledon is a digestive structure (pl. 96G). In many seeds, however, the embryo has no specialized digestive tissues and appears to depend on transfer of materials through its epidermis when, during germination, it takes up food from the endosperm.

REFERENCES

Aleksandrov, V. G., and O. G. Aleksandrova. Ob otmiranii i razrushenii ïader v kletkakh endosperma zlakov kak odnom iz vazhneïshikh faktorov, obuslavlivaïushchikh naliv zernovki. [Death and disintegration of nuclei in cells of cereal endosperm as one of the most important factors determining the filling of the grain.] *Izvest. Akad. Nauk SSSR Ser. Biol.* 1954:88–103. 1954.

Allen, G. S. Embryogeny and the development of the apical meristems of *Pseudotsuga*. II. Late embryogeny. *Amer. Jour. Bot.* 34:73–80. 1947a. III. Development of the apical meristems. *Amer. Jour. Bot.* 34:204–211. 1947b.

Arnott, H. J. The seed, germination, and seedling of *Yucca*. *Calif. Univ., Publs., Bot.* 35: 1–164. 1962.

Artschwager, E. Development of flowers and seed in the sugar beet. *Jour. Agr. Res.* 34: 1–25. 1927.

Avdulov, N. P. Kario-systematicheskoe issledovanie semeïstva zlakov. [Kario-systematic investigation of the grass family.] *Bul. Appl. Bot., Genet., and Plant Breeding, Suppl.* 44. 1931.

Bagley, B. W., J. H. Cherry, M. L. Rollins, and A. M. Altschul. A study of protein bodies during germination of peanut (*Arachis hypogaea*) seed. *Amer. Jour. Bot.* 50:523–532. 1963.

Baude, E. Die Embryoentwicklung von *Stratiotes aloides* L. *Planta* 46:649–671. 1956.

Bennett, C. W., and K. Esau. Further studies on the relation of the curly-top virus to plant tissues. *Jour. Agr. Res.* 53:595–620. 1936.

Berg, R. Y. Seed dispersal, morphology, and phylogeny of *Trillium*. *Norske Vidensk. Akad. i Oslo Math. Nat. Kl. Skr.* 1958:1–36. 1958.

Bhadurim, P. N. Studies on the embryogeny of the Solanaceae. I. *Bot. Gaz.* 98:283–295. 1936.

Blagoveshchenskiï, A. V. Osobennosti belkovykh veshchestv semïan razlichnykh predstaviteleï Leguminosae. [Characteristics of proteinaceous substances in seeds of different representatives of Leguminosae.] In: *Problemy Botaniki*. III. Leningrad, Izdatel'stvo Akademii Nauk SSSR. 1958.

Borthwick, H. A. Development of the macrogametophyte and embryo of *Daucus carota*. *Bot. Gaz.* 92:23–44. 1931.

Brink, R. A., and D. C. Cooper. The endosperm in seed development. *Bot. Rev.* 13:423–477, 479–541. 1947.

Brown, W. V. The morphology of the grass embryo. *Phytomorphology* 10:215–223. 1960.

Buell, K. M. Developmental morphology in *Dianthus*. I. Structure of the pistil and seed development. *Amer. Jour. Bot.* 39:194–210. 1952a. II. Starch accumulation in ovule and seed. *Amer. Jour. Bot.* 39:458–467. 1952b.

Butterfass, T. Fluoroskopische Untersuchungen über Anatomie und Funktion von Koleoptilen. *Deut. Bot. Gesell. Ber.* 69:245–254. 1956.

Buttrose, M. S. Submicroscopic development and structure of starch granules in cereal endosperms. *Jour. Ultrastruct. Res.* 4:231–257. 1960.

Cave, M. S., H. J. Arnott, and S. A. Cook. Embryogeny in the California peonies with reference to their taxonomic position. *Amer. Jour. Bot.* 48:397–404. 1961.

Chopra, R. N. Some observations on endosperm development in the Cucurbitaceae. *Phytomorphology* 5:219–230. 1955.

Coe, G. E. Distribution of carbon 14 in ovules of *Zephyranthes drummondii*. *Bot. Gaz.* 115:342–346. 1954.

Cooper, D. C., and R. A. Brink. The endosperm-embryo relationship in an autonomous apomict, *Taraxacum officinale*. *Bot. Gaz.* 111:139–153. 1949.

Corner, E. J. H. The leguminous seed. *Phytomorphology* 1:117–150. 1951.

Crété, P. L'application de certaines données embryologiques à la systématique des Orobanchacées et de quelques familles voisines. *Phytomorphology* 5:422–435. 1955.

Crocker, W., and L. V. Barton, eds. *Physiology of seeds. An introduction to the experimental study of seed and germination problems.* Vol. 29. Waltham, Mass., Chronica Botanica Company. 1953.

Cutter, V. M., Jr., K. S. Wilson, and B. Freeman. Nuclear behavior and cell formation in the developing endosperm of *Cocos nucifera*. *Amer. Jour. Bot.* 42:109–115. 1955.

Czaja, A. T. Neue Untersuchungen an der Testa der Tomatensamen. *Planta* 59:262–279. 1963.

Dore, W. G. Some grass genera with liquid endosperm. *Torrey Bot. Club Bul.* 83:335–337. 1956.

Esau, K. Morphology and reproduction in guayule and certain other species of *Parthenium*. *Hilgardia* 17:61–120. 1946.

Esau, K. *Anatomy of seed plants.* New York, John Wiley and Sons. 1960.

Flint, L. H., and C. F. Moreland. Note on photosynthetic activity in seeds of the spider lily. *Amer. Jour. Bot.* 30:315–317. 1943.

Foard, D. E., and A. H. Haber. Use of growth characteristics in studies of morphologic relations. I. Similarities between epiblast and coleorhiza. *Amer. Jour. Bot.* 49:520–523. 1962.

Freytag, K. Quellende Schleimzellen der Samenepidermis von *Sinapis alba* im polarizierten Licht. *Protoplasma* 50:166–172. 1958.

Frey-Wyssling, A. *Die pflanzliche Zellwand.* Berlin, Springer-Verlag. 1959.

Guignard, J. L. Recherches sur l'embryogénie des Graminées; rapports des Graminées avec l'autres Monocotylédones. *Ann. des Sci. Nat., Bot.* Ser. 12. 2:491–610. 1961.

Guttenberg, H. von. Grundzüge der Histogenese höherer Pflanzen. I. Die Angiospermen. In: *Handbuch der Pflanzenanatomie.* Band 8. Teil 3. Berlin, Gebrüder Borntraeger. 1960.

Haccius, B. Die Embryoentwicklung bei *Ottelia alismoides* und das Problem des terminalen Monokotylen-Keimblattes. *Planta* 40:443–460. 1952a.

Haccius, B. Verbreitung und Ausbildung der Einkeimblättrigkeit bei den Umbelliferen. *Österr. Bot. Ztschr.* 99:483–505. 1952b.

Haccius, B. Histogenetische Untersuchungen an Wurzelhaube und Kotyledonarscheide geophiler Keimpflanzen (*Podophyllum* und *Eranthis*). *Planta* 41:439–458. 1953.

Haccius, B. Embryologische und histogenetische Studien an "monocotylen Dikotylen". I. *Claytonia virginica* L. *Österr. Bot. Ztschr.* 101:285–303. 1954.

Haskell, G. Pleiocotyly and differentiation within angiosperms. *Phytomorphology* 4:140–152. 1954.

Hyde, E. O. C. The function of the hilum in some Papilionaceae in relation to ripening of the seed and the permeability of the testa. *Ann. Bot.* 18:241–256. 1954.

Iakovlev, M. S. Struktura endosperma i zarodysha zlakov kak systematicheskiĭ priznak. [Structure of endosperm and embryo of cereals as a systematic character.] In: *Morfologiĭa i Anatimiĭa Rastenii.* I. Leningrad, Izdatel'stvo Akademii Nauk SSSR. 1950.

Iakovlev, M. S. Printsipy vydeleniĭa osnovnykh embrional'nykh tipov i ikh znachenie dlĭa filogenii pokrytosemennykh. [Principles of delimitation of basic embryo types and their significance for the phylogeny of angiosperms.] In: *Problemy Botaniki.* III. Leningrad, Izdatel'stvo Akademii Nauk SSSR. 1958.

Ioffe, M. D. Razvitie zarodysha i endosperma u pshenitsy, konskikh bobov i redisa. [Development of embryo and endosperm in wheat, horse beans, and radish.] In: *Morfologĭa i Anatomiĭa Rastenii.* IV. Leningrad, Izdatel'stvo Akademii Nauk SSSR. 1957.

Jacques-Felix, H. Sur une interprétation nouvelle de l'embryon des Graminées. Conséquences terminologiques et rapports avec les autres types d'embryons. *Acad. des Sci. Compt. Rend.* 246:150–153. 1958.

Johansen, D. A. *Plant embryology.* Waltham, Mass., Chronica Botanica Company. 1950.

Kadry, A. E. R. The development of endosperm and embryo in *Cistanche tinctoria* (Forssk.) G. Beck. *Bot. Notiser* 108:231–243. 1955.

Kantor, T. S. Ob activnosti chloroplastov zarodysha l'na. [On the activity of chloroplasts of flax embryo.] *Moskov. Glav. Bot. Sad Bul.* 23:61–67. 1955.

Kooiman, P. On the occurrence of amyloids in plant seeds. *Acta Bot. Neerland.* 9:208–219. 1960.

Lebègue, A. Recherches embryogéniques sur quelques Dicotylédones Dialypétales. *Ann. des Sci. Nat., Bot.* Ser. 11. 13:1–160. 1952.

Maheshwari, P. *An introduction to the embryology of angiosperms.* New York, McGraw-Hill Book Company. 1950.

Mahlberg, P. G. Embryogeny and histogenesis in *Nerium oleander* L.—I. Organization of primary meristematic tissues. *Phytomorphology* 10:118–131. 1960.

Martin, A. C. The comparative internal morphology of seeds. *Amer. Midland Nat.* 36:513–660. 1946.

Matlakówna, M. Ueber Gramineenfrüchte mit weichem Fettendosperm. *Acad. Sci. Cracovie Bul. Ser. B.* 1912:405–416. 1912.

McClure, D. S. Seed characters of selected plant families. *Iowa State Coll. Jour. Sci.* 31:649–682. 1957.

Meeuse, A. D., and E. C. J. Ott. The occurrence of chlorophyll in *Nelumbo* seeds. *Acta Bot. Neerland.* 11:228. 1962.

Meier, H. On the structure of cell walls and cell wall mannans from ivory nuts and from dates. *Biochem. et Biophys. Acta* 28:229–240. 1958.

Meyer, C. F. Cell patterns in early embryogeny of the McIntosh apple. *Amer. Jour. Bot.* 45:341–349. 1958.

Miller, E. V. The accumulation of carbohydrates by seeds and fruits. *Handb. der Pflanzenphysiol.* 6:871–880. 1958.

Miller, H. A., and R. H. Wetmore. Studies in the developmental anatomy of *Phlox drummondii* Hook. I. The embryo. *Amer. Jour. Bot.* 32:588–599. 1945.

Müller, D. Tote Speichergewebe in lebenden Samen. *Planta* 33:721–727. 1943.

Nast, C. G. The embryogeny and seedling morphology of *Juglans regia* L. *Lilloa* 6:163–205. 1941.

Negbi, M., and D. Koller. Homologies in the grass embryo—a re-evaluation. *Phytomorphology* 12:289–296. 1962.

Netolitzky, F. Anatomie der Angiospermen-Samen. In: K. Linsbauer. *Handbuch der Pflanzenanatomie.* Band 10. Lief. 14. 1926.

Norstog, K. The growth and differentiation of cultured barley embryos. *Amer. Jour. Bot.* 48:876–884. 1961.

Pankow, H., and H. von Guttenberg. Vergleichende Studien über die Entwicklung monocotyler Embryonen und Keimpflanzen. *Bot. Stud.* Heft 7:1–39. 1957.

Pant, D. D., and D. D. Nautiyal. Seed cuticles in some modern cycads. *Current Sci.* [India] 31:75–76. 1962.

Periasamy, K. Studies on seeds with ruminate endosperm. II. Development of rumination in the Vitaceae. *Indian Acad. Sci. Proc. Sect. B.* 56:13–26. 1962.

Poddubnaĩa-Arnol'di, V. A. Issledovanie zarodysheĩ u pokrytosemennykh rasteniĩ v zhivom sostoĩanii. [Study of angiosperm embryos in living state.] *Moskov. Glav. Bot. Sad Bul.* 14:3–12. 1952.

Rao, V. S. 'Nuclear endosperm' or 'non-cellular endosperm'? *Ann. Bot.* 23:364. 1959.

Reeder, J. R. The embryo of *Streptochaeta* and its bearing on the homology of the coleoptile. *Amer. Jour. Bot.* 40:77–80. 1953.

Reeder, J. R. The embryo in grass systematics. *Amer. Jour. Bot.* 44:756–768. 1957.

Reeve, R. M. Structural composition of the sclereids in the integument of *Pisum sativum* L. *Amer. Jour. Bot.* 33:191–204. 1946*a*.

Reeve, R. M. Ontogeny of the sclereids in the integument of *Pisum sativum* L. *Amer. Jour. Bot.* 33:806–816. 1946*b*.

Reeve, R. M. Late embryogeny and histogenesis in *Pisum*. *Amer. Jour. Bot.* 35:591–602. 1948.

Robbins, W. W., and H. A. Borthwick. Development of the seed of *Asparagus officinalis*. *Bot. Gaz.* 80:426–438. 1925.

Roth, I. Zur morphologischen Deutung des Grasembryos und verwandter Embryotypen. *Flora* 142:564–600. 1955.

Roth, I. Histogenese und Entwicklungsgeschichte des *Triticum*-Embryos. *Flora* 144:163–212. 1957.

Rutishauser, A. Entwicklungserregung der Eizelle bei pseudogamen Arten der Gattung *Ranunculus*. *Schweiz. Akad. Wiss. Bul.* 10:491–512. 1954.

Sarfatti, G. Studies on the membrane of the almond endosperm haustorium. *Ann. Bot.* 24: 451–457. 1960.

Schnarf, K. Embryologie der Angiospermen. In: K. Linsbauer. *Handbuch der Pflanzenanatomie*. Band 10. Lief. 21. 1927; Lief 23. 1928; Lief 24. 1929.

Schnarf, K. *Vergleichende Embryologie der Angiospermen*. Berlin, Gebrüder Borntraeger. 1931.

Scott, F. M., B. G. Bystrom, and E. Bowler. *Cercidium floridum* seed coat, light and electron microscopic study. *Amer. Jour. Bot.* 49:821–833. 1962.

Singh, B. Studies on the structure and development of seeds of Cucurbitaceae. *Phytomorphology* 3:224–239. 1953.

Smith, O. Pollination and life-history studies of the tomato (*Lycopersicon esculentum* Mill.). *N.Y.* (*Cornell*) *Agr. Expt. Sta. Mem.* 184. 1935.

Souèges, R. *Embryogenie et classification*. Paris, Hermann. 1938, 1939, 1948, 1951.

Souèges, R. L'origine du cône végétatif de la tige et la question de la "terminalité" du cotylédon des Monocotylédones. *Ann. des Sci. Nat., Bot.* Ser. 11. 15:1–20. 1954.

Souèges, R. Essai d'embryogénie comparée dans les limites des Hédysarées. *Ann. des Sci. Nat., Bot.* Ser. 11. 17:325–352. 1956.

Spurr, A. R. Histogenesis and organization of the embryo in *Pinus Strobus* L. *Amer. Jour. Bot.* 36:629–641. 1949.

Spurr, A. R. Organization of the procambium and development of the secretory cells in the embryo of *Pinus Strobus* L. *Amer. Jour. Bot.* 37:185–197. 1950.

Steffen, K. Die Embryoentwicklung von *Impatiens glanduligera* Lidl. *Flora* 139:394–461. 1952.

Steiner, M., and I. Janke. Sind die Malpighischen Zellen die Epidermis der Leguminosentesta? *Österr. Bot. Ztschr.* 102:542–550. 1955.

Swamy, B. G. L., and K. K. Lakshmanan. The origin of epicotylary meristem and cotyledon in *Halophila ovata* Gaudich. *Ann. Bot.* 26:243–249. 1962.

Swamy, B. G. L., and D. Padmanabhan. A reconnaissance of angiosperm embryogenies. *Indian Bot. Sci. Jour.* 41:422–439. 1962.

Swamy, B. G. L., and N. Parameswaran. The helobial endosperm. *Cambridge Phil. Soc. Biol. Rev.* 38:1–50. 1963.

Takhtajan, A. *Die Evolution der Angiospermen*. Jena, Gustav Fischer. 1959.

Tateoka, T. Starch grains in grass systematics. *Bot. Mag. Tokyo* 75:377–383. 1962.

Varner, J. E., and G. Schidlovsky. Intracellular distribution of proteins in pea cotyledons. *Plant Physiol.* 38:139–144. 1963.

Wardlaw, C. W. *Embryogenesis in plants.* New York, John Wiley and Sons. 1955.

Wunderlich, R. Zur Frage der Phylogenie der Endospermtypen bei den Angiospermen. *Österr. Bot. Ztschr.* 106:203–293. 1959.

Yarbrough, J. A. *Arachis hypogaea.* The seedling, its epicotyl and foliar organs. *Amer. Jour. Bot.* 44:19–30. 1957.

Yatsu, L., and A. M. Altschul. Lipid-protein particles: isolation from seeds of *Gossypium hirsutum. Science* 142:1062–1064. 1963.

The following pages, plates 1 to 96, have been printed by the letterpress method to assure the greatest clarity of reproduction.

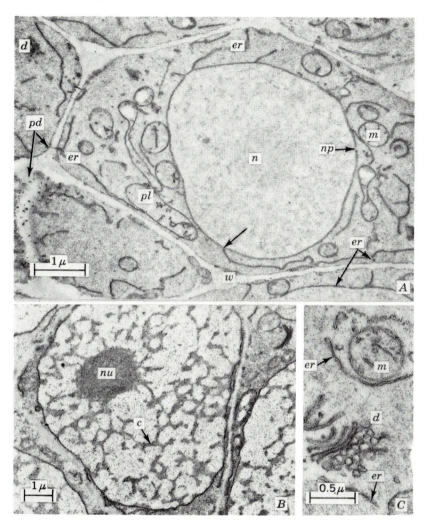

PLATE 1. Ultrastructural details of cells. *A*, meristematic cell from *Vitis* root tip. *B*, nucleus from young cell of *Elodea* shoot. *C*, part of parenchyma cell of *Cucurbita* petiole. Details: *c*, chromatin; *d*, dictyosome; *er*, endoplasmic reticulum; *m*, mitochondrion; *n*, nucleus; *np*, nuclear pore; *nu*, nucleolus (appears to be connected to chromatin); *pl*, plastid; *pd*, plasmodesmata; *w*, cell wall; unlabeled arrow, connection of *er* and nuclear envelope. (*C*, Esau and Cheadle, *Bot. Gaz.* 124, 1962.)

631

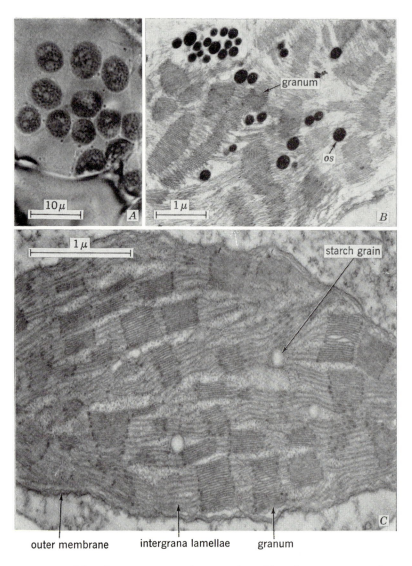

outer membrane   intergrana lamellae   granum

PLATE 2. Chloroplast structure. Grana within chloroplasts appear as dots in light microscope (A), as stacks of membranes in electron microscope (B, C). A, from cotyledon of *Lycopersicon;* B and C, from leaves of *Aspidistra* (B) and *Zea* (C). Detail: *os,* osmiophilic body. (A, Hagemann, *Biol. Ztbl.* 79, 1960; B, courtesy of T. E. Weier; C, Lehninger, *Sci. Amer.* Sept. 1961, and courtesy of A. E. Vatter.)

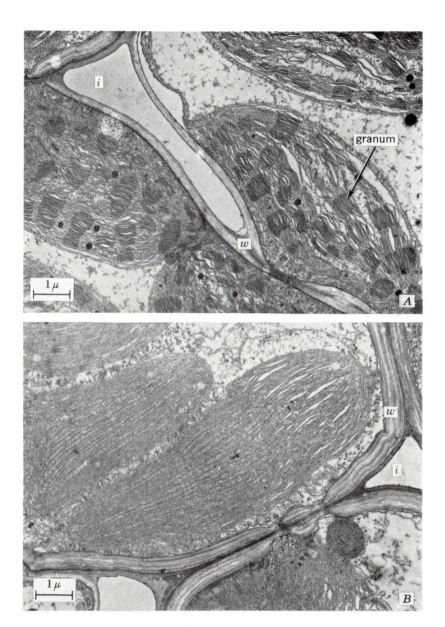

PLATE 3. Ultrastructure of chloroplasts of leaf of *Anthephora* (Gramineae). A, with grana, in mesophyll cells; B, without grana, in bundle sheath cells. Details: *i*, intercellular space; *w*, cell wall. (Courtesy of W. V. Brown.)

633

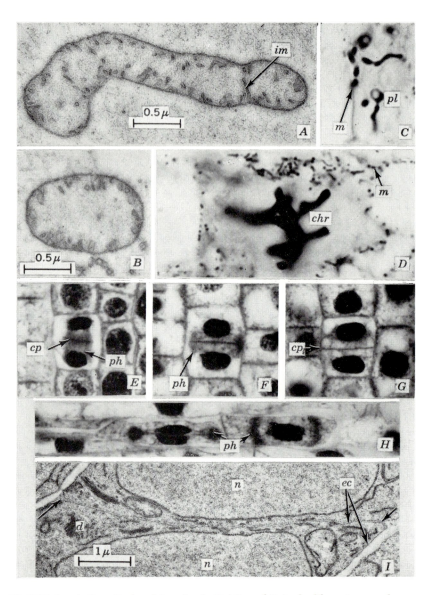

PLATE 4. *A–D*, mitochondria. In *A*, *B* (*Cucurbita*), double outer membranes and tubular inner membranes. *C*, *Zea* scutellum. *D*, *Narcissus* root. *E–I*, cytokinesis. Three successive stages in *E–G* (*Allium* root). Sectional and face views in *H* (*Nicotiana* leaf). *I*, *Ajuga* nectary, unlabeled arrows mark new cell wall. Details: *chr*, chromosomes; *cp*, cell plate; *d*, dictyosome; *ec*, ectoplast; *im*, inner membrane; *m*, mitochondrion; *n*, nucleus; *ph*, phragmoplast; *pl*, plastid with starch; *A*, *B*, *I*, electron micrographs. (*A*, *B*, Esau and Cheadle, *Bot. Gaz.* 124, 1962; *C*, *D*, Newcomer, *Amer. Jour. Bot.* 33, 1946, ×1,800; *E–G*, *Encyclopaedia Britannica*, copyright 1945, ×760; *H*, ×720.)

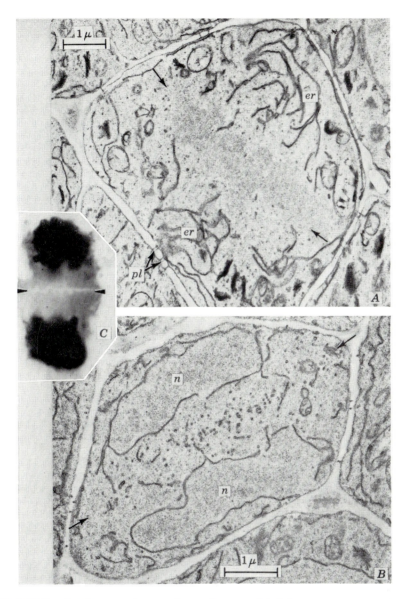

PLATE 5. Cytokinesis. *A*, metaphase plate at unlabeled arrows, concentrations of endoplasmic reticulum (*er*) at poles. *B*, telophase, beginning of cell plate formation at arrows in equatorial plane (circles, possibly vesicles concerned with formation of cell plate); endoplasmic reticulum organizing the envelopes of nuclei (*n*). Plasmodesmata at *pl* in *A*. *A*, *B*, *Ajuga* nectary, electron micrographs. *C* (*Narcissus* root, courtesy of E. H. Newcomer, ×2,500), telophase with cell plate between arrowheads.

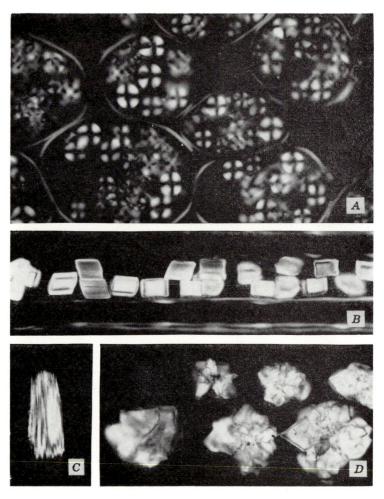

PLATE 6. Doubly refractive ergastic bodies seen in polarized light. A, starch grains in cells of *Convolvulus* root showing the polarization Maltese cross indicating radial or circumferential molecular structure. B, prismatic crystals in phloem parenchyma of root of *Abies*. C, raphides in leaf of *Vitis*. D, druses in cortex of stem of *Tilia*. Double refraction of cell walls also is evident in A, B. (A, C, D, ×750; B, ×500.)

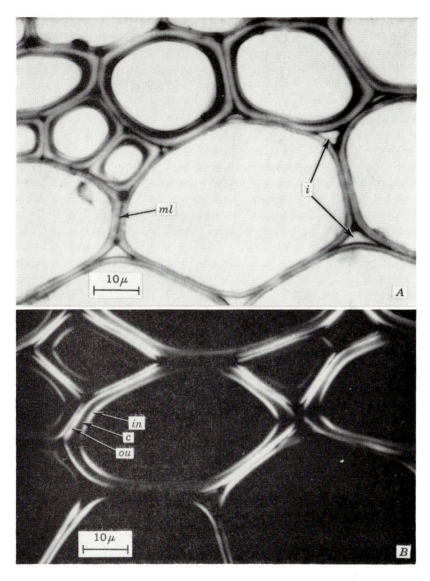

PLATE 7. Cell walls seen in ordinary (*A*) and polarized (*B*) light. Xylem cells (above in each figure) and parenchyma cells from petioles of *Nicotiana*. All cells have secondary walls. In parenchyma cells, primary and secondary walls are indistinguishable; in xylem cells, primary wall is fused with outer layer of secondary wall. Details: *in*, inner, *c*, central, and *ou*, outer layers of secondary wall (*c* in extinction position); *i*, intercellular spaces lined with intercellular material; *ml*, middle lamella.

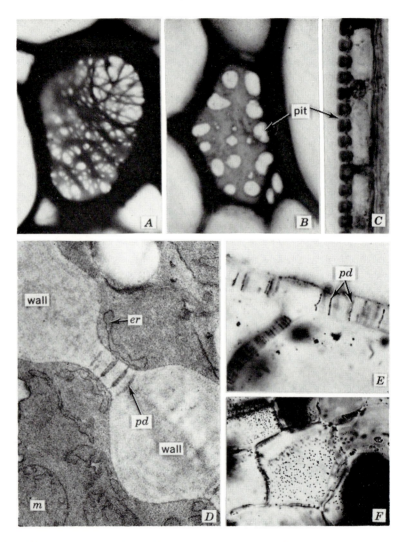

PLATE 8. Pits and plasmodesmata. Parenchyma cells from: root cortex of *Abies* (*A*, ×750); xylem of *Nicotiana* (*B*, ×1,000) and *Vitis* (*C*, ×750); phloem of *Robinia* (*D*, ×15,700, electron micrograph); stem cortex (*E*, ×750) and tuber (*F*, ×325) of *Solanum*. *A*, surface view of reticulum of cellulose; unstained meshes are thin places penetrated by plasmodesmata (not visible). *B*, pits in surface view and, *C*, in section. In *C*, pitted wall between parenchyma cell and vessel. *D*, plasmodesmata (*pd*) in pit membrane, connected to endoplasmic reticulum (*er*). *E*, plasmodesmata (*pd*) in sectional and, *F*, surface (dots) views. (*E*, *F*, Crafts, *Plant Physiol.* 8, 1933.)

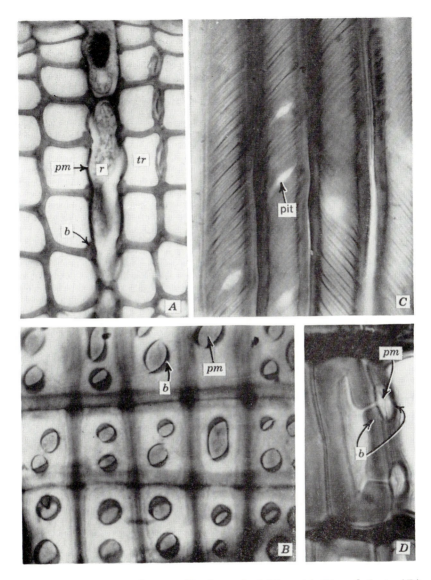

PLATE 9. Pits in tracheary cells of wood of *Pinus* (*A–C*) and *Larix* (*D*). Half-bordered pit-pairs between tracheids (*tr*) and ray cells (*r*) in sectional (*A*, transection of wood) and surface (*B*, radial section) views. The border (*b*) is on the tracheid side. It surrounds the wide pit membrane (*pm*). *C*, compression wood with helical bands of aggregated macrofibrils in inner layer of secondary wall and pits with slit-like apertures. *D*, late-wood tracheid in transection with pits in tangential wall. Border (*b*) is thin on early-wood side, thick on the late-wood side. Pit membrane (*pm*) without torus. (*A–C*, ×750; *D*, ×1,200.)

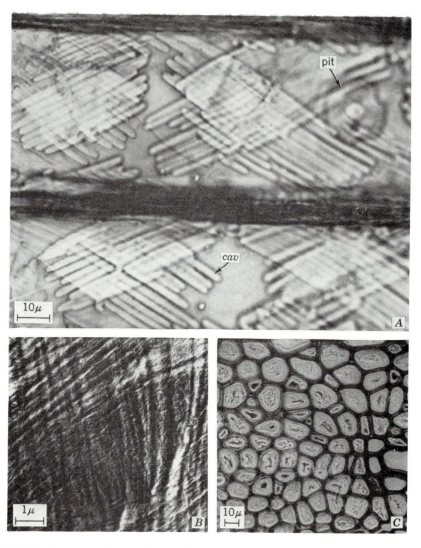

PLATE 10. Secondary wall in xylem. Longitudinal sections of *Sequoia* (*A*) and *Fraxinus* (*B*) wood and transection (*C*) of tension wood of *Grevillea*. In *A*, rotted heartwood has cylindrical cavities (*cav*) produced by fungal enzymes. Cavities are oriented helically in conformity with orientation of microfibrils in central wall layer of tracheid. *B* illustrates parallel orientation of aggregations of microfibrils in the outer layer of fiber wall (electron micrograph). *C*, fibers have gelatinous secondary wall layers (light staining). (*A*, courtesy of L. Bonar; *B*, Bosshard, *Schweiz. Ztschr. f. Forstw.* 107, 1956; *C*, Scurfield and Wardrop, *Austral. Jour. Bot.* 10, 1962.)

640

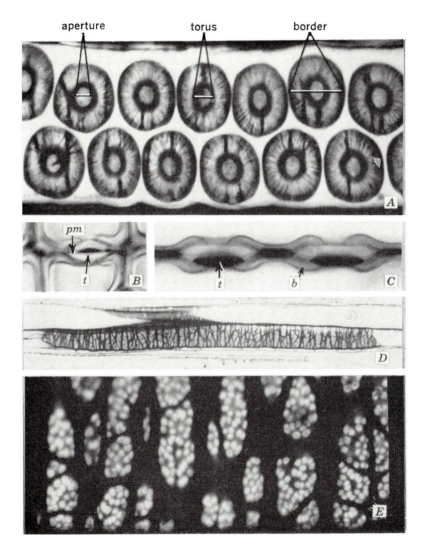

PLATE 11. Bordered pits of conifers (A, Tsuga; B, Abies; C, Pinus) and sieve areas of a monocotyledon (D, E, Cocos). A, surface view of pits with thickenings on pit membranes. B and C, pit-pairs in sectional views with torus (t) on pit membrane (pm) in median position (B) and appressed to the border (b in C; aspirated pit-pair). D, surface view of compound sieve plate with sieve areas in reticulate arrangement. E, part of similar sieve plate. Light spots are callose cylinders. (A, Bannan, Torrey Bot. Club Bul. 68, 1941, ×900; B, C, ×1,200; D, E, Cheadle and Whitford, Amer. Jour. Bot. 28, 1941; D, ×128; E, ×1,000.)

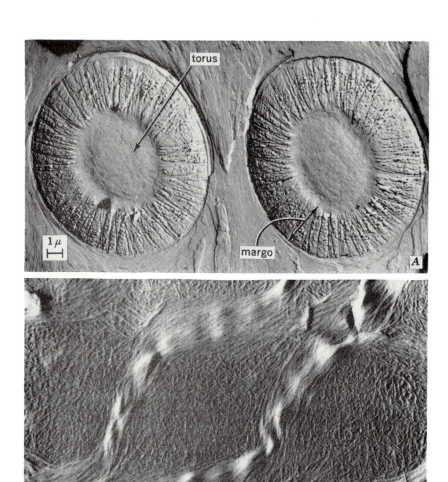

PLATE 12.   Electron micrographs (replication method in *A*) of walls of xylem
cells.   *A*, two pit closing membranes with tori of bordered pits of *Pinus*.   In
margo microfibrils predominantly in radial arrangement.   *B*, vessel wall of
Fraxinus during growth of secondary wall.   The band-like thickenings delimit
future pits.   (*A*, Liese, *Allg. Forstztschr.* 1961; *B*, Bosshard, *Schweiz. Ztschr. f.
Forstw.* 107, 1956).

642

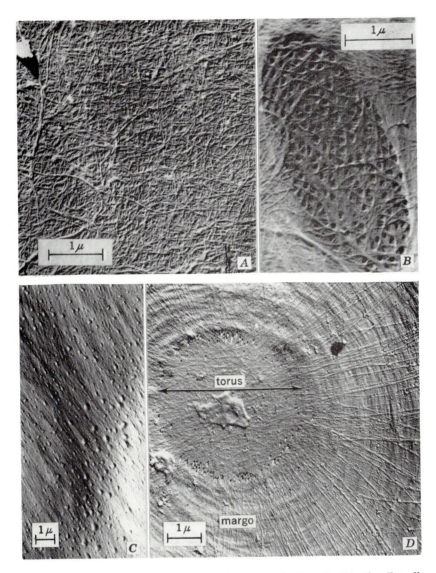

PLATE 13. Electron micrographs (replication method in *C*, *D*) of cell walls. *A*, primary wall of *Linum* fiber with scattered reticulate microfibril arrangement. *B*, primary pit-field with plasmodesmatal openings, from *Zea* coleoptile. *C*, inner side of wall of *Pinus* tracheid with wart-like structures. *D*, part of bordered pit membrane of *Thuja* with poorly developed torus and a margo composed of mainly radially oriented bands of microfibrils. (*A*, Frey-Wyssling et al., *Experientia* 4, 1948; *B*, Mühlethaler, *Biochem. Biophys. Acta* 5, 1950; *C*, Liese, *Allg. Forstztschr.*, 1961; *D*, courtesy of W. Liese.)

643

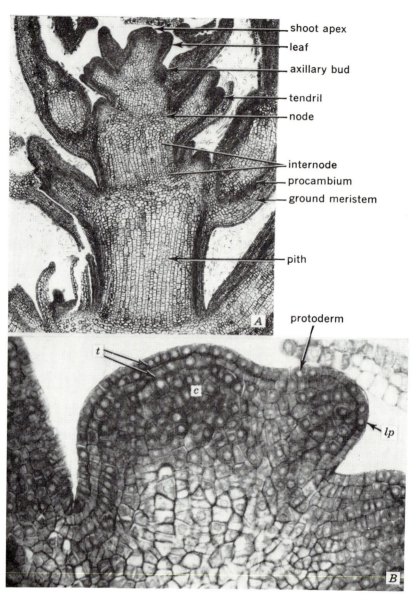

PLATE 14. Shoot (A) and shoot tip (B) of Vitis (grapevine). The apical
meristem has two layers of tunica (t) and a *corpus* (c) several cells deep. Leaf
primordia (one in B, *lp*), tendrils, and increments of stem originate in the apical
meristem; stem elongation occurs chiefly through internodal elongation (rib-
meristem growth, A). (Pratt, *Amer. Jour. Bot.* 46, 1959; courtesy of Brookhaven
National Laboratory; A, ×60; B, ×390.)

644

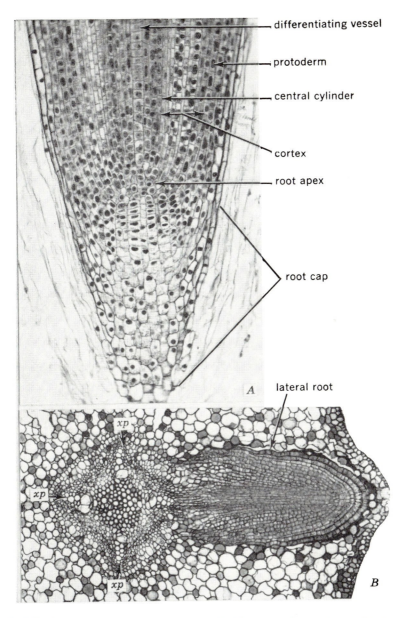

differentiating vessel

protoderm

central cylinder

cortex

root apex

root cap

A   lateral root

xp

xp

xp

B

PLATE 15. Origin of roots. *A*, root tip of *Allium* (garlic) in longitudinal section. *B*, lateral root of *Salix* (willow) that originated opposite one of the four protoxylem poles (*xp*) of the parent root. The latter is seen in transection. (*A*, courtesy of L. K. Mann, ×158; *B*, slide by P. L. Conant, ×82.)

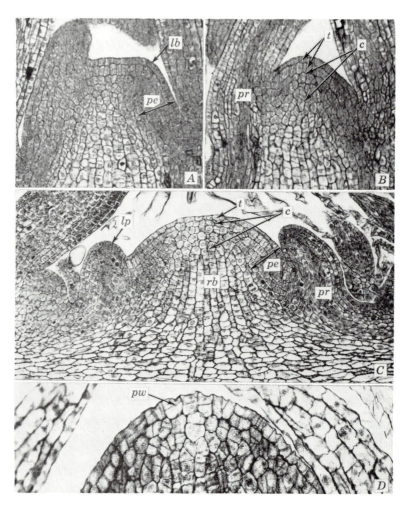

PLATE 16.   Apical meristems in median longitudinal sections of *Salix* (*A*, *B*, ×180), *Opuntia* (*C*, ×126), and *Torreya* (*D*, ×216).   Tunica (*t*) of two layers in *A*, *B*, of one layer in *C*, and none in *D* (periclinal divisions occur in outermost layer).   *A*, flat apex during formation of leaf buttress (*lb*).   *B*, conical apex between periods of leaf initiation.   Details: *c*, corpus; *lb*, leaf buttress; *pe*, peripheral zone; *pr*, procambium; *pw*, periclinal wall; *rb*, rib meristem; *t*, tunica. (*A*, *B*, Reeve, *Amer. Jour. Bot.* 35, 1948; *C*, Boke, *Amer. Jour. Bot.* 31, 1944; *D*, Kemp, *Amer. Jour. Bot.* 30, 1943.)

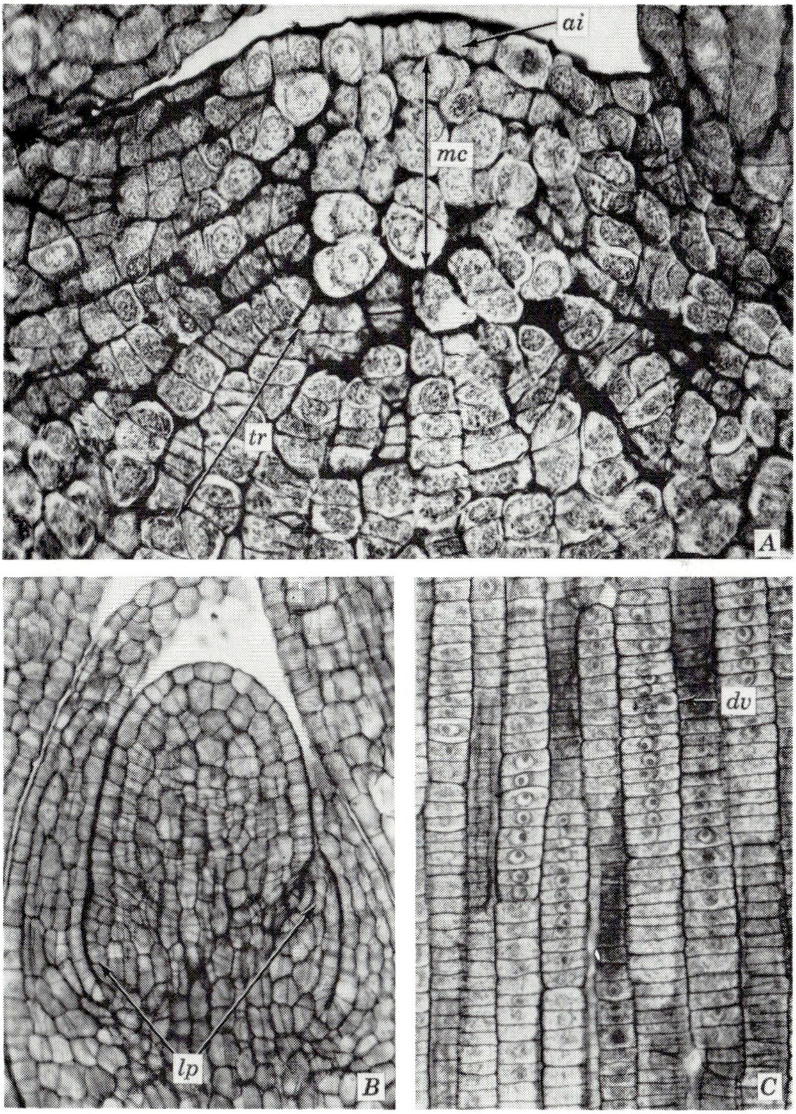

PLATE 17. Longitudinal sections of shoot apices of *Ginkgo* (*A*, ×370) and *Zea* (*B*, ×240) and of rib meristem from root cortex of *Zea* (*C*, ×240). Details: *ai*, apical initial group; *dv*, cell division; *lp*, leaf primordium (two sides of one); *mc*, mother cell zone; *tr*, transition zone. (*A*, Foster, *Torrey Bot. Club Bul.* 65, 1938; slides by: G. I. Patel, *B*; E. M. Gifford, *C*.)

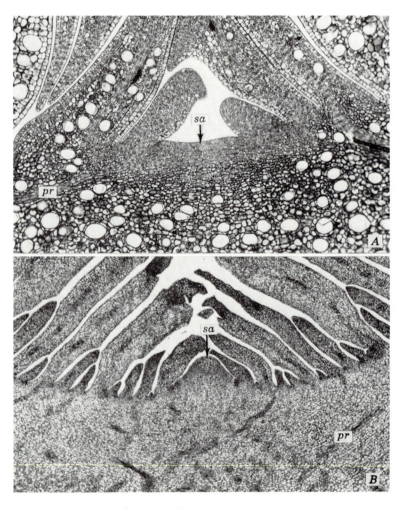

PLATE 18. Distinctive forms of shoot apices (*sa*): flat or slightly concave in *Drimys* (*A*, ×90) and conical but inserted on broad base bearing leaf primordia in *Washingtonia* palm (*B*, ×19). Longitudinal sections. Procambium at *pr*. Oil cells (large cavities) in *A*. (*A*, slide by E. M. Gifford; *B*, Ball, *Amer. Jour. Bot.* 28, 1941.)

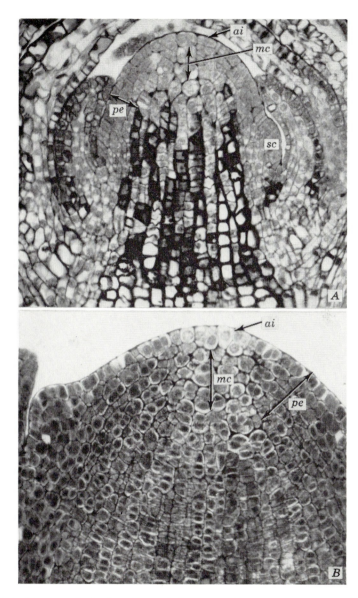

PLATE 19. Longitudinal sections of shoot tips of *Abies* during first phase of seasonal growth (*A*, ×270) and during the winter rest phase (*B*, ×350). In *A*, scale (*sc*) primordia were being initiated, and tannin content in pith distinguishes this region from apex and peripheral zone (*pe*). Results of recent divisions are evident in apex. *B*, zonation less distinctive than in *A*. Apical initial group at *ai*, mother cells at *mc*, peripheral zone at *pe*. (Parke, *Amer. Jour. Bot.* 46, 1959.)

649

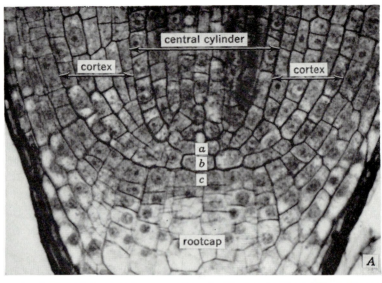

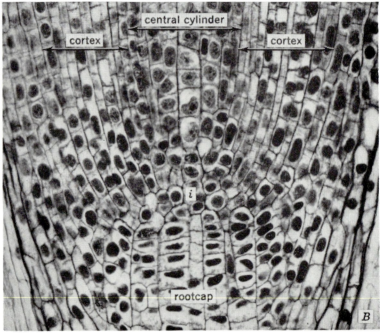

PLATE 20. Longitudinal sections of root tips of *Nicotiana tabacum* (A, ×455) and *Allium sativa* (B, ×600) showing contrasting organizations of apical meristem. In A, tissue regions were initiated in separate cell layers: *a*, central cylinder, *b*, cortex, *c*, rootcap together with epidermis. In B, tissue regions merge in a common initial group (*i*) of cells. (B, Mann, *Hilgardia*, 21, 1952.)

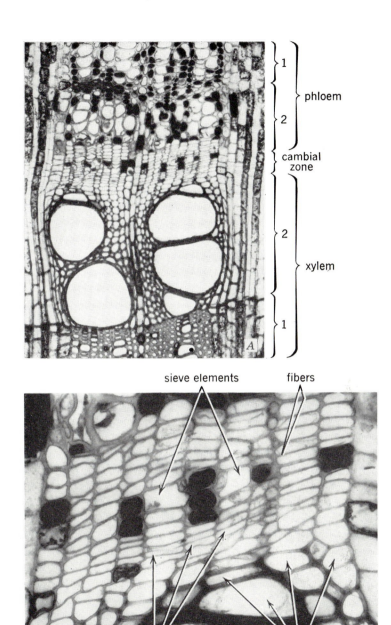

PLATE 21. Transections of vascular cambium and secondary xylem and phloem of *Vitis vinifera* (grapevine). *A*, structural differences between latest part of one seasonal increment (1) and earliest part of another (2) in xylem and phloem. Vessels have disturbed radial seriation of cells in xylem. *B*, cambial zone from *A*. Recently formed tangential walls occur in some cambial initials. Black cell contents, tannin. (*A*, ×112; *B*, ×400. Esau, *Hilgardia* 18, 1948.)

651

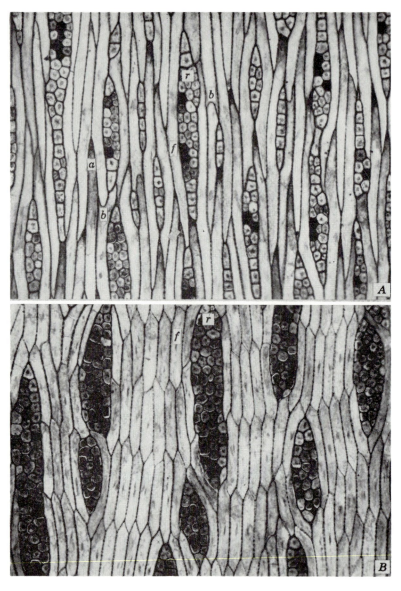

PLATE 22. Tangential longitudinal sections of vascular cambium of *Juglans* (A) and *Robinia* (B). In A, long fusiform initials (*f*) overlap each other (nonstoried cambium). In B, short fusiform initials (*f*) are in horizontal tiers (storied cambium). In both cambia ray initials (*r*) appear in groups, lenticular in outline. Evidence of apical intrusive growth in fusiform cells in A: dense cytoplasm in ends of cells (*a*) and forking (*b*). (Both, ×155. Slides by V. I. Cheadle.)

652

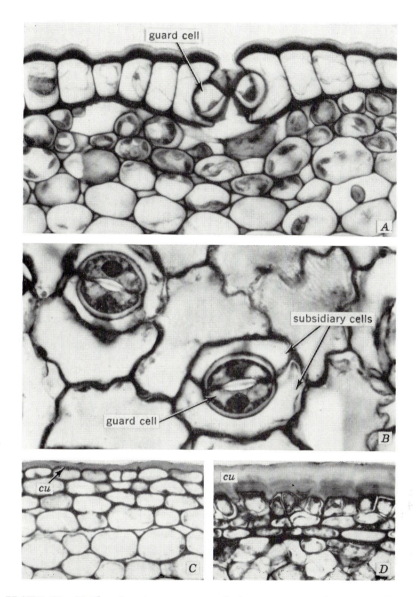

PLATE 23. Epidermis. *A,* transection of *Asparagus* stem showing epidermis and some cortex. Chlorenchyma beneath epidermis and substomatal chamber beneath guard cells. *B,* surface view of *Convolvulus* epidermis with undulate anticlinal walls and stomata. *C, D,* cuticular layers (*cu*) in transections of underground stems of *Menispermum*. In *D* (older stem) cutin has occluded some cells. (*A, B,* ×760; *C, D,* ×700; *A, B, Encyclopaedia Britannica,* copyright 1945.)

653

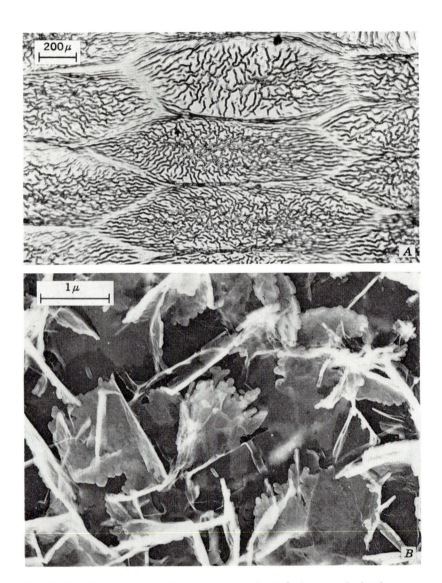

PLATE 24. Plant surfaces. *A*, surface view of cuticle from petal of *Pelargonium* (from vein on abaxial side). Hexagonal areas indicate limits of individual cells. *B*, surface view of adaxial side of *Pisum* leaf illustrating wax projections presumably extruded through cuticle. Electron micrograph. (*A*, from strip preparation by G. Girolami; *B*, Juniper, *Endeavour* 18, 1959.)

654

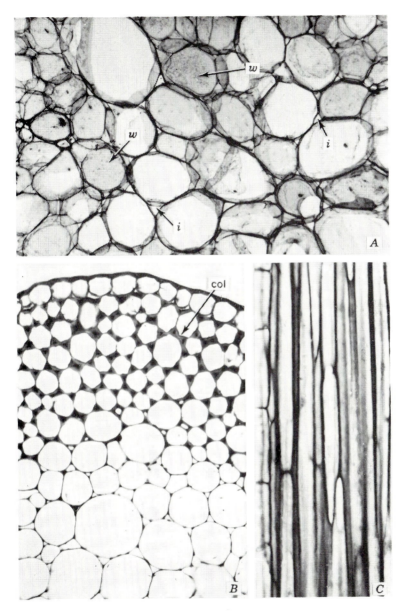

PLATE 25. A, parenchyma from *Lycopersicon* (tomato) stem. Intercellular spaces at *i*, walls in surface views at *w*. Collenchyma (*col*) of *Beta* (sugar beet) petiole in transection (B) and of *Vitis* (grapevine) stem in longitudinal section (C). (A, ×49; B, C, ×285; B, *Encyclopaedia Britannica*, copyright 1945.)

655

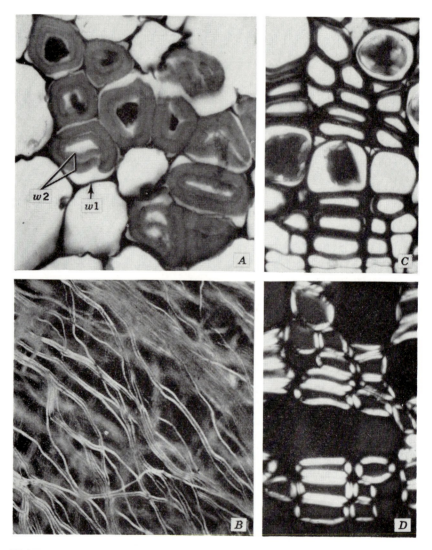

PLATE 26. *A*, transection of immature phloem fibers of *Cannabis* (hemp) stem. Layered secondary wall (*w2*) more or less infolded and separated from primary wall (*w*)—probably an artifact. *B*, filiform sclereids of *Olea* (olive), doubly refractive in polarized light, as seen in cleared leaf. *C*, *D*, transections of secondary phloem of *Abies* in nonpolarized (*C*) and polarized (*D*) light. In *D*, doubly refractive walls (probably secondary) identify sieve cells. (*A*, ×750; *B*, ×57; *C*, *D*, slide by H. E. Wilcox, ×500.)

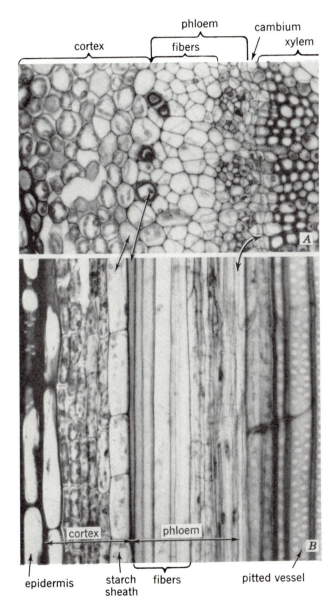

cortex   phloem   cambium
          fibers   xylem

epidermis   starch   fibers   pitted vessel
             sheath

cortex   phloem

PLATE 27. Development of primary phloem fibers in *Linum perenne*. *A*, transverse and, *B*, radial longitudinal sections of stems. Differentiating fibers have wide lumina, scanty contents, first layers of secondary walls in cells nearest cortex. Fibers are much longer than adjacent cortical cells as shown in *B*. (Both, ×385. *A*, Esau, *Amer. Jour. Bot.* 30, 1943.)

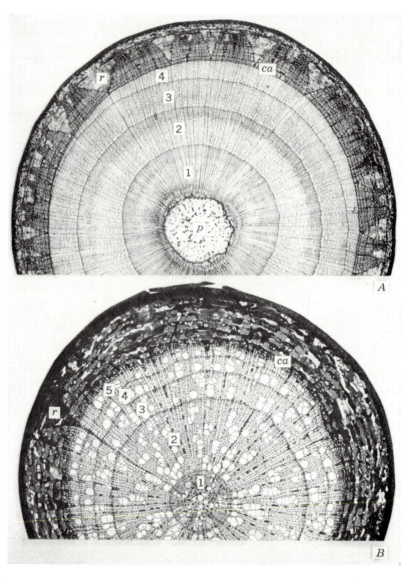

PLATE 28. Stem (*A*, ×12) and root (*B*, ×33) of *Tilia* in transections. Numbers indicate growth increments of secondary xylem. Primary xylem in center in *B*, surrounding pith (*p*) in *A*. Vascular cambium at *ca*, phloem with fibers and dilated rays (*r*) outside cambium. Periderm on surface.

658

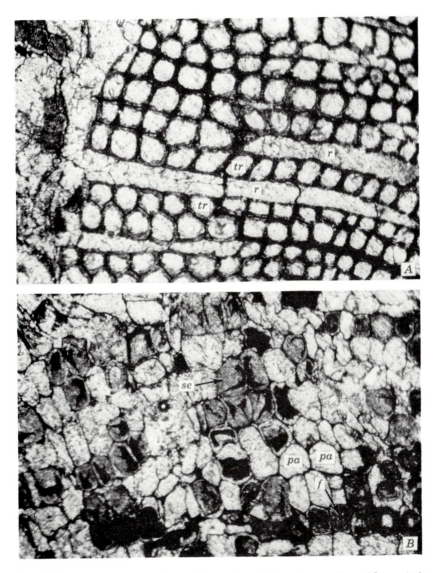

PLATE 29. Transections of secondary xylem (A) and secondary phloem (B) of a fossil plant, *Tetraxylopteris schmidtii*. Tracheids (*tr*) and rays (*r*) in xylem. Parenchyma cells (*pa*), fibers (*f*), and presumed sieve elements (*se*) in phloem. (Beck, *Amer. Jour. Bot.* 44, 1957, ×140.)

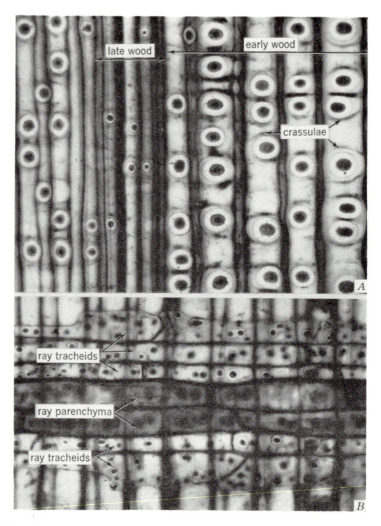

PLATE 30. Longitudinal radial sections of wood of *Pinus*. In *A*, tracheids and related elements of axial system, all with bordered pits. In late wood pits are considerably smaller than in early wood. Crassulae (or bars of Sanio) are thicker portions of intercellular layer and primary walls. *B*, part of ray with two rows of ray parenchyma cells and five rows of ray tracheids. Small bordered pits in ray tracheids. Large simple pits of ray parenchyma cells appear as dark spots in background. (Both, ×255.)

660

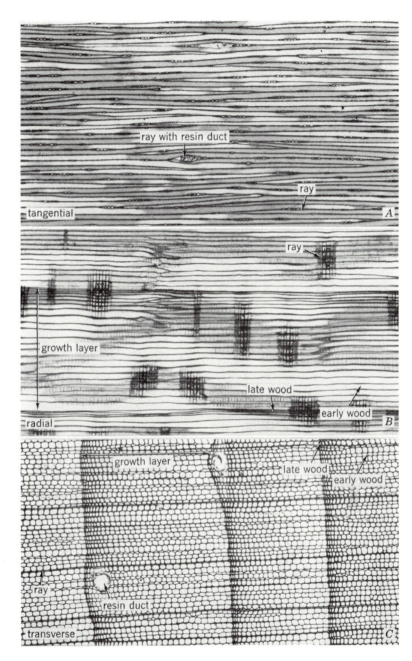

PLATE 31. Secondary xylem of *Pinus strobus* (White Pine) in tangential (*A*), radial (*B*), and transverse (*C*) sections. Conifer wood. (All, ×35.)

661

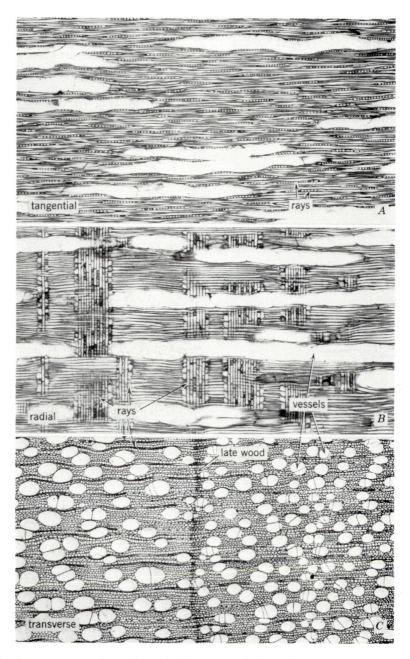

PLATE 32. Secondary xylem of *Salix nigra* (Black Willow) in tangential (*A*), radial (*B*), and transverse (*C*) sections. Diffuse-porous nonstoried dicotyledon wood with uniseriate heterocellular rays. (All, ×35.)

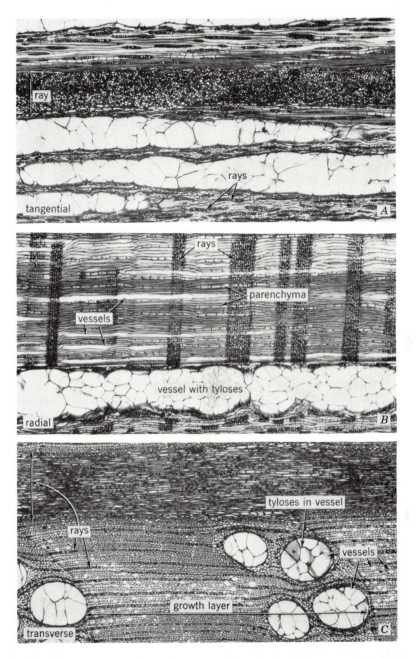

PLATE 33. Secondary xylem of *Quercus alba* (White Oak) in tangential (A), radial (B), and transverse (C) sections. Ring-porous nonstoried dicotyledon wood with high multiseriate and low uniseriate rays. (All, ×35.)

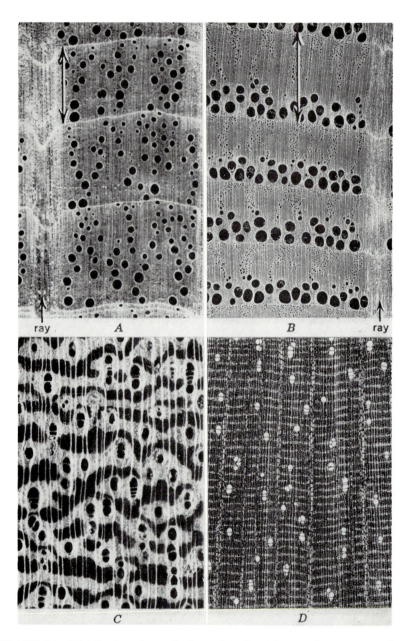

PLATE 34. Distribution of vessels (pores) and axial parenchyma in secondary xylem as seen in transections. A, semi-ring-porous wood of *Quercus virginiana*. B, ring-porous wood of *Quercus bicolor*. Arrows delimit single growth layers. Paratracheal parenchyma in *Andira* (C), apotracheal in *Cymbopetalum* (D). (A, B, ×10; courtesy of H. P. Brown. C, ×10; D, ×20; Record and Hess, *Timbers of the New World*, Yale University Press, 1943.)

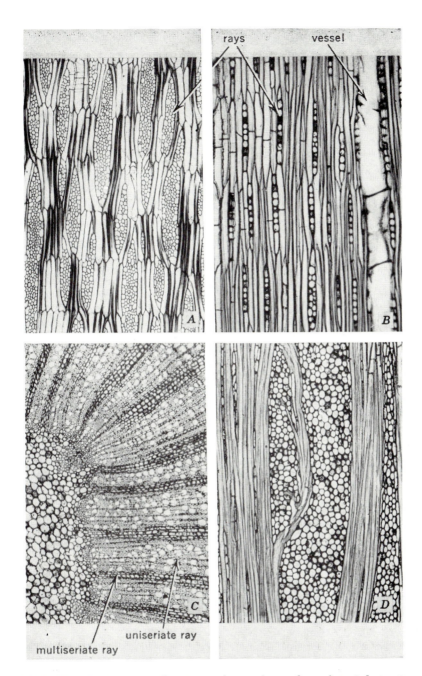

PLATE 35. *A, B,* tangential sections of secondary xylem of storied structure. *A,* high multiseriate rays extend through more than one horizontal tier (*Triplochiton*). *B,* low uniseriate rays, each limited to one horizontal tier (*Canavalia*). *C,* transection of xylem and pith (lower left) of *Villaresia.* Multiseriate rays are continuous with pith. *D,* tangential section of *Crossostyles* xylem. Multiseriate ray dissected by change of ray initials into fusiform initials. (*A, C, D,* ×50; *B,* ×100. Barghoorn, *Amer. Jour. Bot.* 27, 1940; 28, 1941.)

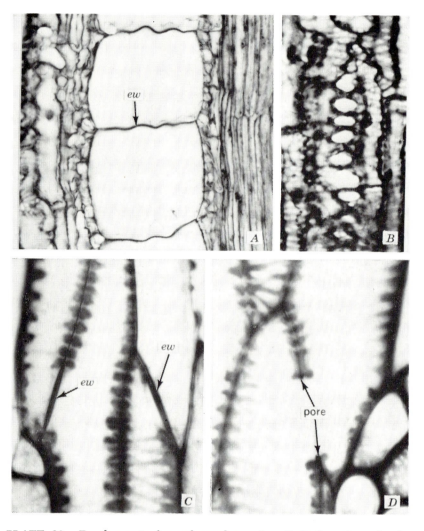

PLATE 36. Development of vessel members. Longitudinal sections of xylem of *Cucurbita* (*A, B*) and *Asimina* (*C, D*). *A,* series of expanded vessel members with end walls (*ew*) still intact and no secondary thickenings on side walls. *B,* cells occurring in contact with vessel members. Some were partly pulled apart because of lateral expansion of vessel members. *C, D,* development of perforation plates in vessel members. Thickened but entirely primary end walls (*ew*) present in *C,* absent in *D.* Secondary thickening present on side walls and rim in *C, D,* except in upper element at right in *C.* (*A,* ×250; *B,* ×400; *C, D,* ×850. *B,* Esau and Hewitt, *Hilgardia* 13, 1940; *C, D,* slide by V. I. Cheadle.)

666

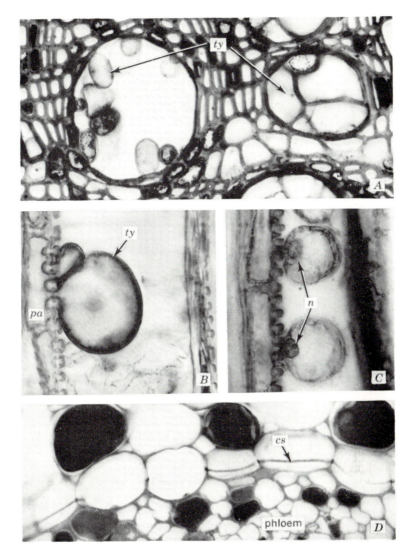

PLATE 37. *A–C*, tyloses (*ty*) in *Vitis* (grapevine) vessels as seen in transverse (*A*) and longitudinal (*B, C*) sections of xylem. *A*, left, young tyloses, right, vessel filled with tyloses. *B* shows continuity between lumina of tyloses and parenchyma cell (*pa*). *C*, nuclei (*n*) have migrated from parenchyma cells to tyloses. *D*, transection of *Myriophyllum* rhizome illustrating endodermis with casparian strips (*cs*). (*A*, ×290; *B, C*, ×750; Esau, *Hilgardia* 18, 1948; *D*, ×290; *Encyclopaedia Britannica*, copyright 1945.)

667

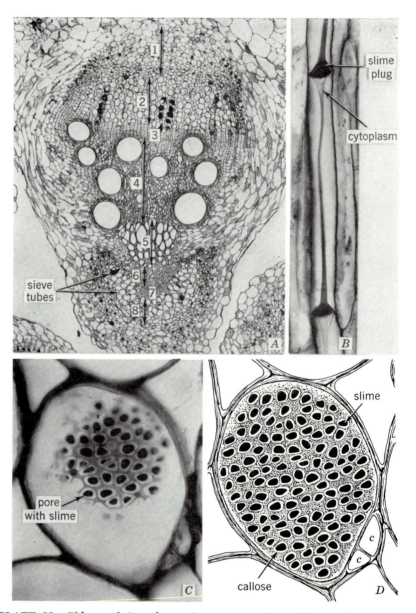

PLATE 38. Phloem of *Cucurbita.* *A,* transection of vascular bundle. Details: 1, external primary phloem; 2, secondary phloem; 3, vascular cambium; 4, secondary xylem; 5, metaxylem; 6, protoxylem; 7, incompletely developed vascular cambium; 8, internal phloem, mostly primary. Slime (black) in some sieve tubes. *B,* part of a sieve tube, in longitudinal section, with one complete member. *C,* surface view of part of sieve plate. *D,* reconstruction, from two successive sections, of sieve plate as in *C.* Companion cells at *c.* (*A,* ×21; *B,* ×220; *C, D,* ×590.)

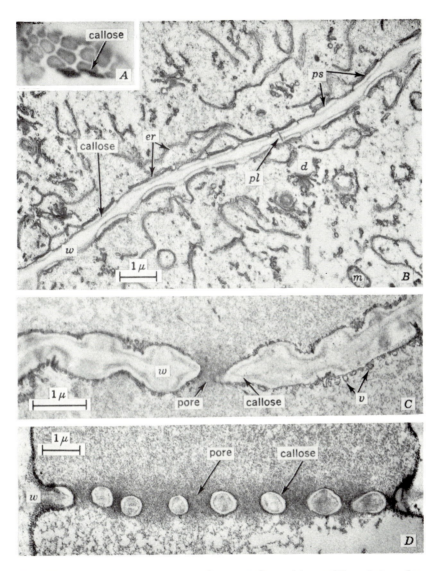

PLATE 39. Development of sieve plate in *Robinia* (*A*, ×970) and *Cucurbita* (*B–D*, electron micrographs of longitudinal sections). *A*, surface view of sieve plate with pore sites covered by callose. *B*, young sieve plate with pore sites (*ps*) covered with callose platelets and endoplasmic reticulum (*er*). Single plasmodesma (*pl*) in one pore site. *C*, recently opened pore lined with callose. In cytoplasm, vesicles (*v*) have replaced *er*. *D*, mature sieve plate. Some slime accumulation above it and in the pores. Sieve plate wall marked *w*. (*A*, Esau et al., *Bot. Gaz.* 123, 1962.)

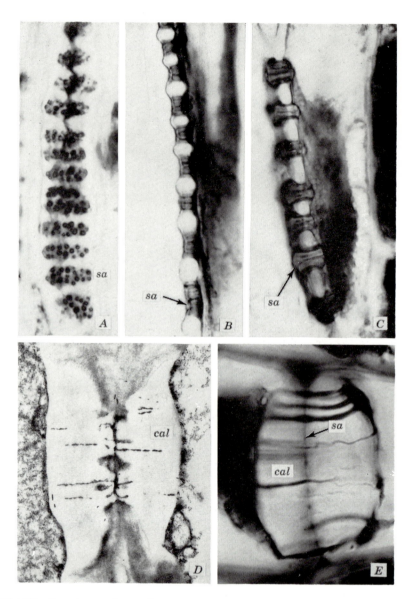

PLATE 40. Sieve plates of *Vitis* (grapevine) in radial section (*A*, surface view) and in tangential sections (*B–E*). *A–C*, compound sieve plates with numerous sieve areas (*sa*) penetrated by strands of slime and cytoplasm. In *C*, sieve areas thickened by callose deposition. Single sieve areas in *D* (electron micrograph), during dormancy, and *E*, during spring reactivation. In *E*, *sa* points to sieve-area wall imbedded in callose. Massive dormancy callose (*cal*) in *D* penetrated by ultra-fine presumably cytoplasmic strands; in *E*, by strands containing slime. (*A–C*, Esau, *Hilgardia* 18, 1948; ×750; *D*, ×14,000; *E*, ×1,200.)

670

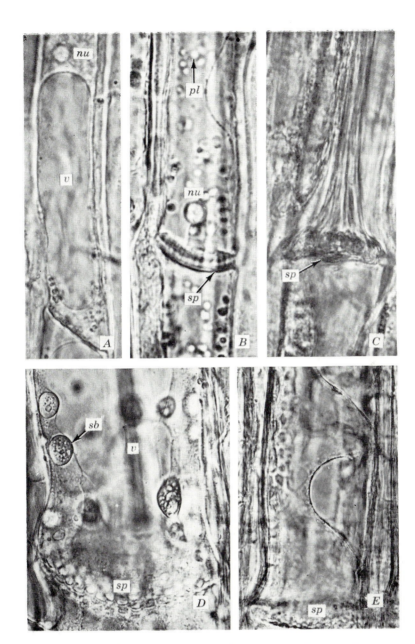

PLATE 41. Sieve elements in longitudinal sections of phloem of *Passiflora* (*A,* *B*), *Bryonia* (*C*), and *Cucurbita* (*D, E*). *A, B, D, E,* from living tissue; *C,* treated with potassium iodide. *A,* young cell with vacuole (*v*) and extruded nucleolus (*nu*). *B,* mature cell, plasmolyzed. Plastids with starch at *pl. C,* contracted stringy protoplast attached to sieve plate (*sp*). *D,* young sieve element with vacuole (*v*) and slime bodies (*sl*) in cytoplasm. *E,* mature cell, plasmolyzed. (*A,* ×1,000; *B. C,* ×1,260; *D,* ×800; *E,* ×500; *A–C,* Kollmann, *Planta* 54, 1960; *D, E,* Huber and Rouschal, *Deut. Bot. Gesell. Ber.* 56, 1938.)

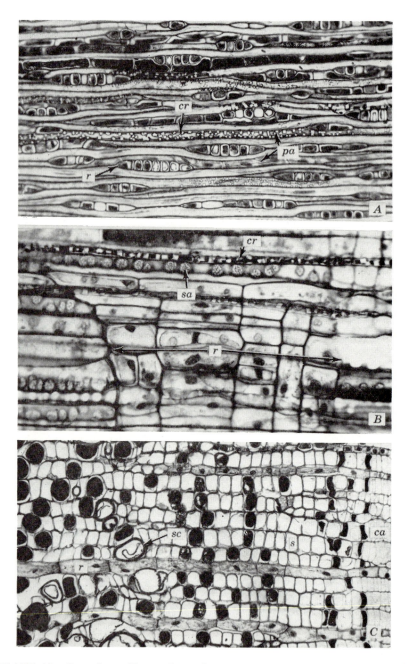

PLATE 42. Secondary phloem of *Pseudotsuga menziesii*, a conifer, in tangential (A), radial (B), and transverse (C) sections. Details: *ca*, cambium; *cr*, crystals; *pa*, parenchyma cells; *s*, sieve cell; *sa*, sieve area; *sc*, sclereid; *r*, ray. (Grillos and Smith, *Forest Sci.* 5, 1959, ×140.)

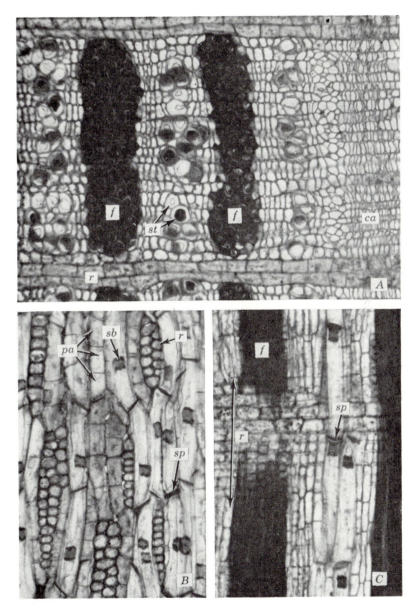

PLATE 43. Secondary phloem of *Robinia pseudoacacia*, dicotyledon, in transverse (*A*), tangential (*B*), and radial (*C*) sections. Details: *ca*, cambium; *f*, fibers; *pa*, parenchyma cells; *r*, ray; *sb*, slime body; *sp*, sieve plate; *st*, sieve tubes. (Courtesy of W. F. Derr, ×180.)

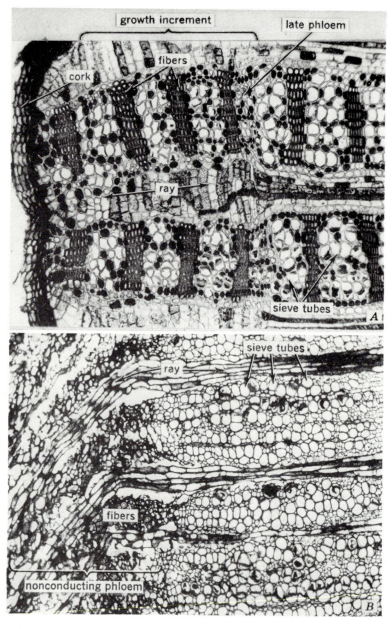

PLATE 44. Transections of secondary phloem of A, *Vitis vinifera* (grapevine) and, B, *Prunus avium* (sweet cherry). Vascular cambium was to the right in both sections. In A, tangential bands of fibers alternate with tangential bands containing sieve tubes, companion cells, and phloem parenchyma cells. Rays are somewhat dilated in older phloem (left). In B, fibers occur in nonconducting phloem (left) where sieve tubes are collapsed; rays are bent in old phloem. (A, Esau, *Hilgardia* 18, 1948, ×90; B, Schneider, *Torrey Bot. Club Bul.* 72, 1945, ×88.)

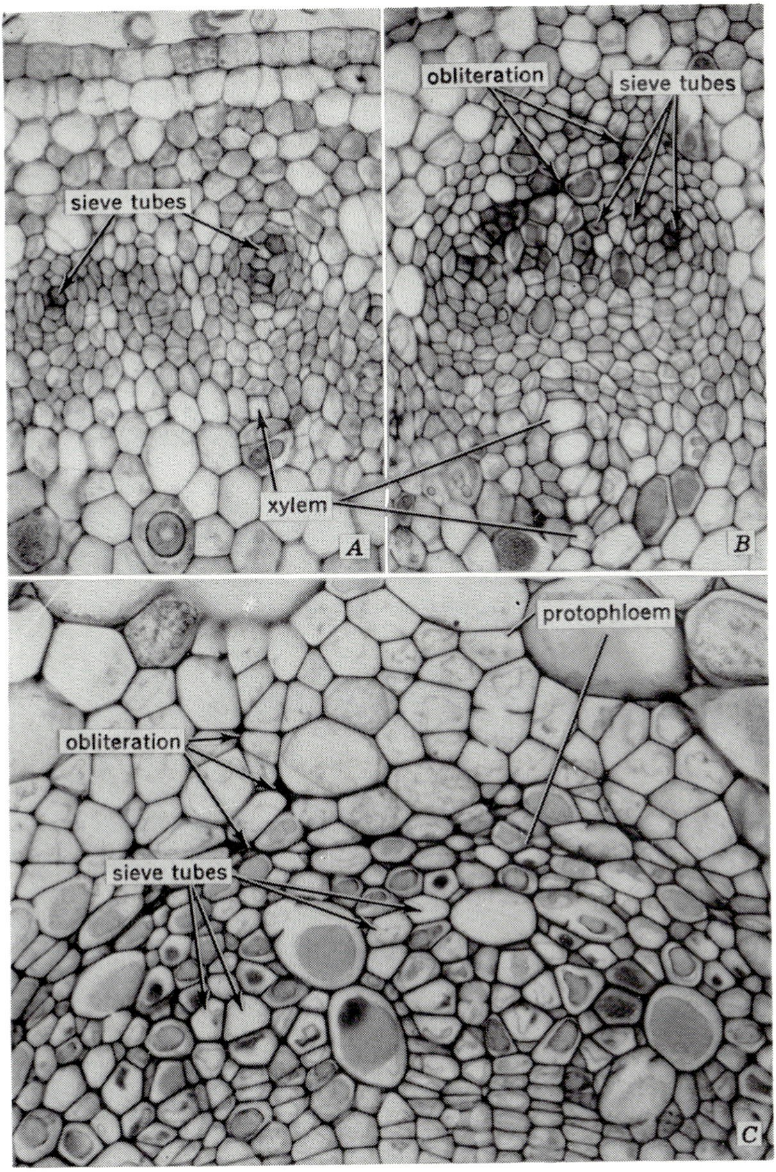

PLATE 45. Differentiation of primary phloem as seen in transection of shoot of *Vitis vinifera. A,* two procambial bundles, one with one sieve tube, the other with several. *B,* vascular bundle with many protophloem sieve tubes. Some of these are obliterated. Protoxylem present in *A, B. C,* protophloem sieve tubes obliterated and metaphloem is differentiated (lower half of figure). Protophloem represented by primordia of fibers. Metaphloem consists of sieve tubes, companion cells, phloem parenchyma, and much enlarged tannin-containing parenchyma cells. (Esau, *Hilgardia* 18, 1948, ×500.)

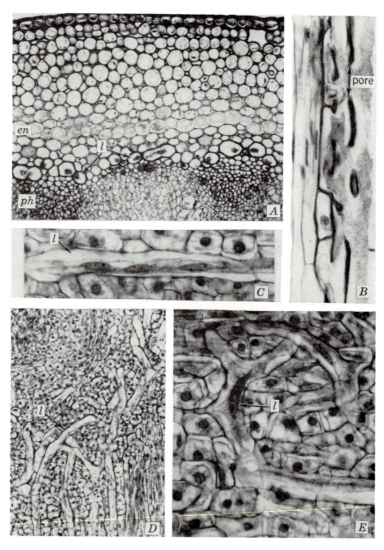

PLATE 46. Laticifers. Transverse (*A*) and longitudinal (*B*) sections of *Lactuca scariola* stem. In *A*, lacunose collenchyma beneath epidermis; endodermis (starch sheath) at *en*, laticifers at *l*, phloem at *ph*. *B*, articulated laticifers with extensively perforated walls. *C–E*, longitudinal sections of *Nerium oleander* stem with nonarticulated laticifers (*l*). *C*, multinucleate condition and, *D*, *E*, branching of laticifers. (*A*, ×150; *B*, ×415; *C*, *E*, ×620; *D*, ×240.)

676

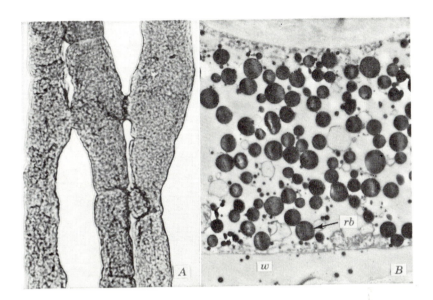

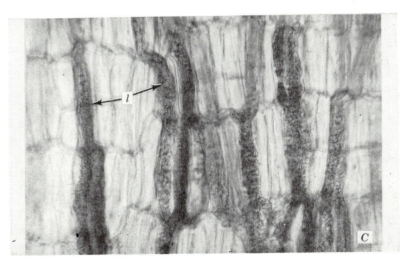

PLATE 47. Laticifers. *A*, articulated laticifers of *Hevea brasiliensis* with inter-connections (from a maceration). *B*, electron micrograph of laticifer of *Hevea* with rubber particles (*rb*) and wall (*w*). *C*, articulated anastomosing laticifers (*l*) of *Taraxacum kok-saghyz* in longitudinal section of secondary phloem of root. *A*, ×400; *B*, ×6,500; *C*, ×280; *A*, *B*, courtesy of W. A. Southorn, Rubber Research Institute of Malaya; *C*, Artschwager and McGuire, *U.S.D.A. Tech. Bul.* 843, 1943.)

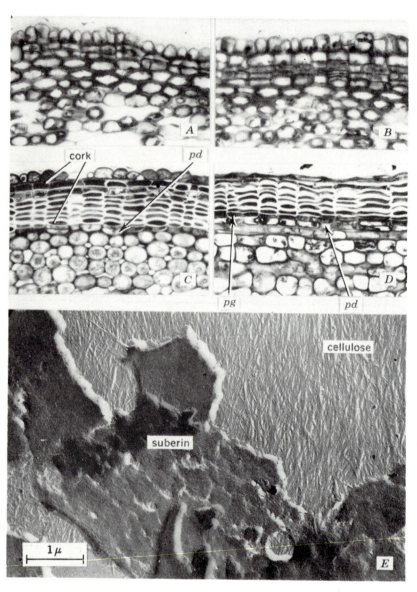

PLATE 48. Earlier (*A*) and later (*B*) stages of development of periderm by periclinal divisions in subepidermal layer. Transections of *Prunus* stem. Periderm in transverse (*C*) and longitudinal (*D*) sections of dormant *Betula* twig. Phellogen at *pg*, phelloderm at *pd*. *E*, electron micrograph of *Quercus suber* cork, which was partially saponified. Amorphous suberin distinct from fibrillar cellulose. (*A–D*, ×280; *E*, Sitte, *Mikroskopie* 10, 1955.)

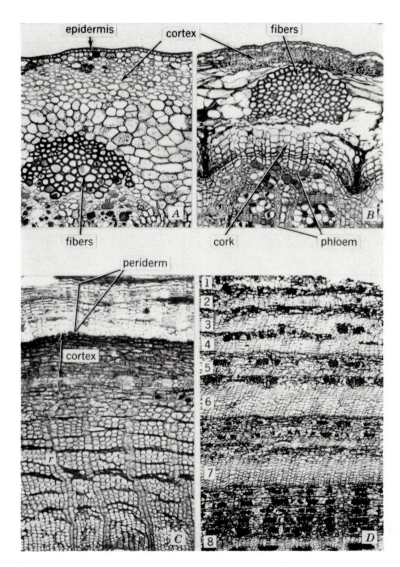

PLATE 49. *A, B,* formation of first periderm in transections of stem of *Vitis* (grapevine). *A,* periderm absent. *B,* periderm arose in primary phloem; tissues outside periderm had died and nonsclerified cells had collapsed. *C, D,* transections of *Robinia* stem with first periderm (*C*) and rhytidome (*D*). In *C,* tangentially oriented layers of crushed cells, mainly sieve-tube elements, in secondary phloem. In rhytidome in *D,* periderm layers are numbered. Tissue layers alternating with periderm layers are dead portions of secondary phloem, the dark patches within which are fibers. (*A–C,* ×75; *D,* ×36; *A, B,* Esau, *Hilgardia* 18, 1948.)

679

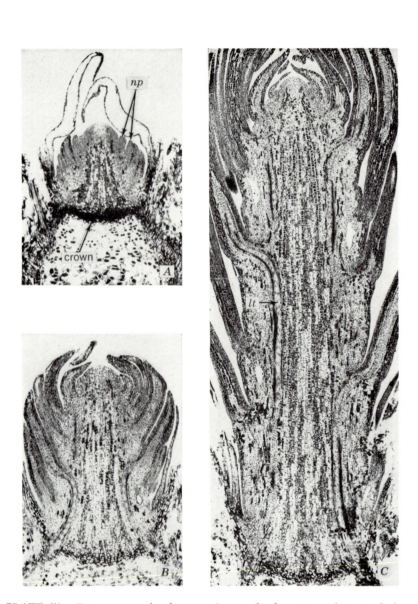

PLATE 50. Primary growth of stem. Longitudinal sections of terminal shoot of *Abies concolor* in three stages of seasonal development. *A*, dormant bud with needle primordia (*np*). *B*, after early internodal elongation and some vascular differentiation above crown. Apical meristem still inactive. *C*, after further internodal elongation and at beginning of formation of scale primordia. Leaf traces (one at *lt*) are conspicuous. (Parke, *Amer. Jour. Bot.* 46, 1959; *A*, *C*, ×33; *B*, ×40.)

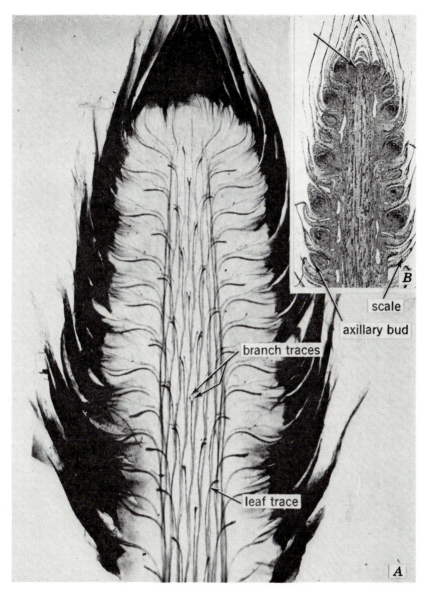

PLATE 51. Young shoot with unexpanded internodes of *Pinus strobus* in longitudinal views. *A,* cleared section showing the primary vascular system composed of leaf traces in sympodial linkages and of branch traces. *B,* median section through shoot showing scales (leaf traces in *A* constitute the vascular supply of these scales) and axillary buds (branch traces in *A* diverge into such buds) subtended by scales. Branch traces occur in pairs. Each branch trace of a pair is connected to a different leaf-trace sympodium. (*A,* ×16; *B,* ×11; Slides by A. R. Spurr.)

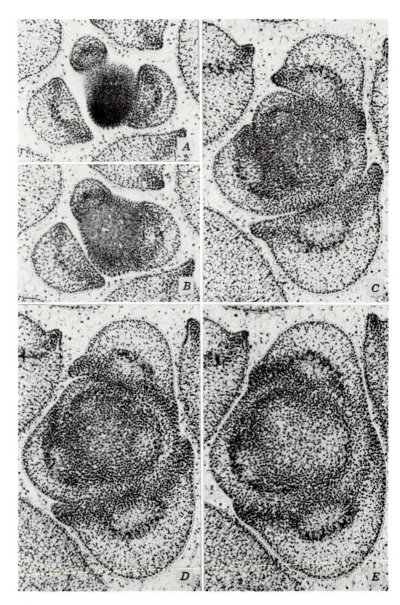

PLATE 52. Initial stages in differentiation of vascular system in *Nicotiana* shoot (including stem and leaf primordia) as seen in transections. *A*, section at level of shoot apex; other sections the following numbers of microns below apex: *B*, 20; *C*, 50; *D*, 70; *E*, 90. At successive levels, gradual decrease in stainability of ground-meristem cells delimits the more densely stained prospective vascular region. Compare with plate 53. (All, ×75.)

682

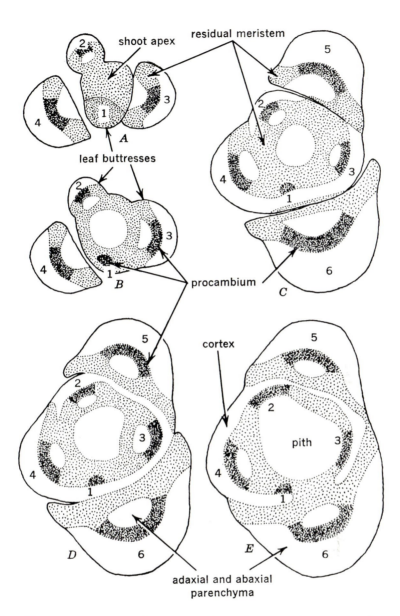

PLATE 53. Diagrams explaining initial vascularization depicted in plate 52. Leaf primordia and their traces are numbered 1–6, beginning with the youngest. Parenchymatic differentiation in cortex, pith, and adaxial and abaxial parts of leaf primordia (all left blank in drawings) delimits as a unit the prospective vascular system of stem and leaves. At early stage of differentiation (here depicted) constituent tissues of vascular system are (1) procambium (densely stippled) with a few mature vascular elements (not indicated) and (2) some less differentiated meristem, residual meristem (lightly stippled), which is the source of additional procambium and of interfascicular and leaf-gap parenchyma. (All, ×75.)

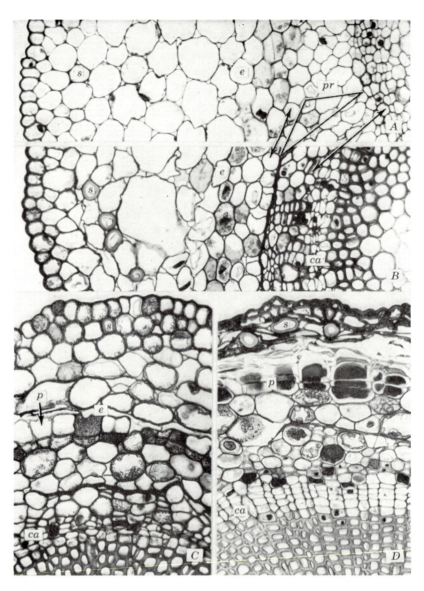

PLATE 54. Development in hypocotyl of *Pseudotsuga menziesii* as seen in longitudinal sections. Precursory phloem (*pr*), intact in *A*, crushed in *B*. Endodermis (*e*) is crushed in *C*. Cambium (*ca*) is progressively better defined from *B* to *D*. Periderm (*p*) is initiated in pericycle beneath endodermis (*e*) in *C*. Tissues outside periderm (*p*) are collapsed in *D*. Sclereids (*s*), without secondary wall in *A*, with such wall in *B* to *D*. (Smith, *Forest Sci.* 4, 1958, ×250.)

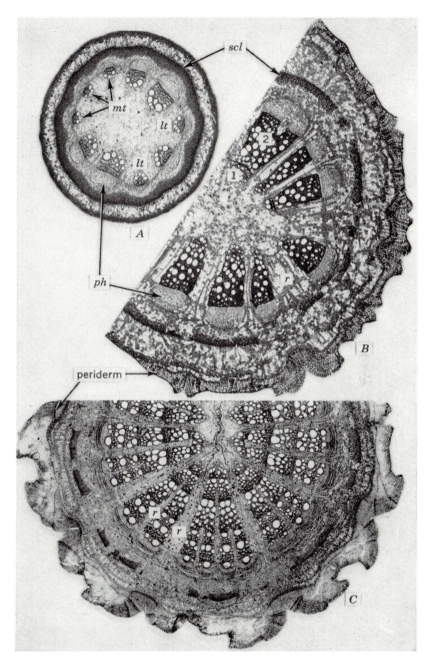

PLATE 55. Transections of stem of *Aristolochia,* one (*A*), two (*B*), and six (*C*) years old. Perivascular sclerenchyma (*scl*) continuous in *A,* disrupted in *B, C.* 1, 2, in *B,* growth increments in xylem. Phloem (*ph*) without fibers. Wide rays (*r*). In *A,* three traces to nearest leaf above: *mt,* median (three bundles) and *lt,* lateral. (*A, B,* ×19; *C,* ×10.)

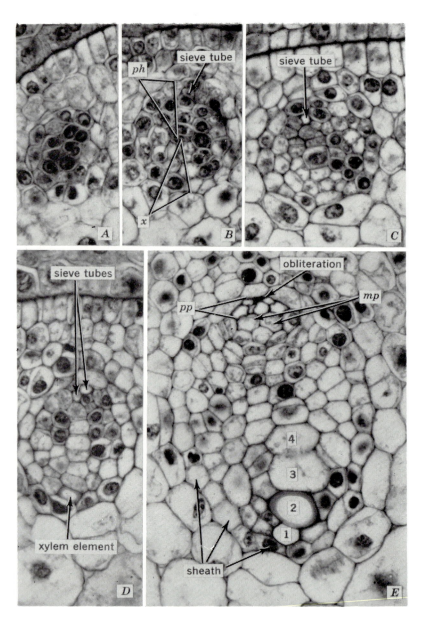

PLATE 56. Differentiation of vascular bundle of *Zea mays* as seen in transsections of leaves. *A–E*, successively older bundles. *A*, procambial strand. *B*, phloic (*ph*) and xylary (*x*) parts of procambium are discernible. First sieve element, still immature, at phloem pole. *C*, first sieve element is mature. *D*, two sieve elements and one xylem element (at xylem pole). *E*, protophloem (*pp*) is completely developed. It consists of sieve tubes only; the earliest are undergoing obliteration. First metaphloem cells (*mp*) are evident. Of the 1–4 protoxylem elements, 1 has annular secondary thickenings, 2 has annular or helical thickenings, 3 and 4 are still without secondary walls. (Esau, *Hilgardia* 15, 1943, ×750.)

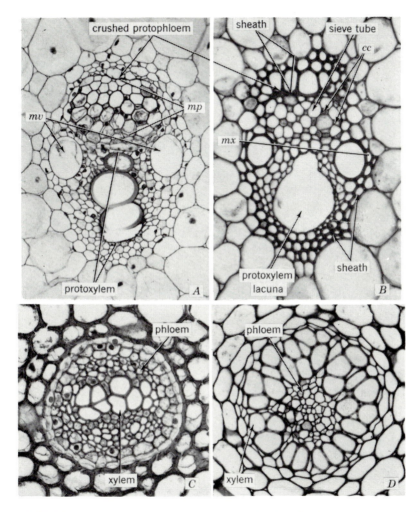

PLATE 57. A, immature, and B, mature vascular bundles from transections of
Zea mays stem. Details: cc, companion cells; mp, metaphloem; mv, large metaxylem vessels, still without secondary wall; mx, metaxylem. Protoxylem disrupted
in B; replaced by lacuna. C, D, concentric vascular bundles, amphicribral in C
(Polypodium), amphivasal in D (Cordyline). (A, ×190; B, ×310; C, D,
×180; D, slide by V. I. Cheadle.)

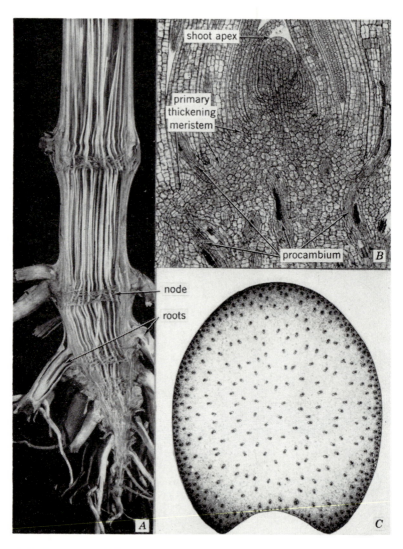

PLATE 58. Stem of *Zea mays* (maize). *A*, mature stem split longitudinally, then partially retted to expose vascular system. Stem increases in width from base upward. *B*, shoot apex, part of subjacent axis, and bases of youngest leaf primordia. *C*, transection of immature internode showing vascular bundles scattered within the ground parenchyma. Indentation in stem at base of figure indicates former location of axillary bud. (*B*, ×90; *C*, ×4.5; *A*, Sharman, *Ann. Bot.* 6, 1942; *B*, slide by G. I. Patel.)

688

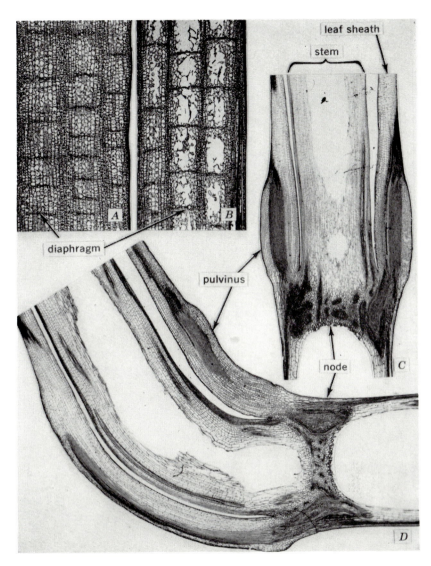

PLATE 59. Details of leaf and stem in a grass. Longitudinal sections. *A, B,*
two stages in rhexigenous lacunae development in midribs of leaf sheath of
Oryza (rice). Diaphragms remain intact between lacunae. *C, D,* intercalary
growth regions (pulvini, joints) of *Hordeum* (barley). *C,* from upright stem,
D, from stem that was rising after it had lodged. In *D,* leaf sheath and stem
became extended on the side next to the ground. Collenchymatous tissue instead
of sclerenchyma in pulvinus region. (*A, B,* Kaufman, *Phytomorph.* 9, 1959,
×140; *C, D,* ×9.)

689

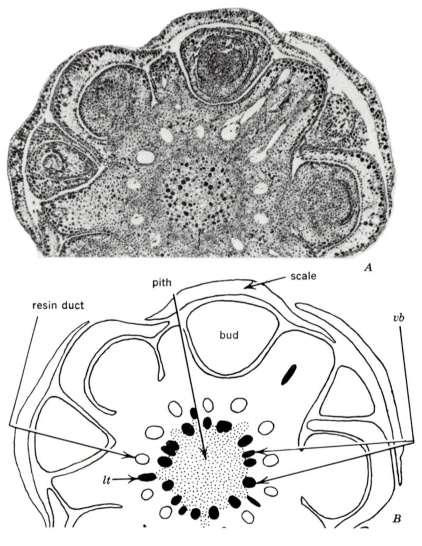

PLATE 60. Transection of *Pinus* stem in primary state of growth, with scales, axillary buds, discrete vascular bundles (*vb*), and leaf traces (one at *lt*). (×46.)

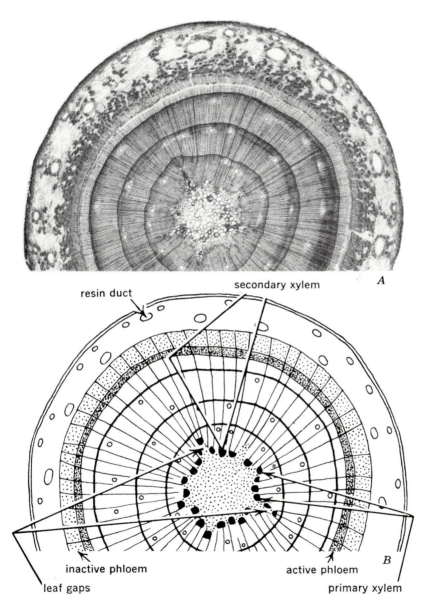

resin duct

secondary xylem

A

inactive phloem

leaf gaps

active phloem

primary xylem

B

PLATE 61. Transection of *Pinus* stem in fourth year of secondary growth. (×21.)

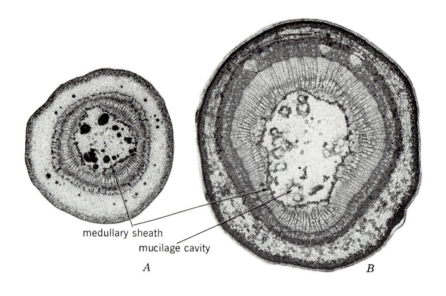

medullary sheath
mucilage cavity

A

B

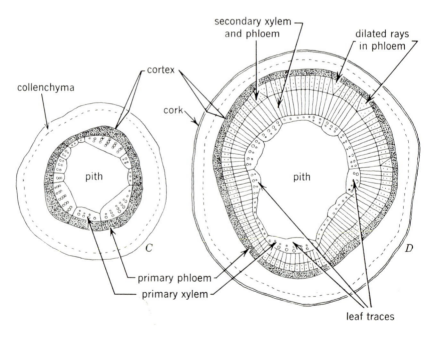

secondary xylem
and phloem

dilated rays
in phloem

cortex

collenchyma

cork

pith

pith

C

D

primary phloem

primary xylem

leaf traces

PLATE 62. Transections of *Tilia* stem made before (*A*, *C*) and after (*B*, *D*) secondary growth was initiated. (All, ×23.)

692

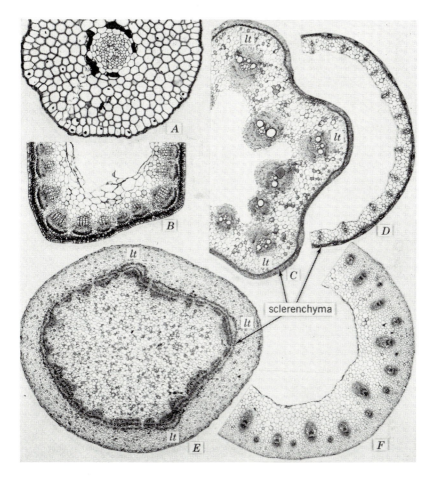

PLATE 63. Comparative structure of stems. Transverse sections. A, *Tmesip-teris*, with clear separation between fundamental and vascular systems (×52). Solid vascular cylinder surrounded by one to two layers of pericycle. B, *Medicago* (alfalfa), herbaceous dicotyledon with discrete vascular bundles (×34). C, *Cucurbita*, dicotyledonous vine with secondary growth restricted to vascular bundles (×8). D, *Secale* (rye), monocotyledon, grass, without secondary growth (×14). E, *Pelargonium*, dicotyledon, at beginning of secondary growth (×7). F, *Ranunculus* dicotyledonous herb without secondary growth (×13). Leaf traces (*lt*) are indicated in C and E. (A, B, *Encyclopaedia Britannica*, copyright 1945.)

693

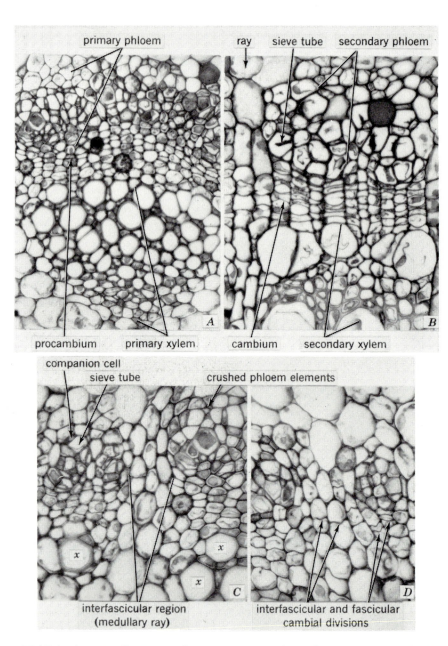

PLATE 64. Vascular tissues from transections of *Sambucus* stem, at end of primary growth (*A*, *C*, *D*, all from same section) and during secondary growth (*B*). In *A*, protoxylem below, with partly obliterated tracheary elements. Metaxylem above protoxylem. In *B*, fascicular cambial zone with expanding vessels below it. Left, ray with interfascicular cambium in line with fascicular to the right. *C*, primary phloem and metaxylem elements (*x*); *D*, only phloem. Results of first cambial divisions in *D*. (*A*, *B*, ×280; *C*, *D*, ×430; *A*, *C*, *D*, Esau, *Amer. Jour. Bot.* 32, 1945; *B*, *Encyclopaedia Britannica*, copyright 1945.)

694

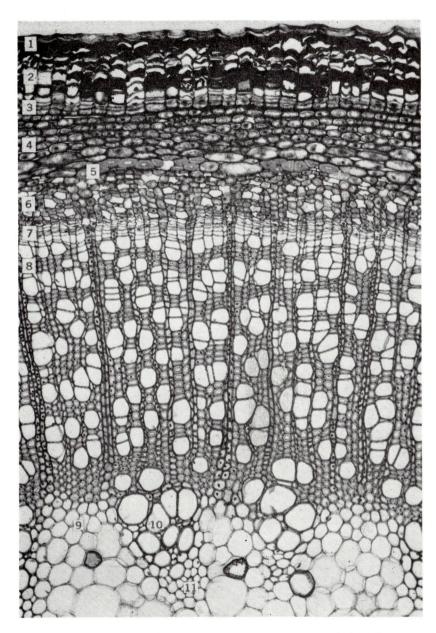

PLATE 65. Transection of *Sambucus* stem. Details: 1, crushed epidermis; 2, cork; 3, phellogen and phelloderm (one layer); 4, cortex with collenchyma; 5, primary phloem fibers; 6, secondary phloem (sieve tubes have wide, clear lumina); 7, vascular cambium; 8, secondary xylem; 9, interfascicular region; 10, metaxylem; 11, protoxylem. (×127.)

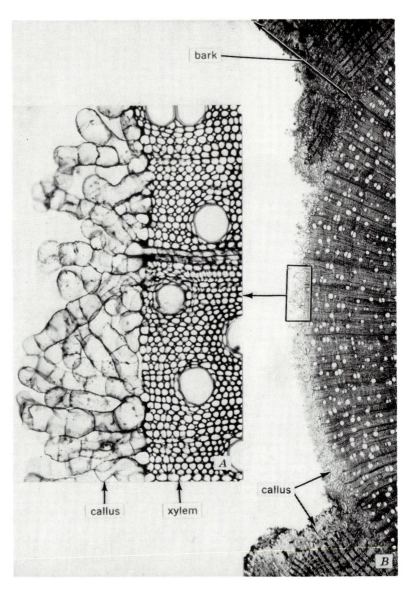

PLATE 66. Wound healing in *Hibiscus* stem. Transections. Callus tissue on the exposed surfaces of wood (*A*, *B*) and bark (*B*). Callus over xylem originated from incompletely differentiated cells of xylem, mainly ray cells. The cells elongated outward and divided by periclinal walls. (*A*, ×120; *B*, ×14; slide courtesy of H. Gunnery; see also Sharples and Gunnery, *Ann. Bot.* 47, 1933.)

696

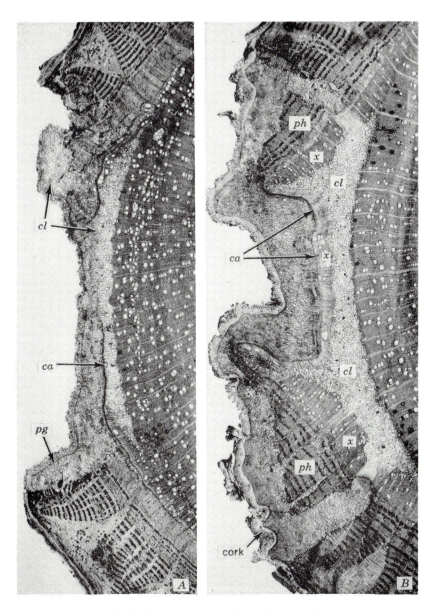

PLATE 67. Wound healing—stages 2 and 3 after the first in plate 66. A, phellogen (*pg*) has appeared beneath the surface of callus (*cl*). Some vascular cambium (*ca*) has appeared in callus in continuity with original cambium. B, regeneration of missing part of stem is complete. Vascular cambium (*ca*) is continuous across callus and has formed secondary xylem (*x*) and phloem (*ph*). Some callus tissue (*cl*) has been imbedded beneath the new xylem. (Both, ×14; credit as for plate 66.)

697

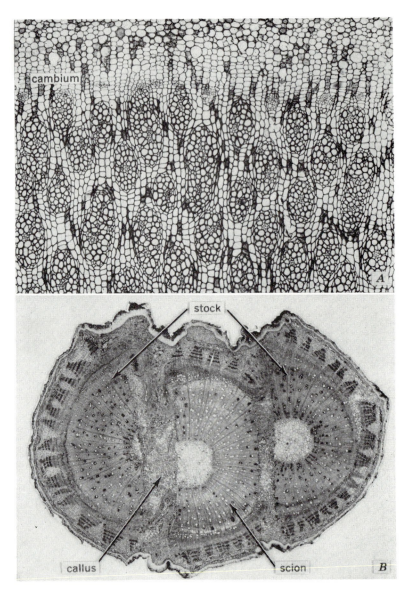

PLATE 68. A, transection of *Cordyline* stem. Cambial secondary growth in a monocotyledon. In secondary tissue below cambium are amphivasal vascular bundles and parenchyma. Outward, cambium has formed some parenchyma (radially seriated). B, transection of *Hibiscus* stem illustrating cleft-graft union. Cambial activity within callus tissue has joined the vascular tissues of stock and scion. (A, ×50; B, ×10; A, Cheadle, *Amer. Jour. Bot.* 30, 1943; B, credit as for plate 66.)

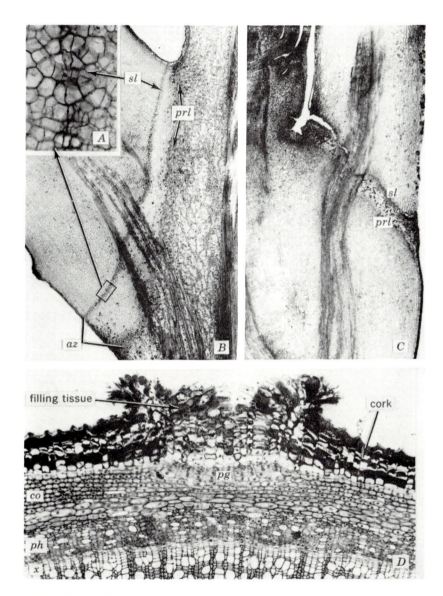

PLATE 69. *A–C*, leaf abscission zones in *Juglans*, walnut (*A, B*), and *Prunus*, cherry (*C*), in longitudinal sections through leaf bases. Abscission zone (*az*) has a separation layer (*sl*) and a protective layer (*prl*) of suberized cells. In *A*, indication of recent cell division in separation layer. *D*, lenticel of *Sambucus*. Details: *co*, cortex; *pg*, phellogen; *ph*, phloem; *x*, xylem. (*A*, ×98; *B*, ×13; *C*, ×17; *D*, ×75.)

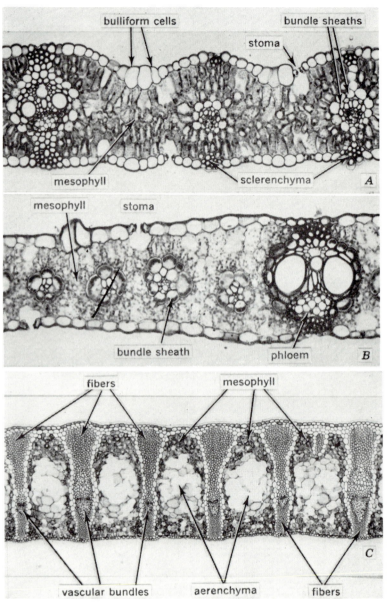

PLATE 70. Transections of monocotyledonous leaves. In *A*, *Triticum* (wheat), adaxial epidermis bears bulliform cells in grooved parts of blade. Subepidermal cells elongated like palisade cells. There is an inner thick-walled and an outer thin-walled bundle sheath. Sclerenchyma in ribs, connected with bundle sheaths. In *B*, *Zea* (maize), mesophyll is relatively undifferentiated. Single-layered thin-walled sheaths enclose vascular bundles. In *C*, *Phormium tenax*, vascular bundles accompanied, above and below, by massive strands of fibers. (*A*, *B*, *Encyclopaedia Britannica, copyright 1945, ×140; C, ×55.*)

700

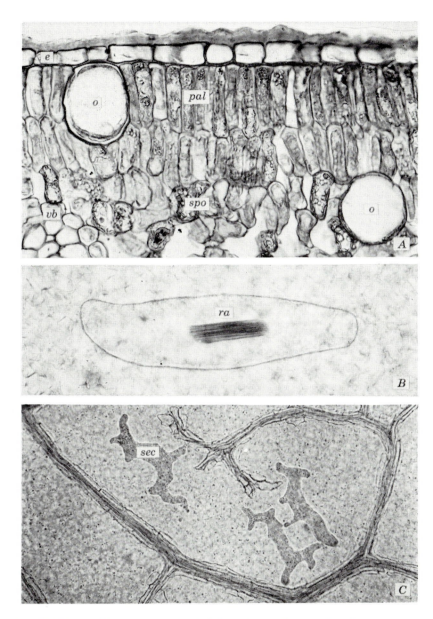

PLATE 71. A Transection of *Umbellularia* leaf with two oil cells (*o*), one in palisade tissue (*pal*), the other in spongy tissue (*spo*). Epidermis at *e*, vascular bundle at *vb*. *B*, idioblast with raphides (*ra*) in cleared petal of *Impatiens*. *C*, three secretory (*sec*) idioblasts in cleared leaf of *Tetracentron*. (Foster, *Protoplasma* 46, 1956; A, ×380; B, ×230; C, ×89.)

701

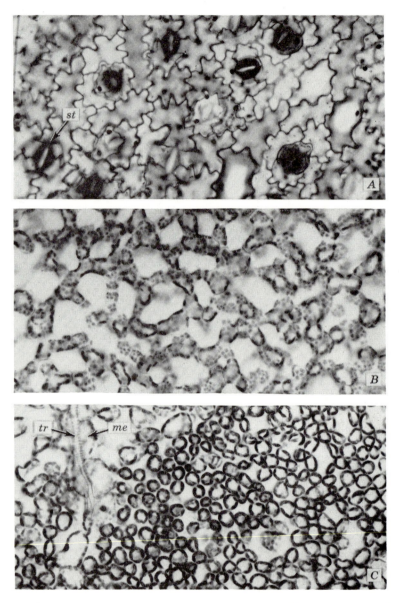

PLATE 72. Leaf of *Nicotiana tabacum*. Sections parallel with surface show abaxial epidermis (*A*) with stomata (one at *st*), spongy parenchyma (*B*), and palisade parenchyma (*C*). Tracheids (*tr*) bordered by mesophyll cells (*me*) in *C*. (All, ×280.)

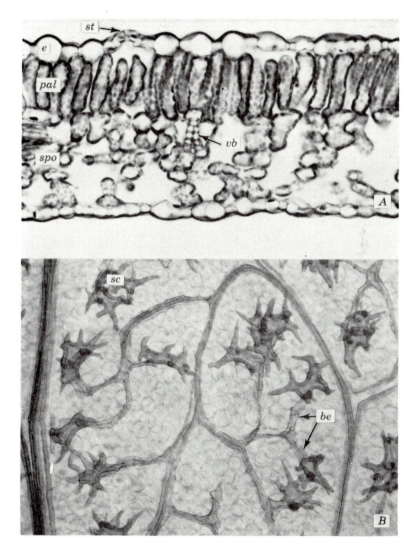

PLATE 73. *A*, transection of leaf of *Nicotiana tabacum*. Details: *e*, epidermis, *pal*, palisade parenchyma; *spo*, spongy parenchyma; *st*, stoma; *vb*, vascular bundle. *B*, cleared *Boronia* leaf. Sclereids (*sc*) at bundle ends (*be*). (*A*, ×280; *B*, Foster, *Amer. Jour. Bot.* 42, 1955, ×93.)

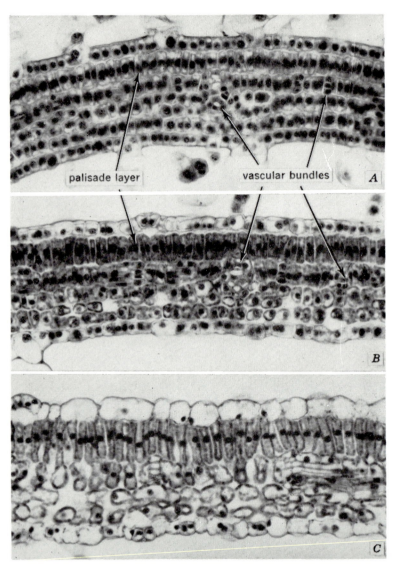

PLATE 74. Development of *Nicotiana tabacum* leaf. Transections showing three stages of development of mesophyll and epidermis. (Fourth stage in plate 73A.) Palisade is just becoming distinct in *A* because of repeated anticlinal divisions and slight elongation of cells at right angles to surface. Extension of cells in spongy parenchyma mainly parallel with leaf surface. Difference between epidermal and palisade cells in tangential dimension increases with age. (All, ×280.)

704

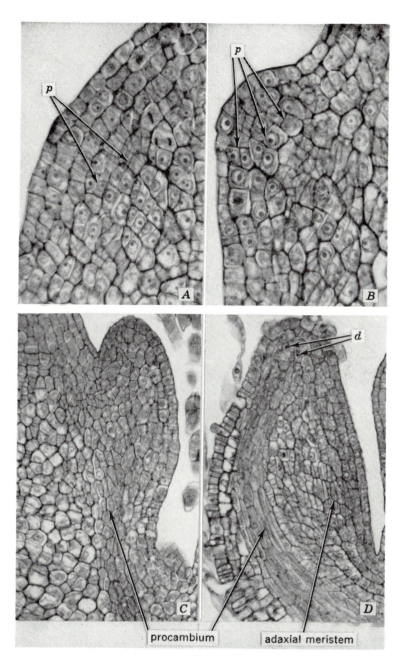

procambium

adaxial meristem

PLATE 75. Origin and development of *Acacia* phyllode (foliar organ) as seen in longitudinal sections. *A*, periclinal divisions occur in corpus at *p*, and then, *B*, also in second and third layers of tunica. Leaf buttress appears on side of shoot apex (*B*). *C*, phyllode primordium 45 microns long has grown upward from buttress. *D*, primordium 234 microns long, with procambial strand and conspicuous vacuolation on abaxial side. Adaxial meristem increases thickness of primordium. Primordia in *C*, *D*, were still growing at their apices (division figures at *d*). (*A*, *B*, ×600; *C*, *D*, ×300; Boke, *Amer. Jour. Bot.* 27, 1940.)

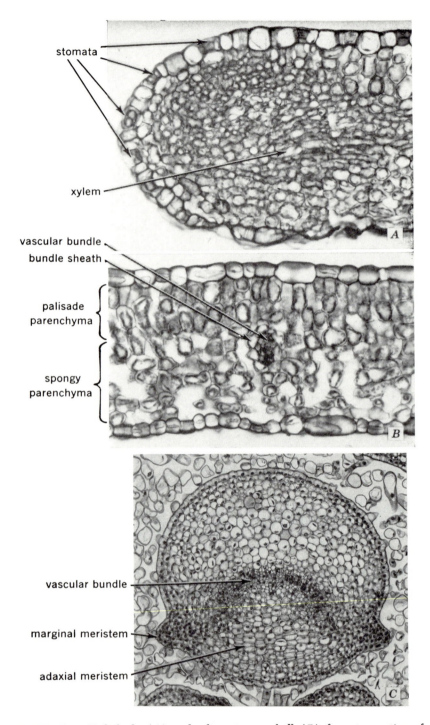

stomata

xylem

vascular bundle
bundle sheath

palisade
parenchyma

spongy
parenchyma

A

B

vascular bundle

marginal meristem

adaxial meristem

C

PLATE 76.   Hydathode (*A*) and adjacent mesophyll (*B*) from transection of *Brassica* leaf.   Xylem in *A* terminates in lacunose epithem tissue.   *C*, transection of *Nicotiana* leaf: initiation of lamina (marginal meristem) and growth of midvein in thickness (adaxial meristem).   (*A*, *B*, ×190; *C*, ×100.)

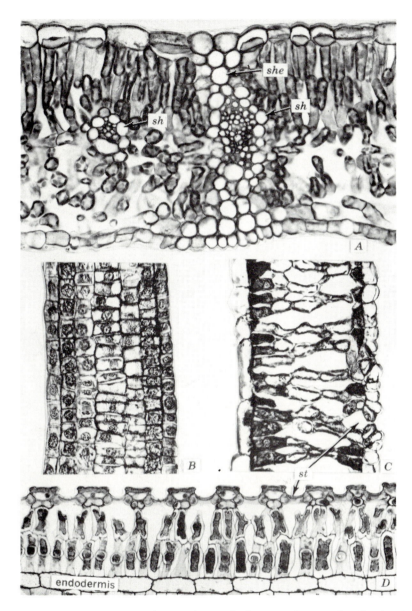

PLATE 77. *A*, transection of *Pyrus* (pear) leaf. Parenchymatic bundle sheaths (*sh*). Bundle-sheath extensions (one at *she*) connect sheath with both epidermal layers. *B*, younger, and *C*, older leaves of *Taxodium* in longitudinal sections. *C*, anticlinal orientation of intercellular spaces. *D*, radial longitudinal section of *Pinus* leaf showing relation of intercellular spaces to stomata (*st*; one guard cell in view in each stoma). (*A*, ×280; *B*, *C*, Cross, *Amer. Jour. Bot.* 27, 1940, ×386; *D*, courtesy of J. A. Sacher, ×150.)

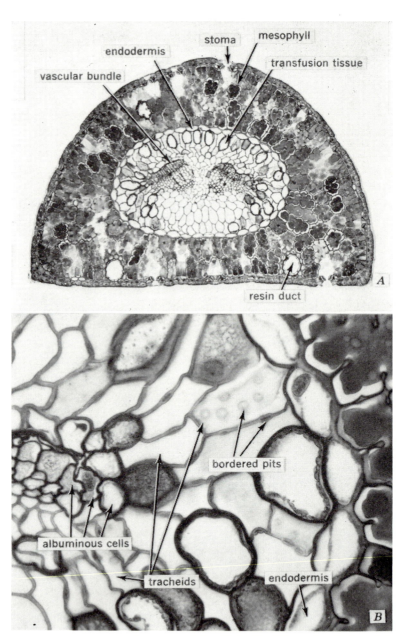

PLATE 78. Conifer leaf, *Pinus resinosa.* Transections. *A*, entire section; *B*, parts of vascular bundle (left), transfusion tissue (middle), and endodermis (right). (*A*, ×78; *B*, ×490.)

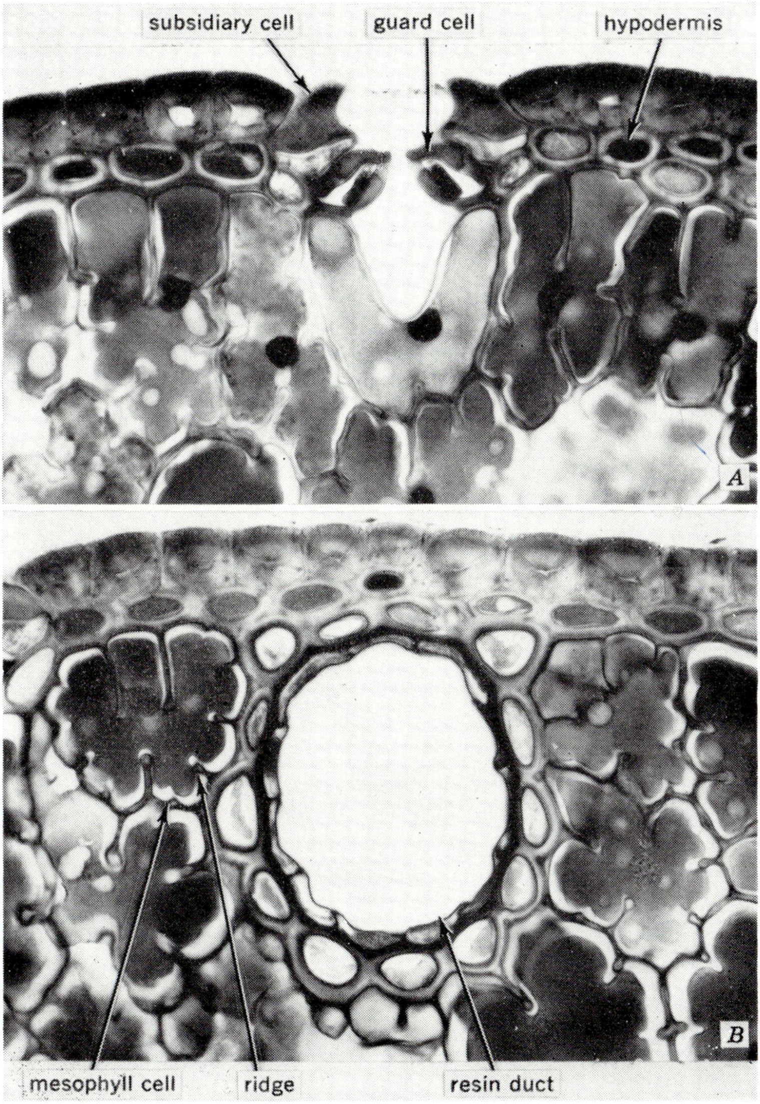

subsidiary cell   guard cell   hypodermis

mesophyll cell   ridge   resin duct

PLATE 79. Conifer leaf, *Pinus resinosa*. Transections through outer parts of a needle showing stoma (*A*) and resin duct (*B*), both with associated mesophyll, epidermis, and hypodermis. (Both, ×490.)

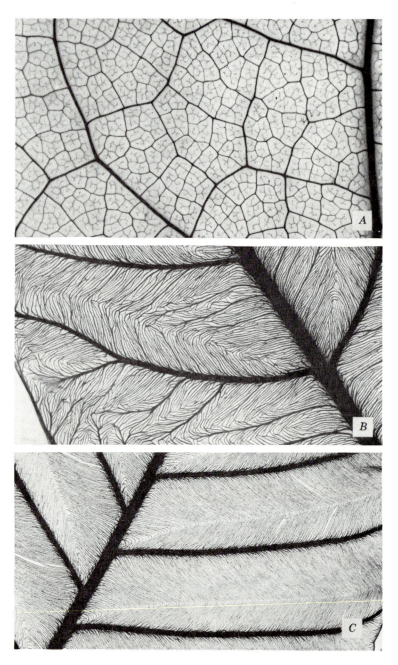

PLATE 80. Venation patterns (cleared leaves): *A, Liriodendron tulipifera,* reticulate, with free endings in mesophyll; *B, Quiina acutangula,* anastomosing plumose; *C, Touroulia guianensis,* anastomosing arcuate. (*A, B,* ×8; *C,* ×9. *A,* Pray, *Amer. Jour. Bot.* 41, 1954; *B, C,* Foster, *Amer. Jour. Bot.* 37, 1950.)

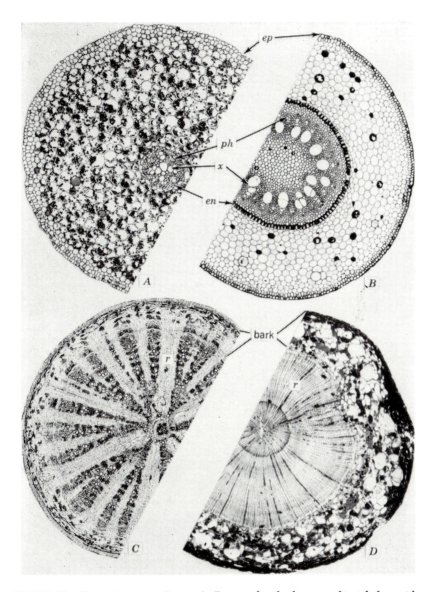

PLATE 81. Roots in transections. *A, Ranunculus,* herbaceous dicotyledon without secondary growth (×52). *B, Smilax,* herbaceous monocotyledon (×39). *C, Medicago* (alfalfa), herbaceous dicotyledon with secondary growth (×18). *D, Abies,* conifer with two increments of secondary tissues (×19). Details: *en,* endodermis; *ep,* epidermis; *ph,* phloem; *r,* ray; *x,* xylem, tetrarch in *A,* polyarch in *B.*

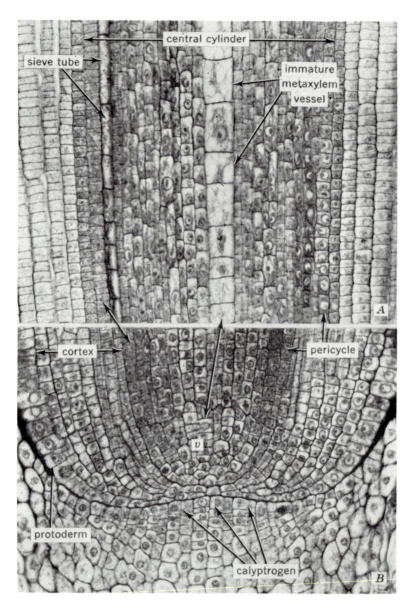

PLATE 82. Longitudinal sections of *Zea* (maize) root tip. *A*, center of root with partly differentiated first sieve tube and a series of metaxylem vessel-member primordia. *B*, apical meristem and recently formed regions of root. These are derivatives of (1) initials of central cylinder, (2) initials of cortex and epidermis, and (3) calyptrogen. Primordia of late-metaxylem vessel members (*v*) occur close to apical initials. Series at *v* is not continued upward because of oblique cut. (Both, ×280. Slides by E. M. Gifford.)

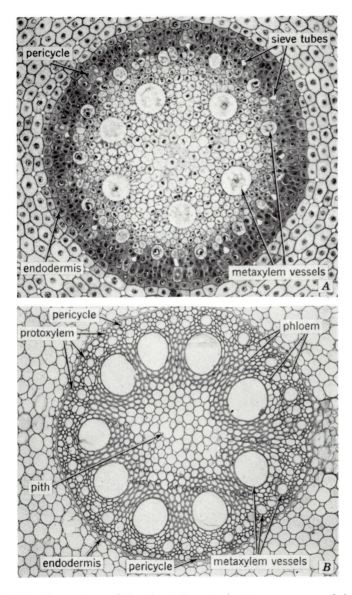

PLATE 83. Transections of *Zea* (maize) roots showing two stages of development of central cylinder. In *A*, first sieve tubes are mature. All cells are differentiated in *B*, vascular parenchyma is sclerified, and endodermis is in third stage of development. The late metaxylem vessels are the widest. (*A*, ×170; *B*, ×102.)

713

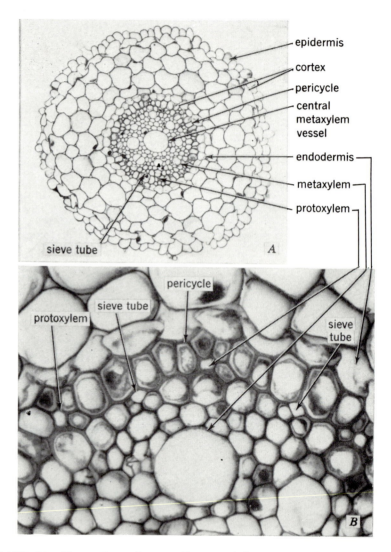

PLATE 84. Transections of young *Triticum* (wheat) root. *A*, entire section. *B*, part of vascular cylinder. The metaxylem (still without secondary walls) includes a circle of relatively narrow vessels and a wide one in center. Protoxylem elements were derived from pericyclic cells. At each phloem pole one sieve tube is mature. Each is flanked by two companion cells. (*A*, ×130; *B*, ×600.)

714

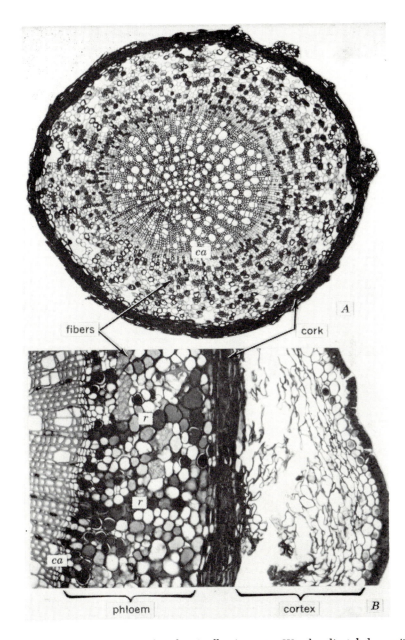

PLATE 85. Transections of *Salix* (willow) root. Woody dicotyledon with secondary growth. *A*, two indistinct growth increments, vascular cambium at *ca*, fibrous secondary phloem, periderm on surface. *B*, abscisson of cortex outside the periderm, and secondary phloem with rays (*r*). (*A*, ×36; *B*, ×145. Slides courtesy of P. L. Conant.)

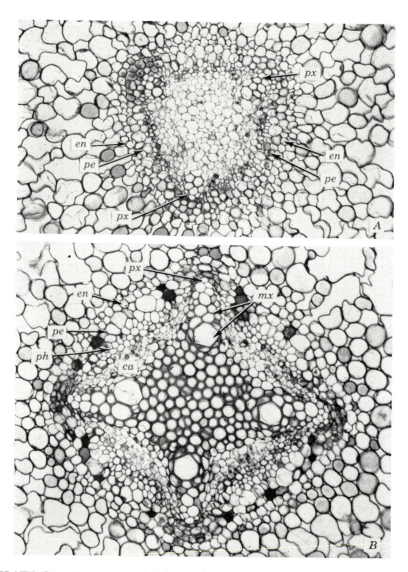

PLATE 86. Two stages of *Salix* (willow) root development in transections. *A*, triarch root without cambial activity and with immature metaxylem. Dense cytoplasm in pericycle at upper left xylem pole: first stage in branch root formation. *B*, tetrarch root with cambial activity initiated and with almost mature tracheary metaxylem elements (*mx*) and sclerenchyma in the center. Details: *ca*, vascular cambium; *en*, endodermis; *pe*, pericycle; *ph*, phloem; *px*, protoxylem *mx*, metaxylem. (Both, ×200. Slides courtesy of P. L. Conant.)

716

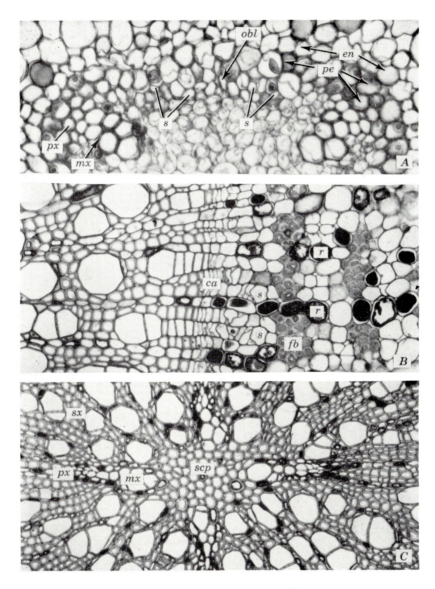

PLATE 87. *Salix* (willow) root in transections. *A*, phloem pole (center above) and two xylem poles from young root (plate 86*A*). *B*, secondary xylem (left of *ca*) and secondary phloem (right of *ca*). *C*, central part of tetrarch root with primary and some secondary xylem. Details: *ca*, vascular cambium; *en* endodermis; *fb*, fibers; *mx*, metaxylem (tracheary elements); *obl*, obliterated phloem cells; *pe*, pericycle; *px*, protoxylem; *r*, ray; *s*, sieve elements; *scp*, sclerified parenchyma; *sx*, secondary xylem. (*A*, ×440; *B*, *C*, ×200. Slides courtesy of P. L. Conant.)

717

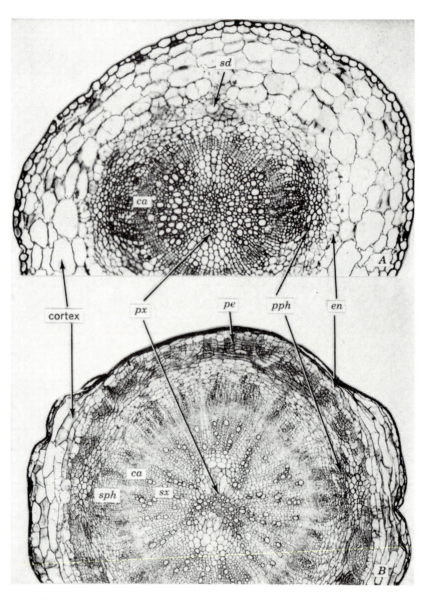

PLATE 88. *Daucus* (carrot) root in transections. *A*, younger root with intact cortex (×92). *B*, older root with cortex partly collapsed (×50). Details: *ca*, vascular cambium; *en*, endodermis; *pe*, pericycle (initiation of periderm); *pph*, primary phloem; *px*, primary xylem; *sd*, secretory duct; *sph*, secondary phloem; *sx*, secondary xylem. (Esau, *Hilgardia* 13, 1940.)

718

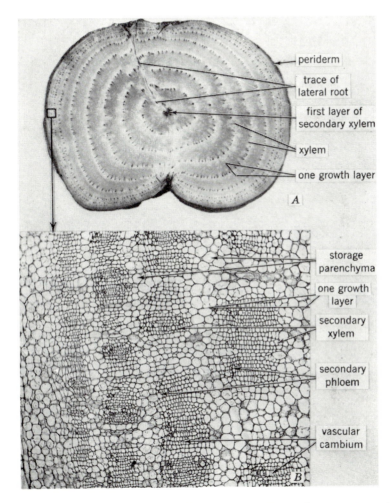

PLATE 89. Transections of root of sugar beet (*Beta vulgaris*). Anomalous secondary growth results from formation of many cambial layers outside the first cambium, each of which gives rise to strands composed of xylem and phloem and to storage parenchyma. (A, ×½; B, ×60; Artschwager, *Jour. Agr. Res.* 33, 1926.)

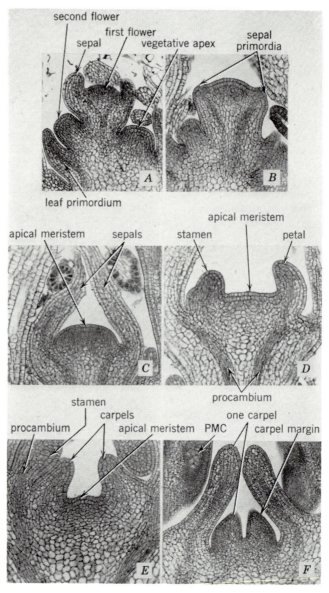

PLATE 90. Development of *Vinca* flower. Longitudinal sections. *A,* young inflorescence with vegetative and floral apices. Both kinds have a two-layered tunica. First flower has sepals. *B,* flower with sepal primordia. *C,* flower with somewhat larger sepals. *D,* median section of flower showing the apical meristem with a two-layered tunica and primordia of petal and stamen. *E, F,* two stages in development of carpels. In *F,* apices of carpels are touching each other and their ventral margins appear below in the center (see plate 91A). (*A, F,* ×106; *B, C,* ×112; *D, E,* ×126; Boke, *Amer. Jour. Bot.* 34, 1947; 35, 1948; 36, 1949.) (PMC, pollen mother cells.)

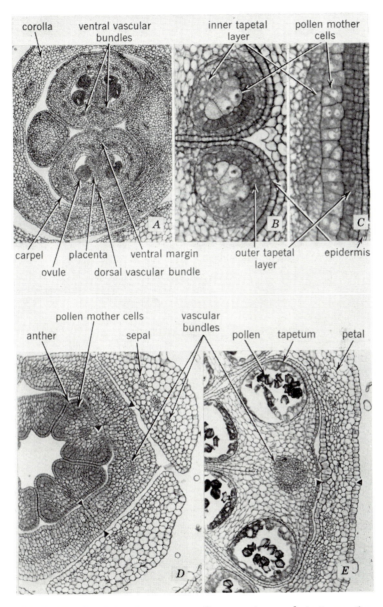

corolla   ventral vascular   inner tapetal   pollen mother
bundles   layer   cells

A   *B*   *C*

carpel   placenta   ventral margin   outer tapetal   epidermis
ovule   dorsal vascular bundle   layer

pollen mother cells   vascular
                       bundles
anther   sepal   pollen   tapetum   petal

D   *E*

PLATE 91. Details of developing *Vinca* flowers. *A,* carpels in transection with ventral margins fused. Placental ridges bear ovules. *B,* transverse and *C,* longitudinal sections of young anthers of *Vinca* with pollen mother cells and wall layers. Outer tapetal layer and wall layers between it and epidermis are derived from archesporial cells. Inner tapetal layer originates in ground tissue. *D, E,* transections showing earlier (*D*) and later (*E*) stages in ontogenetic fusion (arrow heads) of petal margins during formation of corolla tube. (*A,* ×60; *B,* ×247; *C,* ×275; *D, E,* ×110; Boke, *Amer. Jour. Bot.* 35, 1948; 36, 1949.)

721

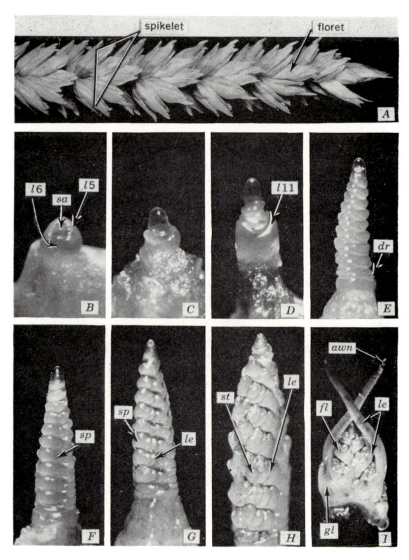

PLATE 92. Development of *Triticum* (wheat) inflorescence. *A*, mature spike.
B–D, shoot apex and associated leaf primordia during vegetative growth. *E, F*,
early stages of spikelet development. The upper member of a double ridge
(*dr*) becomes a spikelet (*sp*). *G, H*. floral parts differentiating in the spikelets.
I, single immature spikelet of awned variety. Details: *dr*, double ridge; *fl*,
floret; *gl* glume; *l*, leaf; *le*, lemma; *sa*, shoot apex; *sp*, spikelet; *st*, stamen. (*A*,
slightly enlarged. *B*, ×35; *C–F*, ×30; *G–I*, ×20. Courtesy of O. T. Bonnett;
B, D, Jour. Agr. Res. 53, 1936.)

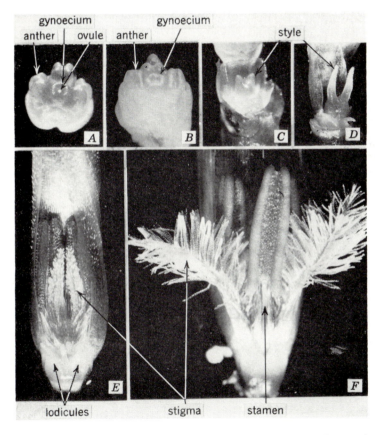

PLATE 93. Development of gynoecium in Gramineae. *A, C–F,* florets of *Avena* (oat); *B,* floret of *Triticum* (wheat). Gynoecium appears as a semicircular ridge in *A,* not enclosing the ovule. In *B,* ovule is completely surrounded by the ridge. *C–F,* stages in development of styles. (*A,* ×40; *B,* ×36; *C,* ×28; *D, E,* ×16; *F,* ×12. Courtesy of O. T. Bonnett. *A, D, F, Jour. Agr. Res.* 54, 1937; *C, E, Univ. Illinois Agr. Expt. St. Bul.* 672, 1961.)

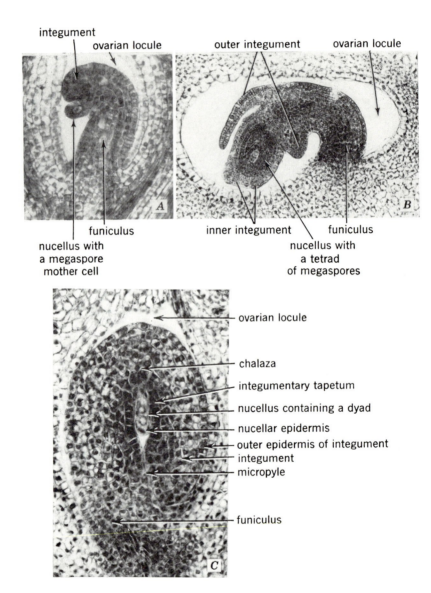

integument
ovarian locule
outer integument
ovarian locule

A

B

funiculus
nucellus with
a megaspore
mother cell

inner integument
nucellus with
a tetrad
of megaspores
funiculus

ovarian locule

chalaza

integumentary tapetum

nucellus containing a dyad

nucellar epidermis

outer epidermis of integument
integument
micropyle

funiculus

C

PLATE 94. Longitudinal sections through ovaries of *Parthenium* (*A, C*) and *Beta* (*B*) showing young ovules. Ovules partly inverted in *A, B*, completely so in *C*. Ovule in *Parthenium* has one integument (*A, C*); that in *Beta* has two (*B*). (*A, C,* ×230; *B,* ×150. *A, C,* Esau, *Hilgardia* 17, 1946.)

724

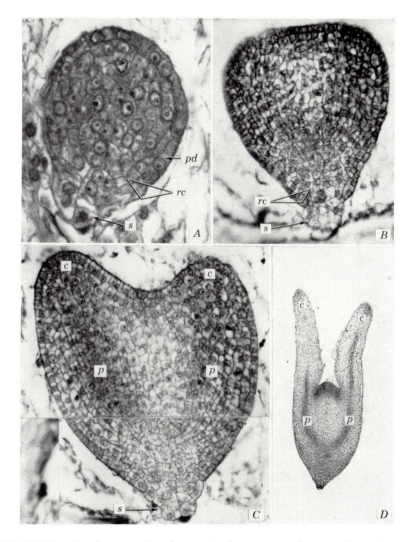

PLATE 95. Development of embryo in *Juglans regia* (walnut). The embryo is spheroidal in *A*, has a flattened apex in *B*, and shows initiation of cotyledons (*c*) in *C*. Oldest embryo in *D* bears the epicotylary meristem between the cotyledons. Details: *c*, cotyledon; *p*, procambium; *pd*, protoderm; *rc*, rootcap; *s*, suspensor. (*A*, ×540; *B, C*, ×240; *D*, ×48; Nast, *Lilloa* 6, 1941.)

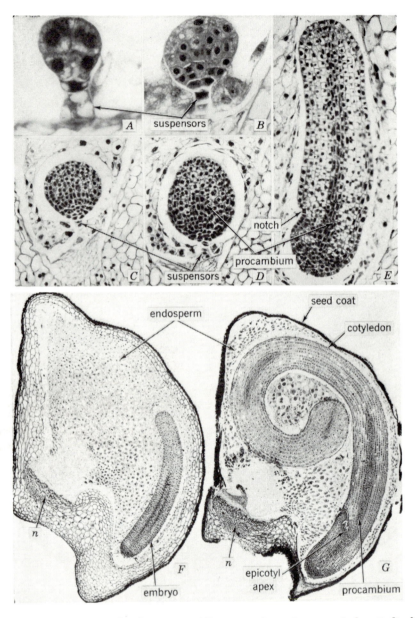

PLATE 96. Embryo development in *Allium cepa* (onion) as seen in longitudinal sections. Parts of ovules with embryos in *A–E*, entire seeds in *F*, *G*. Suspensor in embryos in *A–D*. The embryo in *E* shows a notch below the cotyledon. Epicotyl meristem is organized in the notch region (*G*). Procambium in embryos in *D–G*. Enlarged apex of cotyledon in *G* is a digestive structure. At *n* in *F*, *G*, remnants of nucellar tissue. The ages of the embryos in days after anthesis were: *A*, 12; *B*, 14; *C*, 16; *D*, 18; *E*, *F*, 20; *G*, 30. The seed in *G* is fully developed. (*A*, ×320; *B*, ×195; *C–E*, ×94; *F*, *G*, ×22.)

Author Index

(Bold-face type indicates bibliographic references.)

Abagon, M. A., 160
Abbe, E. C., 106, 108, **124, 125**, 458, **473**
Abbe, L. B., 281, 296, 303, **304**
Addicott, F. T., 471, **473**
Agerter, S. R., **267**
Agthe, C., 313, **334**
Ajello, L., 176, **176**
Aldaba, V. C., 209, 215, **223**
Aleksandrov, V. G., 550, 578, 616, **625**
Aleksandrova, O. G., 616, **625**
Aleshina, L. A., 553, **578**
Allen, E. K., 485, **531**
Allen, G. S., 119, 122, **125**, 147, **176**, 481, **531**, 612, **625**
Allen, O. N., 485, **531**
Allsopp, A., 470, **473**
Al-Talib, K. H., 222, **223**
Altschul, A. M., 617, **625**, 629
Ambronn, H., 198–200, **202**
Anderson, D., 197, **202**
Anderson, D. B., 53, 55, **63**, 170, 171, **176**, 215, 218, **223**
Andrews, E. H., 323, **334**
Andrews, H. N., Jr., 236, **265**, 271, **304**, 540, **578**
Arber, A., 1, **10**, 353, 368, **414**, 423, 424, **473**, 481, 519, 528, **531**, 540, 541, 559, **578**
Armacost, R. R., 440, **473**
Arney, S. E., 470, **473**
Arnold, A., 517, **531**
Arnold, C. A., 1, **10**
Arnott, H. J., 434, **475**, 528, **531**, 612, **625**
Arora, N., 485, **531**
Arreguín, B., 322, **334**
Arslan, N., 444, **473**
Artiushenko, Z. T., 451, **473**
Artschwager, E. F., 72, 73, **86**, 152, 293, 297, **304**, 327, 329, 331, **334**, 411, **415**, 518, **531**, 619, **625**, 677, 719
Arzee, T., 218, 222, 223, **223**
Ash, A. L., 217, 218, **223**
Ashby, E., 467, 469, 470, **473**
Ashton, M., 500, 501, **534**

Askenasy, E., 106, **125**
Audia, W. V., 346, **350**
Audus, L. J., 112, **125**
Avdulov, N. P., 617, **625**
Avers, C. J., 84, **86**, 173, **176**, 498, **533**
Avery, G. S., Jr., 150, **176**, 433, 435, 452, 453, 454, 464, 465, 469, **473**, 528, **531**, **532**

Badenhuizen, N. P., 26, **30**, 233, **265**
Baer, D. F., **124**
Bagley, B. W., 617, **625**
Bailey, I. W., 34–40, 42, 52, 54–57, 59, **63**, 64, 65, 74, 80, **86**, 133–135, 142, **144**, 145, 187, **190**, 204, 210–212, 219–221, **223**, 225, 229, 235–239, 245, 246, 249, 252–255, 258, 259, **265, 268**, 288, 289, **304**, 306, 364, 366, 406, **415**, 425, 440, **473**, 478, 525, **531**, 541, 548, 554–558, **578**
Baillaud, L., 374, **416**
Bain, H. F., 90, 91, **125**
Bain, J. M., 602, **604**
Baker, E., 65, **179**
Baker, G., 153, **176**
Baker, K. C., 597, **606**
Bakshi, T. S., 515, **531**
Bal, A. K., 553, **578**
Balfour, E. E., 363, **419**
Ball, E., 99, **125**, 186, **190**, 400, **415**, 648
Bamber, R. K., 340, **351**
Bancher, E., 27, **31**, 157, **176**, 309, **334**
Bannan, M. W., 80, 81, **86**, 134–136, 138, 139, 140, 142, 144, **145**, 253–255, **266**, 513, 523, **532**, 641
Baranova, E. A., 328, 329, **334**, 483, 513, **532**
Barber, D. A., 188, **191**
Barghoorn, E. S., Jr., 259, **266**, 665
Barker, W. G., 109, 112, **125**, 182, **190**, 404, **415**
Barnard, C., 540, 567, 572, 574–576, **578**, **579**
Barnell, E., 603, **604**

727

Subject Index